AF895728

CyberKnife NeuroRadiosurgery

Alfredo Conti • Pantaleo Romanelli
Evangelos Pantelis • Scott G. Soltys
Young Hyun Cho • Michael Lim
Editors

CyberKnife NeuroRadiosurgery

A practical Guide

Springer

Editors
Alfredo Conti
Associate Professor of Neurosurgery
Alma Mater Studiorum University
of Bologna
Bologna, Italy

IRCCS ISNB Istituto delle Scienze
Neurologiche di Bologna
Bologna, Italy

Resident Research Scientist
Charité University
Berlin, Germany

Evangelos Pantelis
Assistant Professor, Medical Physics
Laboratory, Medical School
National and Kapodistrian
University of Athens
Mikras Asias 75
Athens, Greece

Young Hyun Cho
Associate Professor, Department of
Neurosurgery and Radiosurgery Center
Asan Medical Center, University of
Ulsan, College of Medicine
Seoul, Republic of Korea

Pantaleo Romanelli
Scientific Director
AB Medica
Milano, Italy

Consultant and Scientific Director
Neurosurgery and Brain Radiosurgery
CyberKnife Center, CDI
Milano, Italy

Visiting Scientist
European Synchrotron Radiation
Facility (ESRF)
Grenoble, France

CyberKnife Center, CDI
Via Saint Bon 20
Milano, Italy

Scott G. Soltys
Associate Professor, Department
of Radiation Oncology
Stanford University Cancer Center
875 Blake Wilbur Drive
Stanford, CA
USA

Michael Lim
Department of Neurosurgery
The Johns Hopkins University School
of Medicine
Baltimore, MD
USA

ISBN 978-3-030-50667-4 ISBN 978-3-030-50668-1 (eBook)
https://doi.org/10.1007/978-3-030-50668-1

© Springer Nature Switzerland AG 2020
This work is subject to copyright. All rights are reserved by the Publisher, whether the whole or part of the material is concerned, specifically the rights of translation, reprinting, reuse of illustrations, recitation, broadcasting, reproduction on microfilms or in any other physical way, and transmission or information storage and retrieval, electronic adaptation, computer software, or by similar or dissimilar methodology now known or hereafter developed.
The use of general descriptive names, registered names, trademarks, service marks, etc. in this publication does not imply, even in the absence of a specific statement, that such names are exempt from the relevant protective laws and regulations and therefore free for general use.
The publisher, the authors and the editors are safe to assume that the advice and information in this book are believed to be true and accurate at the date of publication. Neither the publisher nor the authors or the editors give a warranty, express or implied, with respect to the material contained herein or for any errors or omissions that may have been made. The publisher remains neutral with regard to jurisdictional claims in published maps and institutional affiliations.

This Springer imprint is published by the registered company Springer Nature Switzerland AG
The registered company address is: Gewerbestrasse 11, 6330 Cham, Switzerland

"Beautiful things have dents and scratches too"

This book is dedicated to Nathalie Chadeau for her immense passion.

Foreword 1

I am very delighted to write a foreword for this comprehensive book on image-guided robotic (CyberKnife) neuro-radiosurgery for at least three important reasons.

The first reason is that single and hypofractionated stereotactic radiation therapy represents, nowadays, a well-established adjunct to our armamentarium for the treatment of several neurosurgical diseases. I am sure that the neurosurgical community will find a thorough assessment of appropriate indications, clinical benefits, risks, and pitfalls related to this technology useful.

The second reason is related to the fact that I was proud to introduce the CyberKnife at the University Hospital of Messina, Italy, in 2006. At that time, I realized how important it was to make this novel technology available and to evaluate its role and limits in the neurosurgical practice. Since the beginning, Dr. Alfredo Conti, a gifted investigator and neurosurgeon working at that time with me in Messina, was able to bridge his knowledge and experience of image-guided surgery of brain and spine lesions to a new robotic radiosurgical facility.

The third reason is that I consider it extremely important to have a balanced view of the potential and limits of this technology. Sometimes, radiosurgery is proposed as primary option for neurosurgical diseases which could be definitively cured by microsurgical procedures. Thus, it is extremely important to neurosurgeons and radiation oncologists to cooperate in the patient selection and decision-making process and share common experiences and cultural background. Only this factual collaboration will result in the best treatment option for each individual patient.

The occasional overindication for Stereotactic Radiosurgery (SRS) is due, on the one hand, to an understandable enthusiasm and familiarity of radiation oncologists with this technique. On the other hand, neurosurgeons with direct responsibility in SRS should reconsider the definition of SRS as a minimally invasive treatment that can be, to some extent, misleading to patients. As a matter of fact, although radiation is image-guided on targets with a sufficient safety, we cannot underestimate the fact that long-term adverse effects, potentially threatening the quality of life of patients, may occur. Nonetheless, SRS remains a strategic tool for the treatment of hazardous lesions, remnants, and recurrences for fragile patients.

These are all good reasons to welcome this book reporting the perspectives of clinicians and physicists with a large experience in the field of SRS gained through years spent in outstanding international institutions.

The rationale for the use of radiosurgery is highlighted by each contributor in a very systematic discussion of personal data, an extensive literature review with special reference to safety-effectiveness, and a final summary of indications and contraindications. The chapters cover different issues, an historical review, physics, imaging, and a thorough essay of brain and spine tumors, cerebral vascular lesions, and functional disorders. The structure of each chapter is perfectly organized in order to provide a practical guide and to make this book an excellent resource for residents, fellows, and practicing neurosurgeons.

Definitely, I am impressed by the broad and deep insight into applications of this novel robotic tool with different radiation doses and modalities. I am confident that the invaluable scientific information delivered by this book will benefit the international neurosurgical community. The editors and contributors are to be commended for their effort in making us aware of the current possibilities and limitations of radiosurgery for the treatment of neurosurgical disorders.

Francesco Tomasello, MD
University of Messina
Messina, Italy

Foreword 2

The concept of radiosurgery is, and always has been, a genuine neurosurgical concept. The ability to treat and control neurosurgical pathologies with an (invisible) sharp knife, at low risk, has been a success story since its introduction by neurosurgical pioneers. In addition, the therapeutic approach, decision-making, perception of perilesional surgical anatomy, and the pre-interventional risk assessment strongly benefit from decades of lessons learned from open neurosurgical operations. While radiosurgery was initially reserved for benign intracranial tumors, the indications have been broadened over the years into vascular, functional, pediatric, and spinal arenas. This is the reason why skull base, stereotactic, functional, and so on hybrid neurosurgeons, active in radiosurgical and open neurosurgical fields, seem to have an easier career pathway than endovascular/open vascular neurosurgical hybrids, for unknown reasons (although I have my ideas). Being a hybrid neurosurgeon or working in an interdisciplinary setting allows us to better appreciate the advantages and disadvantages of each therapeutic modality that we are able to offer to our patients. This will finally eliminate therapeutic bias from the decision-making process. The patients will benefit most from this development, and this is what counts. I am also a strong supporter of hybrid neurosurgeons since a thorough dual training is the most effective strategy to prevent therapeutic extremism, as we have observed in the past, also in the field of radiosurgery. By this, not everything will look like a nail since one has not only a single hammer. One reason for this smooth and successful development in radiosurgery is the trustful collaboration with our friends and colleagues from radiooncology, to whom we are, and always will be, grateful for their support and friendship.

This view, however, as appealing as it may sound, is currently regarded as "romantic," "unrealistic," or "useless" by critical neurosurgeons and radiation oncologists, even among my friends. This just demonstrates that many still need to be convinced by science- and evidence-based success and progress in the field, and maybe also by overcoming classical borders when it comes to neurointerventions. The single most important aspect that will consolidate the hybrid neurosurgical concepts is the thrive for high-end clinical and academic training in both fields, in order to tackle the argument that one person cannot be good at both therapies. Admittedly, there is still room for improvement in this field for the next years or so.

Having said all this, it is a pleasure to hold this textbook that comprises all aspects of radiosurgery, traditional and innovative ones in my hands. The leaders in the field are discussing these aspects in a very balanced and scientific way to highlight areas of opportunities, challenges, controversy, and common sense. The real value of this book, however, is that all authors involved fulfill my introductory remarks about hybrid neurosurgeons or experienced interdisciplinary teams. They stand for high-quality concepts and balanced views, always struggling for the best outcome in our patients.

It is a pleasure to see Alfredo Conti, a role model of an academic hybrid neurosurgeon and to whom the Charité CyberKnife Program is grateful for his collaboration and expertise, as the editor of this book. I applaud him for this masterpiece. In addition, it is a pleasure to have Francesco Tomasello as author of the second foreword. Francesco, as a very prominent representative of the Italian Neurosurgical Society, has always been a mentor and close friend for me over many years, for what I am truly grateful. However, in the midst of the COVID-19 pandemic and all the associated suffering (when these words are printed), I cannot resist to acknowledge and congratulate the entire Italian Neurosurgical Society for their dedication to all the positive aspects of our neurosurgical life: friendship, positivity, humbleness, elegance, innovation, family, and strong shoulders that carry on and come up with excellent books like this.

Peter Vajkoczy, MD
Department of Neurosurgery
Charité-Universitätsmedizin Berlin
Berlin, Germany

Preface

This book aims to represent a practical guide for image-guided stereotactic radiosurgery and hypofractionated stereotactic radiotherapy of the brain and spine. Leading physicians and scientists from four continents will provide the readers with basic concepts, current evidence, and guidelines for the treatment of neoplastic and non-neoplastic disease of the central nervous system.

Radiosurgery is one of the mainstreams of modern neurosurgery. Indeed, it perfectly complies with the current requirements of a minimally invasive neurosurgery and preservation of health-related quality of life.

Neurosurgery is a formidable challenge that, during its 100-year history, has expanded the opportunities for healing many human diseases, previously invariably mortal. Nonetheless, despite the amazing evolution of techniques and technologies, it remains a substantially invasive and pervasive discipline that often produces a negative impact on the cognitive performance of patients and eventually dramatic effects on their quality of life.

Image-guided radiosurgery represents a refinement of a revolution initiated almost 40 years ago through the introduction of frame-based radiosurgery. As a matter of fact, it has introduced several advantages for both the patient and the physician.

Image-guided radiosurgery offers greater comfort for the patient who does not need to be invasively attached to a stereotaxic frame. Treatment is usually delivered as an outpatient procedure; imaging and treatment should not be performed in a few hours, providing more time for physicians and physicists to work out the best possible treatment for the patient.

Above all, image-guidance has introduced the concept of multisession radiosurgery or hypofractionation, a term that indicates a radiotherapy treatment with a shorter course than conventional radiotherapy with a dose distribution that cannot be significantly different to that of single fraction radiosurgery. This approach has significantly changed the horizon of radiosurgery by expanding its boundaries of curative potential. Actually, by multisession radiosurgery it is now possible to treat larger brain tumors and tumors close to the most critical structures of the brain and spine.

The great value of these features introduced through the first image-guided radiosurgery device, the CyberKnife (Accuray Inc., Sunnyvale, CA), is testified by the adoption of image-guidance by other systems (i.e., Gamma Knife Icon).

The increasingly recognized value of image-guidance urges the widespread diffusion of the knowledge gathered after the treatments of thousands of patients over a period of 20 years.

The purpose of this book is to present the potential of image-guidance in the treatment of neurosurgical diseases, including neuro-oncological, vascular, and functional disorders, to the radiosurgical community.

This text gathers the experience of different centers and professionals with a long and renowned experience in the use of image-guided radiosurgery.

In each chapter, the literature on the topic is critically reviewed.

Several aspects are analyzed in order to present the reader with a critical analysis of single and hypofractionated treatments.

Each chapter provides all the basic radiobiological parameters and risks associated to the treatment. These data represent a summary of the authors' experiences together with the available literature on the topic and a practical guide for the selection of dose, fractions, isodose line, margins, imaging, and other parameters as well as to evaluate the risks associated to each treatment.

To date, literature on image-guided radiosurgery is scarce, and there are currently no texts with a similar practical approach available. Twenty years after the introduction of image-guided radiosurgery, it is time to present a textbook to summarize all the evidence on its great effectiveness and a practical guide for new users to quickly and easily manage this powerful tool.

Bologna, Italy — Alfredo Conti
Milan, Italy — Pantaleo Romanelli
Athens, Greece — Evangelos Pantelis
Stanford, CA, USA — Scott G. Soltys
Seoul, Republic of Korea — Young Hyun Cho
Baltimore, MD, USA — Michael Lim

Contents

Part I

Historical Perspectives

Creating the Future

1

John R. Adler Jr.

During the earliest days of developing the CyberKnife, it was never obvious to me how the technology might impact neuroradiosurgery. Embracing the innovative spirit of Silicon Valley, my Stanford team and I were simply embarking on a mission to free stereotactic radiosurgery from "stereotactic frame"-based immobilization and targeting, primarily with the goal of enabling this procedure to be performed almost anywhere in the human body. However, in time, the late 1980s and early 1990s, it became clear that the practice of neuroradiosurgery was too often dictated by the practical realities of stereotactic frames and not radiobiological or clinical considerations. Moreover, reimbursement at the time (and still largely today) reinforced the existing paradigms of therapeutic radiation, whether single-fraction Gamma Knife SRS or the fractionation schema of conventional radiation therapy. How ironic that two fields of clinical medicine that pride themselves on their scientific roots, and who proudly proclaim to embrace innovation, were so heavily influenced (or blinded) by dogma and money. Regardless, these practical realities have formed the headwinds into which the modern practice of neuroradiosurgery with the CyberKnife has evolved over the past two decades.

J. R. Adler Jr. (✉)
Stanford University, Stanford, CA, USA
e-mail: jra@stanford.edu

© Springer Nature Switzerland AG 2020
A. Conti et al. (eds.), *CyberKnife NeuroRadiosurgery*,
https://doi.org/10.1007/978-3-030-50668-1_1

At the end of the twentieth century, the radiobiologic foundations of radiation therapy, as embodied by the proverbial "4 Rs," provided a powerful intellectual framework for treating malignancies such as Hodgkin's disease and seminoma. However, the complex theory tended to breakdown in the treatment of most solid tumors, which of course represent the vast majority of all cancer. In contrast, the more empirically grounded foundation of radiosurgery emerged from the treatment, or perhaps more appropriately described as the ablation, of non-malignant brain "conditions," such as functional brain disorders or other pathologies such as arteriovenous malformations or benign brain tumors.

To what extent Lars Leksell, the creator of radiosurgery, was aware of the radiobiologic theory that formed the foundation of modern radiation therapy, I am not aware. Over the course of my year fellowship at the Karolinska, my office was immediately adjacent to his, both of us being isolated from the main neurosurgery department. As a consequence, when the semi-retired Leksell was not busy with other work, he would sometimes pop into my office to "entertain" himself. Although quite interested in the radiobiology of vascular injury, the objective of my research at the time, not once did Leksell ever discuss the 4Rs. In hindsight I conclude that Leksell either was unaware or frankly rejected the theory of most radiation therapy has having little relevance to his radiosurgical, or radioablation, pursuits. The reason I mention this fact here is that it nicely illustrates how distinct the foundations of radiosurgery and radiation therapy are from one another. The two are almost parallel disciplines and modes of thought, being forced together not by virtue of common intellectual foundations but by the practical realities of medical specialization and governmental licensing of radiation equipment. Forced into an arranged marriage, the two clinical practices remain uneasy bedfellows today, and the resulting cultural conflict is yet another headwind buffeting CyberKnife neuroradiosurgery.

At Stanford, the decision in the 1990s to utilize hypofractionation to manage some more "difficult" CyberKnife cases was driven ironically, by me, a neurosurgeon. By blending the concept of "normal tissue repair" with radiosurgery's "volume effect," I was seeking to extend radiosurgery to larger brain lesions as well as those immediately adjacent to critical anatomy such as the optic apparatus, other cranial nerves, and the spinal cord. My hope has always been that head-to-head long-term clinical trials might validate the selective benefits of hypofractionated radiosurgery before institutional structures such as government licensing and reimbursement could entomb clinical practice in concrete. Given the rapidity with which modern government and insurance regulations tend to move, I am not optimistic that medical science will ever outrun dogma, and I fear we will see incomplete knowledge being used to freeze clinical practices, much like what happened with radiation therapy.

Despite the "institutional headwinds" confronted by all practitioners of radiosurgery, it is hard, especially for a champion like me, to not view the field as one of the greatest triumphs in all of modern medicine. The ability to replace a very complex and dangerous open operation with a painless, less expensive, and oftentimes even more effective outpatient procedure like CyberKnife radiosurgery truly borders on the miraculous. Meanwhile I am convinced that radiosurgery is still in its infancy. The recent emergence of immunotherapeutic treatments for cancer only amplifies the benefits of simultaneously using radio-ablation as a tool for local control of tumors. Surely, in the years to come, immunologic treatments will get better and better and at every step radiosurgery can provide additional benefits. Meanwhile, there are ample opportunities for radiosurgery technology, even a highly advanced image-guided platform like the CyberKnife, to improve, driven by key radiosurgical metrics such as even greater spatial accuracy, steeper dose gradients, and less complexity, and yes, why not ultimately do all of this at a much lower financial cost? Instead of general-purpose devices, why not utilize dedicated and optimized neuroradiosurgical instruments, perhaps someday even utilize a more exotic radiation source, but only after the latter have been made affordable!! Having treated more than a million patients, the CyberKnife has clearly demonstrated what is possible today with advanced radiosurgical approaches. It is long overdue that such benefits be more widely recognized by all of humankind and, in doing so, be made available to the millions of patients who today go untreated each year. Practitioners of radiosurgery should not rest until we collectively meet this challenge.

2 CyberKnife Warfare in America: Battles at the Border Between Neurosurgery and Radiation Oncology

Cole A. Giller

2.1 Introduction

This chapter focuses on some aspects of CyberKnife history in America that have bothered me for a long time: first, the resistance to the CyberKnife when it was introduced into clinical practice and, second, the subsequent development of conflicts between neurosurgeons and radiation oncologists over matters of finances and treatment standards. These difficulties are rarely discussed openly, and almost never explored in print. But I believe they are real, at least in America. Although I have personally witnessed and even participated in these struggles, I never fully understood their origin or what could have been done to avoid them. My goal here is to approach that understanding not only by sharing my experiences but also by placing my observations in a broader historical, economic, political, and cultural context. For that reason, this chapter will dwell on the history of American medicine more than is ordinarily appropriate for a book of this kind. My hope is that an awareness of the origins of these unfortunate conflicts will be a starting point for their eventual resolution.

A standard narrative of the history of CyberKnife radiosurgery goes something like this: after years of technical inspiration and endless fundraising, the Stanford neurosurgeon John Adler and his team combined the latest advances in robotics and imaging algorithms to create a frameless radiosurgery device with precision rivaling the gold standard of the Gamma Knife [1]. It was a game changer, a "disruptive technology" [2], with potential to offer less invasive radiosurgery at lower cost, to extend treatment to sites other than the brain, to treat otherwise untreatable patients such as infants [3], and to permit exploration of the then novel idea of hypofractionation [4]. These possibilities attracted the attention of neurosurgeons already attuned to the success of Gamma Knife radiosurgery, as well as radiation oncologists already familiar with the CyberKnife planning paradigm due to its similarity to that of standard radiation therapy. Based on the Gamma Knife model, the first CyberKnife centers created multidisciplinary teams consisting of radiation oncologists, neurosurgeons, and medical physicists who established the unique advantages of this new technology. Adler's ideas for frameless, stereotactically delivered radiation therapy are now fully accepted and supported by a wide spectrum of treatment approaches, technologies, and manufacturers, as evidenced throughout this book.

This narration is true enough but is nonetheless incomplete. It omits accounts of resistance to the CyberKnife technology and neglects reports of the conflicts that mold the use of any disruptive technology as its stakeholder's jockey for

C. A. Giller (✉)
Department of Neurosurgery, Augusta University, Augusta, GA, USA
e-mail: cgiller@augusta.edu

© Springer Nature Switzerland AG 2020
A. Conti et al. (eds.), *CyberKnife NeuroRadiosurgery*,
https://doi.org/10.1007/978-3-030-50668-1_2

control, professional standing and financial reward. Early resistance to the CyberKnife was based on doubts about its radiobiological basis and on skepticism of its technical feasibility, even though the concept of radiosurgery had already been vetted by the Gamma Knife. But later, as the CyberKnife method was validated and its potential recognized, resistance arose from competitive, economic, and political issues, spawning clashes between the two interested specialties: neurosurgery and radiation oncology. Although viable partnerships between these groups already existed for Gamma Knife radiosurgery, the relationship had "always been an uneasy one" [5], and it was not surprising to see a steady escalation of conflict. As an example, I recall a group of physicians inviting me to give a talk at their cancer center about the CyberKnife because they were interested in radiosurgery, only to suddenly rescind their invitation without explanation, when they discovered that I worked for a hospital system different from their own. A second example occurred when a colleague from another radiosurgery center confided to me that his neurosurgeons and radiation oncologists frequently "almost came to fistfights" over issues of radiation dose and treatment planning ([6], 154).

How could this happen? How could colleagues who share a mission of excellent patient care allow their behavior to degenerate into petty squabbles and interactions approaching physical violence? How did these conflicts become so vehement, so personal, so *uncivil*, and so out of proportion to the ordinary issues of financial competition? Were these conflicts a fluke, dependent on factors unique to the field of radiosurgery at the time? Or did they arise from long-standing historical differences between the fields of neurosurgery and radiation oncology, or from more generally felt economic, political, and cultural forces? And would answers to these questions lend insight into the destructive consequences of the conflicts or suggest how to mitigate conflicts in the future?

In this chapter, I will share my experiences at several centers over many years, along with numerous conversations with colleagues, to argue that conflict between neurosurgery and radiation oncology has been widespread and long-standing and played a significant role in how CyberKnife radiosurgery is practiced. I believe it is important to understand the evolution of this behavior and to understand the role played by historical, economic, and political forces and even moral factors, in order to inform the future of CyberKnife practice and soothe interdisciplinary tensions that can damage careers and injure patients.

Rancor between medical specialties did not arise overnight, and so I will begin with a discussion of the history of medical specialties before addressing interactions between modern neurosurgeons and radiation oncologists. In particular, I will argue that as medical specialties developed, they took on characteristics that could be called "tribal," each with distinct cultures, each at times viewing themselves as heroes engaged in moral struggles against the villains of competing specialties. I will discuss "turf wars" and other interdisciplinary reactions to the disruptive technology of the CyberKnife in terms of what sociologists have called "border work," i.e., the negotiation of boundaries between competing groups. In this context I will then interpret my own observations of physician behavior relating to the CyberKnife practice. Finally, I will relate this behavior to a newly recognized phenomenon of *incivility* within the medical profession characterized by rudeness and hostility. These are sensitive issues, but my goal is not to assign blame. Instead, I hope to examine the origin of current CyberKnife practice and to explore what the history of CyberKnife development can teach us about managing conflicts between medical specialties.

I have a few caveats. First, my account does not represent all CyberKnife centers. Some centers suffer from physician conflict, but others enjoy a productive and pleasant environment. Conflict is not unavoidable. Second, I will restrict my discussion to radiosurgery of the brain and spinal cord, and not address the dynamics of body radiosurgery. Third, although I argue that medical specialties can take on competitive, tribalistic characteristics, there has also been a historical thread of collegiate cooperation among the specialties. This professional goodwill was crucial for the establishment of medical schools, the founding of productive private practices, and the strong relationships needed to earn the public trust. My goal is not to denigrate these remarkable relationships, but to focus on the competitive—and often unfortunate—aspects of medical specialties in order to better understand their effects on CyberKnife practice.

Fourth, I am limited by the delicacy required by any narration of human conflict. I will be discussing behavior of physicians (including myself) that is not likely to be universally admired, and I will therefore not reveal the identities of the people, the radiosurgery centers, or the time the behavior occurred. This means that I cannot document what I claim to have observed or heard and that my personal observations are more anecdotal than academic. I hope the reader will agree that this is the lesser of two evils. Finally, my aim is to position the nature of conflict involving CyberKnife radiosurgery as dependent on historical antecedents, differences in cultural norms between the "tribes" of neurosurgery and radiation oncology, and the collision of different moral systems. I will also use the sociological framework of boundary work to move the discussion of conflict between physicians beyond conventional discussions of economic and political competition. These arguments require excursions into the fields of history, anthropology, and sociology that may seem out of place in a book focusing on radiosurgery but are needed in order to offer credible arguments for new interpretations rather than unsupported claims. I hope the reader will forgive me for these detours and even find themselves enjoying the view.

2.2 The Beginnings of American Medical Specialization

Here we focus on three historical factors that shaped American medical specialization, setting the stage for the conflicts we are discussing. The first was the formation of specialties as isolated groups with separate beliefs and skills, the second was a change in hospital architecture that spatially isolated the specialists, and the third was the rise of a resident education tradition whose rigor encouraged an intense loyalty of physicians to their own specialties. These resulted in specialties taking on characteristics that could be described as "tribal," each with its own brand of medical morality.

American specialties were slow to develop. Most American physicians were general practitioners rather than specialists until the late 1800s ([7], 1), and even surgery was slow to become a separate field. As late as 1876, the eminent surgeon Samuel Gross said "there is not a medical man on this continent who devotes himself exclusively to the practice of surgery" ([8], 115–215 and 117–118). By the end of the nineteenth century, however, physicians faced an increasingly "competitive urban [medical] marketplace" ([9], 114). In their search for a competitive edge, they began to view specialty training as a pathway to patients, money, and a comfortable lifestyle ([10], 129). But American medicine was young, Europe was virtually the only place to obtain such training, and so American physicians travelled in droves to become specialists ([11], 1351). Specialties arose "with dizzying speed" ([10], 68) as these newly trained specialists returned home, and by the late 1890s, fields such as surgery, ophthalmology, radiology, and pathology were well established.

These changes represented a splintering of the medical workforce, driven by several factors. First, different specialties focused on diseases with different needs. For example, surgeons commonly faced problems requiring urgent action, whereas internists faced chronic problems addressed over days or weeks ([12], 35). Specialties thus diverged in their interests and skill sets, and interaction between specialties became less common as each held their hospital rounds at different times. Furthermore, by 1920, physicians were leaving comments in the patient's chart rather than speaking directly [13], thus communicating in an asynchronous fashion encouraging further isolation. Another factor was financial or political competition over patient volume and hospital control.

This professional compartmentalization of physicians was mirrored by changes in hospital design resulting in a physical compartmentalization of the physician environment. Until the late 1800s, hospital architecture was determined by the belief that diseases spread through contact with "bad air" ([14], 23). Accordingly, most hospitals consisted of large, rectangular "pavilion" wards, spaced far apart to prevent the air from one ward contaminating the others ([14], 41–42). But with the introduction of germ theory, and after Lister's demonstration that aseptic methods could prevent the spread of disease ([14], 98), wide separation was no longer necessary ([14], 91). Hospitals could be built as arrangements of closely spaced units, allowing new economies of scale and accommodating the growing need for specialty spaces. By 1900, separate areas were devoted to operating rooms, x-ray suites, pathology laboratories, and separate clinics for specialties such as dermatology, ophthalmology, gynecology, otology, and orthopedics ([14], 130–1; [9], 172 and 182).

This new physical isolation fostered a culture of exclusivity as each specialty staked out its own "home base." For example, although surgical amphitheaters in the nineteenth century were open to observers, operating rooms were kept locked and made difficult to find based on issues of asepsis by the late 1800s ([15], 14–16 and 126–7). Surgeons thus became monarchs of their new domains, furthering the family type of bonds between them and further enhancing a sense of exclusivity. The surgical theater thereby became a "source of emotionally resonant collective experience" ([9], 182–3). One can speculate that similar factors encouraged a sense of exclusivity in other specialties, including radiation oncology within their defined space of the radiation oncology center.

Exclusivity has a strong presence in modern times, as articulated by the recent statement that "pediatricians almost never enter the ophthalmology suite, internists do not operate the CT machines…emergency physicians do not often wander up to the intensive care units, and psychiatrists are rarely found in the operating rooms" [16]. Exclusivity of specialties was and has become the daily rule.

Changes in medical education magnified the sense of isolation among specialties. The growing demand for inexpensive labor to fuel the rapidly expanding hospital systems was met through the use of medical trainees who were required to live in the hospital with little or no pay. Training was grueling, but the "residents" believed in the process. For example, although a resident complained in 1883 that he had never "seen a class of men so [harshly] worked" ([17], 38), he concluded that the difficult conditions transformed his class into a "united family" ([17], 65). In later years, rigorous training continued to inspire a sense of purpose and cohesion. William Halsted, an early and powerful advocate of arduous training in surgery at the turn of the nineteenth century, "inspired fierce loyalty from his resident staff" ([7], 29). And in the 1950s, a program demanding that residents remain on call 24 hours a day, 7 days a week was reported to "encompass…in the house staff an esprit de corps of such fervor, that everyone eagerly worked until he dropped" [18].

These training conditions evoked a cultural identity within the residents of specialty training programs, a "sense of common purpose" ([7], 43), and feelings of inclusion by a "figurative family" usually led by the chief of service. ([7], 83 and 110–11). Residents developed strong bonds with their fellow residents and their mentors, similar to the bonds that form between soldiers during military training ([19], 1–19). And in the 1930s, new requirements to pass rigorous exams and obtain specialty certification intensified the difficulty of training and the esprit de corps ([7], 11). The claim that arduous training and a difficult certification process can result in group cohesion is supported by the work of anthropologist Naomi Quinn, who showed that "emotionally arousing" experiences, such as initiation rites, are crucial to the formation of a cultural self and, if shared, result in a cultural identity [20].

These three movements—the rise of specialization, the partitioning of hospital space, and the creation of powerful bonds between resident trainees—came to full fruition in the 1930s. Specialists had become isolated groups, intensely loyal to their specialty and bonded together by rites of passage. And as each group developed its own skill sets, traditions, and vocabulary, they developed an "us vs. them" mindset that seemed tribalistic.

2.3 The Emergence of Medical Tribalism

Here I will argue that the forces discussed in the last section influenced American medical specialties such as neurosurgery and radiation oncology to take on *tribal* characteristics, i.e., to act as groups with different cultures, beliefs, rules, and systems of morality. Some would object that the word "tribal" has connotations of primitivity and prejudice, and does not describe all behavior of all specialties. And it is true that the word has fallen out of favor among anthropologists. But I will show how a specific definition of "tribal" captures the intuitive meaning of the word and describes key aspects of medical specialties. Furthermore, I will show how the concept of tribalism aids the understanding of the conflicts between neurosurgery and radiation oncology as well as their individual cultures. Importantly, viewing specialties as tribes does not require us to take sides, but lets us see that neither group is right or wrong—there are only differences that must be understood and respected if there is to be peace.

I will use the definition of "tribal" that was nicely articulated by an independent article sponsored by the RAND Center. Members of tribal groups are bonded together either by kinship or by strong relationships exemplified by brotherhoods or gangs [21]. There are usually strict rules for group inclusion, and the groups thereby exclude outsiders. Relationships between members tend to be egalitarian, and there is often an absence of ruling groups other than those arising from persuasion. Wrongdoing is punished by ostracism or banishment, and a strong emphasis is placed on honor, rituals, pride, and dignity. The exclusivity of these groups is accompanied by an "us vs. them" mindset consonant with exclusivity and a tendency to yield preference to one's own group [22].

It is striking how well these characteristics describe modern medical specialties. For example, members of specialty groups typically share bonds of kinship based on a shared experience of rigorous training and exclusive knowledge. Admission to a specialty is regulated by strict rules, and physicians often consider each other to be peers. Physician leaders do not usually hold legal power over other physicians, although punishment can include banishment from professional societies [23], and mission statements of every specialty contain promises of trust and duty. The tribal characteristics of medical specialties have been recognized by the anthropologist Joan Cassel, who studied surgical teams in the 1980s and concluded that a surgical team is a "savage, exotic, secluded tribe" ([12], xxi).

An "us vs. them" mentality is another tribal characteristic that has been found in medical specialties. The historian George Weisz described such a mindset in the late nineteenth to mid-twentieth century among medical specialists when he described "a belief that specialized expertise was the *highest* form of scientific knowledge [italics mine]" and that the specialties "continued to insist on [their] centrality…to medicine" ([10], 132–133). A more visceral example of tribalism was William Osler's report in 1892 that general physicians referred to gynecological specialists as "butchers and belly-rippers" [24]. And in modern times, Cassell argues that an us vs. them mindset is a strong part of the field of surgery when she remarks that surgeons seem to be members of a "fellowship" ([12], 7) and "join together when threatened…by members of other specialties" ([12], 60). In her studies, surgeons perceived themselves as "a group set off from…internists…[who] they describe by a series of jokes and disparaging remarks" ([12], 61). Others have described healthcare organizations as "tribal confederacies of professional affinity groups" [25] and make reference to a "dysfunctional tribalism in medicine" [26].

It should come as no surprise that medical specialties often harbor tribalistic behavior, for several reasons. First, tribalism is common in other settings such as business organizations [27], politics [28], and sports clubs [29]. Furthermore, elements of tribalism are central to human behavior, and the sociobiologist Edward O. Wilson calls tribalism one of the "absolute universals of human nature" ([30], 57). Studies show that children spontaneously display tribal behavior [22], and tribal behavior can be evoked merely by separation and perceived threats [31]. There are also are numerous other discussions of tribalism in medicine [25, 26, 32–34].

The physician-historian Kenneth Ludmerer observed that every specialty develops "its own temperament, routines and traditions" ([7], 95), and I have argued further that the specialty then becomes a forme fruste for tribalism. Isolated within their specialty, spaces, and times, bonded by intense shared experiences, imbued with a sense of exclusivity, and subject to the natural tendency to become tribal, the specialties arising in the late nineteenth and early twentieth centuries inexorably developed a culture of tribalism. As they did so, they naturally developed their own sense of medical morality.

2.4 Moral Purpose, Heroes, and Interdisciplinary Conflict

The feelings of moral commitment that physicians have always shared were amplified by the new specialty training programs established in America in the 1890s and were transformed into a central component of medical culture. According to Ludmerer, the goal of these programs was to "internalize a moral responsibility for...their patients" and to give the house officers "a professional identity with an ethical core of service and responsibility." Medicine was seen as "a calling [placing] the welfare of ...patients above all else," and residents "zealously guard[ed] the welfare" of their patients ([7], 74–78). The ascetic lifestyle of a resident became associated with a proper moral compass, and in the 1950s, "the underpaid and overworked residents… assumed without question that they were doing good" ([12], 185).

One can speculate that each specialty developed its own brand of morality due to their isolation and to the tribalistic boundaries between them. The role of morality has been particularly well studied for the field of surgery. A study of surgical teams in the 1970s by sociologist Charles Bosk reported that morality was the basis for handling surgical errors. *Technical* errors made by residents due to inexperience were forgiven because the residents were in the process of learning. But errors arising from negligence, dishonesty, or a failure "to discharge…obligations conscientiously" were more serious breaches and represented *moral* errors that were punished ([23], 50). Bosk concludes, "above all, residency training is designed as a moral education" ([23], xvi). In the 1980s, Cassell also concluded that moral issues drove physician behavior. She likened the surgical team to a morality play, in which members represented stereotypes of moral behavior. Because the entire team knew the performance of each member, surgeon behavior was molded by the perceived assignments of morality ([12], 71–8). Like Bosk, Cassell observed that surgeons judge each other on moral grounds, either on the basis of normative errors or on the basis of "surgical sins" such as arrogance and veniality, and affirmed that morality was the basis for every aspect of the team's behavior ([12], 153–81).

With this emphasis on medical morality, it was natural that each specialty felt morally justified in their medical decisions. Surgeons, for example, felt a moral obligation to consider surgical treatment, and as one observer said, physicians became convinced of the "utter rightness of [their] cause" and believed that "they were fighting the good war against death and disease" ([12], 185). But morality was complicated because physicians from different specialties could have contradictory opinions of the best course of action, each believing that theirs was the moral high ground. Arguments for surgery vs. observation, for example, have raged for centuries, and in modern times, the controversy between radiosurgery and conventional radiotherapy continues to thrive.

Collision of different medical moral systems can lead to conflict, thereby promoting an image of heroism in the minds of physicians. If the fate of the patient depends on making the correct decision, and if the decision can only be chosen through a moral battle with another specialty, then those who fight the good fight—the physicians—take the role of a hero. This helps explain why these conflicts often seem to be personal and why disagreements between specialties become vehement, frequently irrational, and mysteriously independent of financial concerns or issues of power [35–37].

2.5 Boundaries and Tension Between Neurosurgery and Radiation Oncology

I have argued that specialties such as neurosurgery and radiation oncology became exclusive and isolated as they developed their own treatment philosophies, spaces, and loyalties. But interactions could not be avoided because the two fields naturally shared patients with intracranial tumors. Prior to the advent of radiosurgery, these interactions were mostly collegial, with boundaries between the two professions defined by the tools they used: surgeons performed surgery, and radiation oncologists delivered radiation. The status quo was disturbed with the introduction of the Gamma Knife, which threatened to allow neurosurgeons to prescribe therapeutic radiation independently from radiation oncologists. But rules were quickly adopted, mandating that Gamma Knife treatments require participation by both neurosurgeons and radiation oncologists. Neurosurgeons would manage placement of the stereotactic frame, focus on anatomical issues, and usually were responsible for follow-up evaluations [38]. Radiation oncologists would concentrate on dose prescription and delivery. Boundaries were preserved and tribal warfare was prevented.

But the peace was uneasy, characterized by struggles for ownership and "turf war statements of radiosurgery as surgery or radiosurgery as radiation therapy" [39]. Some neurosurgeons encroached into radiation oncology territory by insisting on constructing their own treatment plans and by attempting "to control the treatment decisions, in the extreme" ([6], 158). An example of these tensions is seen in an interview reporting a Canadian Gamma Knife center in which neurosurgeons "felt they had to sign off" on Gamma Knife plans, but were not allowed to do so because "there's no need," resulting in "a lot of push and pull from them to have ownership in this way" [40]. In the same interview, the radiation oncologist David Larson comments, "It was sometimes pretty tense." And as late as 2011, Lunsford observed "there are still frequent skirmishes" [41].

Radiation oncologists tolerated this neurosurgical invasion, perhaps because of radiobiological differences between Gamma Knife treatment and radiation therapy. The Gamma Knife ablated its target by destroying organelles as well as damaging DNA, whereas radiation therapy selectively damaged tumor cells by exploiting their sensitivity to radiation. This allowed Gamma Knife treatment to be viewed as a surgical resection rather than radiotherapy, even though the resection was achieved through radiation rather than a scalpel. Gamma Knife treatment seemed more like surgery, making it easier to accept a dominant role of the neurosurgeon.

2.6 Morality and Boundary Work: CyberKnife in the Middle

Conflicts between neurosurgery and radiation oncology were of course motivated by issues of finance and control but were also powerfully driven by differences in moral definitions that commonly occur between isolated tribes. For example, many surgeons did not hesitate to treat multiple metastatic brain tumors with radiosurgery, even though studies had only addressed treatment for fewer than three lesions. They did not require a randomized study any more than they would require a study to resect an extra lesion when encountered in the operating room, and to limit treatment to three lesions in those cases violated their moral obligation to the patient. On the other hand, many radiation oncologists felt it would be scientifically invalid—and morally wrong—to treat more than three lesions because of the uncertainty of efficacy, the potential dangers of radiation injury, and the availability of the tried-and-true conventional radiotherapy and chemotherapy ([6], 154). To them, the accretion of careful studies over years was morally essential. I am not advocating either of these views, but rather suggesting that, as for many tribal beliefs, they are arguably neither right nor wrong. Radiosurgical conflicts can therefore be looked upon as disagreements between two tribes separated by an uncertain boundary, each following their own moral compass.

Interactions at the boundary between two fields have been studied by sociologists with the concept of *boundary work*, defined to be a "purposeful…effort to influence the social, symbolic, material or temporal boundaries…affecting groups, occupations and organizations" [42]. This definition includes turf wars and represents how different occupations compete or cooperate by adjusting the boundaries of "power, social position and status" [42]. Some neurosurgeons did boundary work to promote the goal of "expanding authority into other domains," while some radiation oncologists did boundary work to promote the goal of "monopolizing professional authority by excluding rivals" [42]. One should note that the competition was widespread but by no means universal.

Boundary work can be competitive, cooperative, or configurative [42], and our focus—competitive boundary work—has been classified into three categories. The first category is the work of *defending existing boundaries*, often triggered by an event such as the introduction of the CyberKnife. Defenses include discursive claims of competence or moral superiority (such as claims of superior anatomic knowledge), appeals to normative rules (such as the established rules for radiotherapy), or the use of materiality (as when each group insists that the CyberKnife be located within their department). The second type of work is that of *contesting existing boundaries*, which can occur as a struggle between "challengers" and "incumbents" (as when neurosurgery attempts to legitimize its use of hypofractionation) or achieved with the support of outside relationships (as when each group appeals to their referring physicians for favoritism). The third type of work is that of *creating new boundaries*, which can occur as groups construct a niche for their new activity by arguing for their superior resources or skill (as when neurosurgeons entered the market after inventing the technology) [42]. The value of the concept of boundary work is that it facilitates an understanding of what might otherwise seem to be a series of randomly occurring battles.

2.7 An Example of Boundary Work: Imaging

An example of boundary work may be helpful [43]. When CT/MRI technology was introduced, skills of interpreting anatomic slices and responses to magnetic fields were foreign to both radiologists and non-radiologists, allowing specialties such as cardiology to claim ownership of this new branch of imaging. They did so through boundary work, first claiming superior anatomical knowledge of their body system; second, working materially to demand that the scanners be placed in their own departments; and finally, using their "ownership" of patients to direct the course of radiological studies. Radiology responded by pushing the boundary to their favor, claiming superior visual skills and more intimate familiarity with imaging. And they worked materially to argue that placing the devices in their department would boost throughput and provide an economy of scale, thereby gaining support of the hospital system by promising greater profits.

The result was a long series of struggles. One MRI engineer said, "The relations among radiologists and cardiologists are a huge problem… there are big fights…about prestige and money." A radiologist explained, "The machines are power," and described "a turf-battle…who owns the garden and where is the fence." A radiologist claimed, "surgeons see what they want to in the images," and another advised "internists should leave it to radiologists." The discussions became disrespectful. One radiologist affirmed that the "surgeon fumbles for something and blindly stabs into the body. The radiologist, however, looks for it." The sociologist Burri paraphrases: "The radiologist views the surgeon [as acting] without looking at the patient's body before operating...while [the radiologist] imagines himself...as someone who reflects before acting" [43]. I will argue later that the same dynamics of competitive boundary work can also lead to incivility between neurosurgeons and radiation oncologists.

2.8 Precursors of the CyberKnife: The Stage Is Set for Conflict

The history of the CyberKnife cannot be considered without that of radiosurgery. In this section I will review these histories and argue that conflicts surrounding CyberKnife practice were antedated by skirmishes over radiosurgical hegemony that triggered years of competitive boundary work and, unfortunately, a degree of incivility.

In the early twentieth century, the idea of *hypofractionation*—the use of a small number of fractions—was discredited due to a high rate of tissue damage [44–47]. As one radiation oncologist from that time wrote, "we are forced to fractionate" [47]. Nevertheless, the concept was later explored with the use of helium ions and proton beams beginning in the 1950s [48–51] and resurrected for the brain by Leksell in 1951 [52]. Gamma Knife treatments, using a single fraction, were first delivered in 1967 and subsequently developed into a radiosurgical standard [53, 54]. Despite the violation of conventional principles of radiation oncology, these early efforts were accomplished by collegiate cooperation between neurosurgeons, radiation oncologists, and physicists.

Experience with Gamma Knife was obtained slowly and carefully, limited to Sweden until 1983 ([54], 85) and only brought to North America in 1987 [39]. In the meantime, radiation oncology rapidly expanded, primarily using linear accelerator technology (LINAC) to deliver "conventional" dose schedules of 10–30 fractions. By the 1980s, perhaps inspired by the success of the Gamma Knife, radiation oncology centers were exploring the use of LINAC technology for radiosurgery. Most of the early efforts restricted treatments to a single fraction [55–59], but a few centers included "hypofractionated" regimens, i.e., fractionated treatment typically using 2–7 fractions rather than the higher numbers of conventional radiotherapy [60–62]. By 1995, radiosurgical procedures were offered in America by at least 100 LINAC centers [63]. Radiosurgery was no longer limited to neurosurgically dominated Gamma Knife centers and had found a strong foothold within radiation oncology territory. The boundary defining radiosurgical dominion had shifted.

These early efforts at LINAC radiosurgery were viewed favorably by a growing number of radiation oncologists. David Larson reports that in the 1980s, there was "great individual clinician interest," and educational courses for this "new thing" of stereotactic radiosurgery (SRS) were heavily attended at ASTRO (American Society for Radiation Oncology) meetings [40]. But there was also resistance. Some neurosurgeons opposed the concept of radiosurgery, feeling that craniotomy was a standard of care and that noninvasive therapy was "misguided and bound to fail" [64, 65]. Furthermore, radiosurgery threatened to siphon patients away from neurosurgeons, thereby threatening finances and control. Negative responses from the neurosurgical community can be viewed as efforts at defensive boundary work. Resistance also came from some radiation oncologists for similar reasons. The divergence of radiosurgery from conventional fractionation was not fully accepted, and radiation oncologists who did not have access to radiosurgery feared they would lose income and control [64]. The resistance to radiosurgery represents boundary work by many physicians in both fields to maintain the status quo.

2.9 The History of the CyberKnife and Hypofractionation

The CyberKnife was born into this environment of simultaneous excitement and resistance, propelled by several forces. The first was a perception (not completely deserved) that LINAC technology lacked the precision required for radiosurgery [66, 67]. The resulting motivation to develop new LINAC technology satisfying these requirements was likely an important driver for the invention of the CyberKnife. Furthermore, the interest created by the realization that the frameless platform and precision of the CyberKnife would allow treatment of extracranial targets represented a second favorable driving force.

A third force driving the CyberKnife was the changing perception of hypofractionation. As mentioned, hypofractionation was explored by a small number of LINAC centers in the 1980s, although the majority of radiation therapy was delivered with "conventional" schedule of 10–30 fractions. But support for hypofractionation was voiced in 1991 when radiobiologists argued that hypofractionation would be "more effective at killing" radioresistant tumors than single-fraction regimens and could do so with less damage to a normal tissue [68, 69]. Dose equivalents were calculated, e.g., a single fraction of 20 Gy was calculated as equivalent to 5 fractions of 7.4 Gy [68], and "around five or six" fractions were recommended for malignant brain tumors [69]. In 1993, another group reported similar equivalent doses for early and late responding tissues. Although they argued that fractionation was not likely to confer advantages over single-fraction treatment for small tumors, their results also supported the choice of 5–6 fractions for malignant tumors [70].

In the meantime, clinical data for hypofractionation became available. In 1991, 15 patients with intracranial tumors were treated with six fractions of 7 Gy, although the authors thought that hypofractionation was "still…experimental" [71]. In 1992, a report of 21 patients with intracranial tumors concluded that treatment with 42 Gy delivered in 6 fractions over 2 weeks "appears to be a worthwhile procedure" [71, 72], and the following year, a report of the use of 2 or 3 fractions of 6 Gy for treatment of brain metastases concluded that stereotactic hypofractionation "may be the method of choice" in some conditions [73]. A dose escalation trial of 22 patients with recurrent glioma treated with 4–10 fractions of 5 Gy concluded that stereotactic fractionation "may be a suitable alternative to interstitial therapy" [74], and a report of 49 patients treated with 6 fractions of 6 Gy supported the current "move from dose delivery in a single session to fractionated dose regimens" for selected brain tumors [75].

At this time, "radiosurgery" denoted treatment with a single fraction, while "hypofractionation" denoted the use of a small number of fractions, typically between 1 and 7. Neither of these dose schedules was universally accepted, despite the theoretical advances and clinical experience mentioned above. One report from 1994 stated that radiosurgery should not "become a standard treatment [for] brain metastases" [76], and another referred to some users of radiosurgery as "cavalier cowboys" [77]. Furthermore, many studies from LINAC centers limited their delivery to single fractions [78–86], as did early studies using the CyberKnife for intracranial tumors as late as 1998 [87, 88]. In addition to the disagreement between the radiosurgical and nonradiosurgical communities, there were hints of tension between radiosurgical neurosurgeons and radiation oncologists. The Gamma Knife, for example, was developed without "evidence of any enthusiasm…by radiation oncologists," although it is not clear whether they were excluded or disinterested (Lindquist quoted in [63]). One reviewer noted later that because most radiation oncologists used LINAC technology and most neurosurgeons used the Gamma Knife, the two had very different approaches to radiosurgery that triggered "misunderstandings, differences and antagonisms" [67].

An indication of these conflicts appeared in 1996 when the radiation oncologists David Rosenthal and Eli Glatstein published a critique of radiosurgery and hypofractionation for the treatment of brain tumors [89]. They stated that "well-done clinical studies" had already shown that hypofractionation leads to "increased complications [and] a decrease in local cancer control" when compared to conventional fractionation. They argued further that radiobiological principles were being ignored in order to pursue "a 'fatal attraction' between…new technology and its immediate application." And because comparisons of radiosurgery vs. whole brain radiotherapy were unexplored, the authors believed that radiosurgery for metastatic brain tumors was "at the very least, questionable." Furthermore, the authors believed that radiosurgery at that time represented research that was not labeled as such, because "third-party carriers would not pay [and the] treatment would come to a screeching halt." They summarized by saying that single-fraction radiosurgery "is really stereotactic radiotherapy [and] is suboptimal radiation oncology. [It] is virtually predicated on the ability to perform another craniotomy to remove focal necrosis."

The radiosurgery community responded in a letter to the same journal, critical that the editorial declared radiosurgery to be ineffective and dangerous, despite the "tremendous body of historical and clinical literature relative to outcomes" [90]. For example, by 1993, more than 18,000 patients had been treated with radiosurgery [91], and by the time the editorial appeared, more than 700 articles on stereotactic radiosurgery had been published (a PubMed search under "Gamma Knife" between 1968 and 1994 reveals 762 papers, and searching under "stereotactic radiosurgery" reveals 734 papers; accessed March 12, 2020). They speculated that the editorial may have been motivated by fears that radiosurgery was intended "to impact upon the turf of the radiation oncologist" or by the recent "purchase of an expensive device for fractionated frameless radiosurgery" at the authors' institution (author's note: I was present at the time, and although I never met the authors of the editorial, I was convinced that there were significant tensions between the neurosurgeons and radiation oncologists).

The editorial and the response letter illustrate four levels of conflict relevant to the history of the CyberKnife. The first level is that of science, in which the data opposing the use of single fractions was countered with the more current clinical experience. Scientific discussions like these may represent conflict but are frequently fruitful. The second level is that of turf and tribalism, suggested by the editorial's assertion that "radiosurgery is really stereotactic radiotherapy," its complaint of a "virtual exclusion" of radiation oncologists, its claim that radiosurgery arose "with far more alacrity than scientific thought," and its concern about a "fatal attraction" between technology leading to an "unbridled use" of radiosurgery. The response letter addresses turf issues directly, suggesting that radiation oncologists worried that radiosurgery "does not in all cases rely" upon their talents, and (as quoted above) by asserting that the goal of radiosurgery was "not to impact upon the turf of the radiation oncologist" [90]. These conflicts represent competitive boundary work in action.

A third level of conflict is that between different medical cultures and moralities. The editorial seemed to argue that it would be wrong to offer treatment without many years of careful experimentation, especially when prior data predict complications. The letter seemed to respond that it would be wrong to deprive patients of this important therapy while waiting years for randomized studies, based on the preponderance of new and favorable data. Neither of these approaches was objectively right or wrong but instead represented different viewpoints from different systems of medical morality. The final level of discussion is, unfortunately, that of incivility between specialties. In discussions of sensitive topics evoking strongly held beliefs and layers of professional, economic, and moral issues, the line between the appropriate use of powerful language and frank incivility can become blurred. The reader can decide where the line falls in the essay and letter, but phrases such as "another craniotomy," "fatal attraction," "mid-life" anxieties, and "atavistic" are suggestive. Incivility in the medical profession is unfortunately quite common, as discussed in Sect. 2.13. My intention here is to argue that these layers of conflict, combined with issues of money and ownership, have in many cases led to unfortunate and destructive rifts between the cultures of neurosurgery and radiation oncology.

2.10 A Major Boundary Shift

A major step in boundary work impacting CyberKnife practice was a reassignment of radiosurgical authority by the Nuclear Regulatory Commission (NRC) in the early 2000s. Prior to that time, neurosurgeons and radiation oncologists coexisted in Gamma Knife centers as *authorized users* (AU), an official designation conferring authority during radiosurgery cases. This joint assignment of power stabilized most centers for years despite tensions between the specialties, and the inclusion of neurosurgeons made sense at first because of their crucial role in radiosurgical development. But neurosurgeons did not technically qualify for AU status because of the NRC requirement that an AU be responsible for "ensuring that radioactive materials are handled and used safely" [92]—a task meant for radiation oncologists. Tensions continued, and the NRC eventually approved regulations stripping neurosurgeons from their AU status by requiring that an AU be "a specialist in Radiation Oncology" and be board certified by a radiation/radiology board [93]. Neurosurgeons had been demoted.

These changes were reflected in the Code of Federal Regulations [94] and were "based on recommendations submitted by the Advisory Committee on the Medical Uses of Isotopes [ACMUI]," a standing committee of the NR [93]. The crucial meeting of the ACMUI was held on June 21, 2002, and included representatives for radiation oncologists, health physicists, interventional and nuclear cardiologists, radiologists, nuclear medicine physicians, and nuclear pharmacists—but not for neurosurgeons. Prior to finalizing these changes, on December 4, 2003, the NRC called for comments regarding these changes as they do for all such decisions, specifically for rule RIN 3150-AH19 [95]. One of the comments came as a joint letter on February 23, 2004, from the neurosurgeon L. Dade Lunsford and the executive director of the International RadioSurgery Association (IRSA), Rebecca Emerick. The letter stated:

> *Authorized user for gamma stereotactic radiosurgery should specifically include the neurosurgeon*…gamma radiosurgery...requires a *unique understanding and education in the neuroanatomy of the brain...A neurosurgeon is the only trained physician with this understanding*…IRSA strongly recommends that neurosurgeons be a required authorized user of the gamma stereotactic radiosurgery treatment. To operate without the neurosurgeon as an authorized user is negligent…*Written directives for gamma stereotactic radiosurgery, should require both the signature of the treating neurosurgeon and the radiation oncologist…The neurosurgeon should be required to be physically present throughout all patient treatments involving the gamma unit. Both the neurosurgeon and the radiation oncologist should be required to be present during the initiation of the treatment.* (italics mine) [96]

Responses to comments were published in the Federal Register [97], which summarized one comment as follows:

> One commenter stated that *AUs should be required to be neurosurgeons for use of gamma stereotactic radiosurgery treatments because a neurosurgeon is the only trained physician who has the knowledge unique to understanding the neuroanatomy of the brain.* The commenter also suggested other changes to regulations, including [the requirement that] *WDs for gamma stereotactic radiosurgery be signed by both a treating neurosurgeon and radiation oncologist and that a neurosurgeon should be required to be physically present during treatments involving the gamma unit with the radiation oncologist also present during the initiation of treatment.* (italics mine)

Comparison of the italicized phrases in these two excerpts suggests that the comment discussed in the Federal Register was that of Dr. Lunsford and Ms. Emerick. Importantly, although the letter requested that neurosurgeons should be AUs, the response in the Federal Register was to incorrectly interpret the request to mean, "AUs should be required to be neurosurgeons." In other words, the Federal Register got it backward, stating "it would be an unwarranted intrusion into the practice of medicine to specify that only neurosurgeons may serve as AUs" [97]. It is surprising that entities engaging in details as thoroughly as the Federal Register and the NRC could overlook an interpretation resulting in an important comment remaining unanswered.

This decision was a major blow to the boundary work of neurosurgeons, because it bestowed authority during every Gamma Knife case to the radiation oncologist and eliminated any chance of neurosurgeons being authorized users for CyberKnife cases. The irony of the situation was not missed by neurosurgeon Jason Sheehan who wrote "This meant that my colleague Ladislau Steiner could not serve as the authorized user of the Gamma Knife when treating a patient with an arteriovenous malformation despite having done so first and exhibiting more than 30 years of experience" [98]. Dr. Steiner was a colleague of Leksell and a key figure in the development of Gamma Knife radiosurgery. But regardless, the radiosurgical boundary had been irrevocably pushed toward the advantage of radiation oncology, and the decision marked an ongoing trend to marginalize the role of neurosurgeons in radiosurgery.

2.11 The Promise and Threat of Hypofractionation

By 2004, clinical experience and technical advances had led to the acceptance of hypofractionation as a valid option by both neurosurgeons and radiation oncologists. This conceptual shift was important to LINAC centers because they could easily incorporate the new dose schedules but was particularly favorable to the CyberKnife effort for several reasons. Because of its reputation for frameless precision, the CyberKnife offered the comfort and flexibility of LINAC technology with a precision arguably exceeding LINAC devices and rivaling that of the Gamma Knife [64]. Its use for hypofractionation therefore seemed like a natural choice. Furthermore, the CyberKnife experience with high-dose radiosurgery for intracranial lesions was favorable, lessening the threat of morbidity due to hypofractionation Finally, many LINAC centers did not aggressively offer hypofractionation because of fears of disruption of their throughput and because they were "frequently…discouraged by the limited scope of conditions…resulting small volume of procedures" [67].

On the other hand, the acceptance of hypofractionation disrupted the uneasy peace that had been forged between neurosurgeons and radiation oncologists for Gamma Knife procedures. Extending treatments beyond a single fraction meant the neurosurgical use of stereotactic radiation could no longer be viewed as a type of surgery done with radiation and could no longer be seen as qualitatively different from conventional radiotherapy. Instead, the adoption of hypofractionation by neurosurgeons represented a serious invasion into radiation oncology territory. Would neurosurgeons stop at 3 or 5 fractions, or would their dosage schedules creep toward conventional fractionation? And would hypofractionation serve to siphon patients away from conventional radiotherapy, and if so, what would be the financial and political effects? Did hypofractionation represent a slippery slope injurious to radiation oncology?

These issues placed the CyberKnife at the center of an intense controversy because of the major role it would play in hypofractionated treatment. The conflict came to be embodied in disagreements about definitions. The use of the word "radiosurgery" had been reserved for single-fraction treatment, and continuing to do so would not threaten existing boundaries. But the use of "stereotactic radiosurgery" was disturbing, not only because it was a misnomer but also because it could be interpreted as giving neurosurgeons an open license for fractionation. How activities were defined had enormous implications for the roles of neurosurgeons and radiation oncologists, reimbursement, and control. There was an urgent need for definitions that would determine exact limits of practice, even if those limits were arbitrarily chosen.

Several neurosurgical articles address these controversies [99–101]. One author noted the concern for "the potential erosion of neurosurgery's…role in radiosurgery" and feared that developments "will soon make conformal radiation the exclusive tool of radiation therapy departments" [100]. These discussions resulted in a second consensus statement by ASTRO, the American Association of Neurological Surgeons, and the Congress of Neurological Surgeons, approved in 2006 [102, 103] that again required a team including a neurosurgeon, radiation oncologist, and medical physicist, but extended applications to include spinal lesions and defined "stereotactic radiosurgery" to be treatment delivered "in a limited number of sessions, up to a maximum of five." In theory, the difference between stereotactic radiosurgery and radiation oncology was to be based on whether the intent of treatment was ablation or the use of the "differential sensitivity of [tumors] to fractionated ionizing radiation" [102], but in practice it often became based on the number of fractions. One can argue that the choice of five fractions was arbitrary, but in any case, was likely influenced by the CyberKnife experience and protocols.

By 2010, radiation oncologists had become concerned that neurosurgical involvement and the new definition allowing radiosurgery to have up to five fractions was "usurping the radiation oncologist's traditional role in radiation delivery" [5]. One can speculate that these concerns drove radiation oncologists toward boundary work to their advantage. For example, through the work of powerful groups such as the NRC, boundaries were created as we have seen that excluded neurosurgeons by eliminating their permission to be authorized users. Furthermore, development of new technology such the CyberKnife and IMRT allowed the practice of radiosurgery by radiation oncologists to be more independent of neurosurgical involvement. Material boundary work was also in play against neurosurgeons because of the physical barriers imposed by the placement of the LINAC and CyberKnife devices within radiation oncology departments. Another example is a report of a patient brochure about radiosurgery created in 2008 by ASTRO that omitted any mention of neurosurgical participation [65]. Although later versions of the brochure included neurosurgeons due to efforts of neurosurgery leadership, this is further testimony to the energy devoted to competitive boundary work.

There was also a type of financial boundary work arising from changing practice conditions in the presence of a variety of "turf wars" [104]. When Gamma Knife was the prevailing modality, neurosurgical involvement was very similar from case to case and could be appropriately covered with a single CPT (61793). But as the CyberKnife and other LINAC systems became available, neurosurgical involvement became more intricate and variable. For the CyberKnife, for example, targets were more complex, requiring more time and effort for inverse planning and the possibility of more than one fraction or more than one separate stage. Neurosurgeons realized these changes mandated a new system of unbundled codes and worked to submit recommendations that were approved by the Relative Value Scale Update Committee (RUC). The Centers for Medicare and Medicaid Services, however, discarded the recommendations and reduced the reimbursement for the neurosurgical codes [105]. Soon after this decision, reimbursement to neurosurgeons dropped precipitously, putting at risk the entire body of neurosurgical participation in radiosurgery in general and CyberKnife in particular [105, 106]. Fortunately, neurosurgical reimbursement for radiosurgery was rectified in 2010 due to the efforts of organized neurosurgery, but the specter of competitive boundary work remains.

The neurosurgeons Peter Heilbrun and John Adler articulated these neurosurgical concerns in 2010, speaking of a "devaluation of neurosurgery" threatening to strain relationships with radiation oncologists and "irreparably undermine the practice of radiosurgery." They reported that by 2009 the radiation oncology community acquired favorable CPT codes that considerably increased their reimbursement but that the rejection of the neurosurgical recommendations by the RUC produced a significant decrease in neurosurgical reimbursement [5]. They pointed out that the net effects of financial and other types of boundary work raised the question of whether the "ongoing lobbying efforts" of radiation oncologists will "result in the elimination of all surgeons as partners" and whether rules will change so that "neurosurgeons are not really needed to perform radiosurgery."

Subsequent guidelines published in the radiosurgery literature described the neurosurgeon as "an integral member of the team" but assigned the supervisory duty of "overseeing radiation therapy management of the patient" solely to the radiation oncologist [107]. These statements indicate that the role of the neurosurgeon has changed from that of the group leader to that of a participant and is a good example of boundary work that alters the boundary to benefit one group while excluding the other. These changes are reflected by a recent trend for radiation oncologists to perform stereotactic radiosurgery for intracranial lesions without neurosurgical involvement. For example, a survey of 567 radiation oncologists who treated brain tumors with stereotactic surgery showed that only 44% of these physicians always included neurosurgeons, 36% involved neurosurgeons selectively, and 20% never included neurosurgeons in their cases [108].

These changes represent a marginalization of neurosurgeons from radiosurgery that is also driven by other factors. Increasing demand for surgical productivity impairs the neurosurgeon's interest and ability to spend time in the radiosurgical suite and in many cases prevents their participation in the complex inverse planning often required for CyberKnife treatment. Radiation oncologists naturally fill the void, and the boundary shifts away from neurosurgery. Another example is the importance to the hospital of the technical fees charged for CyberKnife treatment. In some cases, this interest promotes a bond between the hospital and radiation oncologists, who are seen as commanding the radiosurgery facility, leading to an additional boundary shift.

I have argued that neurosurgery and radiation oncology at times behave as tribes and that the political, financial, and cultural events surrounding the CyberKnife can be understood as competitive boundary work. A complementary way to understand this history is to examine individual exchanges between physicians in practice. In the next section, I will discuss examples of these "micro-interactions" and their relationship to conflict and boundary work.

2.12 Personal Experience and Anecdotes

In this section I describe a selection of unfortunate interactions among neurosurgeons and radiation oncologists, based on my experiences and the credible reports of others. They are not always pleasant to read, but they must be understood if the battles they represent are to be avoided. For reasons outlined earlier, I have kept the details anonymous.

Although the Gamma Knife and the CyberKnife were championed by neurosurgeons, both technologies ironically encountered resistance from the neurosurgical community when they were introduced into clinical work. In the 1980s, I remember how neurosurgeons snickered in private after presentations of Gamma Knife radiosurgery for arteriovenous malformations. And during the mid-1990s, when the CyberKnife was being unveiled as a clinical tool, I recall John Adler meeting palpable resistance as he addressed a neurosurgical audience. Although the Gamma Knife had been accepted by that time, neurosurgeons were skeptical that CyberKnife technology would be feasible in practice and concerned about costs. The impact that hypofractionation would have on the treatment of brain and spine tumors was not yet appreciated, nor was it understood that the CyberKnife would enable neurosurgery to push into radiation oncology territory (neurosurgeons were not the only community to voice resistance, and I recall how a radiation oncologist literally stood up at a conference to declare that radiosurgery was "malpractice"). Resistance of this type from neurosurgeons can be viewed as boundary work designed to prevent radiotherapy from replacing craniotomies, thereby reducing neurosurgical case volume [64]. Likewise, resistance from radiation oncologists can be viewed as boundary work designed to prevent the loss of new technology to another specialty.

Resistance was occasionally due to misinformation, as when one hospital committee opposed a CyberKnife instillation because "what if, God forbid, that someone has to crack a chest down there?" Eventually, however, boundaries shifted, and as one neurosurgical review summarized, "after many years of initial resistance by many in our specialty, radiosurgery methods are now essential [for] neurosurgical practice" [65].

Disagreements over turf have strongly colored the interactions between neurosurgeons and radiation oncologists. In a historical narrative of the Gamma Knife, the physician Dan Leksell, who has played a major role in the development of the Gamma Knife, labeled these disputes as a "World War of SRS [stereotactic radiosurgery]" and reported they "became the fiercest battles in the history of SRS politics" [109]. Perhaps the most disruptive "turf war" occurred when hospitals began building radiosurgery facilities at the urging of neurosurgeons. Although these centers promised higher revenue to the neurosurgeons and lucrative technical fees to the hospital, many radiation oncologists saw these new centers as a threat to their practice. Radiation oncologists were of course needed, but at times their participation was all but forced upon them. Later in the process, salt may have been rubbed into these wounds if neurosurgeons were perceived as believing that radiosurgery was "easy", or if they were seen as desirous of control. In the language of boundary work, the establishment of a radiosurgery facility in many cases represented a broach of the already uneasy boundary between neurosurgery and radiation oncology. The reaction of the radiation oncology community, not unexpectedly, was to fiercely defend their boundaries. It is from this struggle that I drew many of the experiences discussed in this section.

The boundary is occasionally explicitly named and discussed, as when a group of radiation oncologists complained to a group of radiosurgical physicians that they were extremely upset because their "book of work" had been violated. But the boundary is more commonly defined by actions, as in the anecdote I described earlier in which my scheduled radiology talk was cancelled by a group of radiation oncologists when they perceived that the talk might position me as an expert, and thereby threaten their authority over radiation therapy. Another such example discussed earlier was the ASTRO pamphlet that initially did not mention neurosurgeons.

A more complex example is a report of a neurosurgeon who organized an IRB-approved study of CyberKnife, in collaboration with radiation oncologists, physicists, and neurologists. Without speaking to these investigators, a radiation oncologist wrote to a handful of colleagues nationwide to ask their opinion of the study. The radiation oncologist then sent a formal letter to the investigators and their chairmen, reporting that the unanimous opinions voiced in this unsolicited survey was that the study was too dangerous even if approved by the IRB. The investigators were shocked, and could not recall a time when responsible, multidisciplinary research was deemed too dangerous to even consider. The study was nevertheless completed, demonstrated that the treatments were effective and safe, presented at international meetings, and published in peer-reviewed journals. Some believed the letter was boundary work intended to shift control of CyberKnife practice to the radiation oncology department at that institution.

Another example of boundary defense was seen when a clinical study written by neurosurgeons, radiation oncologists, and neurologists was submitted for publication. A radiation oncologist, different from the physician mentioned above and not involved with the study, contacted the authors. The radiation oncologist told them that the study did not meet the standards of the radiation oncology department and demanded that the study be retracted. The issue was complex because some (but not all) of the authors had left the institution at which the data had been gathered. When the authors refused to retract the paper, the radiation oncologist responded using language that could be interpreted as a threat to file a lawsuit. The threat was ignored, the paper was published in a peer-reviewed journal, and no further action occurred. Regardless of one's opinion of what was or was not appropriate, it is clear that this interaction was less about clinical disagreements than it was about boundary work over issues of control.

Another example of boundary defense was displayed in a discussion between a neurosurgeon and two radiation oncologists about the dose plan for a CyberKnife treatment of a malignant, circumscribed brain tumor. The neurosurgeon wanted to extend the treatment isodose line a few millimeters into the adjacent bone, a common technique to boost dosage to the tumor itself. In this case the volume of bone was small and late effects not likely due to the patient's expected life span. The radiation oncologists refused—and would not budge—repeatedly stating "we give as little dose as possible to normal tissue" as their sole explanation. The tension in the room was palpable. Regardless of which opinion one favors, this event was a clear boundary defense even if layered upon clinical concern.

Similar observations were reported by the medical physicist Jeff Fiedler, when he wrote that neurosurgeons and radiation oncologists "occasionally aggressively compete among themselves to establish who's top doc" ([6], 159). And another example of boundary defense occurred when a radiation oncologist treated a patient with the CyberKnife without neurosurgical involvement shortly after the CyberKnife was installed (a case done by a neurosurgeon without radiation oncology involvement has been described [40], but this is an example of an attempt at boundary creation).

Other examples include attempts to limit referrals to a CyberKnife center in favor of a competing traditional radiotherapy center without regard for clinical benefits, to refer selectively to a CyberKnife center in the setting of a financial conflict of interest, or to argue for or against CyberKnife treatment at tumor boards based on non-clinical issues. It is important for me to say that I have seen these occurrences committed by both neurosurgeons and by radiation oncologists.

Boundary work is also seen when a CyberKnife is placed within a radiation oncology center distant from the neurosurgical offices, defending the radiation oncology boundary by making it difficult for neurosurgeons to participate in the construction of complicated treatment plans requiring several iterations. An exception was seen at a center in which a large number of patients were referred for radiosurgery by the neurosurgeon. This improved the interactions enough to balance the effects of the location of the radiosurgery center and ensure a peaceful coexistence [42].

Contesting a boundary was evident during a discussion of a radiosurgery case, when a radiation oncologist reminded a neurosurgeon that "you can never, ever take radiation back" once it is given. When the neurosurgeon did not respond, the radiation oncologist replied that of course, neurosurgeons know about that kind of risk. I have witnessed other examples of contesting boundaries, including many tense, angry, and prolonged discussions between radiation oncologists and neurosurgeons regarding dose (sometimes "almost to…fistfights" as mentioned at the beginning of this chapter). And I have seen equally tense arguments over dose increments as small as 0.5 Gy. The issue here was clearly not that of clinical science as much as it was over who was in command.

Another example of contesting the boundary occurred when a radiation oncologist began a study comparing the anatomical outlines drawn for radiosurgical planning by the radiation oncologists to those drawn by the neurosurgeons. The study was never completed, but it is possible that its intent was to eliminate the need for neurosurgeons by showing that their anatomic outlines were not substantially different than those of the radiation oncologists. A final example of contesting boundaries is the report of a neurosurgeon who was ostracized from a radiosurgery center based on the neurosurgeon's desire to be involved with treatment planning as well as dose selection.

These vignettes seem at first glance to be examples of egregious behavior of physicians, battling over money and power with little thought for patient care. But this is not the case. Rather than random hostilities, these stories represent boundary work between two distinct medical specialties, each admittedly motivated by competitive tribal forces to protect their turf. More importantly, each is driven, as I have argued, by a unique moral system arising from their tribal beliefs and culture. These systems are different, but neither is objectively superior to the other. Each group sees themselves as heroes, upholding the sanctity of their profession and their commitment to patient care in different ways. Unfortunately, it is not uncommon for each side to be oblivious to the different moral compass guiding their opposing colleagues. The radiation oncologist who refuses to allow the treatment isodose line to overlap into the bone by even 1 mm is not merely provoking the neurosurgeon, but, rather, is preserving the boundaries of his or her specialty and providing the safest treatment as defined by the beliefs and morals of the field of radiation oncology. Likewise, the neurosurgeon who treats four lesions without support from the literature is providing a lifesaving, aggressive plan as defined by the beliefs and morals of the field of neurosurgery. My aim of providing these stories is to suggest that understanding the opinions of a colleague as rooted in the culture and morals of that colleague's field goes a long way toward defusing what would otherwise continue as a hostile interaction.

Finally, here I must apologize for my bias as a neurosurgeon. An observant reader will notice that my experiences and recollections of others' anecdotes often paint the radiation oncologist as the bad guy. It is not always so in reality, and I know my radiation oncology colleagues have credible experiences implicating me and other neurosurgeons as the villains.

2.13 The Problem of Incivility

As illustrated in the last sections, interactions between members of the CyberKnife team can be intense and fraught with meaning. Unfortunately, these conditions are conducive to a belligerence and hostility that can be described as incivility. Defined as "rude or unsociable speech or behavior," incivility is not limited to radiosurgery. It is ubiquitous among medical professionals, has been well studied, and is destructive to patient care. Examples include surgeons who accuse internists of "hand holding" [12] and internists denigrating orthopedic surgeons when they refer to patients on the orthopedics ward as FOOBA—"found on orthopedics barely alive" [110]. Origins of this behavior include issues of communication and stress [35], empowerment [111], authority [112], professional cultures [34], differences of medical opinion [113], and diversity [114]. One can speculate that medical incivility is driven in part by the perception of medical decisions as moral battles fought by medical heroes, as discussed in previous sections. Discriminating between incivility and appropriate but serious discourse is subjective, and so rather than pointing out what I believe are examples in the stories of the last few sections, I invite the readers to judge for themselves and assure them that examples abound.

2.14 Quo Vadis?

Neurosurgeons have been concerned for many years that their role in radiosurgery could eventually vanish due to influence wielded by radiation oncology [100] and from pressure to expand their conventional neurosurgical practices. I recall Ladislau Steiner sharing his concerns for this trend, showing me letters he had received from radiosurgeons around the world who agreed that there was a problem. The many responses to an article addressing the "devaluation of radiosurgery and its impact on the neurosurgery-oncology partnership" are further testimony to these worries [5]. Likewise, many radiation oncologists have been concerned from the beginning of radiosurgery that their field was being invaded and that their livelihood and professional standing were threatened. I have argued that for decades, both fields have been actively pursuing boundary work to enhance their position within an ongoing contest. One can ask, which field was more effective? Where has this brought us, and what are the consequences?

There is no clear answer, but my opinion is that the boundary work of radiation oncology has been more effective than that of neurosurgery. For example, the exclusion of neurosurgeons from the role as authorized users shifted power to radiation oncologists, demoting neurosurgeons to participants who are being increasingly marginalized [108]. Furthermore, new radiation devices capable of stereotactic precision are available that reside in radiology departments, under the control of radiation oncologists, moving the boundary even further from neurosurgery. And while many neurosurgeons are dedicated to radiosurgery, I have observed how many participate in less meaningful ways, arriving for the procedure solely to agree with an already existing plan and dose before vanishing to the operating room. The study discussed earlier designed to show that the anatomical boundaries drawn by neurosurgeons are no different than those drawn by radiation oncologists was never completed but is also a somber warning of diminution to come.

Perhaps this trend is inevitable. But as a neurosurgeon, I don't think it is for the best. It is often said that neurosurgeons are crucial to radiosurgery because of their expertise in anatomy. And many believe that neurosurgeons bring an approach to risk—an assessment of efficacy vs. safety when facing a lethal disease—that is different from, but counterpoint to, that of the radiation oncologist. I think both are true, but my opinion is that the neurosurgical talent that matters most is the feel for how the patient will respond to injury to specific areas in the brain. What deficits will arise from ablation of region around the radiosurgical target? How likely are they to be permanent or serious? This knowledge determines the value of the anatomical outlines and informs the degree of risk of the radiosurgical plan. It is the neurosurgeon who best understands the limits, accrued through years of resecting such tissue, touching it, following its changes through imaging, watching how postoperative lesions affect neurological function, observing the patient's degree of recovery, getting a feel for how much the tissue can take and at what cost. It is my (biased!) opinion that only neurosurgeons have this knowledge, and that their marginalization from the CyberKnife suite would be tragic for patients and for the field of radiosurgery.

2.15 Conclusion

My interest in the history of the CyberKnife and issues of conflict was triggered by my experience with numerous turf wars, skirmishes, battles, petty incivilities, and personal attacks over CyberKnife practice that I witnessed, heard about, and—to be honest—participated in, over many years. It would be easy to believe these contentious events arose from misguided or malevolent personalities, or from financial greed and political avarice. But I argue instead that these interactions are part of a more general and less personal process of boundary work between neurosurgery and radiation oncology. Although this work finds motivation from understandable issues of finances and control, I argue for the importance of a more powerful but largely ignored factor: the divergence of the two fields as their development focused on different domains of practice, resulting in two isolated groups with unique cultures and priorities. With isolation came beliefs that can be described as tribal. With tribal thinking came a moral system unique to each specialty, and with a moral system came an obligation to champion a hero's cause. That cause, fueled by the difference in perceived moral obligations, has in many cases lent an unfortunate intensity, vehemence, and emotional valence to the disagreements between specialties. My hope is that understanding these disappointing interactions as products of differences in culture and morality, rather than as hostilities based on absolute definitions of right and wrong, will promote the collegiate and nurturing relationships so essential for patient care and so necessary for our own peace of mind.

References

1. Adler JR Jr. Accuracy, Inc.: a neurosurgical business case study. Cures. 2009;1(9):e1. https://doi.org/10.7759/cureus.1.
2. Slavin KV. Disruptive innovation concept. Stereo tact Funct Neurosurg. 2012;121:10. https://doi.org/10.1159/000336475.
3. Giller CA, Berger BD, Pistenmaa DA, Sklar F, Weprin B, Shapiro K, et al. Robotically guided

radiosurgery for children. Pediatr Blood Cancer. 2005;45:304–10.
4. Mould RF, editor. Robotic radiosurgery, vol. 1. Sunnyvale, CA: CyberKnife Society; 2005.
5. Heilbrun MP, Adler JR Jr. The 2009 devaluation of radiosurgery and its impact on the neurosurgery-radiation oncology partnership. J Neurosurg. 2010a;113:10–5.
6. Giller CA, Fiedler JA, Gagnon GJ, Paddick I. Radiosurgical planning: gamma tricks and cyber picks. Hoboken, NJ: Wiley-Blackwell; 2009.
7. Ludmerer KM. Let me heal: the opportunity to preserve health in American medicine. Oxford: Oxford University Press; 2014.
8. Gross SD. Surgery. In: Clarke EH, Bigelow HJ, Gross SD, Thomas TG, Billings JS, editors. A century of American medicine: 1776–1876. Philadelphia: Henry C. Lea; 1876.
9. Rosenberg CE. The care of strangers: the rise of America's hospital system. Baltimore: The Johns Hopkins University Press; 1987.
10. Weisz G. Divide and conquer. Oxford: Oxford University Press; 2006.
11. Wunderlich KRA, quoted in Jan Goldstein. Psychiatry. In: Bynum WF, Porter R, editors. Companion encyclopedia of the history of medicine, vol. 2. London and New York: Routledge; 1993.
12. Cassell J. Expected miracles: surgeons at work. Philadelphia: Temple University Press; 1991.
13. Bartlett W. Records and charts. In: Bartlet W, editor. The after-treatment of surgical patients, vol. 1. St. Louis: CV Mosby; 1921. p. 7–14.
14. Kisacky J. Rise of the modern hospital: an architectural history of health and healing. Pittsburgh: Pittsburgh University Press; 2017.
15. Adams A. Medicine by design: the architect and the modern hospital 1893–1943. Minneapolis: University of Minnesota Press; 2008.
16. McMahon LF Jr, Howell JD. The hospital: still the doctors' workplace(s): a cautionary note for approaches to safety and value improvement. Health Serv Res. 2018;53(2):601–7.
17. Bliss AA. Blockley days: memories and impressions of a resident physician 1883–1884. Private printing;1916.
18. Rankin JS. William Stewart Halsted: a lecture by Dr. Peter D. Olch. Ann Surg. 2006;243:418–25.
19. Manning FJ. Morale and cohesion in military psychiatry. In: Military Psychiatry: Preparing in Peace for War. chapter 1, ed. Franklin D. Jones, Linette R. Sparacino, Victoria L. Wilcox, Joseph M. Rothberg. Washington, DC: Office of the Surgeon General, Department of the Army; 1994.
20. Quinn N, Mathews HF. Emotional arousal in the making of cultural selves. Anthropological Theory. 2016;16:359–89.
21. Rondfeldt D. In search of how societies work: tribes—the first and forever form. Prepared for Rand Pardee Center, WR-433-RPC. December, 2006:29–39. https://www.rand.org/pubs/working_papers/WR433.html.
22. Tajfel H. Experiments in intergroup discrimination. Sci Am. 1970;223:96–103.
23. Bosk C. Forgive and remember: managing medical failure. 2nd ed. Chicago: University of Chicago Press; 2003.
24. Osler W. Remarks on specialism. Boston Med Surg J. 1892;126:457–9.
25. Woodbridge P. Tribalism can trump healthcare improvement. Ind Syst Eng Mag Work. 2018;50(4):22.
26. Hock LK. Tribalism and organised [sic] medicine. College Mirror. 2013;39:4–23.
27. Cova B, Cova V. Tribal aspects of postmodern consumption research: the case of French in-line roller skates. J Consum Behav. 2001;1(1):67–76.
28. Mason L, Wronski J. One tribe to bind them all: how our social group attachments strengthen partisanship. Adv Political Psychol. 2018;39(S1):257–77.
29. Dionísio P, Leal C, Moutinho L. Fandom affiliation and tribal behaviour: a sports marketing application. Qual Mark Res Int J. 2008;11(1):17–39.
30. Wilson EO. The social conquest of earth. New York: Liveright Publishing Corporation; 2012. p. 57.
31. Bohm R, Rusch H, Baron J. The psychology of intergroup conflict: a review of theories. J Econ Behav Organiz. 2018; https://doi.org/10.1016/j.jebo.2018.01.020.
32. Levy P. Healthcare's tribalism—and how to combat it. Athenainsight. 2017:1–7. https://www.athenahealth.com/insight/paul-levy-healthcare-tribalism-and-how-combat-it.
33. Purdy E. Professional socialization, tribalism and career trajectories. 2018. https://onthewards.org/professional-socialization-tribalism-and-career-trajectories/.
34. Braithwaite J, et al. The basis of clinical tribalism, hierarchy and stereotyping: a laboratory controlled teamwork experiment. BMJ Open. 2016:1–10. https://doi.org/10.1136/bmjopen-2016-12467.
35. Azoulay E, Timsit JF, Sprung CL, Soares M, Risinava K, Lafabrie A. Prevalence and factors of intensive care unit conflicts: The Conflicus Study. Am J Respir Crit Care Med. 2009;180:853–60.
36. Bradley V, Liddle S, Shaw R, Savage E, Rabbitts R, Trim C, et al. Sticks and stones: investigating rude, dismissive and aggressive communication between doctors. Clin Med. 2015;15(6):541–5.
37. Pattani R, Ginsburg S, Mascarenhas JA, Moore JE, Jassemi S, Straus SE. Organizational factors contributing to incivility at an academic medical center and systems-based solutions: a qualitative study. Acad Med. 2018;93(4):1569–75.
38. Larson DA, Flickinger JC, Loeffler JS. The radiobiology of radiosurgery. Int J Radiat Oncol Biol Phys. 1993a;25:557–61.
39. Lunsford DL, Niranjan A, Flickinger JC. The first North American clinical Gamma Knife center. Prog Neurol Surg Basel. 2019;34:9–18.

40. Phillips T, Sahgal A. An interview with David Larson. In About-ASTRO/History. 2016. https://www.astro.org/About-ASTRO/History/David-Larson. Accessed 20 April 2020.
41. Lunsford LD. The International Stereotactic Radiosurgery Society is a particular organization. J Radiosurgery BRT. 2011;1:77–83.
42. Langley A, Lindberg K, Mørk BE, Nicolini D, Raviola E, Walter L. Boundary work among groups, occupations, and organizations: from cartography to process. Acad Manag Ann. 2019;13:704–36.
43. Burri RV. Boundary work and symbolic capital in radiology. Soc Stud Sci. 2008;38:35–62.
44. Coutard H. Principles of X ray therapy of malignant diseases. Lancet. 1934;224:1–8.
45. Regato AJ. Claudius Regaud. Int J Radiat Oncol Biol Phys. 1976;1:993–1001.
46. Cox JD. Large-dose fractionation (hypofractionation). Cancer. 1989;55:2105–11.
47. Marcial VA. Time-dose-fractionation relationships in radiation therapy. Natl Cancer Inst Monogr. 1967;24:1887–203.
48. Lawrence JH. Proton irradiation of the pituitary. Cancer. 1957;10:795–8.
49. Larrson B, Leksell L, Rexed B, Sourander P, Mair W, Andersson B. The high-energy proton beam as a neurosurgical tool. Nature. 1958;182:1222–12233.
50. Kjellberg RN, Koehler AM, Preston WM, Sweet WH. Stereotactic instrument for use with the Bragg peak of a proton beam. Confin Neurol. 1962;22:183–9.
51. Fabrikant JI, Lyman JT, Frankel KA. Heavy charged-particle Bragg peak radiosurgery for intracranial vascular disorders. Radiat Res. 1985;104(Suppl 9):S244–58.
52. Leksell L. The stereotactic method and radiosurgery of the brain. Acta Chir Scand. 1951;102:316–9.
53. Leksell L. Stereotaxis and radiosurgery. An operative system. Springfield, IL: Charles Thomas; 1971.
54. Ganz JC. The history of the Gamma Knife. Prog Brain Res. 2014;215:2–136.
55. Betti OO, Derechinsky YE. Irradiations stéréotaxiques multifaisceaux. Neurochirugie. 1983;29: 295–98.
56. Lutz W, Winston KR, Maleki N. A system for stereotactic radiosurgery with a linear accelerator. Int J Radiat Oncol Biol Phys. 1988;14:373–81.
57. Hartmann GH, Schlegel W, Sturm V, Kober B, Pastyr O, Lorenz WJ. Cerebral radiation surgery using moving field irradiation at a linear accelerator facility. Int J Radiat Oncol Biol Phys. 1985;11:1185–92.
58. Larson DA, Gutin PH, Leibel SA, Phillips TL, Sneed PK, Wara WM. Stereotaxic irradiation of brain tumors. Cancer. 1990;65(3 Suppl):792–9.
59. Friedman WA, Bova FJ. The University of Florida radiosurgery system. Surg Neurol. 1989;32:334–42.
60. Colombo F, Benedetti A, Pozza F, Avanzo RC, Marchetti C, et al. External stereotactic irradiation by linear accererator. Neurosurgery. 1985;16:154–60.
61. Podgorsak EB, Olivier A, Pla M, Lefebvre PY, Hazel J. Dynamic stereotactic radiosurgery. Int J Radiat Oncol Biol Phys. 1988;14:115–26.
62. Hariz MI, Henrikkson R, Löfroth PO, Laitinen LV, Säterborg NE. A non-invasive method for fractionated stereotactic irradiation of brain tumors with linear accelerator. Radiother Oncol. 1990;17:57–72.
63. Schwade JG. Introduction. Semin Radiat Oncol. 1995;5:173–4.
64. Knisely JPS, Apuzzo MLJ. Historical aspects of stereotactic radiosurgery: concepts, people, and devices. World Neurosurg. 2019;130:593–607.
65. Heilbrun MP, Adler JR Jr. Radiosurgery and radiation oncology. J Neurosurg. 2010;113:9.
66. Solberg TD, Siddon RL, Kavanagh B. Historical development of stereotactic ablative radiotherapy. In: Lo SS, The BS, Lu JJ, Schefter TE, editors. Stereotactic body radiation therapy. Berlin: Springer-Verlag; 2012. p. 9–35.
67. Schwade JG, Wolf AL. Future trends in radiosurgery. Semin Radiat Oncol. 1995;5:246–50.
68. Brenner DJ, Martel MK, Hall EJ. Fractionated regimens for stereotactic radiotherapy of recurrent tumors in the brain. J Int Radiat Oncol Biol Phys. 1991;21:819–24.
69. Hall EJ, Brenner DJ. The radiobiology of radiosurgery: rationale for different treatment regimes for AVMs and malignancies. Int J Radiat Oncol Biol Phys. 1993;25:381–5.
70. Larson DA, Bova F, Eisert D, Kline R, Loeffler J, Lutz W, et al. Current radiosurgery practice: results of an ASTRO survey. J Radiation Oncology Biol Phys. 1993b;28:523–6.
71. Souhami L, Olivier A, Podgorsak EB, Villemure JG, Pla M, Sadikot AF. Fractionated stereotactic radiation therapy for intracranial tumors. Cancer. 1991;68:2101–8.
72. Olivier A, Sadikopt AF, Villemire JG, Pokrupa R, Souhami L, Podgorsak EB, Hazel J. Fractionated stereotactic radiotherapy for intracranial neoplasms. Stereo tact Funk Neurosurg. 1992;59:193–8.
73. De Salles AA, Hariz M, Bajada CL, Goetsch S, Bergenheim T, Selch M, Holly FE, Solberg T, Becker DP. Comparison between radiosurgery and stereotactic fractionated radiation for the treatment of brain metastases. Acta Neurochir Suppl (Wein). 1993;58:115–8.
74. Laing RW, Warrington AP, Graham J, Britton J, Hines F, Brada M. Efficacy and toxicity of fractionated stereotactic radiotherapy in the treatment of recurrent gliomas (phase I/II study). Radiother Oncol. 1993;1:22–9.
75. Podgorsak EB, Souhami L, Caron J-L, Pla M, Clark B, Pla C, Cadman P. A technique for fractionated stereotactic radiotherapy in the treatment of intracranial tumors. Int J Radiat Oncol Biol. Physics. 1993;27:1225–30.
76. Marks LB, Halperin EC. Radiosurgery is not standard of care for solitary brain metastases. Letter. Int J Radiat Oncol Biol Phys 1995;32:557–58.

77. Sperduto PW, Hall WA. Radiosurgery, cost-effectiveness, gold standards, the scientific method, cavalier cowboys and the cost of hope. Int J Radiat Oncol Biol Phys. 1996;36:511–3.
78. Mehta MP, Rozental JM, Levin AB, Mackie TR, Kubsad SS, Gehring MA, Kinsell TJ. Defining the role of radiosurgery in the management of brain metastases. Int J Radiat Oncol Biol Phys. 1992;24:619–25.
79. Adler JR, Cox RS, Kaplan I, Partin DP. Stereotactic radiosurgical treatment of brain metastases. J Neuro-Oncol. 1992;76:444–9.
80. Adler JR Jr. Indications and limitations of stereotactic radiosurgery. West J Med. 1993;158(1):66.
81. Alexander E III, Moriarty TM, Davis RB, Wen PY, Fine HA, Black PM, Kooy HM, Loeffler JS. Stereotactic radiosurgery for the definitive, noninvasive treatment of brain metastases. J Natl Cancer Inst. 1995;87(1):34–40.
82. Hamilton AJ, Lulu BA, Fosmire H, Stea B, Cassady JR. Preliminary clinical experience with linear accelerator-based spinal stereotactic radiosurgery. Neurosurgery. 1995;36(2):311–9.
83. Loeffler JS, Shrieve DC, Wen PY, Fine HA, Kooy HM, Addesa AE, Black PM, Alexander E III. Radiosurgery for Intracranial malignancies. Semin Radiat Oncol. 1995;5(3):225–34.
84. Sarkaria JN, Mehta MP, Loeffler JS, Buatti JM, Chappell RJ, Levin AB, Alexander E 3rd, Friedman WA, Kinsella TJ. Radiosurgery in the initial management of malignant gliomas: survival comparison with the TROG recursive partitioning analysis. Radiation Therapy Oncology Group. Int J Radiat Oncol Biol Phys. 1995;23(4):931–41.
85. Shaw E, Scott C, Souhami L, Dinapoli R, Bahary J-P, Kline R, Wharam M, Schultz C, Davey P, Loeffler J, del Rowe J, Marks L, Fisher B, Shin K. Radiosurgery for the treatment of previously irradiated recurrent primary brain tumors and brain metastases: initial report of radiation therapy oncology group protocol 90-05. Int J Radiol Oncol Biol Phys. 1996;34(3):647–54.
86. Auchter RM, Lamond JP, Alexander E, Buatti JM, Chappell R, Friedman WA, Kinsell TJ, Levin AB, Noyes WR, Schultz CJ, Loeffler JS, Mehta MP. A multi-institutional outcome and prognostic factor analysis of radiosurgery for resectable single brain metastasis. In J Radiat Oncol Biol Phys. 1996;35(1):278–35.
87. Adler JR Jr, Chang SD, Murphy MJ, Doty J, Geis P, Hancock SL. The CyberKnife: a frameless robotic system for radiosurgery. Stereot Funct Neurosurg. 1997;69(1–4 Pt2):124–8.
88. Chang SD, Murphy M, Geis P, Martin DP, Hancock SL, Doty JR, Adler JR Jr. Clinical experience with image-guided robotic radiosurgery (the CyberKnife) in the treatment of brain and spinal cord tumors. Neuro Med Chir (Tokyo). 1998;38:780–3.
89. Rosenthal DI, Glatstein E. We've got a treatment, but what's the disease? or A brief history of hypofractionation and its relationship to stereotactic radiosurgery. Oncologist. 1996;1:1–7.
90. Lunsford LD, Flickinger JC, Larson D. Letter. Oncologist. 1997;2:59–61.
91. Anonymous. Consensus statement on stereotactic radiosurgery: quality improvement. Neurosurgery. 1994;34(1):193–5.
92. NRCa. Authorized user's supervision of medical programs. https://www.nrc.gov/about-nrc/radiation/protects-you/hppos/hppos145.html. Accessed 20 April 2020.
93. NRCb. Rulemaking issue notation vote SECY-02-0194, October 30, 2002. Attachment 1, Table 8 and Attachment 2. https://www.nrc.gov/reading-rm/doc-collections/commission/secys/2002/secy2002-0194/2002-0194scy.pdf#pagemode=bookmarks. Accessed 20 April 2020.
94. Nuclear Regulatory Commission. Medical use of byproduct material; final rule. 10 CFR Part 35. Federal Register. 2002;67(79):20371–20397. https://www.govinfo.gov/content/pkg/FR-2002-04-24/pdf/02-9663.pdf. Accessed 20 April 2020.
95. U.S. Nuclear Regulatory Commission. News Releases. NRC seeks public comment of proposed rule changes to specialty medical board certification criteria. No. 03-158. 2003;23(49):3–4. https://www.nrc.gov/docs/ML0333/ML033380825.pdf. Accessed 20 April 2020.
96. NRCc. Comment re: (RIN 3150-AH19) medical use of byproduct material—recognition of specialty boards. 2004. https://www.nrc.gov/docs/ML0405/ML040550440.pdf. Accessed 20 April 2020.
97. Nuclear Regulatory Commission. Medical use of byproduct material; final rule. 10 CFR Part 35. Federal Register. 2005;70(60):16354.
98. Sheehan J. Radiation oncology partnership and its impact on neurosurgery. J Neurosurg. 2010;113:3–4.
99. Kondziolka D, Lunsford LD, Loeffler JS, Friedman WA. Radiosurgery and radiotherapy: observations and clarifications. J Neurosurg. 2004;101:585–9.
100. Roberts DW. Radiosurgery and radiotherapy. J Neurosurg. 2004;101:573.
101. Pollock BE, Lunsford LD. A call to define stereotactic radiosurgery. Neurosurgery. 2004;55(6):1371–3.
102. Barnett GH, LInsky ME, Adler JR, Cozzens JW, Friedman WA, Heilbrun MP, et al. Stereotactic radiosurgery—an organized neurosurgery-sanctioned definition. J Neurosurg. 2007;106:1–5.
103. American Society for Therapeutic Radiology and Oncology: ASTRO News. Stereotactic radiosurgery definition. Int J Radiat Oncol Biol Phys. 2007;67:1280.
104. Sheehan J. Radiation oncology partnership and its impact on neurosurgery. J Neurosurg. 2020;113:3–4.
105. Wilson JA, Jacob RP. Stereotactic radiosurgery coding and reimbursement. J Neurosurg. 2010;113:6–8.
106. Kondziolka D. Toward the revaluation of radiosurgery. J Neurosurg. 2010;113:1–2.
107. Seung SK, Larson A, Galvin JM, Mehta MP, Potters L, Schultz CJ, et al. American College of Radiology

(ACR) and American Society for Radiation Oncology (ASTRO) practice guideline for the performance of stereotactic radiosurgery (SRS). Am J Clin Oncol. 2013;36:310–5.
108. Sandler KA, Shaverdian N, Cook RR, Kishan AU, King CR, Yang I, et al. Treatment trends for patients with brain metastases: does practice reflect the data? Cancer. 2017;123:2274–82.
109. Leksell D. The origins and development of radiosurgery and the Leksell Gamma Knife. Prog Neurol Surg Basel. 2019;34:1–8.
110. Goldman B. The secret language of doctors: cracking the code of hospital culture. Chicago: Triumph; 2014. p. 4.
111. Altaker KW, Howie-Esquivel J, Cataldo JK. Relationships among palliative care ethical climate, empowerment, and moral distress in intensive care unit nurses. Am J Clin Care. 2018;27:295–302.
112. Apresoa-Varano EC. A study of boundary work in the hospital. Sociol Perspect. 2013;86:1351–7.
113. Eiser AR, Eiser BJA. How physician leaders can successfully resolve team conflicts. Phys Leadership J. 2016;3:44–7.
114. Fortin S, Maynard S. Diversity, conflict and the recognition of hospital medical practice. Cult Med Psychiatry. 2018;42:32–49.

Part II

Physics

The CyberKnife Robotic Radiosurgery System

3

Argyris Moutsatsos and Evangelos Pantelis

3.1 Introduction

The CyberKnife® (CK) system (Accuray Inc., Sunnyvale, CA, USA) is a dedicated stereotactic radiosurgery (SRS) and stereotactic body radiotherapy (SBRT) unit conceived by Dr. John R Adler [1–4]. It was the first system to obviate the need of a mechanical fixation frame for intracranial SRS and SBRT treatments by utilizing a near real-time image-guided targeting system consisting of two cross-firing X-ray tubes [1, 3–5]. Treatment delivery is facilitated by a lightweight linear accelerator (LINAC) mounted on a robotic manipulator capable of performing movements with 6 degrees of freedom (6-DOF) to deliver many independently targeted (non-isocentric) and non-coplanar treatment beams with high precision [6–9]. This configuration coupled with intrafraction X-ray image guidance allows for the delivery of frameless SRS/SBRT treatments anywhere in the body upon indications with submillimetre accuracy [10–13].

A. Moutsatsos
CyberKnife and TomoTherapy Department, IATROPOLIS Private Special Clinic, Athens, Greece
e-mail: armouts@phys.uoa.gr

E. Pantelis (✉)
Medical Physics Laboratory, Medical School, National and Kapodistrian University of Athens, Athens, Greece

CyberKnife and TomoTherapy Radiotherapy Department, IATROPOLIS Private Special Clinic, Athens, Greece
e-mail: vpantelis@phys.uoa.gr

© Springer Nature Switzerland AG 2020
A. Conti et al. (eds.), *CyberKnife NeuroRadiosurgery*,
https://doi.org/10.1007/978-3-030-50668-1_3

3.2 A Brief CyberKnife Model History

CyberKnife was approved by the US Food and Drug Administration (FDA) for intracranial applications in 1999 and received clearance for radiosurgical treatment of lesions anywhere in the body, where radiation is indicated, in 2001. Since the first commercially available model, the CyberKnife system has undergone substantial technical developments and software upgrades which have enhanced its targeting and tracking accuracy, optimized treatment planning, improved dose calculation accuracy and extended the applicability of CyberKnife treatments to lesions throughout the body. A detailed description of the changes that have been applied to the system over the time is beyond the scope of this chapter and the reader is referred to relevant excellent reviews by Kilby et al. [6, 7]. This section presents a brief history of the CyberKnife system, mentioning the major advancements that came with each model (Fig. 3.1).

Replacing the prototype "Neurotron 1000" (Fig. 3.1) the second generation CyberKnife, released in 1997, introduced a new robotic arm (KUKA Roboter GmbH, Augsburg, Germany) to

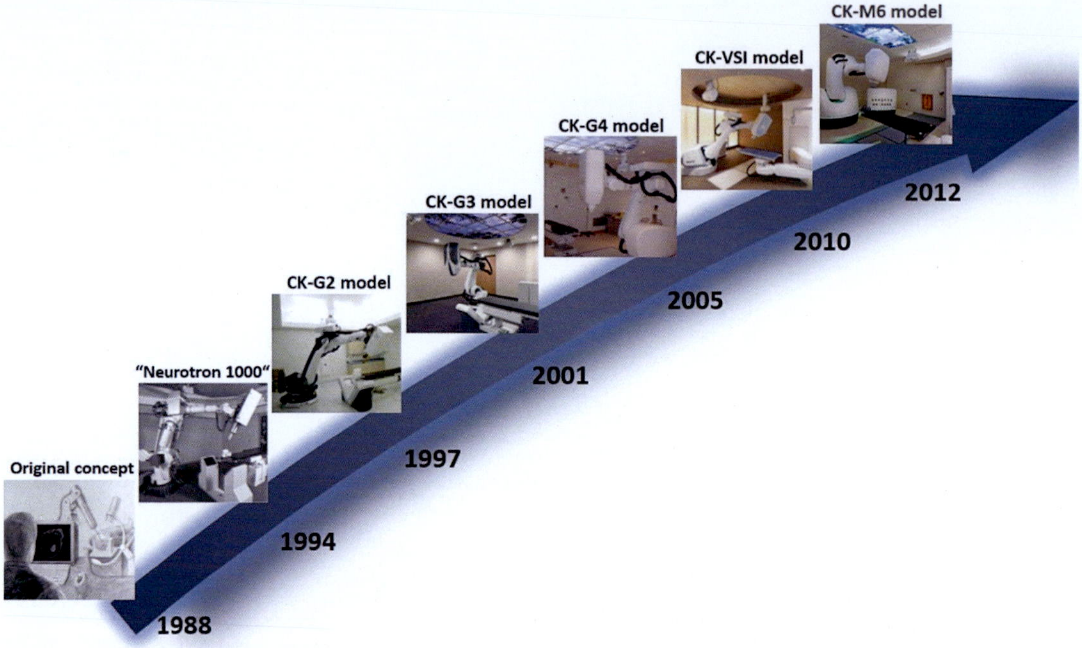

Fig. 3.1 Timeline arrow showing the evolution of the CyberKnife system from its inception on 1988 and its first name "Neurotron 1000" to the CyberKnife M6 model introduced in 2012

manoeuver the LINAC around the patient, a 400 MU/min LINAC, and replaced the fluoroscopic screen camera with high-resolution flat-panel amorphous silicon detectors. Image guidance was based on either three-dimensional (3D) tracking of the patient's skull (for intracranial treatments) or the apparent locations of fiducial markers implanted within or close to the tumour anatomy (for extracranial treatments). In 2001, the G3 model was launched, introducing a lot of advanced image-tracking algorithms that improved the system's delivery accuracy [14]. These included the "6D Skull Tracking" (6D stands for six degrees of freedom) enabling tracking of patient's skull in all six translational and rotational axes, the "XSight® Spine Tracking", facilitating fiducial-free spine tracking to treat lesions located within the spine or those fixed relative to it and the "Synchrony®" module, which coupled with fiducial tracking allowed for dynamic tracking of targets affected by the respiratory motion using an optical camera array to monitor the position of optical markers attached to the patient during all phases of the breathing cycle.

The major changes that came with the G4 model launched in 2005 were the upgraded LINAC with an MU rate of 600 MU/min and the replacement of the floor mounted flat panel detectors with new larger ones mounted flush to the floor, increasing the useful space around the patient. In the following years, until 2009, several system hardware upgrades were released including i) a redesigned LINAC with 800 MU/min rate containing an extra shielding ring to reduce leakage radiation, ii) the Iris™ variable aperture collimator together with the Xchange™ table (a pedestal lying close to the treatment manipulator which allows for automated exchange of the available secondary collimator systems via a pneumatic tool-changing mechanism attached to the LINAC), iii) new oil cooled X-ray tubes with X-ray generators allowing for accelerating potentials of up to 150 kV, and iv) a treatment couch guided by a 6-DOF robotic manipulator, referred to as Robocouch® (see Sect. 3.3). New tracking algorithms enabling spine lesion treatments with the patient in prone position and fiducial-less lung lesions (Lung Optimized Tracking-LOT) were also released [6].

All the above mentioned upgrades, together with a new 1000 MU/min LINAC and a second generation Iris collimator, were made available with the VSI model of the CyberKnife System released in 2010 [6]. At the time these lines are written, the latest member of the CyberKnife family is the M6 model released in 2012, while a new model, under the brand-name "S7", is anticipated on June 2020. Substantial differences relative to the VSI model include a new treatment manipulator robot with larger payload capability, a redesigned room layout to optimize the robot workspace and the introduction of a third secondary collimation system option consisting of a micro multi-leaf collimator (MLC) [15] to improve treatment efficiency and address larger treatment targets. The primary collimator and the beam monitoring system of the LINAC were also redesigned to support the larger primary beam needed for the MLC. The following sections present the major subsystems of the M6 CyberKnife model along with an overview of the treatment planning and delivery procedures.

Finally, besides hardware and robotics engineering advances, the vendor supplying treatment planning software (TPS) has also undergone substantial improvements from the original OnTarget™ system to the Multiplan® (2005) and the most recent Accuray Precision® platform, which have simplified treatment planning and improved dose calculation accuracy. A significant boost of the latter has been the implementation of a Monte Carlo (MC)-based dose calculation engine into both MultiPlan and Precision TPS platforms [16].

3.3 Major CyberKnife Subsystems

The CyberKnife system comprises a diverse combination of advanced technologies, including robotics, linear accelerator, medical imaging and software engineering to deliver frameless radiosurgery anywhere in the body. Figure 3.2 presents the M6 configuration of the CyberKnife system depicting the major subsystems enrolled in treatment delivery, which are analysed as follows.

3.3.1 The Treatment Head

The treatment head is mounted on the robotic manipulator and consists of the LINAC producing the X-ray treatment beam and the assembly of one of the three secondary beam collimation systems available. Already from the conception of the CyberKnife system, the LINAC should be able to deliver irradiation beams throughout the patient's body with high precision. This was addressed employing a compact lightweight LINAC design not requiring a bending magnet coupled with robotic manipulation. The CyberKnife M6 LINAC is powered by an X-band cavity magnetron using a standing wave, side-coupled accelerating waveguide to produce a 6 MV X-ray treatment beam delivering a dose rate of 1000 cGy/min at 15 mm depth inside water and 800 mm away from the source. The M6 LINAC is also flattening filter free (FFF), in favour of the total weight and the delivered dose rate provided the large number of beams comprising the treatment plan. This has an impact on the treatment beam spectrum, by increasing the low energy component, which in turn affects the beam quality specifier, k_{Q,Q_0}, of the reference dosimetry field defined by the fixed secondary collimator of 60 mm nominal diameter at 800 mm distance from the source [17] (for further information refer to Chaps. 6 and 7). Within the LINAC housing, a large tungsten enclosure comprises the primary collimator accommodating the X-ray target (i.e. beam source) and uses it to minimize radiation leakage in all directions apart from a fixed rectangular aperture defining the maximum possible treatment field size. Beneath lies a sealed, gas-filled ion chamber which monitors the treatment beam—in terms of MU delivered per beam, dose rate, beam uniformity and beam symmetry—and controls the dose delivered to the patient by signaling termination of irradiation when the planned monitor units have been reached. A laser-mirror assembly is also accommodated within the LINAC housing to direct a low-power optical (red) laser along a direction coinciding with the treatment beam central axis. This laser is employed for quality

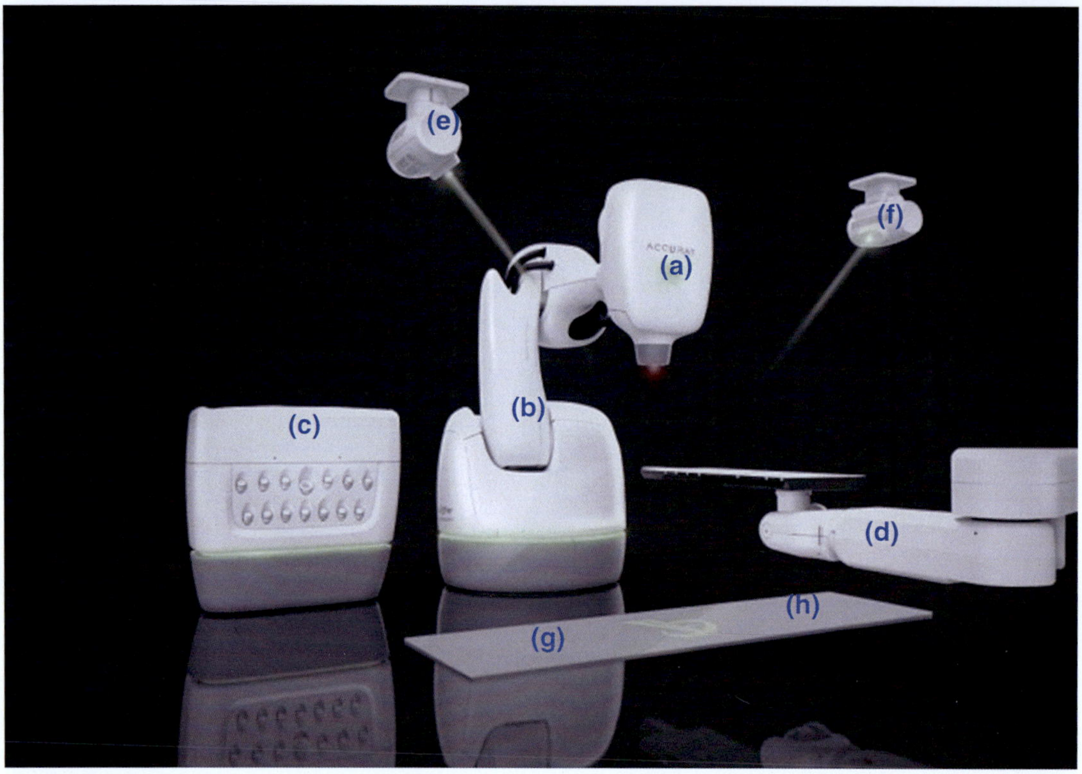

Fig. 3.2 The latest commercially available CyberKnife™ model known with the brand name CyberKnife M6™. (**a**) The lightweight X band 6MV linear accelerator, (**b**) The six-joint robotic manipulator capable of performing movements with 6 degrees of freedom, (**c**) The Xchange™ table used for automated exchange and storage of the secondary collimator assemblies, (**d**) The six-joint robotic couch (RoboCouch™) exhibiting the ability to move in all six translational and rotational axes, (**e** and **f**) The kV X-ray tubes used for patient setup and intrafraction imaging, (**g** and **h**) The floor mounted high resolution amorphous silicon detectors used for image detection. The infrared stereo camera system used for the Synchrony™ treatment delivery mode is not shown

assurance and geometric calibration procedures (see Chap. 7) while also being used by the automated exchange collimator system.

Downstream of the LINAC (i.e. approaching the couch) is the detachable part of the treatment head containing the secondary collimation system, which defines the geometric characteristics of the treatment beam delivered to the patient. There are three collimating systems available in the M6 model (Fig. 3.3). At the top of each one is an intermediate collimator made of tungsten and designed to reduce the field size exiting the primary collimator down to the maximum field size achievable by the selected secondary collimation system. The bottom portion of each collimator housing is encased in a touch sensor used to trigger an interlock terminating robot movement and irradiation in the case of collision or if the proximity between the robot and the patient's body drops below a predefined distance limit. Following is a description of these collimating systems.

1. Fixed collimators (Fig. 3.3a). A set of 12 conical collimators made of tungsten with a circular aperture. The nominal diameter of the beams defined by these collimators ranges from 5 mm to 60 mm (5, 7.5, 10, 12.5, 15, 20, 25, 30, 35, 40, 50 and 60 mm) at the reference distance of 800 mm. To minimize beam penumbra, the collimator apertures are focused to the X-ray target except for the two smallest sizes which have straight apertures. The fixed collimators fit on a corresponding collimator assembly attached on the LINAC head that can be installed automatically (see Sect. 3.3).

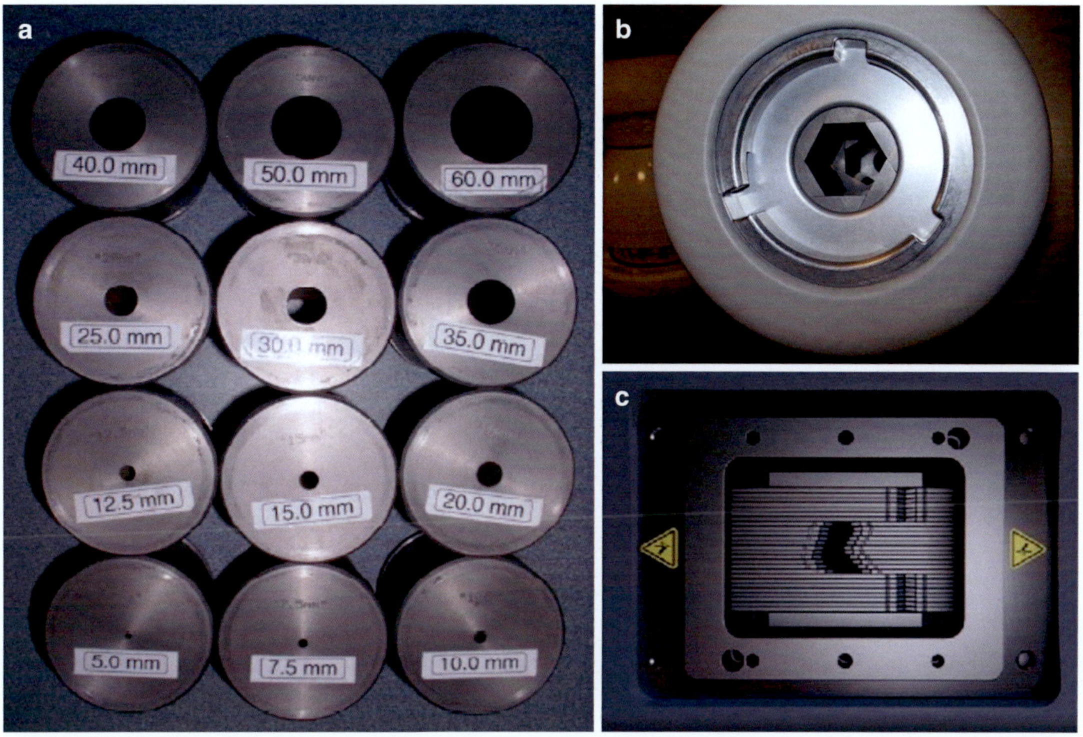

Fig. 3.3 The three secondary beam collimation systems available on the CyberKnife system: (**a**) A set of 12 fixed circular collimator cones with nominal diameters ranging from 5 to 60 mm at 800 mm distance from the source, (**b**) the Iris® variable aperture collimator capable of achieving the same nominal field sizes with those of fixed collimators and (**c**) the Incise™ multi leaf collimator exhibiting 3.85 mm leaf width and maximum achievable field size of 100 × 115 mm^2 at 800 mm distance from the source

The automatic fixed collimator exchange option available in the previous CK models has been removed, and the fixed collimators in the M6 model can only be changed manually. The collimator size is monitored by the system, and an interlock prevents treatment delivery unless this matches the size assigned in the treatment plan to each beam. It is noted that for each fixed collimator size the treatment manipulator traverses a separate path of nodes (see Sect. 3.2.1). Fixed collimators are generally preferred to treat small targets, especially small lesions lying in the central nervous system (CNS), since they deliver the sharpest beam penumbra with respect to the other collimation systems while avoiding aperture size uncertainties.

2. The Iris™ Variable Aperture Collimators [18] are capable of achieving the same set of 12 circular field sizes as those obtained with the fixed collimators using a single variable aperture and therefore provide the flexibility to apply any field size at any beam position without the need to swap collimators during treatment. This combination contributes to both treatment time reduction and dosimetric benefits over fixed collimators in many cases, especially to extracranial targets [18]. The variable aperture is created by two banks of six triangular tungsten segments, each creating a hexagonal aperture. All 12 segments are driven from a single motor. The two banks are rotated by 30 degrees with respect to each other, resulting in a dodecahedral aperture which is virtually circular (Fig. 3.3b). This rotational offset also minimizes the radiation leakage between the segments. The upper bank (closer to the LINAC) has a smaller aperture than the lower one to approximate the focusing of the fixed collimators.

3. The InCise™ multi leaf collimator (MLC) [15], which has recently been updated to its second generation [InCise™ 2, Fig. 3.3c] consists of 26 tungsten leaf pairs with a leaf width of 3.85 mm projected at 800 mm distance from the source. The maximum field size is 115 mm (in leaf motion direction) by 100 mm (in the vertical direction) at 800 mm distance from the source. Each leaf is driven independently and is capable of unlimited interdigitation and overtravel. The sides and tips of the leaves collimate the edges of the treatment field and are machined such as to minimize the beam penumbra width (<3.5 mm for a field size of 10 mm × 10 mm at 800 mm distance from the source and 50 mm depth in water). However, to minimize interleaf radiation leakage, the entire leaf assembly is tilted by 0.5 degrees, which makes a compromise between penumbra and leakage. The MLC collimation system is the most flexible of all options available in the CyberKnife system, since it allows for the delivery of irregularly shaped (noncircular) and larger radiation fields using fewer beams and lower total monitor units compared to (non-isocentric) fixed or Iris variable aperture collimated fields exploiting the non-coplanar, non-isocentric workspace. Therefore, a treatment time reduction of the order of 30–35% has been reported along with a better dose gradient in the low dose region [19–21]. It is noted though, that MLC treatment delivery is associated with larger radiation leakage (<0.5% maximum relative to a 100 mm × 100 mm field size at 800 mm distance from the source) and mechanical positioning tolerances (better than ±0.95 mm at 800 mm distance from the source) than the other collimation systems.

3.3.2 The Treatment Manipulator

The CyberKnife M6 system utilizes a KUKA QUANTEC KR300 R2500 Ultra robot. It is a six-joint robot allowing for 6-DOF movements with a maximum payload of 300 kg, 2496 mm reach and position repeatability of ±0.06 mm. The main difference of this robot relative to that used in the VSI system is the payload increased by 60 kg required for the addition of the MLC secondary collimation system. The increase in payload came at the expense of the overall reach, which is reduced by almost 200 mm relative to the VSI robot. To compensate for this, the treatment room layout had to be amended in two ways. First, the treatment robot was moved from superior-right or superior-left (both configurations were supported by the vendor and referred to as "normal" or "mirror", respectively) relative to the patient to the head of the couch being in alignment with its longitudinal axis (see Fig. 3.2). Second, the robot is situated upon a custom designed pedestal of 412 mm height, allowing the system to reach over the patient and maximizing the robot workspace for treatment delivery [7].

The robot workspace within the treatment room is defined by three primary coordinate frames. The first, referred to as "the robot world frame", has its origin at the centre of the pedestal the robot is mounted upon, while the second, referred to as "the robot tool frame", is aligned with the laser that is mounted inside the LINAC (see Sect. 3.1) and is centred at the LINAC target (i.e. the X-ray beam source). Finally, the "robot user frame" has its origin at the machine centre, with rotations aligned to the robot world frame. The machine centre in CyberKnife terminology is the point where the central axes of the imaging X-ray beams intersect. This also defines the origin of the imaging system coordinate frame (see Sect. 3.5.1) and the centre of the CyberKnife treatment volume. Calibration of the treatment robot relative to the machine centre is mandatory and through this procedure, the correlation between robot and imaging system coordinates is defined. Briefly, a calibration post is inserted into a floor frame with a photodetector located at its tip, and the LINAC laser is instructed to scan across this point from positions throughout the robot workspace. Using the scanning data, a least-squares minimization relationship is found which provides the calibrated robot tool frame with respect to the origin of the robot user and imaging system frames. More details are provided in Chap. 7 and elsewhere [6, 8, 9].

3.3.2.1 Nodes and Treatment Paths

The points that the robot can reach during treatment in the three-dimensional (3D) space inside the treatment room are referred to as nodes. At each node, the robot positions the centre of the LINAC X-ray target; therefore, nodes represent the source positions of the treatment beams. Nodes are preselected by the vendor from a set of points located on concentric spheres about the machine centre (with radii ranging from 650 to 1200 mm), on the grounds of maximizing robot reachability during real-time tracking of patient motion as well as to provide flexibility for robot transversals to other nodes while avoiding collisions in the room and ensuring that the cable management is not stretched or compacted too much [7].

The nodes are grouped to larger sets, referred to as treatment paths. Among the properties associated with each node of a treatment path are the maximum rotational and translational tracking corrections, X-ray imaging status (i.e. whether placing the X-ray target at the node causes the LINAC or the robot to obstruct one of the X-ray imaging cameras), as well as properties to facilitate the safe and efficient movement of the LINAC between these nodes. During treatment planning, the TPS is based on this information combined with other treatment parameters, such as the target anatomy and the collimator type, to define the proper set of nodes that (1) allows for full range of the tracking corrections needed (e.g. in prostate treatments the robot must support ±5 degrees of pitch correction, instead of the standard ±1.5 degrees), (2) involves a sufficient number of "imaging nodes" (nodes where the robot or the LINAC is not blocking either of the X-ray cameras) and (3) minimizes the time spent moving between "dose nodes", without deteriorating imaging opportunities.

There are different treatment paths depending on the treated anatomy and the secondary collimation type used. The two primary paths are the "head" and "body" paths used for treating intracranial and extracranial lesions, respectively. The "head" path involves nodes with shorter distances from the treated target in the range of 650–900 mm to maximize the effective dose rate. The "body" path involves nodes extending throughout the superior-inferior axis, distributed at longer distances (800–120 mm) to allow for larger patient clearance due to, for example, respiratory motion tracking, prostate pitch tracking, and the larger range of possible patient body alignment positions. It is noted that while the fixed and the Iris collimation systems share the same housing geometry, the MLC systems is different, thus requiring different "head" and "body" paths. This results in four primary treatment paths, neglecting those employed for quality assurance purposes. This flexibility necessitates larger installation vaults and primary beam barriers.

Each primary treatment path needs to be calibrated independently following a procedure where the robot moves to each node and performs a scan of the calibration post using the LINAC laser resulting in a list of offsets relative to their nominal position which is stored and applied during delivery. Potential residual systematic offset owing, for example, to discrepancies between the laser and treatment beam axes, is possible and its translational component is compensated by a final correction referred to as "DeltaMan". This correction is derived from a sequence of phantom based end-to-end (E2E) tests, which estimate the total system error (TSE) using radiochromic films. Further details are analysed in Chap. 7 and elsewhere [8]. With this combination of calibrations, the CyberKnife system achieves submillimeter accuracy in treatment delivery which can be sustained long-term following a comprehensive quality assurance program [13].

3.3.3 The Xchange Table

The CyberKnife system offers the feature of automated exchange of the secondary collimator systems via a pneumatic tool-changing mechanism. This is facilitated by a custom designed pedestal, referred to as Xchange table in the CyberKnife terminology, where the secondary collimator assemblies are stored when not in use. A separate storage well is assigned to each type of collimator. At the centre of each storage well lies a sensor that the laser LINAC scans to instruct the robot to drop off or mount a collimator housing. It is, therefore, of mandatory importance that the robot, accommodating the tool-changing mechanism, knows the exact position of the table and the sensors. This is addressed by defining a rigid 6-DOF calibration of the table coordinate frame and the sensor positions. The table has three spots to hold calibration posts similar (smaller) to those used for primary system calibration. Using the maximum reading from successive vertical and horizontal scans of the laser LINAC, a set of three points in space is provided giving a full 6-DOF representation of the table-top plane relative to the robot. For the storage well sensor positions, a similar procedure is performed, and, in turn, the robot picks up and drops off each housing to ensure that the calibration scans successfully found the centre positions. Finally, in addition to automated collimator exchange, the table also facilitates a laser alignment check prior to each treatment. Provided that the LINAC laser is aligned with the treatment beam, this check ensures that the treatment beam is also aligned consistently (see Chap. 7).

3.3.4 The Patient Positioning Systems

Two types of patient positioning systems are available with the CyberKnife M6 system: The RoboCouch® patient positioning system (provided optionally) and the Standard Treatment Couch, which are analysed as follows.

3.3.4.1 The RoboCouch

RoboCouch® comprises a six-joint serial link robotic system manipulating the treatment couch. The robot is custom designed, while utilizing the same KUKA controller and wrist as those of the treatment robot. RoboCouch allows for patient alignment in all 6-DOF eliminating the need of manual adjustments by the therapist inside the treatment room. Moreover, the required patient positioning adjustments are performed by correcting all degrees of freedom simultaneously. RoboCouch exhibits a payload of 227 kg (500 lb) retaining submillimetre mechanical accuracy and precision better than 0.1 mm and 0.1 degrees for translational and rotational movements, respectively. The workspace involves a maximum travel of 100 cm along the patient's inferior-superior direction, ±18 cm in the patient's left-right direction and a maximum travel of 37 cm posterior to the machine centre (i.e. there is a minimum load height of 55 cm from the floor). Regarding rotations, a range of ±5 degrees about each axis is achievable, although these limits are reduced at the boundaries of translation limits.

3.3.4.2 The Standard Treatment Couch

Relative to the RoboCouch, the standard treatment couch has a reduced payload of 159 kg and slightly limited translational travel limits (91 cm along the inferior-superior axis, ±15 cm in the right-left direction and 28 cm posterior to the machine centre). Rotational motion is automated only for the head up/head down (pitch) and right/left (roll) directions, within the same limits as the RoboCouch (±5 degrees), while rotations about the patient's anterior/posterior central axis (yaw) need to be manually adjusted by the therapist inside the room. The mechanical accuracy of the standard couch is well below 1 mm, while its movement precision is better than 0.3 mm and 0.3 degrees for translational and rotational adjustments, respectively. In contrast to RoboCouch, the required corrections are performed serially.

3.3.4.3 Calibration of Patient Positioning Systems

Irrespective of the type, the treatment couch is calibrated to the rest of the CyberKnife treatment delivery subsystems. This is accomplished by performing a specific sequence of couch movements and tracking calibration targets using the X-ray imaging system. Correlation of the couch coordinate frame to the imaging system coordinates allows for accurate alignment of the patient to the treatment position by instructing the couch to move based on registration results between live X-ray images prior to treatment and the DRRs calculated during treatment planning.

3.3.5 The Image Guidance System

Image guidance in the CyberKnife system is facilitated by both X-ray and optical imaging systems, which are described briefly in the following subsections. A more detailed description of the image guidance system including the accompanied software options of the CyberKnife is given in Chap. 4.

3.3.5.1 The X-Ray Imaging System

The X-ray imaging system comprises two ceiling-mounted kilovoltage (kV) X-ray tubes coupled with corresponding image detectors mounted at floor level under load-bearing covers. The central axis of each X-ray beam is tilted at 45 degrees with respect to the vertical, such that the two beams intersect orthogonally providing a stereo image pair from which the three-dimensional (3D) location of an imaged object can be determined. The X-ray tubes are operated at voltages ranging from 40 to 150 kV using one of the two available focal spot sizes (0.6 mm and 1.2 mm). The generated beams are collimated by a fixed aperture projecting a field size of approximately 19 × 19 cm^2 at the machine centre. The distance from each X-ray tube to the machine centre (the point in the room at which the centres of the two X-ray beams intersect, which is coincident with the center of the treatment robot workspace) is nominally 2.2 m, and the distance from machine centre to the image detector is nominally 1.42 m, giving an image magnification factor of 1.6. Each detector is an amorphous silicon photodiode beneath a cesium iodide scintillator. The detector has a total sensitive area of 43 cm × 43 cm and pixel size of 278 μm × 278 μm at the detector. The entrance surface dose at the machine centre (i.e. the air kerma at isocenter in the absent of backscatter) is analogue to the used kV and mAs settings, and for a typical exposure (120 kV, 10 mAs) is equal to 0.18 mGy [22]. For an intracranial single session treatment acquiring 100 x-ray image pairs, the aforementioned entrance dose is associated with an effective dose of 0.4 mSv [23, 24]. After images are acquired for initial patient alignment to treatment position, the user specifies the interval between subsequent image acquisitions which are used to detect and correct for intrafraction target motion.

3.3.5.2 The Optical Camera System

The optical imaging system consists of three cameras in a ceiling-mounted retractable boom-arm and is used to track targets that are affected by respiratory motion in combination with intrafraction X-ray images. This camera array detects the position of three optical markers (red LEDs communicating with the system via optical fibres) attached to the patient surface. The LEDs are pulsed sequentially (i.e. marker 1, then 2, then 3) so that they can be differentiated by the camera, which reads their positions at about 100 Hz. The camera controller calculates each marker position in a 3D camera frame, which is later reduced to a scalar measurement along the marker principal axis of motion. These measurements are used to create a correlation model between the patient's breathing pattern, determined by the external markers, and the precise location of the tumour (or a tumour surrogate, e.g. fiducial markers) determined by the X-ray intrafraction images at various points of a normal respiration cycle. This way, the treatment is delivered during all phases of patient breathing, unlike gating or breath-hold methods, since the robotic manipulator, uniquely featured in the CyberKnife system, allows for the treatment head to follow the 3D tumour trajectory nearly in real time.

3.4 Treatment Planning and Delivery Overview

3.4.1 Treatment Planning

Treatment planning is performed using a vendor-provided software suite (currently, the Accuray Precision®) and aims to determine the optimum geometric arrangement of treatment beams and radiation fluence per beam. A prerequisite to start the planning process is the acquisition of a 3D patient CT scan, transferred to the TPS via a dedicated database server. This CT image series, referred to as the primary image set, is mandatory since it allows for the construction of the digitalized reconstructed radiography (DRRs) used for tracking during treatment delivery and determines a 3D patient coordinate system where the electron density and other tissue properties (e.g. mass density) required by the dose calculation algorithms can be calculated on a voxel basis. Target volumes and organs at risk (OARs) are segmented onto the primary CT images using a combination of automatic (atlas- and/or model-based) and manual methods. Since CT images are characterized by poor soft tissue contrast, the user is enabled to import up to five (or 15 in the case of multiple respiration phases of a 4D CT scan) secondary 3D image sets (e.g. MRI, PET/CT etc.) to aid the segmentation process by registering each secondary image set to the primary CT series using rigid (normalized mutual information) or deformable methods [25, 26], not precluding manual adjustments.

In the next step, the generated 3D patient model is virtually aligned (through the corresponding TPS platform, referred to as *Align*) to the imaging system, such that the treatment target lies close to the machine centre. This process correlates the patient coordinate frame with that of the imaging system, which allows for a set of feasible (in terms of robot reachability) treatment beams to be defined onto the patient model since the transformation between imaging and robot coordinate systems is already known. Typically, more than 100 (reaching to several thousand for non-isocentric circular collimator plans) of these candidate beams are generated. During this step, a set of DRRs is constructed based on the simulated orientation of the X-ray imaging system relative to the 3D CT-based patient model.

The planner then needs to determine the optimum geometric beam arrangement and radiation fluence per beam that best matches the plan objectives including target coverage goals and OAR sparing constraints. This is performed in the plan optimization step facilitated by two dose optimization algorithms integrated in the TPS: Sequential Optimization (SO) [27] and VOLO™ [28]. In the SO algorithm, plan objectives are specified either as hard constraints (i.e. cannot be violated) or goals (i.e. can be violated). Goals are sequentially optimized one at a time based on a priority list defined by the user, and the results are converted into new hard constraints. In VOLO, all plan objectives are specified as goals, each one associated with a user defined weight, and optimized simultaneously by minimizing a cost function using a quasi-Newton gradient search algorithm. Another difference constitutes the optimization of MLC-based plans. The SO relies on shape heuristics, parameterized by the user and applied to each node, to create the set of beams comprising the solution space. Then optimization is facilitated by weighting the MUs delivered by each generated beam. In VOLO, the projection of the MLC is divided into beamlets, that create a fluence map projecting from each node to each target volume. This way, the fluence map is initially optimized by optimizing individual beamlet weights and is subsequently converted into deliverable MLC apertures used to calculate the dose distribution corresponding to this fluence map.

Dose calculation is performed using either type A and/or type B algorithms [29], both available at the CyberKnife TPS platforms. Type A algorithms include ray-tracing [6] for circular collimator beams and a finite size pencil beam (FSPB) algorithm developed for the MLC fields [30]. Both algorithms account for relative electron density variation effects on photon energy loss using the equivalent path length estimation. The actual depth defined by the ray-line linking the X-ray source (node) to the centre of the dose calculation voxel is converted to an equivalent depth in water calculated as the integral of relative electron densities along the ray, i.e. the ratio of electron density at each voxel relative to the electron density of water. The type B algorithm offered in CyberKnife TPS comprises a Monte Carlo (MC)-based dose calculation engine, where the interactions of primary photons with atomic electrons and the generated secondary electrons, as well as scattered photons are explicitly simulated. Photon interaction with matter is stochastic, and, therefore, simulation is based on density probability distributions, requiring a large number of simulated incident photons to result in dose calculations of acceptable statistical uncertainty (typically, better than 1.5%). In view of this, MC-based dose optimization is slower (of the order of minutes) compared to type A algorithms. Usually, MC is preferred for lung cancer and some head cases, where the associated dosimetric accuracy is superior compared to type A algorithms, due to the presence of tissue heterogeneities.

From a technical point of view, circular collimators enable both isocentric and non-isocentric treatment delivery techniques. In the isocentric mode, the planner selects the position of one or more pseudo-isocenters within the patient model and a beam linking each node to each pseudo-isocenter is automatically generated. The resulting dose distribution comprises approximately spherical dose clouds around each pseudo-isocenter in a similar fashion to other radiosurgery systems using circular collimators. The non-isocentric mode takes advantage of the ability of the robotic manipulator to direct each beam at a unique point within the target volume, eliminating the need to reposition the patient between beam delivery. The delivered dose distributions are characterized by complex shapes highly conformed to the target volume and steep dose gradients approximating—far more efficiently—those obtained using multiple pencil beams. The MLC collimation system is the most flexible of all options available, since it allows for the delivery of irregularly shaped (noncircular) and larger radiation fields using fewer beams and lower total MUs compared to (non-isocentric) fixed or Iris variable aperture collimated fields exploiting the non-coplanar, non-isocentric workspace [12, 13].

Once a treatment plan is approved (by a radiation oncologist and/or a neurosurgeon), a set of machine commands, including treatment delivery instructions, is generated and transferred to the treatment delivery computer via an integrated Data Management System (iDMS™), along with the generated set of DRRs. Details of the CyberKnife treatment planning procedures are analysed in depth in Chap. 5.

3.4.2 Treatment Delivery

Prior to treatment delivery, patient alignment to treatment position is performed by registering live stereoscopic X-ray images to the DDRs using bony anatomy (skull or spine), implanted fiducial markers or soft tissue anatomy for lung tumours, with the aid of dedicated software installed on the treatment delivery computer. Once this is accomplished, the couch remains static during treatment delivery, and all fine alignment corrections are achieved by adjusting the treatment manipulator position and orientation based on the target pose deviation from the settings stored in the treatment plan. This modus operandi removes the need for the patient to be considered as a rigid object statically attached to the couch [7].

During treatment delivery, intrafraction motion is tracked by comparing the live X-ray images to corresponding DRRs using dedicated image-guided tracking algorithms. The employed tracking algorithm and the temporal resolution of live image acquisition (user-selected or semi-automatically determined for specific treatment paths, e.g. in-tempo prostate path) depends on the target anatomy and the respiration motion effect—if existent. Retrospective analysis of a large body of CyberKnife intracranial and spine image tracking data suggested that submillimeter accuracy is achieved with image acquisition at a temporal resolution of 60–90 s [13, 31, 32].

The 6D skull tracking method is mostly used for intracranial lesions, also facilitating tracking of some upper cervical and head and neck targets. This method utilizes a rigid 2D to 3D image registration algorithm based on the skeletal features of the patient skull depicted in the live images and the corresponding DRRs. For spinal lesions, the Xsight® spine tracking (XST) method is used. The XST algorithm relies on the high image contrast of the spine, which is further enhanced in the DRRs by ray tracing only through a spine tracking volume (STV) defined in the planning CT. The apparent locations of the bony structures are compared between live images and DRRs using a grid of nodes distributed over the STV portion along the treated and the two adjacent vertebrae. A displacement vector of each tracking node is independently determined using an intensity-based similarity measure, and the node displacement vectors are combined to calculate a rigid transformation of the target in each projection applying smoothness constraints. This registration algorithm accounts for local deformations of the vertebrae, e.g. due to different spine flexing between the time of planning CT acquisition and beam delivery. Finally, the 2D registration results, corresponding to each projection, are combined by back projection to calculate a global 3D transformation as in 6D skull tracking.

For soft tissue lesions that are not fixed relative to the spine or skull (e.g. prostate, liver, pancreas), a set of fiducial markers (typically gold seeds) are implanted inside or near the treated lesions, and the fiducial tracking method is used. The fiducial markers are usually implanted percutaneously under image guidance, and a minimum of three of them are required to enable rotational tracking. During treatment planning, the user identifies the fiducial locations within the CT images and tracking is performed by registering these known features in the DRRs with their apparent location in the live images. To mitigate the risk of fiducial migration between implantation and planning CT acquisition, a typical time period of 1 week is interleaved between these processes.

For lung lesions within the lung parenchyma of adequate density, size and position to be visualized in the acquired X-ray images, the Xsight® Lung (XSL) tracking algorithm is commonly used. In this algorithm, tumour tracking is performed by image registration of the tumour region in the DRRs to the corresponding region in the treatment X-ray images. Specifically, the image intensity pattern of the tumour region in the DRR is matched to the most similar region in the X-ray image. A matching window for the tumour is defined based on the tumour silhouette in each projection. As with the other target locating methods, the registration process is conducted separately for each projection, resulting in 2D translations for each projection. The 3D tumour translation is determined by backprojection of the 2D translations [6].

For targets that move with respiration, the real-time, dynamic capabilities of the CyberKnife are used by means of the Synchrony® Respiratory Motion Tracking system. Synchrony is utilized in combination with the fiducial or Xsight® Lung [33] tracking methods. The latter enables fiducial-less tracking of a lung tumour when the tumour is visible in the images of both or at least one (1-vew tracking) of the X-ray imaging tubes during the (majority of) respiration motion phases [34]. The Synchrony system utilizes the optical camera system, described in Sect. 3.5.2, to monitor the position of three markers emitting pulsed, visible (red) light in real time (approximately every 10 ms). The markers are placed on the patient surface, usually attached on a special vest. Prior to treatment, the position and time stamp data of the external markers are correlated to the internal target positions using a series of X-ray images acquired at multiple phases of the breathing cycle. During treatment, the system uses the optical signal of the external markers combined with this correlation model to determine the treatment manipulator corrections needed to track the target in real time and maintain the same static beam-target orientation that was simulated in the treatment plan [7]. Additional intrafraction X-ray images are used to verify and adapt the correlation model throughout treatment. The CyberKnife tracking algorithms and treatment delivery process are analysed in detail in Chap. 4.

References

1. Adler JR. Accuray, Inc.: a neurosurgical business case study. Cureus. 2009;1:1–15.
2. Guthrie BL, Adler JR. Computer-assisted preoperative planning, interactive surgery, and frameless stereotaxy. Clin Neurosurg. 1992;38:112–31.
3. Adler JR Jr, Chang SD, Murphy MJ, et al. The Cyberknife: a frameless robotic system for radiosurgery. Stereotact Funct Neurosurg. 1997;69:124–8.
4. Adler JR, Murphy MJ, Chang SD, Hancock SL. Image-guided robotic radiosurgery. Neurosurgery. 1999;44:1299–306.

5. Murphy MJ, Cox RS. The accuracy of dose localization for an image-guided frameless radiosurgery system. Med Phys. 1996;23:2043–9.
6. Kilby W, Dooley JR, Kuduvalli G, et al. The CyberKnife robotic radiosurgery system in 2010. Technol Cancer Res Treat. 2010;9:433–52.
7. Kilby W, Naylor M, Dooley JR, et al. A technical overview of the CyberKnife system. In: Handbook of robotic and image-guided surgery: Elsevier; 2020. p. 15–38.
8. Dieterich S, Cavedon C, Chuang CF, et al. Report of AAPM TG 135: quality assurance for robotic radiosurgery. Med Phys. 2011;38:2914–36.
9. Antypas C, Pantelis E. Performance evaluation of a CyberKnife G4 image-guided robotic stereotactic radiosurgery system. Phys Med Biol. 2008;53:4697–718.
10. Chang SD, Veeravagu A. Cyberknife stereotactic radiosurgery. In: Brain, vol. 1. 1st ed. Stanford, CA: Nova Science Publishers, Stanford University Medical Center; 2014.
11. Chang SD, Veeravagu A. Cyberknife stereotactic radiosurgery. In: Spine, vol. 2. 1st ed. Stanford, CA: Nova Science Publishers, Stanford University Medical Center; 2014.
12. Urschel HC, Kresl JJ, Luketich JD, et al. Treating tumors that move with respiration. 1st ed. Berlin Heidelberg, Berlin, Heidelberg: Springer; 2007.
13. Pantelis E, Moutsatsos A, Antypas C, et al. On the total system error of a robotic radiosurgery system: phantom measurements, clinical evaluation and long-term analysis. Phys Med Biol. 2018;63:165015.
14. Fu D, Kuduvalli G. A fast, accurate, and automatic 2D–3D image registration for image-guided cranial radiosurgery. Med Phys. 2008;35:2180.
15. Asmerom G, Bourne D, Chappelow J, et al. The design and physical characterization of a multileaf collimator for robotic radiosurgery. Biomed Phys Eng Express. 2016;2:017003.
16. Ma C-M, Li JS, Deng J, Fan J. Implementation of Monte Carlo Dose calculation for CyberKnife treatment planning. J Phys Conf Ser. 2008;102:012016.
17. Kawachi T, Saitoh H, Inoue M, et al. Reference dosimetry condition and beam quality correction factor for CyberKnife beam. Med Phys. 2008;35(10):4591–8.
18. Echner GG, Kilby W, Lee M, et al. The design, physical properties and clinical utility of an iris collimator for robotic radiosurgery. Phys Med Biol. 2009;54:5359–80.
19. McGuinness CM, Gottschalk AR, Lessard E, et al. Investigating the clinical advantages of a robotic linac equipped with a multileaf collimator in the treatment of brain and prostate cancer patients. J Appl Clin Med Phys. 2015;16:284–95.
20. Jin L, Price RA, Wang L, et al. Dosimetric and delivery efficiency investigation for treating hepatic lesions with a MLC-equipped robotic radiosurgery-radiotherapy combined system. Med Phys. 2016;43:727–33.
21. Kathriarachchi V, Shang C, Evans G, et al. Dosimetric and radiobiological comparison of CyberKnife M6™ InCise multileaf collimator over IRIS™ variable collimator in prostate stereotactic body radiation therapy. J Med Phys. 2016;41:135–43.
22. Accuray. CyberKnife robotic radio. Surgery system physics essentials guide. 2017.
23. Le Heron JC. Estimation of effective dose to the patient during medical X-ray examinations from measurements of the dose-area product. Phys Med Biol. 1992;37:2117–26.
24. Murphy MJ, Balter J, Balter S, et al. The management of imaging dose during image-guided radiotherapy: report of the AAPM Task Group 75. Med Phys. 2007;34:4041.
25. Jordan P, Myronenko A, Gorczowski K, et al. Deformable image registration: description and evaluation. White Papper: Accuray Precision; 2017.
26. Gupta V, Wang Y, Méndez Romero A, et al. Fast and robust adaptation of organs-at-risk delineations from planning scans to match daily anatomy in pre-treatment scans for online-adaptive radiotherapy of abdominal tumors. Radiother Oncol. 2018;127:332–8.
27. Schlaefer A, Schweikard A. Stepwise multi-criteria optimization for robotic radiosurgery. Med Phys. 2008;35:2094–103.
28. Zeverino M, Marguet M, Zulliger C, et al. Novel inverse planning optimization algorithm for robotic radiosurgery: first clinical implementation and dosimetric evaluation. Phys Med. 2019;64:230–7.
29. Knöös T, Wieslander E, Cozzi L, et al. Comparison of dose calculation algorithms for treatment planning in external photon beam therapy for clinical situations. Phys Med Biol. 2006;51:5785–807.
30. Jeleń U, Söhn M, Alber M. A finite size pencil beam for IMRT dose optimization. Phys Med Biol. 2005;50:1747–66.
31. Fürweger C, Drexler C, Kufeld M, et al. Patient motion and targeting accuracy in robotic spinal radiosurgery: 260 single-fraction fiducial-free cases. Int J Radiat Oncol Biol Phys. 2010;78:937–45.
32. Murphy MJ. Intrafraction geometric uncertainties in frameless image-guided radiosurgery. Int J Radiat Oncol Biol Phys. 2009;73:1364–8.
33. Fu D, Kahn R, Wang B, et al. Xsight lung tracking system: a fiducial-less method for respiratory motion tracking. In: Urschel HC, Kresl JJ, Luketich JD, et al., editors. Treating tumors that move with respiration. Berlin Heidelberg, Berlin, Heidelberg: Springer; 2007. p. 265–82.
34. Jordan P, West J, Sharda A, Maurer C. SU-GG-J-24: retrospective clinical data analysis of fiducial-free lung tracking. Med Phys. 2010;37:3150.

The Target Locating System for CyberKnife Neuroradiosurgery

4

Warren Kilby

4.1 Introduction

The CyberKnife® System is designed to deliver treatment using precisely targeted beams of ionizing radiation. High precision beam targeting is achieved using multiple imaging systems within the treatment room combined with algorithms operating within the treatment planning and delivery software, referred to collectively as the target locating system (TLS), which forms the topic of this chapter. CyberKnife is designed for a wide range of clinical applications, including both central nervous system (CNS) and non-CNS targets. Given the nature of this textbook, the scope of this chapter is limited to a discussion of the TLS for CNS applications. For a complete description of the CyberKnife TLS applied to treatment sites throughout the entire body the reader is referred to [1].

Essentially, the TLS uses in-room imaging to localize the target position and orientation continually during each treatment session. This information is used to adjust the treatment couch and the treatment delivery robot throughout treatment such that the alignment of each treatment beam relative to the target is maintained, and that it exactly matches the optimal alignment defined in the patient-specific treatment plan. There is no assumption that the target is static during treatment, and the sequence of image-localize-align occurs continually throughout every treatment session. The entire process is fully automated, but feedback is provided to the user who can interrupt treatment at any time.

It is important to understand that image-based stereotactic beam alignment is an integral part of every CyberKnife treatment, and the TLS is fully integrated within the CyberKnife System. Therefore, before reading this chapter, it is advisable to review the overall CyberKnife System description provided in Chap. 3, and for an overview that is not solely focused on CNS applications, the reader is referred to [1]. In addition, as with any other innovative technology, this description is only complete at the time of writing. This chapter describes the CyberKnife M6 System version 11.1, which was current in January 2020. The reader is advised to review the more recent literature and query the vendor regarding technical advances subsequent to this date and system version.

Section 4.2 of this chapter will describe the major TLS components within the CyberKnife treatment suite and the overall TLS function. Sections 4.3 and 4.4 will describe in more detail aspects of the TLS that are specific to neuroradiosurgical targets within the skull and those inside the spinal cord or close to the spinal vertebrae, respectively. Section 4.5 will describe methods that have been used to measure the geometric treatment delivery accuracy and review the results of these tests.

W. Kilby (✉)
Accuray Incorporated, Sunnyvale, CA, USA
e-mail: wkilby@accuray.com

© Springer Nature Switzerland AG 2020
A. Conti et al. (eds.), *CyberKnife NeuroRadiosurgery*,
https://doi.org/10.1007/978-3-030-50668-1_4

4.2 Target Locating System Overview

4.2.1 TLS Layout

The treatment suite layout including the TLS is shown in Chap. 3. The major TLS hardware components include a stereoscopic X-ray imaging system and a stereoscopic optical camera array. The latter is combined with the X-ray system for targets affected by respiratory motion, which in the CNS occurs only in the thoracic and lumbar spine if the patient is treated in the prone position.

The X-ray imaging system layout is shown in Fig. 4.1. This system is comprised of two ceiling mounted X-ray tubes, and two X-ray image detectors positioned just beneath the floor. The central axis of each imaging beam is at 45° from vertical, and therefore the two imaging beams are separated by 90°. This provides an orthogonal stereo image pair which allows the three-dimensional (3D) position and rotation of image features to be calculated. The distance from each X-ray target to the machine center (the point in the room at which the centers of the two X-ray beams intersect, which is coincident with the center of the treatment robot workspace) is nominally 2.3 m, and the distance from the machine center to the image detector is nominally 1.4 m, giving an image magnification factor of 1.6.

The optical camera array contains three cameras mounted on a moveable boom arm fixed to the ceiling (Fig. 4.2). When tracking a target affected by respiratory motion this arm is normally positioned near the foot of the couch, from where it observes three optical markers that are placed on the patient surface close to the treatment site. When not in use the arm can be retracted.

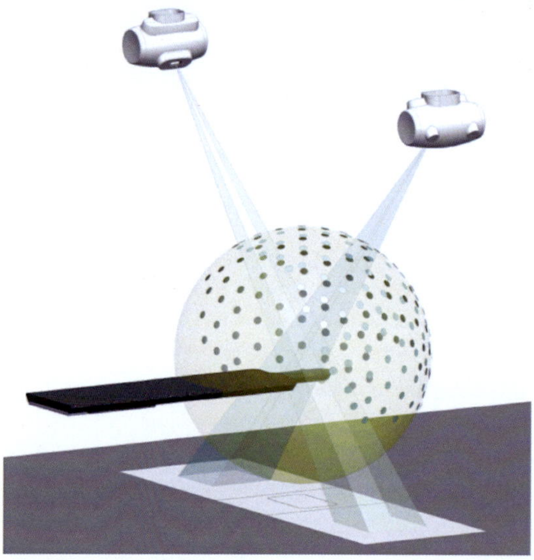

Fig. 4.1 The in-room X-ray imaging system includes two ceiling mounted X-ray tubes, each generating an imaging beam (shown in blue) directed at an image detector placed beneath the floor. The beams are oriented at 45° to the vertical and intersect at the machine center, which is at the center of the treatment delivery robot workspace. This workspace is shown by the yellow sphere, the dots on which illustrate positions from which treatment beams can be delivered to the target (note that the workspace shown is for head and upper spine treatment, different workspaces are provided for the rest of the body, as described in Chap. 3)

4.2.2 Major TLS Hardware Components

High-voltage power to the ceiling mounted X-ray tubes is provided by generators that may be situated in the treatment room or an adjacent equipment room. They provide a range of X-ray techniques from 40 to 150 kVp with settings determined by the user via the treatment delivery PC. The oil-cooled X-ray sources are mounted via struts to the ceiling. Each imaging field is limited by a variable rectangular collimator. This is adjusted during system installation to provide a field of view matched to the detector sensitive area. The image detectors are recessed into the floor with load bearing covers sitting flush with the finished floor level. Each detector is an amorphous silicon photodiode beneath a cesium iodide scintillator. The detector has a total sensitive area of 43 cm × 43 cm and pixel size of 278 μm × 278 μm at the detector, projecting to 169 μm × 169 μm at machine center. The exact alignment of the imaging system with respect to the treatment robot is carefully calibrated during system installation and is checked routinely as part of the recommended quality assurance program.

The optical tracking components, which are strictly part of the motion tracking system (MTS) rather than the TLS, include an array of three linear cameras mounted on a retractable boom arm

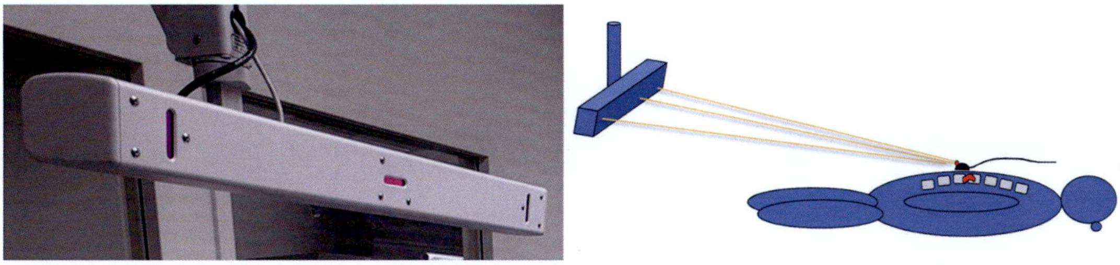

Fig. 4.2 Left: The optical camera array, showing the retractable ceiling mounted boom arm. The three linear cameras are positioned behind the purple slots in the cover. Right: An illustration of the camera array positioned to track optical markers during a prone spine treatment. In this case one optical marker is shown on the skin surface close to the target location (indicated in red). Yellow rays illustrate the optical signals detected by the three cameras from which the marker position is measured about one hundred times per second, which is fast enough to capture respiratory motion. The internal target position is measured in real-time by combining the 100 Hz marker position measurements with intermittent X-ray image measurements of the corresponding spinal vertebrae positions, as described in Sect. 4.4.2

near the foot of the couch, a tight-fitting vest worn by the patient during treatment, three optical makers that are attached to the vest and placed close to the treatment target, optical fibers and light sources for the markers, and control electronics. Each marker is formed by the tip of an optical fiber that is attached below the couch to a red optical LED pulsing at about 100 Hz. The three markers are pulsed in sequence so that the camera can temporally discriminate between them and measure the position of each marker at 100 Hz. Each camera can measure the marker positions perpendicular to its line of sight, and the combination of camera images provides the depth perception by triangulation needed to measure the third position component. During treatment the 3D marker positions are reduced to a scalar quantity, indicating the marker position along its principal axis of motion. The TLS does not require the boom arm to be positioned repeatably between treatments, and its position is not calibrated.

4.2.3 TLS Operation

All CyberKnife System treatments follow the general sequence described in Table 4.1.

During treatment, X-ray image acquisition is controlled from the treatment delivery PC at the operator station. X-ray image pairs are acquired using radiographic mode (i.e., fixed kVp, mA,

Table 4.1 The main steps in the sequence of planning and delivering treatment using the CyberKnife System

Step	Description
1	Prior to treatment a virtual 3D model of the patient is constructed from multimodality medical image sets. As a minimum, this includes a CT scan of the patient in the same pose will be used during treatment. The anatomical structures and other volumes of interest within this model, including the target volume(s) for treatment, are segmented
2	A treatment plan is constructed in which a simulation of the CyberKnife delivery robot workspace and the X-ray imaging system geometry is aligned with the patient model, placing the target volume(s) close to the simulated machine center (see Fig. 4.1). This allows the position of treatment beams, together with their shapes and radiation doses to be optimized according to clinical goals and objectives specified by the user. The resulting treatment plan includes a set of machine instructions informing the treatment delivery system how to reproduce the planned set of beam positions, orientations, shapes, and radiation doses
3	Once the treatment plan is finalized and approved for treatment, the simulated X-ray system geometry is used to generate a library of digitally reconstructed radiographs (DRRs) by ray casting from the simulated X-ray sources through the CT-based patient model and onto the simulated X-ray detectors
4	At the start of each treatment session, the patient is manually positioned on the couch in the same pose as they were in for the planning CT scan, and the couch is translated and rotated to place the target volume approximately at the machine center

(continued)

Table 4.1 (continued)

Step	Description
5	This rough alignment is fine-tuned by acquiring X-ray image pairs which are automatically registered to the DRRs to calculate the precise 6D corrections (3 translations and 3 rotations) needed to bring the target volume into the same alignment with the imaging system as was simulated in the treatment plan. These corrections are performed using the couch. This process repeats until the residual offsets are typically <1 mm and <1°. At this point the couch is not adjusted further and treatment delivery can commence (see Fig. 4.3)
6	The residual alignment corrections are applied by adjusting the position and orientation of the treatment delivery robot on which the linear accelerator is mounted, such that the alignment of each beam relative to the target volume corresponds precisely with that simulated in the treatment plan
7	Because the target volume is not static throughout treatment (either because of gross patient movement or internal organ motion), the sequence of acquire X-ray images-register live images to DRR's–calculate alignment offsets (translations and rotations)–perform beam alignment corrections using the treatment delivery robot, repeats throughout every treatment session. This process is augmented by more rapid imaging using the optical camera system if the target is affected by respiratory motion

and exposure time). The two images can be acquired simultaneously or sequentially, with the second image acquired immediately after the first. The simultaneous option is used for targets affected by respiratory motion, so for most CNS targets the sequential option is used. This slightly improves the image quality by removing scattered X-rays from the other imaging beam.

Intra-treatment images are acquired automatically by the control system, although the user specifies the image acquisition interval (typically 30s–60s) and has the ability to pause the treatment at any time to acquire additional images. The system also provides a method to intelligently adjust the imaging interval based on the observed stability of the target pose (position and orientation). Essentially, if the changes in target position and rotation calculated between the most recent image pair and the pair acquired immediately before exceeds user defined thresholds, then the inter-imaging time interval is automatically reduced to a minimum setting, or treatment can be paused (at the user's discretion). If individual treatment beams take longer than this interval to deliver, they will be interrupted in order to acquire images. Once the target position and rotation variations are consistently smaller than these thresholds then the imaging interval increases back to the initial value defined by the user.

As the treatment robot moves around the patient during treatment, it can obstruct one or both imaging beams. These positions are identified during treatment planning and are referred to as "blocked nodes." If images are required while the treatment robot is at a blocked node, the robot is automatically moved to a nearby unblocked node and then back to the blocked node after the images are acquired.

If the treatment target is in the thoracic or lumbar spine and the patient is treated in the prone position, then target motion caused by breathing can be significant. This cannot be managed using the approach described above because the breathing motion is more rapid than X-ray imaging and alignment correction at ≥15 s intervals. In this situation the real-time optical imaging components are combined with the X-ray imaging system to allow real-time tracking of the respiratory motion.

The details of this approach specific to skull and spine tracking, using the 6D skull tracking and Xsight® Spine Tracking systems, respectively, are described in Sects. 4.3 and 4.4.

4.3 6D Skull Tracking

In this TLS mode, the skull is used as a surrogate for the target volume, which requires a fixed rigid relationship between the target volume and the skull. It can be used for intra-cranial and some upper cervical spine and head and neck targets. The patient is usually positioned supine and fitted with a thermoplastic mask and simple head and neck immobilization (Fig. 4.4). This is noninvasive and is not relied upon to align the patient for

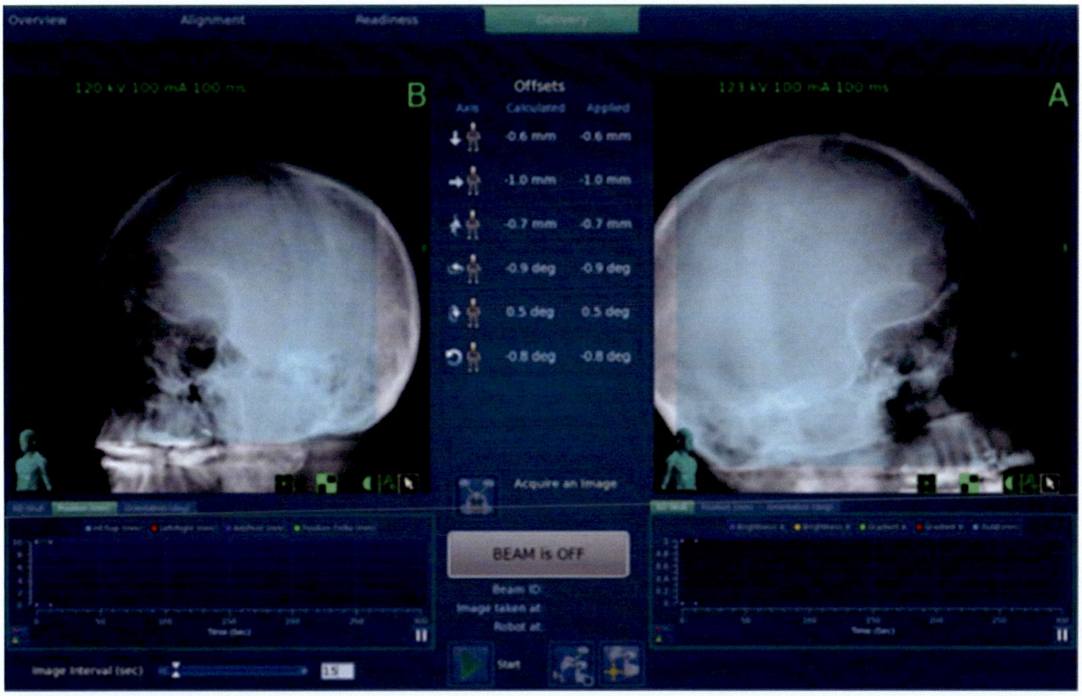

Fig. 4.3 The treatment delivery PC screen during treatment of an intracranial target. X-ray image pairs are automatically acquired throughout treatment, and the target location and orientation (i.e., 6D pose, including three translational and three rotational offsets) calculated from the most recent image pair is used to apply corrections to the treatment delivery robot so that each treatment beam remains precisely aligned. In this screen the offsets calculated from the most recent image pair are shown in the central area. The image interval, which is set by the user, is shown in the lower left (in this tracking mode the available range is 5s–150s). Automatic adjustment of the image interval is not enabled in this example. The control to acquire additional images manually is in the central bar

treatment, only to make it quicker to manually position them in the approximate pose at the start of treatment (step 4 in Table 4.1) and to reduce intra-treatment motion. During treatment planning, the simulated machine center is placed near the center of the skull such that the whole of the skull is within the simulated X-ray field of view (Fig. 4.3). A library of DRR's are calculated by ray-casting through the 3D patient model, from which X-ray attenuation is estimated using the CT number in each voxel and interpolation along each ray. High-resolution CT and sufficiently large scan field of view to capture all the relevant bony anatomy is essential to this process. A 512 × 512 pixel resolution and maximum slice separation of 1.5 mm, with scan limits that extend from 10 mm above the head to the suprasternal notch are recommended by the vendor. Multiple

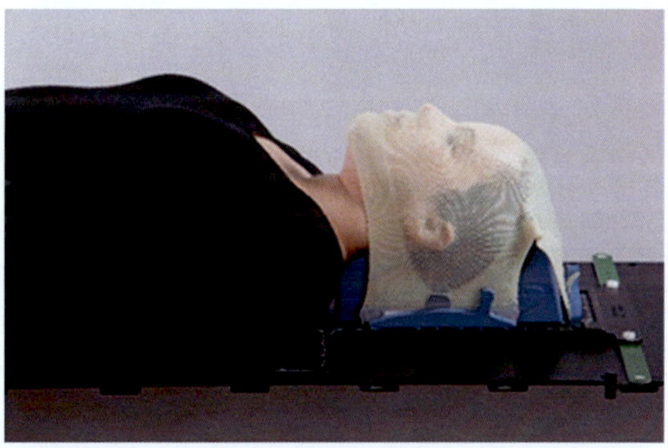

Fig. 4.4 Noninvasive patient immobilization using a head and neck support, thermoplastic mask, and couch indexing device. The immobilization device is only used to make initial patient set-up faster and to reduce the amount of intra-treatment motion. Treatment alignment is based entirely on X-ray images of skeletal features within the skull and registration of these images with pre-treatment DRR's. This process repeats throughout treatment to compensate for any patient motion

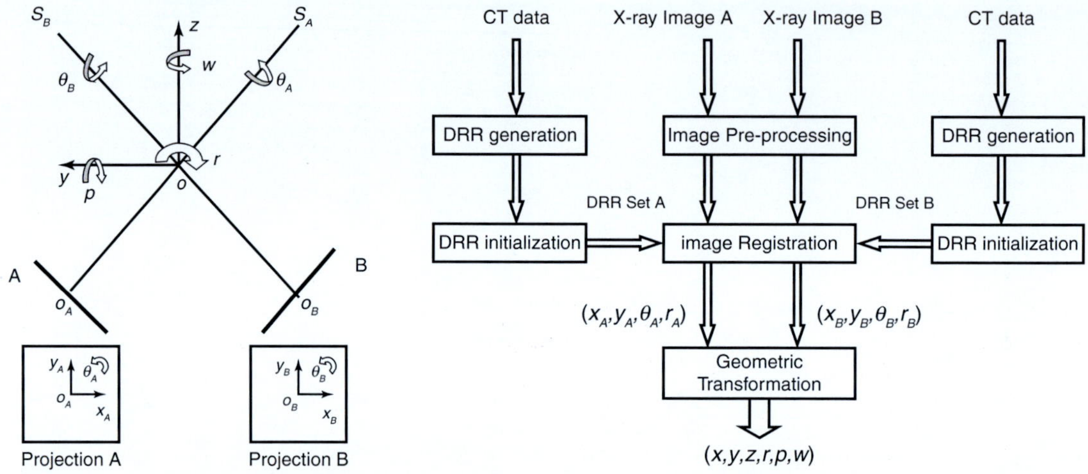

Fig. 4.5 Overview of the offset calculation geometry and method. Left: the coordinate system of each X-ray projection A and B is shown relative to the room co-ordinate system. Right: The main steps involved in generating the three rigid-body translational and three rotational alignment offsets in the room system from each X-ray image pair

DRRs are calculated for each X-ray beam in order to include multiple patient roll angles (i.e., the 3D model is rotated about the superior-inferior axis for each DRR calculation). Immediately after X-ray images are acquired during treatment delivery, they are automatically registered to this DRR library in order to calculate the 6D alignment offsets, which will be applied to subsequent treatment beams. The process is described in detail in [2]. An overview is shown in Fig. 4.5 and is described below:

- Each live image is registered to the corresponding zero roll-angle DRR using an image intensity-based similarity measure. This registration process calculates the in-plane 2D translations and in-plane rotation needed to optimally align the DRR to the live image in that projection.
- For that same X-ray projection, this initial 2D transformation is applied as a starting point estimate to register the live image with the library of DRR's simulating different out of plane (roll) rotations, initially sampled at 1° resolution. At each roll angle, the in-plane registration is fine-tuned using another image intensity-based similarity measure. The result provides the initial estimate of the out-of-plane rotation.
- The previous two steps are iterated, with the results of each previous step providing the initial estimate for the roll-angle and in-plane transformation applied in the next step. These iterations use progressively more accurate methods and finer DRR roll-angle resolution. This iterative search converges on a stable 2D transformation (2 translations and one in-plane rotation) and roll-angle for each of the two X-ray projections.
- Finally, the two 2D transformations are back-projected to provide most of the final 3D transformation (i.e., all three translations, and the rotations about the patient left-right and anterior-superior axes). The rotation about the inferior-superior axis is taken as the average of the roll angles calculated using the two projections.

During treatment delivery, the user is presented with feedback on the quality of the offset calculation process which can be used to assess the accuracy of the tracking result. Some of these metrics compare the variation in pixel intensities between the DRR and live images. The most direct quality metric is the difference in the calculated translational offset along the patient inferior-superior axis between the two X-ray projections, since this can be measured independently in each projection (i.e., the difference between x_A and x_B in Fig. 4.5). Maximum thresholds can be set on these metrics to automatically interrupt treatment. If this occurs, the problem resolution is usually to make adjustments to the X-ray technique and re-acquire images. Treatment will also automatically pause if the calculated offsets become unexpectedly large.

4.4 Xsight Spine Tracking

4.4.1 Patient Supine

This tracking method relies upon a fixed rigid relationship between the target volume and skeletal features within the spinal vertebrae closest to the target. If the patient is treated in the supine position the general approach is similar to 6D skull tracking, although tracking vertebral features introduces new challenges:

- Other bony features may overlay the vertebrae in each X-ray projection (e.g., ribs, clavicle), and these may also deform relative to the vertebrae (e.g., due to arm position changing between CT imaging and treatment).
- The spinal column is non-rigid, with the potential for vertebral bodies to deform relative to each.

The problem of overlying bony anatomy is mitigated by calculating the DRR library using only a sub-volume of the 3D patient model, such that the spine is included but overlying structures such as ribs are excluded (see Fig. 4.6). This sub-volume is defined during treatment planning.

The second problem is addressed by positioning the simulated machine center at the vertebral level closest to the target volume during treatment planning and limiting the tracking offset calculations to a small volume of the bony anatomy around this position. The size of this sub-region is user-adjustable and typically contains the three closest vertebral bodies to the target (see Fig. 4.7). Within this region, there is still the potential for the skeletal features to deform relative to each other, and therefore the DRR to live image registration problem is solved using a non-rigid method. In each projection the tracking region is divided into a mesh (Fig. 4.7). Around each vertex of this mesh, or node, a small region or "matching block" is defined in one image and an image intensity-based similarity measure is used to identify the optimum translational offset for that matching block within a small search window in the other image. In this way, the translational offset of each node is calculated independently, enabling the mesh to deform. The algorithm proceeds iteratively, with the mesh resolution increasing and the search window size decreasing at each iteration. At each step, the registration results from the previous step provide the starting point estimate for node translations using linear interpolation. In the final iteration, the tracking region is divided into a 9 × 9 node-mesh. A 2D displacement field in each projection is generated from the final set of node translations combined with a smoothness constraint, with a typical result shown in Fig. 4.8. From this displacement field, rigid translations and in-plane rotation of the target are calculated in each projection, and the 2D results are backprojected to obtain the 3D target translations and two rotations (yaw and pitch). As with skull tracking, the out-of-plane (roll) rotation is calculated by comparing each live image to a library of DRR's generated with varying roll angle applied,

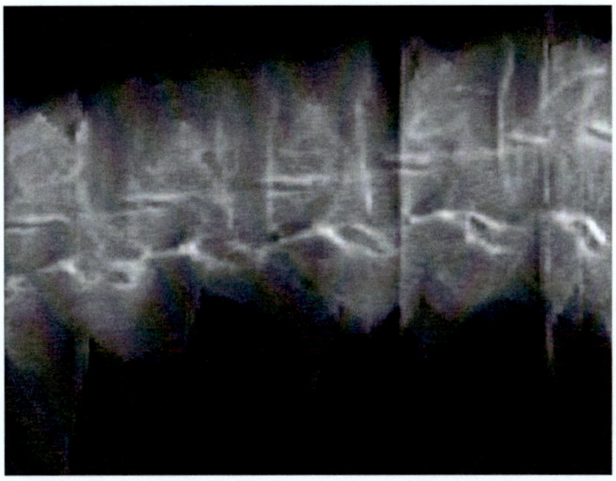

Fig. 4.6 Example of a spine tracking DRR. The DRR is generated using only the CT image data within a segmented spine region, so that overlaying anatomical structures do not influence the tracking result

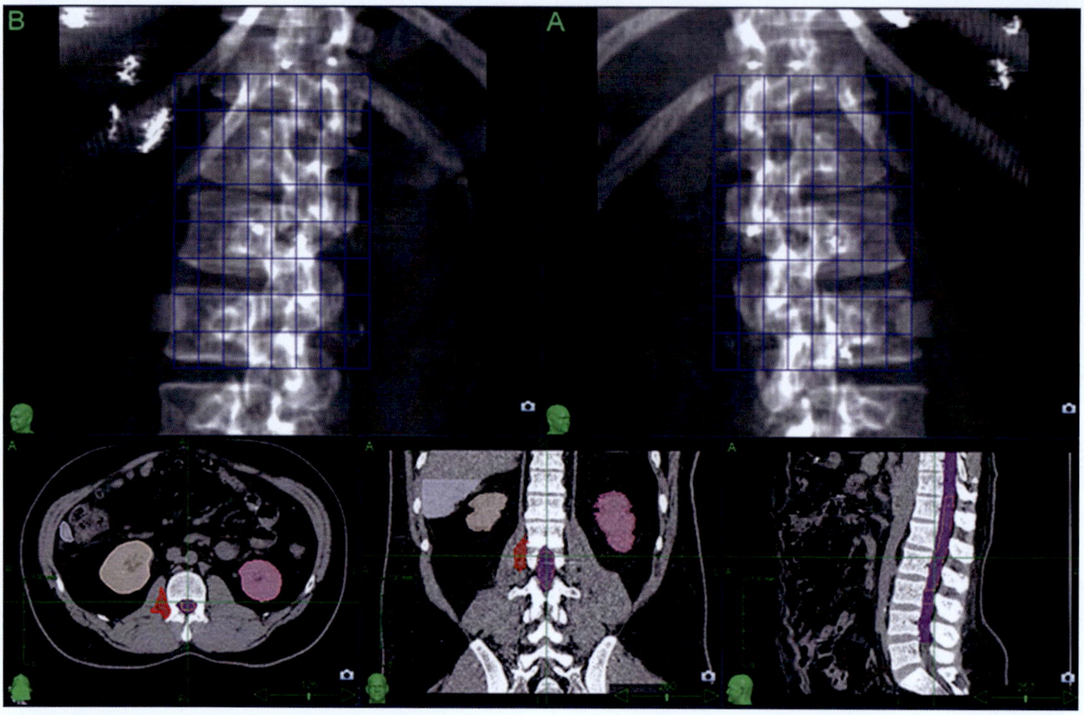

Fig. 4.7 A treatment planning system screenshot showing the set-up for a target volume (shown in red) to be tracked using spine tracking. In the three cardinal image planes at the bottom of the image the simulated machine center, shown by the green crosshairs, is positioned at the vertebral level closest to the target volume. The top images show the corresponding DRRs, one for each of the X-ray beams, zoomed in to the machine center. The blue box indicates the sub-region of each image that will be used to register the DRR to the corresponding live X-ray images. This region covers approximately the three vertebral bodies closest to the target volume. The vertices of the blue mesh are the nodes used to solve the non-rigid registration problem

although the methods used are slightly different to skull tracking. Details of this algorithm are provided in [3–5].

The user is presented with feedback describing the quality of the tracking offset calculations. This includes the difference in superior-inferior translation calculated using the two orthogonal images (as with the 6D skull algorithm), and the difference in roll angle (about the inferior-superior axis) resulting from the registrations performed in the two projections. In addition, the reliability of the registration result at each node is estimated using the difference between its translation result and the median result of the surrounding nodes. If this exceeds a threshold, that node is considered "false" and its translations are replaced with the median of the surrounding nodes. The proportion of nodes considered "false nodes" is reported as one of the tracking quality metrics, and treatment is interrupted if this exceeds a user-defined threshold. As with 6D skull tracking, treatment is also interrupted if the calculated alignment offsets are unexpectedly large.

4.4.2 Patient Prone

The description of spine tracking in the supine position is equally applicable in the prone position but in addition, the impact of respiratory motion on targets within the thorax and abdomen must be considered and managed. All patients can be treated supine, which has the advantage of avoiding respiratory motion and reducing the treatment complexity, and usually of maximizing patient comfort. However, because the treatment beams are delivered predominantly from above

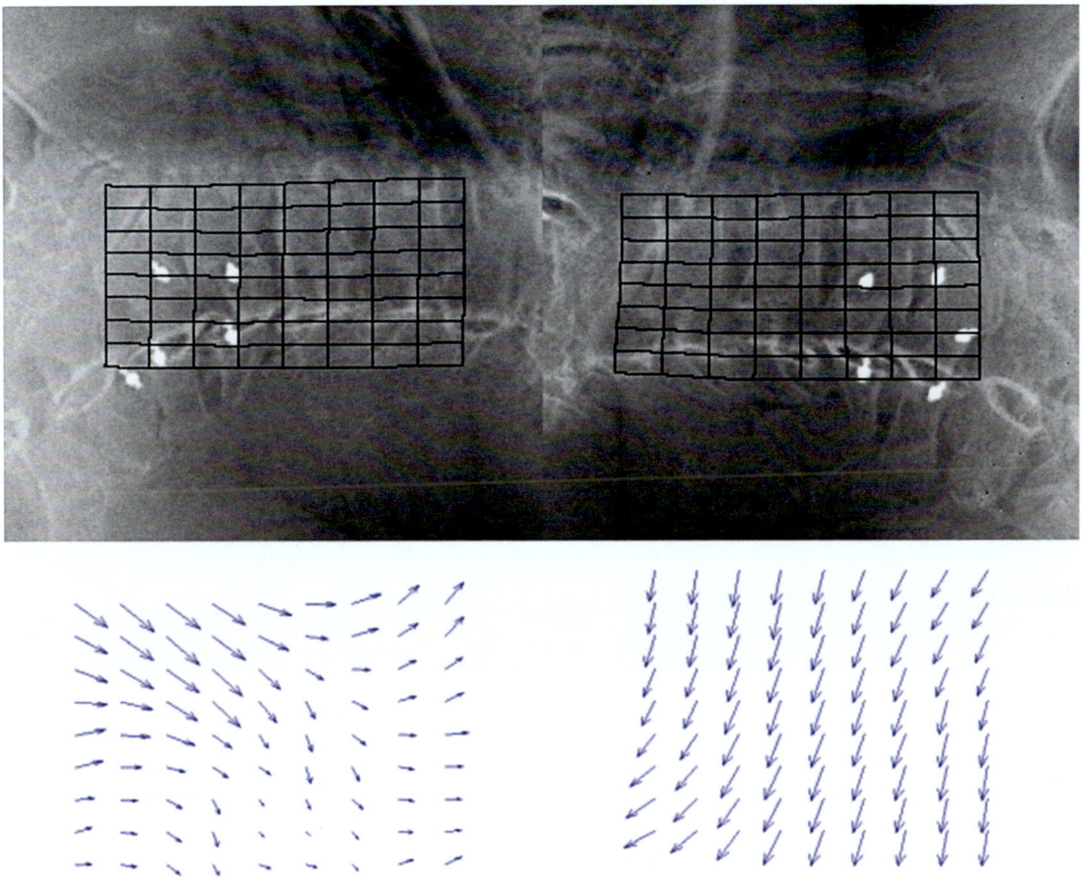

Fig. 4.8 Top: The tracking region divided into the final 9 × 9 node-mesh in projection A (left) and B (right) showing the translational offsets at each node calculated during the matching block step. Bottom: the corresponding 2D displacement fields calculated from the node translation results combined with the smoothness constraint

the couch (see Fig. 4.1), there can be dosimetric advantages to treating patients in the prone position because this minimizes the volume of normal tissue traversed by each beam before the target volume. The choice of patient position is made on a case by case basis.

When the prone position is chosen respiratory motion must be assessed. If it is not significant (e.g., in the neck or upper thorax), then treatment can proceed exactly as described in Sect. 4.4.1. If respiratory motion is significant, this is managed by combining the Xsight Spine method in with the Synchrony® Respiratory Motion Tracking System, which is described in greater detail elsewhere [1, 6]. This system uses the correlation between the target position (measured each time an X-ray image pair is acquired using the Xsight Spine tracking method described in Sect. 4.4.1, typically every 30s–60s) and external marker positions measured at about 100 Hz by the optical camera system described in Sect. 4.2.2. During treatment the patient breaths normally, and treatment is delivered through the entire breathing cycle (i.e., there is no breath-holding or gating). At the start of each treatment session, X-ray images are acquired at multiple phases of the breathing cycle, and the positions of the external markers at the corresponding times are recorded. Up to 15 of these data points are used to construct a correlation model, which describes the position of the target volume as a function of the external marker position (Fig. 4.9). Once this model is constructed, it can be used to generate alignment offsets in real-time from the external

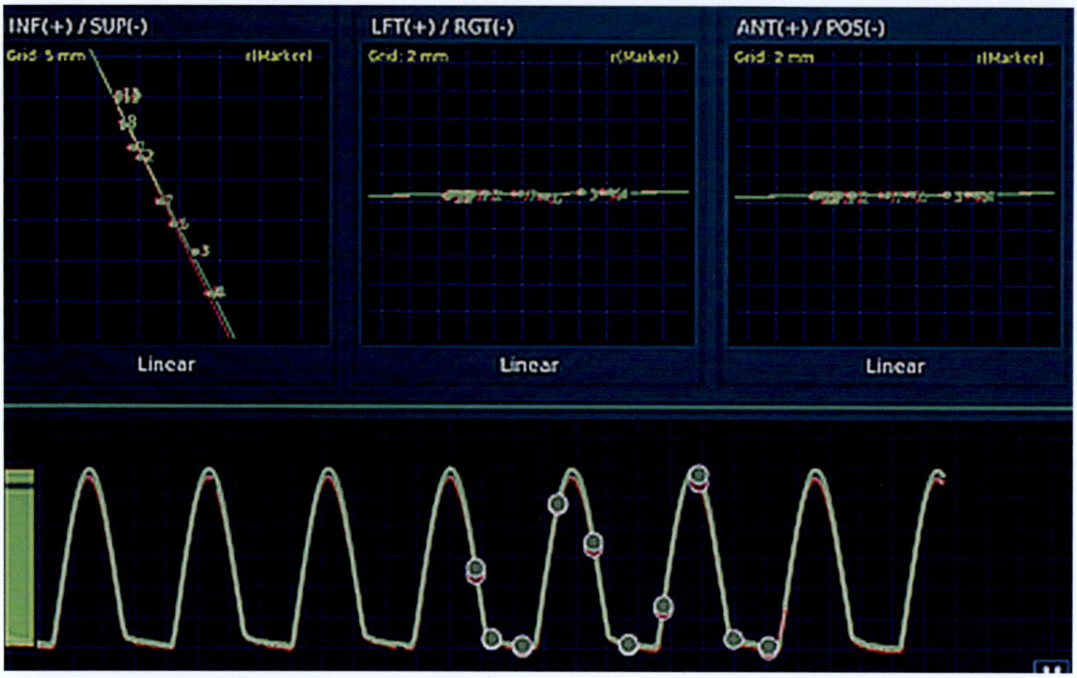

Fig. 4.9 A screenshot from the treatment delivery system showing a Synchrony correlation model. The lower plot shows external marker position as a function of time (green line), with the instants at which X-ray image pairs were acquired superimposed as green circles. The upper three plots display the corresponding model points as dots, showing the correlation between marker position (X-axis) and target position (Y-axis). From left to right, the Y-axis in these three plots corresponds to inferior-superior, left-right, and anterior-posterior. In this case almost all the target motion is in the inferior-superior direction, and the correlation model fit is essentially linear. More complex non-linear fitting is provided for situations where the target motion exhibits hysteresis (i.e., the trajectory during inhalation is different to that during exhalation) or where there is a non-zero phase relationship between target motion and marker motion

marker signal to be fed to the treatment delivery robot, enabling real-time tracking of the target as it moves due to respiration. Additional X-ray images are acquired throughout treatment, which provide additional model points that are used to automatically adapt the correlation model to any changes in the respiratory motion pattern.

In addition to the spine tracking accuracy feedback described in Sect. 4.4.1, further metrics and interruption mechanisms are provided with respiratory motion tracking. These include (a) treatment is not possible unless the correlation model includes data-points that cover the full range of respiratory motion, (b) treatment is automatically interrupted if the target position measured with a new X-ray image pair is not within a user-defined maximum distance from the position predicted by the previous correlation model, and (c) treatment is automatically interrupted if the external marker position deviates significantly from the range of positions included in the correlation model (e.g., if the patient takes a much deeper breath or coughs).

4.5 TLS Accuracy

The image registration accuracy of the TLS has been tested by comparing registration results of test objects with corresponding values measured manually using a caliper and a digital level [7]. TLS registration accuracy of 0.2 mm for translations and 0.2° for rotations was reported for 6D skull and Xsight Spine methods [7]. Fu and Kuduvalli (2008) evaluated the accuracy of the 6D skull algorithm by moving an anthropomorphic head and neck phantom to predefined positions inside the imaging field of view with the aid of the robotic manipulator [2]. The authors reported mean registration errors of 0.33 mm in translations and 0.29° in rotations. Using the same technique, Fürweger et al. (2011) reported a mean deviation of 0.2 mm from the nominal translational offset and a maximum root mean square (RMS) error of less than 0.4 mm for the Xsight Spine registration algorithm [8].

The most meaningful measurement of TLS performance is the total system error (TSE), which describes the total end-to-end (E2E) geometric error in the entire process of pre-treatment imaging, treatment planning, and treatment delivery. This is measured as the radial offset between the centroid of a delivered dose distribution and the intended dose centroid from the corresponding treatment plan. Methods for measuring TSE using anthropomorphic phantoms are described in Chap. 7. The vendors' tolerance on these tests is ≤0.95 mm for all tracking methods. Test results obtained for both 6D Skull and Xsight Spine tracking have been reported by multiple CyberKnife users [9–11] and have previously been summarized [12]. All of these results show TSE within the vendors' tolerance. Most recently, a retrospective analysis of E2E test results acquired on a single system over a period of 11 years demonstrated TSE (mean ± s.d.) of 0.40 mm ± 0.18 mm and 0.55 mm ± 0.20 mm for 6D Skull and Xsight Spine tests, with no degradation observed over time [13]. It should be noted that this result was obtained only after following a regular program of quality assurance and equipment servicing, which included TLS recalibrations after certain system updates. A limitation of this test method is that the target volume must be positioned very close to the machine center because of the phantom construction. Since this is almost identical to the geometry used to calibrate the delivery system, the systems' ability to deliver treatment accurately for targets positioned away from the machine center, and therefore further from the calibration condition, is not tested. This limitation was addressed by an alternate E2E test method using polymer gel dosimetry to measure the TSE for treatment of seven separate intracranial targets distributed across the brain of a 3D-printed head phantom generated from patient CT. The TSE for those targets ranged between 0.29 mm and 0.66 mm, demonstrating that the vendors' tolerance is met for targets up to 8 cm away from the machine center [13].

All of these tests were performed using static phantoms, neglecting the possibility of intra-treatment target motion. As described previously, when using 6D Skull and Xsight Spine (supine) tracking intra-treatment motion is corrected at the frequency of X-ray image acquisition, typically once every 30s–60s (minimum 15 s). The impact of residual intra-treatment motion between these corrections can be evaluated from the stability of target positions reported during treatment (i.e., the changes between successive treatment offsets). Evaluations of this kind have been performed for skull and spine tracking and have been previously summarized [12]. The reported systematic TSE increase associated with residual motion (i.e. the offset between mean actual target position and TLS identified target position, due to uncorrected motion between images) was typically <0.5 mm for an imaging interval of 60s–120s [14, 15]. There was a clear trend for this offset to decrease with decreasing imaging interval, and so with corrections applied every 30s–60s the expected errors would be smaller. The most recent of these studies evaluated the residual motion for 260 spinal radiosurgery patients with a mean imaging interval of 90s. The median change in target offset between successive image pairs, which combines both systematic and random inter-image position changes, was 0.48 mm and remained <1 mm for 95% of all cervical, thoracic, and lumbar treatments. The authors concluded that their study "provided technical and clinical evidence that submillimeter targeting can be achieved in single-session spinal treatments despite intrafraction patient motion when using a noninvasive fiducial-free tracking technique" [16].

The most direct measurement of TSE performed to-date used the clinical data of a patient treated with intracranial functional radiosurgery [13]. In this treatment 120 Gy was prescribed in a

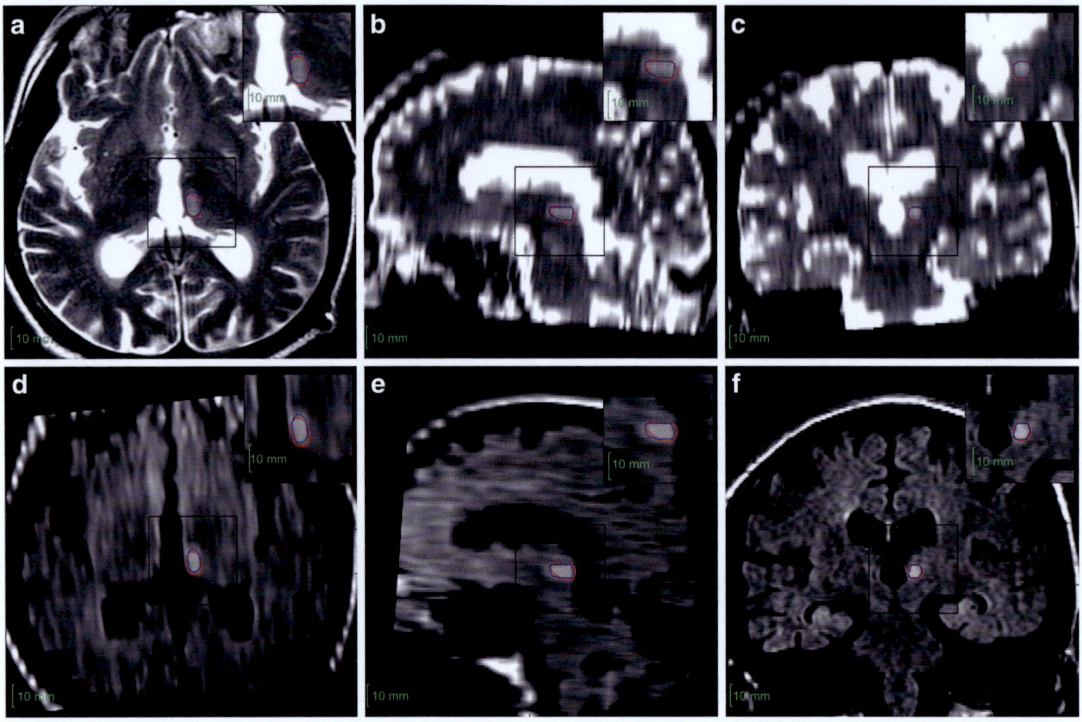

Fig. 4.10 Axial, sagittal and coronal T2 (top) and FLAIR (bottom) MR images of a patient treated for neuropathic pain, acquired 6 months post-treatment. The prescription dose is depicted with red line and the delineated target using the T2 and FLAIR images is shown with blue line. (Reprinted from [13] with the kind permission of Dr. Pantelis)

single fraction for medial thalamotomy. The TSE was measured by comparing the dose centroid visualized in the treatment plan with the centroid of the radiation induced lesion observed in post-treatment MR imaging acquired using two different pulse sequences. The average TSE obtained from those two sequences was 0.87 mm ± 0.25 mm (see Fig. 4.10).

4.6 Summary

The CyberKnife System uses a noninvasive target locating system (TLS) to enable stereotactic treatment alignment to be performed anywhere in the body. The major hardware component is an in-room orthogonal X-ray imaging system that acquires radiographic images of the patient anatomy close to the treatment target. In addition, an optical camera system is used to monitor the position of markers on the patient surface in real-time for targets that are affected by respiratory motion. Prior to treatment, a virtual patient model is generated using CT images, and the target and other structures are localized within this model. During treatment planning, this model is registered to a simulation of the in-room X-ray imaging system, defining a reference pose (translation and orientation) of the treatment target in the imaging coordinate system and allowing a set of treatment beams to be defined in this same coordinate system. The final stage of pre-treatment preparation is to calculate a set of digitally reconstructed radiographs by ray-casting from the simulated X-ray sources, through the virtual patient model, onto the simulated X-ray detectors. During neuroradiosurgical treatment delivery, live X-ray images are automatically registered to these DRR's using either the bony features of the skull (for intracranial targets) or vertebral bodies (for spinal targets). This image registration allows the translational and rotational offsets between the target pose during treatment planning and treatment delivery to be calculated. Initially, these offsets are applied to adjust the target pose using the robotic treatment couch. Once these coarse corrections are applied, fine adjustments are made using the treatment robot on which the linear accelerator is mounted. The sequence of acquire images-register images to DRR's–determine target offsets–apply corrections, continues throughout treatment in order to manage intra-treatment motion with new images usually acquired every 30s–60s. The exception to this method is for spinal treatments if the patient is prone and respiratory target motion is significant. In this case the more rapid respiratory motion is managed using the real-time optical camera system in combination with X-ray imaging. The overall geometric accuracy of this approach is quantified by the Total System Error (TSE), which is the treatment accuracy combining all aspects of pre-treatment imaging, treatment planning, and treatment delivery. Multiple phantom-based and clinical methods have been developed to measure TSE for neuroradiosurgical targets. All of these studies demonstrate that TSE <1 mm is consistently achieved using the TLS described in this chapter.

Acknowledgments The author would like to acknowledge his colleagues Kolos Lugosi and Jonathan McCoy in the Accuray Research & Development team for helpful discussions during the preparation of this chapter.

References

1. Kilby W, Naylor M, Dooley JR, Maurer CR Jr, Sayeh S. A technical overview of the CyberKnife system. In: Abedin-Nasab MH, editor. Handbook of robotic and image-guided surgery. Amsterdam: Elsevier; 2019. p. 15–38.
2. Fu D, Kuduvalli G. A fast, accurate, and automatic 2D-3D image registration for image-guided cranial radiosurgery. Med Phys. 2008;35:2180–94.
3. Fu D, Kuduvalli G. Enhancing skeletal features in digitally reconstructed radiographs. Proc SPIE. 2006;6144:846–51.
4. Fu D, Kuduvalli G, Maurer CR Jr, Allison JW, Adler JR Jr. 3D target localization using 2D local displacements of skeletal structures in orthogonal x-ray images for image-guided spinal radiosurgery. Int J Comput Assist Radiol Surg. 2006;1:198–200.
5. Fu D, Wang H, Maurer CR Jr, Kuduvalli G. Fiducial-less 2D-3D spine image registration using spine region segmented in CT image. Proc SPIE. 2007;6509:650935.
6. Sayeh S, Wang J, Main WT, Kilby W, Maurer CR Jr. Respiratory motion tracking for robotic radiosurgery. In: Urschel Jr HC, Kresl JJ, Luketich JD, Papiez L, Timmerman RD, editors. Robotic radiosurgery: treating tumors that move with respiration. Berlin: Springer-Verlag; 2007. p. 15–29.
7. Pantelis E, Petrokokkinos L, Antypas C. Image guidance quality assurance of a G4 CyberKnife robotic stereotactic radiosurgery system. J Instrum. 2009;4:P05009.
8. Furweger C, Drexler C, Kufeld M, Muacevic A, Wowra B. Advances in fiducial-free image-guidance for spinal radiosurgery with CyberKnife—a phantom study. JACMP. 2011;12:20–8.
9. Antypas C, Pantelis E. Performance evaluation of a CyberKnife G4 image-guided robotic stereotactic radiosurgery system. Phys Med Biol. 2008;53:4697–718.
10. Muacevic A, Staehler M, Drexler C, Wowra B, Reiser M, Tonn JC. Technical description, phantom accuracy, and clinical feasibility for fiducial-free frameless real-time image-guided spinal radiosurgery. J Neurosurg Spine. 2006;5:303–12.
11. Ho AK, Fu D, Cotrutz C, Hancock SL, Chang SD, Gibbs IC, Maurer CR Jr, Adler JR Jr. A study of the accuracy of Cyberknife spinal radiosurgery using skeletal structure tracking. Neurosurgery. 2007;60:147–56.
12. Kilby W, Dooley JR, Kuduvalli G, Sayeh S, Maurer CR Jr. The CyberKnife® robotic radiosurgery system in 2010. Technol Cancer Res Treat. 2010;9:433–52.
13. Pantelis E, Moutsatsos A, Antypas C, Zoros E, Pantelakos P, Lekas L, Romanelli P, Zourari K, Hourdakis CJ. On the total system error of a robotic radiosurgery system: phantom measurements, clinical evaluation and long-term analysis. Phys Med Biol. 2018;63:165015.
14. Hoogeman MS, Nuyttens JJ, Levendag PC, Heijmen BJ. Time dependence of intrafraction patient motion assessed by repeat stereoscopic imaging. Int J Radiat Oncol Biol Phys. 2008;70:609–18.
15. Murphy MJ. Intrafraction geometric uncertainties in frameless image-guided radiosurgery. Int J Radiat Oncol Biol Phys. 2009;73:1364–8.
16. Fürweger C, Drexler C, Kufeld M, Muacevic A, Wowra B, Schlaefer A. Patient motion and targeting accuracy in robotic spinal radiosurgery: 260 single-fraction fiducial-free cases. Int J Radiat Oncol Biol Phys. 2010;78:937–45.

Treatment Planning

5

Matthias Schlüter, Daniela Schmitt, Christoph Fürweger, Achim Schweikard, and Alexander Schlaefer

5.1 Introduction

Treatment planning involves a variety of steps. Starting with image data of the patient, three-dimensional volumes of interest (VOIs) are contoured. Margins around the volumes account for uncertainties, e.g., due to patient positioning or organ motion. Extending the tumor or target structure and the organs at risk (OARs) leads to the planning target volume (PTV) and the planning organ at risk volumes (PRVs) as the key input for the treatment planning [1, 2]. The other inputs are goals regarding the desired doses for each VOI, and the result of planning is a set of treatment beams which deliver a dose distribution best fulfilling the goals. The parameters that can be varied depend on the actual treatment system and for the CyberKnife typically include the aperture, orientation, and number of monitor units of the treatment beams.

While the CyberKnife was first used for intracranial stereotactic radiosurgery [3, 4], it has evolved into a versatile tool for the treatment of targets in the whole body [5, 6].

Hence, treatment planning needs to be flexible and to allow for different priorities, e.g., conformal dose distributions for small cranial targets and efficient dose delivery for larger targets in thorax or abdomen. Over time, this has been reflected in the CyberKnife's beam delivery. For example, while initially there were only a few predefined paths available for moving the linear accelerator (linac) during the treatment, the trajectory is now more flexible and based on a larger set of possible beam positions, also called beam nodes. In addition to a set of 12 circular collimators the more flexible Iris™-based collimator (a 12-sided polygonal cross section that is effectively circular) [7] and, more recently, a multileaf collimator (MLC) [8] allow shaping beams of variable aperture at any position. Three exemplary cases are shown in Fig. 5.1.

M. Schlüter · A. Schlaefer (✉)
Institute of Medical Technology and Intelligent Systems, Hamburg University of Technology, Hamburg, Germany
e-mail: matthias.schlueter@tuhh.de; schlaefer@tuhh.de

D. Schmitt
Department of Radiation Oncology, Heidelberg University Hospital, Heidelberg, Germany
e-mail: Daniela.Schmitt@med.uni-heidelberg.de

C. Fürweger
Europäisches Cyberknife Zentrum München-Großhadern, Munich, Germany

Department of Stereotaxy and Functional Neurosurgery, University of Cologne, Faculty of Medicine and University Hospital Cologne, Cologne, Germany
e-mail: christoph.fuerweger@cyber-knife.net

A. Schweikard
Institute for Robotics and Cognitive Systems, Universität zu Lübeck, Lübeck, Germany
e-mail: schweikard@rob.uni-luebeck.de

© Springer Nature Switzerland AG 2020
A. Conti et al. (eds.), *CyberKnife NeuroRadiosurgery*,
https://doi.org/10.1007/978-3-030-50668-1_5

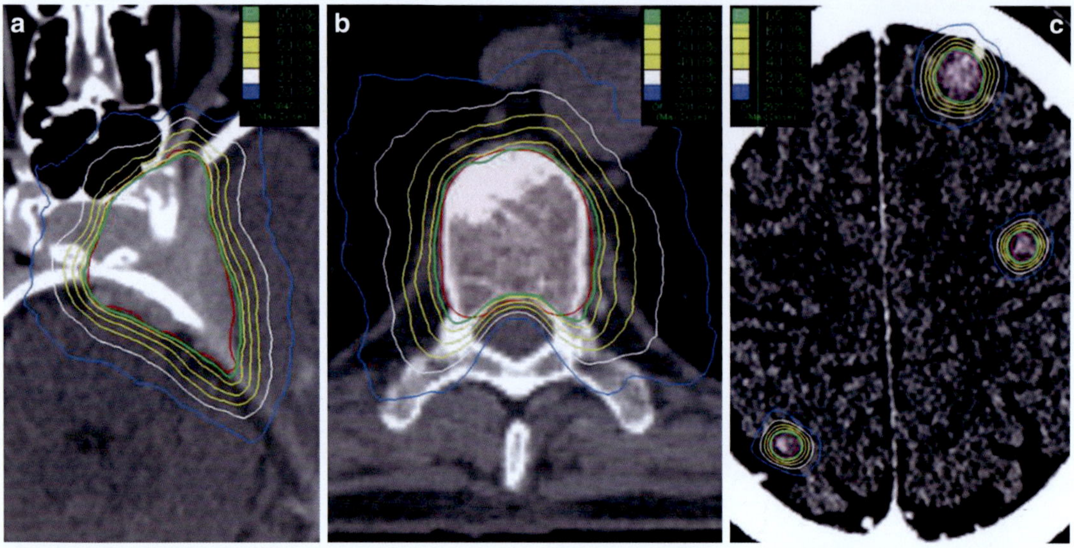

Fig. 5.1 Characteristics of treatment plans in robotic radiosurgery: The dose distribution can be shaped to closely match targets of irregular shape (**a**: Meningioma, 5 × 5 Gy@65%, MLC with Monte Carlo), to selectively spare a radiosensitive organ at risk in close proximity (**b**: Thoracic spine metastasis adjacent to the spinal cord, 1 × 21 Gy@65%, MLC with MC), or to cover multiple lesions with minimal exposure of the intermediate healthy tissue (**c**: Multiple brain metastases, 1 × 19 Gy@65%, Iris with RT)

This has been also reflected in the planning software, which provides several interfaces and underlying mathematical models to facilitate the search for acceptable and ideally optimal treatment plans.

While delivery is currently limited to a step-and-shoot mode, the MLC has recently also been proposed for experimental continuous irradiation [9, 10].

Inherent to treatment planning is the conflicting nature of the goals. As high-energy photon beams pass through the whole body and deliver dose to all structures in between, there is no way to avoid healthy tissue when delivering dose to a target. Note, that the choice of margins may further complicate the planning task, e.g., when a PTV is defined such that it overlaps with an OAR. Typically, beams will vary with regard to efficiency, e.g., passing through less or less dense tissue before reaching the target region. However, generally only a larger number of overlapping beams will shape a dose distribution conforming to the PTV shape. The non-isocentric and highly non-coplanar geometry of the CyberKnife provides particular freedom to shape the dose distribution (Fig. 5.2). Understanding the intricate relationship between many individual beams and the resulting dose distribution is generally impossible and beams and their respective monitor units result from inverse planning, i.e., from solving an optimization problem. On a more

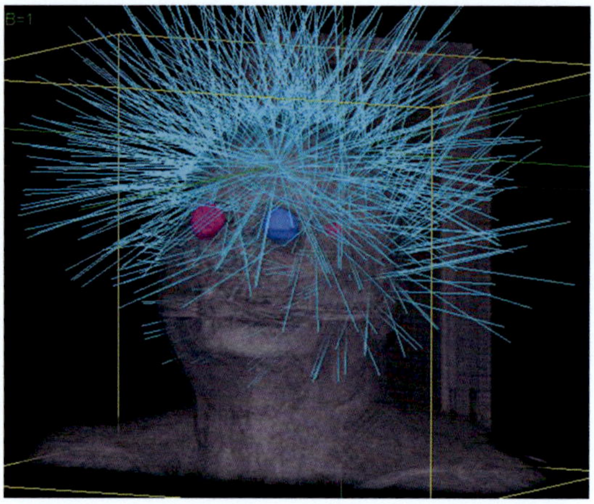

Fig. 5.2 Example of the highly non-coplanar and non-isocentric approach: 3D view of the 756 beams in light blue for a case with 9 brain metastases. The spared eyes can be seen as well

abstract level, different possible dose distributions need to be evaluated and treatment planning essentially means searching for acceptable trade-offs between conflicting goals. These include the coverage, i.e., how much of the PTV actually receives at least the prescribed dose; the homogeneity, i.e., how equally distributed the dose in the target is; and the conformity, i.e., how well the prescribed dose follows the PTV shape.

In this chapter, we will provide an overview of treatment planning for the CyberKnife. Starting with an introduction to define the relevant volumes of interest and clinical goals, we will then describe the different collimators and beam orientations which are available. Afterward, we define the mathematical treatment planning problem and introduce the two optimization strategies of the CyberKnife system to approach this problem. Finally, we provide a practical guide with general considerations for planning of intra and extracranial treatments.

5.2 Basic Clinical Aspects of Treatment Planning

The basis of all treatment planning procedures is volumetric image data of the involved region of the patient. For dose calculation purposes a computed tomography (CT) data set is used as primary data set and depending on treatment site and indication additional secondary image data sets are necessary for precise definition of target volumes and OARs. Examples include contrast-enhanced CT, MRI, PET, and angiography image data. These secondary image sets have to be registered to the primary CT in the treatment planning system.

Contouring follows the principles given by the ICRU reports, with the most recent versions accounting for intensity modulated [1] and stereotactic [2] treatments, while the indication specific volume definition is described in the clinical chapters of this book. Generally, all VOIs are contoured on the appropriate image data. After defining the target volume, for tumorous diseases mostly the gross tumor volume (GTV), a clinical target volume (CTV) and a planning target volume (PTV) need to be defined. The CTV contains all structures which should receive a certain therapeutic dose, and the PTV is a geometric expansion to ensure that the target receives this dose under consideration of all uncertainties of the treatment (margin concept). In general, the clinical goal will be to deliver a sufficient therapeutic dose to the target volume, while the organs at risk should be spared from critical doses based on ALARA (As Low As Reasonably Achievable) principle for healthy tissue. Of course, there are cases, in which these goals are conflicting, e.g., if the target volume is very close to an OAR. Then the PTV will potentially even overlap with the OAR, and it is a clinical decision (under inclusion of the patient) which of the conflicting goals should be prioritized. Depending on the target volume, an inhomogeneous dose distribution may be desirable inside the target volume to achieve local high doses inside the GTV and at the same time keep the dose at the border between healthy and tumorous tissue acceptable (e.g., for brain metastases). For OARs there are in principle two concepts of critical doses, an absolute maximum dose in a small volume for serial OARs (e.g., the spinal cord) and dose volume limits for more parallel organized OARs, like the common average dose constraint for the cochlea. To ease the potential conflict of target and OAR dose, the most important method is the distribution of the radiation dose to more than one portion (i.e., fractions) on several days. In the context of this book, this may range up to approximately 7 fractions.

To safely deliver the high radiation doses typical for stereotactic treatments, the dose distribution has to fulfill the following properties: the shape of the prescribed isodose surface has to be highly conformal with respect to the target volume, the dose gradient toward OARs needs to be very steep, and generally the dose gradient around the target should be as steep as possible to spare surrounding healthy tissue and the dose outside the direct proximity of the target should be very low. This can be achieved by delivering a large number of treatment beams distributed over a large solid angle combined with a very precise beam delivery technique, continuously correcting for small changes in the tumor's position. Quantitative parameters to determine and compare these dose distribution properties will be defined in the next sections.

5.3 Basic Physical Aspects of Treatment Planning

The fundamental idea of stereotactic treatments is that beams from many different directions are used to shape a highly conformal dose distribution with steep gradients. The CyberKnife's robotic linac allows delivering beams from a large solid angle around the patient and with virtually arbitrary orientation. Each treatment beam starts at one of currently up to about 180 positions or beam nodes. The actual number of available nodes depends on the selected targets, paths, and the type of collimators. While ideally these nodes would be broadly distributed around the patient, they are practically limited because there has to be a safe path for the robot to reach all of them efficiently [11]. Further restrictions limiting the feasible beam directions can be defined, e.g., like exit-only areas that may only be passed by a beam after it has already passed the target. Such restrictions account for practical aspects, e.g., if the image data does not cover the full beam and hence the dose calculations would be infeasible for the respective area. As an alternative to the non-isocentric setup it is also possible to define one or more virtual isocenters in which beams intersect to mimic isocentric treatments.

The shape of a beam is determined by the collimator mounted to the linac. In the original version of the CyberKnife system, 12 fixed-sized cylinder collimators with field sizes from 5 to 60 mm were available to form conical beams. While these collimators have favorable dosimetric properties, re-running the chosen path with each collimator is time-consuming. As an alternative, the Iris™ collimator was introduced [7]. It shapes the 12 circular apertures using an electromechanical iris and hence beams of different diameters can be quickly realized.

More recently, the InCise 2 multi-leaf collimator (MLC) has been introduced as an alternative [8, 12]. It consists of 26 pairs of motorized leaves with a width of 3.85 mm, which can be moved individually to form apertures with arbitrary shape. Depending on the target and the clinical goals, using MLCs can substantially reduce the treatment time while providing similar plan quality [13, 14].

In principle, two different approaches can be used to derive the beam apertures during optimization. In beamlet-based inverse planning, the beams or fields are discretized into a typically regular grid, and each grid element or beamlet is individually weighted during plan optimization. The advantage is that the beam apertures can be readily obtained by searching for connected regions of similar weight within the grid. However, there may be impractically many such regions and the actual aperture area will be different from that of the individual beamlets, resulting in differences in the dose calculations.

Hence, another approach is to first define a number of candidate beams with known aperture, to perform the actual dose calculations, and to use inverse planning to select a subset of beams that will be used for treatment. For example, the projection of the PTV from a node's eye view can serve as a promising start for MLC apertures. Since the actual apertures are known before optimization, this approach is also called Direct Aperture Optimization (DAO).

Firstly, note that the way the openings are modeled—using a circular, Iris™ or multi-leaf collimators—makes no difference from a conceptual point of view. Secondly, both approaches are incomplete: the beamlet-based method ignores that the actual dose depends on the area of the aperture, and therefore the resulting beams are not physically optimal and the direct aperture method simply cannot consider all possible beam openings and therefore the truly optimal set of beams may be missing.

Thirdly, the lack of completeness is somewhat mitigated by the fact that typically many different beam sets exist which result in similar dose dis-

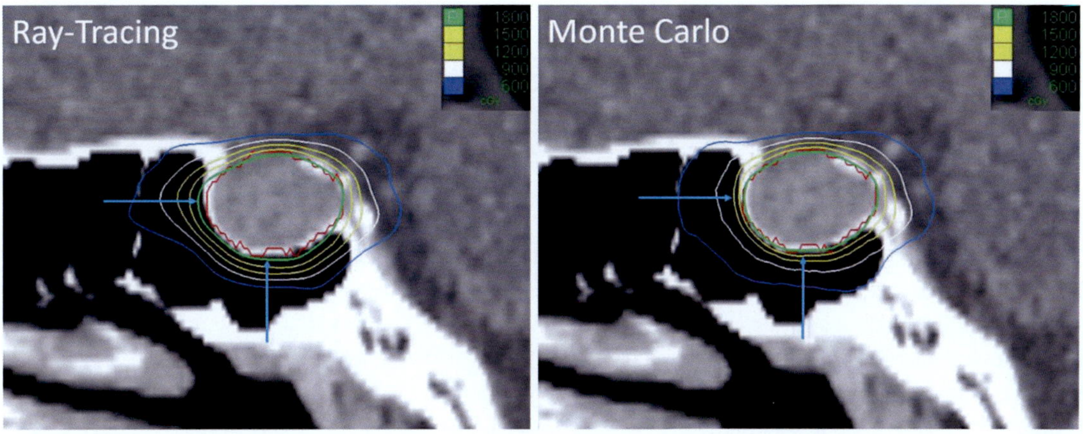

Fig. 5.3 Ray-tracing vs. Monte-Carlo calculated dose distributions for a pituitary adenoma case with a prescription dose of 1 × 18 Gy (green) in sagittal view. In comparison to Monte-Carlo, ray-tracing overestimates the dose along the border to the nasal cavities (blue arrows)

tributions, particularly with respect to clinical planning goals. This also allows considering more elaborate heuristics, e.g., based on machine learning methods [15]. Finally, a notable fact is that the CyberKnife does not use a flattening filter allowing for higher dose rates [16]. Hence, the dose profile of a beam is not constant but shows a continuous decay from a center point, making direct aperture optimization preferable.

The actual dose calculations for the current CyberKnife treatment planning are realized with different algorithms. The most elaborate is based on Monte Carlo simulations and typically provides the highest accuracy [17, 18] but also requires extensive computational effort. For faster estimates a ray-tracing algorithm is used for circular apertures (Fig. 5.3) and a finite-size pencil-beam (FSPB) algorithm for MLC apertures [19]. Monte Carlo-based algorithms are especially needed in geometries with large density inhomogeneities, e.g., in or near the lung. All algorithms estimate the dose in Gray per monitor unit that a beam delivers at a point inside the patient. Typically, the dose is estimated for each CT voxel. The resulting dose coefficients per beam and voxel are stored in a matrix D. Multiplying D with a vector of monitor units $\boldsymbol{x}$ yields a vector $\boldsymbol{d}$ containing the delivered dose for every voxel. Note that for optimization only a subset of the voxels is considered, e.g., voxels inside relevant VOIs.

5.4 Basic Mathematical Aspects of Treatment Planning

On an abstract level, treatment planning represents the problem to search for a set of beams that deliver an acceptable dose distribution that best fulfills the clinical goals. To find these beams, the clinical problem is typically represented as a mathematical optimization problem. The dose at a specific point inside the patient depends on all the tissue passed by the beams before reaching that point, and therefore there is no simple analytic form to compute it. Instead, the dose is estimated at discrete points, e.g., the centroids of the CT voxels. Also, as discussed before, the CyberKnife can in principle generate infinitely many different beams, and hence not all beams can be considered when modeling the optimization problem. A typical approach to address this for CyberKnife planning is DAO with a finite set of N candidate beams, e.g., $N = 6000$. Considering a column vector $\boldsymbol{x} = (x_1, \ldots, x_N)^{\mathrm{T}}$ of monitor units per beam, M points where the dose is estimated, and the dose deposition matrix D with elements d_{ij} representing the dose per monitor unit, point i, and beam j, the discrete dose distribution at the points is given as a vector $\boldsymbol{d} = D\boldsymbol{x}$. For a specific VOI v, its dose vector $\boldsymbol{d}_v$ is obtained from the corresponding submatrix D_v, i.e., the rows of the full matrix corresponding to voxels of this VOI.

To guide the actual search for suitable beams, the clinical goals need to be expressed as objective functions. A simple approach is to minimize the difference between the clinically desired dose distribution $\boldsymbol{d}$ and the actually realized dose distribution $\boldsymbol{d}$. Note that mathematically it is convenient to minimize the squared difference, as the resulting optimization problem is convex and efficiently solved using some variant of gradient descent. In practice, however, not all differences will be zero and different deviations for different VOI have different clinical importance. This can be reflected by coefficients c_v in the objective function, often called importance factors, leading to $\sum_{v=1}^{V} c_v \left(\tilde{\boldsymbol{d}}_v - \boldsymbol{d}_v\right)^{\mathrm{T}} \left(\tilde{\boldsymbol{d}}_v - \boldsymbol{d}_v\right)$, where v ranges over all V relevant VOIs. One disadvantage of this approach is the lack of strict bounds on the deviation from the desired dose, particularly, if a term penalizing the total monitor units is added. An alternative approach is to define lower and upper bounds on the dose in the PTV and upper bounds on the dose in OARs and to minimize $\sum_{i=1}^{N} x_i$. The resulting linear program can be solved efficiently. Note that all bounds for a VOI v can be expressed as $D_v\boldsymbol{x}_v \geq \boldsymbol{b}_v$ or $D_v\boldsymbol{x}_v \leq \boldsymbol{b}_v$ and that in this simple form the problem may be infeasible.

While some clinical goals with respect to the dose in voxels or VOIs can be expressed as convex objective functions, this is not universally true. For example, dose-volume constraints require that the volume exceeding a certain dose is limited. Considering the discrete representation of VOIs by sets of voxels, this is equivalent to asking that at most k out of the M_v voxel representing VOI v have a dose larger than some bound b_{DVC}. Technically, any subset of k voxels could receive a larger dose, which clearly shows that this is a combinatorial problem. Note that the objective to optimize the PTV coverage at a desired dose level has a similar structure. Likewise, a clinical goal to limit the number of beams used in a treatment to at most l would require the optimization method to consider all possible combinations of l beams.

While in principle it would be straightforward to extend the linear program illustrated before into a mixed-integer program to account for combinatorics, the run-time is typically prohibitive when solving practical planning problems to optimality.

Instead, approximations and heuristics are used to identify locally optimal solutions. Note that, as discussed before, global optimality is virtually impossible to achieve due to the finite discretization with respect to the beams and typically many local solutions have similar objective values, i.e., the resulting treatment plans are often still near optimal.

To further shape the dose distribution without limiting each individual voxel, it is typical to define so called virtual VOIs. This is particularly useful to control the gradient of the dose around the PTV by setting up a shell-like structure surrounding the PTV (Fig. 5.4). Considering all the VOIs and the respective clinical goals, it is clear that treatment planning represents a multi-criteria optimization problem. Typically, different solutions to the optimization problem have different advantages, e.g., while the PTV coverage for one resulting plan may be preferable, the OAR sparing may be preferable for another plan. All plans that cannot be improved with respect to any single objective term without compromising at least one other objective are called Pareto-efficient and form the Pareto-frontier. These plans represent the reasonable trade-offs and the choice among these plans depends on the clinical preference. Note that in principle the level of abstraction for representing clinical goals can be more fine grained, e.g., by adding substructures as VOIs and approaches to consider voxel-level trade-offs have been proposed [20]. However, only clinical goals reflected by mathematical objective terms should be compared. For example, studying the dose gradient outside the PTV is only reasonable if actual or virtual VOIs that map goals with

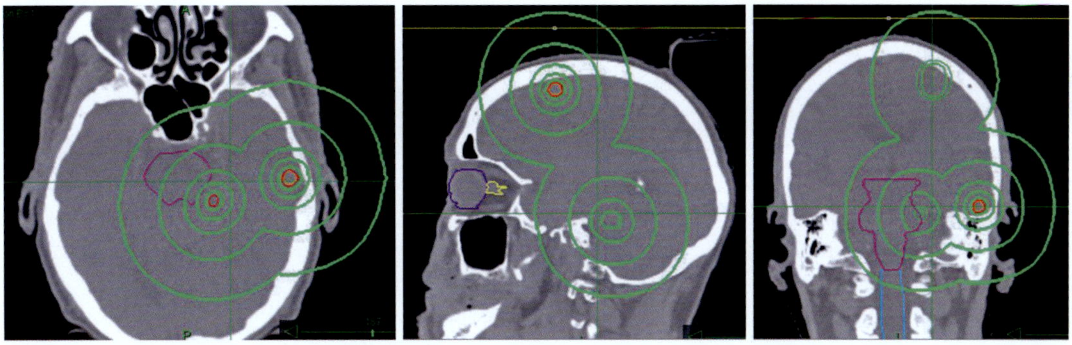

Fig. 5.4 GTVs (orange) and PTVs (red) surrounded by shells (green) to control the dose gradients for a case with three brain metastases

respect to the shape of the dose distribution outside the PTV to mathematical objectives have been defined.

The current planning system for the CyberKnife primarily offers two ways to optimize treatment plans. The sequential optimization method approaches the search for an optimal trade-off as a sequence of steps that optimize with respect to a single objective at a time. The key advantage is that plan quality can be preserved by hard constraints after each step, which guarantees that no unintended degradation occurs subsequently. Another approach called VOLO implements a more conventional gradient descent-based optimization which is typically fast but offers no constraints. We will present both methods in more detail.

5.5 Sequential Optimization

The sequential optimization approach employs linear programming to realize step-wise way of multi-criteria optimization [21, 22]. It considers several objective functions but only optimizes one at a time while maintaining strict bounds on all other values. Thereby, no weighting factors between different objectives have to be tuned. Objective functions can be defined with respect to:

- Maximization of the minimum dose (OMI) in a target VOI, e.g., the GTV
- Maximization of the coverage (OCO) of the PTV
- Optimization of the conformity (OCI) by minimizing the maximum dose in a shell
- Minimization of the maximum dose (OMA) in an OAR
- Minimization of the mean dose (OME) in an OAR
- Minimization of the total monitor units (OMU)

All these objectives share the property that they are linear functions or can be approximated by linear functions. In addition to the vector of variables $\boldsymbol{x}$ denoting the beams' monitor units, a vector of auxiliary variables $\boldsymbol{s}$ measuring how far the dose in a voxel deviates from a dose bound is introduced. Thereby the constraints for the dose in each voxel of the optimization problem can be written as $D\boldsymbol{x} + \boldsymbol{s} = \boldsymbol{b}$. It is clear that for any value that some component s_i takes it is possible to set s_i to zero by subtracting its value from the corresponding right-hand-side entry b_i, i.e., $b_i' = b_i - s_i$, without changing the constraint. Hence, it is also possible to only allow components of $\boldsymbol{s}$ that are related to one clinical goal to take values larger than zero and to minimize their sum. For example, if $\boldsymbol{s}_{\mathrm{PTV}}$ is related to the subset of constraints for PTV voxels then by setting the respective $\boldsymbol{b}_{\mathrm{PTV}}$ to the desired PTV dose, converting all other constraints into the form $D\boldsymbol{x} = \boldsymbol{b}'$, and minimizing the sum of the components of $\boldsymbol{s}_{\mathrm{PTV}}$, the total deviation to the desired PTV dose will be minimized while strictly maintaining all constraints defined for other voxels. Note, that the variables $\boldsymbol{x}$ are not part of the objective, but they are part of the constraints, including constraints containing the variables s_{PTV}. Minimizing the sum of the components of s_{PTV} implies that the delivered dose has to increase to maintain equality, i.e., the x will reflect a different set of active beams. But writing $Dx = b'$ and minimizing the sum of all monitor units is also possible and results in the set of beams with lowest total monitor units delivering the doses described by b'. Note that the monitor units for each individual beam or the sum over all beams can also be limited, e.g., as $\sum_{i=1}^{N} x_i \leq b_{MU}$.

While optimizing coverage is generally not convex, it can be approximated by minimizing the total deviation from the desired PTV dose as outlined before. When the desired coverage approaches 100%, the deviation will approach zero and the approximation is accurate. The search for the best clinical trade-off can now be approached by defining reasonable upper bounds on the dose and the total monitor units and starting to optimize PTV coverage. If coverage is (almost) 100%, further clinical goals, e.g., regarding OAR sparing or total monitor units can be addressed in subsequent steps. However, given the multi-criteria nature of the problem, the resulting coverage may not be acceptable. In this case it is clear that some of the constraints need to be relaxed to further increase the coverage. Any set of constraints, e.g., regarding some OAR, and in principle constraints for any individual voxel can be deliberately set to new values and the optimization of PTV coverage can be repeated. While the constraints guarantee that the new bounds will not be violated, there is no guarantee that the amount of relaxation was sufficient, i.e., this sequence of relaxation and optimization steps may need to be repeated until the trade-off is acceptable. Note that the steps of the sequential optimization approach can be stored and applied to futures cases, often resulting in good initial plans if the treatment geometry is comparable.

5.6 VOLO and MLC Optimization

The VOLO algorithm represents a conventional approach to optimize treatment plans. It does not maintain strict bounds and simply solves for the weighted least squares deviation between desired and the actual doses, i.e., using an objective function of the form $\sum_{i=1}^{O} c_i \left(\tilde{d}_i - d_i \right)^2$ consisting of O individual objectives. However, VOLO only considers single-sided deviations. This allows to specify, for example, desired upper doses and only voxels exceeding the desired limits contribute to the objective. Further generalization of this structure also allows for dose-volume objectives.

Optimization of such objective functions using a gradient descent-based optimizer is typically fast, and deviations are small for many voxels. However, maintaining dose bounds and finding appropriate trade-offs requires varying the coefficients c_i. Depending on the complexity of the planning problem multiple iterations of adapting the coefficients may be required.

VOLO allows employing beamlet-based optimization for MLC treatments in an iterative fashion. In a first fluence-map optimization, the monitor units of neighboring beams are smoothed by adding an appropriate term to the objective function. From these smoothed maps, segments are assembled, which are meaningful with respect to practical realization and subject to acceptable dosimetric uncertainty. Subsequently, the monitor units of the segments are optimized, and their shape is further fine-tuned by loosening the grid structure.

Like other similar implementations, VOLO realizes a fast general-purpose search for treatment plans which will often produce acceptable plans quickly [23]. However, the lack of explicit dose bounds and the side effects of altering one coefficient in the objective function on all other individual objectives can make it tricky to use in more complex cases. To find a plan actually satisfying all clinical goals appropriately might require extensive and potentially unintuitive tuning of the coefficients.

5.7 Practical Treatment Planning for Intra- and Extracranial Lesions and Plan Evaluation

The neuro-radiosurgical targets addressed in this book will almost always require a static treatment planning CT and, as breathing motion is often negligible, four-dimensional or respiratory phase-dependent imaging is typically not considered. However, depending on treatment site and target size, the CT has to fulfill some requirements: The slice thickness should not exceed 1 mm for intracranial or small spinal targets, while a maximum slice thickness of 1.5 mm is suitable for larger extracranial targets. The treatment planning CT for intracranial cases has to include the complete head and head rest plus a minimum of 1 cm air in superior and anterior direction to enable good automatic image registration with the X-ray images during treatment delivery. For extracranial treatments, the field of view has to be extended by 15 cm from the target in both longitudinal directions to cover the full path of beams through the patient and hence to allow correct dose calculations for non-coplanar beams. For the same reason, all patient setup devices like thermoplastic head masks or extracranial vacuum cushions, which will be used during treatment, have to be present in the same way when acquiring the treatment planning CT. Note that regions not covering all material along a potential beam path and regions where artifacts are visible need to be contoured and blocked with a so-called "exit only" blocking structure. For example, this is often the case for the arms. Generally, the smallest field of view which includes the whole geometry should be used in order to maximize the resolution, i.e., to realize the smallest in-plane pixel size.

Secondary imaging also has to include the region necessary for target or OAR delineation and enough information (e.g., field of view) to perform a reliable image registration with the planning CT in the treatment planning system. For automatic registration a region of interest within a nearly static region around the target should be chosen to avoid effects of irrelevant structures on the registration, e.g., the cervical spine for an intracranial treatment. After registration of the secondary images, all VOIs should be contoured on the image set best suited in terms of contrast or resolution. Contours are automatically available in the CT data set for dose calculation and optimization. Further geometric operations can be applied, e.g., adding a margin to form the PTV, subtracting overlapping regions to avoid conflicting goals during treatment planning, or combining multiple disjoint volumes into a single logical PTV_{total}. The latter is particularly useful for multiple targets. Since the introduction of Precision treatment planning system for CyberKnife treatment planning, it is also possible to import VOIs and isodoses from previous treatments using standard DICOM files, which is particularly helpful for patients returning with additional brain or bone metastases. For example, this information can be used to block pre-treated structures or to define OAR regions with lower dose constraints.

Comparing intra- and extracranial treatments, the geometric uncertainty is typically larger for the latter, as reproducibility of patient position is complicated by less rigidity of setup devices and a larger deformability around the target volume. Such deformations may degrade the accuracy of the automatic image registration between DRRs derived from the planning CT and the X-ray images acquired during the treatment. This should be considered when choosing PTV margins, which typically range from 0 to 2 mm. Also, besides the standard supine position, for extracranial treatments using the "XSight Spine" tracking mode, it is possible to position the patient in prone position. This has the advantage that more beams with a shorter path length before reaching the target can be used for spinal treatments, typically resulting in less dose in the bowels and thorax. A disadvantage of prone positioning is respiratory motion affecting the spine. While this may be compensated using the "Synchrony Respiratory Tracking System", motion tracking is usually less accurate than static spine tracking and superior beam access must be carefully weighed against the need for larger margins because of higher uncertainty during beam delivery [24].

Another step in the treatment planning is the selection of the alignment center position within the treatment planning CT. The alignment center is the point in the treatment room where the central axes of the X-ray sources intersect and defines the virtual isocenter for dose calculations. This means, the choice of the alignment center determines the required absolute position of the patient inside the treatment room and the patient region inside the field of view of the X-ray sources during treatment. For intracranial treatments, the skull should be visible with most information included and about 1 cm air in superior and anterior direction for the automatic registration. For extracranial treatments (here always based on "XSight Spine" tracking mode), the aim is to position the alignment center as near as possible to the treated lesion with the whole vertebras inside the grid. In the cervical or lumbar spine or the sacrum, problems can occur due to overlapping of the spine with the jaw or the pelvic bones in the DRR and the X-ray images or due to a larger distance between alignment center and treated lesion, potentially decreasing tracking accuracy. In these cases, the alignment center should be chosen under consideration of all these aspects, possibly taking into account larger PTV margins [25].

The actual beams that can be generated are defined by a so-called path. A path includes the collimator, the body region, and the number and distribution of beam nodes that can be used for beam generation. Often it is preferable to choose the "full path" to provide maximum flexibility for the subsequent optimization. Moreover, typically circular fields are used for small nearly spherical targets, like most brain metastases, or irregular shaped targets with small extensions like some AVMs, while the MLC may be preferable in larger, irregular shaped targets like resection cavities and many extracranial targets.

In the following, a short introduction in practical treatment planning with Precision version 2.0 will be given, providing examples for planning with circular fields and the sequential optimization algorithm and for planning MLC based treatments using the VOLO algorithm.

5.7.1 Practical Optimization with the Sequential Optimization Method

The principle of sequential optimization is covered in Sect. 5.6 and in [22], and a step-by-step manual on treatment planning using the sequential optimization for intra- and extracranial targets is given in the discussion and supplementary material of [26]. Here, an overview will be given with additional suggestions for frequent cases. The main user steps during optimization are the following in an iterative process, while the number of iterations depends on the complexity of the case:

- Definition of maximum or dose volume constraints for OARs and targets
- Run of a sequential optimization with a target structure as first step and potentially other target or shell or OAR steps (see below)
- Adaption of constraints
- Re-run of sequential optimization with the same or other steps

The whole process is organized as follows:

First, up to three corresponding collimator sizes between 5 and 60 mm have to be chosen per target volume to generate beams targeting at the respective target volume. Typically, it is sufficient to consider the orientations toward the PTV, as the large number of beams implies that there is flexibility to shape the dose inside the PTV, e.g., to boost the dose in the GTV. While the best collimator sizes depend on size, shape, and location of the PTV and the surrounding OARs, a rule of thumb is to choose one collimator to be at the same size or slightly smaller than the diameter of the lesion, while smaller collimators can be used in addition to shape the dose distribution to irregular formed parts of the target. If multiple targets are to be treated, the collimators should be chosen accordingly for each target.

As mentioned before, shell structures are defined to shape the dose conformity to the target and the dose fall-off outside the target region. To avoid hotspots due to beam entrance, additional shells in larger distance to the target can be defined and the number of MU per node may be limited. For the treatment of multiple targets, shell structures for each target have to be defined, completed by middle-sized and large shells around the PTV_{total} to reduce hotspots caused by crossing beams targeting at different targets and to reduce skin dose.

The run-time of the initial "dose volume generation" operation can be influenced by the definition of the so-called dose optimization box around, which should cover the target volume, the relevant OARs, and regions where the dose gradient is critical. Only for voxels inside this box the dose will be calculated and optimized, i.e., setting a smaller box reduces the number of voxels in the optimization problem. To guarantee low dose in OARs in larger distance from the target, VOIs can be blocked with status "never," meaning that beams will not pass through these structures and therefore no constraints, objectives, or dose volume histogram calculations need to be considered.

In the sequential optimization, hard constraints are used. This means, if, for example, the brain stem maximum dose is set to 2 Gy, this will always be met, if the optimization is performed in high resolution, i.e., CT resolution, and all voxels of the OAR are included in the dose optimization box. Note, that a dose-cut-off per beam and per voxel of 0.01 cGy/MU will result in a small deviation in final dose calculation in the evaluate tab, i.e., the final dose will always be a bit higher relative to the target dose than during optimization, making a max dose constraint about 0.5–1 Gy lower than finally desired necessary.

In addition to OAR constraints, a maximum dose for the target structures has to be set to shape the dose-volume-histogram and to define the prescription isodose. For example, a brain metastasis may be treated with 20 Gy to the 70% isodose line. The PTV maximum dose has to be set to 100% = 28.57 Gy, and the first optimization step can be OCO w.r.t. the PTV with the objective to cover the PTV at 20.25 Gy. Note the optional extra 0.25 Gy can be considered extra slack, i.e., the constraint can be relaxed to 20 Gy later in the optimization process. If the OAR constraints are compatible with the target dose goals, they can be maintained throughout the optimization process. If the OAR dose is in conflict with the target dose, the easiest way to quantify this conflict is to adjust the OAR constraint and repeat the optimization with an objective for a target structure (e.g., OCO w.r.t. the PTV). Once a satisfying OAR dose can be reached while preserving target coverage, additional steps can be added to the sequential optimization. Typically, these steps include optimization to minimize the shell dose (OCI with 0.01Gy objective, 0 is not possible, and, e.g., 0.2 Gy as relaxation value) in the order of increasing shell distance. For the treatment of multiple targets, the PTV_{total} is optimized in the first step followed by the shells in increasing order per target beginning with the smallest target. The results of these optimizations can be transferred to a maximum constraint of the shells (and deletion of the shell optimization steps) and can eventually be further optimized manually under preservation of target coverage. For the treatment of multiple targets, it may be necessary to include single targets in additional sequential steps in the progress of optimization. Additional dose constraints on tuning structures between target volumes may be necessary to avoid dose bridges in between. The mean dose optimization (OME) can be used to "push" low-to-middle dose outside an OAR. For example, if the brain stem has an 8 Gy maximum because of a nearby target volume, it is possible to significantly reduce the brainstem volume getting 4 Gy by the OME optimization.

If two levels of prescription doses are aimed at different targets or as an integrated boost concept, these targets must be optimized as the first two steps, while the optimal order must be tested for the specific case. In principle, the order of the sequential steps reflects the priority of the goals. For complex cases, this priority must be explored in combination with the objectives and their dose values.

A possible problem when optimizing conformity with collimators smaller than the lesion diameter is the occurrence of so-called donut-shaped dose distributions with the highest doses near the boundary and lower dose region in the center of the lesion. This is typically related to an unfavorable choice of collimators, and there are different ways to address this problem: larger and/or smaller collimators can be chosen, an inner region of the target (e.g., PTV minus 3 mm) can be defined and optimized with a higher dose, or the maximum total MU constraint can be modified.

After a satisfactory dose distribution has been reached, it may be necessary to reduce the treatment time. There are two possibilities to reach that goal: the time reduction and the node/beam reduction process. In the time reduction process, the desired treatment time can be defined and via several iterations of beam generation and optimization the treatment time is decreased heuristically. Subsequently, the different solutions can be evaluated and a decision on trade-offs between plan quality and treatment time can be made.

Additionally, and for multiple targets with a better result, a node or beam reduction can be performed by defining a lower MU limit per node or beam. All nodes or beams with less MU will be deleted, and the weightings of the residual beams will be re-optimized. This can be done repeatedly. At the end, a final beam reduction shall be performed by deleting all beams with less than 5–10 MU per fraction to guarantee a high-quality delivery due to the accelerator's limited dose-to-MU linearity.

5.7.2 Practical Optimization with the VOLO Method

Whereas sequential optimization uses "hard" dose bounds and a hierarchical order for planning goals, VOLO is based on a single multi-criteria cost function that includes all individual "soft" objectives at the same time, which are to be prioritized by different penalty weights. As a consequence, the practical approach to create a plan with VOLO is so different from sequential optimization that previous experience of the planner with sequential optimization is of very limited use. In particular, MLC optimization with VOLO introduces a new two-step process—fluence optimization followed by segment optimization—which shall be described here in detail from a practical point of view.

As a first step of VOLO optimization, all clinical goals need to be formulated by assigning up to five objectives to each VOI, which can include maximum dose, minimum dose (targets only), and dose-volume objectives. These numerical objectives can be entered manually or via click, drag, and drop in the dose-volume histogram (DVH) view. A penalty must be applied to each objective on a weight scale of 1–1000, representing the importance of the corresponding clinical goal. For most cases, the list of objectives will at least include:

- The intended prescription dose (e.g., a minimum dose of 20 Gy covering 98% of the PTV)
- The maximum dose to the target (to meet a specific prescription isodose level, e.g., to the 70%)
- A maximum dose and/or DVH constraint to each OAR in proximity to the target
- A maximum dose to an inner tuning shell (to achieve a conformal dose distribution)
- A maximum dose to an outer tuning shell (to shape the dose fall-off)
- A maximum dose to a skin ring structure (to avoid dose hot spots in beam entry zones)

In addition, the number of nodes available to create the initial beamset needs to be specified. Choosing the maximum value is often a reasonable starting point as it enables highest flexibility for beam access, potentially improving plan quality at the expense of some robot travel time. In case the final treatment time turns out to be unacceptably long, optimization may be rerun with a lower number of nodes.

At this point, fluence map optimization can be initialized. The progress and current state of the solution is represented by means of DVHs only. The process is interactive and fast, i.e., goals can be changed during run-time and are immediately reflected by an adaptation of the solution. For simplicity, it is often advisable to start with equal penalty weights of 1 for all objectives and then continue with iteratively adjusting weights and dose goals while the fluence map optimization is ongoing. The complexity of the individual fluence maps can be influenced by altering the fluence smoothness penalty as another objective of the cost function.

When the DVHs are considered satisfactory, the transition from ideal fluence maps to actually deliverable MLC segments needs to be made. For this purpose, a set of parameters related to assembling segments (maximum and minimum permissible number of MUs per segment, maximum MUs per beamlet, maximum numbers of beams, segments per beam and nodes to create the initial beamset) is required.

These input parameters are rigorous in the sense that the final solution is guaranteed to satisfy the given values by iteratively adjusting the list of segments and the MU weights during the next segment optimization step.

The aim of segment optimization is to find optimal weights for the segments derived from the fluence maps and to fine-tune the segment shapes.

For this purpose, the list of objectives used during fluence optimization is reused either unchanged or can be adapted at the planner's discretion. Frequently, adaptation is necessary because the fluence map DVHs will deteriorate to some extent due to inherent limitations of deliverable segments. As a consequence, less ambitious goals on OARs or shells as well as an increased weight on the intended prescription dose will likely be required to maintain an acceptable target coverage after segment optimization. Finally, a three-dimensional dose distribution and DVH metrics are presented to the planner to allow for assessment of plan quality.

The absence of hard bounds in VOLO can impact the planning approach. Importantly, while a selected penalty of 1000 puts maximum weight on a particular objective such as a maximum dose limit, this definition still does not represent an absolute, "hard" dose bound. The solution may yet turn out to violate the constraint to some extent, which is an inherent characteristic of optimizing a single cost function with relative objective weights. The practical consequence is that it can be difficult with VOLO to follow clinical protocols relying heavily on strict maximum dose limits to OARs. In order to fulfill such specific dose constraints, several iterations of parameter adjustments and subsequent fluence- and segment optimizations may be required. However, due to the fast optimization process of VOLO, it is quite feasible to explore multiple solutions in a short time and to identify a clinically preferable plan. This is reflected in a recent report of better quality as well as reduced delivery and optimization times for plans with VOLO compared to sequential optimization [23], although it must be noted that an unbiased comparison is difficult, given that VOLO cannot prevent side effects to the dose distribution.

5.7.3 Plan Evaluation

A final step during treatment planning is the plan evaluation, which often considers metrics that are not directly reflected in the underlying optimization problem. These include target and OAR-specific metrics, dose gradients, the dose distribution outside the proximity of the target, and parameters regarding treatment efficiency like treatment time and—related—the number of MU, nodes, or beams.

For the target, two quantitative values are important, the conformity of the prescription isodose to the target and the coverage of the target by the prescribed dose. In the Precision treatment planning system, conformity is expressed by two metrics, the conformity index CI and the new conformity index nCI [27]. In the following formulas, dose conformity is expressed with respect to the PTV. The nCI is defined as the inverse Paddick conformity index [28], where

$$\mathrm{nCI} = \frac{V_{\mathrm{P}}}{V_{\mathrm{PTV,P}}} \frac{V_{\mathrm{PTV}}}{V_{\mathrm{PTV,P}}},$$

with the PTV volume V_{PTV}, the volume covered by the prescription dose V_{P}, and the volume of the PTV covered by the prescription dose $V_{\mathrm{PTV,P}}$. The coverage C is defined as the fraction of the PTV covered by the prescribed dose, i.e.,

$$C = \frac{V_{\mathrm{PTV,P}}}{V_{\mathrm{PTV}}}.$$

The nCI would ideally be 1 and should be as low as possible. For spherical targets, an nCI of <1.1 is often possible. For very irregular targets, values of 1.2–1.4 can be acceptable. An nCI very close to 1 will be accompanied with a high coverage for the respective volume. The coverage is important for all types of target volumes. Typically, 100% coverage will be requested for GTV and CTV, while the goal for the PTV will often be at least 98% coverage.

The high dose region inside the target volume should always be evaluated. Preferably, the high dose region (e.g., 95% isodose line) would be an uncompromised volume in the center of the lesion. There are some cases where sensitive structures are inside the target volume, and these structures should be outside the high dose region. The evaluation of OAR doses depends on the type of OAR, and typical metrics are the maximum dose, the near maximum dose, the mean dose, and dose volume constraints. For example, it would be typical to consider the brainstem volume subjected to a dose of 10 Gy or to ensure a mean dose of at most 8 Gy for the cochlea. For intracranial treatments, often the volume of the healthy brain receiving 12 Gy (i.e., V_{12Gy}) is evaluated to estimate the risk of necrosis [29]. Outside the defined VOIs, the dose gradient may be evaluated, for example, defined as a high dose and a low dose gradient following [30]. Additionally, the low-dose isodoses, e.g., down to the 10% isodose line, should be evaluated to estimate the "dose bath," with the aim of having no 20 or 30% isodose islands outside the direct target vicinity for intracranial targets and extracranial targets, respectively. The whole CT should be evaluated regarding unnecessary hotspots far away from the target region.

Treatment efficiency can be assessed on the basis of treatment time. The acceptable treatment time depends heavily on the complexity of the case and the general condition of the patient. For radiation protection reasons, a lower number of MUs should be preferred for the same prescription dose, if it is possible to reduce the MUs without compromising the quality of the plan. Due to the dosimetric precision, a lower number of beams with a higher number of MU each are preferable.

References

1. ICRU. Report 83: prescribing, recording, and reporting photon-beam intensity-modulated radiation therapy (IMRT). J ICRU. 2010;10(1).
2. ICRU. Report 91: prescribing, recording, and reporting of stereotactic treatments with small photon beams. J ICRU. (2014);14(2).
3. Adler JR, Chang SD, Murphy MJ, Doty J, Geis P, Hancock SL. The Cyberknife: a frameless robotic system for radiosurgery. Stereotact Funct Neurosurg. 1997;69:124–8. https://doi.org/10.1159/000099863.
4. Adler JR, Murphy MJ, Chang SD, Hancock SL. Image-guided robotic radiosurgery. Neurosurgery. 1999;44:1299–306. https://doi.org/10.1097/00006123-199906000-00079.
5. Kilby W, Dooley JR, Kuduvalli G, Sayeh S, Maurer CR. The CyberKnife® robotic radiosurgery system in 2010. Technol Cancer Res Treat. 2010;9:433–52. https://doi.org/10.1177/153303461000900502.
6. Kilby W, Naylor M, Dooley JR, Maurer CR, Sayeh S. A Technical Overview of the {CyberKnife} System. In: Abedin-Nasab MH. editor, Handbook of Robotic and Image-Guided Surgery, Elsevier; 2020. p. 15—38. https://doi.org/10.1016/b978-0-12-814245-5.00002-5.
7. Echner GG, Kilby W, Lee M, Earnst E, Sayeh S, Schlaefer A, et al. The design, physical properties and clinical utility of an iris collimator for robotic radiosurgery. Phys Med Biol. 2009;54:5359. https://doi.org/10.1088/0031-9155/54/18/001.
8. Asmerom G, Bourne D, Chappelow J, Goggin LM, Heitz R, Jordan P, et al. The design and physical characterization of a multileaf collimator for robotic radiosurgery. Biomed Phys Eng Express. 2016;2:017003. https://doi.org/10.1088/2057-1976/2/1/017003.
9. Bedford J, Tsang H, Nill S, Oelfke U. Treatment planning optimization with beam motion modeling for dynamic arc delivery of SBRT using Cyberknife with multileaf collimation. Med Phys. 2019;46(12):5421–33. https://doi.org/10.1002/mp.13848.
10. Kearney V, Descovich M, Sudhyadhom A, Cheung JP, McGuinness C, Solberg TD. A continuous arc delivery optimization algorithm for CyberKnife M6. Med Phys. 2018;45:3861–70. https://doi.org/10.1002/mp.13022.
11. Dieterich S, Gibbs IC. The CyberKnife in clinical use: current roles, future expectations. In: IMRT, IGRT, SBRT—advances in the treatment planning and delivery of radiotherapy, vol. 43: Karger Publishers; 2011. p. 181–94. https://doi.org/10.1159/000322423.
12. Fürweger C, Prins P, Coskan H, Heijmen BJ. Characteristics and performance of the first commercial multileaf collimator for a robotic radiosurgery system. Med Phys. 2016;43:2063–71. https://doi.org/10.1118/1.4944740.
13. Jang SY, Lalonde R, Ozhasoglu C, Burton S, Heron D, Huq MS. Dosimetric comparison between cone/Iris-based and InCise MLC-based CyberKnife plans for single and multiple brain metastases. J Appl Clin Med Phys. 2016;17:184–99. https://doi.org/10.1120/jacmp.v17i5.6260.
14. Kim N, Lee H, Kim JS, Baek JG, Lee CG, Chang SK, Koom WS. Clinical outcomes of multileaf collimator-based CyberKnife for spine stereotactic body radiation therapy. Br J Radiol. 2017;90:20170523. https://doi.org/10.1259/bjr.20170523.
15. Schlaefer A, Dieterich S. Feasibility of case-based beam generation for robotic radiosurgery. Artif Intell Med. 2011;52(2):67–75.

16. Xiao Y, Kry SF, Popple R, Yorke E, Papanikolaou N, Stathakis S, et al. Flattening filter-free accelerators: a report from the AAPM Therapy Emerging Technology Assessment Work Group. J Appl Clin Med Phys. 2015;16:12–29. https://doi.org/10.1120/jacmp.v16i3.5219.
17. Okoye CC, Patel RB, Hasan S, Podder T, Khouri A, Fabien J, et al. Comparison of ray tracing and Monte Carlo calculation algorithms for thoracic spine lesions treated with CyberKnife-based stereotactic body radiation therapy. Technol Cancer Res Treat. 2015;15:196–202. https://doi.org/10.1177/1533034614568026.
18. Sharma SC, Ott JT, Williams JB, Dickow D. Clinical implications of adopting Monte Carlo treatment planning for CyberKnife. J Appl Clin Med Phys. 2010;11:170–5. https://doi.org/10.1120/jacmp.v11i1.3142.
19. Jeleń U, Söhn M, Alber M. A finite size pencil beam for IMRT dose optimization. Phys Med Biol. 2005;50:1747–66. https://doi.org/10.1088/0031-9155/50/8/009.
20. Schlaefer A, Viulet T, Muacevic A, Fürweger C. Multicriteria optimization of the spatial dose distribution. Med Phys. 2013;40(12):121720. https://doi.org/10.1118/1.4828840.
21. Lessard E, Kilby W, Dooley J, Sims C, Schlaefer A, Blanck O, Maurer CR. Sequential optimization scripts to facilitate treatment planning for robotic radiosurgery clinical studies for prostate and lung cancers. In: IFMBE proceedings. Berlin Heidelberg: Springer; 2009. p. 1031–4. https://doi.org/10.1007/978-3-642-03474-9_290.
22. Schlaefer A, Schweikard A. Stepwise multi-criteria optimization for robotic radiosurgery. Med Phys. 2008;35:2094–103. https://doi.org/10.1118/1.2900716.
23. Zeverino M, Marguet M, Zulliger C, Durham A, Jumeau R, Herrera F, et al. Novel inverse planning optimization algorithm for robotic radiosurgery: first clinical implementation and dosimetric evaluation. Phys Med. 2019;64:230–7. https://doi.org/10.1016/j.ejmp.2019.07.020.
24. Fürweger C, Drexler C, Muacevic A, Wowra B, de Klerck E, Hoogeman M. CyberKnife robotic spinal radiosurgery in prone position: dosimetric advantage due to posterior radiation access. J Appl Clin Med Phys. 2014;15(4):4427. https://doi.org/10.1120/jacmp.v15i4.4427.
25. Muacevic A, Drexler C, Kufeld M, Romanelli P, Duerr H, Wowra B. Fiducial-free real-time image-guided robotic radiosurgery for tumors of the sacrum/pelvis. Radiother Oncol. 2009;93(1):37–44. https://doi.org/10.1016/j.radonc.2009.05.023.
26. Blanck O, Wang L, Baus W, Grimm J, Lacornerie T, Nilsson J, et al. Inverse treatment planning for spinal robotic radiosurgery: an international multi-institutional benchmark trial. J Appl Clin Med Phys. 2016;17(3):313–30. https://doi.org/10.1120/jacmp.v17i3.6151.
27. Nakamura J, Verhey L, Smith V, Petti P, Lamborn K, Larson D, et al. Dose conformity of gamma knife radiosurgery and risk factors for complications. Int J Radiat Oncol Biol Phys. 2001;51(5):1313–9.
28. Paddick I. A simple scoring ratio to index the conformity of radiosurgical treatment plans. Technical note. J Neurosurg. 2000;93(Suppl 3):219–22. https://doi.org/10.3171/jns.2000.93.supplement.
29. Flickinger J, Kondziolka D, Lunsford L, Kassam A, Phuong L, Liscak R, Pollock B. Development of a model to predict permanent symptomatic postradiosurgery injury for arteriovenous malformation patients. Arteriovenous Malformation Radiosurgery Study Group. Int J Radiat Oncol Biol Phys. 2000;46(5):1143–8. https://doi.org/10.1016/S0360-3016(99)00513-1.
30. Paddick I, Lippitz B. A simple dose gradient measurement tool to complement the conformity index. J Neurosurg. 2006;105(Suppl):194–201. https://doi.org/10.3171/sup.2006.105.7.194.

Small Field Dosimetry

6

Evangelos Pantelis and Argyris Moutsatsos

6.1 Introduction

In stereotactic radiosurgery (SRS), an increased dose of radiation is delivered to a well-defined lesion of the body in one or few fractions. This ablative dose is delivered safely using systems that enable localization of the lesion with stereotactic accuracy and delivery of radiation fields with decreased penumbra. Accurate localization of the lesion involves its delineation on anatomical images (i.e., CT and/or MRI) of increased spatial resolution and accuracy, as well as precise detection of the target inside the treatment room. Radiation field penumbra depends on the field size, as well as beam energy and type (i.e., photons, charge particles). It is known to increase with photon beam energy, and, therefore, generators with nominal accelerating potential of less than 10 MV are typically used in SRS platforms [1, 2]. Decreasing beam energy, on the other hand, affects inversely the dose received by tissues situated at smaller depths relative to the target. This is resolved by using multiple radiation beams focusing on the target but emerging from a set of diverse orientations with respect to the patient body. The use of multiple beams increases also the spatial dose gradient of the delivered dose distributions, thus sparing the surrounding organs at risk. The shape of SRS radiation beams is configured by cylindrical collimators or micro-multi-leaf collimators (micro-MLCs) specifically designed to decrease field penumbra [2]. The need to treat small target volumes with steep spatial dose gradients dictates also the use of small field sizes, which comes at the expense of beam output and, hence, treatment time (beam on). Beam output is reduced in small fields due to the reduction of phantom and head photon scatter component combined with partial occlusion of the primary photon source for the smaller field sizes [2–6]. To compensate beam output reduction and its effect on treatment time, SRS systems use beam generators achieving higher dose rates, up to 2400 MU/min [7, 8], by omitting the flattening filter in the treatment beam generation chain and/or reducing the source to target distance.

As field size decreases, Bragg-Gray cavity theory conditions break down, and the specific geometrical and constructional details of the employed detectors affect their response, increasing the uncertainty of dosimetry measurements

E. Pantelis (✉)
Medical Physics Laboratory, Medical School, National and Kapodistrian University of Athens, Athens, Greece

CyberKnife and TomoTherapy Radiotherapy Department, IATROPOLIS Private Special Clinic, Athens, Greece
e-mail: vpantelis@phys.uoa.gr

A. Moutsatsos
CyberKnife and TomoTherapy Radiotherapy Department, IATROPOLIS Private Special Clinic, Athens, Greece
e-mail: armouts@phys.uoa.gr

© Springer Nature Switzerland AG 2020
A. Conti et al. (eds.), *CyberKnife NeuroRadiosurgery*,
https://doi.org/10.1007/978-3-030-50668-1_6

[5, 9, 10]. Furthermore, the sensitive volume dimensions of a detector relative to the measured field size, as well as the exact positioning of its sensitive volume at the measuring point, could introduce additional dosimetric uncertainties [5, 10]. For example, the use of a Farmer-type chamber having a cavity length of 24 mm for output factor measurements of small beams defined by a micro-MLC without further corrections has been reported as the main cause of an accidental overdosage of 145 patients [11].

To reduce dosimetric uncertainties, the International Atomic Energy Agency (IAEA) and the American Association of Physicists in Medicine (AAPM) have developed international, standardized recommendations for the dosimetry of small and non-standard megavoltage photon fields used in external beam radiotherapy. These recommendations are included in the IAEA Technical Report Series No. 483 (TRS-483) along with a dosimetric formalism for relevant dosimetry measurements [5]. TRS-483 is considered an extension of the IAEA TRS-398 code of practice (CoP) introducing a set of additional detector and field size specific correction factors that should be applied in both reference and relative dosimetry measurements [9].

TRS-483 is relevant to the CyberKnife system and should be used for dosimetry measurements during commissioning and quality assurance procedures. However, it must be noted that the CyberKnife factors given in TRS-483 were not obtained from corresponding Monte Carlo (MC) simulation or measurement studies reported in the literature [12–23], but they were derived from data from other linacs and measurement conditions which are then adjusted to fit to the CyberKnife radiation beams [9]. For example, the reported chamber-specific $k_{Q_{msr},Q_o}^{f_{msr},f_{ref}}$ factors for CyberKnife reference dosimetry measurements were obtained by adjusting previous linac-based k_{Q,Q_o} data to account for beam quality differences and volume averaging effects due to the absence of flattening filter in the CyberKnife system. As a result, the $k_{Q_{msr},Q_o}^{f_{msr},f_{ref}}$ correction factors given in TRS-483 may have some limitations when applied in dosimetry measurements in different generations of the CyberKnife system, due to corresponding differences in the primary beam collimator. Additionally, there are some concerns about the suggested TRS-483 correction factor data for the air-filled microchambers and the PTW-60019 synthetic microdiamond detector for output factor measurements in the small fields of all CyberKnife system generations [24]. In this chapter the challenges encountered in small field dosimetry measurements are discussed, and practical guidelines are given for the application of the TRS-483 CoP in small field dosimetry measurements for the CyberKnife system.

6.2 Definitions and Challenges

6.2.1 Radiation Fields

The CyberKnife offers 3 secondary collimation systems, which include (a) 12 fixed collimators producing circular radiation fields with nominal diameters ranging from 5 to 60 mm defined at 800 mm distance away from the source (source to axis distance, SAD) [25], (b) an Iris variable aperture collimator capable of replicating the same set of 12 circular field sizes [26], and (c) the InCise MLC capable of creating arbitrary field shapes with nominal dimensions ranging from (7.7 × 7.6) mm^2 up to (115 × 100.1) mm^2 [27], defined at 800 mm SAD. More details on the collimation systems can be found in Chap. 3.

6.2.1.1 Field Size

The field size in small field dosimetry is defined as the full width at half maximum (FWHM) of the lateral beam profile determined at a measurement depth sufficient to eliminate the contribution of contamination electrons [5]. Field size is used as a synonym of the irradiation field size defined by the International Electrotechnical Commission [28]. It must be noted that, while the measured radiation field sizes should be used for calculating the corresponding detector-specific correction factors for output factor measurements (see Sect. 6.4.2.1), their corresponding nominal values defined at 800 mm SAD are displayed by the treatment planning system, as well as the electronic patient records of the CyberKnife system.

6.2.1.2 Equivalent Square Field

The concept of equivalent square field is defined for arbitrary rectangular or circular fields as the square field having the same area. Therefore, for a circular field with diameter, D, the corresponding equivalent square field, S_{clin}, is given by:

$$S_{clin} = \frac{\sqrt{\pi} * D}{2} \quad (6.1)$$

For rectangular fields with unequal in-plane and cross-plane dimensions, the size of the equivalent square field is given by:

$$S_{clin} = \sqrt{AB} \quad (6.2)$$

where A and B correspond to the in-plane and cross-plane irradiation field widths, defined as the FWHM at the measurement depth. This equation should be used for radiation fields with dimensions in the range $0.7 < A/B < 1.4$.

It is noted that the above guidance for calculating equivalent square fields is used by the TRS-483 to obtain correction factors for output factor measurements of non-square small fields based on corresponding published data for square fields. For broad photon beams, the equivalent square field size is not calculated on the assumption of the same area; rather it is based on ensuring equal photon scatter contributions (see also Sect. 6.2.3).

6.2.2 Definition of Small Fields

An external photon beam is characterized "small" if at least one of the following three physical conditions is met: (a) there is a loss of lateral charge particle equilibrium on the beam axis, (b) there is partial occlusion of the primary photon source by the collimating devices on the beam axis, and (c) the size of the detector is similar or larger than the cross-sectional beam dimensions at the depth of measurement [2, 5]. The first two conditions are beam related, while the third one is detector related for a given field size. All three of these conditions result in an overlap between the field penumbrae and the detector volume [2–5].

Lateral charge particle equilibrium (LCPE) occurs on the central beam axis if the beam half-width is smaller than the maximum range of the majority of secondary electrons. This width can be viewed as a "lateral charged particle equilibrium range," r_{LCPE}, and can be used to determine quantitatively if a radiation field is small. Monte Carlo calculations have shown that r_{LCPE} can be expressed as a function of photon beam quality specifier $TPR_{20,10}(10)$ or $\%dd(10,10)_X$ suggested by the IAEA TRS-398 [29] or AAPM TG-51 and its Addendum [30, 31], respectively. Specifically, when $TPR_{20,10}(10)$ beam quality specifier is used, the r_{LCPE} (in cm) can be determined by

$$r_{LCPE} = 8.369 \ TPR_{20,10}(10) - 4.382 \quad (6.3)$$

while when $\%dd(10,10)_X$ is used as beam quality specifier, r_{LCPE} (in cm) can be determined by

$$r_{LCPE} = 77.97 \times 10^{-3} \ \%dd(10)x - 4.112 \quad (6.4)$$

where $TPR_{20,10}(10)$ is the tissue-phantom ratio in water at depths of 20 and 10 g cm^2 for a 10 cm square field size at the depth of the detector and a source-to-detector distance (SDD) of 100 cm [29] and $\%dd(10,10)_X$ is the percentage depth dose, due to photons only (designated by the symbol "X"), at 10 cm depth in a water phantom, determined for a field size of (10 × 10) cm^2 at the phantom surface and a source-to-surface distance (SSD) of 100 cm [30, 31].

Applying Eq. (6.3) for a nominal 6 MV linac having a typical $TPR_{20,10}(10)$ beam quality of 0.68 results in a r_{LCPE} of 1.31 cm showing that on points laying on beam axis, LCPE breaks down for field sizes smaller than 2.6 cm. This field is much smaller than the 6 cm field size used for calibrating the output of the CyberKnife system (see also Sect. 6.2.3).

Partial occlusion of the primary photon source is related to the finite size of the primary photon beam source. Its size is usually defined as the full width at half maximum (FWHM) of the bremsstrahlung photon fluence distribution exiting the target. A small field shaped by a collimator that shields part of the finite primary photon source will produce a lower beam output on the beam axis compared to field sizes where the source is not partially blocked. The FWHM of the primary photon source of modern linacs is less than 5 mm. For the CyberKnife, direct focal spot measurements have not been performed, but MC studies have suggested a primary photon source with a FWHM of 2.2 mm [32]. Source occlusion usually occurs at field sizes smaller than those where lateral electron disequilibrium starts [4]. Partial occlusion of the primary photon source influences the particle spectrum and is a source of steep local absorbed dose gradients, both of which can have a large effect on the detector response. The loss of LCPE and the primary photon source occlusion effect are both responsible for a sharp drop in beam output with decreasing field size at very small sizes. This drop becomes more pronounced when the photon beam energy increases or the density of the medium decreases (since in both cases, the electron ranges increase) [10].

The third condition that characterizes a small field is the size of the detector relative to the size of the radiation field. Detector response is affected by the dose gradient over its sensitive volume an effect that is usually referred to as volume averaging effect. The presence of a non-water-equivalent detector perturbs additionally the charged particle fluence. In the presence of large dose gradients and in the absence of lateral charged particle equilibrium conditions, fluence perturbations become large and difficult to model since they can depend on minor variations of the detector design, even within engineering tolerances [5, 10]. The dosimetric difficulties that this causes start to show up as soon as the effects of lateral absorbed dose gradients and charged particle disequilibrium reach the detector volume. Small field conditions can thus be assumed to exist when the external edge of the detector volume is at a distance from the field edge smaller than the lateral charged particle equilibrium range in the medium (r_{LCPE}) [5, 10].

6.2.3 Machine-Specific Reference Field

For the radiation beam generators that cannot establish the conventional (10 × 10) cm^2 uniform reference field, the concept of the machine-specific reference (msr) field, f_{msr}, has been introduced [5, 9]. The f_{msr} is defined as the radiation field closest to the conventional reference field and should extend at least a distance r_{LCPE} beyond the outer boundaries of the used reference ionization chamber to avoid presence of small field conditions. If the greatest distance between two points on the outer boundary of the used detector is d, the FWHM of the field in the direction of d has to fulfill the condition:

$$\mathrm{FWHM} \geq 2\ r_{\mathrm{LCPE}} + d \qquad (6.5)$$

It should be noted that the stem and cable are normally not considered part of the detector volume, but to account for the strong disequilibrium conditions at the field edges, part of the stem starting from the air cavity, with a length equal to the maximum wall thickness (including the sleeve) in the other directions, should be considered in the detector size.

While the latest M6 version of the CyberKnife equipped with the InCise MLC, is capable of establishing a nominal field of (10 × 10.1) cm^2 which is similar to the conventional reference field size, this collimator is not available to all CyberKnife generations, and therefore the f_{msr} used for reference dosimetry measurements is the circular field with nominal diameter of 60 mm defined at 800 mm distance from the source established by the fixed collimator [33]. In Fig. 6.1 the off-axis profile of the msr field of the M6 and prior to M6 (referred to as pre-M6 hereafter) CyberKnife models is shown. A flatter off-axis profile in the f_{msr} of the M6 version can be observed which is attributed to corresponding differences of the primary beam collimator.

The equivalent uniform square msr field can be calculated based on equal scatter contributions [34] and is equal to 51.6 mm and 50.4 mm for the M6 and pre-M6 versions, respectively. This difference is attributed to the flatter 60 mm off-axis beam profile of the M6 system relative to the pre-M6 systems (see Fig. 6.1).

Applying Eq. (6.5) for a classical Farmer-type ionization chamber with a cavity length of 24 mm and an external dimension of about 30 mm (including also part of the stem), the minimum field that fulfills small field conditions is equal to

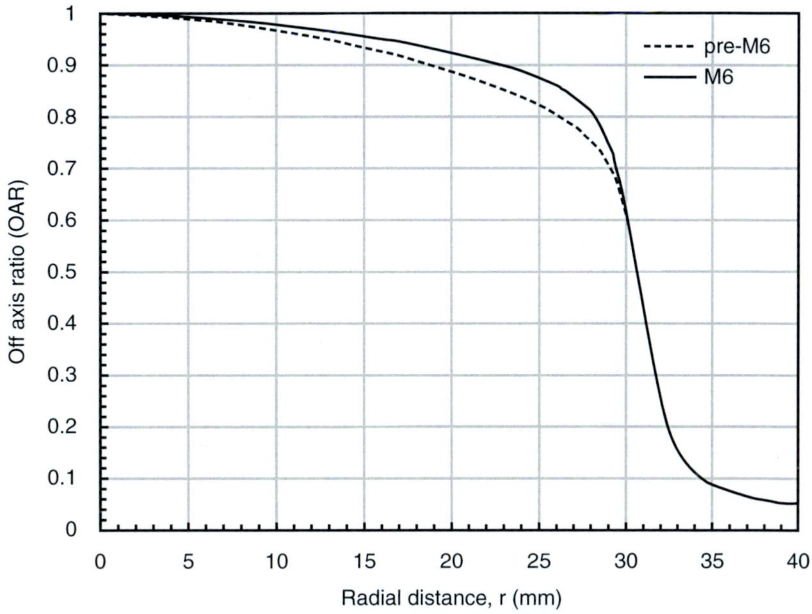

Fig. 6.1 The off-axis ratio (OAR) of the circular 60 mm nominal diameter msr field of pre-M6 and M6 CyberKnife models. The presented data correspond to average measured values from multiple CyberKnife systems and were extracted from data given in Ref. [24]

55 mm. This indicates that small field conditions are not present even when a classical Farmer chamber is used for reference dosimetry measurements in the 60 mm f_{msr} field of the CyberKnife. However, caution is required when choosing ionization chamber for reference dosimetry measurements, since the unflattened lateral beam profile of the f_{msr} may induce considerable volume averaging effects (see Sect. 6.2.4) [20, 22, 24].

6.2.4 Volume Averaging

A detector produces a signal that is proportional to the mean absorbed dose over its sensitive volume, and this signal is affected by the uniformity of the absorbed dose over the detection volume (volume averaging). For the CyberKnife the beam uniformity of the msr field (see Fig. 6.1) over the detection volume of a classical Farmer chamber changes from 95.5 to 97.0% for the pre-M6 and M6 CyberKnife systems, respectively [24]. To account for this effect, a volume averaging correction factor, k_{vol}, has been introduced defined as the ratio of the absorbed dose to water at the reference point in the water phantom in the absence of the detector and the mean absorbed dose to water over the sensitive volume of the detector (still in the absence of the detector) [5]. Volume averaging correction factors can be calculated by integrating the 3D dose distribution in the water phantom over the sensitive volume of the detector [5]. The expression to calculate the volume averaging correction factor is:

$$k_{vol} = \frac{\iint_A w(x,y)\,\mathrm{d}x\mathrm{d}y}{\iint_A w(x,y)\mathrm{OAR}(x,y)\,\mathrm{d}x\mathrm{d}y} \tag{6.6}$$

where x and y are the coordinates orthogonal to the beam central axis, A is the area of the projection of the sensitive volume of the detector on a plane orthogonal to the beam axis, OAR(x, y) is the off-axis ratio which is the 2D lateral beam profile at the measurement depth normalized to unity on the central axis, and $w(x, y)$ is a weighting function representing the extensions of the detector's sensitive volume along the beam axis as function of the beam lateral coordinates. Different expressions of the weighting function are given in page 148 of TRS-483 based on the detector design and required accuracy.

The accuracy of the calculated volume averaging correction factors depends on the exact knowledge of the sensitive volume dimensions of the used detector and the off-axis beam profile data. Detector dimensions are commonly stated by the vendors, and a corresponding list for several detectors is included in TRS-483. Regarding the lateral beam profile data, it is important to note that these must be measured using a high-resolution detector with increased signal-to-noise ratio and following the guidelines given in TRS-483 for detector positioning. Unshielded diodes or synthetic diamond dosimeters should be preferred [32].

Since volume averaging correction factor increases with cavity length due to the unflattened lateral beam profile of the CyberKnife, shorter cavity length (<10 mm) ion chambers should be preferred for reference dosimetry measurements. It is also important to select a chamber that meets the other requirements for reference dosimetry in small beams, as described in TRS-483.

6.2.5 Beam Quality

Photon beam quality (Q) is classically determined using the $TPR_{20,10}$ and $\%dd(10)_X$ beam quality specifiers calculated based on measured depth dose data [29–31]. Depth dose data should be acquired in a water phantom of (30 × 30 × 30) cm^3 dimensions using a 10 cm square reference field and 100 cm source-to-detector distance (SDD) for the TPR [29] and source to surface distance (SSD) for the %dd [30, 31]. The primary beam of the CyberKnife is unflattened, and the 60 mm in diameter msr field differs from the classical 10 cm^2 reference field (see Fig. 6.1). Moreover, output calibration is performed at 800 mm SAD.

For the CyberKnife, TRS-483 suggests that beam quality should be determined for the msr field using the closest achievable to the reference measurement conditions. Beam quality index for a hypothetical (10 × 10) cm^2 reference field, $TPR_{20,10}(10)$ or $\%dd(10{,}10)_X$, can be calculated using measured $TPR_{20,10}$ or $\%dd(10)_X$ values, respectively, in the msr field and applying the following equations proposed by Palmans [35]:

$$TPR_{20,10}(10) = \frac{TPR_{20,10} + c(10 - S)}{1 + c(10 - S)} \tag{6.7}$$

and

$$\%dd(10{,}10) = \frac{\%dd(10)_X + 80c(10 - S)}{1 + c(10 - S)} \tag{6.8}$$

where S is the equivalent square field of the msr field at 100 cm distance from the source (S is equal to 6.45 cm and 6.3 cm for the M6 and pre-M6 CyberKnife systems, respectively), $c = (16.15 \pm 0.12) \times 10^{-3}$ when $TPR_{20,10}$ is used, and $c = (53.4 \pm 1.1) \times 10^{-3}$ when %dd is used as beam quality specifier. It must be noted that the detector reference point (located for cylindrical chambers, on the central axis at the center of the cavity volume) should be positioned 100 cm away from the source for $TPR_{20,10}$ measurements. Similarly, percentage depth dose measurements should be performed using an SSD of 100 cm with the effective point of measurement of the chamber (located 0.6 times the cavity radius from the chamber axis toward the photon source) positioned at 10 cm depth [5].

The beam quality specifier can be used to determine ion chamber beam quality correction factors for reference dosimetry measurements (see Sect. 6.3.1). TRS-483 assumed that the beam quality varies slightly between the installed CyberKnife systems, and they used published CyberKnife beam data to calculate beam quality correction factors for an extended list of ion chambers appropriate for reference dosimetry measurements in the msr field of the CyberKnife system (see Sect. 6.3.1) [5]. It is crucial to note that the given correction factors include the volume averaging correction which as shown in Sect. 6.2.4 depend on the shape of the lateral profile of the msr field. Therefore, the TRS-483 correction factors are applicable only to pre-M6 CyberKnife versions (see Sect. 6.3.1).

Literature findings support the assumption that beam quality deviations between the different CyberKnife systems are minor and have been further improved in the newest M6 generation due to the use of a fully digital advanced magnetron modulator and linac control system [24]. Despite the above findings, it is advised to measure and control the beam quality for quality assurance and verification with other CyberKnife systems.

6.3 Dosimetric Formalism

6.3.1 Reference Dosimetry

A reference class ionization chamber is advised to be used for beam output calibration measurements [29–31]. The dose response of the ionization chamber should be determined in terms of absorbed dose to water, $N_{D,w}$, in a standard (10 × 10) cm² field (f_{ref}) of beam quality Q_o produced by a ^{60}Co generator in a primary or secondary standard laboratory. The dose at reference depth inside water for the msr field of the CyberKnife, $D_{w,Q_{\mathrm{msr}}}^{f_{\mathrm{msr}}}$, is given by:

$$D_{w,Q_{\mathrm{msr}}}^{f_{\mathrm{msr}}} = M_{Q_{\mathrm{msr}}}^{f_{\mathrm{msr}}} N_{D,w,Q_o}^{f_{\mathrm{ref}}} k_{Q_{\mathrm{msr}},Q_o}^{f_{\mathrm{msr}},f_{\mathrm{ref}}} \qquad (6.9)$$

where $M_{Q_{\mathrm{msr}}}^{f_{\mathrm{msr}}}$ is the reading of the chamber in the msr field f_{msr} corrected for influence quantities, such as pressure, temperature, incomplete charge collection, polarity effects, etc. (the unfamiliar reader is advised to refer to TRS-483 [5], TRS-398 [29], or TG-51 [30, 31] dosimetric protocols for determining the correction of the response of the chamber for these influence quantities); $N_{D,w,Q_o}^{f_{\mathrm{ref}}}$ is the calibration coefficient in terms of absorbed dose to water of the ionization chamber measured at the standard laboratory using a conventional standard (10 × 10) cm² reference calibration field f_{ref} with beam quality Q_o; and $k_{Q_{\mathrm{msr}},Q_o}^{f_{\mathrm{msr}},f_{\mathrm{ref}}}$ is the beam quality correction factor to account for the different response of the chamber in the conventional reference field f_{ref} with beam quality Q_o and in the f_{msr} with beam quality Q_{msr} [5].

Beam quality correction factors $k_{Q_{\mathrm{msr}},Q_o}^{f_{\mathrm{msr}},f_{\mathrm{ref}}}$ are given in TRS-483 for the CyberKnife and for a list of reference class ionization chambers (see Table 13 in TRS-483) [5]. These correction factors were obtained using corresponding beam quality correction factor data given in previous dosimetric protocols, such as the TRS-398 [29] or TG-51 and its Addendum [30, 31], and correcting for (a) the volume averaging effect using k_{vol} values calculated based on the cavity length of each detector and the f_{msr} off-axis profile, given by Antypas and Pantelis [36], and (b) the different water-to-air stopping power ratios between the WFF and the FFF beams [37]. Differences in the fluence perturbations between WFF and FFF beams with the same beam quality specifier were assumed negligible, and their contribution to the beam quality correction factor was therefore ignored.

As also Stated in Sect. 6.2.5, TRS-483 assumes that beam quality and off-axis profiles are minor between CyberKnife systems. The impact of inter-unit variations on $k_{Q_{\mathrm{msr}},Q_o}^{f_{\mathrm{msr}},f_{\mathrm{ref}}}$ correction factors was further studied by Francescon et al. [24] using beam commissioning data from 139 different CyberKnife systems including pre-M6 and M6 versions of the system with and without a lead beam filter. The authors found that:

(a) The $k_{Q_{\mathrm{msr}},Q_o}^{f_{\mathrm{msr}},f_{\mathrm{ref}}}$ correction factor for output calibration is independent of the presence or absence of the lead beam filter.
(b) The flatter primary beam found with the M6 system (see Fig. 6.1) is associated with a systematic decrease in $k_{Q_{\mathrm{msr}},Q_o}^{f_{\mathrm{msr}},f_{\mathrm{ref}}}$ versus pre-M6 systems. The magnitude of this change increases with cavity size, and is approximately 0.4% for a Farmer chamber (cavity length 20–25 mm), decreasing to 0.1% for lengths <10 mm. The CyberKnife $k_{Q_{\mathrm{msr}},Q_o}^{f_{\mathrm{msr}},f_{\mathrm{ref}}}$ data in TRS-483 are systematically higher than the $k_{Q_{\mathrm{msr}},Q_o}^{f_{\mathrm{msr}},f_{\mathrm{ref}}}$ data for the M6 (by up to 0.6%), which is attributed to the fact that TRS-483 data were generated using an analytic correction for volume averaging based on pre-M6 CyberKnife 60 mm collimator measured off-axis ratio values. Therefore the $k_{Q_{\mathrm{msr}},Q_o}^{f_{\mathrm{msr}},f_{\mathrm{ref}}}$ correction factors given by Francescon et al. [24] should be preferred for reference dosimetry measurements in the M6 CyberKnife system. If TRS-483 $k_{Q_{\mathrm{msr}},Q_o}^{f_{\mathrm{msr}},f_{\mathrm{ref}}}$ data are used, they should be adjusted to account for the different volume averaging correction factor in the pre-M6 and M6 systems as described in [24] (see also Sect. 6.4.1).
(c) The variations in $k_{Q_{\mathrm{msr}},Q_o}^{f_{\mathrm{msr}},f_{\mathrm{ref}}}$ due to other inter-unit changes in beam profile increase with

cavity length and are larger for the pre-M6 system than the M6 system owing to the flatter and more consistent primary beam observed with M6. For a Farmer chamber, this uncertainty is estimated to be ±0.8% for pre-M6 reducing to ±0.4% for M6. When a short chamber with cavity length <10 mm is used, this uncertainty reduces to ±0.5% and ±0.2%, respectively.

6.3.2 Relative Dosimetry Measurements

The dose at the reference depth in water on central axis for a clinical field f_{clin} of beam quality Q_{clin} is given by:

$$D^{f_{\text{clin}}}_{w,Q_{\text{clin}}} = \Omega^{f_{\text{clin}},f_{\text{msr}}}_{Q_{\text{clin}},Q_{\text{msr}}} * D^{f_{\text{msr}}}_{w,Q_{\text{msr}}} \tag{6.10}$$

where $\Omega^{f_{\text{clin}},f_{\text{msr}}}_{Q_{\text{clin}},Q_{\text{msr}}}$ is the output factor for the specific clinical field and generator (also known as total scatter factor [4]). The field output factor can be determined by:

$$\Omega^{f_{\text{clin}},f_{\text{msr}}}_{Q_{\text{clin}},Q_{\text{msr}}} = \frac{M^{f_{\text{clin}}}_{Q_{\text{clin}}}}{M^{f_{\text{msr}}}_{Q_{\text{msr}}}} k^{f_{\text{clin}},f_{\text{msr}}}_{Q_{\text{clin}},Q_{\text{msr}}} \tag{6.11}$$

where $M^{f_{\text{clin}}}_{Q_{\text{clin}}}$ and $M^{f_{\text{msr}}}_{Q_{\text{msr}}}$ are the measured signal in the measurement depth for the clinical field and msr field, respectively, and $k^{f_{\text{clin}},f_{\text{msr}}}_{Q_{\text{clin}},Q_{\text{msr}}}$ is the field output correction factor that accounts for the differences between the response of the detector in the clinical field and that in the msr reference field [5].

$k^{f_{\text{clin}},f_{\text{msr}}}_{Q_{\text{clin}},Q_{\text{msr}}}$ correction factors are given in TRS-483 (see Table 23 in TRS-483) for a number of detectors including shielded and unshielded diodes, a diamond detector, and cylindrical micro-ionization chambers. Diode and diamond detectors are suggested to be positioned with their stem parallel to beam axis, and the ion chambers are suggested to be positioned with their stem perpendicular to beam axis [5]. For the above detectors and orientations, $k^{f_{\text{clin}},f_{\text{msr}}}_{Q_{\text{clin}},Q_{\text{msr}}}$ were calculated based on detector-specific fitted functions over published $k^{f_{\text{clin}},f_{\text{msr}}}_{Q_{\text{clin}},Q_{\text{msr}}}$ data versus square (or equivalent square) field sizes. Generic volume averaging correction factors were also implemented for the flattening filter-free linacs.

It is important to note that the published data used by TRS-483 for calculating the $k^{f_{\text{clin}},f_{\text{msr}}}_{Q_{\text{clin}},Q_{\text{msr}}}$ values did not include any of the published MC and experimental CyberKnife-specific data. Moreover, the suggested perpendicular to beam axis orientation of the micro-ionization chambers, suggested by TRS-483 for small field output factor measurements, is not commonly used by the CyberKnife users [12, 14, 19, 23]. In most of the CyberKnife published studies, the ion chambers are positioned so that the smallest cavity dimension to be positioned perpendicular to beam axis in order to reduce volume averaging effects [17]. Since in the used ion chambers cavity diameter was smaller than cavity length, the chambers were positioned with their stem parallel to beam axis.

Francescon et al. [24] studied the inter-unit variations on the $k^{f_{\text{clin}},f_{\text{msr}}}_{Q_{\text{clin}},Q_{\text{msr}}}$ correction factor data and found that:

(a) For small field OF measurement, preference should be given to diode or the synthetic micro diamond detector [24, 38]. These have smaller corrections than micro-ionization chambers and are less sensitive to inter-unit variations in beam profiles [24]. If using a microdiamond, the CyberKnife-specific correction factors provided in reference [32] should be preferred to TRS-483 [24].

(b) If using a micro-ionization chamber, it should be positioned with its stem parallel to the beam axis. This minimizes the correction factor magnitude and its sensitivity to inter-unit beam profile variations.

(c) For micro-ionization chambers in the perpendicular orientation, large differences of up to 9% have been found between the CyberKnife-specific $k^{f_{\text{clin}},f_{\text{msr}}}_{Q_{\text{clin}},Q_{\text{msr}}}$ data and those reported in TRS-483, and therefore the latter should not be used without careful prior validation.

6.4 Practical Implementation

6.4.1 Reference Dosimetry Measurements

The following components are advised to be used for reference dosimetry measurements:

- One or more reference class ionization chambers, including a permanently attached cable and connector. It is advised that the ionization chambers are calibrated in terms of $N_{D,w}$ at reference beam quality (Q_o) and are chosen specifically for the CyberKnife (see Sect. 6.2.4).
- A (30 × 30 × 30) cm^3 or larger water phantom with movable mechanisms and holder for attaching the detector at measurement point.
- An electrometer often separately calibrated in terms of charge or current per scale division.
- One or more stability check sources specifically designed for the chosen ionization chamber.
- Calibrated thermometer and barometer.

Equation (6.9) should be used to calculate the absorbed dose in water. The output of the CyberKnife is calibrated so that 1 cGy is equal to 1 MU at 15 mm depth inside water for the 60 mm in diameter circular field established with the fixed collimator at 80 cm distance from the source. However, calibration measurements should be performed at 10 cm depth in water according to TRS-483 and transferred to 15 mm using the corresponding measured TPR data. The details of the experimental setup that should be established for calibration dosimetry measurements are given in Table 6.1.

The pressure and temperature conditions should be stable and recorded throughout the measurements. The raw ionization reading $M_{Q_{msr}}^{f_{msr}}$ must be corrected for temperature and pressure, polarity, ion recombination effects, and electrometer calibration (if calibrated separately) according to TRS-483.

Beam quality correction factors $k_{Q_{msr},Q_o}^{f_{msr},f_{ref}}$ are given in TRS-483 for pre-M6 CyberKnife systems and for a list of reference class ionization chambers (see Table 13 in TRS-483) [5]. For the newest M6 CyberKnife system, the user can either:

Table 6.1 Reference conditions for beam output calibration of the CyberKnife system

Influence quantity	Reference value/characteristics
Phantom material	Water
Phantom shape and size	At least 30 cm × 30 cm × 30 cm
Chamber type	Cylindrical
Measurement depth	10 cm
Reference point of chamber	On the central axis at the center of the cavity volume
Position of reference point of chamber	At the measurement depth
Source-to-detector distance	80 cm
Field shape and size	Circular fixed collimator 60 mm nominal field size
Radiation beam axis	Vertical to the water phantom surface
Chamber stem orientation	Perpendicular to beam axis

1. Use the $k_{Q_{msr},Q_o}^{f_{msr},f_{ref}}$ correction factor values given in Table 2 of reference [24]
2. Use the TRS-483 $k_{Q_{msr},Q_o}^{f_{msr},f_{ref}}$ reported values and correct them by an amount based on Figure 3 of reference [24] (approximately 0.4% for a Farmer chamber, 0.1% for cavity lengths <10 mm) to account for the change in primary beam profile between pre-M6 and M6 systems
3. Follow the methodology presented in TRS-483 [5] to calculate the $k_{Q_{msr},Q_o}^{f_{msr},f_{ref}}$ correction factor of the used chamber, i.e.:
   (a) Calculate the beam quality specifier ($TPR_{20,10}(10)$ or %dd(10,10)) of the used CyberKnife beam following the methodology described in Sect. 6.2.5 and Eq. (6.7) or (6.8).
   (b) Obtain the beam quality correction factor based on the calculated beam quality specifier and corresponding beam quality correction factor data given in Tables 28 and 29 of TRS-483, or in the previous TRS-398 [29] and TG-51 [30, 31] dosimetric protocols.

(c) Correct the obtained beam quality correction factor for volume averaging effects. The volume averaging correction factor k_{vol} (see Sect. 6.2.4) can be calculated using the measured off-axis profile of the msr field, the geometrical details of the ion-chamber collection cavity, and Eq. (6.6) (or an approximated equation given in TRS-483).
(d) Apply an additional correction factor beam to account for the different water-to-air stopping power ratios between the standard WFF beam used for obtaining beam quality correction factors and the FFF beam of the CyberKnife (see Table 30 of TRS-483).

The users are advised to apply all the above methods (if possible) for calculating the beam quality correction factor of the used chamber and compare their results.

6.4.2 Relative Dosimetry Measurements

6.4.2.1 Output Factor Measurements

Guidelines for choosing detectors for small field output factor measurements are given in TRS-483 [5]. An ideal detector for small field dosimetry should (a) sample the fluence at a point, (b) be water equivalent, and (c) have a linear dose response which is energy and absorbed dose rate independent. In practice, an easily handled real-time detector with increased absorbed dose sensitivity, small active volume dimensions, and well-known and close to unity (within ±5% [5]) $k_{Q_{clin},Q_{msr}}^{f_{clin},f_{msr}}$ correction factor should be preferred.

Air-filled small and micro-ionization chambers have been used in small field dosimetry measurements. However, their use for small field output factor measurements in the CyberKnife system may lead to increased uncertainties. These uncertainties are associated with (a) the increased volume averaging effects due to limitations on the minimum active volume dimensions required to have an increased measured signal compared to the background signal created from other chamber components such as the stem and the cable, (b) electron fluence perturbations especially if a metal central electrode is used, (c) increased polarity effects [23], (d) increased leakage, and (e) their dependence on the exact dimensions on the focal spot dimensions [12]. TRS-483 suggests that the volume averaging effect is one of the limiting issues for the choice of a detector [5]. The detector size should be such that the volume averaging correction factor k_{vol} (calculated using Eq. (6.6) and the small field off-axis profile) for the small field of interest is less than 1.05.

Among the available detectors currently in the market, the unshielded silicon diodes and a synthetic microdiamond present favorable characteristics and should be preferred. These characteristics include their increased absorbed dose sensitivity and small sensitive volume dimensions resulting to minor or small volume averaging effects. The unshielded diodes have been well studied, and their $k_{Q_{clin},Q_{msr}}^{f_{clin},f_{msr}}$ for the CyberKnife small fields are well-known [18, 24, 32], and don't depend on the source focal spot dimensions [12]. The PTW-6009 synthetic microdiamond presents smaller compared to the unshielded diodes $k_{Q_{clin},Q_{msr}}^{f_{clin},f_{msr}}$ correction factor values [32] and has been successfully used for CyberKnife small field output factor measurements [38].

As noted in Sect. 6.3.2, the TRS-483 $k_{Q_{clin},Q_{msr}}^{f_{clin},f_{msr}}$ reported values for the CyberKnife system were calculated using published data that didn't include any of the published MC and experimental CyberKnife-specific data. Therefore, for the unshielded diodes, the TRS-483 reported values should be compared with the corresponding CyberKnife-specific MC-derived data prior to their use to establish the level of uncertainty of the measured small field output factor values. For the microdiamond detector, the CyberKnife-specific correction factors provided in reference [32] should be preferred to TRS-483 [24].

Both the unshielded diodes and the microdiamond detector are used without bias voltage and should be positioned with their stem parallel to beam axis, except from the EDGE diode detector

(Sun Nuclear Corp., Melbourne, FL, USA) which is positioned with its stem perpendicular to beam axis due its specific constructional design. The microdiamond detector should be irradiated prior to its use, while diodes should be checked for radiation damage [5]. Off-axis profiles should be performed using small steps (e.g., 0.1 mm) to align the detector reference point with beam axis. The reference point of the detector should be positioned at 15 mm depth inside water where the output factors of the CyberKnife system are defined. At least five measurements should be acquired for each field size, and results should be averaged. For the output factor measurements of the small fields established with the Iris or the InCise multi-leaf collimators, it is suggested to instruct the collimator to fully open or fully close and re-establish the measured field in between measurements.

Measurement sequences are often long, and the atmospheric conditions can vary substantially affecting the response of the used detectors. Therefore, the environmental conditions should be monitored during the measurements.

Finally, it must be also noted that while the unshielded diodes and the microdiamond detector are preferred for small field output factor measurements, they are constructed from high density and atomic number materials perturbing the electron fluence of the measured fields. While these effects are corrected using corresponding MC calculated correction factors, their accuracy depends on the exact knowledge of the geometrical and physical characteristics of the used detector as well as on the reproducibility of the manufacturing procedure. Therefore, it is advised that the obtained small field output factor values be compared with corresponding data measured using a water-equivalent dosimeter such as a plastic scintillator and/or GafChromic films.

6.4.2.2 Off-Axis Profile and Depth Dose Measurements

Guidelines for choosing detectors for small field lateral profile measurements are given in TRS-483. Sensitive volume dimensions are crucial to avoid volume averaging effects. Therefore, preference should be given to unshielded diodes and the synthetic microdiamond detector, with the latter to have the advantage of almost uniform directional response. Detector positioning is important to avoid stem effects. Therefore the detector should not be positioned with its stem parallel with the scanning direction [5]. In detail, for the small volume ionization chambers, the stem should be positioned parallel to beam axis or perpendicular to beam axis and scanning direction. For the solid state detectors (diodes and the microdiamond), the stem should be positioned parallel to beam axis except for the EDGE diode detector which should be positioned perpendicular to beam axis and scanning direction due to its constructional characteristics [17].

The correction factor suggested for small field output factor measurement has been extended for lateral off-axis and depth dose profile measurements. PDD corrections at depths higher than 15 mm have been found to be less than 2% for all detectors except for the IBA Razor (IBA SA, Louvain-la-Neuve, Belgium) where a maximum 4% correction has been observed at 300 mm depth [32]. Off-axis ratio (OAR) corrections have been found smaller inside the field than outside. At the beam edge micro-ionization chamber OAR corrections of up to 15% have been reported caused mainly by density perturbation effects [32]. With larger beams and depths, correction factors outside the beam have been found to increase reaching up to 20% for the PTW and IBA diodes. These effects are most noticeable for large field size and depth, where they are dominated by electron fluence perturbations and stopping power differences. For the microdiamond OAR corrections have been found to be less than 3% outside the beam.

Due to the variation of the correction factors with depth and off-axis distance, their application in dosimetry measurements is not suggested, and the PDD and OAR corrections published by Francescon et al. [32] should be used to guide detector selection and inform the evaluation of results.

References

1. Chin LS, Regine WF. Principles and practice of stereotactic radiosurgery. New York, NY: Springer; 2015.
2. International Commission on Radiation Units and Measurements (ICRU) Repot 91. Prescribing, recording and reporting of stereotactic treatments with small photon beams. J Int Comm Radiat Units Meas. 2014;14(2):1–160.
3. Das IJ, Ding GX, Ahnesjö A. Small fields: nonequilibrium radiation dosimetry. Med Phys. 2007;35:206–15.
4. Aspradakis MM, Byrne JP, Palmans H, et al. IPEM report 103: small field MV photon dosimetry. 2010.
5. International Atomic Energy Agency (IAEA) Technical Report Series (TRS) No. 483. Dosimetry of small static fields used in external beam radiotherapy: an IAEA-AAPM International Code of Practice for reference and relative dose determination. Vienna, Austria; 2017.
6. Zhu TC, Ahnesjö A, Lam KL, et al. Report of AAPM Therapy Physics Committee Task Group 74: in-air output ratio, Sc, for megavoltage photon beams. Med Phys. 2009;36:5261–91.
7. Georg D, Knöös T, McClean B. Current status and future perspective of flattening filter free photon beams. Med Phys. 2011;38:1280.
8. Xiao Y, Kry SF, Popple R, et al. Flattening filter-free accelerators: a report from the AAPM Therapy Emerging Technology Assessment Work Group. J Appl Clin Med Phys. 2015;16:12–29.
9. Alfonso R, Andreo P, Capote R, et al. A new formalism for reference dosimetry of small and nonstandard fields. Med Phys. 2008;35:5179.
10. Palmans H, Andreo P, Huq MS, et al. Dosimetry of small static fields used in external photon beam radiotherapy: summary of TRS-483, the IAEA–AAPM International Code of Practice for reference and relative dose determination. Med Phys. 2018;45:e1123–45.
11. Derreumaux S, Etard C, Huet C, et al. Lessons from recent accidents in radiation therapy in France. Radiat Prot Dosim. 2008;131:130–5.
12. Francescon P, Cora S, Cavedon C. Total scatter factors of small beams: a multidetector and Monte Carlo study. Med Phys. 2008;35:504–13.
13. Francescon P, Cora S, Satariano N. Calculation of k(Q(clin),Q(msr)) (f(clin),f(msr)) for several small detectors and for two linear accelerators using Monte Carlo simulations. Med Phys. 2011;38:6513–27.
14. Bassinet C, Huet C, Derreumaux S, et al. Small fields output factors measurements and correction factors determination for several detectors for a CyberKnife® and linear accelerators equipped with microMLC and circular cones. Med Phys. 2013;40:071725.
15. Moignier C, Huet C, Makovicka L. Determination of the kQclin,Qmsrfclin,fmsr correction factors for detectors used with an 800 MU/min CyberKnife® system equipped with fixed collimators and a study of detector response to small photon beams using a Monte Carlo method. Med Phys. 2014;41:071702.
16. Francescon P, Kilby W, Satariano N, Cora S. Monte Carlo simulated correction factors for machine specific reference field dose calibration and output factor measurement using fixed and iris collimators on the CyberKnife system. Phys Med Biol. 2012;57:3741–58.
17. Francescon P, Kilby W, Satariano N. Monte Carlo simulated correction factors for output factor measurement with the CyberKnife system-results for new detectors and correction factor dependence on measurement distance and detector orientation. Phys Med Biol. 2014;59:N11–7.
18. Francescon P, Beddar S, Satariano N, Das IJ. Variation of kQclin,Qmsrfclin,fmsr for the small-field dosimetric parameters percentage depth dose, tissue-maximum ratio, and off-axis ratio. Med Phys. 2014;41:101708.
19. Pantelis E, Antypas C, Petrokokkinos L, et al. Dosimetric characterization of CyberKnife radiosurgical photon beams using polymer gels. Med Phys. 2008;35:2312–20.
20. Kawachi T, Saitoh H, Inoue M, et al. Reference dosimetry condition and beam quality correction factor for CyberKnife beam. Med Phys. 2008;35:4591–8.
21. Araki F. Monte Carlo study of a Cyberknife stereotactic radiosurgery system. Med Phys. 2006;33:2955–63.
22. Pantelis E, Moutsatsos A, Zourari K, et al. On the implementation of a recently proposed dosimetric formalism to a robotic radiosurgery system. Med Phys. 2010;37:2369–79.
23. Pantelis E, Moutsatsos A, Zourari K, et al. On the output factor measurements of the CyberKnife iris collimator small fields: experimental determination of the kQclin,Qmsrfclin,fmsr correction factors for microchamber and diode detectors. Med Phys. 2012;39:4875–85.
24. Francescon P, Kilby W, Satariano N, et al. The impact of inter-unit variations on small field dosimetry correction factors, with application to the CyberKnife system. Phys Med Biol. 2019;64:035006.
25. Kilby W, Dooley JR, Kuduvalli G, et al. The CyberKnife robotic radiosurgery system in 2010. Technol Cancer Res Treat. 2010;9:433–52.
26. Echner GG, Kilby W, Lee M, et al. The design, physical properties and clinical utility of an iris collimator for robotic radiosurgery. Phys Med Biol. 2009;54:5359–80.
27. Asmerom G, Bourne D, Chappelow J, et al. The design and physical characterization of a multileaf collimator for robotic radiosurgery. Biomed Phys Eng Express. 2016;2:017003.
28. International Electrotechnical Commission (IEC). Medical electrical equipment—Glossary of defined terms. 2004.
29. International Atomic Energy Agency (IAEA) Technical Report Series (TRS) No. 398. Absorbed dose determination in external beam radiotherapy: an international code of practice for dosimetry based on standards of absorbed dose to water. Vienna, Austria; 2000.

30. Almond PR, Biggs PJ, Coursey BM, et al. AAPM's TG-51 protocol for clinical reference dosimetry of high-energy photon and electron beams. Med Phys. 1999;26:1847–70.
31. McEwen M, Dewerd L, Ibbott G, et al. Addendum to the AAPM's TG-51 protocol for clinical reference dosimetry of high-energy photon beams. Med Phys. 2014;41:1–20.
32. Francescon P, Kilby W, Noll JM, et al. Monte Carlo simulated corrections for beam commissioning measurements with circular and MLC shaped fields on the CyberKnife M6 system: a study including diode, microchamber, point scintillator, and synthetic microdiamond detectors. Phys Med Biol. 2017;62:1076–95.
33. Physics Essentials Guide (P/N 100722). Accuray Inc.
34. Sauer OA. Determination of the quality index (Q) for photon beams at arbitrary field sizes. Med Phys. 2009;36:4168–72.
35. Palmans H. Determination of the beam quality index of high-energy photon beams. Med Phys. 2012;39:5513–9.
36. Antypas C, Pantelis E. Performance evaluation of a CyberKnife G4 image-guided robotic stereotactic radiosurgery system. Phys Med Biol. 2008;53:4697–718.
37. Dalaryd M, Knöös T, Ceberg C. Combining tissue-phantom ratios to provide a beam-quality specifier for flattening filter free photon beams. Med Phys. 2014;41:111716.
38. Russo S, Masi L, Francescon P, et al. Multicenter evaluation of a synthetic single-crystal diamond detector for CyberKnife small field size output factors. Phys Med. 2016;32:575–81.

7 Quality Control

Evangelos Pantelis and Argyris Moutsatsos

7.1 Introduction

A wide range of different technologies spanning from image guidance and artificial intelligence to increased dose rate flattening filter free (FFF) linacs are incorporated into modern radiotherapy devices [1]. Specifically, the CyberKnife system (Accuray Inc., Sunnyvale, CA, USA) combines robotic, image guidance and multi-leaf collimator (MLC) technologies with inverse treatment planning to create and deliver highly conformal three-dimensional (3D) dose distributions with stereotactic accuracy to lesions throughout the body [2, 3]. These dose distributions are characterized by steep spatial dose gradients, which combined with the complexity of the system, renders quality assurance (QA) of the CyberKnife a challenging task to perform.

The Task Group 135 (TG-135) of the American Association of Physicists in Medicine (AAPM) has suggested a number of procedures for the QA of the CyberKnife system [4]. However, as new technologies are introduced into the system (Iris™ and InCise™ collimators, Xchange™ collimator table, Monte Carlo dose calculation algorithms, lung optimized tracking algorithm, etc.), additional QA procedures have been suggested by the vendor [5]. In the following sections, the QA procedures that should be followed to ensure performance and safe use of the CyberKnife system are described. Tables with daily, monthly, and annual QA procedures are given in Sect. 7.8 along with corresponding tolerances.

7.2 Patient Safety Checks

The CyberKnife uses a robot to support and position the linear accelerator. A detailed description of the system is given in Chap. 3. During treatment delivery the robot moves between beam delivery positions. There are no inherent mechanical restrictions placed on the robot's movement, except for the collimator assembly collision detector. Therefore, checking the collimator assembly collision detector as part of the daily QA is suggested [4, 5].

Besides the collimator collision detector, a robot proximity detection program (PDP) is also executed during treatment delivery. This program restricts any part of the robot from entering predefined restriction zones. There are two zones; the first is fixed and is associated with the

E. Pantelis (✉)
Medical Physics Laboratory, Medical School, National and Kapodistrian University of Athens, Athens, Greece

CyberKnife and TomoTherapy Department, IATROPOLIS Private Special Clinic, Athens, Greece
e-mail: vpantelis@phys.uoa.gr

A. Moutsatsos
CyberKnife and TomoTherapy Department, IATROPOLIS Private Special Clinic, Athens, Greece
e-mail: armouts@phys.uoa.gr

© Springer Nature Switzerland AG 2020
A. Conti et al. (eds.), *CyberKnife NeuroRadiosurgery*,
https://doi.org/10.1007/978-3-030-50668-1_7

x-ray tubes, the Xchange™ table, the floor, the walls, and the ceiling of the treatment room that are fixed with respect to the robot base. The second zone is the patient safety zone and is defined relative to the treatment couch and thus must be tested at various couch locations within the range of couch motions. Both fixed and patient safety zones should be tested prior to the first clinical use of the system and after any major software upgrade. A testing procedure is provided by the manufacturer during installation but requires the assistance of a field service engineer.

All patient safety systems incorporated into the facility design must be also verified. These systems include audio and visual monitors, emergency interruption for robot movement, emergency power off, and door interlocks. These systems must be checked at installation, periodically as part of daily and monthly QA, and each time they may have been disabled or disconnected during maintenance work. Interlocks must occur immediately upon activation and remain engaged until the generating condition is reversed and acknowledged by the operator.

7.3 Quality Assurance of CyberKnife Subsystems

7.3.1 Robot Mechanical Accuracy

During installation, system upgrade, or maintenance, the robot is calibrated, in terms of spatial coordinates, relative to a "geometrical center" which is physically represented by a small crystal ("isocrystal") situated on the top of a floor-mounted post, the so-called isopost [4]. Mechanical calibration of the robot is performed using a laser beam aligned with the treatment beam. During calibration the robot is instructed to scan through all the nodes comprising each treatment path, and, at each node, the position and direction of the laser beam producing the maximum intensity signal on the isocrystal is recorded. Quality assurance of the mechanical accuracy of the robot can be performed in three levels.

In the first level, a light sensor on the Xchange™ table is used. Provided that the sensor and the Xchange table have not been moved since calibration, the sensor serves as a reference point inside the treatment room. In this test the robot is instructed to move and point the laser beam on the light sensor (Fig. 7.1a). The signal intensity of the sensor is recorded and compared to the corresponding calibrated value. If the intensity value is within 80% of the calibrated value, the check passes. This test is called "laser alignment check" and should be performed daily after system warm-up. It must be noted that while a procedure is given by the manufacturer to adjust the light intensity baseline value of the central sensor for possible degradation of laser output over time, most of CyberKnife users mark the laser beam spot on the treatment room floor when the treatment robot is in the perch position to verify that the laser beam does not shift from day to day.

The mechanical accuracy of the robot in placing each treatment beam should be also tested. For this second level test, the anthropomorphic phantom supplied by the manufacturer is commonly used. Computed tomography (CT) images of the phantom are acquired and imported in the treatment planning system. A single pseudo isocentric plan is created with all beams aiming a reference point situated at the surface of the phantom (e.g., tip of the nose) (Fig. 7.1b). Delivery of the created plan in "Demonstration" mode (a special mode where the plan is executed but instead of using the x-ray beam the laser beam is used) allows for a qualitative evaluation of robot mechanical accuracy to a level of approximately ±1.5 mm. This test is called "BB test" and should be performed monthly (one path set per month).

The third level test of robot's mechanical accuracy is a rigorous repeat of path calibration process using the isopost (Fig. 7.1c) [5]. The results are quantitative and produce a detailed list of node-by-node deviations that can be evaluated individually or in combination. The user should record the node-by-node results and verify that no individual node exceeds 0.5 mm deviation or that the total RMS deviation does not exceed

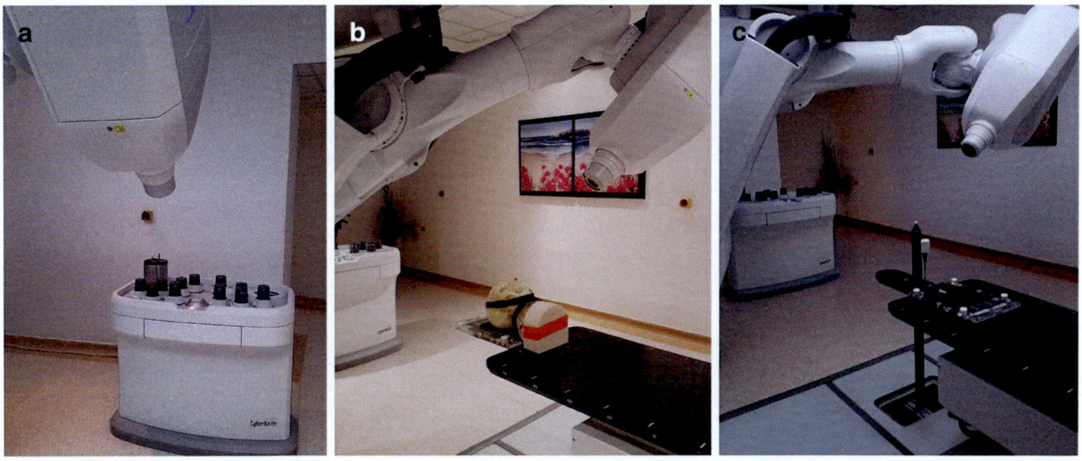

Fig. 7.1 Pictures of a CyberKnife system performing robot mechanical accuracy tests. (**a**) In the laser alignment test (first level), the robot is instructed to aim the laser beam to a constant sensor on the collimator Xchange table. (**b**) The second level test includes the delivery of a pseudo isocentric plan on a reference point on a phantom using the "Demonstration" mode on the operation console. In this mode the radiation beam is not used, and it is replaced by the laser beam (BB test). (**c**) The third level test is the path verification test. In this test the robot is instructed to pass through all nodes of a treatment path, scan with the laser beam to find the node coordinates producing the maximum signal on the isocrystal at the tip of the isopost, and compare them with the corresponding coordinates obtained during path calibration

0.3 mm [4]. It is noted that these deviations have been suggested for the KR240-2 (series 2000) six degrees of freedom (6DOF) robotic manipulator (KUKA Roboter GmbH, Augsburg, Germany) that has a manufacturer specification for position repeatability of ±0.12 mm. Stricter tolerances may be employed for the CyberKnife M6 model which uses the KUKA QUANTEC KR300 R2500 Ultra robot having improved position repeatability of ±0.06 mm [3].

The aforementioned test procedures use the laser beam for evaluating the mechanical accuracy of the robot. Therefore, test failure may be also associated with a possible laser beam position shift relative to its calibration position. Use of the radiation beam for checking robot mechanical accuracy would be more appropriate. Procedures involving the irradiation of diode arrays or aSi flat panels with one or more radiation beams have been suggested [6]. A plan with multiple beams from a single node can be created and delivered on the two-dimensional (2D) detector. The position of the radiation beam axes can be identified on the measured 2D dose maps and compared with the corresponding positions from the treatment plan for testing robot mechanical accuracy. A procedure based on a cylindrical phantom containing diode detectors in a cylindrical configuration has been also suggested for testing the treatment beam mechanical accuracy [7].

7.3.2 Couch Mechanical Accuracy

Quality control testing of the treatment couch demonstrates whether the mechanical accuracy of the couch and the safety interlocks are within the manufacturer's specifications and work properly. The mechanical accuracy of individual table movements can be evaluated using a mechanical ruler and a digital level. The test should be performed on a monthly basis. The accuracy of couch movements for the standard AXUM couch should be less than ±5% in translations and ±0.3° in rotations [5]. Stricter tolerances may be employed for the robotic version of the treatment couch (RoboCouch™).

7.3.3 Image Guidance Subsystem Performance

The image guidance subsystem is responsible for measuring patient and target location in stereotactic space and guiding the robot to deliver each beam to its position and direction obtained during treatment planning [2]. Accuracy of the image guidance subsystem depends on the spatial resolution and contrast of the radiographs recorded by the flat panel detectors and of the digital reconstructed radiographs (DRRs) calculated based on patient's simulation CT, as well as on the used target locating algorithm [2, 8, 9]. Quality control of the imaging guidance subsystem includes tests for the mechanical alignment of the x-ray tubes and flat panel detectors and the performance of the x-ray generators and the digital detectors as well as for the accuracy of the target locating algorithms [4].

7.3.3.1 Mechanical Alignment of the Imaging System

The mechanical alignment of the image guidance subsystem is evaluated by mounting the isopost on the flat panel frame and acquiring two x-ray images of its position using 60 kVp and 2.5 mAs exposure settings. Using the treatment delivery computer tools, the coordinates of the projected crystal are obtained which should be in the central pixel of both x-rays within a maximum tolerance of ±1 mm in both imaging directions [4, 5]. Mechanical alignment should be tested during installation and system maintenance and quarterly thereafter.

7.3.3.2 Quality Assurance of the X-Ray Generators

The accuracy and precision of kVp and time settings should be checked. A noninvasive multifunction meter is used to measure kVp values and exposure times. The multifunction meter should be positioned at the isocenter facing the x-ray tube being tested, and a set of measurements are acquired for each generator using clinically used nominal kVp values and exposure times. Kilovoltage accuracy and reproducibility should be better than 10% and 5%, respectively [10–12].

Imaging x-ray beam exposure output reproducibility and linearity should be also checked. A high-precision diagnostic electrometer connected to a detector with a flat energy response and calibrated in kV diagnostic beams should be used. The detector must be positioned at the isocenter facing the tested x-ray tube. At clinically used kV settings (e.g., 80–150 kV) and constant exposure time (e.g., 100 ms), a set of exposures are acquired for mAs values ranging from 5 to 30 mAs by independently changing the mA setting. The kV beam output reproducibility and exposure linearity between all mAs settings shall be within 5% and 10%, respectively [10–12]. Quality assurance of the kV-imaging subsystem parameters should be checked during acceptance of the system, x-ray tube or generator replacement and yearly thereafter.

7.3.3.3 Quality Assurance of the Flat Panel Detectors

Quality assurance of the flat panel detectors is performed in terms of high-contrast resolution, low-contrast resolution, and geometrical distortion. An appropriate multipurpose diagnostic QA tool should be used to quantitatively evaluate the above characteristics (Fig. 7.2) [4, 8, 13]. Each detector should be tested individually with the QA tool facing each camera. The raw x-ray images are distorted due to the oblique irradiation geometry (Fig. 7.2b). High-contrast and low-contrast resolution should be measured on the raw x-rays using the corresponding features of the QA tool. Geometric distortion should be also evaluated either qualitatively or quantitatively by measuring the ratio of known phantom dimensions, for example, length/width, on the digitally aligned x-ray images (Fig. 7.2c). X-ray images should be also tested for artifacts and bad pixels. There are several spatial resolution and contrast detail phantoms available on the market today for the QA of the imaging subsystem of the CyberKnife. Baseline values of the imaging

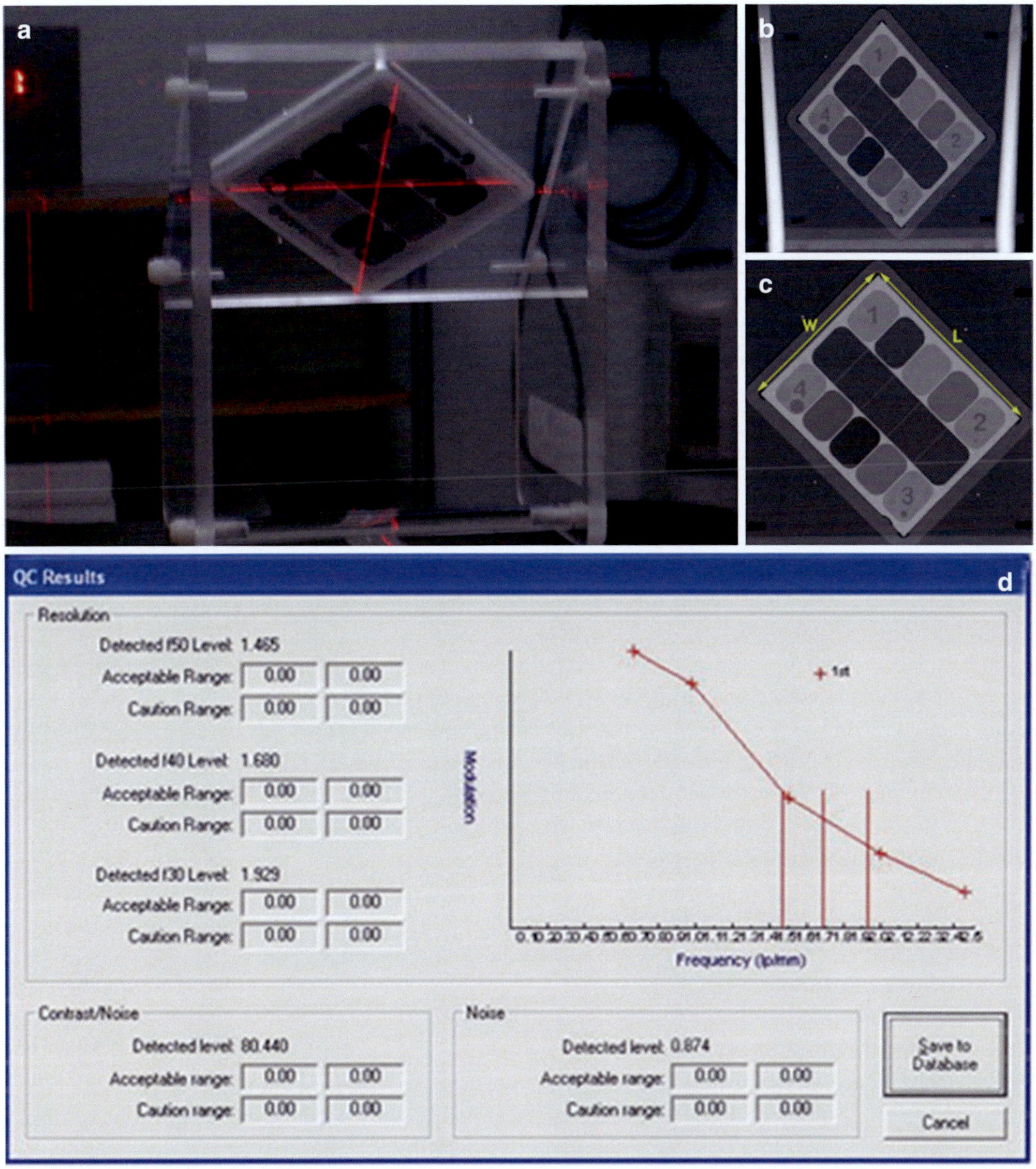

Fig. 7.2 Quality control of the imaging characteristics of the digital flat panels of the CyberKnife using a commercially available imaging QA phantom (**a**) and corresponding image analysis software (**d**). Since the flat panels lay horizontal in the treatment room floor and the central kV x-ray axis hits them with an angle of 45°, the raw acquired images are distorted (**b**). Prior their use by the tracking algorithm, they are digitally corrected for geometrical distortion (**c**)

characteristics should be determined during installation of the system and checked annuallly or after system maintenance. X-ray images is advised to be tested quarterly for bad pixels and image artifacts [4].

7.3.3.4 Quality Assurance of the Tracking Algorithms

The accuracy of the target locating algorithms should be verified during initial acceptance testing or major image guidance system upgrades. At installation, the vendors' engineers use an anthropomorphic head and neck phantom containing hidden targets with an automated test tool to evaluate the accuracy of each image guidance algorithm [14, 15]. The phantom is attached directly to the robotic manipulator arm. The robot then moves the phantom such that all three degrees of positional and three degrees of angular freedom are tested throughout the range of clinical significance. The translational accuracy should be within 0.2 mm, and the rotational accuracy within 0.2° below 2° rotation from setup, and 0.5° at more than 2° rotation [4].

Alternatively, a procedure using the treatment couch to move the phantom can be used. This procedure requires development of a phantom treatment plan for each tracking algorithm. The phantom is aligned on the treatment couch using the corresponding tracking algorithm to get couch offset close to zero [13, 16]. Using couch digital readouts, the phantom can be moved away to several positions in both translational and rotational axes (with the standard couch, the yaw cannot be tested using this method). At each position, x-ray images are acquired, and the tracking algorithm results are recorded and compared with the actual offsets and rotations.

7.3.4 Quality Assurance of the Linear Accelerator

7.3.4.1 Laser and Radiation Beam Axis Coincidence

The CyberKnife doesn't use a light field to represent the radiation beam, but a laser beam is used to depict radiation beam axis. Prior to measuring beam data or performing robot mechanical accuracy test (see Sect. 7.3.1), the laser and radiation beam axes must be checked for coincidence [4, 5]. Axes coincidence can be checked using radiographic or radiochromic films, irradiated using at least two collimators (e.g., 20 mm and 40 mm) and from two source to film distances (e.g., 800 mm and 1200 mm). Care must be taken to align vertically the beam axis with film's plane and to mark accurately laser spot on the film for this test. Radiation and laser beam axes coincidence can be quantitatively evaluated using freeware film analysis software tools and should be less than 1 mm [4, 5].

7.3.4.2 Radiation Beam Data

General recommendations on linac QA and beam data commissioning should be followed for the quality control of the CyberKnife radiation beams [17–23]. Dosimetric measurements to be used also for treatment planning beam data input include beam quality specification, off-axis and depth-dose (PDD/TPR) profile measurements, output factors, output calibration, reproducibility, linearity, output constancy versus linac orientation, collimator transmission, leakage radiation, and end effect.

Beam energy, flatness, symmetry, and penumbra are measured during commissioning to establish baseline values and are monitored on a monthly basis. These measurements are performed using a water tank, but since they are time-consuming, acrylic phantom combined with appropriate detectors can be used. The $TPR_{20,10}$ beam quality index is used for monitoring beam energy [17]. The 60 mm fixed collimator is used for this test to irradiate an ion-chamber positioned inside acrylic phantom slabs. Beam symmetry, flatness, and penumbra can be measured using film dosimetry or other 2D detectors (e.g., diode arrays or flat panel detectors).

Beam output is calibrated so that 1 cGy is equal to 1MU at 15 mm inside water for the 60 mm fixed collimator defined at 800 mm distance from the x-ray source [5]. A flattening filter is not used which affects the shape of the off-axis beam profile (i.e., the profile has a conical shape; see Fig. 6.1 in Chap. 6) and the emitting photon spectrum [24, 25]. Volume averaging effects have been reported during reference dosimetry measurements due to the unflattened beam profile [17, 25, 26]. Caution is required for the proper choice of the ion chamber used for calibration dosimetry measurements. Ion chambers with a cavity length less than 10 mm are recommended [17, 25, 26]. More details on reference dosimetry measurements are given in Chap. 6.

Diode detectors feature the highest response per volume of all common detector types. Hence their sensitive volume is usually small enough to avoid dose volume effects down to very small fields. However, their directional response and their response to low-energy scattered photons are not ideal. To reduce the effect of low-energy photons, diodes exist in a shielded design where the shield reduces the signal from these photons. In small fields the low-energy scatter contribution is low, hence diode shielding is not needed, and unshielded diodes are recommended for small fields [23, 27]. Alternatively, a synthetic diamond detector has been proposed [28–30]. Diamond detectors present lower sensitivity compared to diodes but are characterized by near water equivalence and independence from the energy of photons. Both unshielded diodes and synthetic diamond detectors can be used for output factor, relative off-axis, and depth-dose profile small field measurements provided that appropriate correction factors are applied [17, 30].

7.3.4.3 Beam Output Constancy

Beam output constancy should be checked daily. For this check, the birdcage provided by the vendor is attached on the linac nozzle to support an ion chamber at 800 mm from the source with a buildup material (Fig. 7.3). The chamber is irradiated with 100 MUs using the 60 mm fixed collimator. A dosimetric protocol can be used to calculate the dose at calibration conditions (i.e., 15 mm depth in water) using an additional correction factor to account for the different experimental setup [17]. This factor can be determined experimentally by irradiating the chamber with the same MUs in the two experimental setups (i.e., at 15 mm depth inside the water phantom and in the birdcage) and comparing the measured charge values corrected for temperature and pressure.

7.3.4.4 Iris Collimator Performance

The Iris variable aperture collimator uses two stacked hexagonal banks of tungsten segments to produce a 12-sided aperture. The Iris apertures

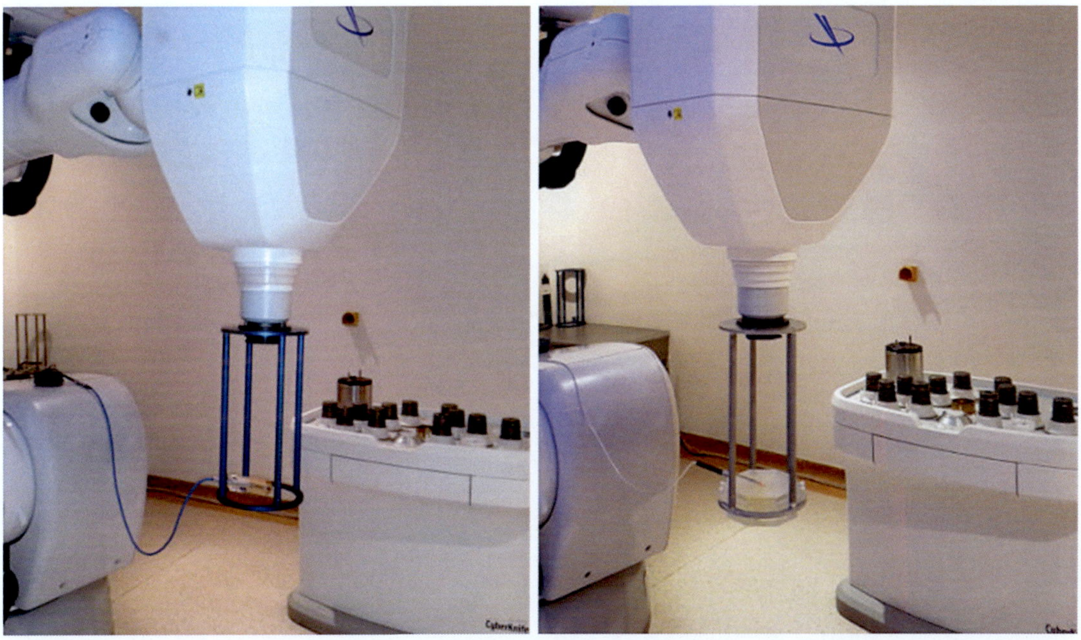

Fig. 7.3 Ion-chamber setups for daily beam output constancy measurements. On the left a classical Farmer-type ion chamber is positioned on the birdcage using a 15-mm-thick buildup cap. On the right a short cavity length (7.5 mm) thimble ion chamber is positioned at 15 mm depth inside a custom-made 5-cm-thick PMMA phantom attached on the birdcage. In both setups the fixed 60 mm collimator is used

replicate the fixed collimator apertures with an accuracy and reproducibility of ±0.2 mm at 800 mm SAD [31]. During commissioning of the Iris collimator, measurements of all 12 clinical available field sizes are performed and compared to a fixed collimator aperture in order to establish a baseline for further periodic quality assurance measurements. While water tank measurements using a scanning detector are closer to the beam data used by the treatment planning system, these are time-consuming for periodic QA, and therefore alternative methods using acrylic phantoms have been developed.

The vendor recommends the use of radiochromic films for the quality assurance of the Iris collimator [5]. A dedicated film-based tool (Iris QA tool) is provided with the system for this test to assure measurement setup reproducibility (Fig. 7.4). The Iris QA tool consists of a 5-cm-thick acrylic base plate with a 15-mm-thick buildup plate that fits into the birdcage assembly. The birdcage assembly is mounted on the linac head assuring that the film is positioned at 800 mm distance from the source during measurement. During Iris collimator commissioning, a series of measurements are performed by irradiating radiochromic films with all 12 Iris collimator apertures and the 30 mm fixed collimator which is used for reference. The films are scanned in a flatbed optical-transparency scanner using a resolution higher than 300 dpi. The film images are analyzed with the aid of the Iris QA software as shown in Fig. 7.4, and the diameter of each radiation field is measured. These results serve as a baseline dataset during quality assurance measurements.

Film measurements do not provide direct results, and therefore other solutions have been suggested [6, 32–34]. Heidorn et al. [32] used a large area parallel plate ion chamber to measure

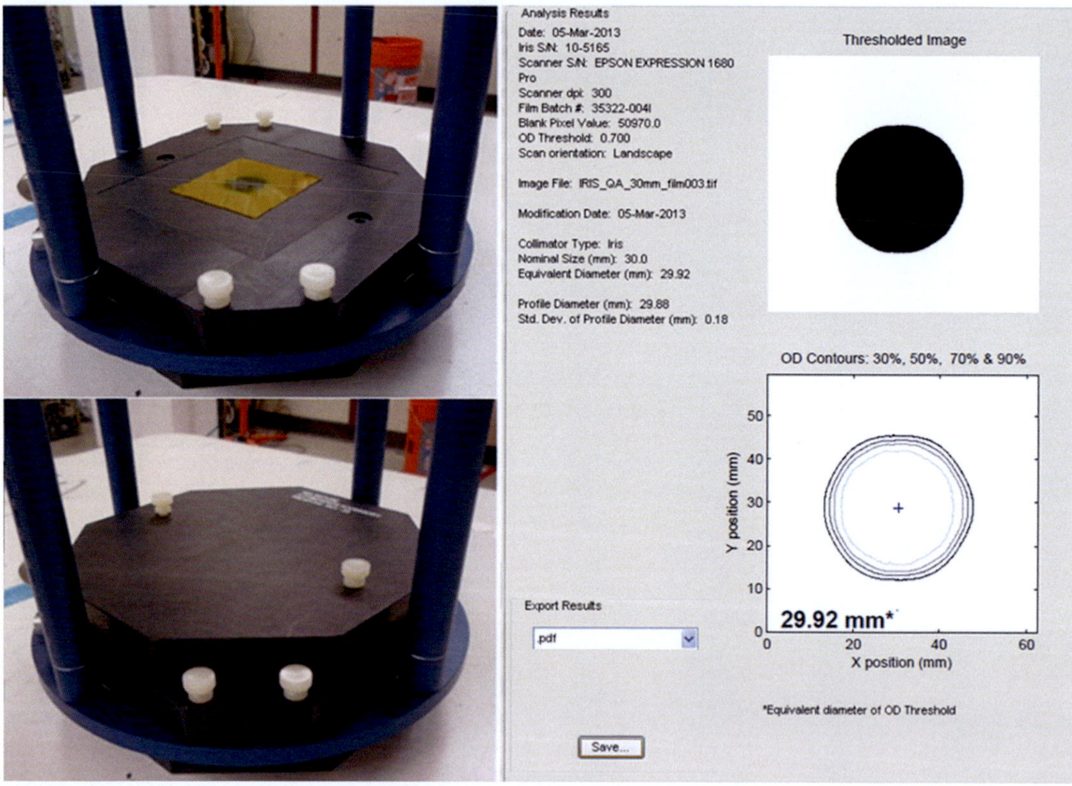

Fig. 7.4 The Iris QA tool mounted on birdcage assembly with film (upper left) and buildup plate on film (lower left). A screenshot of the Iris QA software used to analyze the exposed films and calculate the field size is shown on the right

the dose area product (DAP) of all Iris apertures and the 60 mm fixed reference aperture. Baseline values were established during commissioning and used to evaluate the stability of the Iris collimation system [32]. Other commercially available solutions have been proposed replacing the radiochromic films with diode arrays or amorphous silicon (aSi) flat panels [6, 33, 34]. Software is also provided to analyze the measured charge values from the Iris collimator apertures and give direct results for the field size, flatness, and penumbra.

The stability of the Iris collimator should be checked during monthly QA of the system and after an upgrade, recalibration, or major service.

7.3.4.5 InCise MLC Performance

The general guidelines of the TG-142 [19] and TG-50 [35] AAPM reports should be followed for the quality control of the InCise MLC. The TG-50, or as also referred to as Picket Fence, pattern is used for a quick qualitative daily QA check of leaf-positioning accuracy due to its increased sensitivity [35]. The Bayouth or Garden Fence test is used for a quantitative measurement of leaf-positioning accuracy [36]. Both methods use multiple fields so the leaf positions can be measured at multiple positions across the field. The Garden Fence test is used for leaf calibration by a service engineer and for measuring the leaf-positioning accuracy during acceptance and periodic QA.

A film-based tool (MLC QA tool) that mounts directly to the accessory mounting points is provided by the vendor for performing the Garden Fence and Picket Fence tests (Fig. 7.5). This tool holds a radiochromic film and two radiopaque markers that allow for very precise film marking to determine the beam center on the film and rotation of the film relative to the X/Y MLC axes. The irradiated film is scanned using an optical-transparency scanner with a 600 dpi resolution, and software is used to automatically determine the center and rotation of the film based on the radiopaque marks. The variation in measured dose (or OD) on the film is used to evaluate the leaf positions and perform Picket Fence and Garden Fence tests.

For a Garden Fence test run at perch position, the following criteria should be met: The mean bank offset from the expected positions should be less than or equal to 0.2 mm for each bank, and at least 90% of the measured leaf positions should have an offset of less than or equal to 0.5 mm from the expected position for each bank. For a single leaf on each bank, there can be no more than 1 deviation from the expected position that is greater than 0.5 mm. For Garden Fence tests run at any linac orientation, all measured leaf positions for each bank must be less than 0.95 mm. The above tolerance values should be expressed at 800 SAD [5].

Indicative Picket Fence test patterns created using EBT2 films are presented in Fig. 7.5. The uniform exposure of the film shown on the left indicates a correctly calibrated MLC. Areas of underexposure on the central film image of Fig. 7.5 show that one bank is underextending (by 0.5 mm). Overextending of one bank would be manifested as areas of overexposure such as those observed on the right film image of Fig. 7.5. Tilt errors can also be detected by Picket Fence tests by the variation in the thickness of under- and overexposed film areas as seen in Fig. 7.5.

Besides Picket and Garden Fence test patterns, MLC leaf transmission and leakage should also be measured. The MLC leakage and transmission should be less than 0.5% with a mean value of less than 0.3%. Commercially available film-less solutions using diode arrays or aSi flat panels have been suggested for direct measurement of the MLC performance [6, 34].

7.3.5 Treatment Planning Quality Assurance

CyberKnife treatment plans are developed using the Precision™ treatment planning software (TPS) (Accuray Inc., Sunnyvale, CA, USA). During commissioning, maintenance or upgrade the transfer of patient imaging data from the used imaging modalities (CT, MRI, etc.) to the TPS should be tested for geometrical accuracy. The transfer of imaging data must preserve the pixel values and magnification and linearity of the

Fig. 7.5 Top row: Experimental film setup using the MLC QA tool provided by the vendor for the quality assurance of the MLC. Lower row: Picket Fence test patterns created using EBT2 film with a correctly calibrated InCise MLC (left), with one bank underextending shown by the underexposed areas (middle), and one bank overextending shown by the overexposed areas (right)

image. The (manual or automatic) definition of volumes of interest (geometrically simple objects) should correspond to their real dimensions and positions. The TPS should be also tested for accurately showing the used beams on patient 3D model.

Three dose calculation algorithms are provided by the Precision TPS. The "ray tracing" algorithm provides a fast dose calculation method based on measured beam data look-up and can only be used with the fixed and Iris collimators. For the InCise MLC, a finite size pencil beam (FSPB) algorithm is used. This algorithm generates the dose distribution by the weighted summation of kernels distributed across an arbitrary 2D beam aperture that each represents the dose distribution delivered by a narrow pencil beam in an infinite water phantom [37]. Heterogeneity corrections are performed in both algorithms using equivalent path length (EPL), where the effective depth along a ray-line linking the x-ray source to the dose calculation point is calculated as the integral of the relative electron density of the voxels along the ray. The ray tracing algorithm accuracy can be improved using an obliquity correction by casting multiple rays within each beam which is implemented as an option [2]. The FSPB algorithm accuracy can be further improved in low-density materials by scaling the kernels laterally based on local density to simulate the extended electron range [38], which is implemented as an option. A fast Monte Carlo (MC) dose calculation algorithm is also provided for increased dose accuracy in low-density organs (such as lungs) or at the interfaces between soft tissues and low-density materials [3, 39]. Studies have shown that EPL correction-based algorithms overestimate dose relative to more accurate MC methods by up to 30% for lung lesions. A modest overestimation of less than 5% has been found for brain and spinal lesions [40, 41].

During commissioning, system maintenance, or upgrade, the accuracy of the used dose calculation algorithms should be evaluated. Single-beam irradiation geometries can be applied using previously CT-scanned solid water phantoms for this purpose. Calculated dose distributions can be exported in DICOM RTDOSE format and compared with corresponding measured values [5]. TPS dose calculation accuracy can be repeated using solid water phantoms containing tissue heterogeneities [42]. Independent dose calculation systems can be also used for the evaluation of the accuracy of TPS dosimetry calculations [43, 44].

7.4 The Automated Quality Assurance (AQA) Test

The automated quality assurance (AQA) test is a daily test to check target reproducibility, robot mastering, and mechanical stability of the image guidance system. The AQA test is analogous to the Winston–Lutz gantry linac stereotactic QA technique of placing a radiopaque ball at the treatment isocenter and observing the concentricity of the beam and shadow of the ball [45]. For a non-isocentric image-guided system such as the CyberKnife, the technique has been modified, i.e., the target ball is not mechanically placed precisely at the room isocenter but inserted into a specifically designed phantom as shown in Fig. 7.6.

The fiducials that are inserted in the phantom allow the image guidance system to direct the radiation beam at the ball. As the ball is spherically symmetric and is located at the center of the phantom, this technique can determine the translational targeting error (TTE), which is a combination of image guidance and treatment delivery error. Two radiochromic films (RCFs) are inserted in the coronal and sagittal planes of the AQA phantom and irradiated by a two-beam plan (one anterior and one lateral beam) as shown in Fig. 7.6. The shadow of the radiopaque ball is exposed on the films. Films are scanned in a flatbed transparency scanner and analyzed for concentricity of the beam and shadow using a dedicated software tool provided by the vendor. Targeting errors should deviate less than 1 mm from the baseline value set at time of calibration [4].

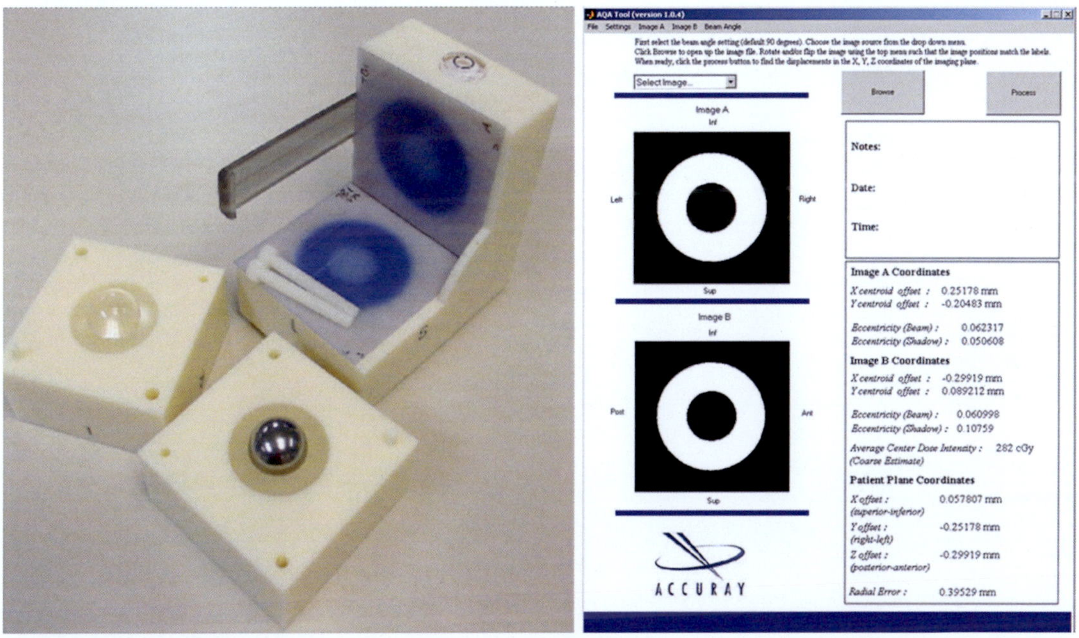

Fig. 7.6 An AQA phantom (left) and vendor-provided software to analyze concentricity of radiation beam and shadow caused by metal ball (right)

7.5 Total System Error Assessment

While the accuracy of each component affecting treatment delivery can be tested independently [13, 14, 46–48], it is more meaningful to measure the total system error (TSE) using the planned dose distribution to the patient. The so-called end-to-end (E2E) test integrates all components of the therapeutic procedure including CT scan, treatment planning (CT data import, contouring, dose calculation), software generating digitally reconstructed radiographs, tracking algorithm, and treatment delivery using the robot. The vendor provides an anthropomorphic head and neck phantom for TSE measurements [4]. This phantom resembles the x-ray attenuation properties and radiographic appearance of the corresponding human anatomy and consists of a ball cube in which a pair of orthogonal radiochromic films can be placed (Fig. 7.7). A large ball cube situated close to the center of the head phantom or a smaller one residing on a region close to the spine at the longitudinal height of neck is provided by the vendor to test 6D skull and fiducial or XST tracking methods, respectively.

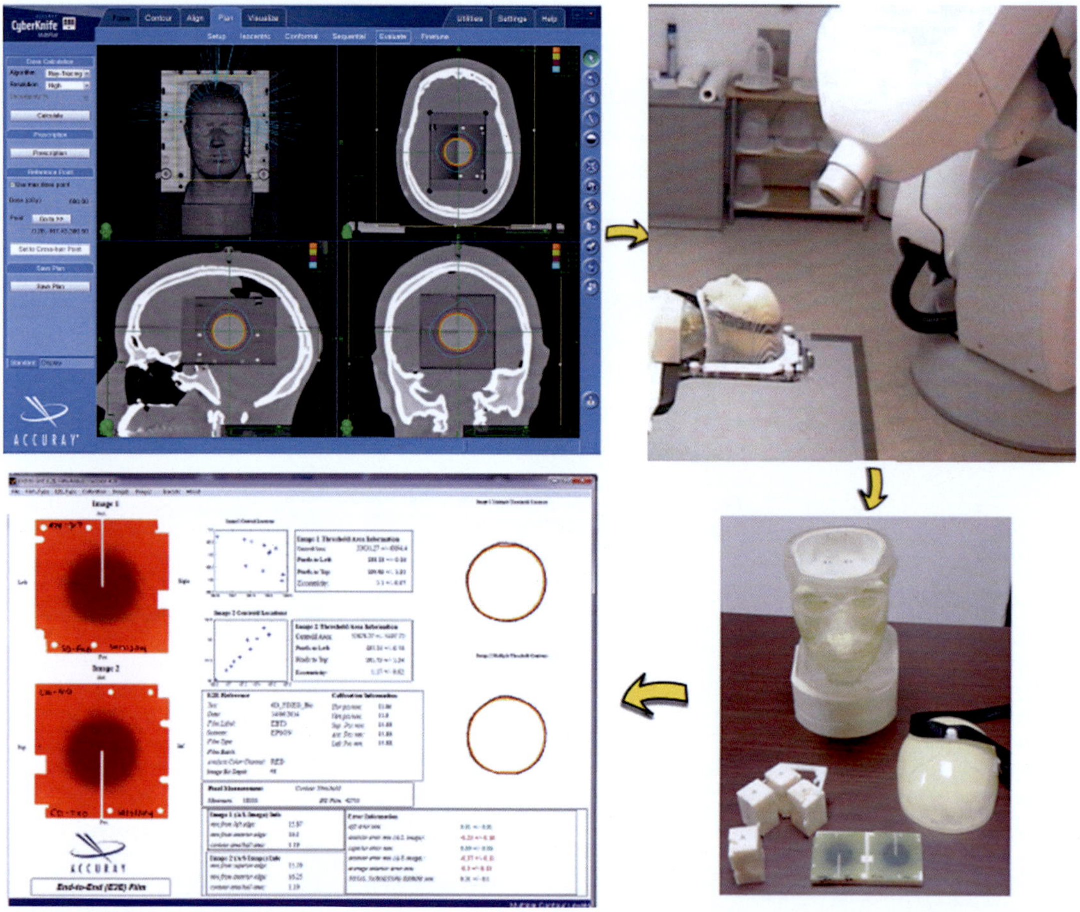

Fig. 7.7 Quality assurance of the total targeting error of the CyberKnife system using an anthropomorphic head and neck phantom and radiochromic films. Computed tomography images of the phantom are acquired and transferred into the treatment planning system. The radiopaque sphere located at the center of the ball cube is delineated, and an isocentric plan is created prescribing 420 cGy at the 70% isodose encompassing the spherical target. The plan is delivered, and the irradiated films are scanned using an optical transparent scanner. The coordinates of the centroid of the irradiated area are calculated in each film and compared with their nominal values using a software tool provided by the vendor

The phantom is CT scanned using the clinical imaging protocol (e.g., 1 mm slice thickness, 300 mm field of view, 512 × 512 matrix). CT images are imported in the TPS where the radiopaque sphere is delineated. After selecting the tracking method, an isocentric conformal treatment plan is prepared, so that the 70% isodose line to conform with the spherical contour of the (Fig. 7.7). After plan delivery, the films are scanned in an optical-transparency scanner using 300 dpi resolution. Analysis of the exposed films is performed using a film analysis software provided by the vendor and includes alignment of the films, calculation of the optical densities (ODs), and corresponding relative dosimetry values (Fig. 7.7). The TSE of the system is assessed by comparing the center of mass (CoM) coordinates of the area encompassed by the 70% isodose line measured on each film with the known coordinates of the geometrical center of the radiopaque sphere inside the ball cube. Due to the orientation of the films inside each ball cube, one measurement of the TSE was performed for each test along the left-right and superior-inferior directions and two measurements for the anterior-posterior direction which are compared for consistency and averaged.

E2E tests should be performed during commissioning to establish accuracy of the system for each image guidance algorithm and on a monthly basis to document constancy of the system delivery accuracy. On an annual basis, a complete set of E2E tests for all available tracking algorithms should be conducted including new CT scans, planning, and delivery. Indicative total targeting error results are given in Fig. 7.8 using box plots and correspond to E2E measurements performed in a single G4 CyberKnife system for a period of 10 years using the 6D skull and XST tracking algorithms, ball cube versions, linac designs, and collimation systems (i.e., Fixed and Iris) [9].

7.6 Dose Delivery Verification

The E2E test allows the user to quantify the total system error of the CyberKnife system for each tracking algorithm. This test however is based on isocentric beam delivery, and it does not give any information as to the dose delivery accuracy to complex targets that are treated using non-isocentric treatment plans, which are most of the plans delivered with CyberKnife. It is suggested to perform patient-specific dosimetry measurements using high spatial resolution dosimeters on a phantom [4]. The acceptance criterion for the dosimetry quality assurance (DQA) tests should be at least 90% pass rate using 2%/2 mm dose difference and distance agreement criteria, respectively, for the tumor and critical structures and in the high-dose region down to the 50% isodose level [4].

Radiochromic films are water equivalent, have excellent spatial response, and have been proposed as the optimum dosimeters for dose verification measurements [4, 49]. Indicative film-based

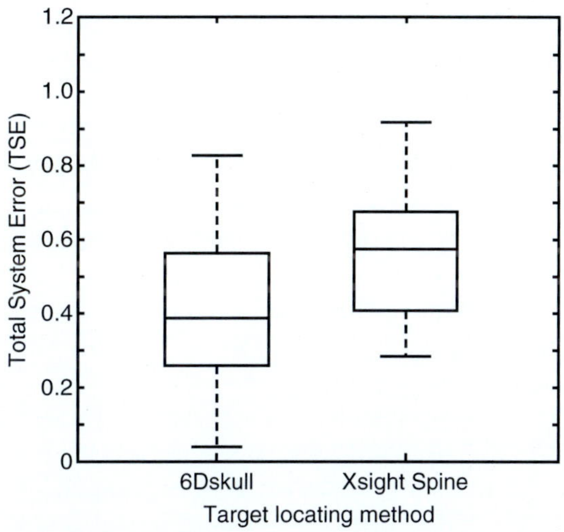

Fig. 7.8 Box plots of the total targeting error of a G4 CyberKnife system using the 6D skull and XST tracking algorithms, measured for a period of more than 10 years

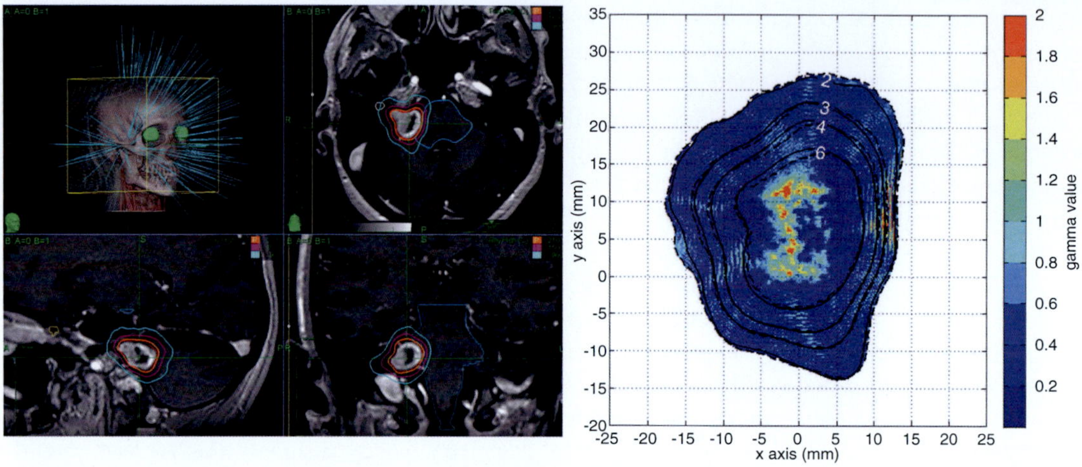

Fig. 7.9 Verified treatment plan for vestibular schwannoma (left). Film and corresponding treatment planning system dosimetry values (right)

DQA results are presented in Fig. 7.9 which correspond to irradiation of a 3.1 cm^3 vestibular schwannoma. The treatment plan was created using the ray tracing dose calculation algorithm and consisted of 204 beams with the 5 mm and 10 mm fixed diameter collimators. The prescribed dose of 21 Gy at the 75% isodose line encompassing the 98% of the target volume was delivered in three fractions. Prior to treatment the treatment plan was overlaid on the CT images of a spherical water-equivalent phantom containing an EBT2 radiochromic films using the corresponding TPS tools. A dose of 6 Gy at 75% was prescribed to the film, and the verification plan was saved as deliverable.

The phantom contained metal pins to create small holes on the film surface and aid the spatial registration of film results with corresponding TPS dose calculations exported in DICOM RTDOSE format. Four fiducials were also positioned on the surface of the phantom to allow for the verification plan to be delivered using the fiducial tracking method of the system. One day post-verification plan delivery, the film was scanned in an Epson Expression 1680Pro flatbed optical scanner using 150 dpi and 48 bit RGB color mode. A triple channel-based protocol was used to convert measured pixel values to absorbed dose [50]. The results are presented in Fig. 7.9 for doses greater than 2 Gy, and a good agreement between the film and the TPS data can be observed. The presented dose distributions were also compared using a gamma analysis criterion of 2% dose difference and 2 mm distance to agreement [4]. The proportion of pixels meeting this criterion was 95%.

Besides film-based DQA tests, other more sophisticated methods using 3D printing technologies, polymer gel 3D dosimeters, and MRI for radiation-induced polymerization signal readout have been suggested in the literature [51, 52]. While these techniques have been used for verifying both dose and spatial accuracy in SRS applications [9, 52], they are still time-consuming especially if an increased spatial resolution down to 1 mm isotropic voxels is required.

7.7 Conclusions

The CyberKnife is a radiation treatment system optimized for frameless radiosurgery anywhere in the body that it is clinically indicated. It provides a unique combination of robotic and image-guided technologies. Establishment of a QA program with daily, monthly, annual, and upgrade performance tests is of paramount importance for the safe use of the CyberKnife system. These QA tests aim to monitor the performance of the robotic manipulator, the image guidance subsystem, the linac producing the x-ray treatment beam, and the secondary collimation systems. The mechanical stability of the robotic manipulator and image guidance subsystem should be measured daily using the AQA test. The total targeting error of the system should be monitored on a monthly basis. Provided that the total targeting system error, the constancy of beam output, and the secondary collimator systems are monitored, patient-specific DQA tests can be performed on a monthly basis.

7.8 Practical Guide

7.8.1 Daily QA

Test	Tolerance
Safety interlocks (doors, console, CCTV cameras and monitors, audio system, collimator assembly collision detector)	Functional
Accelerator warm-up: 3000 MU	N/A
X-ray tubes warm-up	N/A
Linac output constancy	<2%
Detection of incorrect and missing secondary collimator	Functional
Laser alignment test	Pass. If failed and laser hits the sensor, adjust baseline value for laser intensity, and repeat. If check continues to fail, perform AQA test
Laser floor spot check	<1 mm
AQA test	<1 mm from baseline If >1 mm, perform E2E test to verify TSE

7.8.2 Monthly QA

Test	Tolerance
Beam energy, symmetry, flatness, and penumbra constancy	<2% from commissioning
Image guidance system mechanical accuracy	<1 mm
Imaging artifacts	Check for artifacts and bad pixels
Laser beam and radiation beam coincidence	<1 mm
BB test for checking robot mechanical accuracy	Visually check isocentric plan to verify beam laser illuminates the reference point on the phantom
Iris collimator aperture accuracy	≤0.2 mm
InCise MLC QA. Deliver Picket Fence and Garden Fence test patterns	Mean bank offset ≤0.2 mm from expected position
At least 90% of leaf positions should deviate ≤0.5 mm from their expected position	
Only one leaf can deviate more than 0.5 mm from the expected position	
All measured leaf positions must deviate ≤0.95 mm from their expected positions.	
E2E test	<0.95 mm. If failed create a new plan using a new phantom CT, check laser coincidence and robot mechanical accuracy, and beam symmetry before recalibrating the system
Dose delivery verification	>90% pass rate using 2 mm/2% DTA and dose difference criteria

7.8.3 Quarterly QA

Test	Tolerance
Imaging contrast, noise, geometrical accuracy, and spatial resolution of flat panel detectors	Contrast: baseline ±2 groups of lines Noise: baseline ±5% Spatial resolution: baseline minus 25% Geometrical accuracy: ≤ 2%

Test	Tolerance
Image guidance tracking accuracy	Translations: ≤0.2 mm Rotations: ≤0.2° for rotations below 2° and ≤0.5° for greater rotations
Treatment couch accuracy	Translations, <5%; rotations, <0.3°

7.8.4 Annual QA

Test	Tolerance
Daily QA	Update parameters
Monthly QA	Update parameters
Reference dosimetry measurements using the TRS-483 COP	<1%. Adjust calibration if larger difference is found
Beam data checks on at least three collimators including largest and smallest collimator (TPR or PDD, OCR, and output factors)	<2% from commissioning
Dose output linearity to lowest MU/beam used	1%
kVp accuracy	≤10%
kVp reproducibility	≤5%
mA exposure linearity	≤10%
Exposure reproducibility	≤5%
Treatment path verification test	Each node <0.5 mm, RMS < 0.3 mm
E2E test using new CT imaging data and plan	<0.95 mm

References

1. Van Dyk J. The Modern Technology of Radiation Oncology, Volume 3: A compendium for medical physicists and radiation oncologists. Madison, WI: Medical Physics Publishing Corporation; 2014.
2. Kilby W, Dooley JR, Kuduvalli G, Sayeh S, Maurer CR. The CyberKnife robotic radiosurgery system in 2010. Technol Cancer Res Treat. 2010;9:433–52.
3. Kilby W, Naylor M, Dooley JR, Maurer CR, Sayeh S. A technical overview of the CyberKnife system. In: Abedin-Nasab M H (editor). Handbook of robotic and image-guided surgery: Elsevier; 2020. p. 15–38.
4. Dieterich S, Cavedon C, Chuang CF, Cohen AB, Garrett JA, Lee CL, et al. Report of AAPM TG 135: quality assurance for robotic radiosurgery. Med Phys. 2011;38:2914–36.
5. Physics Essentials Guide (P/N 100722). Accuray Inc.
6. Palmgren J-E. QA for CyberKnife radiosurgery. Moving towards filmless QA. ISRS Educ. course, Kuopio, Finland; 2018.
7. Yang B, Wong WKR, Lam WW, Geng H, Kong CW, Cheung KY, et al. A novel method for monitoring the constancy of beam path accuracy in CyberKnife. J Appl Clin Med Phys. 2019;20:109–19.
8. Antypas C, Pantelis E. Performance evaluation of a CyberKnife G4 image-guided robotic stereotactic radiosurgery system. Phys Med Biol. 2008;53:4697–718.
9. Pantelis E, Moutsatsos A, Antypas C, Zoros E, Pantelakos P, Lekas L, et al. On the total system error of a robotic radiosurgery system: phantom measurements, clinical evaluation and long-term analysis. Phys Med Biol. 2018;63:165015.
10. Shepard SJ, Lin PP, Boone JM, Cody D, Fisher JR, Frey GD, et al. AAPM report 74. Quality control in diagnostic radiology. 2002.
11. IPEM Report 91. Recommended standards for the routine performance testing of diagnostic x-ray imaging systems. 2005.
12. Rossi RP, Lin PP, Rauch PL, Strauss KJ. AAPM report 14. Performance specifications and x-ray generators and automatic exposure control devices. 1985.
13. Pantelis E, Petrokokkinos L, Antypas C. Image guidance quality assurance of a G4 CyberKnife robotic stereotactic radiosurgery system. J Instrum. 2009;4:P05009.
14. Fu D, Kuduvalli G. A fast, accurate, and automatic 2D–3D image registration for image-guided cranial radiosurgery. Med Phys. 2008;35:2180.
15. Fürweger C, Drexler C, Kufeld M, Muacevic A, Wowra B, Schlaefer A. Patient motion and targeting accuracy in robotic spinal radiosurgery: 260 single-fraction fiducial-free cases. Int J Radiat Oncol Biol Phys. 2010;78:937–45.
16. Ho AK, Fu D, Cotrutz C, Hancock SL, Chang SD, Gibbs IC, et al. A study of the accuracy of cyberknife spinal radiosurgery using skeletal structure tracking. Neurosurgery. 2007;60:ONS147–56; discussion ONS156.
17. IAEA TRS 483. Dosimetry of small static fields used in external beam radiotherapy: An IAEA-AAPM International Code of Practice for reference and relative dose determination. Technical Report Series No. 483. Vienna, Austria; 2016.
18. Schell MC, Bova FJ, Larson DA, Leavitt DD, Lutz WR, Podgorsak EB, et al. AAPM report No. 54: Stereotactic radiosurgery. 1995.
19. Klein EE, Hanley J, Bayouth J, Yin F-FF, Simon W, Dresser S, et al. Task group 142 report: quality assurance of medical accelerators. Med Phys. 2009;36:4197–212.
20. Smith K, Balter P, Duhon J, White GA, Vassy DL, Miller RA, et al. AAPM medical physics practice guideline 8.a.: linear accelerator performance tests. J Appl Clin Med Phys. 2017;18:23–39.
21. Palmans H, Andreo P, Huq S, Seuntjens J, Christaki K E, Meghzifene A. Dosimetry of small static fields used

in external beam radiotherapy: Summary of TRS-483, the IAEA-AAPM International Code of Practice for reference and relative dose determination. Med Phys. 2018;45(11):e1123–45.
22. Pawlicki T. Quality and safety in radiotherapy: Taylor & Francis; 2011.
23. Aspradakis MM, Byrne JP, Palmans H, Duane S, Conway J, Warrington AP, et al. IPEM report 103: small field MV photon dosimetry. 2010.
24. Francescon P, Kilby W, Noll JM, Masi L, Satariano N, Russo S. Monte Carlo simulated corrections for beam commissioning measurements with circular and MLC shaped fields on the CyberKnife M6 system: a study including diode, microchamber, point scintillator, and synthetic microdiamond detectors. Phys Med Biol. 2017;62(3):1076–95.
25. Kawachi T, Saitoh H, Inoue M, Katayose T, Myojoyama A, Hatano K. Reference dosimetry condition and beam quality correction factor for CyberKnife beam. Med Phys. 2008;35(10):4591–8.
26. Pantelis E, Moutsatsos A, Zourari K, Kilby W, Antypas C, Papagiannis P, et al. On the implementation of a recently proposed dosimetric formalism to a robotic radiosurgery system. Med Phys. 2010;37:2369.
27. Seuntjens J. 2. Small field dosimetry. J Int Comm Radiat Units Meas. 2014;14:31–53.
28. Russo S, Reggiori G, Cagni E, Clemente S, Esposito M, Falco MD, et al. Small field output factors evaluation with a microDiamond detector over 30 Italian centers. Phys Med. 2016;32:1644–50.
29. De Coste V, Francescon P, Marinelli M, Masi L, Paganini L, Pimpinella M, et al. Is the PTW 60019 microDiamond a suitable candidate for small field reference dosimetry? Phys Med Biol. 2017;62:7036–55.
30. Francescon P, Kilby W, Satariano N, Orlandi C, Elshamndy S. The impact of inter-unit variations on small field dosimetry correction factors, with application to the CyberKnife system. Phys Med Biol. 2019;64
31. Echner GG, Kilby W, Lee M, Earnst E, Sayeh S, Schlaefer A, et al. The design, physical properties and clinical utility of an iris collimator for robotic radiosurgery. Phys Med Biol. 2009;54:5359–80.
32. Heidorn S, Kremer N, Fürweger C. A novel method for quality assurance of the Cyberknife iris variable aperture collimator. Cureus. 2016;8:1–12.
33. Sibata CH, Thongphiew D, Copenhaver J, Allison RR. Reproducibility of the iris collimator for Cyberknife treatments. Int J Radiat Oncol. 2010;78:S789.
34. Gersh J. WE-AB-BRB-10: filmless QA of CyberKnife MLC-collimated and Iris-collimated fields. Med Phys. 2015;42:3651–2.
35. Boyer A, Biggs P, Galvin J, Klein E, et al. AAPM TG50 report: basic applications of multileaf collimators. 2001.
36. Bayouth JE, Wendt D, Morrill SM. MLC quality assurance techniques for IMRT applications. Med Phys. 2003;30:743–50.
37. Jeleń U, Söhn M, Alber M. A finite size pencil beam for IMRT dose optimization. Phys Med Biol. 2005;50:1747–66.
38. Jeleń U, Alber M. A finite size pencil beam algorithm for IMRT dose optimization: density corrections. Phys Med Biol. 2007;52:617–33.
39. Ma C-M, Li JS, Deng J, Fan J. Implementation of Monte Carlo dose calculation for CyberKnife treatment planning. J Phys Conf Ser. 2008;102:012016.
40. Ho A, Lo AT, Dieterich S, Soltys SG, Gibbs IC, Chang SG, et al. Trigeminal neuralgia treatment dosimetry of the Cyberknife. Med Dosim. 2012;37:42–6.
41. Wilcox EE, Daskalov GM, Lincoln H. Stereotactic radiosurgery-radiotherapy: should Monte Carlo treatment planning be used for all sites? Pract Radiat Oncol. 2011;1:251–60.
42. Wilcox EE, Daskalov GM. Evaluation of GAFCHROMIC® EBT film for CyberKnife® dosimetry. Med Phys. 2007;34:1967.
43. Mackeprang PH, Vuong D, Volken W, Henzen D, Schmidhalter D, Malthaner M, et al. Benchmarking Monte-Carlo dose calculation for MLC CyberKnife treatments. Radiat Oncol. 2019;14:1–11.
44. Mackeprang PH, Vuong D, Volken W, Henzen D, Schmidhalter D, Malthaner M, et al. Independent Monte-Carlo dose calculation for MLC based CyberKnife radiotherapy. Phys Med Biol. 2018;63(1):015015.
45. Lutz W, Winston KR, Maleki N. A system for stereotactic radiosurgery with a linear accelerator. Int J Radiat Oncol Biol Phys. 1988;14:373–81.
46. Suh Y, Dieterich S, Keall PJ. Geometric uncertainty of 2D projection imaging in monitoring 3D tumor motion. Phys Med Biol. 2007;52:3439–54.
47. Wong KH, Dieterich S, Tang J, Cleary K. Quantitative measurement of CyberKnife robotic arm steering. Technol Cancer Res Treat. 2007;6:589–94.
48. Fürweger C, Drexler C, Kufeld M, Muacevic A, Wowra B. Advances in fiducial-free image-guidance for spinal radiosurgery with CyberKnife—a phantom study. J Appl Clin Med Phys. 2011;12:20–8.
49. Pantelis E, Niroomand-Rad A. Use of Radiochromic films in commissioning and quality assurance of CyberKnife. In: Das IJ, editor. Radiochromic film: role and application in radiation dosimetry: CRC Press; 2018.
50. Micke A, Lewis DF, Yu X. Multichannel film dosimetry with nonuniformity correction. Med Phys. 2011;38:2523–34.
51. Makris DN, Pappas EP, Zoros E, Papanikolaou N, Saenz DL, Kalaitzakis G, et al. Characterization of a novel 3D printed patient specific phantom for quality assurance in cranial stereotactic radiosurgery applications. Phys Med Biol. 2019;64:105009.
52. Saenz D, Rasmussen K, Pappas E, Kirby N, Stathakis S, Shi Z, et al. QA for SBRT of spine lesions: introducing a novel 3D gel dosimeter for spatial and dosimetric end-to-end testing. Int J Radiat Oncol. 2018;102:e517.

Part III

Imaging

Morphological Imaging

8

Eleftherios P. Pappas and Evangelos Pantelis

8.1 Introduction

Volumetric imaging is in the heart of modern radiotherapy techniques. Computed tomography (CT) remains the main imaging modality for radiotherapy treatment planning, mainly due it its geometrical accuracy and its capability to provide an estimate of the electron density distribution in the patient. The latter is required by the dose calculation algorithms of the treatment planning systems to account for the different scattering and absorption properties of the human tissues. Magnetic resonance imaging (MRI) is also employed in order to take advantage of the superior soft tissue contrast it exhibits, which is necessary for accurate tumor and soft tissue delineation [1]. Especially for lesions of the central nervous system (CNS), MRI provides unsurpassed soft tissue contrast, following administration of appropriate contrast agents.

E. P. Pappas
CyberKnife and TomoTherapy Department,
IATROPOLIS Private Special Clinic, Athens, Greece
e-mail: elepappas@phys.uoa.gr

E. Pantelis (✉)
Medical Physics Laboratory, Medical School,
National and Kapodistrian University of Athens,
Athens, Greece

CyberKnife and TomoTherapy Radiotherapy
Department, IATROPOLIS Private Special Clinic,
Athens, Greece
e-mail: vpantelis@phys.uoa.gr

Multiple anatomical (or morphological) and functional volumetric image studies can be imported in the CyberKnife system. In this chapter, only the morphological imaging techniques used in CyberKnife treatment are discussed. The use of functional imaging is presented in Chaps. 9 and 10. A prerequisite for a CyberKnife treatment is to acquire a CT scan of the patient in treatment position. Therefore, Section 8.2 is focused on the role of CT in CyberKnife, followed by image acquisition guidelines and the use of contrast agents. Section 8.3 gives details on the use of MRI and describes the most frequent MR imaging sequences used in CyberKnife treatments for CNS lesions.

8.2 Computed Tomography

8.2.1 The Role of CT

One volumetric CT study of the patient is the minimum requirement to perform a CyberKnife radiosurgery procedure. This CT study is imported into the CyberKnife data management system (iDMS), and when loaded in the treatment planning system (TPS), it automatically creates a 3D model of the patient within which the contours of the target(s) and organs at risk (OARs) as well as the radiation beams and the dose distribution are defined (Fig. 8.1). Tissue heterogeneities are considered by all available

© Springer Nature Switzerland AG 2020
A. Conti et al. (eds.), *CyberKnife NeuroRadiosurgery*,
https://doi.org/10.1007/978-3-030-50668-1_8

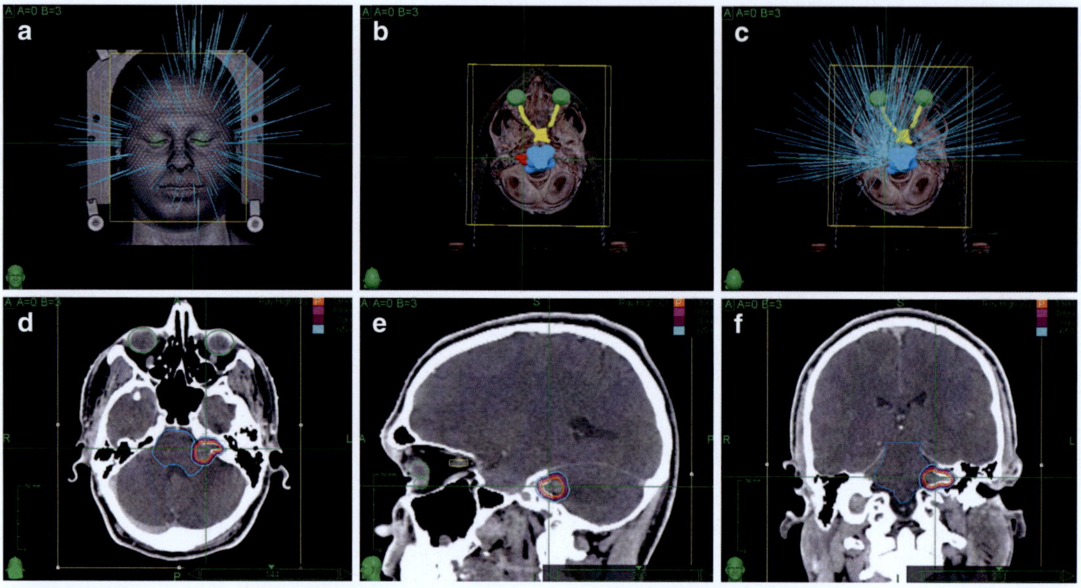

Fig. 8.1 (**a–c**) A three-dimensional representation of a patient model created by the treatment planning system using the volumetric CT scan of the patient. (**d–f**) Axial, sagittal, and coronal two-dimensional representations of the patient anatomy generated from the acquired CT scan. The presented patient was treated for a vestibular schwannoma. The delineated schwannoma, optic apparatus, brain stem, and ipsilateral cochlea can be seen along with the radiation beams and the corresponding dose distribution of the delivered treatment plan

dose calculation algorithms (i.e., ray tracing, finite size pencil beam (FSPB), and Monte Carlo) during dosimetry calculations using the electron density relative to that of water (ray tracing and FSPB), or the mass density (Monte Carlo) of each voxel of the patient model, both determined using the corresponding measured Hounsfield units (HU) and the CT density calibration model of the used scanner (Sect. 8.2.4).

CT is characterized by high geometric accuracy which is of paramount importance in radiosurgery applications. It also offers volumetric imaging at excellent spatial resolution, achieving sub-millimeter voxel sizes even for large field of views (FoVs) and extended scan lengths, as well as scanning times of the order of few seconds, thus avoiding patient motion artifacts in the acquired images. The CyberKnife is a frameless image guidance radiosurgery system. The developed treatment plan dose distributions are registered on the treated lesion using a sophisticated target locating system (TLS) based on kV x-ray stereoscopic imaging. A pair of orthogonal x-ray images of the patient in treatment position is acquired and compared using intensity similarity-based algorithms with corresponding digitally reconstructed radiographs (DRRs). These DRRs are calculated using a ray tracing algorithm on the patient model created using the primary CT of the patient (Fig. 8.2). More details on the target locating system and the used image registration algorithms can be found in Chap. 4.

Ideally, besides dose calculation and image-guided treatment delivery, the target(s) and organs at risk (OARs) should be delineated on the acquired planning CT scan of the patient to avoid registration uncertainties between the planning CT and other volumetric studies (e.g., MRI). However, the CT contrast depends on the density differences of the imaging tissues. As a result, CT exhibits excellent bone-soft tissue contrast, which is particularly useful for the delineation of the spinal canal and skull. Other tissues with different relative electron density compared to their environment (e.g., eyes and eye lenses, cochlea using a bone reconstruction filter) are also delineated on CT images. When soft tissue contrast is

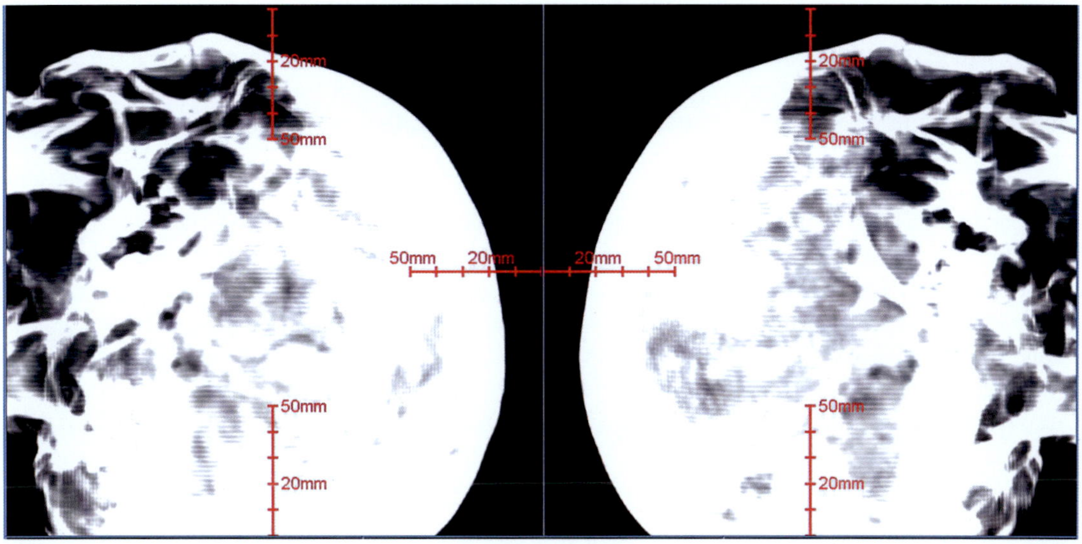

Fig. 8.2 An example of digitally reconstructed radiographs (DRRs) of a patient treated for an intracranial lesion using the CyberKnife frameless radiosurgery system. The DRRs are calculated using a ray tracing algorithm over the patient model created based on the planning CT scan of the patient

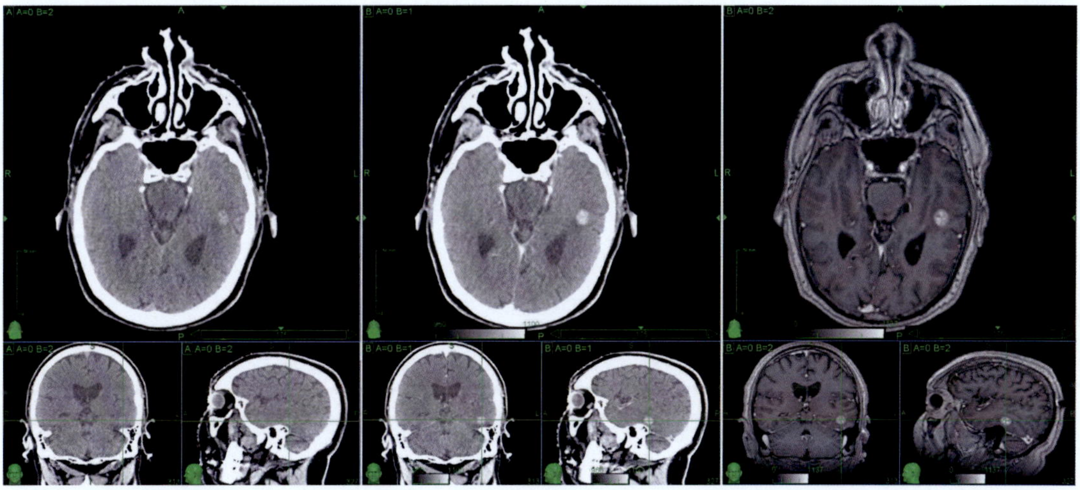

Fig. 8.3 Axial, sagittal, and coronal slices of a patient presented with a single brain metastasis in the left temporal lobe and treated with the CyberKnife system. Starting from the left set of axial, coronal, and sagittal slices, the CT studies without and with the administration of intravenous contrast agent, as well as the corresponding contrast-enhanced T1-weighted MRI study of the patient are presented. The improvement of target identification on the CT with contrast agent compared to the CT study of native contrast is evident. Excellent lesion identification in the contrast-enhanced T1-weighted MRI study can be observed

limited, administration of intravenous contrast agents can be exploited. In Fig. 8.3 an indicative case of a patient treated for a single brain metastasis in the left temporal lobe is presented. As can be seen, target identification is restricted in the non-enhanced CT scan of the patient. When intravenous iodine-based contrast agent is given to the patient, the target can be easily identified. Nevertheless, both CT studies cannot reach the soft tissue contrast of the MRI study of the patient. Other lesions showing improved target identification when contrast agent is administered are meningiomas, arteriovenous malformations (AVMs), and vestibular schwannomas [2].

CT contrast agents are commonly based on iodine material due to its enhanced x-ray attenuation. Several iodine-based contrast agents have been proposed and are clinically available for CT imaging. Clinically approved contrast media exhibit high water solubility, low binding to biological receptors, low osmolality, and toxicity [3]. Commercially available contrast media include iopamidol (Isovue, Bracco Imaging), iohexol (Omnipaque, GE Healthcare), iopromide (Ultravist, Bayer Healthcare), and iodixanol (Visipaque, GE Healthcare).

8.2.2 Image Quality Specifications

In a CT scan, several factors, scanning conditions, and parameters may affect image quality [4]. In some cases, a suboptimal image will be acquired with no direct consequences in CyberKnife treatment delivery, while in other cases target definition and/or dosimetric calculations may be considerably affected, potentially compromising treatment efficiency. In this section the key factors that may impact image quality to a degree that could be considered unacceptable for stereotactic radiosurgery (SRS) treatment planning purposes are discussed. Tolerance criteria, wherever referred, are strongly associated with the specifications and construction limitations of the phantom employed and method used for the evaluation.

8.2.2.1 Spatial Resolution and Total Number of Slices

Modern CT scanners offer the option for submillimeter spatial resolution. Images can be reconstructed at a matrix of 512 × 512 pixels and a slice thickness of <1 mm. The vendor suggests that slice thickness of more than 1.5 mm should not be used for the primary planning CT image [5]. Using a slice thickness of 1.0–1.5 mm seems a reasonable choice. Further reducing the slice thickness and using the finest spatial resolution available is tempting, but such an approach could also be suboptimal for treatment planning purposes. First of all, reducing the voxel size comes at the expense of increased noise in the image [4]. Moreover, rendering unnecessarily large image volumes can affect the performance of the dose calculation algorithm resulting in prolonged dose calculation and optimization times by the treatment planning system.

Care should be taken on the FoV used to reconstruct the acquired CT images. The FoV should include the entire anatomy since it is used for dose calculation and image tracking but should not be larger than necessary in order to minimize in plane voxel dimensions. For intracranial lesions, the FoV should be defined so that at least 1 cm of air gap is anterior of the patient and 1 cm of air gap superior of the patient (Fig. 8.4). For extracranial lesions, care should be taken that the entire cross section of the patient is scanned, and an adequate number of slices are acquired inferior and superior to the target since the beam delivery is non-coplanar. A general guideline is that 15 cm above and below the target are enough for planning purposes.

8.2.2.2 Contrast Agents

As already discussed, concentration of contrast agents locally increases HUs. As a result, the native HUs and density information of the imaged tissue are lost. Specifically, following intravenous injection of iodine-based contrast agent, the HUs of brain veins and brain parenchyma increased on average by 103 ± 29 and 7 ± 4, respectively [6]. For conventional whole-brain radiotherapy, this resulted in an increase in monitor units by less than 1%. For arteriovenous malformations, a HU difference of 152 between enhanced and unenhanced images has been reported. Still, the induced mean and maximum dose deviation in the calculated SRS treatment plans were 0.67% and 1.8%, respectively [7]. Specifically for extracranial CyberKnife applications, Kim et al. [8] assessed changes in HUs due to the presence of contrast agent as well as the related dosimetric effect for a variety of treatment sites. Detected dose deviations at the reference point depended on the target location as well as the calculation algorithm employed. Indicatively, for spine lesions differences reached 2.1% and 1.9% if the ray tracing or the Monte

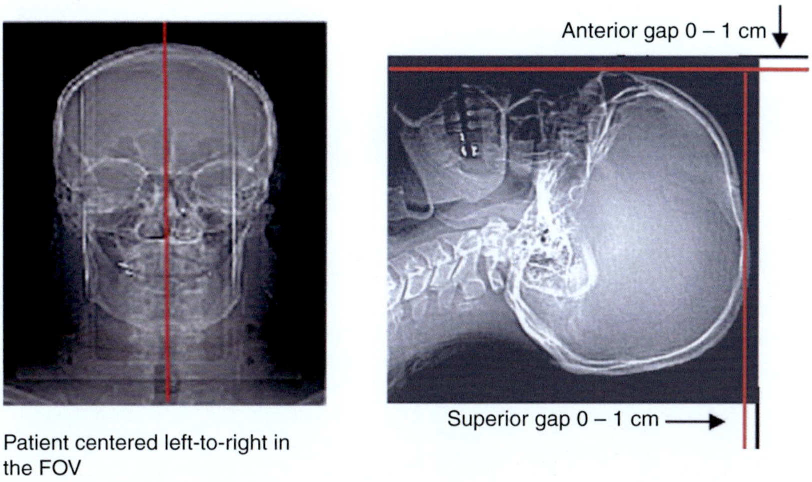

Fig. 8.4 Field of View (FoV) adjustment in CT imaging for cranial applications. The patient is centered laterally within the FoV, as shown on the left panel. The FoV should extend by 1 cm from the anterior and superior tips of the patient, as shown on the right panel. (Image taken from the CyberKnife user's manual version 11.1, revised 2018-06 [5]). The reader should always refer to the most recent version available for updated recommendations

Carlo dose engine is used, respectively [8]. Larger deviations were reported for lung tumors.

It should be noted that the CyberKnife manufacturer recommends that contrast-enhanced CT images are not used as the primary CT volumes for dose calculations for all applications, irrespective of the treatment site [5]. This conservative approach is driven by the desire to minimize dose calculation uncertainties as well as the fact that presence of contrast agent could affect DRR quality.

8.2.2.3 Artifacts and Implants

Several artifacts can cause image quality degradation. For SRS treatment planning, the most important ones are those caused by the presence of artificial implants. High Z materials induce streaking artifacts in the image which are often not limited to the vicinity of the implant. On the contrary, signal voids can span through the entire patient geometry. Moreover, artifacts are often demonstrated by excessive signal around the high Z material. Figures 8.5 and 8.6 present a cranial and an extracranial neuro-radiosurgery case, respectively, with severe artifacts, in and around the target location.

In addition to lost anatomical information, HUs are also affected. Consequently, accuracy of dose calculations is affected if the effect is not taken into account. Excessive HUs and signal voids need to be identified and contoured at least in the areas where primary photon beams are expected to pass through. Density override within the contoured regions can mitigate erroneous dose calculations.

8.2.3 Image Acquisition Protocol

Based on the above remarks, a specific imaging protocol should be established and followed in all CT imaging procedures referred to for CyberKnife treatment planning. It is crucial that determined imaging parameters do not vary between patients. For instance, a change in the kV or filtering parameters can jeopardize treatment outcome. The vendor provides general guidelines and specifications for the primary CT imaging procedure that need to be adopted [5]. In Table 8.1, a CT image acquisition protocol for treatment planning and dose calculation purposes in CyberKnife neurosurgical applications is given.

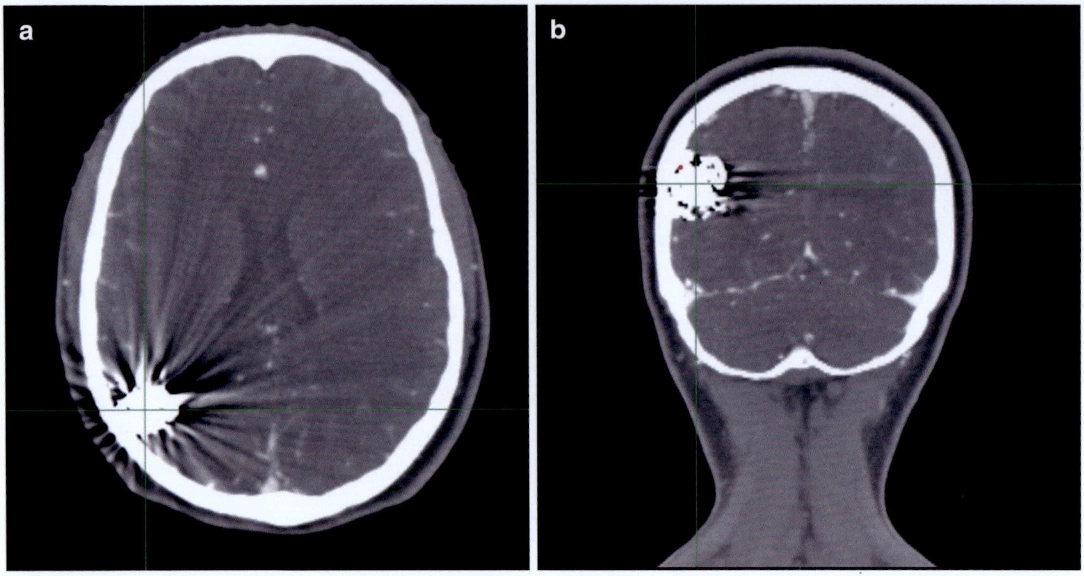

Fig. 8.5 (**a**) Axial and (**b**) coronal slices of a CT image stack of a patient with embolized arteriovenous malformation. The emboli induce severe streaking artifact in and around the target. The density model cannot be directly applied to determine density distribution. Density override at signal voids and high HU areas should be performed to mitigate erroneous dose calculations

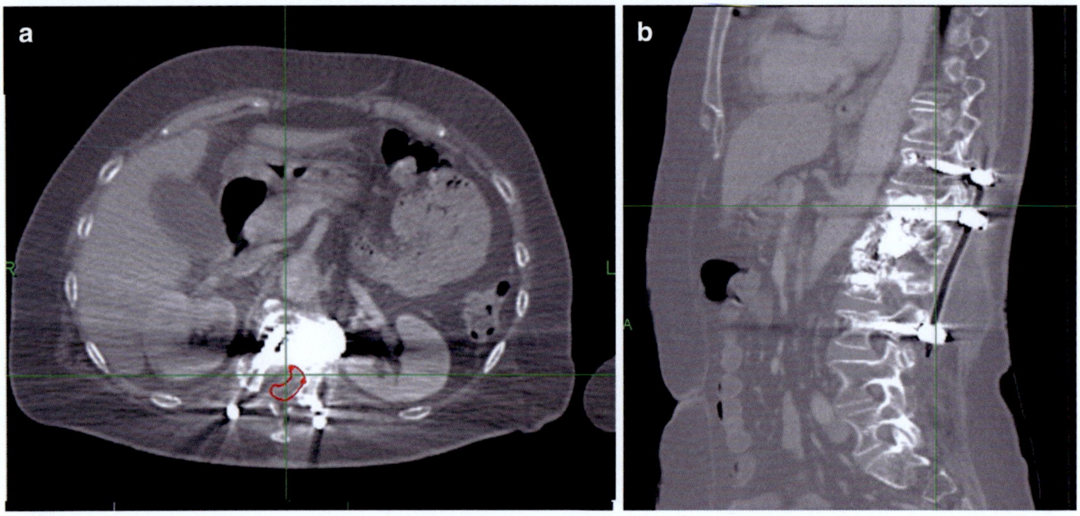

Fig. 8.6 (**a**) Axial and (**b**) sagittal slices of a primary planning CT image stack of a patient having a spine-supporting metal implant causing streaking artifacts, in and around the target location (depicted by the red contour). On the primary images, density override at signal voids and high HU areas should be performed to mitigate erroneous dose calculations

Table 8.1 A CT imaging protocol for CyberKnife neuro-radiosurgery clinical applications

Imaging parameter/condition	Recommendation
Patient positioning	The patient should be aligned on the CT couch with his craniocaudal axis parallel to the CT scanner's longitudinal axis For intracranial and intracanal spine lesions, the patient should be positioned supine. Prone positions could be explored for extra-canal spine lesions
Immobilization devices	For intracranial lesions, the use of a head rest and a thermoplastic mask is suggested. These devices could also be used for cervical spine lesions For spinal lesions, the use of a vacuum foam could be explored In all cases, the used immobilization devices should make the patient feel comfortable on treatment couch For cervical and upper thoracic spine lesion, alignment of the spine horizontal (if possible) on treatment couch should be considered to aid image guidance
CT tabletops	A flat tabletop should be used without embedded guide wires. If guide wires are present, position patient in a way that the tracked volume of interest (e.g., the skull) does not intersect with guide wires
Intravenous injection of contrast agent	Not to be used in primary CT images. Can be injected for secondary image studies
Imaging mode	Helical scan with a pitch equal to 1
Gantry angle	Always at 0°
Tube voltage	120 kV
Tube charge	400 mAs or scanner's maximum
Reconstruction matrix	Up to 512 × 512 × 512 matrix size supported
Slice thickness	1 mm is the typical value (not more than 1.5 mm). No gaps between slices
Variable slice thickness	Not supported
Pixel size	Only square pixel sizes are supported. Use the minimum suggested field of view (FOV) for best in-plane resolution
FoV for cranial cases	The patient should be at the center (laterally) within the FoV. The thermoplastic mask and part of the head rest should be included. The entire volume where beams are expected to intersect with the patient should be included in the images Extend the FoV by 1 cm anteriorly (from the nose tip) and superiorly (top of skull), as shown in Fig. 8.4
FoV for extracranial cases	The patient should be at the center (laterally) within the FoV The target should be centered in the inferior-superior direction with a minimum of 15 cm scan length in either direction
Reconstruction kernel or filter	Use a medium smooth reconstruction kernel. High-pass reconstruction kernels could be additionally used in specific cases to aid delineation of specific structures like the cochlea in a vestibular schwannoma case and the embolization material (if present) in an arteriovenous malformation case

8.2.4 CT Density Calibration Model

As already discussed, the primary CT image volume contains the material density distribution, necessary to perform dosimetric calculations. More specifically, the ray tracing and finite size pencil beam (FSPB) dose calculation algorithms require the relative electron density information within the patient anatomy. On the other hand, the Monte Carlo-based dose calculation algorithm uses the mass density. Both quantities can be estimated by the HU distribution acquired from a CT scan, provided that a HU-to-electron density or a HU-to-mass density calibration curve has been determined for the specific scanner, scanning protocol, and image reconstruction parameters.

The CT calibration curve of the scanner (i.e., the CT density model) can be derived by scanning a phantom containing reference materials of known mass and relative electron densities (Fig. 8.7). The phantom should incorporate inserts filled with common materials equivalent to the ones encountered during a CT scan of a patient. However, a CT density model derived using the above reference materials often fails to

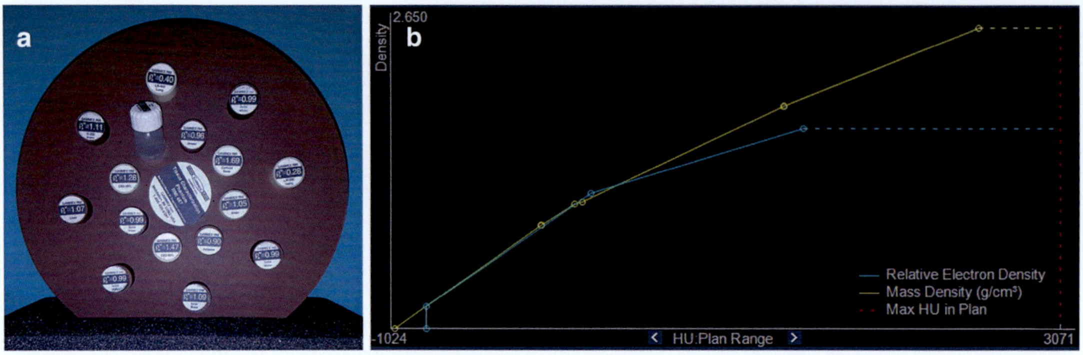

Fig. 8.7 (**a**) A photograph of a commercially available tissue characterization phantom used to determine the CT density calibration curve. (**b**) An example of a CT density calibration curve used to assign measured HU to relative electron densities (used by the ray tracing and FSPB dose calculation algorithms) and mass densities (used by the Monte Carlo dose calculation algorithm)

accurately predict densities of other non-standard high Z materials. In these cases an appropriate extrapolation method should be used, with the "flat" option (i.e., all voxels with HU higher than the maximum HU value of the CT density model are mapped to densities equal to the maximum density of the model) to be the safer method since it avoids mapping to erroneous densities. Nevertheless, the user can delineate the high Z material and assign its density (if known) to the delineated structure and override the automatic density assignment method.

Most importantly, the CT density model is strongly dependent on the kilovoltage potential (kV) used for image acquisition. A change of the kV by the radiographers without proper communication to the medical physicists can result in erroneous dose calculations [9]. To avoid confusion, it is suggested that only one fixed kV is always used for CT imaging for treatment planning purposes, irrespective of the treatment site considered [9].

8.3 Magnetic Resonance Imaging

8.3.1 An Introduction to MR Contrast Weighting

MRI offers unsurpassed multi-contrast capabilities, mainly stemming from the inherent T1 and T2 relaxation times and proton density (PD) which are characteristics of each tissue. Moreover, by taking advantage of changes in the local magnetic field, other sources of contrast can be exploited for reconstructing images based on T2* relaxation time, accumulation of a contrast agent, the chemical shift effect, local magnetic susceptibility variations, diffusion of water, or even brain activation [10–12].

Selecting the contrast, i.e., the preferred weighting of the source of image signal, starts by choosing the most appropriate pulse sequence and is followed by optimizing its main parameters. The following sections present the underlying key concepts related to these two steps.

8.3.1.1 Sequence Selection

There are basically two groups of sequences [13, 14]: the spin echo (SE) and the gradient echo (GE). All sequences employ a radiofrequency (RF) excitation pulse to disturb the proton spins lying within the volume of interest and a set of gradients for space encoding. However, SE sequences also use a second RF pulse to regain the lost phase coherence associated with the local microscopic magnetic field variations. Therefore, SE sequences can be used to acquire T1-, T2-, and PD-weighted (T1w, T2w, and PDw, respectively) images, but not T2*. On the other hand, a GE sequence uses a single RF excitation pulse, and gradient fields are applied to rephase only those spins who were deliberately de-phased by a preceding gradient field. Since local magnetic field inhomogeneities are not accounted for, GE sequences produce T2*-weighted (T2*w) images, in addition to T1w and PDw. This characteristic makes GE images sensitive to local magnetic field inhomogeneities. However, using a single excitation pulse allows for much faster image acquisitions, as compared to SE sequences.

Each group contains numerous sequences developed and implemented in clinical practice either as faster or more optimal scanning protocols or to answer different clinical questions. A very brief and incomplete overview follows, covering only the most common ones with emphasis to brain imaging. For a complete description of all available sequences and more technical considerations, the reader should refer to MRI-dedicated publications [10–12].

Turbo spin echo is one of the most common sequences in the SE group and is usually labeled as *TSE* or *FSE* (turbo spin echo or fast spin echo, respectively) depending on the scanner manufacturer. A conventional SE sequence contains one RF refocusing pulse, and only one signal can be collected, following each excitation. In a TSE sequence, however, multiple acquisitions can be performed by adjusting the number of refocusing pulses, a parameter often labeled as "turbo factor" or "echo train length." Therefore, TSE is routinely used for faster T2w image acquisitions. A TSE sequence can be employed in either 2D or 3D imaging mode. In the latter case, the refocusing pulses can also have variable flip angles, depending on the scanner manufacturer. Examples of such sequences are *SPACE*, *3D-VIEW*, *CUBE*, and *VISTA*.

Inversion recovery (IR) sequences are special variants of the SE group. Briefly, an extra inversed RF excitation pulse is first applied to turn all proton spins upside down. By the T1 relaxation mechanism, protons begin to return to equilibrium. The main excitation pulse is then emitted at a time point when the magnetization of a specific tissue is zero. In other words, specific tissues are not excited, i.e., they are nulled. If the signal from fat is nulled, the corresponding pulse sequence is known as *STIR* (short TI inversion recovery). Similarly, in brain scans, signal from the cerebrospinal fluid can be nulled by employing the *FLAIR* (fluid-attenuated inversion recovery) sequence. The sequence names vary among scanner manufacturers.

Spoiled gradient echo: In a conventional GE sequence, following signal acquisition, residual T2-related (transverse) magnetization still remains often after applying the next RF excitation pulse. Spoiling of the transverse magnetization can occur by using gradients of variable amplitude just before the next excitation or by employing additional RF spoiling pulses. Sequences applying such techniques are known as spoiled gradient echo (SGE) sequences [13, 15]. They are primarily used to acquire T1w images, but with proper selection of parameters, T2*w and PDw images can be reconstructed. Examples of SGE sequences are *SPGR*, *T1-FFE*, *FLASH*, and *FE*.

Ultrafast gradient echo: Conventional SGE sequences are very slow GE image acquisition techniques. Rapid versions of SGEs are known with the generic name ultrafast gradient echo (UGE). In order to speed up image acquisition, very short TRs and flip angles (see Sect. 8.3.1.2) are used, resulting in poor T1 contrast. To mitigate for the latter, pre-pulses are applied in combination with spoiling gradients and a time delay before the main excitation pulse. For 2D imaging, common commercial names for UGE sequences are *FSPGR*, *TFE*, *Turbo-FLASH*, and *Fast FE*, often exhibiting technical differences. Accordingly, 3D UGE sequences are usually labeled as *BRAVO*, VIBE, THRIVE, *3D TFE*, *MPRAGE, MP2RAGE*, and *3D Fast FE*.

Time-of-flight (TOF) sequences are typically GE-based and are used in MR angiography. Briefly, they take advantage of the so called inflow effect (also referred to as the time-of-flight effect), according to which blood that has just flown in a slice excited by preceding RF pulses is in equilibrium (in contrast to stationary surrounding tissues which are partially saturated) and therefore can produce high signal if a new excitation pulse is applied. In TOF images, blood vessels appear bright against dark saturated stationary tissues.

Echo planar imaging (EPI) is a family of sequences that can be either SE- or GE-based. The main characteristic of EPI is the extreme acquisition times, typically less than 100 ms per slice. This is achieved by applying rapidly oscillating frequency encoding gradients in addition to low amplitude bleep-type phase encoding gradients. Data for an entire slice can be obtained following a single or a few RF excitation pulses, sequences termed as single- and multi-shot EPIs, respectively. Although fast, EPIs exhibit low spatial resolution and suffer from image artifacts and increased geometric distortion. In clinical practice, SE-EPI is the sequence of choice for diffusion-weighted imaging, while GE-EPI is implemented in functional MRI (fMRI; see Chap. 9).

8.3.1.2 Sequence Optimization

Implementation of a pulse sequence comes with the adjustment of a great deal of parameters, affecting contrast weighting, spatial resolution, noise levels, signal nulling, geometric distortion, acquisition time, etc. The main parameters are matrix size, slice thickness, echo time (TE) or effective echo time (TE_{eff}) for TSE sequences, repetition time (TR), inversion time (IT), echo train length (ETL) or turbo factor (TF), flip angle (FA), bandwidth (BW) or bandwidth per pixel (PBW), and number of signal acquisitions (NSA or NEX). It should be noted that parameters' terms, definitions, and units may vary between scanner manufacturers.

The first step is to decide on the main contrast weighting, i.e., T1w, T2w, T2*w, or PDw. In SE sequences this is mainly affected by TE and TR selection, while FA is typically set to 90°. For GE sequences FA is variable and fundamentally impacts contrast weighting. Typically, T1 weighting is related to short TRs and TEs, while T2 or T2* weighting is achieved by using long TRs and TEs. PDw arises from long TRs and short TEs. Quantitatively, "long" and "short" TEs and TRs do not hold the same meaning among sequences. In SE imaging a long TE and TR would typically be >60 ms and >2000 ms, with corresponding "short" values ranging between 10–25 ms and 250–700 ms, respectively [14]. On the other hand, long TEs and TRs in a GE sequence would typically mean TE > 10 ms and TR > 100 ms [14].

Other parameters such as the matrix size (determining the number of frequency and phase encoding steps), slice thickness, NSA, PBW, and TF are interdependent affecting noise levels, spatial resolution, and accuracy as well as image acquisition time. The matrix size affects the number of frequency and phase encoding steps N_{FE} and N_{PE}, respectively. In a specific 2D sequence, changing PBW = BW/N_{FE}, NSA, and N_{FE} noise levels can be relatively estimated [11]:

$$\text{noise} \sim \frac{\sqrt{\text{PBW}}}{\sqrt{\text{NSA} \cdot N_{\text{PE}}}} \tag{8.1}$$

Regarding acquisition time for a particular 2D sequence, it is affected by N_{PE} and NSA according to [11]:

$$\text{Acquisition Time} = \text{NSA} \cdot \text{TR} \cdot N_{\text{PE}} \tag{8.2}$$

It should be noted that increasing the spatial resolution in the frequency encoding direction, i.e., increasing N_{FE}, will not burden scan time but will degrade signal-to-noise ratio. More specifically, the achievable signal-to-noise ratio is directly proportional to the voxel size if the scanning time is held constant.

Regarding 3D scanning protocols, an additional phase encoding gradient is applied to the through-plane direction, introducing extra phase encoding steps, $N_{PE,Z}$. Corresponding noise levels and acquisition time estimations in 3D sequences can be performed if $N_{PE,Z}$ is introduced as an extra term in Eqs. (8.1) and (8.2).

A compromise among the above should be determined to meet the requirements of the specific clinical application. In SRS treatment planning, spatial resolution and accuracy are of paramount concern.

8.3.2 Role and Mechanism of Gd-Based Contrast Agents

Although MRI offers multi-contrast capabilities, especially for soft tissues, contrast agents are routinely administered, enhancing signal contrast at specific tissues of interest. More specifically, paramagnetic ions can act both on T1 and T2 relaxation times, if accumulated locally in the tissue. Gadolinium (Gd^{3+}) is the most commonly used one, but, as being an ion, it is administered chelated by a molecule to avoid the toxicity of free ions [16], forming the group of the Gd-based contrast agents. Such contrast agents will mainly result in shortening T1 relaxation time in the tissues accumulated and, thus, significantly increasing the brightness in a T1w image. Gd is very routinely used in brain lesion detection, such as brain metastases. Following injection into the body, it is distributed to all perfused tissues but, chelated in a large molecule, cannot cross the blood-brain barrier quickly. In tumors, however, the barrier is disrupted which results in the Gd-based contrast agent leaking into the interstitial space, a mechanism which results in a significant increase in T1w signal from the tumor. Gd also reduces T2 relaxation time but the normal rate is still the dominant one [11]. The underlying mechanism that reduces relaxation times is related to the paramagnetic nature of Gd. Its magnetic susceptibility changes the local magnetic field in the vicinity of the molecule, acting as a local field inhomogeneity.

Dosage varies depending on the formulation, imaging application, and body weight of the patient. Typical doses for brain lesion localization are 0.1–0.2 mmol/kg of body mass, administered dissolved in saline. The main contraindications for administration of such contrast agents are poor renal function and pregnancy. Gd-based contrast agents have been linked to nephrogenic systemic fibrosis as a side effect [17, 18] which has led to regulatory recommendations by the authorities [11]. Moreover, Gd-based contrast agents have been reported to be deposited in the brain after repeated administrations, although the clinical significance and risks associated with this finding are still unknown [19, 20].

8.3.3 Applications in Neuro-radiosurgery

MRI is the most common secondary imaging modality used in SRS treatment planning. In cranial cases such as brain metastases, vestibular schwannomas, meningiomas, arteriovenous malformations, trigeminal neuralgia, and others, appropriate MR scanning protocols can provide excellent contrast between the lesion and surrounding healthy tissues, in high 3D spatial resolution.

In a typical application, a T1w image offers better contrast between gray and white matter. Following intravenous injection of a Gd-based contrast agent, brain lesions (wherever the blood-brain barrier is disrupted, such as in tumors) will appear brighter providing excellent contrast with surrounding normal brain parenchyma (Figs. 8.3 and 8.8a). T2w images are generally sensitive to fluid collections. In addition to brain tumors, they can be used for nerve identification in trigeminal neuralgia (Fig. 8.8d) or vestibular schwannoma cases (Fig. 8.8b). An IR sequence might be useful to null the signal from the cerebrospinal fluid (CSF) and enable detection of small lesions in the periventricular area. For extracranial neuro-radiosurgery cases, T2*w images are more appropriate to reveal the internal anatomy of the spinal cord, while T2w images are very useful for tracing the nerve roots.

In any case, it should be noted that MRI is fundamental for accurate lesion identification, localization, and delineation in SRS treatment planning. The exact boundaries of the target(s) and adjacent critical organs will mainly—if not solely—rely on the MR image quality and particularly the spatial resolution, spatial accuracy, artifacts, and signal-to-noise ratio of the set of MR images acquired.

8.3.4 Image Acquisition Protocol

The CyberKnife vendor provides general guidelines and specifications for the MR images used for treatment planning [5]. In Table 8.2, a protocol for acquiring patient MR images appropriate for treatment planning with the CyberKnife system is presented.

8.3.5 Artifacts

MR images can exhibit a great deal of artifacts associated with the imaging parameters selected, hardware performance, patient/organ motion,

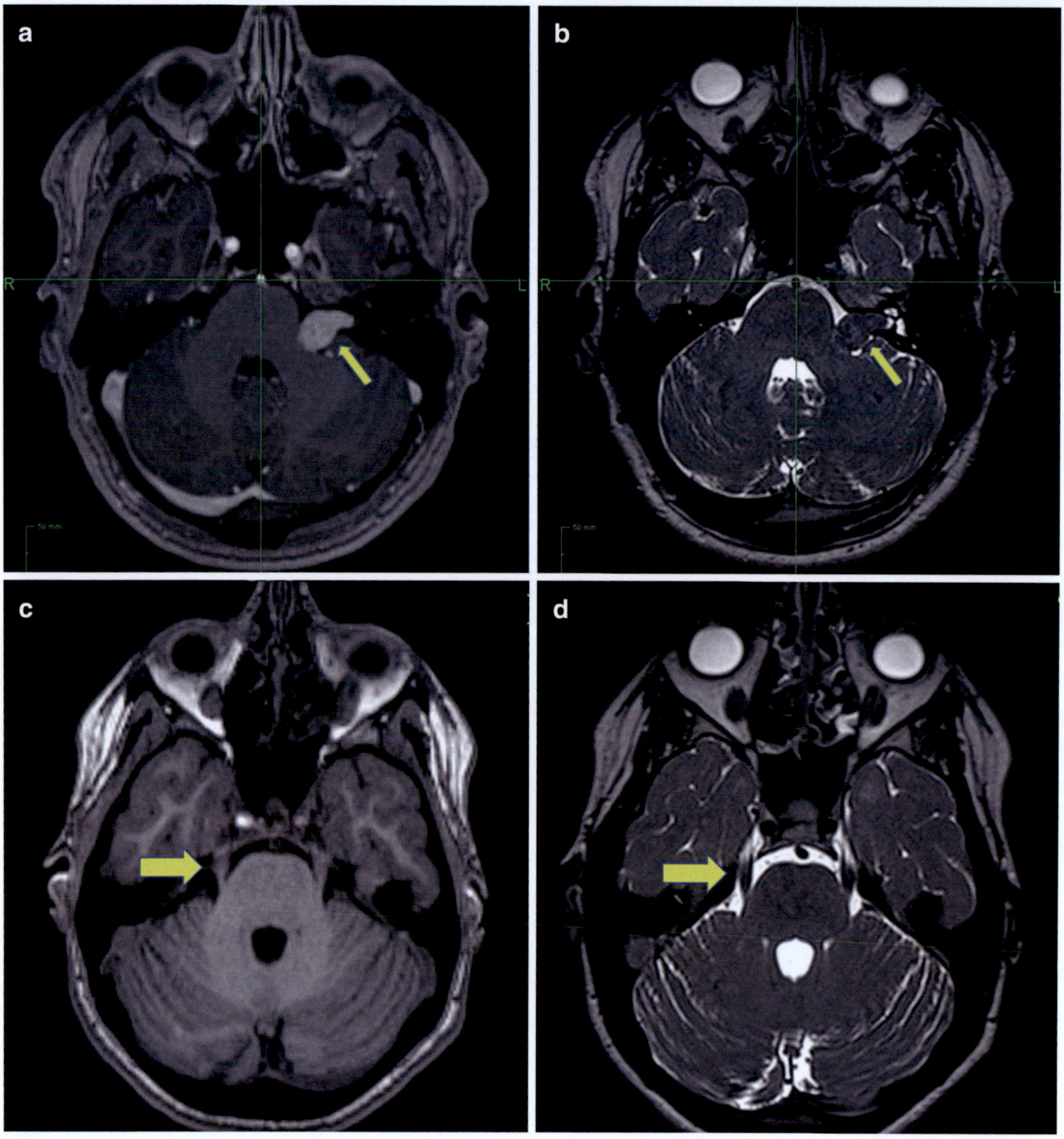

Fig. 8.8 (**a**) Contrast-enhanced T1w and (**b**) T2w axial slices of a patient suffering from a vestibular schwannoma (indicated with the yellow arrow). (**c**) T1w images without administration of contrast agent and (**d**) corresponding T2w axial slices of a patient suffering from trigeminal neuralgia of the right trigeminal nerve indicated with the yellow arrow

magnetic properties of the image volume, or presence of implants. General guidelines to avoid or mitigate artifacts in an MR image have been described in the literature [21, 22].

Artifacts caused by implants are very common in CNS imaging. As an instance, for the case shown in Fig. 8.9a, an artificial CSF pump severely distorts the image and causes loss of signal in the brain parenchyma. Figure 8.9b presents an extracranial case with metal implants supporting the spine. In both cases, the lesion lies in and around the artifact, and, therefore, target localization and delineation is challenging. Sequences and imaging parameters that are less prone to such artifacts should be considered, in addition to employing other imaging modalities.

Table 8.2 A protocol for acquiring MR images for CyberKnife treatment planning

Imaging parameter/condition	Recommendation
Patient positioning	Ideally the same position as with the CT imaging positioned should be used to minimize image registration uncertainties
Immobilization devices	Ideally the immobilization devices present during CT scanning should be used to ensure same patient position. However, for intracranial cases the base plate used to hold the head rest and the thermoplastic mask do not fit on commonly used MR scanners, and their use can be avoided
Imaging orientation	Axial, coronal, sagittal, or oblique up to 30° are supported
Reconstruction matrix	Up to 1024 × 1024 × 512 matrix size supported
Slice thickness	1 mm. Depending on the case, a value of 2 mm can be also used
Variable slice thickness	Not supported
Pixel size	1 mm or less. Only square pixel sizes are supported
Field of view (FoV) for cranial cases	Center the patient laterally within the FoV. Extend the FoV by 1 cm anteriorly (from the nose tip) and superiorly (top of skull)
FoV for extracranial cases	Center the patient left-right and anterior-posterior within the FoV. The target should be centered in the inferior-superior direction
Acquisition sequence	– Contrast-enhanced T1w sequences are used for the identification of brain metastases, meningiomas, vestibular schwannomas, pituitary adenomas, gliomas, etc. 3D acquisition sequences based on spoiled gradient echo or ultrafast gradient echo sequences are appealing due to their short acquisition times – T2w sequences can be additionally used for vestibular schwannomas, cavernous sinus meningiomas, pituitary adenomas, and trigeminal neuralgia. It is noted that in these cases the corticospinal fluid surrounding the lesion(s) or the nerves appears white. In case that the signal from the corticospinal fluid (CSF) is obscuring lesion identification, a fluid-attenuated inversion recovery (FLAIR) T2w sequence can be used (e.g., to show edema in a recurrent glioma) – For the identification of AVMs, a T1w time-of-flight (TOF) 3D gradient echo sequence should be used
Intravenous injection of contrast agent	Yes, for the identification of brain metastases, meningiomas, vestibular schwannomas, pituitary adenomas, gliomas, arteriovenous malformations, etc.

8.3.6 Spatial Distortion

Besides the common image quality indices (such as signal-to-noise ratio, contrast-to-noise ratio, etc.), an imaging modality's accuracy in localizing in space anatomical structures of interest is of paramount importance, especially in the case images are employed in radiosurgery treatment planning. It is well-known that MR images are inherently distorted [23]. Distortion of a few millimeters is not expected to affect typical diagnostic applications. However, if the images are used to identify and delineate a target or a critical organ with minimum spatial error tolerance, then MR-related geometric distortion might set limitations or raise concerns. Particularly for intracranial SRS, spatial inaccuracies of the order of 1 mm may have a significant dosimetric impact (e.g., a significant reduction to the absorbed dose by the target), in cases where steep dose gradients exist. Therefore, significance of the geometric distortion depends on the application the image will be employed for.

Spatial accuracy degradation is mainly exhibited at the edges of the imaged volume and increases with increasing FoV [24, 25]. Geometric distortions mainly stem from static magnetic field, B_0, inhomogeneity, gradient field nonlinearity, differences in the magnetic susceptibility of the object/subject being imaged, and the chemical shift effect [23]. Other sources of distortions and/or artifacts are related to the eddy currents, temperature drift, aliasing, etc. [26–31] which will not be discussed.

8.3.6.1 Gradient Field Nonlinearity

Signal in MR imaging is tagged with respect to its origin by applying gradient magnetic fields,

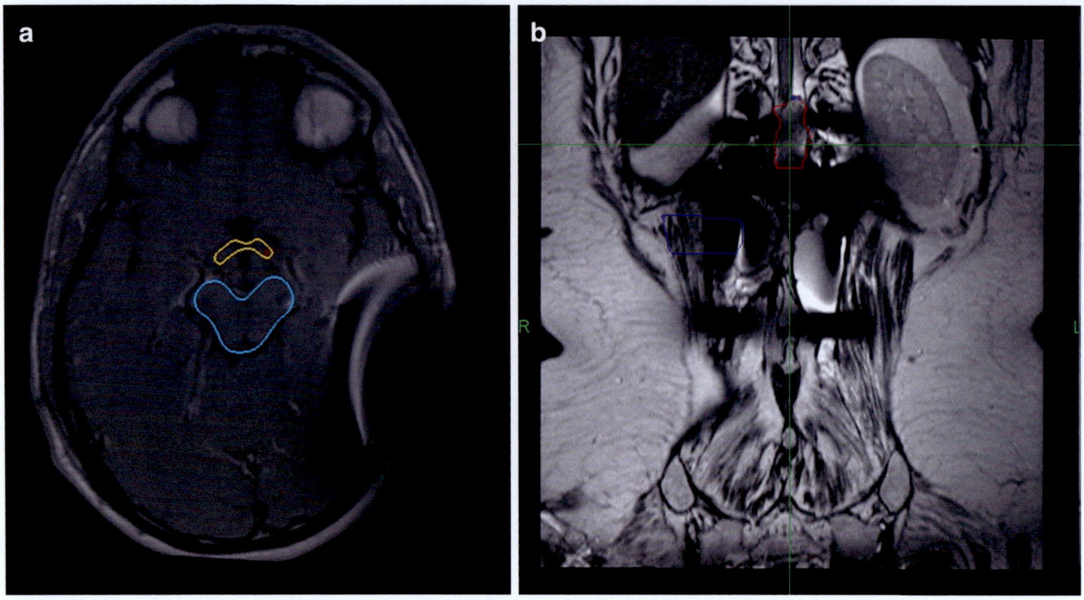

Fig. 8.9 (**a**) Axial MR image of a patient with a CSF pump resulting in an artifact obscuring part of the brain parenchyma. (**b**) A spine case with the target indicated by the red contour. Presence of metal implants for spine support causes signal loss in the vicinity of the target

G_x, G_y, and G_z, on the three dimensions. Gradient fields are enabled either during excitation (slice selection gradient in 2D pulse sequences) or prior to (phase encoding gradient) or during (frequency encoding gradient) the MR signal read-out at TE. Therefore, gradient fields are strongly associated with the coordinates of the voxel being imaged.

The gradient of the magnetic field is supposed to be uniform throughout the volume being imaged. In other words, when a gradient field is enabled on one dimension, magnetic field is supposed to increase or decrease linearly with respect to distance from the isocenter on this dimension. When the gradient field is enabled on x axis, G_x, the local magnetic field at location x should be:

$$B(x) = B_0 + G_x \cdot x \tag{8.3}$$

The MRI systems are designed to apply a constant G_x with respect to x, and, therefore, $G_x \cdot x$ is expected to vary linearly with x location. The same concept applies to the other two dimensions, as well. This is fundamental for encoding the MR signal in space.

However, a considerable deviation from the assumed linearity in space will result to mis-encoding of the signal and, consequently, to a geometric offset for the voxel in the MR image corresponding to the specific location. If no correction is applied, the image will appear heavily warped, with distortion often exceeding 3 mm for a brain scan, exhibited at the edges of the FoV [32, 33]. Thus, all major MR scanner manufacturers have developed post-imaging routines to partly correct for machine-related gradient nonlinearities [32]. Depending on the manufacturer and sequence selected, correction is performed either in 2D or 3D. Using pre-calculated gradient distortion maps [32, 34–36], the image series is corrected by applying a transformation from the distorted image space to the undistorted one. This is actually an interpolation task [23] with several different approaches presented, ranging from spherical harmonics to polynomial and spline interpolations [37–40].

An important remark on vendor-supplied correction algorithms is that they can be applied automatically, if enabled, with a negligible impact on image reconstruction time [1] and no impact on scanning time. Thus, it is highly recommended that 3D distortion correction algorithms are always enabled, irrespective of the intended MR application. Strikingly, this option is not enabled by default in some scanners. Figure 8.10 presents an axial T1w slice of a multimodality MR distortion phantom acquired at 3.0T without and with the distortion correction option enabled, fused with the corresponding CT image study. The significant spatial distortions observed (mainly at the edges of the FoV) are minimized when built-in distortion correction algorithms are enabled.

Still, residual distortion should not be considered negligible for SRS treatment planning purposes. Mean distortion within a typical FoV for brain scans was reported to reach 0.53 mm with maximum detected distortion exceeding 1.0 mm [41]. It should be noted, however, that gradient nonlinearity induced distortion is system-specific, while its magnitude levels are affected by the imaging parameters selected, such as the pixel bandwidth. Moreover, distortion greatly varies within the imaging volume. Therefore, one should evaluate residual distortion levels for the specific MR scanner, sequence, imaging parameters, and volume of interest clinically employed in SRS treatment planning.

8.3.6.2 Static Magnetic Field Inhomogeneity

MRI strongly relies on the application of a static magnetic field, $\vec{B}_0$, constant in magnitude and direction, in order to separate the energy levels of the spins, according to the Zeeman effect. Higher B_0 strength results in enhanced signal-to-noise or shorter scanning times.

A potential local inhomogeneity in the strength of the static magnetic field will result in a strength of $B_0^{'}$, which will directly affect the Larmor precessing frequency of the spins, according to the equation $\omega_0^{'} = \gamma B_0^{'}$. This will also impact the spatial information, as $B_0^{'}$ will be summed with the gradient field to encode the location of the imaged volume, according to Eq. 8.3.

The inhomogeneous $B_0^{'}$ field, summed with a decreasing gradient field, G, results in a volume actually located at r_0 to be imaged at $r_0^{'} < r_0$. However, reversing the polarity of the gradient field will result in a distorted image location at $r_0^{'} > r_0$. In other words, reversing the polarity of the gradient field will change the sign of distortion without affecting the distortion magnitude [24, 42]. This is in contrast to the corresponding remark made for the gradient field nonlinearity-related distortion.

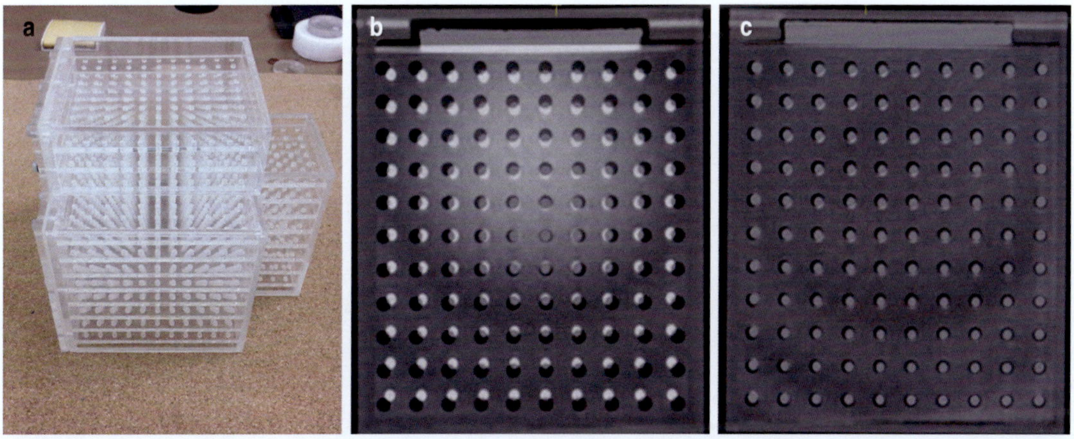

Fig. 8.10 (**a**) An in-house built distortion detection phantom. Axial T1w MR images fused with corresponding CT study of the phantom acquired at 3.0T without (**b**) and with (**c**) having enabled the vendor-supplied distortion correction algorithm. The severe spatial distortion of the MR images observed mainly at the edges of the field of view is minimized when the distortion correction algorithm is enabled

Another key difference between gradient field nonlinearity- and B_0 inhomogeneity-related distortions is that the latter is mainly exhibited on the frequency encoding direction (and the slice selection direction in 2D imaging protocols) in typical (non-EPI) T1w images, while the former affects all dimensions.

8.3.6.3 Susceptibility Differences

The magnetic susceptibility (commonly referred to as volume susceptibility [43]), χ, is an important magnetic property of a material. It indicates whether a material is attracted into or repelled out of a magnetic field. In short, when an external uniform magnetic field $\vec{B}_0 = \mu_0 \vec{H}_0$ is applied inside a (non-permanently magnetized) material, the actual field $\vec{B}$ inside the material is given by [12]:

$$\vec{B} = \mu_0 (1+\chi)\vec{H} \tag{8.4}$$

where $\vec{H}$ field is measured in A/m and μ_0 is the vacuum permeability ($4\pi \cdot 10^7$) with units of Tm/A [12] in the SI system convention. Therefore, magnetic susceptibility, χ, is dimensionless.

Based on the macroscopic behavior under the influence of an external magnetic field, various materials are classified into diamagnetic, paramagnetic, and ferromagnetic materials. According to Eq. 8.4, if the susceptibility $\chi > 0$, the material is considered as paramagnetic, and if $\chi < 0$, the material is diamagnetic. For vacuum, $\chi = 0$ [12, 43]. Superconductors are characterized by the smallest susceptibility value, $\chi = -1$, while for soft ferromagnetic materials $\chi > 10^5$. However, for materials involved in MRI, $|\chi| \ll 1$.

Table 8.3 lists the magnetic susceptibility of various substances or materials found in vivo. Although most of the materials listed are diamagnetic (i.e., $\chi < 0$), significant variations in magnetic susceptibility are observed. According to Eq. 8.4, the local magnetic field inside a substance depends on the local susceptibility, and, consequently, B_0 uniformity is inevitably compromised by the presence of materials. As a result, the Larmor precession frequency of spins inside a substance will also be affected. In an MR image, the center of a uniform material will be mis-encoded in space, resulting in a geometric offset which may or may not be significant depending on the susceptibility value. At material interfaces, due to the abrupt change in susceptibility, geometric distortion and artifacts might be observed in an MR image. More specifically, darker and brighter areas might appear along with surrounding tissues being distorted in the MR space.

Table 8.3 Magnetic susceptibility (volume susceptibility) of various materials or substances found in vivo

Material/tissue	Magnetic susceptibility χ
Pure water (37 °C)	$-9.05 \cdot 10^{-6}$
Air (NTP)	$0.36 \cdot 10^{-6}$
Human tissues	$-11 \cdot 10^{-6}$ to $-7 \cdot 10^{-6}$
Liver	$\sim 0.0 \cdot 10^{-6}$
Whole blood (deoxygenated)	$-7.9 \cdot 10^{-6}$
Red blood cell (deoxygenated)	$-6.52 \cdot 10^{-6}$
Hemoglobin protein (without Fe ions)	$-9.91 \cdot 10^{-6}$
Cortical bone	$-12.82 \cdot 10^{-6}$
Lipids (stearic acid)	$-10 \cdot 10^{-6}$

Values from [45]

For the majority of soft tissues, we may assume that their magnetic susceptibility is equal to that of water. For a typical MRI scan for diagnostic purposes, differences in magnetic susceptibility such that $|\chi - \chi_{water}| < 10^{-5}$ are expected to cause minimum or negligible distortion in the image, even if they lie close to the anatomical site of interest [45]. However, if the MR image is employed in SRS treatment planning, for which spatial accuracy is of paramount importance, more strict tolerances may be needed.

Susceptibility-related distortion depends on the employed MR pulse sequence and parameters selected. Moreover, the resulting geometric offset increases with increasing static magnetic field strengths [23], TEs used, and decreasing bandwidth [33]. To reduce susceptibility-related artifacts, SE and TSE sequences with very short TEs should be preferred [32].

However, it should be noted that this type of distortion appears only in the frequency-encoded direction (for non-EPI sequences) and the slice selection direction (only for 2D sequences). This results to the susceptibility-related distortion being dependent on the relative position of the materials being imaged. If the interface of materials with considerable susceptibility difference is perpendicular to the frequency-encoded direction, the effect will be maximized. On the other hand, if the material interface is parallel to the frequency-encoded direction, no distortion is expected [46, 47].

More specifically for SRS applications, in contrast-enhanced MR images, Gd accumulates in target locations. Due to its paramagnetic nature (see Sect. 8.3.2), the created susceptibility cavity induces distortion, affecting lesion location as well as surrounding tissues. For a sequence and imaging parameters used in SRS clinical practice, the corresponding offset has been reported to reach on average 0.5 mm [48], although the magnitude varies with respect to cavity location, size, and orientation with respect to the main magnetic field [10, 12, 45, 49]. However, Gd-induced distortion could also have a favorable impact; susceptibility-related displacements can either add up or partly counterbalance other sources of MR-related distortion (particularly B_0 inhomogeneity and chemical shift), resulting in severely increased or minimized overall spatial offset on the frequency encoding axis [48].

8.3.6.4 Chemical Shift

The chemical shift effect is related to the time-averaged interaction of the electrons within a molecule (i.e., intramolecular) and/or between neighboring molecules (intermolecular) [12]. Rotating electrons induce magnetic field which is anti-parallel to static external magnetic field. Therefore, a local uniform shift is caused to the magnetic field as experienced by protons which will also affect Larmor precession frequency. This shift is proportional to the applied external field [12].

More specifically, at 1.5T the Larmor frequency of a proton spin in the molecule of water is approximately 63.9 MHz, while for a proton spin in a fat molecule, it is reduced by 210 Hz. These values apply for an object/subject being scanned at a temperature of 37 °C. However, for a phantom at the room temperature of 22 °C, the water-fat Larmor frequency shift is 224 Hz [50]. Therefore, large temperature drifts could also cause additional imaging issues in MRI [50–52].

At 3.0T, the abovementioned water-fat shifts are doubled. Therefore, chemical shift-related distortion is proportional to the applied external main magnetic field strength. In a more in-depth analysis, it should be noted that fat exhibits a more complex NMR spectrum. It comprises of several secondary resonance peaks [50]. However, the secondary peaks' amplitudes are significantly lower than the main peak's amplitude, and, therefore, they are often considered negligible.

Similar to the B_0 inhomogeneity and susceptibility-related distortions, chemical shift has a considerable impact only on the frequency encoding (for non-EPI pulse sequences) and the slice selection (only for 2D imaging protocols) directions.

8.3.6.5 Estimating Spatial Distortion

Spatial distortion can stem either from the MR system used or the subject being imaged, i.e., system-related or patient-induced distortion, respectively. In the former case, distortion arises from B0 inhomogeneity and residual gradient nonlinearity. It can be mapped using specially designed phantoms (Fig. 8.10a) that incorporate distinct points in space capable of detecting image warping, often referred to as control points [43, 44, 53–56]. Reference control point locations are usually mechanically pre-determined or defined by a CT scan. As part of a comprehensive quality assurance protocol, it is common that system-related distortion is monitored regularly using specifically the sequence and parameters used in SRS clinical practice.

On the other hand, patient-induced distortion (i.e., chemical shift or susceptibility-related effects) cannot be predicted. Simulation studies can provide an estimate of the expected spatial offset [38, 49], although the magnetic properties of a patient cannot be predicted and are not constant in time [23].

The read gradient polarity reversal method [57, 58] exploits the fact that B_0 inhomogeneity-, chemical shift-, and susceptibility-related distortions (often collectively referred to as sequence-dependent distortions [27]) change sign with respect to the frequency encoding direction. For a given patient and imaging protocol, one can easily evaluate sequence-dependent distortion magnitude by acquiring two identical image series except for the read gradient polarity. Differences in lesion boundaries identified independently in the two scans offer an estimate of the distortion magnitude, exactly at the target location [25, 48]. Figure 8.11a presents fused forward and reversed read gradient polarity contrast-enhanced T1w MR scans of a patient with brain metastases. Due to the collective effect of all sequence-dependent distortions, there is a considerable spatial offset in the target (as well as vessel) locations as identified in the two MR scans and highlighted in the pixel intensity profiles across the frequency encoding direction (Fig. 8.11b). The undistorted lesion location is supposed to be at the intermediate position. This technique is attractive as being rather straightforward and no image processing is required. However, with this approach patient scan time is doubled.

Field mapping is another technique for sequence-dependent distortion assessment [59]. Briefly, a GE sequence with two echoes is acquired, and the corresponding phase difference distribution is proportional to the distortion magnitude. However, a post-imaging process for phase unwrapping should be preceded, which might be time-consuming and often subject to errors [27, 48]. On the other hand, this method burdens the total acquisition time by an additional short scan of a few minutes. As with the read gradient reversal technique, gradient nonlinearity induced distortions are not taken into account.

Although not new, sequence-dependent (and thus patient-specific) distortion assessment protocols have not gained wide acceptance in clinical practice. Sub-millimeter distortion is expected in most cases for a FoV relative to a brain scan, which could be significant for SRS applications, if added up to the overall distortion and spatial uncertainty budget. Moreover, distortion magnitude varies according to the scanner's specifications (e.g., main magnetic field strength), imaging parameters (e.g., sequence and bandwidth selection), distance from the MR isocenter, contrast agent accumulation, and the magnetic properties

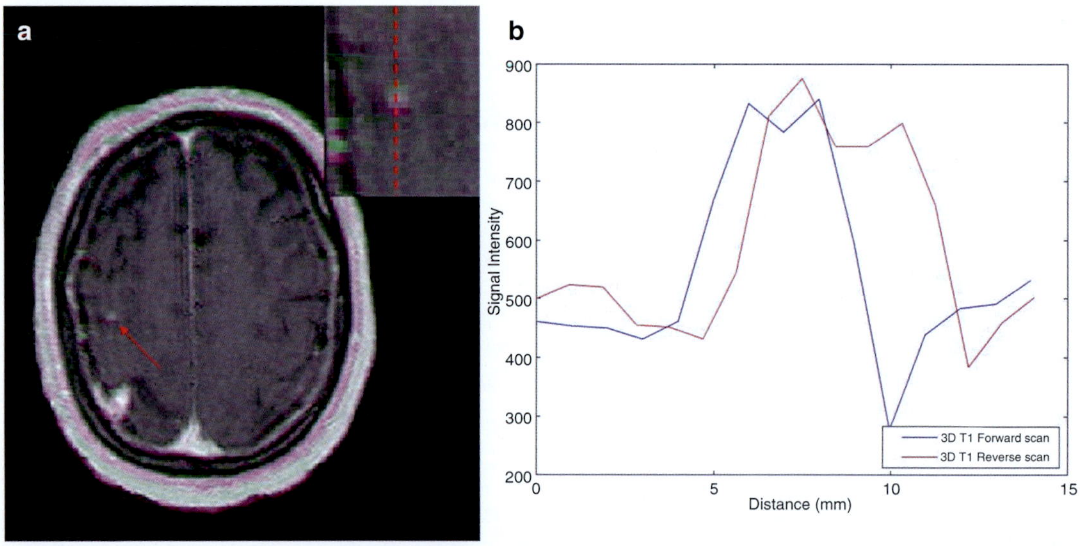

Fig. 8.11 (**a**) Fused forward and reversed polarity T1w MR images of a patient with a small brain metastasis (shown by the red arrow), acquired using a clinically employed protocol for SRS treatment planning. Differences in pixel intensities are highlighted in color (green and purple for higher values in forward and reversed images, respectively). The area around the lesion is shown magnified in the figure insert. (**b**) Pixel intensity profiles for both images, along the red dashed line, coinciding with the frequency encoding direction (anterior-posterior direction)

of the tissue of interest. Therefore, sequence-dependent distortion should not be considered negligible for SRS applications. A thorough sequence optimization to reduce the expected distortion levels while maintaining acceptable signal-to-noise ratio and scan time is a first step toward reassuring treatment efficiency.

References

1. Schmidt MA, Payne GS. Radiotherapy planning using MRI. Phys Med Biol. 2015;60:R323–61.
2. Chin LS, Regine WF. Principles and practice of stereotactic radiosurgery. New York, NY: Springer; 2015.
3. Lusic H, Grinstaff MW. X-ray-computed tomography contrast agents. Chem Rev. 2013;113:1641–66.
4. Verdun FR, Racine D, Ott JG, et al. Image quality in CT: from physical measurements to model observers. Phys Med. 2015;31:823–43.
5. CyberKnife Treatment Delivery Manual (version 1062940-ENG-A). Accuray Inc.
6. Shibamoto Y, Naruse A, Fukuma H, et al. Influence of contrast materials on dose calculation in radiotherapy planning using computed tomography for tumors at various anatomical regions: a prospective study. Radiother Oncol. 2007;84:52–5.
7. Zabel-Du Bois A, Ackermann B, Hauswald H, et al. Influence of intravenous contrast agent on dose calculation in 3-d treatment planning for radiosurgery of cerebral arteriovenous malformations. Strahlentherapie und Onkol. 2009;185:318–24.
8. Kim HJ, Chang AR, Park YK, Ye SJ. Dosimetric effect of CT contrast agent in CyberKnife treatment plans. Radiat Oncol. 2013;8:1.
9. International Atomic Energy Agency (IAEA). Quality assurance programme for computed tomography: diagnostic and therapy applications. IAEA Human Health Series No. 19. Vienna, Austria; 2012.
10. Brown RW, Cheng YCN, Haacke EM, et al. Magnetic resonance imaging: physical principles and sequence design. Second Edi: Wiley; 2014.
11. McRobbie DW, Moore EA, Graves MJ. MRI from picture to proton: Cambridge University Press; 2017.
12. Haacke EM, Reichenbach JR. Susceptibility weighted imaging in MRI: basic concepts and clinical applications: Wiley-Blackwell; 2014.
13. Chavhan GB, Babyn PS, Jankharia BG, et al. Steady-state MR imaging sequences: physics, classification, and clinical applications. Radiographics. 2008;28:1147–60.
14. Bitar R, Leung G, Perng R, et al. MR pulse sequences: what every radiologist wants to know but is afraid to ask. Radiographics. 2006;26:513–37.
15. Hargreaves B. Rapid gradient-echo imaging. J Magn Reson Imaging. 2012;36:1300–13.
16. Gossuin Y, Hocq A, Gillis P, Vuong LQ. Physics of magnetic resonance imaging: from spin to pixel. J Phys D Appl Phys. 2010;43:213001.
17. Ranga A, Agarwal Y, Garg K. Gadolinium based contrast agents in current practice: risks of accumulation and toxicity in patients with normal renal function. Indian J Radiol Imaging. 2017;27:141.
18. Khawaja AZ, Cassidy DB, Al Shakarchi J, et al. Revisiting the risks of MRI with gadolinium based contrast agents—review of literature and guidelines. Insights Imaging. 2015;6:553–8.
19. Kanda T, Nakai Y, Oba H, et al. Gadolinium deposition in the brain. Magn Reson Imaging. 2016;34:1346–50.
20. Gulani V, Calamante F, Shellock FG, et al. Gadolinium deposition in the brain: summary of evidence and recommendations. Lancet Neurol. 2017;16:564–70.
21. Ruan C. MRI artifacts: mechanism and control. Pers Conclus. 2013:1–9.
22. Bennett LH, Wang PS, Donahue MJ. Artifacts in magnetic resonance imaging from metals. J Appl Phys. 1996;79:4712.
23. Weygand J, Fuller CD, Ibbott GS, et al. Spatial precision in magnetic resonance imaging–guided radiation therapy: the role of geometric distortion. Int J Radiat Oncol. 2016;95:1304–16.
24. Seibert TM, White NS, Kim GY, et al. Distortion inherent to magnetic resonance imaging can lead to geometric miss in radiosurgery planning. Pract Radiat Oncol. 2016;6:e319–28.
25. Karaiskos P, Moutsatsos A, Pappas E, et al. A simple and efficient methodology to improve geometric accuracy in gamma knife radiation surgery: implementation in multiple brain metastases. Int J Radiat Oncol Biol Phys. 2014;90:1234–41.
26. Pappas EP, Alshanqity M, Moutsatsos A, et al. MRI-related geometric distortions in stereotactic radiotherapy treatment planning: evaluation and dosimetric impact. Technol Cancer Res Treat. 2017;16:153303461773545.
27. Baldwin LN, Wachowicz K, Fallone BG. A two-step scheme for distortion rectification of magnetic resonance images. Med Phys. 2009;36:3917–26.
28. Le Bihan D, Poupon C, Amadon A, Lethimonnier F. Artifacts and pitfalls in diffusion MRI. J Magn Reson Imaging. 2006;24:478–88.
29. Erasmus LJ, Hurter D, Naudé M, et al. A short overview of MRI artefacts. SA J Radiol. 2004;8:13–7.
30. Smith T, Nayak K. MRI artifacts and correction strategies. Imaging. 2010;2:445–57.
31. Heiland S. From A as in aliasing to Z as in zipper: artifacts in MRI. Clin Neuroradiol. 2008;18:25–36.
32. Stadler A, Schima W, Ba-Ssalamah A, et al. Artifacts in body MR imaging: their appearance and how to eliminate them. Eur Radiol. 2007;17:1242–55.
33. Bernstein MA, Huston J, Ward HA. Imaging artifacts at 3.0T. J Magn Reson Imaging. 2006;24:735–46.
34. Karger CP, Höss A, Bendl R, et al. Accuracy of device-specific 2D and 3D image distortion correction algorithms for magnetic resonance imaging of

the head provided by a manufacturer. Phys Med Biol. 2006;51:N253–61.
35. Pappas EP, Seimenis I, Dellios D, et al. EP-1726: efficacy of vendor supplied distortion correction algorithms for a variety of MRI scanners. Radiother Oncol. 2017;123:S947–8.
36. Price RG, Kadbi M, Kim J, et al. Technical note: characterization and correction of gradient nonlinearity induced distortion on a 1.0 T open bore MR-SIM. Med Phys. 2015;42:5955–60.
37. Price RG, Knight RA, Hwang K-P, et al. Optimization of a novel large field of view distortion phantom for MR-only treatment planning. J Appl Clin Med Phys. 2017;18:51–61.
38. Adjeiwaah M, Bylund M, Lundman JA, et al. Quantifying the effect of 3T magnetic resonance imaging residual system distortions and patient-induced susceptibility distortions on radiation therapy treatment planning for prostate cancer. Int J Radiat Oncol. 2018;100:317–24.
39. Tadic T, Jaffray DA, Stanescu T. Harmonic analysis for the characterization and correction of geometric distortion in MRI. Med Phys. 2014;41:112303.
40. Janke A, Zhao H, Cowin GJ, et al. Use of spherical harmonic deconvolution methods to compensate for nonlinear gradient effects on MRI images. Magn Reson Med. 2004;52:115–22.
41. Caramanos Z, Fonov VS, Francis SJ, et al. Gradient distortions in MRI: characterizing and correcting for their effects on SIENA-generated measures of brain volume change. NeuroImage. 2010;49:1601–11.
42. Maikusa N, Yamashita F, Tanaka K, et al. Improved volumetric measurement of brain structure with a distortion correction procedure using an ADNI phantom. Med Phys. 2013;40:062303.
43. Pappas EP, Seimenis I, Moutsatsos A, et al. Characterization of system-related geometric distortions in MR images employed in Gamma Knife radiosurgery applications. Phys Med Biol. 2016;61:6993–7011.
44. Baldwin LN, Wachowicz K, Thomas SD, et al. Characterization, prediction, and correction of geometric distortion in 3T MR images. Med Phys. 2007;34:388–99.
45. Schenck J. The role of magnetic susceptibility in magnetic resonance imaging: MRI magnetic compatibility of the first and second kinds. Med Phys. 1996;23:815–50.
46. De Deene Y. Fundamentals of MRI measurements for gel dosimetry. J Phys Conf Ser. 2004;3:87–114.
47. De Deene Y. Review of quantitative MRI principles for gel dosimetry. J Phys Conf Ser. 2009;164:012033.
48. Pappas EP, Seimenis I, Dellios D, et al. Assessment of sequence dependent geometric distortion in contrast-enhanced MR images employed in stereotactic radiosurgery treatment planning. Phys Med Biol. 2018;63:135006.
49. Stanescu T, Wachowicz K, Jaffray DA. Characterization of tissue magnetic susceptibility-induced distortions for MRIgRT. Med Phys. 2012;39:7185–93.
50. Bley TA, Wieben O, Francois CJ, et al. Fat and water magnetic resonance imaging. J Magn Reson Imaging. 2010;31:4–18.
51. De Deene Y, De Wagter C. Artefacts in multi-echo T2 imaging for high-precision gel dosimetry: III. Effects of temperature drift during scanning. Phys Med Biol. 2001;46:2697–711.
52. Hijnen NM, Elevelt A, Pikkemaat J, et al. The magnetic susceptibility effect of gadolinium-based contrast agents on PRFS-based MR thermometry during thermal interventions. J Ther Ultrasound. 2013;1:8.
53. Moutsatsos A, Karaiskos P, Petrokokkinos L, et al. Assessment and characterization of the total geometric uncertainty in Gamma Knife radiosurgery using polymer gels. Med Phys. 2013;40:031704.
54. Wang D, Strugnell W, Cowin G, et al. Geometric distortion in clinical MRI systems: part I: evaluation using a 3D phantom. Magn Reson Imaging. 2004;22:1211–21.
55. Stanescu T, Jans HS, Wachowicz K, Gino Fallone B. Investigation of a 3D system distortion correction method for MR images. J Appl Clin Med Phys. 2010;11:200–16.
56. Damyanovich AZ, Rieker M, Zhang B, et al. Design and implementation of a 3D-MR/CT geometric image distortion phantom/analysis system for stereotactic radiosurgery. Phys Med Biol. 2018;63:075010.
57. Chang H, Fitzpatrick JM. A technique for accurate magnetic resonance imaging in the presence of field inhomogeneities. IEEE Trans Med Imaging. 1992;11:319–29.
58. Maurer CR, Aboutanos GB, Dawant BM, et al. Technical note. Effect of geometrical distortion correction in MR on image registration accuracy. J Comput Assist Tomogr. 1996;20:666–79.
59. Jezzard P, Balaban RS. Correction for geometric distortion in echo planar images from B0 field variations. Magn Reson Med. 1995;34:65–73.

Functional Imaging

9

Argyris Moutsatsos and Evangelos Pantelis

9.1 Introduction

Stereotactic radiosurgery (SRS) is a widely acknowledged treatment modality for the management of a variety of intracranial lesions, due, mainly, to its comparable to surgery efficacy sparing the need of heavily invasive procedures [1–3]. Especially for lesions located at critical brain regions not easily accessible via neurosurgical approaches, SRS has almost replaced surgery [4–6]. Despite the benefits of SRS, though, potential radiation-induced complications should not be overlooked. It has been reported that the treatment of arteriovenous malformations (AVMs) located at the motor cortex is associated with a 3% risk of radiation-related complications [7]. This risk reaches 12–19% when lesions located in the thalamus, basal ganglia, or brain stem are treated [2, 8, 9]. To this end, it has been shown that identification of eloquent cortical and subcortical brain regions via advanced neuroimaging techniques can elevate the therapeutic benefit of CyberKnife SRS applications by involving these regions in treatment planning as "functional" organs at risk (fOARs) [10–15]. Functional neuroimaging, including blood oxygenation level-dependent (BOLD) magnetic resonance imaging (MRI) and diffusion tensor imaging (DTI) tractography, can identify the location of fOARs along with white matter pathways [16–18]. Due to their noninvasive nature and widespread availability, both neuroimaging techniques have gained acceptance over the past two decades as important tools for both pre-surgical and SRS planning [19–21]. In terms of CyberKnife SRS, the most investigated fOARs are the sensorimotor cortex, the visual system including the primary visual cortex, and the activation areas of language function. In all studies performed, it was demonstrated that with proper beam delivery optimization, the fOARs neighboring the treating lesion can be spared in terms of both maximum absorbed and integral dose. This chapter focuses on the use and incorporation of BOLD-based functional MRI (fMRI) into the CyberKnife treatment planning, while DTI is analyzed in Chap. 11.

A. Moutsatsos
CyberKnife and TomoTherapy Department, IATROPOLIS Private Special Clinic, Athens, Greece
e-mail: armouts@phys.uoa.gr

E. Pantelis (✉)
Medical Physics Laboratory, Medical School, National and Kapodistrian University of Athens, Athens, Greece

CyberKnife and TomoTherapy Radiotherapy Department, IATROPOLIS Private Special Clinic, Athens, Greece
e-mail: vpantelis@phys.uoa.gr

© Springer Nature Switzerland AG 2020
A. Conti et al. (eds.), *CyberKnife NeuroRadiosurgery*,
https://doi.org/10.1007/978-3-030-50668-1_9

9.2 fMRI Data Acquisition

9.2.1 The BOLD Contrast Signal

The vast majority of studies reporting on the incorporation of fMRI into CyberKnife treatment planning make use of BOLD signal contrast as a surrogate marker of neuronal activity [22–25]. Although good agreement between BOLD activation areas and intraoperative localization (e.g., through direct cortical stimulation, DCS) as well as navigated brain stimulation (NBS) and magnetoencephalography (MEG) corresponding results have been observed [10, 11, 26], the fMRI activation areas trend to be slightly overestimated. This is mainly due to spatial smoothing of the acquired image data necessitated to increase signal-to-noise ratio (SNR) and post-analysis statistics power, as well as to the fact that BOLD signal closely correlates with the extracellular local field potentials (LFPs), reflecting the total activity of regional neural networks rather than the number of "firing" nerve cells. This overestimation results in a safer estimation of the radiation-induced damage risk to fOARs but should be acknowledged when compromises between target coverage and fOAR sparing need to be done.

BOLD contrast stems from regional alterations of the relative concentrations of oxyhemoglobin (diamagnetic) and deoxyhemoglobin (paramagnetic) in response to neural activity. In general, as the relative concentration of deoxyhemoglobin increases, the T2 and T2* relaxation times of the brain decrease (and, hence, so does MR image signal), due to local magnetic field distortions in and around blood vessels induced by the paramagnetic deoxyhemoglobin inside red blood cells which affect:

1. Nearby stationary and/or slowly moving spins by altering their resonance frequencies and inducing phase shifts. This "intravoxel dephasing" is a classic T2* shortening effect most prominent near larger veins and detectable by Gradient Echo (GRE) sequences with echo times (TEs) close to T2*. The effect scales linearly with field strength (B_0) and is the dominant mechanism for BOLD contrast at 1.5T.
2. Protons in water molecules diffusing in and around these vessels. Such protons experience randomly changing frequency offsets and undergo unrecoverable dephasing. This diffusion-related T2-signal loss is best appreciated using spin echo techniques (that reverse phase losses secondary to static field inhomogeneity effects) and is more prominent adjacent to capillaries (than near larger vessels). True-T2 diffusion effects scale with the square of the magnetic field strength (Bo^2) and constitute the dominant mechanism for BOLD contrast at 4.0T and higher, offering better SNR and spatial resolution (hence, higher statistics power) compared to the low field T2*/T2 effect.

At 3T, where most clinical fMRI studies take place, the T2 and T2* effects have comparable contributions to the BOLD contrast.

Surprisingly, while regional cerebral blood flow (CBF) does increase during neuron activation, the metabolic rate of oxygen consumption is not elevated accordingly. Therefore, the T2*/T2 shortening effect of deoxyhemoglobin is obscured, and the apparent location of activated areas in fMR images is demarcated by high MR signal distributions. Briefly, the regional BOLD response following a short peripheral stimulus, such as finger tapping, is known as the hemodynamic response function (HRF; see Fig. 9.1). As seen in Fig. 9.1, HRF typically demonstrates an initial dip, followed by the BOLD effect peak, and then, a post-stimulus undershoot which is variable and most commonly observed for prolonged stimuli. The initial dip, although offering higher spa-

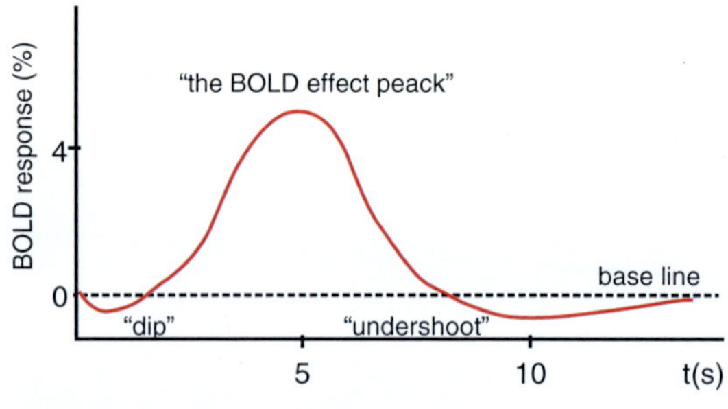

Fig. 9.1 Graphical representation of the HRF function

tial specificity compared to the HRF peak, is also inconsistent and variable and more commonly observed at high fields (≥7T). The most probable sources of its origin are (a) an increased early metabolic extraction of blood oxygen and (b) an early increase of cerebral blood volume. Bulk BOLD response is associated with a positive dominant peak, following the initial dip. It should be noted that a time interval of the order of 5–15 s is typically required between the HRF peak and even a very brief stimulus.

9.2.2 fMRI Pulse Sequence(s)

In view of the discussion of Sect. 9.2.1, a BOLD pulse sequence should have the following characteristics: (a) sensitivity to T2 and/or T2* alterations; (b) low signal detection threshold, since BOLD signal is intrinsically low, typically reaching up to a few percent higher levels than the baseline; and (c) sufficient spatial and temporal resolution to receive readings from the entire brain at multiple closely spaced time points. For static magnetic fields ≤3T, T2*-weighted GRE echo planar imaging (EPI) pulse sequences are typically used to meet those characteristics [22]. At higher imaging fields, T2-weighted spin echo (SE) techniques are preferred due to increased spatial resolution and SNR compared to T2* GRE sequences for a given temporal resolution. Provided that most clinical fMRI acquisitions are nowadays performed at 3T, Table 9.1 provides typical values of some essential fMRI pulse sequence parameters along with a brief rationale of their selection. Please note that the tabulated values reflect the average practice communicated in the literature and are based on a tradeoff between SNR, spatial resolution, and temporal resolution. Amendments and parameter fine-tuning according to the specific functional experiment and the individual being examined may be required.

Three-dimensional (3D) acquisitions have also been proposed in view of their comparative advantages in signal integrity and stability while offering contiguous acquisition without gaps or the need for slice time corrections (see Sect. 9.3.4). Their implementation, however, is difficult and has limitations as TR and readout time may be prolonged and hence unacceptable for many applications [22]. Segmented 3D methods with parallel imaging in two directions may help alleviate these problems. Z-shimming techniques

Table 9.1 Typical GRE EPI MR pulse sequence parameters used for fMRI acquisition at 3T

Parameter	Value	Rationale
Echo time (TE)	30–35 ms	Compromise between T2* of the tissue and susceptibility artifacts/signal dropout induced by prolonged TEs
Repetition time (TR)	1000–1400 ms	Should be less than HRF time course Caution for TEs <1500 ms for saturation effects and blood inflow signal
Slice thickness	2–4 mm	Trade-off between SNR and partial volume averaging
Slice acquisition order	Interleaved (1,3,5,…2,4,6…)	Reduce slice crosstalk artifacts
Matrix acquisition	Matrix: ≤128 × 128 Resulting in (2 × 2) – (3 × 3) mm^2 in-plane resolution	Increasing spatial resolution decreases temporal resolution (by increasing total imaging time), lengthens readout time inducing more artifacts, and reduces SNR
Parallel imaging [27]	Desirable, but low acceleration factors R ≈ 2	Parallel MRI can reduce imaging time and, hence, increase temporal resolution while reducing susceptibility artifacts High acceleration factors may impair SNR severely
Total imaging time	45–60 min in total 10–12 min per individual experiment	Ensure patient compliance

may also be used involving the application of a compensating gradient along the slice reconstruction direction (z-axis) that ensures the k-space trajectory has returned to the origin at time TE [28]. Finally, the (magnetic) excitement of multiple slices simultaneously has been made possible using "multi-band" techniques which, however, comes at the expense of a prolonged EPI readout train [29].

9.2.3 fMRI Paradigm Design

There are three basic categories in which task-based BOLD-fMRI studies can be divided: (1) block, (2) event-related, and (3) mixed [30–33]. Figure 9.2 shows a graphical representation of the functional template used by each BOLD fMRI design.

Block designs are the oldest functional imaging paradigms comprising task periods alternated with periods of rest. After proper pre- and post-processing (disscussed in the following sections), the lower BOLD signal during rest state is digitally "subtracted" from higher BOLD signal during task periods to indicate focal areas of cortical activation. Block designs are the simplest and most straightforward paradigms to implement and the most widely used for clinical fMRI studies to identify eloquent cortical areas prior to surgery or SRS applications. In a classic "finger-tapping" experiment, for example, the patient taps his/her fingers for ~15 s followed by an equal length period of rest. More complex designs, examining two or more activities, are also possible. After correcting for noise and spatial distortions, areas whose net BOLD signals exceed certain statistical thresholds are identified as "activated" regions.

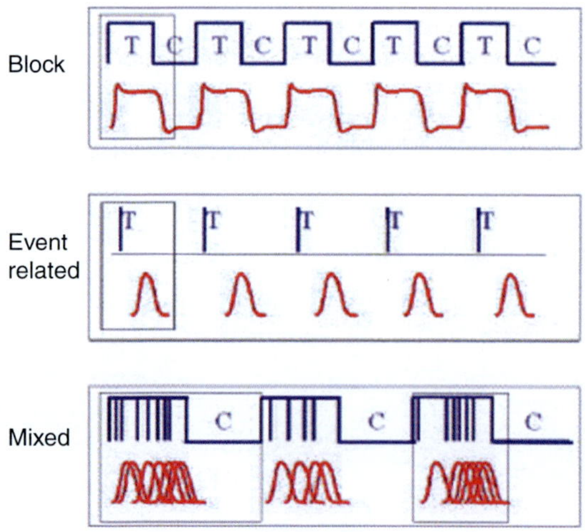

Fig. 9.2 Graphical representation of the functional template used by the different BOLD fMRI designs. (Used with permission from Ref. [30])

The comparative advantages of block designs relative to other fMRI paradigms include higher SNR levels, increased statistical power, and maximal time efficiency. On the other hand, block designs have limited utility, especially within the setting of more complex neuropsychological experiments involving non-binary tasks. Also, even with simple tasks, subjects can anticipate the order or duration of the simple blocks, introducing confounding variables. Finally, because blocks are measured over relatively long periods (10–20 s), information about the hemodynamic response and fMRI signal timing are difficult to measure.

Event-related designs allow for single or multiple tasks and stimuli to take place at short and variable time intervals. Therefore, they provide the increased flexibility required for sophisticated neuropsychological experiments. Events can be randomized, and different types of events can be mixed, so that the subject cannot predict when or what will occur. Event-related designs allow for better temporal resolution and estimation of the HRF time course. All these come at the expense of SNR and statistical power, which are lower compared to block designs thus requiring longer imaging times and more trials per subject. Analysis of the data is significantly more complex and dependent on accurate modeling of the HRF.

Mixed paradigms embody features of blocked and event-related designs. Here, semi-randomized events take place during the task blocks, with rest periods in between. Mixed paradigms thus tend to preserve the favorable signal-to-noise characteristics of blocked methods with the flexibility of event-related ones.

9.2.3.1 Quantifying BOLD Signal

Independent of the paradigm design, raw MR BOLD signal is not absolute but affected by both technical and patient-specific factors. Technical factors include field strength and the employed pulse sequence (SE or GRE) parameters (e.g., TE, TR, slice acquisition order, voxel size), while patient-related factors include hematocrit, respiratory rate, head size, age, gender, hormonal status, and medications. Therefore, the BOLD signal is typically expressed in arbitrary units (AU) or as a percent change from baseline and an "activated" brain region must be differentiated from "non-activated" areas using sophisticated statistical techniques based on changes measured between "on task" and "rest" states.

9.3 fMRI Data Pre-processing

Prior to applying any kind of statistical analysis to raw fMRI data, pre-processing procedures, including the following, must be performed.

9.3.1 Inspection of Raw fMR Images

Inspection of the acquired fMRI data on a slice-by-slice basis is mandatory. It is common that individual slices suffer from random variations in average signal intensity, noise spikes, ghosts, and image data abnormalities, stemming from physiological sources (e.g., patient motion, respiration, cardiac pulsations, and anxiety) or/and technical deficiencies (e.g., magnetic field inhomogeneities and eddy currents). These discrepancies should not pass undetected, since their inclusion in data statistical analysis may induce detrimental effects to the whole experiment. Image data inspection is usually facilitated by visual inspection of all the acquired slices on a montage mode using either manual means of fMRI software platforms along with embedded tools helping to find and exclude aberrant slices.

9.3.2 Distortion Correction

The T2*-weighted EPI sequences, usually employed for BOLD-fMRI studies, acquire gradient echoes and hence are sensitive to magnetic inhomogeneity effects. These, however, induce spatial distortions and signal dropout which are more pronounced in the boundaries of different magnetic susceptibility tissue (e.g., soft tissue-air, bone-air, blood vessels-brain parenchyma) and affect the acquired MR images primarily in the phase encoding and slice reconstruction (z-axis) directions [34, 35]. Intracranially, such distortions and signal discrepancies appear near the skull base, typically affecting the anterior frontal and temporal lobes. Field mapping and "phase unwarping" methods are available to correct these distortions [36] but are time consuming and require advanced skills in medical image processing. Therefore, for basic eloquent cortex mapping performed in clinical fMRI studies, a standard shimming procedure precedes the fMRI acquisition to suppress field inhomogeneities.

9.3.3 Motion Correction

Patient (head) motion is the dominant source of error in fMRI studies, and a variety of strategies have been developed to cope with this problem [37]. Immobilization of the head using padding and straps is essential, while more rigid restrictions using bite bars and masks are also occasionally employed. Proper coaching and training of the subject prior to imaging is important. Usually motion correction is performed retrospectively considering the head as a rigid body with three directions of translation (displacement) and three axes of rotation. A specific (usually the first) fMRI volume corresponding to a single run is chosen as the reference to which all other volumes are aligned. An iterative procedure is then performed in which each volume is rotated and aligned with the reference, with the goal to minimize a cost function (such as the mean-squared difference) until no further improvement can be achieved. Motion correction algorithms are available in almost all fMRI software platforms.

9.3.4 Slice Timing Correction

fMRI studies usually acquire one slice at a time resulting in time offsets that should be accounted for when comparing the signal from different slices. Slice timing correction is not trivial, especially in the case that simultaneous multi-slice acquisition is employed and depends on whether the slices have been acquired in sequential or in interleaved order. Although slice timing effect seems to be nonimportant for simple block design experiments, it can impart considerable errors in rapid, event-related fMRI studies if not accounted for. Two basic strategies have been developed for slice timing correction [38]. Data shifting is the most used one and involves back (in time) projection of the recorded points to reflect their proper offset from the time of stimulus. The second method is model shifting (applied post-hoc), where the expected location of HRF is varied, treating slice location as an additional independent variable in the subsequent statistical analysis.

9.3.5 Spatial Smoothing

Spatial smoothing improves the SNR by averaging the signal of neighboring voxels but decreases spatial resolution, blurs the image, and results in overestimations of the activated volumes. The process can be justified because closely neighboring brain voxels are usually inherently correlated in their function and blood supply. A standard smoothing method is to convolve the fMRI data with a 3D Gaussian kernel that averages signals from neighboring voxels with weights that decrease with increasing distance from the target voxel. The optimal kernel size depends on several factors such as slice thickness, in-plane resolution, and the volume threshold for separate activation regions. In practice, the full width at half maximum (FWHM) value of the Gaussian spatial filter is typically set to about 4–6 mm for single subject studies.

9.3.6 Temporal Filtering

It is common that fMRI data exhibit slow wandering of the baseline signal over time as well as rapid fluctuations due to noise. The removal of low frequency drifts is known as detrending [39]. Detrending may be accomplished using either high-pass filtering after Fourier transformation or by time-domain averaging methods. High-frequency signal fluctuations (i.e., noise) can be removed by low-pass filtering. However, low-pass filtering is generally not recommended for most studies since it may distort estimation of individual HRFs and reduce the fMRI signals of interest.

9.4 Statistical Analysis and Generation of Activation Maps

Statistical analysis of fMRI data can be performed using several methods encoded to software algorithms available either in the form of standalone applications or embedded into comprehensive fMRI analysis platforms. The most commonly used methods are (1) the general linear model (GLM) [40, 41], (2) the independent component analysis (ICA) [42], and (3) the multi-voxel pattern analysis (MVPA) and networks [43, 44]. The GLM model is the one widely used for SRS applications, including CyberKnife studies. Briefly, the GLM model assumes that the fMRI signal [$\mathbf{Y}_t$] for a given imaging voxel at time = t is analogous to the sum of one or more experimental variables [$\mathbf{X}_{i,t}$] each multiplied by a weighting factor ($\boldsymbol{\beta}_i$) plus a random error term [$\boldsymbol{\varepsilon}_t$] (note that residual errors should follow a Gaussian distribution with zero mean value in order to be "random"). The components of the vector $\mathbf{X}_{i,t}$ are the timings (t) corresponding to the (delayed) response of stimulus (i.e., finger tapping). The weighted factor β_i corresponds to each different type of stimulus, while the random error, ε_t, evaluates the discrepancy between Υ_t and $\beta_i{}^*X_{i,t}$ for each time measurement, t.

Using the statistical analysis results of fMRI data, corresponding activation maps are developed by identifying regions of voxel that apparently present statistically significant levels of activation (or correlation) and assigning them bright color tones from a given color map. The size and number of these voxel regions are chosen on a semi-empirical basis and always involve trade-offs between excluding false positives and accepting false negatives. The first criterion used for color assignment to a voxel or brain region is based on the calculation of a statistical test (e.g., a T-, F-, or Z-test and corresponding scores). Under the null hypothesis (H_0) that no true activation has occurred, a *p* value can be determined, representing the probability that the calculated test statistic score or larger has occurred by chance. Whenever the *p* value is less than an arbitrary preselected level of significance, *a* (e.g., $a = 0.05$), it is concluded that the measurement is unlikely to have occurred by chance and the voxel is classified as "activated" (or "correlated"). A common problem affecting the validity of fMRI activation maps stems from the arbitrary nature of the selected confidence level (and, hence, the set *p* value) which is called "the multiple comparison problem" [45, 46], which results in excessive false positives and is partially accounted for in the Bonferroni method [47].

The fMRI activation maps are then co-registered with MR anatomic images acquired, usually, during the same session. The anatomic reference, MRI_{anat}, is typically a 3D MR sequence (e.g., MPRAGE) acquired with high-resolution isotropic voxels (e.g., $1 \times 1 \times 1$ mm^3). The isotropic voxel acquisition allows for the data to be rotated, re-sliced on oblique planes, and manipulated. Many of the same image correction and alignment techniques for fMRI data preprocessing are also used to perform fMRI-MRI_{anat} co-registration. Typically, this procedure begins with image resampling using interpolation techniques followed by rigid body transformations. An iterative optimization protocol is then employed relying usually on the minimization of a cost function which measures the degree of disparity between the fMRI and MRI_{anat} after each iteration [48]. Figure 9.3 shows an fMRI activation map fused with the anatomical MRI data depicting the Broadmann-17 functional structure.

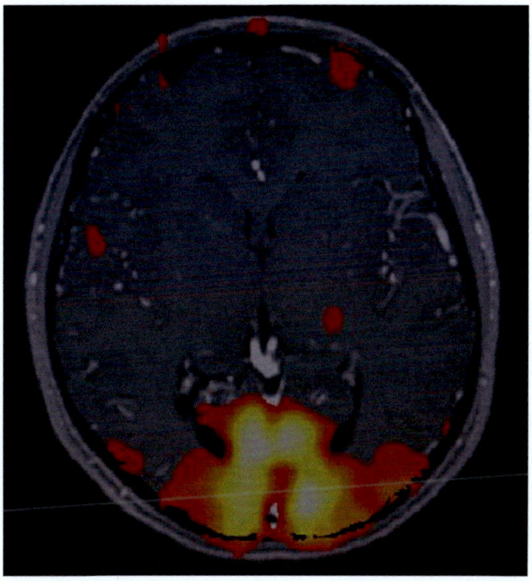

Fig. 9.3 fMRI activation map overlaid to corresponding anatomical MRI data indicating the Broadmann-17 functional structure. (Reproduced from Ref. [12] with permission)

For CyberKnife SRS applications, the motor, visual, and language functions are commonly evaluated. Motor functions are triggered usually by means of hand, foot, and tongue movements. Visual function is commonly evaluated by cuing corresponding signs to the patient via, for example, a flashing checkboard. Language-cognitive functions are probed using category generation, letter generation, simple questions, and verbal generation tasks. For motor studies, functional maps are typically obtained by means of a t-test analysis with a *p* value of 0.05 (false-positive corrected) and a minimum number of 20 adjacent voxels to define an activation cluster. For language-cognitive functions, the corresponding values typically used are $p = 0.001$ (uncorrected for false positives) and 40 neighboring voxels. These differences in activation analysis are required due to the spread of activated areas in language-related tasks.

9.5 Integration to CyberKnife Treatment Planning

9.5.1 Image Registration and Delineation of fOARs

Prior to incorporating fMRI data to the CyberKnife treatment planning system, a sequence of image registration procedures must be performed: (1) registration of the reference anatomical MRI volume and MRI_{anat} to the CT volume and (2) application of transformation obtained from step 1 to the activation maps already registered to the MRI_{anat}. This sequence allows the functional maps to be spatially registered with the CT volume used for treatment planning and dosimetry calculations and, therefore, made it possible to define regions of interest corresponding to functional areas (fOARs), within which dose constraints could be imposed during the optimization process. Figure 9.4 shows the fMRI activation map data presented in Fig. 9.3, registered to the corresponding CT volume used for treatment planning and dose calculations.

9.5.2 Treatment Planning and Dose Optimization

The dosimetric impact to functionally eloquent brain structures in CyberKnife SRS applications is commonly evaluated by comparing two instances of the same treatment plan differentiated by the inclusion or not of those structures in the dose optimization process as fOARs (see Fig. 9.5). As shown by published results, considerable reductions to both maximum dose as well

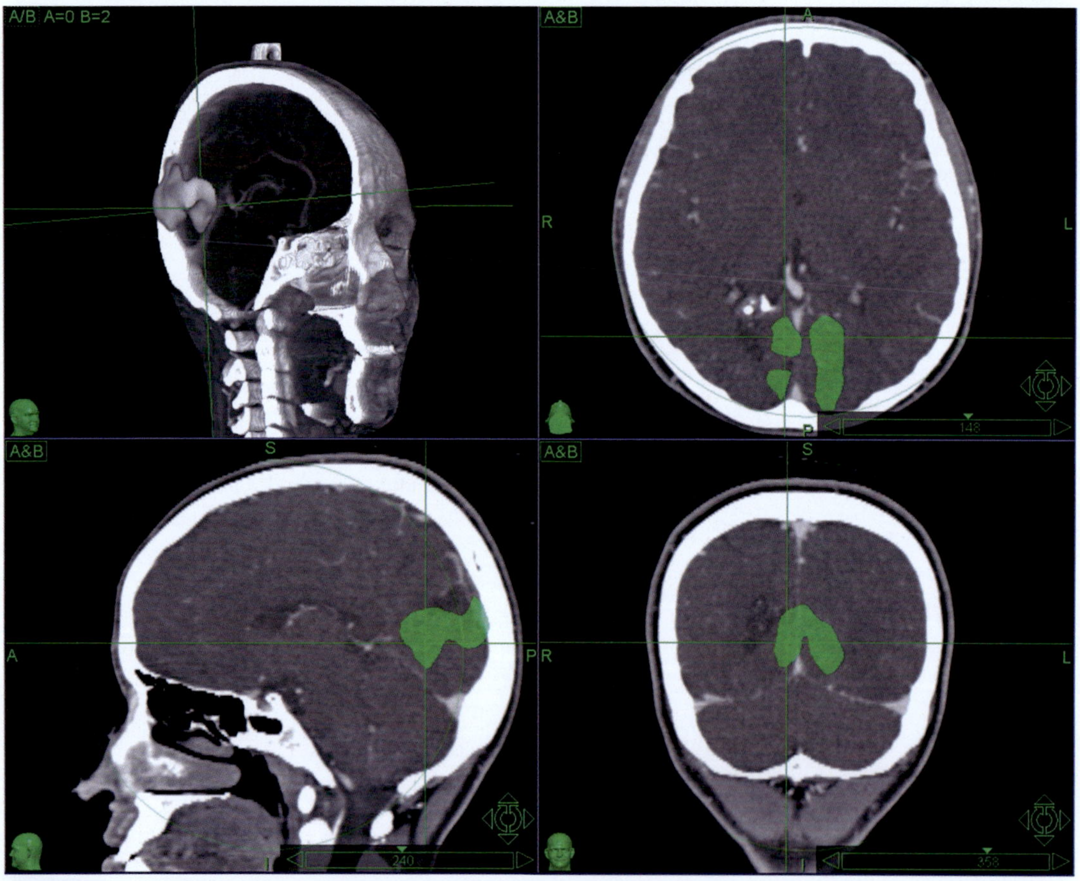

Fig. 9.4 The fMRI activation map data shown in Fig. 9.3, registered to the CT volume used for CyberKnife treatment planning and dose calculations. (Reproduced from Ref. [12] with permission)

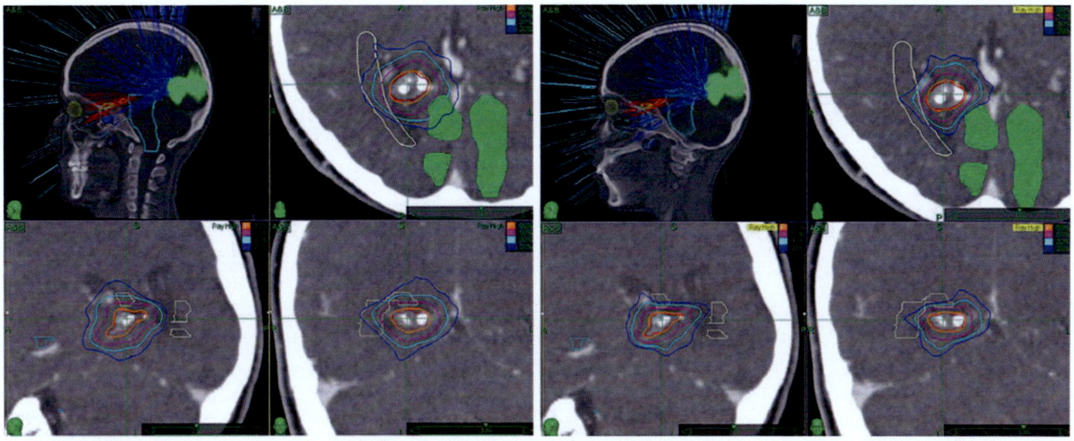

Fig. 9.5 Irradiation plans developed for the same AVM treatment without (left panels) and with (right panels) the introduction of the Broadmann-17 functional structure (light green) and optic tract as fOARs in the optimization process. The configuration of the irradiation beams, the dose distributions in terms of iso-dose lines and the optic tract (light yellow) are also depicted. (Reproduced from Ref. [12] with permission)

as other dose-volume metrics, such as the mean dose, can be achieved for fOARs depending on the size and relative position of each fOAR with respect to the target [10–15]. For the AVM indicative case presented in Fig. 9.5 the maximum dose to the Broadmann-17 functional structure reduced from 1400 cGy to 700 cGy when included in treatment planning. Correspondingly, the maximum dose to the adjacent optic tract was reduced from 2000 cGy to 1200 cGy upon inclusion to dose optimization [12].

Generally, the above dose reductions are attainable without compromising target coverage but come at the expense of delivered monitor units (MU) (and, hence, total treatment time) due to the use of smaller collimator sizes and the non-isocentric irradiation configuration usually required. Also, increases to the integral dose, calculated as the mean dose to all target and critical structures multiplied by the volume of all soft tissues, and the time of treatment have also been reported [13].

References

1. Mehta MP, Tsao MN, Whelan TJ, et al. The American Society for Therapeutic Radiology and Oncology (ASTRO) evidence-based review of the role of radiosurgery for brain metastases. Int J Radiat Oncol. 2005;63:37–46.
2. Pollock BE, Gorman DA, Brown PD. Radiosurgery for arteriovenous malformations of the basal ganglia, thalamus, and brainstem. J Neurosurg. 2004;100:210–4.
3. Chopra R, Kondziolka D, Niranjan A, et al. Long-term follow-up of acoustic Schwannoma radiosurgery with marginal tumor doses of 12 to 13 Gy. Int J Radiat Oncol Biol Phys. 2007;68:845–51.
4. Choi CYH, Chang SD, Gibbs IC, et al. Stereotactic radiosurgery of the postoperative resection cavity for brain metastases: prospective evaluation of target margin on tumor control. Int J Radiat Oncol Biol Phys. 2012;84:336–42.
5. Chao ST, De Salles A, Hayashi M, et al. Stereotactic radiosurgery in the management of limited (1-4) brain metasteses: systematic review and International Stereotactic Radiosurgery Society practice guideline. Clin Neurosurg. 2018;83:345–53.
6. Lee CC, Trifiletti DM, Sahgal A, et al. Stereotactic radiosurgery for benign (World Health Organization Grade I) cavernous sinus meningiomas—International Stereotactic Radiosurgery Society (ISRS) practice guideline: a systematic review. Clin Neurosurg. 2018;83:1128–41.
7. Hadjipanayis CG, Levy EI, Niranjan A, et al. Stereotactic radiosurgery for motor cortex region arteriovenous malformations. Neurosurgery. 2001;48:70–7.
8. Sasaki T, Kurita H, Saito I, et al. Arteriovenous malformations in the basal ganglia and thalamus: management and results in 101 cases. J Neurosurg. 1998;88:285–92.
9. Andrade-Souza YM, Zadeh G, Scora D, et al. Radiosurgery for basal ganglia, internal capsule, and thalamus arteriovenous malformation: clinical outcome. Neurosurgery. 2005;56:56–64.

10. Stancanello J, Cavedon C, Francescon P, et al. BOLD fMRI integration into radiosurgery treatment planning of cerebral vascular malformations. Med Phys. 2007;34:1176.
11. Colombo F, Cavedon C, Casentini L, et al. Early results of CyberKnife radiosurgery for arteriovenous malformations. J Neurosurg. 2009;111:807–19.
12. Pantelis E, Papadakis N, Verigos K, et al. Integration of functional MRI and white matter tractography in stereotactic radiosurgery clinical practice. Int J Radiat Oncol Biol Phys. 2010;78:257–67.
13. Conti A, Pontoriero A, Ricciardi GK, et al. Integration of functional neuroimaging in CyberKnife radiosurgery: feasibility and dosimetric results. Neurosurg Focus. 2013;34:1–8.
14. Sun L, Qu B, Wang J, et al. Integration of functional MRI and white matter tractography in CyberKnife radiosurgery. Technol Cancer Res Treat. 2017;16:850–6.
15. De Martin E, Duran D, Ghielmetti F, et al. Integration of functional magnetic resonance imaging and magnetoencephalography functional maps into a CyberKnife planning system: feasibility study for motor activity localization and dose planning. World Neurosurg. 2017;108:756–62.
16. Faro SH, Mohamed FB. Functional MRI: basic principles and clinical applications: Springer; 2006.
17. Kim PE, Singh M. Functional magnetic resonance imaging for brain mapping in neurosurgery. Neurosurg Focus. 2003;15:1–7.
18. Melhem ER, Mori S, Mukundan G, et al. Diffusion tensor MR imaging of the brain and white matter tractography. Am J Roentgenol. 2002;178:3–16.
19. Petrella JR, Shah LM, Harris KM, et al. Preoperative functional MR imaging localization of language and motor areas: effect on therapeutic decision making in patients with potentially resectable brain tumors. Radiology. 2006;240:793–802.
20. Maruyama K, Kamada K, Shin M, et al. Optic radiation tractography integrated into simulated treatment planning for Gamma Knife surgery. J Neurosurg. 2007;107:721–6.
21. Maruyama K, Kamada K, Ota T, et al. Tolerance of pyramidal tract to gamma knife radiosurgery based on diffusion-tensor tractography. Int J Radiat Oncol Biol Phys. 2008;70:1330–5.
22. Chen JE, Glover GH. Functional magnetic resonance imaging methods. Neuropsychol Rev. 2015;25:289–313.
23. Ogawa S, Tank DW, Menon R, et al. Intrinsic signal changes accompanying sensory stimulation: functional brain mapping with magnetic resonance imaging. Proc Natl Acad Sci U S A. 1992;89:5951–5.
24. Sandrone S, Bacigaluppi M, Galloni MR, et al. Weighing brain activity with the balance: Angelo Mosso's original manuscripts come to light. Brain. 2014;137:621–33.
25. Buxton RB. The physics of functional magnetic resonance imaging (fMRI). Reports Prog Phys. 2013;76:096601.
26. Hall EL, Robson SE, Morris PG, Brookes MJ. The relationship between MEG and fMRI. NeuroImage. 2014;102:80–91.
27. Preibisch C, Wallenhorst T, Heidemann R, et al. Comparison of parallel acquisition techniques generalized autocalibrating partially parallel acquisitions (GRAPPA) and modified sensitivity encoding (mSENSE) in functional MRI (fMRI) at 3T. J Magn Reson Imaging. 2008;27:590–8.
28. Glover GH. 3D z-shim method for reduction of susceptibility effects in BOLD fMRI. Magn Reson Med. 1999;42:290–9.
29. Feinberg DA, Setsompop K. Ultra-fast MRI of the human brain with simultaneous multi-slice imaging. J Magn Reson. 2013;229:90–100.
30. Amaro E, Barker GJ. Study design in fMRI: basic principles. Brain Cogn. 2006;60:220–32.
31. Dale AM. Optimal experimental design for event-related fMRI. Hum Brain Mapp. 1999;8:109–14.
32. Huettel SA. Event-related fMRI in cognition. NeuroImage. 2012;62:1152–6.
33. Maus B, van Breukelen GJP, Goebel R, Berger MPF. Optimization of blocked designs in fMRI studies. Psychometrika. 2010;75:373–90.
34. Baldwin LN, Wachowicz K, Thomas SD, et al. Characterization, prediction, and correction of geometric distortion in 3T MR images. Med Phys. 2007;34:388–99.
35. Baldwin LN, Wachowicz K, Fallone BG. A two-step scheme for distortion rectification of magnetic resonance images. Med Phys. 2009;36:3917–26.
36. Cusack R, Papadakis N. New robust 3-D phase unwrapping algorithms: application to magnetic field mapping and Undistorting Echoplanar images. NeuroImage. 2002;16:754–64.
37. Maclaren J, Herbst M, Speck O, Zaitsev M. Prospective motion correction in brain imaging: a review. Magn Reson Med. 2013;69:621–36.
38. Sladky R, Friston KJ, Tröstl J, et al. Slice-timing effects and their correction in functional MRI. NeuroImage. 2011;58:588–94.
39. Tanabe J, Miller D, Tregellas J, et al. Comparison of Detrending methods for optimal fMRI preprocessing. NeuroImage. 2002;15:902–7.
40. Monti MM. Statistical analysis of fMRI time-series: a critical review of the GLM approach. Front Hum Neurosci. 2011;5:28.
41. Poline J-B, Brett M. The general linear model and fMRI: does love last forever? NeuroImage. 2012;62:871–80.
42. Mckeown MJ, Makeig S, Brown GG, et al. Analysis of fMRI data by blind separation into independent spatial components. Hum Brain Mapp. 1998;6:160–88.

43. Haxby JV, Gobbini MI, Furey ML, et al. Distributed and overlapping representations of faces and objects in ventral temporal cortex. Science. 2001;293:2425–30.
44. Mahmoudi A, Takerkart S, Regragui F, et al. Multivoxel pattern analysis for fMRI data: a review. Comput Math Methods Med. 2012;2012:1–14.
45. Colquhoun D. An investigation of the false discovery rate and the misinterpretation of *p*-values. R Soc Open Sci. 2014;1:140216.
46. Engel SA, Burton PC. Confidence intervals for fMRI activation maps. PLoS One. 2013;8:e82419.
47. Nichols T, Hayasaka S. Controlling the familywise error rate in functional neuroimaging: a comparative review. Stat Methods Med Res. 2003;12:419–46.
48. Klein A, Andersson J, Ardekani BA, et al. Evaluation of 14 nonlinear deformation algorithms applied to human brain MRI registration. NeuroImage. 2009;46:786–802.

Diffusion Tensor Imaging (DTI) Tractography

10

Enmin Wang

10.1 Introduction

Stereotactic radiosurgery (SRS) has been widely recognized as an effective treatment modality in the management of various intracranial lesions [1–5]. Although SRS is known as one of the least invasive treatment modalities for cerebral arteriovenous malformations (AVM), the associated risk of radiation-induced neuropathy occurs in 5–20% of patients [1, 6–9], which is not negligible for patients with AVM in deep-seated eloquent areas, such as the thalamus, the basal ganglia, and precentral gyrus. With the aid of modern magnetic resonance imaging (MRI) techniques, functional brain areas and white matter fiber pathways can be well demarcated in imaging studies [10–12]. These imaging techniques have been implemented in modern neuronavigation systems and used to guide the surgical removal of critically located intracranial lesions [13–16]. The incorporation of this information for radiosurgery planning has also been proposed [17–20]. More specifically, functional MRI (fMRI) and diffusion tensor imaging (DTI) tractography have been used to identify functional structures and white matter pathways of the brain as critical volumes (i.e., volumes to which dose constraint are assigned) in treatment planning optimization strategies [20–23]. Despite the limited number of published studies that utilize these imaging techniques, it has been shown that these techniques could be useful for sparing healthy and sensitive parts of the brain from high doses of radiation [20–28].

E. Wang (✉)
Department of Neurosurgery, Huashan Hospital, Fudan University, Shanghai, China
e-mail: wangem@fudan.edu.cn

10.2 Basic Principles of MR, DTI, and Tractography

Diffusion tensor imaging, a type of MR-based neuroimaging technique, has the capability of delineating the white matter tracts in the brain through estimation of the directional movement of water molecules [29]. The diffusion of water in three dimensions is calculated by fitting a tensor to every voxel in the brain of a diffusion-weighted MR scan. This tensor may be represented as an ellipsoid-shaped mathematical model, defined by three orthogonal eigenvectors and corresponding eigenvalues. The shape of the tensor depends on the average direction and magnitude of water diffusion within a given voxel. Since white matter bundles are elongated and consist of structural barriers such as myelin, axonal membranes, and microtubules, the diffusion of water is greater along the length of the axon than across it. This results in the shape of the tensor being more elongated, representing anisotropic diffusion [30]. Diffusion tensor tractography (DTT) is a robust technique based on diffusion

© Springer Nature Switzerland AG 2020
A. Conti et al. (eds.), *CyberKnife NeuroRadiosurgery*,
https://doi.org/10.1007/978-3-030-50668-1_10

tensor imaging which allows noninvasive in vivo reconstruction of the trajectory of the neuronal fiber tracts [30, 31]. This technique may provide information about the course, integrity, anatomical connectivity, or possible disruption of neural pathways. DTT may be helpful in better visualizing the anatomy of the brain structures, in assisting to design CyberKnife treatment plans to avoid damaging the important structures.

10.3 Image Acquisition and Data for Postprocessing

In our center, DTI was mainly used in patients with AVMs located adjacent to eloquent areas or within eloquent areas. For each patient, five different imaging studies were acquired: a computed tomography (CT) scan for radiosurgery treatment planning and for target tracking during treatment delivery (Fig. 10.1a), an anatomical contrast-enhanced MRA (3D time-of-flight magnetic resonance angiography), MRI T2 weighted images to provide a complete set of morphological MR data (Fig. 10.1b, c), MRI T1 weighted images (Fig. 10.1d), and a DTI to provide white matter tractography data (Fig. 10.1e, f). Axial CT images (1 mm slice thickness) of each patient were acquired using an Aquilion 64-slice scanner (Toshiba Medical Systems, Japan). MRI volumes were acquired using a 3.0 T MAGNETOM Trio scanner (Siemens, Germany), by a 3D magnetization-prepared rapid acquisition of gradient echo sequence (3D MPRAGE).

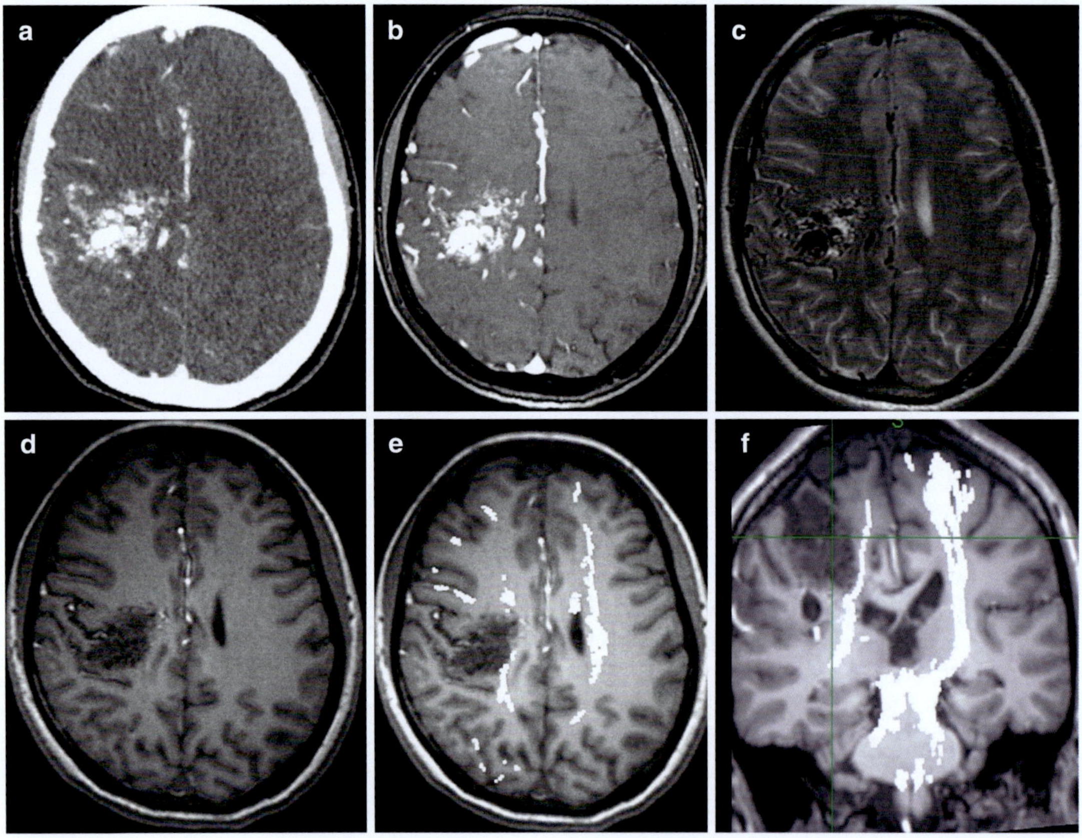

Fig. 10.1 Acquired axial CT and anatomical MRI slices from the AVM patient. DTI fused with the anatomical MRI data showing corticospinal tract in corresponding axial, coronal, MRI plane. (**a**) Contrast-enhanced CT. (**b**) Axial contrast-enhanced MRA. (**c**) T2-wighted image. (**d**) T1-weighted image. (**e** and **f**) DTI in the combined images. All MR images fused with CT in the CyberKnife Multiplan

10.4 Diffusion Tensor Imaging Studies

Diffusion tensor imaging of MRI, including diffusion tensor tractography, is a unique tool to visualize and segment the white matter pathways in vivo. Three-dimensional visualization of the white matter fibers, such as corticospinal (pyramidal) tracts, with relationship to brain lesions (vascular malformations and brain tumors) is extremely helpful for stereotactic radiosurgery. On the day before the CT and MRI localization, patients with AVMs in deep-seated eloquent areas underwent DTI in the intraoperative MRI suite (MAGNETOM Verio, Siemens AG, Germany), which is a 3 T MR imaging scanner. It is equipped with Neuro 3D Analysis (Workstation MR Imaging Software). The anatomical images were obtained by a 3D magnetization-prepared rapid acquisition of gradient echo sequence (3D MPRAGE). First, whole-brain, axial, T1-weighted images were obtained with section thickness, 1 mm; TR, 1900 ms; TE, 2.98 ms; flip angle, 90°; voxel size, 1.0 × 1.0 × 1.0 mm; field of view, 256 mm; and matrix, 256 × 256. To get the DTI, we applied a single-shot multi-slice 2D spin-echo diffusion - sensitized and fat-suppressed echo planar imaging (EPI) sequence (axial, 42 sections; section thickness, 2 mm no gap; TR, 9900 ms; TE, 90 ms; voxel size, 1.5 × 1.5 × 3 mm; flip angle, 90°; field of view, 240 mm; matrix, 128 × 128) of 42 slices covering the entire brain. We usually use four MRI sequences (Table 10.1) for DTI studies.

The whole-brain, axial T1-weighted images and DTI data were transferred to a Neuro 3D workstation (Siemens, Erlangen, Germany) for postprocessing. The Neuro 3D software is an additional component of established intraoperative neuronavigational software already in clinical use for preoperative planning and tract generation for cranial neurosurgery. The eigenvalues and eigenvectors of the anisotropic components of each voxel were determined. Fiber tracking was initiated in accordance with the principle eigenvector and terminated when the fractional anisotropy (FA) value was below 0.2 and the angle threshold was set at 30°. In our study, to reconstruct the corticospinal tract with tractography, two paired region of interests (ROIs) were segmented on symmetrical transverse images as follows: the first was at the level of the posterior limb of the internal capsule, and the second was on the plane of the centrum semiovale. The motor pathway volumes that were derived from DTI tractography integrated the 3D structural MRI by rigid registration. The generated tracts were exported with the navigation examination (T1-weighted) and a hybrid examination (T1-weighted with superimposed/burned 3D tract) in DICOM format using the 3D Object Data dialog. The corticospinal tract studies were burned to DICOM CD for transfer to CyberKnife Multiplan clinical application (Fig. 10.2).

Table 10.1 The fMRI sequences for DTI (3 T MAGTENOM Verio, Siemens)

1	Localizer HEAD
2	T1 MPRAGE_TRA iso1.0 HEAD
3	ep2d_diff_DTI_20_p2_TENSOR HEAD
4	ep2d_diff_DTI_20_p2_ADC HEAD
5	ep2d_diff_DTI_20_p2_TRACEW HEAD
6	ep2d_diff_DTI_20_p2_FA HEAD
7	ep2d_diff_DTI_20_p2_ColFA HEAD

10.5 CyberKnife Radiosurgery and Treatment Planning

The CyberKnife (Accuray, Sunnyvale, California) is a new system for SRS. It uses a real-time, non-invasive, image-guidance system, which can ensure the accuracy of localization. The frameless feature can alleviate the patient's distress to some extent. And its original skull-tracking system ensures it has natural advantage in the treatment of intracranial tumors. The Multiplan system enables us to fuse MRI, fMRI, and DSA to the CT image which was used for target tracking during the treatment delivery. We demonstrated the efficacy of the integration of DTI tractography data into CyberKnife radiosurgery

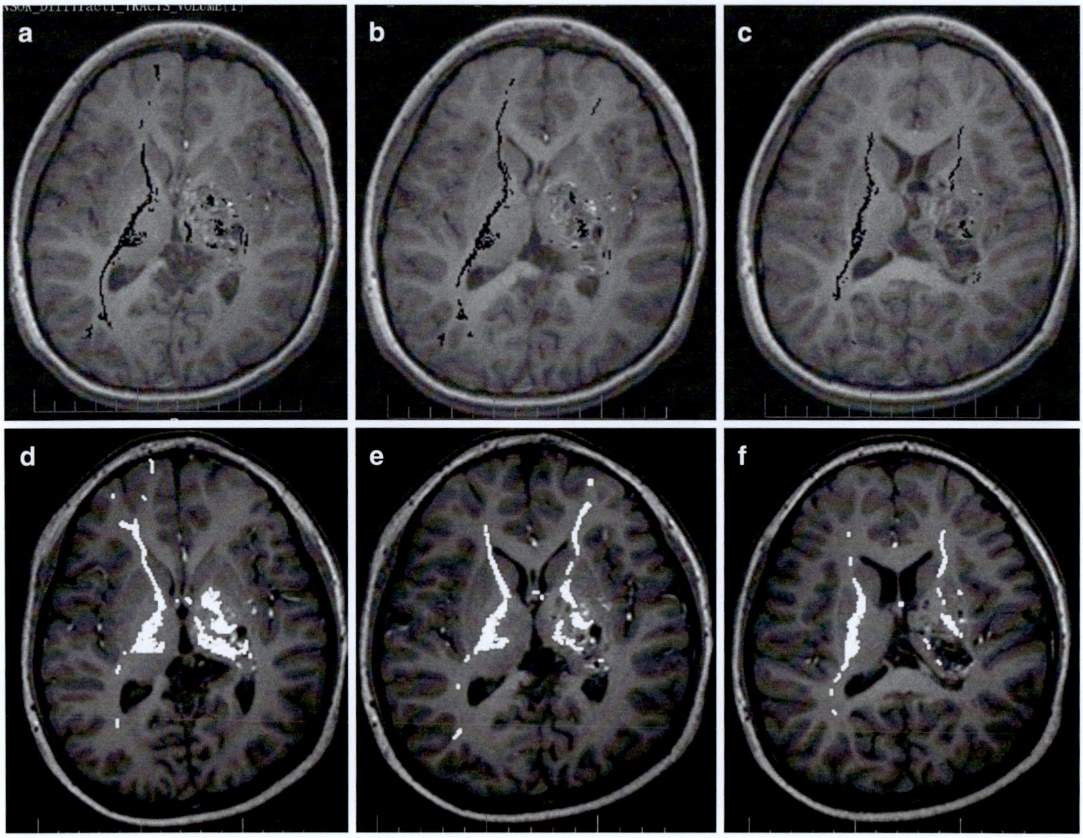

Fig. 10.2 A 12-year-old boy presented right-side hand weakness and unsteady walking. An MRI examination was performed. A brain AVM was found in the left basal ganglia and thalamus. Parts (**a**–**c**) were hybrid MR images (T1-weighted with superimposed/burned 3D nerve tract: corticospinal tract). The nerve tracts on the left hemisphere and basal ganglia decreased in volume and passed through the AVM nidus. Parts (**d**–**f**) were the same patient's hybrid MR images at 10 months post CyberKnife radiosurgery. The corticospinal tract on the left basal ganglia increased in volume

treatment planning to increase the therapeutic potential and safety of AVM radiosurgery.

The fused fMRI activation maps and the white matter tracts overlaid on the anatomical MRI volume were exported as separate grayscale DICOM images to the CyberKnife system and loaded onto the Multiplan treatment planning system (TPS) software version 2.3 before the year of 2016 or version 4.6.1 after 2016. The anatomical MRI images were registered with the CT volume for each patient, using the registration algorithm of the TPS. The registration parameters were then applied to the imported fMRI on the tractography data sets in order to attain the fusion of the tractography images with the CT study. Figure 10.3 shows the treatment plan on the corresponding axial, coronal, and sagittal CT planes for the AVM case. The target of AVM and organs at risk (OARs) were delineated in the axial contrast-enhanced MRA and T1-weighted image. The corticospinal tract was contoured in DTI combined images. We used conformal simplex (Multiplan version 2.3) or the sequential optimization (Multiplan version 4.6.1) to create the treatment planning. The maximum dose to the corticospinal tract in the internal capsule was 21 Gy in three fractions (or 18 Gy in two fractions) and in the precentral and postcentral gyrus was 22.5 Gy in three fractions. When the AVM is a large lesion in volume, two-staged CyberKnife radiosurgery was administrated 10 months apart.

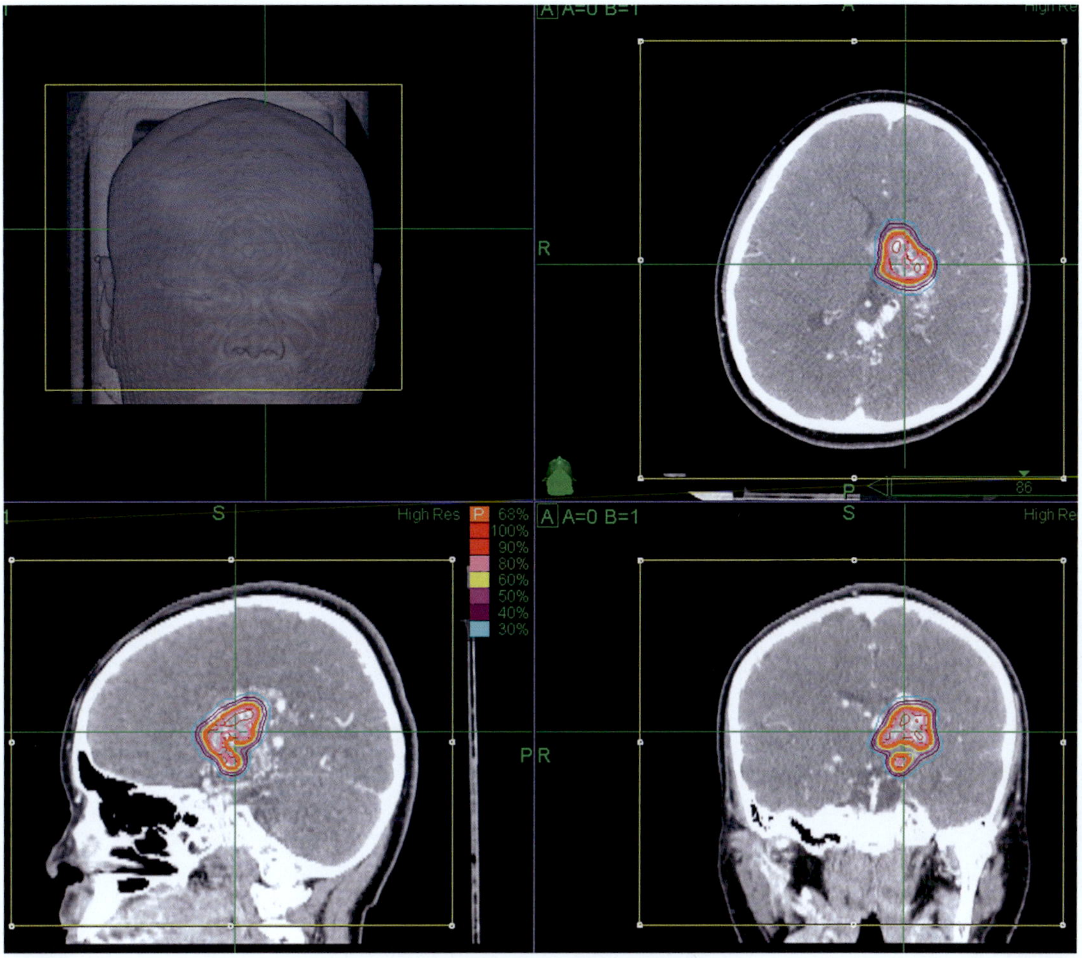

Fig. 10.3 The treatment planning of the AVM in the axial, sagittal, and coronal CT planes

10.6 A Case Demonstration

A 12-year-old boy was referred to our institution complaining of slow and progressive weakness of right hand and unsteady walking for 6 months. He had an otherwise uneventful clinical history. The neurological examination revealed right limbs muscle power grade IV. MRI showed a lesion on the left basal ganglia and thalamus. DSA demonstrated a large complex AVM receiving blood supply from middle cerebral artery and posterior cerebral artery (Fig. 10.7a). This unruptured AVM was located in the basal ganglia and thalamus. The multidisciplinary team (neurologists, neurosurgeons, neuro-anesthesiologists, and neuro-radiologists) discussed the treatment option for the high-grade AVM. Microsurgical resection of the basal ganglia AVM was unfeasible. Embolization is often used as the first-line treatment for this unruptured AVM. On the balance between natural history and the risks associated with the treatment, two-staged CyberKnife radiosurgery (CKRS) was referred as optional treatment. To minimize the complication, we integrated DTI tractography into treatment planning for CKRS. DTI showed the corticospinal tract pass through the AVM nidus. The anterior part of the AVM was treated during the first stage of CKRS. The volume of AVM was 6.2 cm^3. The prescription dose was 21.0 Gy at the periphery in three fractions, with an isodose line at 68% covering the AVM nidus (Figs. 10.3, 10.4, and 10.5).

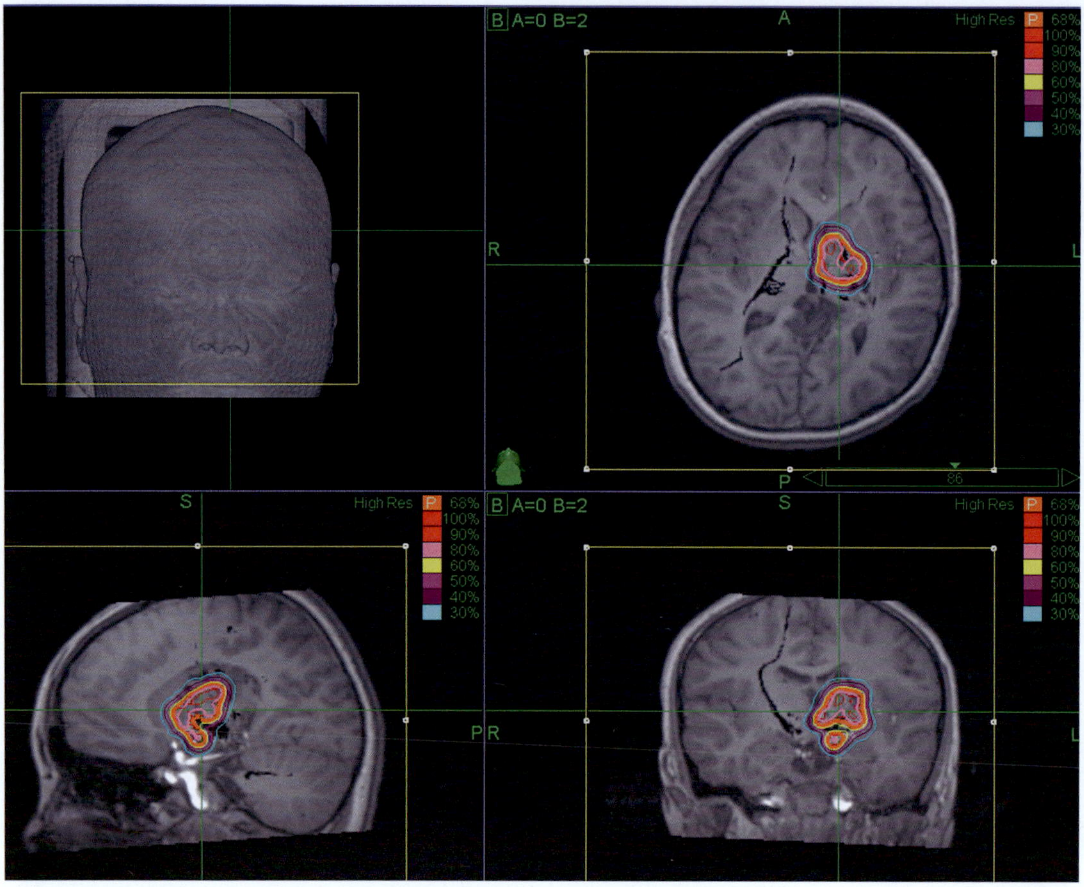

Fig. 10.4 The exact location of corticospinal tract, especially in deep white matter, could not be identified using conventional sequences such as enhanced CT or T1-weighted MR imaging studies. The fused DTI MR images are overlaid on the corresponding axial, sagittal, and coronal MRI planes, showing the position of the corticospinal tracts relative to the treated lesions on the Multiplan

We spared the corticospinal tract during the creation of the treatment plan (Fig. 10.5). The second stage of CKRS was administrated 10 months later for the residual AVM (posterior part of AVM). The volume was 7.4 cm^3. The prescription dose was 21.0 Gy at the periphery in two fractions, with the isodose line at 66% covering the AVM nidus (Fig. 10.6). The follow-up MRI revealed no apparent brain edema post hypofractionated CKRS and AVM diminishing in volume at 6 months, 1 year, 2 years, and 3 years (Fig. 10.6). DSA demonstrated that the AVM was almost obliterated at 3.5 years post CKRS (Fig. 10.7). The patient improved his symptoms progressively and had normal walk and hand power.

In Huashan Hospital Fudan University, the authors have treated 52 patients with deep-seated AVMs, integrating DTI tractography into treatment planning. We contoured the corticospinal tract in MR images directly and set dose constraint for functional organs at risk. The results showed that delineation of the functional structures and fiber tracts is beneficial and could further reduce the doses received to these critical structures and thus decrease the risk of radiation-induced complications. This is especially vital when treating AVM lesions situated in critical areas of the brain (Fig. 10.8). We also integrated DTI tractography into glioma treatment planning, when the patients considered more about their motor function post radiosurgery (Fig. 10.9).

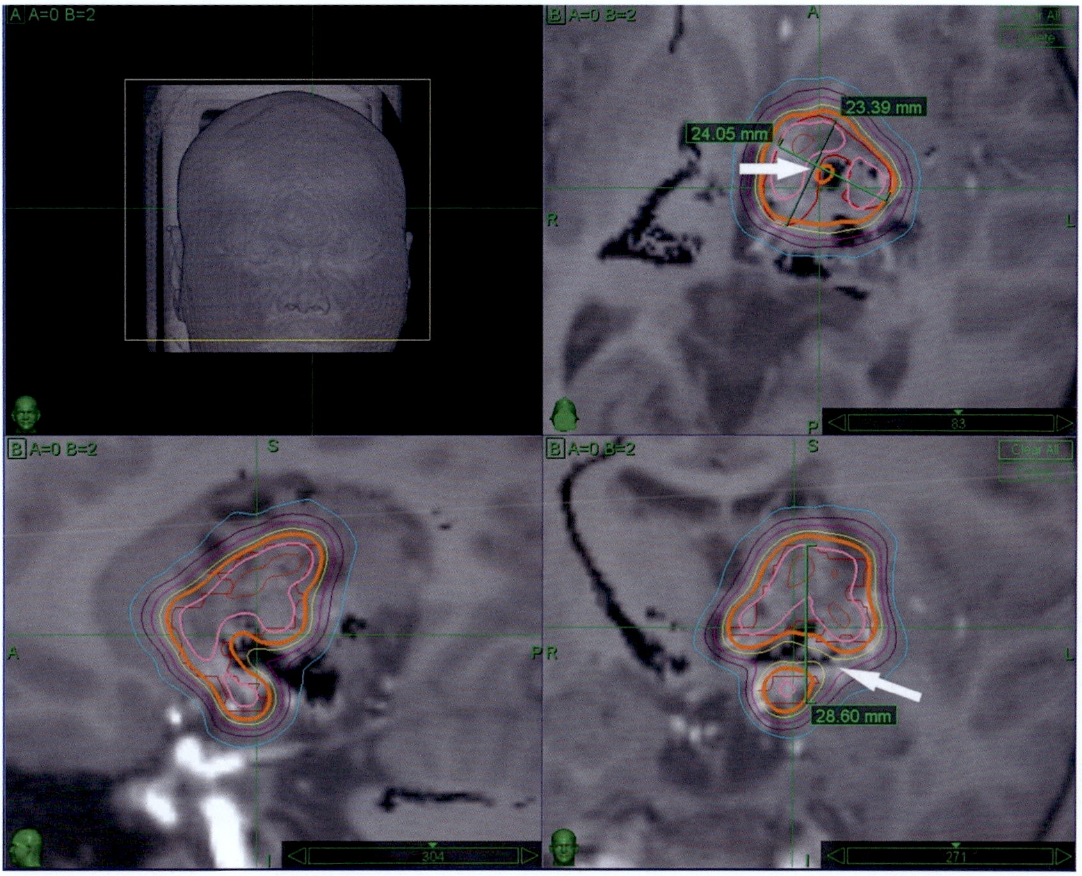

Fig. 10.5 The first stage of CyberKnife treatment planning. The corticospinal tracts pass through the AVM nidus and were spared to avoid high irradiation dose. Sparing of functional structures and fiber tracts was achieved in this planning. The black bundles were pyramidal tracts (corticospinal tracts)

10.7 DTI and Tractography in Stereotactic Radiosurgery Clinical Practice

Maruyama et al. reported integration of three-dimensional corticospinal tractography into treatment planning for Gamma Knife surgery in 2005 [17]. Seven patients with cerebral AVMs located adjacent to the corticospinal tract (CST) underwent this technique. After image fusion of the anatomical MRI and DTI studies, the combined images were transferred to a Gamma Knife treatment-planning workstation. The spatial relationship between the dose distribution and the CST was clearly demonstrated. The univariate logistic regression analysis of transient or permanent motor complications revealed a significant independent correlation with the volume of the CST that received 25 Gy or more and with a maximum dose to the CST [17]. They also reported optic radiation tractography integrated into treatment planning for Gamma Knife surgery. The results demonstrated that a maximum dose to the optic radiation tractography of less than 12 Gy did not cause new visual field deficits. A maximum dose to the optic radiation tractography of 8 Gy or more was significantly related to neurological change ($p < 0.05$), including visual field deficits [19, 20].

Stancanello et al. first reported integrating BOLD functional MRI into CyberKnife radiosurgery treatment planning of cerebral vascular malformations [21]. Five patients affected by AVMs and scheduled to undergo radiosurgery

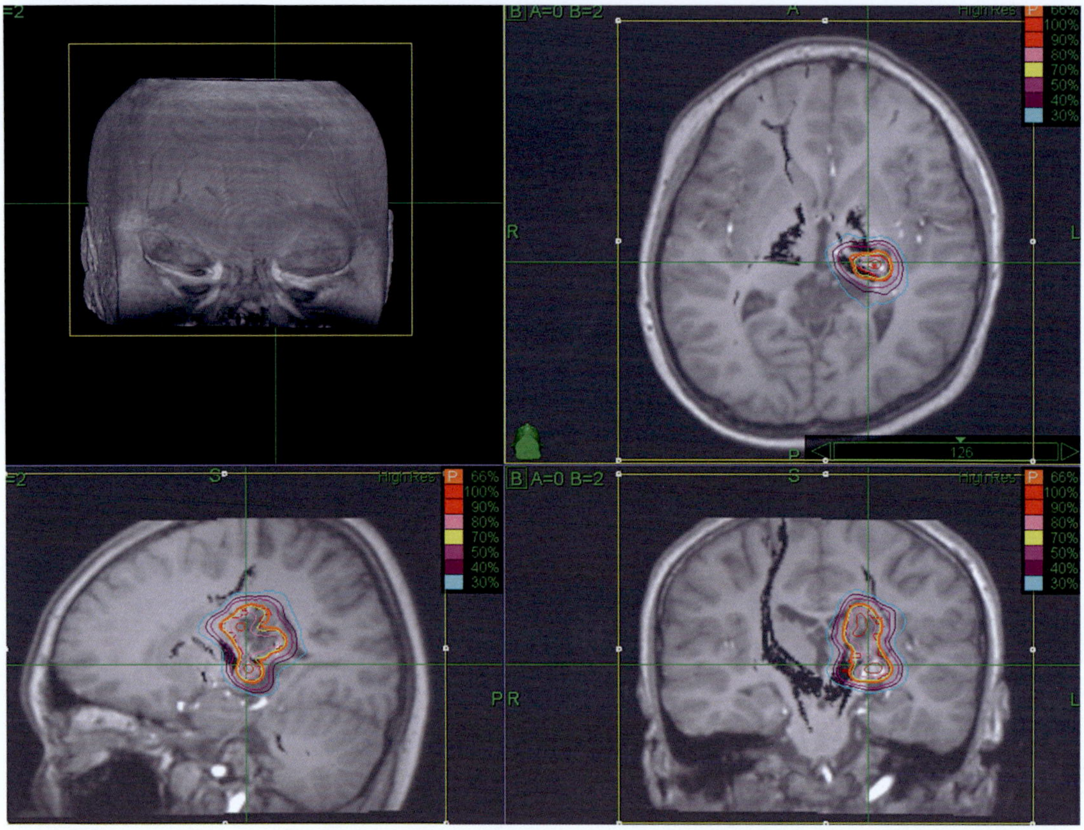

Fig. 10.6 The second stage of CyberKnife treatment planning. The AVM nidus in the thalamus was treated in the second stage of radiosurgery. The black bundles were pyramidal tracts

were scanned with functional MRI. Functional data were superimposed on three-dimensional rotational angiography and CT used for treatment planning. Treatment plans studied with and without considering functional organs at risk were significantly different, in particular with respect to both maximum dose and dose-volume histograms. Consideration of the functional organs at risk allowed quality indices of treatment plans to remain almost constant or to improve in four out of five cases compared to plans with no consideration of functional organs at risk. Pantelis et al. described four cases were treated using CyberKnife, integrating of functional MRI and white matter tractography into treatment planning [23]. Treatment plans with and without the incorporation of the functional structures and the fiber tracts into the optimization process were developed and compared. The first patient, a 25-year-old woman, was suffering from an AVM located adjacent to the posterior part of the visual pathway and near the calcarine sulcus at the right occipital lobe. The functional structures of the brain and the fiber tracts situated near the target were delineated by a neurosurgeon, using the fused activation maps and tractography images. The results showed that in the AVM, the doses received by the Broadmann-17 structure and the optic tract were reduced to 700 cGy from 1400 cGy and to 1200 cGy from 2000 cGy respectively, upon inclusion into the optimization process.

Conti et al. reported integration of functional neuroimaging in CyberKnife radiosurgery [24]. Among patients with brain lesions in critical areas, treatment planning with the integration of functional neuroimaging was performed in 25 patients. Morphological and functional imaging

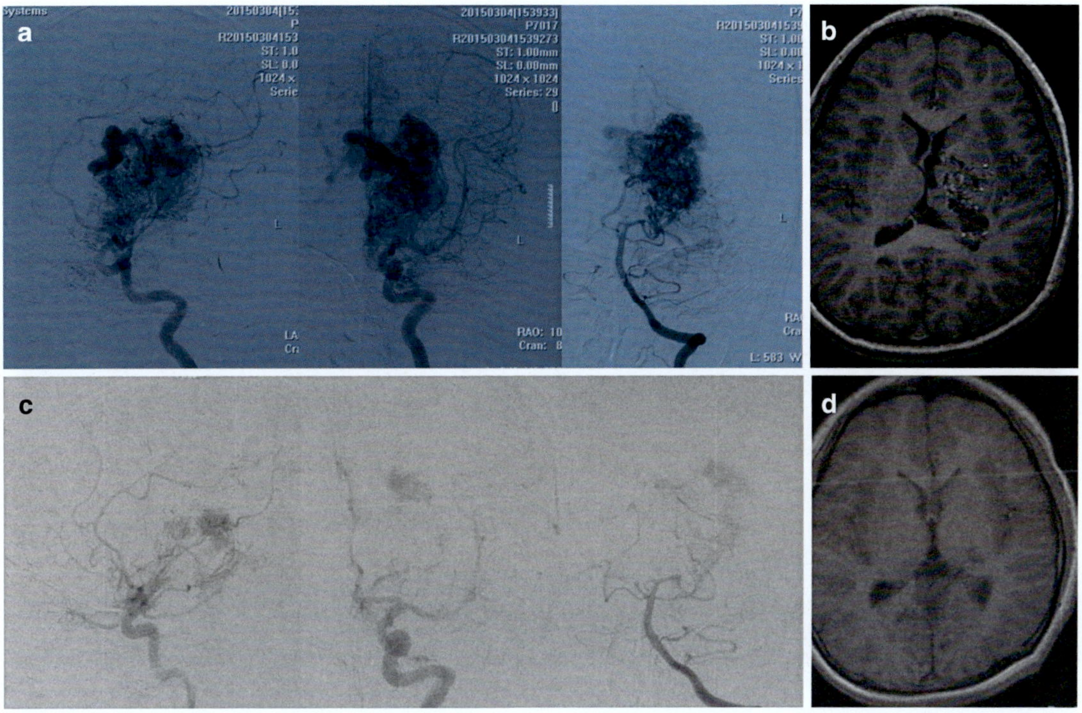

Fig. 10.7 The same patient with AVM in the left side basal ganglia and thalamus (**a**) was treated using two staged CKRS. The integration of DTI tractography into treatment planning has decreased the damage to pyramidal tracts and improved the safety of SRS. (**b**) AVM was in T1-weighted MRI before CyberKnife. DSA demonstrated that AVM was almost obliterated at 3.5 years post two-staged CKRS (**c**). AVM almost disappeared in T1-weighted MRI and no brain edema appeared at 3.5 years post CKRS (**d**)

data sets were co-registered using the Multiplan treatment planning system. The integration of functional data allows a reduction in radiation doses to functional organs at risk, including critical cortical areas, subcortical tracts, and vascular structures. The authors achieved an average of 17% reduction in the radiation dose to functional areas. No neurological deficit due to radiation was recorded at the short-term follow-up. Sun et al. reported on the integration of functional magnetic resonance imaging and diffusion-tensor imaging tractography data into CyberKnife radiosurgery for intracranial tumor management [26]. The authors investigated the data of 16 patients who had undergone CyberKnife to treat brain lesions in critical areas. The lesions included two meningioma, eight brain metastases, and six arteriovenous malformation. Radiation dose distributions with and without the functionally relevant cortical and subcortical areas into the optimization process were developed and compared. The results demonstrated that there were significant differences between the treatment plans with and without the functionally relevant cortical and subcortical areas into the optimization process. An average 22.7% reduction in the maximum dose to functional areas was observed. No neurological complication due to radiation damage was observed in the follow-up period.

Gomes et al. has integrated DTI into Gamma Knife thalamotomy planning. He demonstrated the internal capsule constraint of <15 Gy was safe for pyramidal tract [27]. Kim et al. illustrated the feasibility of LINAC thalamotomies and DTI-based segmentation guided therapies [31]. These two studies showed that DTI tractography has the potential to guide and refine thalamic targeting for improved efficacy in neuromodulation and thalamotomies.

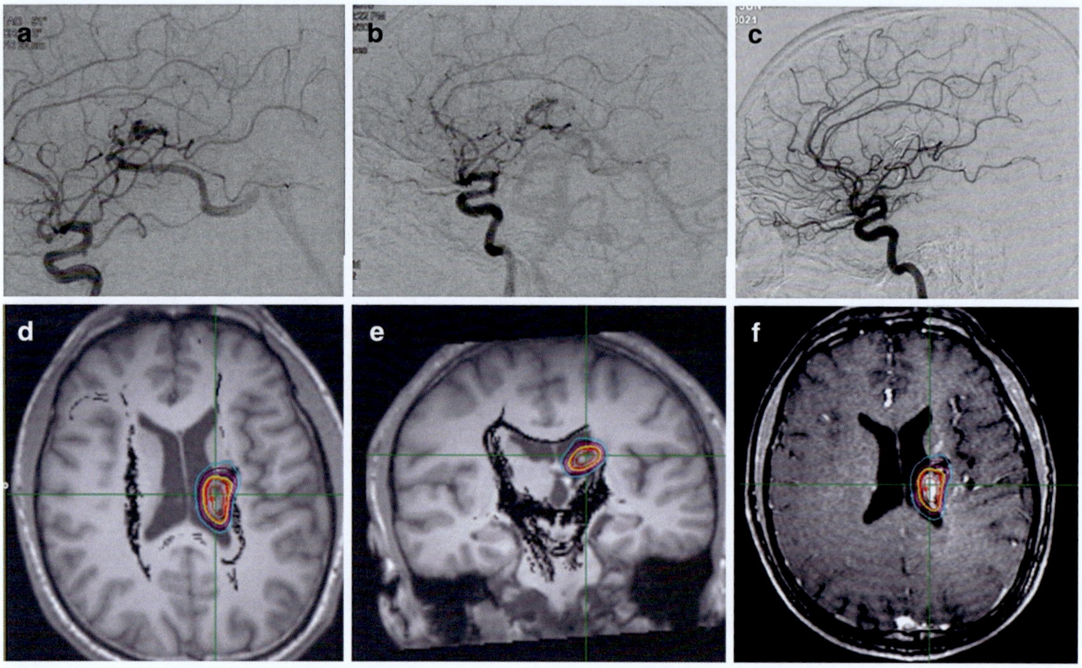

Fig. 10.8 A female patient had intracranial hemorrhage history. DSA showed left side AVM in the deep brain (**a**). Embolization was selected as the first-line treatment. One year later, complete occlusion was not obtained by embolization alone (**b**) and subsequent CKRS was performed. DTI was used to visualize the corticospinal tract and helped reduce radiation doses to the nerve tract (**d**–**f**). Left internal carotid angiogram shows complete cure of the AVM at 3 years after CKRS (**c**)

10.8 Tolerance of Corticospinal Tract

The tolerance of the corticospinal tract to radiosurgery is not well known. Maruyama reported that the risk of a motor complication was estimated to be 50% when 60 mm^3 of the visualized corticospinal tract received more than 25 Gy (single fraction of Gamma Knife) or when the maximum dose to the corticospinal tract was 28 Gy [17], whereas further study suggested that the corticospinal tracts only needed to be kept outside the 20 Gy isodose line, which is the standard margin dose used to treat arteriovenous malformations (by Gamma Knife). The risk of motor complication is less than 5% [22]. Some authors suggested that the tolerance of corticospinal tract in the basal ganglia were less than 16–18 Gy (single fraction SRS) [27, 28, 32]. In our center, patients with an AVM in the deep parietal lobe, precentral gyrus, basal ganglia, and thalamus usually had undergone hypofractionated CyberKnife radiosurgery since 2008. DTI of the corticospinal tracts had been integrated into treatment planning of CyberKnife since 2010, and the maximum dose received by the corticospinal tracts was attempted to be less than 21 Gy in 3 fractions or 18 Gy in 2 fractions. If the AVM located in the precentral gyrus (or in the vicinity of the precentral or postcentral gyrus), the prescription dose was 22.5 Gy in three fractions. No new neurological deficits due to radiation were developed at the long-term follow-up. The tolerance of the white matter tracts is shown in Table 10.2.

10.9 Limitations

With any tractography method, there is the inherent drawback of the estimated white matter pathways not being wholly representative of

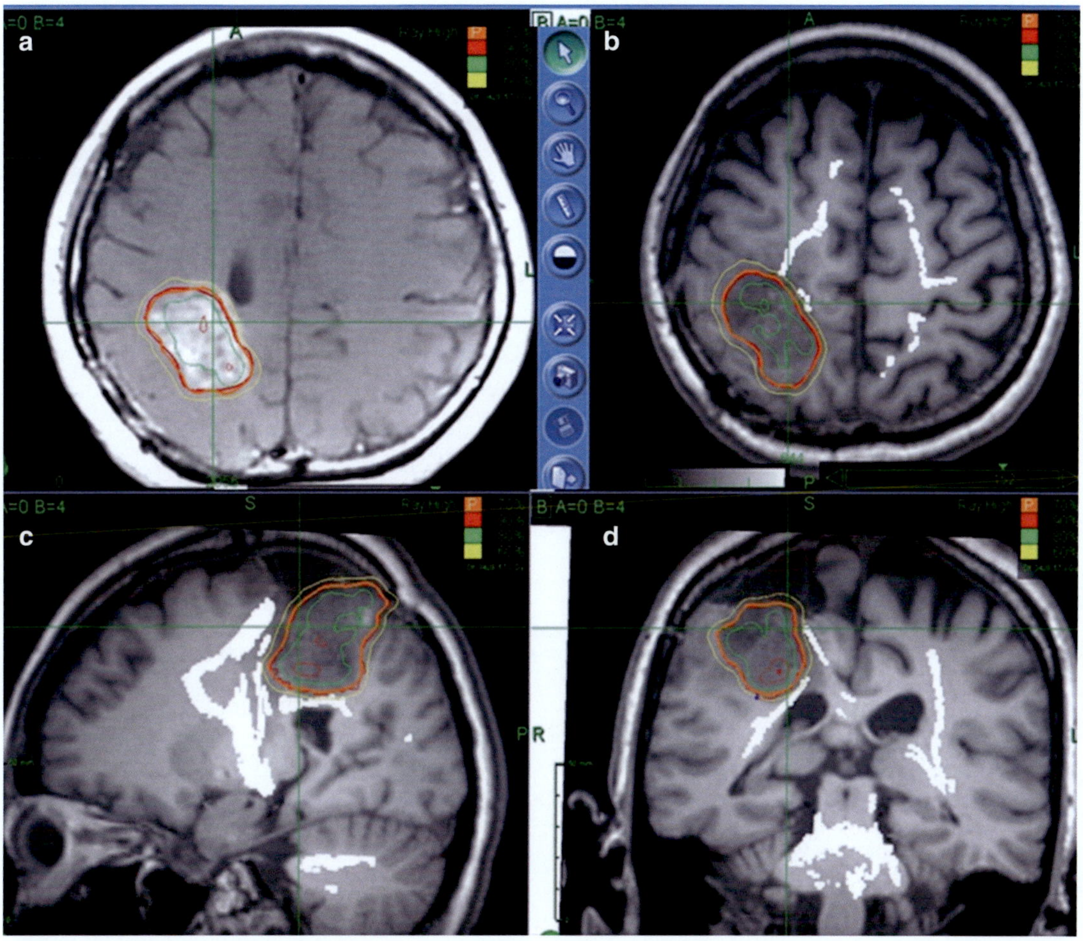

Fig. 10.9 Contrast-enhanced MRI showed the tumor and treatment planning (**a**), and the corticospinal tracts (the white bundles) were visualized in axial (**b**), sagittal (**c**) and coronal (**d**) MR imaging

Table 10.2 Tolerance dose of white matter tracts

Author	Type of radiosurgery	CST	Optic tract	Central gyrus	Arcuate fasciculus
Maruyama [17]	GKS	25 Gy			
Maruyama [20]					10 Gy in the frontal
Maruyama [22]	GKS	23 Gy			
Pantelis [23]	GKS	20 Gy	12 Gy		
Gomes [25]	GKS	15 Gy in basal ganglia			
Koga [28]	GKS	17 Gy in basal ganglia		20 Gy	
Conti [24]	CKS	NR	10 Gy		
Sun [26]	CKS	15 Gy/1F 22.5 Gy/3F			
The author	CKS	21 Gy/3F 18 Gy/2F	14.5 Gy/3F	22.5 Gy/3F	

Abbreviations: *GKS* Gamma Knife radiosurgery, *CKS* CyberKnife radiosurgery, *NR* not reported, *F* fraction, *CST* corticospinal tract actual tissue microstructure or anatomical connectivity. For one, the diameter of individual axons is much smaller (1.0 μm) than a typical DTI voxel (resolutions in the range of 1–3 mm). In addition, many areas of the brain contain multiple population(s) of fibers or fibers that cross each other. Another disadvantages of fMRI as a clinical tool is the time required for postprocessing.

In an era of precision medicine, it is time we move beyond conventional MRI and embrace the integration of these advanced neuroimaging modalities in CyberKnife radiosurgery. In particular, DTI has the ability to better define anatomical structures by allowing detailed visualization of white matter tracts. The integration of DTI tractography into treatment planning has increased the therapeutic potential and safety of stereotactic radiosurgery for AVMs in eloquent areas. With the aid of the above-described techniques, these structures can be marked as critical structures and spared during the treatment planning process. In cases where this is not feasible due to close proximity of the critical volumes to the treated lesions, the biologically equivalent dose could be delivered in more than one fraction to reduce the risk for any radiation induced complications.

References

1. Pollock BE, Gorman DA, Brown PD. Radiosurgery for arteriovenous malformations of the basal ganglia, thalamus, and brainstem. J Neurosurg. 2004;100:210–4.
2. Chopra R, Kondziolka D, Niranjan A, Lunsford LD, Flickinger JC. Long-term follow-up of acoustic schwannoma radiosurgery with marginal tumor doses of 12 to 13 Gy. Int J Radiat Oncol Biol Phys. 2007;68(3):845–51.
3. Sahgal A, Larson D, Knisely J. Stereotactic radiosurgery alone for brain metastases. Lancet Oncol. 2015;16(3):249–50.
4. Pollock BE, Stafford SL, Link MJ, Garces YI, Foote RL. Single-fraction radiosurgery of benign cavernous sinus meningiomas. J Neurosurg. 2013;119(3):675–82.
5. Wang X, Zhu H, Knisely J, Mei G, Liu X, Dai J, Mao Y, Pan L, Qin Z, Wang E. Hypofractionated stereotactic radiosurgery: a new treatment strategy for giant cavernous sinus hemangiomas. J Neurosurg. 2018;128(1):60–7.
6. Maruyama K, Kawahara N, Shin M, et al. The risk of hemorrhage after radiosurgery for cerebral arteriovenous malformations. N Engl J Med. 2005;352:146–53.
7. Andrade-Souza YM, Zadeh G, Scora D, Tsao MN, Schwartz ML. Radiosurgery for basal ganglia, internal capsule, and thalamus arteriovenous malformation: clinical outcome. Neurosurgery. 2005;56:56–64.
8. Flickinger JC, Kondziolka D, Lunsford LD, Pollock BE, Yamamoto M, Gorman DA, Schomberg PJ, Sneed P, Larson D, Smith V, McDermott MW, Miyawaki L, Chilton J, Morantz RA, Young B, Jokura H, Liscak R. A multi-institutional analysis of complication outcomes after arteriovenous malformation radiosurgery. Int J Radiat Oncol Biol Phys. 1999;44:67–74.
9. Sasaki T, Kurita H, Saito I, Kawamoto S, Nemoto S, Terahara A, Kirino T, Takakura K. Arteriovenous malformations in the basal ganglia and thalamus: management and results in 101 cases. J Neurosurg. 1998;88:285–92.
10. Wengenroth M, Blatow M, Guenther J. Diagnostic benefits of presurgical fMRI in patients with brain tumors in the primary sensorimotor cortex. Eur Radiol. 2001;21:1517–25.
11. Kim PE, Singh M. Functional magnetic resonance imaging for brain mapping in neurosurgery. Neurosurg Focus. 2003;15(1):1–7.
12. Kamada K, Todo T, Masutani Y, Aoki S, Ino K, Takano T, Kirino T, Kawahara N, Morita A. Combined use of tractography-integrated functional neuronavigation and direct fiber simulation. J Neurosurg. 2005;102:664–72.
13. Javadi SA, Nabavi A, Giordano M, Faghihzadeh E, Samii A. Evaluation of diffusion tensor imaging-based Tractography of the corticospinal tract: a correlative study with intraoperative magnetic resonance imaging and direct electrical subcortical stimulation. Neurosurgery. 2017;80(2):287–99.
14. Pirotte B, Voordecker P, Neuroschl C, Baleriaux D, Wikler D, Metens T, Denolin V, Joffroy A, Massager N, Brotchi J, Levivier M. Combination of functional magnetic resonance imaging-guided neuronavigation and intraoperative cortical brain mapping improves targeting of motor cortex stimulation in neuropathic pain. Neurosurgery. 2005;56(2 Suppl):344–59.
15. Petrella RJ, Shah LM, Harris KM. Preoperative functional MR imaging localization of language and motor areas: effect on therapeutic decision making in patients with potentially resectable brain tumors. Radiology. 2006;240:793–802.
16. Ellis MJ, Rutka JT, Kulkarni AV, Dirks PB, Widjaja E. Corticospinal tract mapping in children with ruptured arteriovenous malformations using functionally guided diffusion-tensor imaging. J Neurosurg Pediatr. 2012;9(5):505–10.
17. Maruyama K, Kamada K, Shin M, Itoh D, Aoki S, Masutani Y, Tago M, Kirino T. Integration of three-dimensional corticospinal tractography into treatment planning for gamma knife surgery. J Neurosurg. 2005;102(4):673–7.

18. Schad LR. Improved target volume characterization in stereotactic treatment planning of brain lesions by using high-resolution BOLD MR venography. NMR Biomed. 2001;14(7–8):478–83.
19. Maruyama K, Kamada K, Shin M, Itoh D, Masutani Y, Ino K, Tago M, Saito N. Optic radiation tractography integrated into simulated treatment planning for Gamma Knife surgery. J Neurosurg. 2007;107(4):721–6.
20. Maruyama K, Koga T, Kamada K, Ota T, Itoh D, Ino K, Igaki H, Aoki S, Masutani Y, Shin M, Saito N. Arcuate fasciculus tractography integrated into Gamma Knife surgery. J Neurosurg. 2009;111(3):520–6.
21. Stancanello J, Cavedon C, Francescon P, Causin F, Avanzo M, Colombo F, Cerveri P, Ferrigno G, Uggeri F. BOLD fMRI integration into radiosurgery treatment planning of cerebral vascular malformations. Med Phys. 2007;34:1176–84.
22. Maruyama K, Kamada K, Ota T, Koga T, Itoh D, Ino K, Aoki S, Tago M, Masutani Y, Shin M, Saito N. Tolerance of pyramidal tract to gamma knife radiosurgery based on diffusion-tensor tractography. Int J Radiat Oncol Biol Phys. 2008;70(5):1330–5.
23. Pantelis E, Papadakis N, Verigos K, Stathochristopoulou I, Antypas C, Lekas L, Tzouras A, Georgiou E, Salvaras N. Integration of functional MRI and white matter tractography in stereotactic radiosurgery clinical practice. Int J Radiat Oncol Biol Phys. 2010;78(1):257–67.
24. Conti A, Pontoriero A, Ricciardi GK, Granata F, Vinci S, Angileri FF, Pergolizzi S, Alafaci C, Rizzo V, Quartarone A, Germanò A, Foroni RI, De Renzis C, Tomasello F. Integration of functional neuroimaging in CyberKnife radiosurgery: feasibility and dosimetric results. Neurosurg Focus. 2013;34(4):E5.
25. Gomes JG, Gorgulho AA, de Oliveira López A, Saraiva CW, Damiani LP, Pássaro AM, Salvajoli JV, de Oliveira Siqueira L, Salvajoli BP, De Salles AA. The role of diffusion tensor imaging tractography for Gamma Knife thalamotomy planning. J Neurosurg. 2016;125(Suppl 1):129–38.
26. Sun L, Qu B, Wang J, Ju Z, Zhang Z, Cui Z, Jack Y, Ling Z, Yu X, Pan L. Integration of functional MRI and white matter tractography in CyberKnife radiosurgery. Technol Cancer Res Treat. 2017;16(6):850–6.
27. Gavin CG, Ian Sabin H. Stereotactic diffusion tensor imaging tractography for Gamma Knife radiosurgery. J Neurosurg. 2016;125(Suppl 1):139–46.
28. Koga T, Shin M, Maruyama K, Kamada K, Ota T, Itoh D, Kunii N, Ino K, Aoki S, Masutani Y, Igaki H, Onoe T, Saito N. Integration of corticospinal tractography reduces motor complications after radiosurgery. Int J Radiat Oncol Biol Phys. 2012;83(1):129–33.
29. Alexander AL, Lee JE, Lazar M, Field AS. Diffusion tensor imaging of the brain. Neurotherapeutics. 2007;4:316–29.
30. Le Bihan D, Mangin JF, Poupon C, Clark CA, Pappata S, Molko N, Chabriat H. Diffusion tensor imaging: concepts and applications. J Magn Reson Imaging. 2001;13(4):534–4626.
31. Jeurissen B, Descoteaux M, Mori S, Leemans A. Diffusion MRI fiber tractography of the brain. NMR Biomed. 2019;32(4):e3785.
32. Kim W, Sharim J, Tenn S, Kaprealian T, Bordelon Y, Agazaryan N, Pouratian N. Diffusion tractography imaging-guided frameless linear accelerator stereotactic radiosurgical thalamotomy for tremor: case report. J Neurosurg. 2018;128(1):215–21.

11 Metabolic Imaging

Andrea d'Amico

11.1 Introduction

Identifying the boundaries of a tumor lesion is crucial in the planning process of radiation therapy. The development of highly specific radiopharmaceuticals also allows the use of PET in the context of the therapy of intracranial tumors with a CyberKnife.

The integration of molecular imaging methods, firstly, significantly reduces interobserver variability in target volume delineation. Various automatic or semi-automatic methods for the segmentation of molecular images have been proposed. This also allows us to evaluate the intratumoral heterogeneity for applying biologically conformed radiotherapy or the possibility of determining the early response to therapy.

11.2 Technical Considerations

The accurate localization of the tumor and its definition boundaries is the pivotal concept for radiotherapy treatment planning. An imaging technique must detect a sufficient signal difference between the tumor and the surrounding healthy tissues to be suitable for planning radiation therapy.

A. d'Amico (✉)
Diagnostic PET Department, Maria Sklodowska-Curie National Research Institute of Oncology, Gliwice, Poland
e-mail: adamico@io.gliwice.pl

Computed tomography and magnetic resonance imaging are commonly used radiological modalities for preliminary evaluation of patients with intracranial neoplasms, but the role of DTI tractography techniques is also increasing.

Nuclear medical imaging modalities can add important complementary information involving the different biological and metabolic characteristics of healthy and neoplastic brain tissues. The first molecular imaging modality for the evaluation of solid tumor borders is PET with 2-[(18)F]-fluoro-2-deoxy-D-glucose (FDG).

However, the high uptake in healthy gray matter greatly reduces the possibility of applying this radiotracer for the evaluation of primary or secondary intracranial malignant lesions. More importantly, the inflammation induced by radiotherapy treatment is an important non-specific accumulation factor that can simulate the presence of a neoplasm or can mask the actual location of the tumor.

Several more specific PET radiotracers, such as the radiolabeled somatostatin analogues (68Ga)-DOTA-Tyr3-octreotide (68Ga-DOTATOC), 68Ga-DOTA-D-Phe1-Tyr3-octreotate (68Ga-DOTATATE), and 68Ga-DOTA-1-Nal3-octreotide (68Ga-DOTANOC) or the amino acids [18F]-fluorothymidine (FLT) and *O*-(2-[(18)F]-fluoroethyl)-L-tyrosine (FET), have a greater specificity without accumulation in healthy brain tissue. This leads to an easier evaluation of CNS tumors [1–3].

© Springer Nature Switzerland AG 2020
A. Conti et al. (eds.), *CyberKnife NeuroRadiosurgery*,
https://doi.org/10.1007/978-3-030-50668-1_11

The combination of radiological and nuclear medicine imaging creates an increasing range of logistical and technical difficulties when performing examinations. Unlike radiological images, the quality of molecular images varies greatly depending on the patient's preparation for the acquisition protocol and the type of PET device. Numerous recommendations and guidelines have been published for imaging procedures that use FDG and other radiotracers [4–8].

Spatial localization of the tumor must be ensured during radiological and molecular examinations, as well as during radiation therapy treatment. This is achieved by using thermoplastic masks or other means of immobilization. These immobilization devices must also be used during the PET examination, being careful not to place any accessory equipment outside the visual field of the transaxial image to avoid artifacts in correcting the attenuation [9].

The CT component of the PET CT exam can be performed in several ways. In general, a low-dose acquisition is performed to mitigate the correction, but another possibility is to perform a single diagnostic-quality CT scan, which can then be used for treatment planning. In the latter case, image registration is not needed to align the PET with the treatment planning exam.

Due to the wide range of iterative reconstruction algorithms and PET image correction methods currently available, each center should determine its best acquisition and reconstruction method for image analysis according with the available equipment and radiotracers used.

A previous acquisition of a series of images using a phantom for the evaluation of the different reconstruction functions should be recommended. This is especially true in the case of multicentric study protocols, as it will ensure the production of images of comparable quality regardless of the site of acquisition.

PET images must be successively registered with diagnostic MR or CT images for treatment planning. Since the effective visualization of the target lesion largely depends on the accuracy of the registration, the protocols for the acquisition and processing of molecular images must follow principles compatible with those used for the acquisition of diagnostic CT images [10].

One factor to consider is that the injection of the contrast medium can cause an increase in voxel values in the region of high contrast concentration [11]. Such cases require repetition of the acquisition of the low-dose CT to be used for the correction of the attenuation of the PET which will not be used for treatment planning.

In the case of PET-MR equipment, the images used for attenuation correction could potentially be used for planning radiation therapy [12, 13]. However, the intensity value of the resonance signal in a single voxel cannot be uniquely mapped for the definition of the gamma ray attenuation coefficient. Studies are underway to improve the accuracy of pseudo-CT mapping, but have not yet reached the stage of clinical validation [14].

For CNS tumor analysis, the rigid recording algorithms usually included in the commercial software packages are generally sufficient for image registration. Diagnostic CT images performed for radiotherapy planning are first registered to the CT component of the PET-CT image.

The resulting spatial transformation is then successively applied to the PET image, which can subsequently be merged with the CT for radiotherapy planning. One necessary step is to take into account the differences in the size of the PET voxel compared to the CT voxels, which requires up-sampling of the larger PET voxels. However, the resampling operation changes the standardized update value (SUV). Therefore, the quantitative evaluation of the response to the treatment must be performed on the PET images before resampling.

For the registration of MR and PET images, the recommended protocol is to proceed with the use T1-weighted volumetric sequences later during the CT image recording. The MR and CT modalities have intrinsically different signal values, so this operation can be carried out using dedicated software.

PET signal quantification is mainly performed by calculating the SUV of each lesion. That SUV is obtained by normalizing the concentration of radioactivity measured in a voxel for the injected activity and the body weight. Both the maximum and average SUVs can be easily calculated. However, they do not necessarily represent a reliable measure of the underlying biological process. In fact, these values are extremely sensitive to the conditions of acquisition and reconstruction of the images.

A first source of possible errors is the range of technical factors that relate to the acquisition of the molecular image. The different processes of PET acquisition and processing must be carefully evaluated. Test-retest studies on FDG PET images have shown statistical uncertainties that contribute 15% of the SUV max [15, 16]. Generally, lesions with diameters less than 4 or 5 mm are displayed with difficulty due to the reduced spatial resolution of the PET method [17]. This has obvious implications for the segmentation of small tumor localizations and for the evaluation of a small tumor's response to treatment.

Patient preparation is a further factor that strongly influences image quality. In particular, having the patient fast before the administration of radiopharmaceuticals such as FET or FDG is recommended to ensure an optimal level of tracer accumulation in the tumor.

Lastly, the anatomical position of the signal source in different points of the field of view of the detector can cause variations in SUV, with the possibility of underestimating the signal for the more peripheral lesions [18].

A kinetic analysis of the accumulator dynamics of the radiotracer can provide more reliable information than the nature of the pathological process, but this entails a series of logistical difficulties related to the longer acquisition time and more complex image processing [19].

Advances in radiation distribution and dosimetry currently allow precise delivery of the dose at any point determined by the operator. The main problem lies in the clear identification of the location where this dose is to be delivered.

The uncertainty regarding the identification of the target volume with respect to the patient's dosimetry is estimated to be ten times greater than the impact of a patient's movement during radiotherapy [20, 21]. Molecular imaging can play a role in reducing interobserver variability for the delineation of the target volume. Consequently, it allows the attainment of greater conformity between the boundaries of the target volume and the real boundaries of the tumor in the patient's body.

The boundary definition with PET can be performed using different strategies. The most common and easiest method is the manual segmentation process. An important bias of this method is the occurrence of radiotracer accumulations that are not related to the cancer. Very often, the presence of areas of inflammation in highly vascularized or granular tissues near the tumor can lead to focal or diffuse uptake that can be confused with a specific accumulation in the neoplastic site. Assuming a good knowledge of the biological processes that are measured in the molecular images, an expert operator should be able to distinguish the pathological tissue from the non-specific accumulations in the majority of cases.

When performing manual segmentation, the experience of the nuclear medicine specialist is fundamental. Nevertheless, a certain degree of inter- and intra-observer variability is inevitable. The centers that use this method are recommended to establish detailed protocols regarding the display modes, with definition of the window level, the color settings, and all the other parameters that can influence the choice of the contour by the operator [22].

Conversely, automatic segmentation strategies are preferable to reduce variability and increase the speed of execution. The simplest strategy is based on the definition of an SUV threshold value determined a priori, based on the degree of avidity of the tumor being studied.

Automatic segmentation protocols can also be defined by setting a percentage value of the SUV max for each lesion or by using subtraction algorithms which consider the physiological uptake calculated for an organ (i.e., the liver) as constant [23, 24]. The SUVs show a certain degree of variability related to the type of scanner, the reconstruction and acquisition parameters, and the partial volume effect.

Therefore, it is not possible to define an ideal method that suits any clinical context. In general, any automatic segmentation requires a review by an experienced physician. This prompted the proposition of iterative algorithms

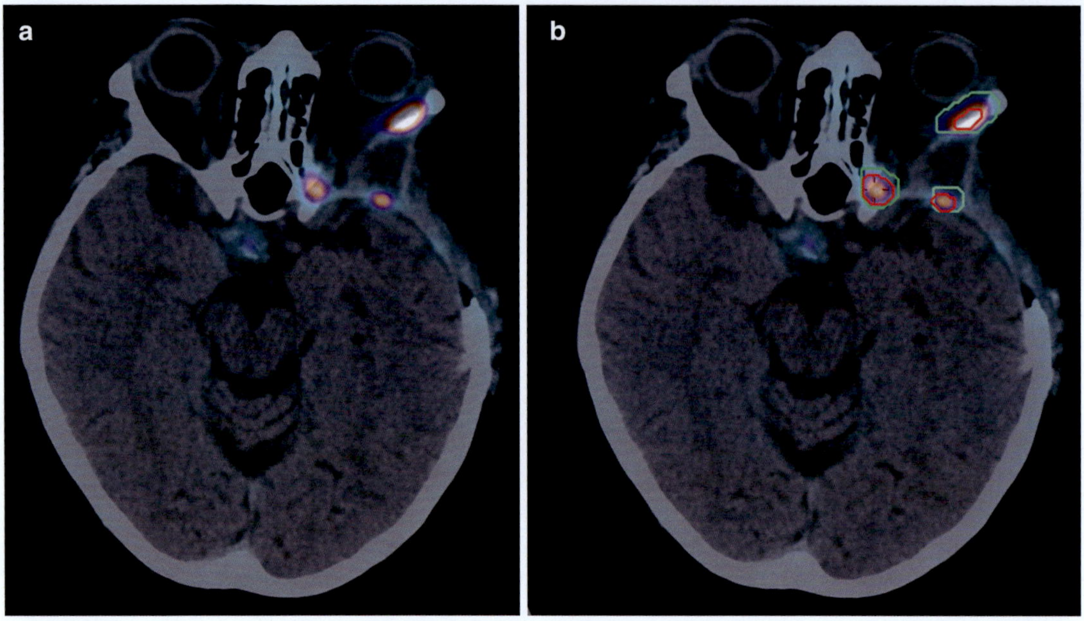

Fig. 11.1 (**a**) PET-CT with Ga-68 Dotatate in a patient after non radical excision of a large left frontal meningioma. Residual tumor areas are clearly visible, with intense radiopharmaceutical uptake. (**b**) The same image after segmentation with different automatic methods: adaptive threshold (green borders) and threshold at 40% SUV max (red borders)

based on machine learning or image filtering based on texture as potential substitutes for this human requirement [25, 26]. However, although these methods are fascinating, no algorithm has yet reached the level of clinical validation.

It has also been shown that alternative radiotracers have much better specific characteristics: those based on amino acids, such as fluorinated Ola methionine labeled with carbon fluoride are able to localize the active tumor thanks to the greater expression of amino acid transporters. Phenylalanine has also shown extremely promising results. Similarly, radioactive gallium-labeled somatostatin analogues have proven useful in the segmentation of neuroendocrine tumors and meningiomas.

11.3 Clinical Applications

A typical application of molecular imaging techniques for the therapeutic application of the CyberKnife is represented by meningiomas located near the neurovascular structures of the cranial base. The volumetric progression of these tumors can be blocked by stereotactic radiosurgery.

However, these cases very often have great limits for the precise definition of the edges of the tumor based exclusively on MR, especially in patients who have undergone neurosurgical interventions [27]. Meningiomas have a high degree of somatostatin receptor expression, which allows for accurate visualization using PET images and somatostatin analogs marked with gallium 68 [28] (Fig. 11.1).

Numerous studies have demonstrated the usefulness of these images in the planning of stereotactic radiosurgery [29–34]. The extremely high tumor/background ratio generally allows manual segmentation of PET images. A study performed in 2013 showed that the addition of PET images in the definition of GTV led to substantial modifications (more than 10%) in over two thirds of the patients when compared to the GTV defined based on the MR images [35].

Early evaluation studies of the efficacy of radiotherapy with the use of molecular markers of apoptosis have been excluded. Fluorine-labeled ML-10 is a known radiopharmaceutical capable of evaluating apoptosis in vivo [36, 37]. A Chinese group in 2018 evaluated the degree of accumulation of this radiopharmaceutical before and after radiation therapy in a series of 29 patients with various types of intracranial tumors undergoing treatment with CyberKnife [38].

The baseline examination has proven effective in determining the localization of the tumor, based on the absence of accumulation in healthy brain tissues. Repetition of the PET exam 48 h after radiation therapy showed a significant increase in tumor accumulation. This accumulation was significantly correlated with subsequent volume reduction and was more evident for lesions with increased malignancy.

FET is a clinically applied radiotracer for the evaluation of tumor recurrence in gliomas after treatment with radiotherapy [39, 40]. The recurrence of brain metastases previously treated with CyberKnife can be effectively assessed with this tracer.

A study conducted by Romagna in 2016 examined 22 patients previously treated with the CyberKnife for brain metastases and showing suspected recurrence on MR images. Kinetic analysis of the FET distribution within the suspected lesions showed sensitivity and specificity values of 93% and 84% respectively, thereby confirming the clinical utility of this radiotracer in the evaluation of recurrence of both primary and metastatic intracranial malignancies [41].

Finally, molecular imaging methods have proven useful in the particular cases of patients with metastatic tumors spread in the spine after surgical treatment. The previous implantation of metal fixation elements hindered the definition of tumor boundaries in diagnostic CT scans.

A series of three patients described by a Korean group in 2006 revealed that PET with FDG was useful both in defining the tumor volume and in evaluating the response to treatment [42]. Notably, in these patients, the standard FDG radiopharmaceutical for PET could be used, due to the absence of high physiological uptake by the vertebral column elements or the paraspinal tissues.

References

1. Suchorska B, Tonn JC, Jansen NL. PET imaging for brain tumor diagnostics. Curr Opin Neurol. 2014;27(6):683–8. https://doi.org/10.1097/WCO.0000000000000143. Review. PubMed PMID: 25333605.
2. Pirotte B, Goldman S, Massager N, David P, Wikler D, Vandesteene A, Salmon I, Brotchi J, Levivier M. Comparison of 18F-FDG and 11C-methionine for PET-guided stereotactic brain biopsy of gliomas. J Nucl Med. 2004;45(8):1293–8. PubMed PMID: 15299051.
3. Heiss P, Mayer S, Herz M, Wester HJ, Schwaiger M, Senekowitsch-Schmidtke R. Investigation of transport mechanism and uptake kinetics of O-(2-[18F]fluoroethyl)-L-tyrosine in vitro and in vivo. J Nucl Med. 1999;40(8):1367–73. PubMed PMID: 10450690.
4. Law I, Albert NL, Arbizu J, Boellaard R, Drzezga A, Galldiks N, la Fougère C, Langen KJ, Lopci E, Lowe V, McConathy J, Quick HH, Sattler B, Schuster DM, Tonn JC, Weller M. Joint EANM/EANO/RANO practice guidelines/SNMMI procedure standards for imaging of gliomas using PET with radiolabelled amino acids and [(18)F]FDG:version 1.0. Eur J Nucl Med Mol Imaging. 2019;46(3):540–57. https://doi.org/10.1007/s00259-018-4207-9. Epub 2018 Dec 5. PubMed PMID: 30519867; PubMed Central PMCID: PMC6351513.
5. Segall G, Delbeke D, Stabin MG, Even-Sapir E, Fair J, Sajdak R, Smith GT, SNM. SNM practice guideline for sodium 18F-fluoride PET/CT bone scans 1.0. J Nucl Med. 2010;51(11):1813–20. https://doi.org/10.2967/jnumed.110.082263. Erratum in: J Nucl Med. 2011;52(3):495. PubMed PMID: 21051652.
6. Vander Borght T, Asenbaum S, Bartenstein P, Halldin C, Kapucu O, Van Laere K, Varrone A, Tatsch K, European Association of Nuclear Medicine (EANM). EANM procedure guidelines for brain tumour imaging using labelled amino acid analogues. Eur J Nucl Med Mol Imaging. 2006;33(11):1374–80. PubMed PMID: 16932934.
7. Delbeke D, Coleman RE, Guiberteau MJ, Brown ML, Royal HD, Siegel BA, Townsend DW, Berland LL, Parker JA, Hubner K, Stabin MG, Zubal G, Kachelriess M, Cronin V, Holbrook S. Procedure guideline for tumor imaging with 18F-FDG PET/CT 1.0. J Nucl Med. 2006;47(5):885–95. Erratum in: J Nucl Med. 2006 Jun;47(6):903. PubMed PMID: 16644760.
8. Boellaard R, O'Doherty MJ, Weber WA, Mottaghy FM, Lonsdale MN, Stroobants SG, Oyen WJ, Kotzerke J, Hoekstra OS, Pruim J, Marsden PK, Tatsch K, Hoekstra CJ, Visser EP, Arends B, Verzijlbergen FJ, Zijlstra JM, Comans EF, Lammertsma AA, Paans AM, Willemsen AT, Beyer T, Bockisch A, Schaefer-Prokop C, Delbeke D, Baum RP, Chiti A, Krause BJ. FDG PET and PET/CT: EANM procedure

guidelines for tumour PET imaging: version 1.0. Eur J Nucl Med Mol Imaging. 2010;37(1):181–200. https://doi.org/10.1007/s00259-009-1297-4. PubMed PMID: 19915839; PubMed Central PMCID: PMC2791475.
9. Mantlik F, Hofmann M, Werner MK, Sauter A, Kupferschläger J, Schölkopf B, Pichler BJ, Beyer T. The effect of patient positioning aids on PET quantification in PET/MR imaging. Eur J Nucl Med Mol Imaging. 2011;38(5):920–9. https://doi.org/10.1007/s00259-010-1721-9. Epub 2011 Feb 10. PubMed PMID: 21308373.
10. Mutic S, Palta JR, Butker EK, Das IJ, Huq MS, Loo LN, Salter BJ, McCollough CH, Van Dyk J, AAPM Radiation Therapy Committee Task Group No. 66. Quality assurance for computed-tomography simulators and the computed-tomography-simulation process: report of the AAPM Radiation Therapy Committee Task Group No. 66. Med Phys. 2003;30(10):2762–92. PubMed PMID: 14596315.
11. Mawlawi O, Erasmus JJ, Munden RF, Pan T, Knight AE, Macapinlac HA, Podoloff DA, Chasen M. Quantifying the effect of IV contrast media on integrated PET/CT: clinical evaluation. AJR Am J Roentgenol. 2006;186(2):308–19. PubMed PMID: 16423932.
12. Paulus DH, Oehmigen M, Grüneisen J, Umutlu L, Quick HH. Whole-body hybrid imaging concept for the integration of PET/MR into radiation therapy treatment planning. Phys Med Biol. 2016;61(9):3504–20. https://doi.org/10.1088/0031-9155/61/9/3504. Epub 2016 Apr 7. PubMed PMID: 27055014.
13. Paulus DH, Thorwath D, Schmidt H, Quick HH. Towards integration of PET/MR hybrid imaging into radiation therapy treatment planning. Med Phys. 2014;41(7):072505. https://doi.org/10.1118/1.4881317. PubMed PMID: 24989408.
14. Hofmann M, Steinke F, Scheel V, Charpiat G, Farquhar J, Aschoff P, Brady M, Schölkopf B, Pichler BJ. MRI-based attenuation correction for PET/MRI: a novel approach combining pattern recognition and atlas registration. J Nucl Med. 2008;49(11):1875–83. https://doi.org/10.2967/jnumed.107.049353. Epub 2008 Oct 16. PubMed PMID: 18927326.
15. Minn H, Zasadny KR, Quint LE, Wahl RL. Lung cancer: reproducibility of quantitative measurements for evaluating 2-[F-18]-fluoro-2-deoxy-D-glucose uptake at PET. Radiology. 1995;196(1):167–73. PubMed PMID: 7784562.
16. EANM Physics Committee, Busemann Sokole E, Płachcínska A, Britten A, EANM Working Group on Nuclear Medicine Instrumentation Quality Control, Lyra Georgosopoulou M, Tindale W, Klett R. Routine quality control recommendations for nuclear medicine instrumentation. Eur J Nucl Med Mol Imaging. 2010;37(3):662–71. https://doi.org/10.1007/s00259-009-1347-y. PubMed PMID: 20130859.
17. Kumar R, Chauhan A, Zhuang H, Chandra P, Schnall M, Alavi A. Clinicopathologic factors associated with false negative FDG-PET in primary breast cancer. Breast Cancer Res Treat. 2006;98(3):267–74. Epub 2006 Mar 23. PubMed PMID: 16555126.
18. McCall KC, Barbee DL, Kissick MW, Jeraj R. PET imaging for the quantification of biologically heterogeneous tumours: measuring the effect of relative position on image-based quantification of dose-painting targets. Phys Med Biol. 2010;55(10):2789–806. https://doi.org/10.1088/0031-9155/55/10/001. Epub 2010 Apr 22. PubMed PMID: 20413832; PubMed Central PMCID: PMC2942022.
19. Jaskowiak CJ, Bianco JA, Perlman SB, Fine JP. Influence of reconstruction iterations on 18F-FDG PET/CT standardized uptake values. J Nucl Med. 2005;46(3):424–8. PubMed PMID: 15750154.
20. Keall PJ, Mageras GS, Balter JM, Emery RS, Forster KM, Jiang SB, Kapatoes JM, Low DA, Murphy MJ, Murray BR, Ramsey CR, Van Herk MB, Vedam SS, Wong JW, Yorke E. The management of respiratory motion in radiation oncology report of AAPM Task Group 76. Med Phys. 2006;33(10):3874–900. PubMed PMID: 17089851.
21. Van de Steene J, Linthout N, de Mey J, Vinh-Hung V, Claassens C, Noppen M, Bel A, Storme G. Definition of gross tumor volume in lung cancer: inter-observer variability. Radiother Oncol. 2002;62(1):37–49. PubMed PMID: 11830311.
22. Berson AM, Stein NF, Riegel AC, Destian S, Ng T, Tena LB, Mitnick RJ, Heiba S. Variability of gross tumor volume delineation in head-and-neck cancer using PET/CT fusion, Part II: the impact of a contouring protocol. Med Dosim. 2009;34(1):30–5. https://doi.org/10.1016/j.meddos.2007.08.003. Epub 2007 Sep 29. PubMed PMID: 19181253.
23. Schakel T, Hoogduin JM, Terhaard CH, Philippens ME. Diffusion weighted MRI in head-and-neck cancer: geometrical accuracy. Radiother Oncol. 2013;109(3):394–7. https://doi.org/10.1016/j.radonc.2013.10.004. Epub 2013 Oct 31. PubMed PMID: 24183864.
24. Wang D, Schultz CJ, Jursinic PA, Bialkowski M, Zhu XR, Brown WD, Rand SD, Michel MA, Campbell BH, Wong S, Li XA, Wilson JF. Initial experience of FDG-PET/CT guided IMRT of head-and-neck carcinoma. Int J Radiat Oncol Biol Phys. 2006;65(1):143–51. PubMed PMID: 16618577.
25. Shepherd T, Teras M, Beichel RR, Boellaard R, Bruynooghe M, Dicken V, Gooding MJ, Julyan PJ, Lee JA, Lefèvre S, Mix M, Naranjo V, Wu X, Zaidi H, Zeng Z, Minn H. Comparative study with new accuracy metrics for target volume contouring in PET image guided radiation therapy. IEEE Trans Med Imaging. 2012;31(11):2006–24. https://doi.org/10.1109/TMI.2012.2202322. Epub 2012 Jun 4. PubMed PMID: 22692898; PubMed Central PMCID: PMC5570440.
26. Xing L, Siebers J, Keall P. Computational challenges for image-guided radiation therapy: framework and current research. Semin Radiat Oncol. 2007;17(4):245–57. Review. PubMed PMID: 17903702.

27. Kaul D, Badakhshi H, Gevaert T, Pasemann D, Budach V, Tuleasca C, Gruen A, Prasad V, Levivier M, Kufeld M. Dosimetric comparison of different treatment modalities for stereotactic radiosurgery of meningioma. Acta Neurochir (Wien). 2015;157(4):559–63. https://doi.org/10.1007/s00701-014-2272-9. discussion 563–4. . Epub 2014 Nov 21. Erratum in: Acta Neurochir (Wien). 2015;157(4):565. PubMed PMID:25413163.
28. Gehler B, Paulsen F, Oksüz MO, Hauser TK, Eschmann SM, Bares R, Pfannenberg C, Bamberg M, Bartenstein P, Belka C, Ganswindt U. [68Ga]-DOTATOC-PET/CT for meningioma IMRT treatment planning. Radiat Oncol. 2009;4:56. https://doi.org/10.1186/1748-717X-4-56. PubMed PMID: 19922642; PubMed Central PMCID: PMC2785827.
29. Graf R, Nyuyki F, Steffen IG, Michel R, Fahdt D, Wust P, Brenner W, Budach V, Wurm R, Plotkin M. Contribution of 68Ga-DOTATOC PET/CT to target volume delineation of skull base meningiomas treated with stereotactic radiation therapy. Int J Radiat Oncol Biol Phys. 2013;85(1):68–73. https://doi.org/10.1016/j.ijrobp.2012.03.021. Epub 2012 May 9. PubMed PMID: 22575489.
30. Milker-Zabel S, Zabel-du Bois A, Henze M, Huber P, Schulz-Ertner D, Hoess A, Haberkorn U, Debus J. Improved target volume definition for fractionated stereotactic radiotherapy in patients with intracranial meningiomas by correlation of CT, MRI, and [68Ga]-DOTATOC-PET. Int J Radiat Oncol Biol Phys. 2006;65(1):222–7. Epub 2006 Feb 20. PubMed PMID: 16488553.
31. Rogers L, Barani I, Chamberlain M, Kaley TJ, McDermott M, Raizer J, Schiff D, Weber DC, Wen PY, Vogelbaum MA. Meningiomas: knowledge base, treatment outcomes, and uncertainties. A RANO review. J Neurosurg. 2015;122(1):4–23. https://doi.org/10.3171/2014.7.JNS131644. Review. PubMed PMID: 25343186; PubMed Central PMCID: PMC5062955.
32. Stade F, Dittmar JO, Jäkel O, Kratochwil C, Haberkorn U, Debus J, et al. Influence of 68Ga-DOTATOC on sparing of normal tissue for radiation therapy of skull base meningioma: differential impact of photon and proton radiotherapy. Radiat Oncol. 2018;13:58.
33. Thorwarth D, Henke G, Müller AC, Reimold M, Beyer T, Boss A, Kolb A, Pichler B, Pfannenberg C. Simultaneous 68Ga-DOTATOC-PET/MRI for IMRT treatment planning for meningioma: first experience. Int J Radiat Oncol Biol Phys. 2011;81(1):277–83. https://doi.org/10.1016/j.ijrobp.2010.10.078. Epub 2011 Feb 6. PubMed PMID:21300465.
34. Acker G, Kluge A, Lukas M, Conti A, Pasemann D, Meinert F, Anh Nguyen PT, Jelgersma C, Loebel F, Budach V, Vajkoczy P, Furth C, Baur ADJ, Senger C. Impact of 68Ga-DOTATOC PET/MRI on robotic radiosurgery treatment planning in meningioma patients: first experiences in a single institution. Neurosurg Focus. 2019;46(6):E9. https://doi.org/10.3171/2019.3.FOCUS1925. PubMed PMID: 31153151.
35. Nyuyki F, Plotkin M, Graf R, Michel R, Steffen I, Denecke T, Geworski L, Fahdt D, Brenner W, Wurm R. Potential impact of (68)Ga-DOTATOC PET/CT on stereotactic radiotherapy planning of meningiomas. Eur J Nucl Med Mol Imaging. 2010;37(2):310–8. https://doi.org/10.1007/s00259-009-1270-2. Epub 2009 Sep 18. PubMed PMID: 19763565.
36. Oborski MJ, Laymon CM, Qian Y, Lieberman FS, Nelson AD, Mountz JM. Challenges and approaches to quantitative therapy response assessment in glioblastoma multiforme using the novel apoptosis positron emission tomography tracer F-18 ML-10. Transl Oncol. 2014;7(1):111–9. https://doi.org/10.1593/tlo.13868.
37. Oborski MJ, Laymon CM, Lieberman FS, Drappatz J, Hamilton RL, Mountz JM. First use of 18F-labeled ML-10 PET to assess apoptosis change in a newly diagnosed glioblastoma multiforme patient before and early after therapy. Brain Behav. 2014;4(2):312–5. https://doi.org/10.1002/brb3.217.
38. Sun L, Zhou K, Wang W, Zhang X, Ju Z, Qu B, Zhang Z, Wang J, Ling Z, Yu X, Zhang J, Pan L. [18F] ML-10 imaging for assessment of apoptosis response of intracranial tumor early after radiosurgery by PET/CT. Contrast Media Mol Imaging. 2018;2018:9365174. https://doi.org/10.1155/2018/9365174. eCollection 2018. Erratum in: Contrast Media Mol Imaging. 2019;2019:4967404. PubMed PMID:29983648; PubMed Central PMCID: PMC6015719.
39. Suchorska B, Albert NL, Tonn JC. Usefulness of PET imaging to guide treatment options in gliomas. Curr Treat Options Neurol. 2016;18:4. https://doi.org/10.1007/s11940-015-0384-z.
40. Galldiks N, Stoffels G, Filss CP, et al. Role of O-(2-(18)F-fluoroethyl)-L-tyrosine PET for differentiation of local recurrent brain metastasis from radiation necrosis. J Nucl Med. 2012;53:1367–74. https://doi.org/10.2967/jnumed.112.103325.
41. Romagna A, Unterrainer M, Schmid-Tannwald C, Brendel M, Tonn JC, Nachbichler SB, Muacevic A, Bartenstein P, Kreth FW, Albert NL. Suspected recurrence of brain metastases after focused high dose radiotherapy: can [(18)F]FET- PET overcome diagnostic uncertainties? Radiat Oncol. 2016;11(1):139. PubMed PMID: 27769279; PubMed Central PMCID: PMC5073742.
42. Gwak HS, Youn SM, Chang U, Lee DH, Cheon GJ, Rhee CH, Kim K, Kim HJ. Usefulness of (18)F-fluorodeoxyglucose PET for radiosurgery planning and response monitoring in patients with recurrent spinal metastasis. Minim Invasive Neurosurg. 2006;49(3):127–34. PubMed PMID: 16921451.

Part IV

Radiobiology

12 Radiobiology of Radiosurgery and Hypofractionated Treatments

Antonio Pontoriero

12.1 Introduction

Stereotactic radiosurgery (SRS) has gained a major role in the treatment of brain tumors. This is based on its ability to precisely and accurately deliver a high dose of radiations to a target, thus effectively ablating all viable tumors while minimizing the dose and preventing damage in surrounding normal tissue [1]. While the high dose per fraction observed in single-fraction SRS may be quite effective in damaging vascularization and enhancing local control, the resulting impaired perfusion could limit the transport of antigens and immune cells, inhibiting the global immunomodulatory effect of radiation [2].

Indeed, interesting evidence is emerging showing that irradiation of tumors may also release antigens stimulating the immune system and leading to improved local control. Also, and perhaps more importantly, this may influence the appearance of new distant disease in the brain and body [3]. It has been suggested that a hypofractionated regimen could still generate antigens without impairing transport and that this treatment strategy would produce a more robust immune response [3, 4]. Such an approach might have an even greater impact when combined with one or more of the immunomodulating drugs that have entered and profoundly changed clinical practice, though much remains to be understood about this relationship. Recent studies on the radiobiological effect of single-fraction high-dose radiotherapy have shed light on the underlying mechanism of radiation damage [5].

The goal of radiotherapy is to achieve local control while minimizing normal tissue toxicity. In standard radiotherapy, dose is deposited in both normal tissue and tumor. Hence, it is important to keep the therapeutic ratio in mind. The therapeutic ratio is defined as in the equation (Fig. 12.1):

Probability of Tumor Cure / Probability of Complications

This is a key concept, because all treatment decisions are based on the therapeutic ratio [6]. When radiation interacts with matter, it causes the ionization of atoms. Ionization of atoms leads to the formation of free radicals that cause DNA damage. What differentiates radiation-induced DNA damage from other types of DNA damage is that in radiation-induced DNA damage, the lesions are clustered in one location, thereby making repair difficult. DNA damage can occur either via direct damage or indirect damage. Direct damage occurs when radiation directly ionizes DNA causing single- or double-stranded breaks which, if not repaired, can lead to cell

A. Pontoriero (✉)
Department of Radiation Oncology A.O.U. "G. Martino", Cyberknife Centre of Messina, Messina, Italy
e-mail: apontoriero@unime.it

© Springer Nature Switzerland AG 2020
A. Conti et al. (eds.), *CyberKnife NeuroRadiosurgery*, https://doi.org/10.1007/978-3-030-50668-1_12

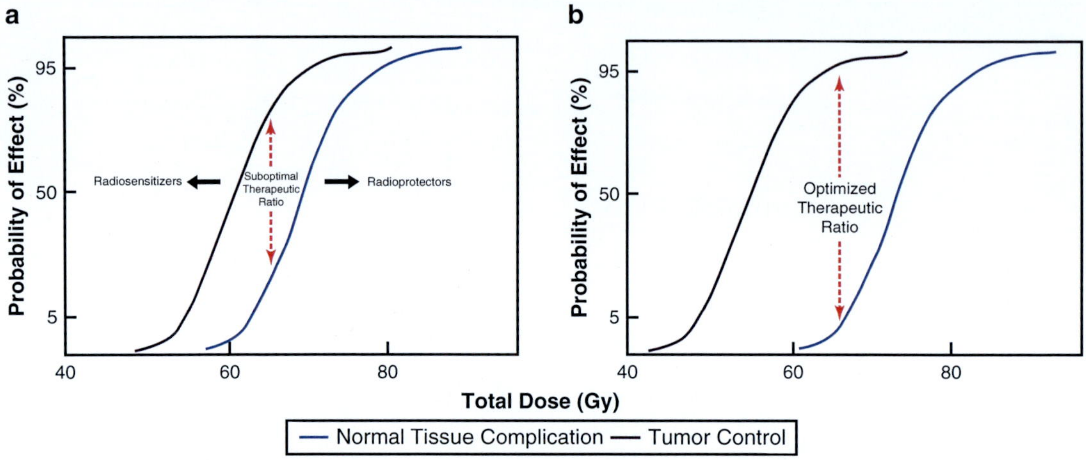

Fig. 12.1 The diagram represents the dose-dose relationship between total dose and tumor control, as well as total dose and normal tissue complications. The therapeutic window at a given dose is the separation between tumor control and normal tissue complication curves. (**a**) shows a suboptimal therapeutic relationship. Radioprotectors, radiosensitizers and advances in radiation transmission increase the separation between normal tissue multiplication and tumor control curves and thus optimize the therapeutic relationship as represented in (**b**). The optimal therapeutic ratio is obtained when the two curves are separated to the maximum [5]

death. Indirect DNA damage, which is more common, is mediated by the formation of hydroxy [7] radicals from water which then interact with DNA to produce DNA strand breaks.

Indirect damage is the predominant way by which DNA is damaged due to the radiation of photons or electrons ("sparsely ionizing radiation").

When alpha particles are used as a radiation source ("densely ionizing radiotherapy"), direct DNA damage predominates [8]. DNA damage can be classified as sublethal damage and potentially lethal damage. Under normal conditions sublethal damage is completely repaired within hours, unless repair mechanisms within cells are overwhelmed by the damage.

Potentially life-threatening damage can only be repaired in conditions that are not optimal for replication and therefore not readily available in vivo. When DNA repair mechanisms fail to repair DNA damage, chromosomal instability occurs and cells undergo post-mitotic cell death. While tumors mostly suffer post-mitotic cell death after irradiation, normal cells undergo apoptosis after irradiation [5].

Radiosurgery is an intriguing field, and our understanding of the molecular response to SRS is rapidly evolving. Research on radiosensitizers has so far been disappointing with very few, if any, radiosensitizers proven to be clinically useful. However, the development of animal models, the development of irradiators for simulating stereotactic body radiation therapy (SBRT) in small animals [9], and future in vitro studies will push the boundaries of our current understanding, thereby jumpstarting the development of targeted radiosensitizing agents [5].

A potential radiobiological disadvantage of SRS, or even any hypofractionated regimen, is the inability to exploit cell cycle redistribution. Single-fraction SRS may not result in improved cell kill compared to a prolonged conventional radiotherapy regimen because cells do not have time to redistribute into more radiosensitive phases (G2 and M) of the cell cycle.

A second potential radiobiological disadvantage of single-fraction SRS is inability to exploit reoxygenation in hypoxic tumor cells, which is another potential advantage of conventional radiotherapy. However, a potential radiobiological advantage of SRS is its ability to limit tumor cell repopulation, a disadvantage seen in conventional radiotherapy [10]. Another potential radiobiological advantage of single-fraction SRS is the increase in antitumor immunity after irradiation of the tumor, commonly known as the abscopal effect [10].

Hypofractionated radiotherapy may provide a different pathway of biological effects either used alone or combined with chemoradiotherapy. A potential advantage of hypofractionated radiation therapy that makes it an attractive approach for the management of advanced cancers is the reduction of time and costs of treatment and the reduction of the burden of frequent and numerous radiotherapy sessions [11]. Hypofractionated radiation therapy can be approached in two different ways. The first is to consider the α/β ratio and biologically effective dose (BED) where the "classical" concepts of repair, reassortment, reoxygenation, and repopulation (4Rs) are applicable [11]. This is a categorical approach for hypofractionated radiotherapy that uses 3–6 Gy dose fractions. The second approach is the hypofractionation schedule that uses over 8 Gy doses/fraction in radiotherapy, in which the biological changes different than the "classical" 4Rs are felt to be applicable, generally known as high-dose hypofractionation radiation therapy (HDHRT) [11]. There are data to suggest that the use of HDHRT radiation is effective as an alternative means of dose escalation with a conventional fractionation treatment schedule.

12.2 "Classical" 4Rs and SRS-HDHRT

There are a number of processes that will be affected by dose size and fractionation that could be exploited, including changes in the "4Rs" (repair, repopulation, redistribution, and reoxygenation), consequences of endothelial damage (which could worsen hypoxia) or tumor shrinkage (which could lessen hypoxia), and impact of the high dose on factors secreted by the tumor [12].

12.2.1 Reoxygenation

The reoxygenation of hypoxic cells in irradiated tumors would occur when the blood flow and therefore the oxygen supplied to the tumor cells increases or oxygen consumption is reduced.

Given the massive vascular destruction in tumors after high-dose irradiation, it is highly unlikely that reoxygenation of hypoxic cells would occur in the tumors within 2–3 days after receiving high-dose hypofractionated SBRT and SRS. However, it is probable that oxygen consumption would drastically diminish after massive death of tumor cells and thus the surviving hypoxic cells may be reoxygenated. The changes in oxygenation status in tumors following high-dose hypofractionated irradiation remain to be elucidated.

12.2.2 Repair

The half-time for the completion of sublethal radiation damage repair in mammalian cells has been reported to be about 30 min [13].

Therefore, in treating tumors with SBRT or SRS, which takes a considerably long irradiation time, repair of sublethal damage may occur during the protracted irradiation. How the deterioration of intratumor environment due to vascular damage affects the repair of sublethal radiation damages after high-dose irradiation still needs to be investigated.

12.2.3 Repopulation

Depletion of cell population by injury, including ionizing radiation, evokes repopulation of cells in both tumors and normal tissues. The time of outset of the compensatory repopulation would vary depending on tissue type and radiation dose. In fractionated radiotherapy, repopulation of tumor cells occurs 2–3 weeks after initiation of radiotherapy. It is conceivable that the repopulation of tumor cells may begin earlier in SBRT and ablative SRS than in fractional radiotherapy.

12.2.4 Redistribution

In general, irradiation with moderate doses slows down the cell cycle progression through the G1 and S phases and arrests the cell in the G2 phase in a dose-dependent manner. Fractions of the cells arrested in the G2 phase may successfully complete mitosis and progress into the G1 phase or die during mitosis.

After irradiation of various cell lines with extremely high doses, i.e., 20 Gy in a single exposure, it was discovered that the progression of the cell cycle was markedly delayed and many cells died in the phases of the cycle in which they were irradiated although some of the irradiated cells slowly progressed to the G2 phase and died [14].

12.3 Radiobiologic Rationale of Radiosurgery

The relationship between radiation dose and tumor cell survival may be represented by the linear quadratic model, at least below 10 Gy per fraction [8]. The probability of cell survival after a single dose of radiation is a function of absorbed dose and is represented by cell survival curves as represented in Fig. 12.2.

The shape of the cell survival curve does not differ much between cell lines and has two characteristic regions: low-dose shoulder region and the high-dose steep region [15]. A simple model for cell death assumes that double-stranded DNA breaks are sufficient to cause cell death. Double-stranded breaks can be achieved by a single electron ("single-hit aberrations") or by two separate electrons ("exchange-type aberrations"). Single-hit aberrations are more likely to result from "heavy radiation" such as protons, whereas exchange-type aberrations are more likely to occur from "light radiation" such as electrons and photons. Such a model can be represented by the linear-quadratic formula:

$$S = e^{-\alpha D - \beta D^2}$$

where S represents the cell survival fraction, D represents the dose, α is the number of log kills from the linear portion of the cell survival curve (single-hit aberrations), and β is the number of log kills from the quadratic portion of the cell survival curve (exchange-type aberrations). An interpretation of this model is that the shoulder region of the cell survival curve represents sublethal damage and lethality from accumulation of several sublethal lesions. The α/β ratio is a measure of the relative contributions of single-hit and exchange-type aberrations and represents the dose at which these two components are equal [6]. Tumors that are radioresistant to conventional small-dose fractionation have a lower α/β, whereas more radiosensitive tumors have a higher α/β [16, 17].

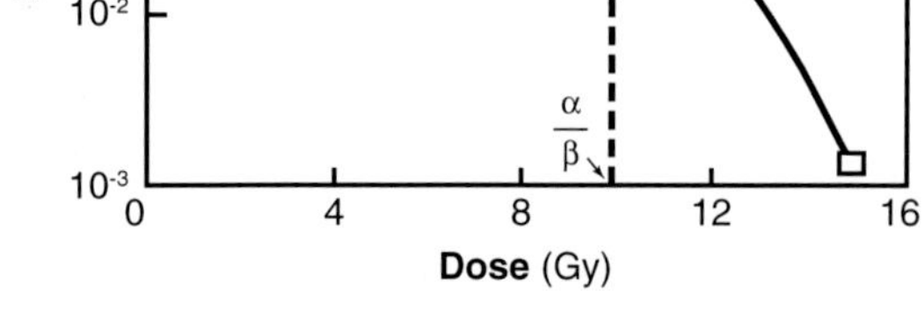

Fig. 12.2 The graph shows the survival of the cell log as a function of the dose (Gy). Referring to the text for a description of the *a*/*b* ratio. The "shoulder" region of the cell survival curve indicates the radiation doses that cause sublethal DNA damage. Fractional radiotherapy in doses within the "shoulder" region will allow normal tissues to repair DNA damage. Since tumor cells cannot repair DNA damage due to faulty repair mechanisms, tumor cells remain sensitive to the next dose of radiation [5]

To understand this, the β component can be considered the cell's repairing ability. Since exchange-type aberrations require damage from two separate electrons, each break in the DNA strand has a greater chance of repair due to the availability of the model strand.

Since radioresistant tumors have better DNA repair mechanisms, their β component is larger, leading to lower α/β ratio. This principle also applies to normal tissues, which have robust repair mechanisms.

The α/β ratio is based on preclinical and clinical data and is 2 for tissue in the CNS, 3 for late-responding tissues, and 10 for early responding tissues. The α/β ratio is also necessary when calculating the LQ equation to determine the dose equivalent to a conventional radiotherapy regimen. These ratios can then be used to help to determine the dose for tumor control while minimizing normal tissue toxicity. In general, malignant tumors such as brain metastases and malignant brain tumors have higher α/β ratios, estimated to be closer to 10 and representative of early responding tissues, while slow-growing benign brain tumors such as pituitary adenomas, arteriovenous malformations (AVMs), and benign meningiomas have a lower α/β ratio, estimated to be closer to 3 and representative of late responding tissues [18, 19]. Regardless of the uncertainties of the true α/β ratio of all tumors, especially in the brain, the overall goal of SRS is to provide highly conformal treatment with radiation to the tumor while sparing normal CNS tissue surrounding the target volume.

In the linear-quadratic model, a plot of surviving cell fraction (SCF) versus radiation dose shows that the log of the SCF is initially linearly proportional to the dose (D, units Gy) with a slope of $-\alpha$ (i.e., SCF = exp[$-\alpha D$]). As the dose increases, SCF decreases even more rapidly, and at moderate doses, SCF depends on dose and dose squared (i.e., SCF = exp[$-\alpha D - \beta D^2$]). Tissue response to radiation is often characterized by the α/β ratio, which tends to be on the order of 2–3 Gy for brain tissue and 10 Gy for many rapidly proliferating tumors. Of course, the response to radiation is also influenced by many other factors, including the microenvironment (e.g., oxygen content) and the capacity of cells to repair, repopulate, and redistribute in the cell cycle [20, 21].

Using the linear-quadratic model, one can calculate a biologically effective dose (BED) for a particular α/β ratio (units Gy), total dose (D), and dose/fraction (d, Gy):

$$\mathrm{BED}_{\alpha/\beta} = \mathrm{D}\left[1 + d/(\alpha/\beta)\right].$$

Thus, the BED for a low α/β tissue will increase much more rapidly with increasing dose per fraction than the BED for a high α/β tissue. Consequently, the difference in α/β ratio between the tumor and normal tissue by fractionating the dose could be potentially exploited, improving the therapeutic ratio. Representative isoeffect plots are presented in Fig. 12.3.

The linear-quadratic (LQ) model has been used to calculate the effects of ionization radiation to normal and neoplastic cells, to calculate isoeffect doses between different therapeutic regimens, and to describe tumor cell kill (through one or two tracks resulting in chromosome breaks) with these five principles in mind. Although these principles are often applied to conventional radiotherapy, the main difference between conventional radiotherapy and SRS is the size of the dose delivered and the target volume.

The radiobiology and application of the LQ model to SRS continue to be a matter of investi-

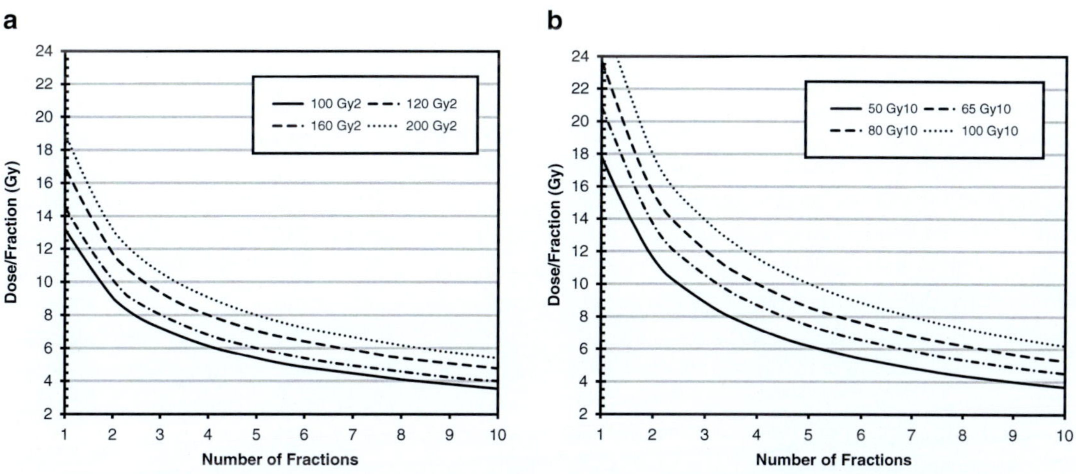

Fig. 12.3 Biologically effective dose (BED) isoeffect plots for dose/fraction and number of fractions administered for (**a**) α/β = 2 Gy and (**b**) α/β = 10 Gy calculated using the linear-quadratic (LQ) model [8]

gation and debate because clinical results have validated applications of the LQ model to radiation doses within the dose range of 1–5 Gy per fraction. Using this dose range, the LQ model has been used to approximate in vitro clonogenic survival. Single doses >5 Gy, which are commonly used in SRS, are considered by some to affect the validity of the LQ model [3, 15].

The use of the LQ model is theorized to underestimate tumor control at the high doses commonly used in SRS and does not reflect other mechanisms involved in killing cancer cells.

In addition to DNA strand breakage mechanisms and chromosomal aberrations by conventional radiation therapy, SRS with doses >10 Gy per fraction is hypothesized to cause vascular damage resulting in reduced blood perfusion and leading to indirect death of cancer cells [3]. Others have proposed the use of alternate models such as the lethal-potentially lethal (LPL) model that may be used for large fractions/acute doses. However, the LPL model is limited by its general applicability to clinical data. Modification of the LQ model, otherwise known as the modified LQ (MLQ) model, introduces a parameter characterized not only by in vitro cell survival data of human tumor cell lines but also by in vivo animal isoeffect curves, which results in closer fitting of isoeffect data than the original LQ model. This has resulted in better approximation of the radiobiological effects of high single doses of irradiation, as used in SRS. The use of this model is not only consistent with the LPL model but also retains the generalizable characteristics of the LQ model [22, 23].

The shape of the dose-response curve above 10 Gy is controversial [3, 15, 24]. Some argue that the linear-quadratic model provides an adequate representation of the dose-response relationship at high doses and that observed clinical outcomes are entirely consistent with the predictions of this model [25–27]. Others assert that radiobiological mechanisms, such as profound vascular damage [28, 29] and antigen expression, different from classic DNA damage, are evoked above a threshold dose of 8–12 Gy and that the high levels of tumor control observed in radiosurgery reflect this "new radiobiology" and enhanced dose-response [27, 30–32] (see Fig. 12.4).

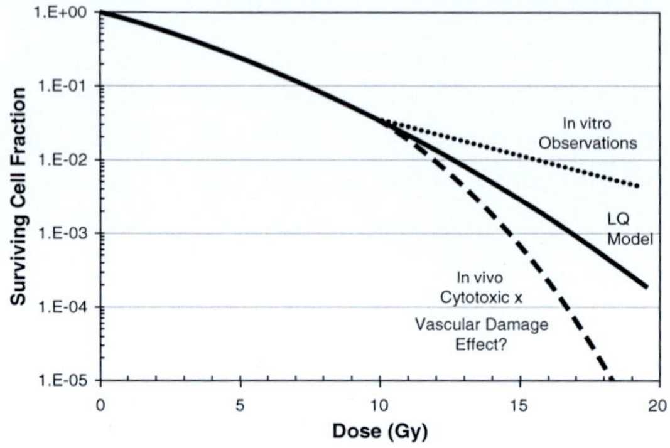

Fig. 12.4 Speculative surviving cell fraction (SCF) versus single-dose irradiation response curves for the linear-quadratic (LQ) model, in vitro cell cultures and in vivo tumors with SCF determined by the product of direct cell kill and indirect vascular damage [1]

If there is actually a gradual change in response above a certain threshold and there is no fundamental reason why this threshold should be the same for tumor compared to normal tissue, it would seem appropriate to design treatment plans and select such dose regimens wherein the dose in the tumor always exceeds this threshold. On the contrary, the plan should be designed in such a way that the dose in the surrounding normal tissue rarely exceeds the threshold. In any case, a better understanding of the dose-response curves in vivo and of the underlying radiobiological mechanisms for tumors and normal tissues (which probably differ) would not only help with the design of rational plans but could open new ways to increase the therapeutic ratio.

The previous dose-response problem does not include the other critical elements in the assessment of the volume toxicity of normal irradiated tissues.

As discussed by Marks et al. in the QUANTEC series of papers [33, 34], normal tissue complications increase as the volume of tissue receiving some minimum dose increases, and this phenomenon is observed in a wide variety of tissues. For example, the volume of brain tissue receiving 12 Gy or more in radiosurgery appears to be correlated with the risk of radionecrosis, particularly when this volume exceeds 10–15 mL. Note, however, that this limitation appears overly restrictive, as it appears that virtually every single-fraction radiosurgery plan would exceed this limit when large lesions were treated to accepted doses [35].

While the linear-quadratic model can be used to convert doses into a uniform basis, the most relevant method for doing so remains unclear [36]. Recognizing these limitations, the fundamental principles of SRS highly conformal treatment plans, the minimum margin around the target, the accurate and precise target localization, the minimization of position deviation, and the solid quality assurance should aid in minimizing the irradiated volume and should always be employed.

The linear-quadratic (LQ) model assumes that DNA double-stranded break is responsible for the radiation-induced clonogenic cell death and that hypoxic cells are fully reoxygenated during the interval of fractionated irradiation [12]. As shown in Fig. 12.2, the LQ survival curve bends downward due to the quadratic component in the formula, and thus the LQ model has been suggested to overestimate cell death with the increase in radiation dose. Interestingly, despite the inherent problem with the LQ model, some investigators reported that the LQ model fits certain clinical outcomes of SBRT and SRS and thus asserted that direct cell death due to DNA damage alone is sufficient to account for the high clinical efficacy of SBRT and SRS [23, 26]. As shown in Fig. 12.5, the radiation survival curve of tumor cells in vivo also bends downward as the radiation dose is increased above approximately 10 Gy due to the secondary cell death caused by vascular damage. Therefore, it is conceivable that in certain clinical situation, the calculated cell death by the LQ formula may incidentally not overestimate but approximate the total cell death by SBRT and SRS, which encompasses not only direct but also indirect cell deaths.

Interestingly, the LQ model may even underestimate the outcome of SBRT and SRS in situations where indirect cell death is extensive, as shown by curve "d" in Fig. 12.6 [30]. This implies

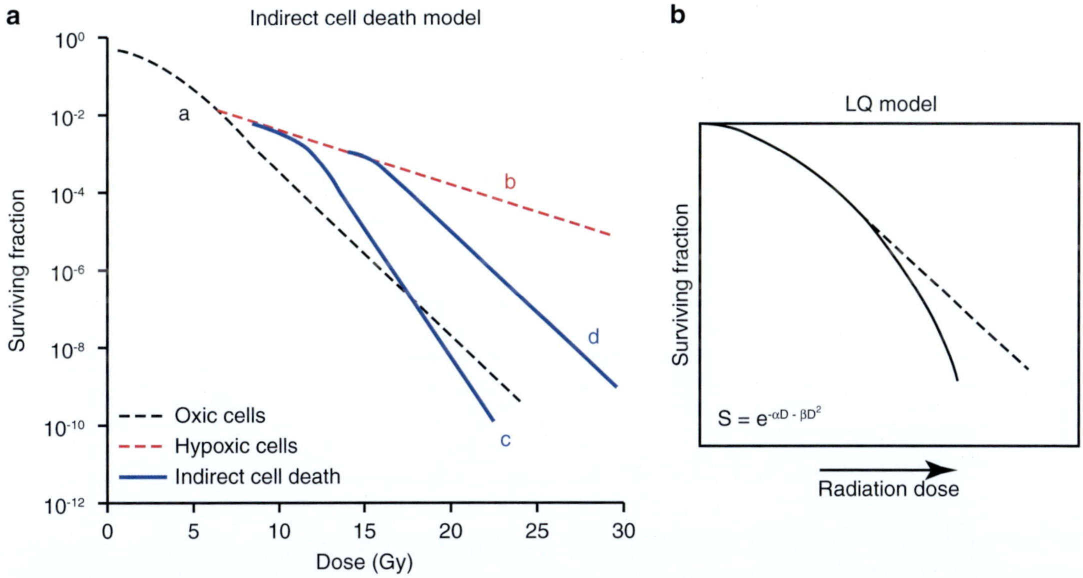

Fig. 12.5 (**a**) Hypothetical radiation survival curve of tumor cells in vivo, assuming about 10% of the tumor cells are radiobiologically hypoxic. The 'a' corresponds the radiation-induced death of hypoxic cells and 'b' indicates the death of hypoxic cells assuming that radiation-induced cell death is due only to direct damage in DNA/chromosomes. The 'c' and 'd' show indirect and additional cell death due to vascular damages at high radiation doses. (**b**) The dotted line indicates decline in cell survival when radiation-induced cell death is linearly related to radiation dose. Solid line is the linear-quadratic (LQ) survival curve which bends downward at high radiation dose indicating that the LQ model overestimates cell death at high radiation doses [12]

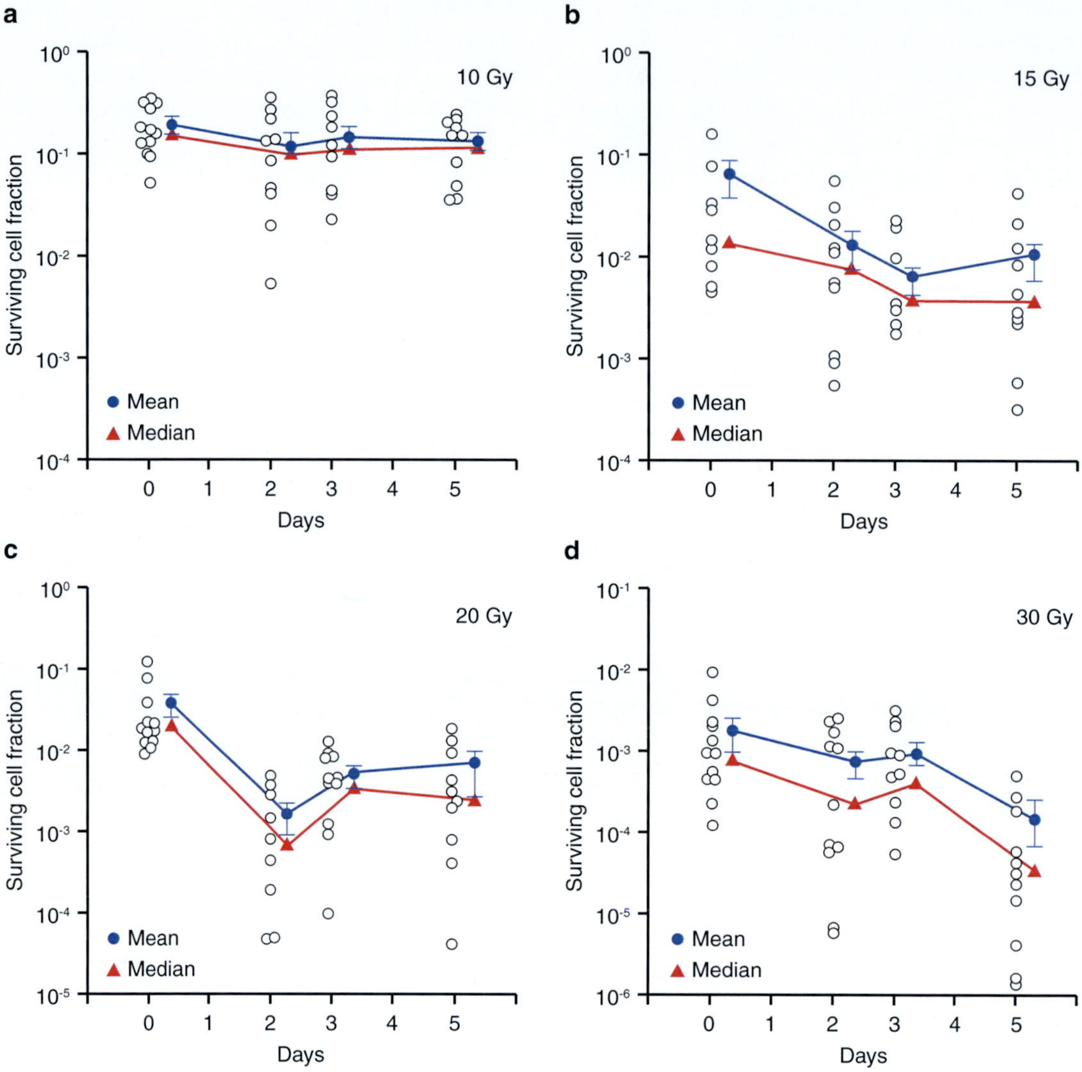

Fig. 12.6 Survival of clonogenic cells (**a–d**) in FSaII tumors grown s.c. in the legs of C3H mice at 0–5 days after 10–30 Gy irradiation in a single dose. Cell survival was determined by the in vitro excision assay method in vivo and cell fractions surviving in irradiated tumors were obtained by normalizing cell survival in irradiated tumors to that of non-irradiated control tumors. Open circles indicate cell survival in each individual tumor. The mean + a standard error (blue) and a median (red) are shown [12]

that the LQ model works for SBRT and SRS in certain clinical situations, not because tumor cells are killed only through a direct effect of radiation, but rather because significant fractions of tumor cells are indirectly killed through secondary mechanisms in addition to direct cell death.

As a result of direct or indirect DNA damage, oxygen forms peroxides with the damaged DNA molecule, stabilizing the damage and preventing repair mechanisms from working effectively. Thus, oxygen serves as a radiosensitizer. Due to the poor vascular supply of most tumors, tumors with large volumes often have a hypoxic core that is more radioresistant due to lack of oxygen sensitization. When the outer part of the tumor is eliminated due to radiotherapy, the inner part receives a higher blood flow leading to sensitization of the remaining tumor cells and making them more radiosensitive.

To identify and select the optimal dose regimen that maximizes tumor kill and minimizes normal tissue damage, time should also be considered [1]. Decreasing the time between fractions and the total length of the treatment course should decrease tumor cell repopulation and thus enhance the efficacy of the regimen [1]. In particular, this should be more beneficial in the more rapidly growing malignant tumors (e.g., metastases, high-grade gliomas) than in the indolent benign tumors (e.g., World Health Organization grade I schwannomas and meningiomas).

However, a too short interval between treatments could lead to less complete repair of normal tissues and more pronounced delayed effects.

While a minimum 8-h interval between treatments was generally considered adequate to allow for repair of normal tissues, the QUANTEC analysis of daily brain treatments compared to those twice a day questioned it. In hypofractionated SRS, treatment may be administered once daily on consecutive days or as infrequently as twice a week.

In this case, the issue still revolves around the optimum timing that permits adequate repair of normal tissues while minimizing the adverse impact of tumor cell repopulation [1].

12.4 Radiobiology of Radiosurgery

High doses administered in a single fraction of SRS and SBRT challenged conventional radiobiological wisdom, and therefore its clinical utility was discussed [5]. Opponents of SRS and SBRT have suggested that single-fraction radiotherapy has not exploited the full potential of reoxygenation and restocking of interactions within cancer cells [5]. Furthermore, they suggested that the highly conformal plans generated for SRS and SBRT treatment would not adequately cover microscopic disease and very high doses (>40 Gy) as predicted by the linear-quadratic model [37]. Interestingly, they found that in the high-dose group (23–24 Gy), tumor histology was not a statistically significant predictor of local control ($p = 0.90$).

These results are in contrast with classical radiobiology because classical radiobiology would predict tumor histology as a significant predictor of local control. The rationale for using higher doses for the fraction in fractional radiotherapy is to overcome the "shoulder" of the cell survival curve for radioresistant tumors such as renal cell carcinoma, melanoma, and sarcoma [5].

However, recent data show that in high-fraction single-dose radiotherapy, it is not only DNA damage that causes cell death, but it is also inflammation of the endothelial cells [38] and apoptosis through the sphingomyelin pathway [39] that causes the subsequent microvascular dysfunction which are the triggering factors for the death of cancer cells [40].

DNA damage continues to be an important insult to the tumor cell, and the interaction between microvascular dysfunction and DNA damage eventually leads to tumor death (two-target model) [41].

Upon exposure to high-dose radiation, endothelial cells in the tumor vasculature undergo a wave of apoptosis which stops by 15–20 h post-irradiation, after which these cells are replaced by fibroblasts.

Apoptosis in endothelial cells is mediated by the sphingomyelin pathway. Following the exposure to high-dose radiation, acid sphingomyelinase (ASMase) is translocated to the plasma membrane of endothelial cells where it plays a role in generating ceramide from sphingomyelin [42].

Deng et al. showed that ceramide biogenesis is required for radiation-induced apoptosis [43]. Ceramide release leads to the activation of the apoptotic protein BAX [44]. BAX is part of the Bcl-2 family of proteins and is an important pro-apoptotic regulator. Activation of BAX leads to the release of mitochondrial cytochrome c which indicates the cell's commitment to apoptosis through the intrinsic pathway [45].

Endothelial apoptosis peaks within 6 h after radiation and causes microvascular dysfunction and hence acutely disrupts tumor perfusion [29]. This model of acute hypoxia and reperfusion injury has been well described in cardiology literature in the context of ischemic heart disease. It is important to note that the above-described molecular responses do not take place until the dose of radiation exceeded the 10 Gy threshold [29]. Thus, it is not surprising that Lu et al. [46] found that distinct signaling pathways are activated in response to high or low doses of radiation. At doses <17 Gy, ceramide is generated via the above-described ASMase-mediated pathway. However, at doses ≥17 Gy, ceramide is also produced by ceramide synthase, and this pathway is modulated by ataxia telangiectasia-mutated (ATM) kinase.

Under normal conditions ATM kinase represses ceramide synthase. Ch'ang et al. showed that in ATM knockout mice, stem cell radiosensitivity increased 3.7-fold without sensitizing the microvascular response [41].

Endothelial cell apoptosis mediated by the sphingomyelin pathway acts synchronously with DNA damage in cancer stem cells to lead to tumor death. Figure 12.7 details this interaction.

With a single high dose of radiation, tumor stem cells undergo DNA damage. This DNA usually repairs itself and stem cells are able to survive, leading to local failure. Rotolo et al. showed that inactivation of DNA damage repair pathways signals the generation of ceramide via the ceramide synthase pathway even at low radiation doses [47]. This suggests that unrepaired double-stranded DNA breaks can cause the activation of the ceramide synthase pathway leading to apoptosis [47]. However, both these models do not consider the emerging understanding of molecular and cellular effects of high-dose single-fraction radiotherapy which will be important in developing models that accurately predict the effect of high-dose radiation on normal tissue as well as tumor cells. Recent radiobiological evi-

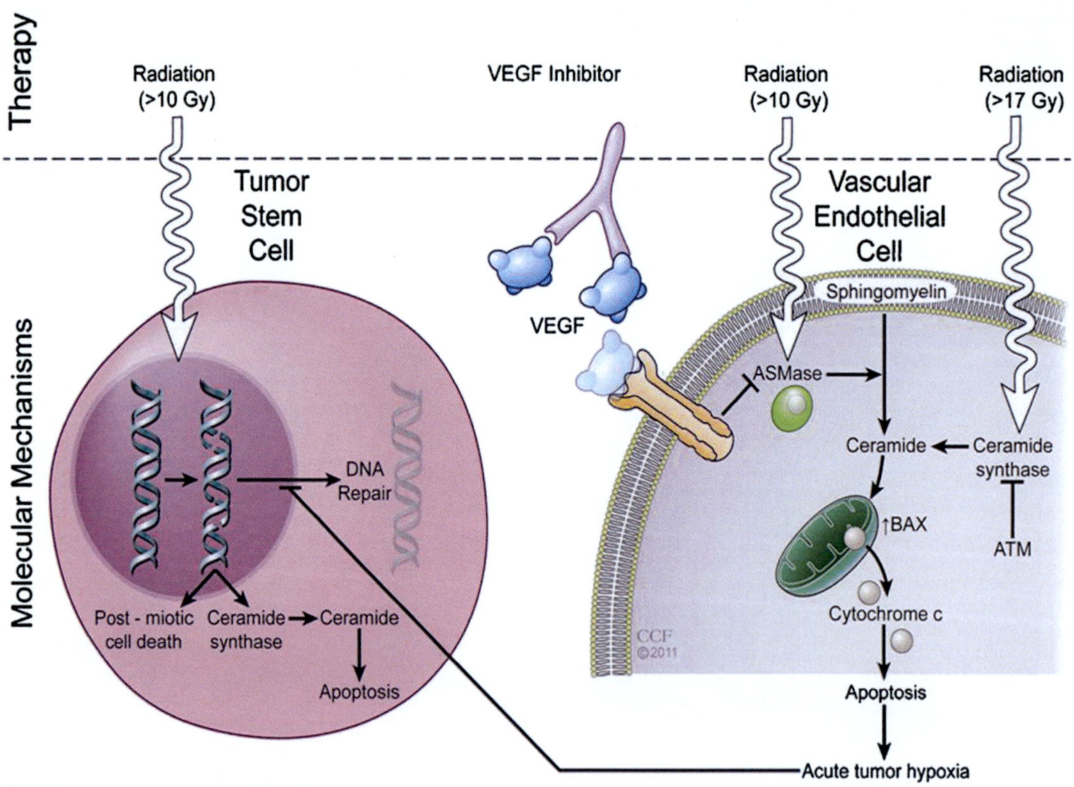

Fig. 12.7 Diagram representing the current understanding of the molecular response to radiosurgery. *VEGF* vascular endothelial growth factor, *ASMase* acid sphingomyelinase, *ATM* ataxia-telangiectasia mutated kinase [5]

Table 12.1 Comparison of 3D conformal radiotherapy (non-IMRT, non-IGRT), stereotactic radiosurgery, and stereotactic body radiotherapy

	Low dose per fraction	High dose per fraction	
	3D conformal radiotherapy	Stereotactic radiosurgery[a]	Stereotactic body radiotherapy[b]
Treatment delivery	Multiple beams Less conformal to target Simpler treatment planning	Multiple non-coplanar beams Highly conformal to target Sophisticated treatment planning	
Dosing and fractionation	Multiple fractions (1.5–5 Gy per fraction)	Typically single fraction (10–35 Gy per fraction)	1–5 fractions (7–34 Gy per fraction)
Treatment delivery duration	Up to 7 weeks	1 day	Up to 5 days
Mechanism of cell death	DNA damage → post-mitotic cell death	Endothelial apoptosis → ischemia and reperfusion injury DNA damage	
Molecular mechanisms	1. Direct DNA ionization 2. Indirect DNA ionization via hydroxyl free radicals	1. ASMase-dependent generation of ceramide → activation of pro-apoptotic BAX → endothelial cell apoptosis 2. DNA damage	

[a]For stereotactic brain and spine radiotherapy, SRS can be delivered in 1–5 fractions
[b]SBRT usually refers to extracranial and stereotactic radiotherapy
Modified by Balagamwala E. H. et al. [5]

dence indicates that the mechanism of radiation damage from SRS and SBRT is different from conventionally fractionated radiation. Table 12.1 summarizes the differences between 3D conformal radiotherapy, SRS, and SBRT.

12.5 Radiobiologic Rationale of Hypofractionated Radiation Therapy

Biologically, new mechanistic insights suggest that hypofractionated radiation therapy (HDHRT) may cause four unique effects that can be further exploited for sensitization. HDHRT can cause non-targeted pharmacodynamic effects (such as intratumoral bystander as well as abscopal effects) mediated by TNF-α, TRAIL, PAR-4, and ceramide [48–50]; it can robustly induce tumor endothelial death at doses above 8–11 Gy [29]; it can increase host immune recognition of radiation-induced enhanced antigen presentation, in such a way that a single fraction may incite an immune response that enhances the effects of radiation [51]; and may cause a better response of those tumors that are heterogeneous with different cell populations whose clonal radiosensitivity differs significantly [52].

The effects of interaction between HDHRT and hypoxia would depend in part on the initial hypoxic fraction, the dose size used, and fractionation, as reoxygenation could occur [11]. Another interesting consideration could be the use of conventional radiation therapy following single high dose or high dose in combination with chemotherapeutic drugs to improve the response of tumors to treatment.

Strong biological data exist that suggest that a large dose of radiation induction preceding conventional fractional radiotherapy causes significantly greater tumor regression [53, 54].

12.6 Radiobiology of Brain Metastases

The radiobiology of this SRS response is not completely understood. It is postulated that the higher biologically equivalent dose (BED) alone may not fully explain the higher local control rates observed with SRS. Additional biological factors and/or cellular pathways are believed to be involved in the pathophysiology of the SRS response [55].

Seminal laboratory data by Fuks et al. has shown activation of the acid sphingomyelinase pathway at fraction sizes above 8 Gy, which in turn serves to activate tumor endothelial cell apoptosis, disrupt tumor vasculature, and increase tumor cell death [44]. This sequence of events is depicted diagrammatically in Fig. 12.8. The secretory form of the enzyme, acid sphingomyelinase, is found in 20-fold higher concentrations in endothelial cells than in any other cell of the body. High doses of radiation ≥8 Gy cause cholesterol-enriched rafts in the cell membrane where it hydrolyzes sphingomyelin to generate pro-apoptotic ceramide. Ceramide, in turn, acts as a second messenger, triggering a mitochondrial-mediated apoptotic response by stimulating the Bax pathway of pro-apoptotic signals with resultant cytochrome c release from the mitochondria.

In addition, ceramide can create membrane rafts to alter extracellular as well as intracellular signaling pathways.

Experimental data from Garcia-Barros et al. has demonstrated ceramide-mediated apoptosis in tumor endothelial cells 1–6 h following receipt of single large radiation doses of 15–20 Gy [29, 57]. In separate experiments involving ASMase and Bax knockout mice, this wave of apoptosis was not observed.

The apparent dose threshold of radiation therapy to induce the ASMase pathway was 8–10 Gy, with a dose-response relationship seen up to 20–25 Gy.

In addition to the acid sphingomyelinase pathway, the immune system, which has long been recognized to play a key role in tumor surveillance and suppression, is an integral component of the SRS response.

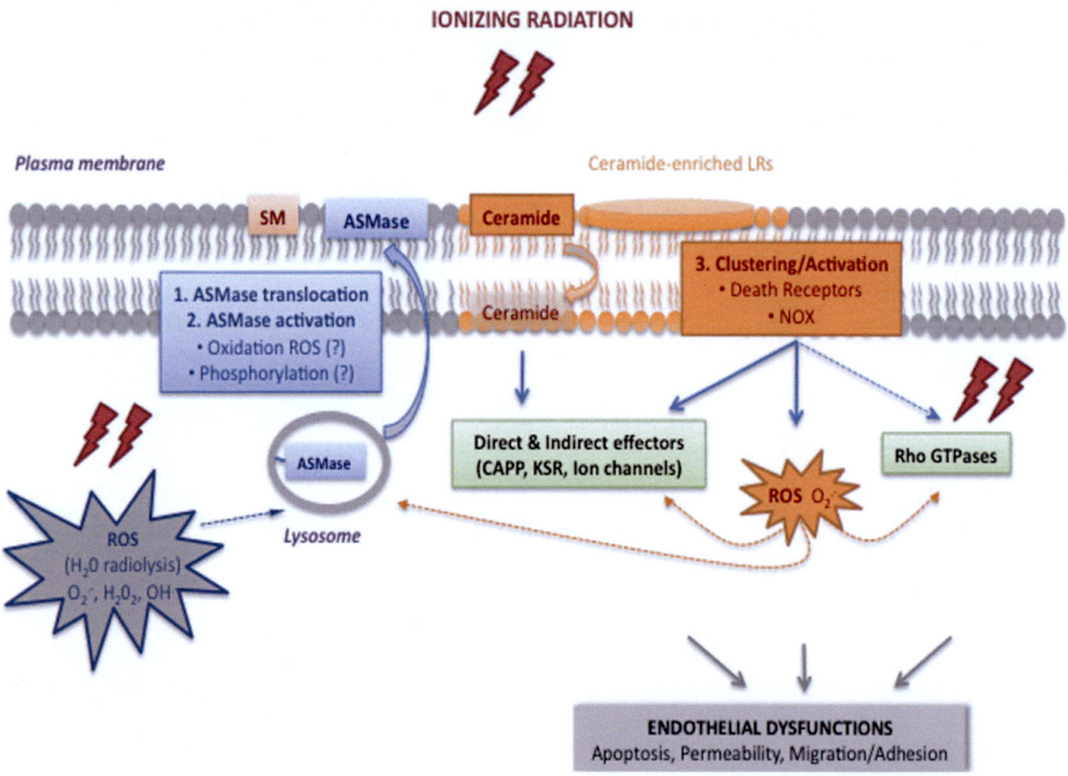

Fig. 12.8 Cell membrane signal pathways in endothelial cells induced by stereotactic ablative radiotherapy. *ASMase* acid sphingomyelinase, *CAPP* ceramide-activated protein phosphatase, *KSR* kinase suppressor of Ras, *LRs* lipid rafts, *NOX* nicotinamide adenine dinucleotide phosphate oxidase, *ROS* reactive oxygen species, *SM* sphingomyelin (adapted from [55, 56])

It is thought that the extreme hypofractionation that characterizes SRS results in the release of tumor-specific antigens, leading to the priming of CD8+ T-cells and a subsequent immune-mediated response, further enhancing tumor cell death (Fig. 12.9). Support for this theory comes from a study examining the effect of ablative radiotherapeutic doses in mouse melanoma models [51].

In this study, mice with B16 melanomas were subject to extreme hypofractionation, receiving a dose of 20 Gy in a single fraction.

Histopathologic examination of the tumor microenvironment and lymphoid tissues 1–2 weeks post-treatment demonstrated tumor regression as well as an influx of T-cells. By contrast, no significant decrease in tumor volume was noted when the experiment was repeated in athymic mice lacking T-cells. Separate experiments utilizing CD 8 depletion strategies in wild-type mice with B16 tumors have documented a diminished response to ablative radiation. Taken in combination, these studies suggest that CD8+ T-cells play a critical role

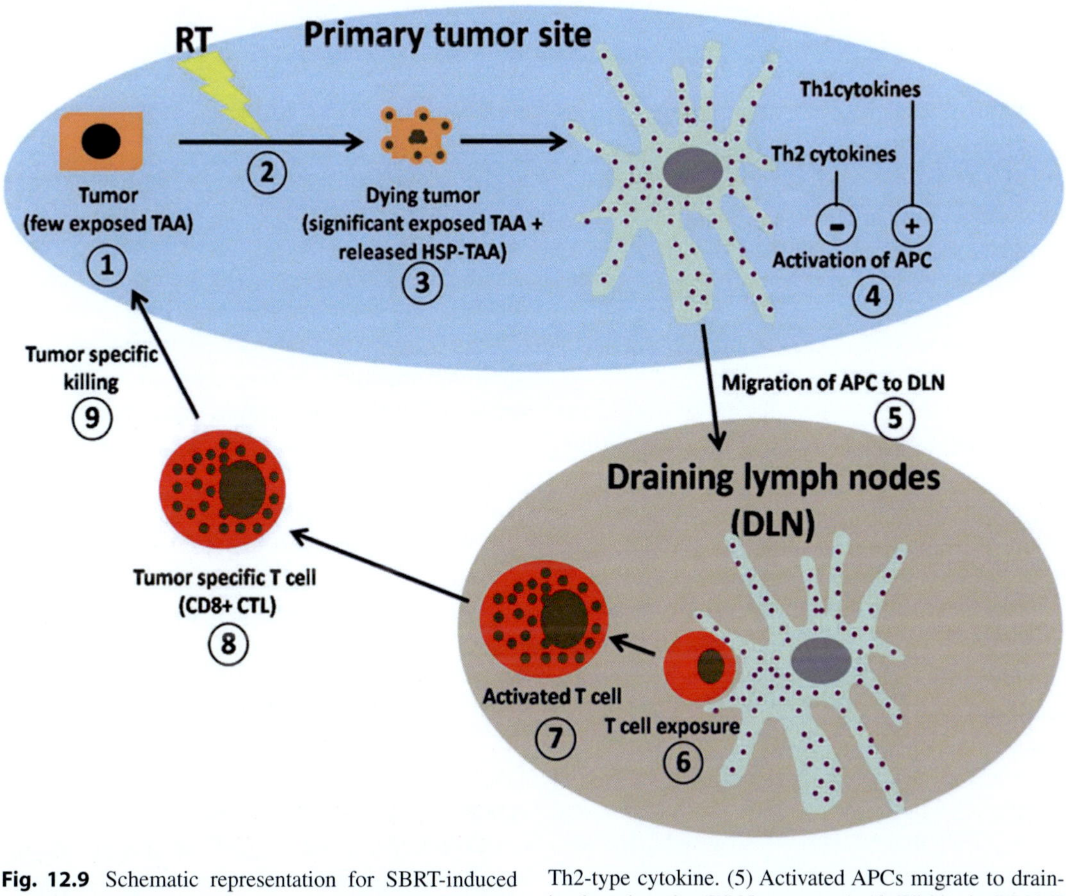

Fig. 12.9 Schematic representation for SBRT-induced anti-tumor immune regulation. (1) Within the primary tumor microenvironment (blue area) untreated tumors express limited exposed tumorassociated antigens (TAAs). (2) Exposure to RT induces the (3) dying tumors to express significantly more TAAs on their surface and to release DAMPS, (4) which are both taken up by antigen-presenting cells (APC) resulting in their activation. Activation of APC is enhanced (+) by the presence of Th1-type cytokines and suppressed (−) by the presence of Th2-type cytokine. (5) Activated APCs migrate to draining lymph nodes (DLN; gray area). (6) Within the DLN, T cell exposure to APC is achieved by direct contact with activated APCs. (7) Activated T cells increase in size and granularity. (8) The activated T cells migrate from the DLN as tumor-specific T cells (CD8+ CTL) into the tumor microenvironment. (9) Within the tumor microenvironment CD8+ CTL perform tumor-specific killing (adapted from [55, 58])

in radiation-induced antitumor immune response following stereotactic ablative RT.

12.7 Radiobiology of Brain Arteriovenous Malformations (BAVM)

SRS represents the only biological therapy for BAVM that avoids the need for invasive treatment. Obliteration is the hallmark of successful radiosurgical treatment of BAVM and is defined by "complete absence of pathological vessels forming the AVM nidus, disappearance or normalization of veins draining the AVM, appearance of normal circulatory kinetics, and absence of visible arteriovenous shunt" [59, 60]. Despite the widespread use of SRS in the management of BAVMs, the exact mechanism of radiosurgical obliteration remains poorly understood [61]. Available data regarding the biology of radiation-induced vascular obliteration result from observations in BAVM tissue resected after radiosurgical treatment [62, 63] and in irradiated arteries in animal models [64–67].

Observations of BAVM tissue [62, 63] suggested that damaged endothelial cells shrink, detach from neighboring endothelial cells and the basement membrane, and allow for platelet infiltration with fibrin and hyaline deposition.

As these endothelial cells slough off over time, the inhibition of the proliferation of smooth muscle cells is lost, and the migration of smooth muscle cells into the subintimal layer causes the deposition of collagen which thickens the subintima and the adventitia, progressively narrowing the lumen and finally occluding it.

Study of irradiated arteries in animal models has suggested that the radiosensitivity of BAVMs originates in endothelial cells [61, 64–67]. Failed mitosis of irradiated endothelial cells damaged by direct interactions with irradiating electrons and indirect free radical by-products results in eventual apoptosis [28] and initiation of radiation-induced arteriopathy.

As such, the latency period of BAVM obliteration after SRS is believed to depend on the turnover rate of endothelial cells, which typically varies from the order of a couple of months to a couple of years, since the onset of arteriopathy occurs only once endothelial cells attempt mitosis [28, 61, 65]. Currently, the major disadvantage of SRS is the latency period before a BAVM might successfully become obliterated, during which time the patient remains unprotected against risk of new hemorrhage [68]. Widespread variation in patient response to SRS treatment of BAVMs may be the result of varying degrees of endothelial cell turnover, which is known to be abnormal in BAVMs [69].

Important progress has recently been made in animal models of SRS-induced arteriopathy [61] providing the basis for future studies in transgenic mice as to the role of genetic variation in modulating response to SRS. Future studies in patients with BAVMs will include proteomic analyses and gene expression profiling of peripheral blood cell populations, which may reflect indirect interactions from circulating through diseased tissue [70] as well as direct interactions in the pathophysiology of BAVM [71–75].

These peripheral blood cells could provide important biofeedback regarding progression to successful BAVM obliteration after initiation of SRS treatment. Such biomarkers of the BAVM response to SRS could not only guide treatment planning but could identify new targets for adjuvant therapies designed to promote obliteration after SRS [59].

12.8 High-Dose Radiation Induces Factors Leading to Bystander and Abscopal Effects

Brooks et al. reported the first observation of radiation-induced non-targeted effects in a hamster model [76]. Although the evidence for these effects has accumulated over time, the exact mechanisms by which they cause tumor regression away from the irradiation site remain somewhat speculative.

A few important mechanistic categories have been proposed to explain the abscopal effects on the basis of studies involving different malignancies: the immune system, cytokines, and the pseudo-abscopal effect [77]. Cell-cell communication appears to play an important role in mediating the bystander effect, and there may also be contributions from transfers of soluble mediators generated in the irradiated medium.

The presence of gap junctions is not essential. Transfer of radiation-conditioned medium (RCM) from confluent cell culture is more effective, a phenomenon that is termed as "indirect radiation effects" [78–81]. Irradiated cells may release clastogenic factors into a serum that will induce chromosomal damage when transferred to cultured cells from unirradiated donors [82–84].

For example, in a rat study, clastogenic activity remained in the circulating plasma of irradiated animals for the duration of the 10-week study and was not abrogated by diluting with non-irradiated serum. Serum irradiated in vitro was not clastogenic, suggesting that these factors were released from the irradiated cells [85]. Although evidence for the presence of these factors has been accumulating over the past decades, their exact nature as well as the mechanisms by which they cause the distant bystander effects (more of an abscopal effect) has proven elusive.

One of these mechanisms could be through the first radiation-induced genes and cytokine induction. In fact, TNF-α and TRAIL are directly involved in apoptosis and are induced by ionizing radiation [86–90].

There is a demonstrated correlation of therapeutic efficacy following GRID therapy, or spatially fractionated radiation therapy (SFGRT), with TNF-α induction in the serum obtained from these patients as well as ceramide production [49, 50]. For SFGRT, the "bystander effect" is within the GRID-irradiated tumor volume that falls directly under shielded regions (low-dose regions) of the GRID. Bystander factors such as TNF-α shown by Sathishkumar et al. [50] and Shareef et al. [48], TRAIL shown by Shareef et al. [48], and ceramide shown by Sathishkumar et al. [49] are induced in cells that are under the open field of the high-dose GRID areas and are hypothesized to be responsible for initiating the cell death cascade both in the epithelial and endothelial compartments of the tumor microenvironment.

Recent reports have demonstrated the presence of radiation-induced signal transduction leading to significant DNA damage and cellular stress [68, 91]. In addition, the baystender effect within the GRID-irradiated tumor, Peters et al. [92] reported that there is a robust "abscopal effect" in distant tumors or metastatic lesions that are not irradiated or treated and has been reported clinically with the use of large doses [93].

In this regard, recently, by using SFGRT, both the bystander and the abscopal effects have been shown in mice carrying adenocarcinoma pulmonary xenograft A549 contralateral tumors [94]. The maximum abscopal effect was observed in non-irradiated right tumor when mice were exposed to 15 Gy SFGRT followed by five fractions of 2 Gy to the left tumor suggesting that the abscopal effect can be amplified by the sequential combination of SFGRT with conventional fractionation.

Lee et al. [51] reported that reduction of tumor burden after ablative radiation depends largely on T-cell responses as it dramatically increases T-cell priming in draining lymphoid tissues, leading to reduction/eradication of the primary tumor or distant metastasis in a CD8(+) T-cell-dependent fashion.

Interestingly, this study observed that immune responses initiated with ablative radiation and tumor reduction are abrogated by conventional fractional RT or adjuvant chemotherapy (if administered after a week of a single ablative dose) but greatly amplified by local immunotherapy.

However, in SFGRT settings, significant enhanced response was demonstrated when the high-dose radiation was followed by fractionated 2 Gy fractions (given after 24 h), implying that spatial fractionation of radiation delivery might activate immune factors that can synergize with the conventional fractionated radiation. These results strongly argue for more detailed investigations to elucidate the role of immune factors in radiation therapy [11].

High-dose radiation induces damage to the endothelium. The engagement of the vascular component in the tumor response to radiation therapy has been a topic of interest in recent literature. However, in addition to a release of cytokines, impaired blood vessel formation and induction of endothelial cell death in tumors not exposed to radiation have been demonstrated to play a role in the abscopal effect [95]. Endothelial cells generate 20-fold more of a unique form of acid sphingomyelinase (ASMase), termed Secretory ASMase, than any other cell type in the body. Secretory ASMase activation is required for ionizing radiation to kill endothelium [96], as endothelia in lung, gut, and brain. They are thoroughly resistant to radiation-induced apoptotic death in the absence of ASMase. Garcia-Barros et al. [29] have postulated that high-dose radiation-induced damage (15 Gy) to the endothelial cells could convert potentially lethal damage (PLD) in tumor cells and cancer stem cells to lethal damage resulting in tumor cell death.

Animal studies have shown that radiation at doses higher than 10 Gy induces endothelial apoptosis by activation of acid sphingomyelinase (ASMase) and ceramide generation [29, 94]; these effects are not observed with conventional radiation doses. The findings of Garcia-Barros et al. [29] suggest that high-dose radiation-induced tumor regression can be entirely dependent on tumor endothelium apoptosis since these effects were abolished in ASMase knockout animals implanted with functional ASMase

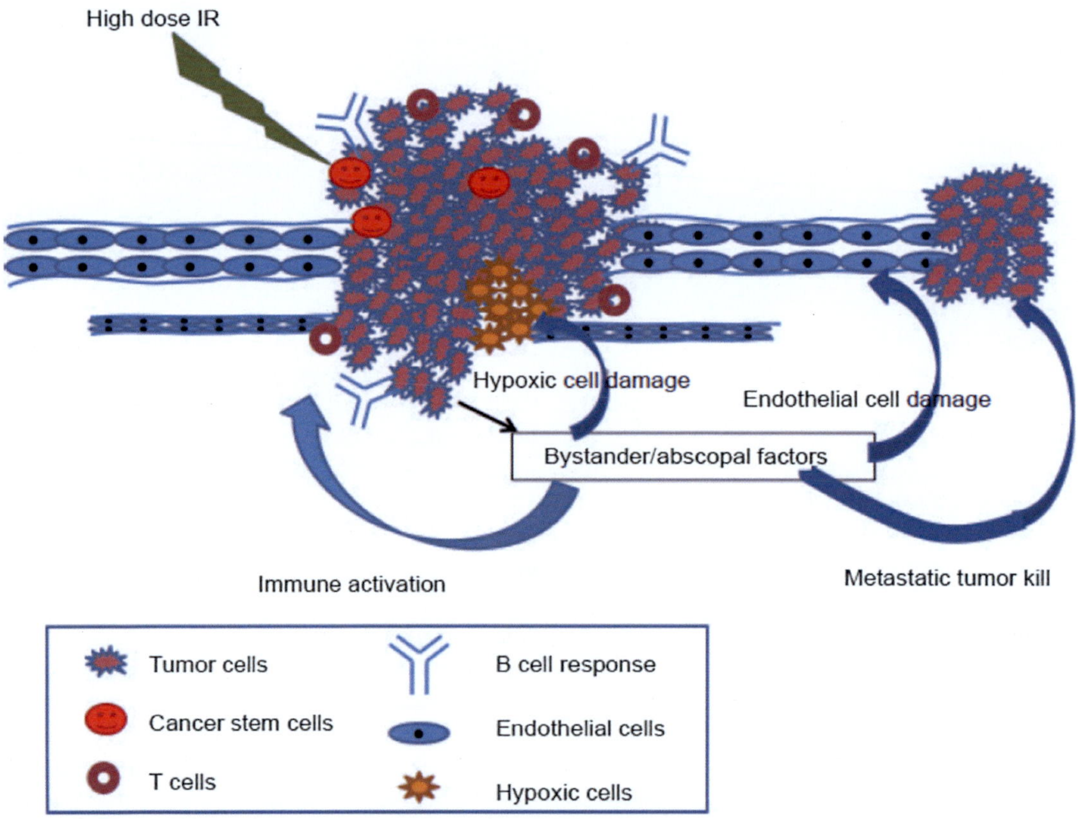

Fig. 12.10 Impact of high-dose ablative RT on the components of the tumor micro-environment. High-dose ablative RT administered in lattice (2 vertices) to the tumor induces bystander/abscopal factors, endothelial cell death coupled with immune activation. The underlying radiobiological mechanisms for an improved outcome obtained by high dose hypofractionated radiation therapy could be multifactorial. Differential effects on tumor endothelium and cancer stem cells could be responsible for this enhanced response. In addition, complex immunological pathways could be linked to high dose radiation-induced mechanisms. All these pathways could be affected by the bystander/abscopal factors released from the tumor following spatially fractionated radiotherapy. An animation of these events can be found at URL: http://youtu.be/KvQ8z91J6A8 [11]

MCA/129 fibrosarcomas and B16F1 melanomas and restored upon bone marrow transplantation of ASMase functional stem cells. Furthermore, elevated sphingomyelinase activity and ceramide concentration in the serum of patients undergoing high-dose spatially fractionated radiation treatment were observed [49]. The latter hypothesis implies that simultaneous strategies (such as hypoxic cytotoxin) aimed directly at hypoxic cells might improve the therapeutic ratio of SABR and allow clinicians to deal with a larger fraction in the patient population. Fractional doses in hypofractionation schemes vary significantly in clinical practice, from 3 Gy/fraction to 20 Gy/fraction.

There are numerous processes that will be performed based on the dose size and fractionation that could be exploited, including changes in the "4Rs" (repair, restocking, redistribution, and reoxygenation), a consequence of endothelial damage (which could worsen hypoxia) or tumor reduction (which could reduce hypoxia), and impact of the high dose on the factors secreted by the tumor [11]. It remains to be determined how to make the most of the effect of the high dose but also not to damage normal tissue. This could include partial treatment of cancer to high dose using a variety of techniques such as the high-dose LATTICE approach. This might have positive effects on damage to the endothelial compartment and/or on immune activation [11].

Another important aspect that is not discussed in detail could be the differential effect of hypofractionation on cancer stem cells. The success of hypofractionated radiation therapy depends on its ability to deliver a markedly higher dose to the target volume without damaging the surrounding normal tissue.

The underlying radiobiological mechanisms for improving the results obtained with high-dose hypofractionated radiation therapy could be multifactorial, including endothelial and cancer stem cell killing, overcoming hypoxic radioresistance, activating complex immunological pathways, and bystander/abscopal tumor effects, with consequent treatment outcome (Fig. 12.10).

12.9 Conclusions

Although an increasing number of cancer patients have been treated with SBRT and SRS in recent years, the biological mechanisms of these new modalities have not been clear [12]. A simple calculation based on the radiobiological principles for the conventional multi-fractionated radiotherapy clearly suggests that tumor cell death caused by DNA damages by direct effect of radiation alone cannot account for the high efficacy of SBRT and SRS [12]. Evidence now indicates that SBRT and SRS with doses higher than about 10 Gy per fraction induce severe vascular damages in tumors, which then cause secondary and additional tumor cell death [12]. The ensuing degradation of tumor cells would then release massive tumor-specific antigens, thereby elevating antitumor immune response leading to the suppression of recurrence of tumors and metastases [12]. The role of the 4Rs and the LQ model is limited in SBRT and SRS. In conclusion, ablative hypofractionation schemes are effective in certain solid tumors that may take advantage of new aspects of radiation biology by involving certain components of tumor microenvironment such as effects on vasculature as well as immunologic modulation [11].

References

1. Kirkpatrick JP, Soltys SG, Lo SS, Beal K, Shrieve DC, Brown PD. The radiosurgery fractionation quandary: single fraction or hypofractionation? Neuro-Oncology. 2017;19(Suppl_2):ii38–49.
2. Demaria S, Golden EB, Formenti SC. Role of local radiation therapy in cancer immunotherapy. JAMA Oncol. 2015;1(9):1325–32.
3. Park HJ, Griffin RJ, Hui S, Levitt SH, Song CW. Radiation-induced vascular damage in tumors: implications of vascular damage in ablative hypofractionated radiotherapy (SBRT and SRS). Radiat Res. 2012;177(3):311–27.
4. Demaria S, Formenti SC. Radiation as an immunological adjuvant: current evidence on dose and fractionation. Front Oncol. 2012;2:153.
5. Balagamwala E, Chao ABST, Suh JH. Principles of radiobiology of stereotactic radiosurgery and Clinical applications in the central nervous system. Technol Cancer Res Treat. 2012;11:3–13.

6. MS LLGM, MD JET. Clinical radiation oncology. Churchill Livingstone; 2006.
7. Mehta MP, Tsao MN, Whelan TJ, Morris DE, Hayman JA, Flickinger JC, Mills M, Rogers CL, Souhami L. The American Society for Therapeutic Radiology and Oncology (ASTRO) evidence-based review of the role of radiosurgery for brain metastases. Int J Radiat Oncol Biol Phys. 2005;63:37–46.
8. Hall E. Radiobiology for the radiologist. Philadelphia: Lippincott Williams & Wilkins; 2006.
9. Cho J, Kodym R, Seliounine S, Richardson JA, Solberg TD, Story MD. High dose-per-fraction irradiation of limited lung volumes using an image-guided, highly focused irradiator: simulating stereotactic body radiotherapy regimens in a small-animal model. Int J Radiat Oncol Biol Phys. 2010;77:895–902.
10. Kondziolka D, Shin SM, Brunswick A, Kim I, Silverman JS. The biology of radiosurgery and its clinical applications for brain tumors. Neuro-Oncology. 2015;17(1):29–44.
11. Prasanna A, Ahmed MM, Mohiuddin M, Norman Coleman C. Exploiting sensitization windows of opportunity in hyper and hypofractionated radiation therapy. J Thorac Dis. 2014;6(4):287–302.
12. Kim M-S, Kim W, In Hwan Park BA, Hee Jong Kim MS, Eunjin Lee MS, Jung J-H, Cho LC, Song CW. Radiobiological mechanisms of stereotactic body radiation therapy and stereotactic radiation surgery. Radiat Oncol J. 2015;33(4):265–75.
13. Fowler JF, Welsh JS, Howard SP. Loss of biological effect in prolonged fraction delivery. Int J Radiat Oncol Biol Phys. 2004;59:242–9.
14. Jeong JH, Park IW, Kang MA, Kim MS, Song CW. Effect of high dose hypofractionated irradiation (SBRT/SRS) on cell cycle progression [abstract]. In: 61th Radiation research society annual meeting, Weston, FL, 19–22 Sep 2015; 2015. Abstract no. PS1-24.
15. Hanin LG, Zaider M. Cell-survival probability at large doses: an alternative to the linear-quadratic model. Phys Med Biol. 2010;55(16):4687–702.
16. Thames HD, Withers HR, Peters LJ, Fletcher GH. Changes in early and late radiation responses with altered dose fractionation: implications for dose-survival relationships. Int J Radiat Oncol Biol Phys. 1982;8:219–26.
17. Withers HR. Biologic basis for altered fractionation schemes. Cancer. 1985;55:2086–95.
18. Santacroce A, Kamp MA, Budach W, et al. Radiobiology of radiosurgery for the central nervous system. Biomed Res Int. 2013;2013:362761.
19. Hall EJ, Brenner DJ. The radiobiology of radiosurgery: rationale for different treatment regimes for AVMs and malignancies. Int J Radiat Oncol Biol Phys. 1993;25(2):381–5.
20. Steel GG, McMillan TJ, Peacock JH. The 5Rs of radiobiology. Int J Radiat Biol. 1989;56(6):1045–8.
21. Withers HR. The four R's of radiotherapy. New York: Academic; 1975.
22. Guerrero M, Li XA. Extending the linear-quadratic model for large fraction doses pertinent to stereotactic radiotherapy. Phys Med Biol. 2004;49(20):4825–35.
23. Brenner DJ. The linear-quadratic model is an appropriate methodology for determining isoeffective doses at large doses per fraction. Semin Radiat Oncol. 2008;18(4):234–9.
24. Park C, Papiez L, Zhang S, Story M, Timmerman RD. Universal survival curve and single fraction equivalent dose: useful tools in understanding potency of ablative radiotherapy. J Radiat Oncol Biol Phys. 2008;70(3):847–52.
25. Brown JM, Carlson DJ, Brenner DJ. Dose escalation, not "new biology," can account for the efficacy of stereotactic body radiation therapy with non-small cell lung cancer. In reply to Rao et al. Int J Radiat Oncol Biol Phys. 2014;89(3):693–4.
26. Brown JM, Carlson DJ, Brenner DJ. The tumor radiobiology of SRS and SBRT: are more than the 5 Rs involved? Int J Radiat Oncol Biol Phys. 2014;88(2):254–62.
27. Kirkpatrick JP, Brenner DJ, Orton CG. Point/counterpoint. The linear-quadratic model is inappropriate to model high dose per fraction effects in radiosurgery. Med Phys. 2009;36(8):3381–4.
28. Fuks Z, Kolesnick R. Engaging the vascular component of the tumor response. Cancer Cell. 2005;8(2):89–91.
29. Garcia-Barros M, Paris F, Cordon-Cardo C, et al. Tumor response to radiotherapy regulated by endothelial cell apoptosis. Science. 2003;300(5622):1155–9.
30. Kirkpatrick JP, Meyer JJ, Marks LB. The linear-quadratic model is inappropriate to model high dose per fraction effects in radiosurgery. Semin Radiat Oncol. 2008;18(4):240–3.
31. Song CW, Lee YJ, Griffin RJ, et al. Indirect tumor cell death after high dose hypofractionated irradiation: implications for stereotactic body radiation therapy and stereotactic radiation surgery. Int J Radiat Oncol Biol Phys. 2015;93(1):166–72.
32. Sperduto PW, Song CW, Kirkpatrick JP, Glatstein E. A hypothesis: indirect cell death in the radiosurgery era. Int J Radiat Oncol Biol Phys. 2015;91(1):11–3.
33. Bentzen SM, Constine LS, Deasy JO, et al. Quantitative analyses of normal tissue effects in the clinic (QUANTEC): an introduction to the scientific issues. Int J Radiat Oncol Biol Phys. 2010;76(Suppl 3):S3–9.
34. Vogelbaum MA, Angelov L, Lee SY, Li L, Barnett GH, Suh JH. Local control of brain metastases by stereotactic radiosurgery in relation to dose to the tumor margin. J Neurosurg. 2006;104(6):907–12.
35. Kirkpatrick JP, Marks LB, Mayo CS, Lawrence YR, Bhandare N, Ryu S. Estimating normal tissue toxicity in radiosurgery of the CNS: application and limitations of QUANTEC. J Radiosurg SBRT. 2011;1:95–102.
36. Blonigen BJ, Steinmetz RD, Levin L, Lamba MA, Warnick RE, Breneman JC. Irradiated volume as a predictor of brain radionecrosis after linear accelera-

tor stereotactic radiosurgery. Int J Radiat Oncol Biol Phys. 2010;77(4):996–1001.
37. Rosenthal DI, Glatstein E. We've got a treatment, but What's the 15. Disease? Or a brief history of Hypofractionation and its relationship to stereotactic. Radiosurgery. Oncologist. 1996;1:1–7.
38. Sharp CD, Jawahar A, Warren AC, Elrod JW, Nanda A, Alexander JS. Gamma knife irradiation increases cerebral endothelial expression of intercellular adhesion molecule 1 and E-selectin. Neurosurgery. 2003;53:154–60. discussion 160–161.
39. Peña LA, Fuks Z, Kolesnick R. Stress-induced apoptosis and the sphingomyelin pathway. Biochem Pharmacol. 1997;53:615–21.
40. Paris F, Fuks Z, Kang A, Capodieci P, Juan G, Ehleiter D, Haimovitz-Friedman A, Cordon-Cardo C, Kolesnick R. Endothelial apoptosis as the primary lesion initiating intestinal radiation damage in mice. Science. 2001;293:293–7.
41. Ch'ang H-J, Maj JG, Paris F, Xing HR, Zhang J, Truman J-P, Cardon-Cardo C, Haimovitz-Friedman A, Kolesnick R, Fuks Z. ATM regulates target switching to escalating doses of radiation in the intestines. Nat Med. 2005;11:484–90.
42. Gulbins E, Kolesnick R. Raft ceramide in molecular medicine. Oncogene. 2003;22:7070–7.
43. Deng X, Yin X, Allan R, Lu DD, Maurer CW, Haimovitz-Friedman A, Fuks Z, Shaham S, Kolesnick R. Ceramide biogenesis is required for radiation-induced apoptosis in the germ line of C. elegans. Science. 2008;322:110–5.
44. Kolesnick R, Fuks Z. Radiation and ceramide-induced apoptosis. Oncogene. 2003;22:5897–906.
45. Danial NN, Korsmeyer SJ. Cell death: critical control points. Cell. 2004;116:205–19.
46. Lu T-P, Lai L-C, Lin B-I, Chen L-H, Hsiao T-H, Liber HL, Cook JA, Mitchell JB, Tsai M-H, Chuang EY. Distinct signaling pathways after higher or lower doses of radiation in three closely related human lymphoblast cell lines. Int J Radiat Oncol Biol Phys. 2010;76:212–9.
47. Rotolo JA, Mesicek J, Maj J, Truman J-P, Haimovitz-Friedman A, Kolesnick R, Fuks Z. Regulation of ceramide synthase-mediated crypt epithelium apoptosis by DNA damage repair enzymes. Cancer Res. 2010;70:957–67.
48. Shareef MM, Cui N, Burikhanov R, et al. Role of tumor necrosis factor alpha and TRAIL in high-dose radiation-induced bystander signaling in lung adenocarcinoma. Cancer Res. 2007;67:11811–20.
49. Sathishkumar S, Boyanovsky B, Karakashian AA, et al. Elevated sphingomyelinase activity and ceramide concentration in serum of patients undergoing high dose spatially fractionated radiation treatment: implications for endothelial apoptosis. Cancer Biol Ther. 2005;4:979–86.
50. Sathishkumar S, Dey S, Meigooni AS, et al. The impact of TNF-alpha induction on therapeutic efficacy following high dose spatially fractionated (GRID) radiation. Technol Cancer Res Treat. 2002;1:141–7.
51. Lee Y, Auh SL, Wang Y, et al. Therapeutic effects of ablative radiation on local tumor require CD8+ T cells: changing strategies for cancer treatment. Blood. 2009;114:589–95.
52. Bao S, Wu Q, McLendon RE, et al. Glioma stem cells promote radioresistance by preferential activation of the DNA damage response. Nature. 2006;444:756–60.
53. Schenken LL, Poulakos L, Hagemann RF. Responses of an experimental solid tumour to irradiation: a comparison of modes of fractionation. Br J Cancer. 1975;31:228–36.
54. Sakamoto K, Sakka M. The effect of bleomycin and its combined effect with radiation on murine squamous carcinoma treated in vivo. Br J Cancer. 1974;30:463–8.
55. Thiagarajan A, Yamada Y. Radiobiology and radiotherapy of brain metastases. Clin Exp Metastasis. 2017;34(6–7):411–9.
56. Corre I, Guillonneau M, Paris F. Membrane signaling induced by high doses of ionizing radiation in the endothelial compartment. Relevance in radiation toxicity. Int J Mol Sci. 2013;14:22678–96.
57. Garcia-Barros M, Lacorazza D, Petrie H, Haimovitz-Friedman A, Cardon-Cardo C, Nimer S, Fuks Z, Kolesnick R. Host acid sphingomyelinase regulates microvascular function not tumor immunity. Cancer Res. 2004;64:8285–91.
58. Kaur P, Asea A. Radiation-induced effects and the immune system in cancer. Front Oncol. 2012;2:191.
59. Achrol AS, Guzman R, Varga M, Adler JR, Steinberg GK, Chang SD. Pathogenesis and radiobiology of brain arteriovenous malformations: implications for risk stratification in natural history and posttreatment course. Neurosurg Focus. 2009;26(5):E9.
60. Lindqvist M, Steiner L, Blomgren H, Arndt J, Berggren BM. Stereotactic radiation therapy of intracranial arteriovenous malformations. Acta Radiol Suppl. 1986;369:610–3.
61. Lawton MT, Arnold CM, Kim YJ, Bogarin EA, Stewart CL, Wulfstat AA, et al. Radiation arteriopathy in the transgenic arteriovenous fistula model. Neurosurgery. 2008;62:1129–38.
62. Chang SD, Shuster DL, Steinberg GK, Levy RP, Frankel K. Stereotactic radiosurgery of arteriovenous malformations: pathologic changes in resected tissue. Clin Neuropathol. 1997;16:111–6.
63. Schneider BF, Eberhard DA, Steiner LE. Histopathology of arteriovenous malformations after gamma knife radiosurgery. J Neurosurg. 1997;87:352–7.
64. Bitzer M, Topka H. Progressive cerebral occlusive disease after radiation therapy. Stroke. 1995;26:131–6.
65. Kamiryo T, Lopes MB, Berr SS, Lee KS, Kassell NF, Steiner L. Occlusion of the anterior cerebral artery after gamma knife irradiation in a rat. Acta Neurochir. 1996;138:983–90.
66. Munter MW, Karger CP, Reith W, Schneider HM, Peschke P, Debus J. Delayed vascular injury after single high-dose irradiation in the rat brain: histologic

immunohistochemical, and angiographic studies. Radiology. 1999;212:475–82.
67. Qi F, Sugihara T, Yamamoto Y, Abe K. Arterial changes following single-dose irradiation. J Reconstr Microsurg. 1998;14:153–9.
68. Asur RS, Sharma S, Chang CW, et al. Spatially fractionated radiation induces cytotoxicity and changes in gene expression in bystander and radiation adjacent murine carcinoma cells. Radiat Res. 2012;177:751–65.
69. Hashimoto T, Mesa-Tejada R, Quick CM, Bollen AW, Joshi S, Pile-Spellman J, et al. Evidence of increased endothelial cell turnover in brain arteriovenous malformations. Neurosurgery. 2001;49:124–31.
70. Sharp FR, Xu H, Lit L, Walker W, Pinter J, Apperson M, et al. Genomic profiles of stroke in blood. Stroke. 2007;38:691–3.
71. Chen Y, Fan Y, Poon KY, Achrol AS, Lawton MT, Zhu Y, et al. MMP-9 expression is associated with leukocytic but not endothelial markers in brain arteriovenous malformations. Front Biosci. 2006;11:3121–8.
72. Chen Y, Pawlikowska L, Yao JS, Shen F, Zhai W, Achrol AS, et al. Interleukin-6 involvement in brain arteriovenous malformations. Ann Neurol. 2006;59:72–80.
73. Chen Y, Zhu W, Bollen AW, Lawton MT, Barbaro NM, Dowd CF, et al. Evidence of inflammatory cell involvement in brain arteriovenous malformations. Neurosurgery. 2008;62:1340–50.
74. Hao Q, Chen Y, Zhu Y, Fan Y, Palmer D, Su H, et al. Neutrophil depletion decreases VEGF-induced focal angiogenesis in the mature mouse brain. J Cereb Blood Flow Metab. 2007;27:1853–60.
75. Nuki Y, Matsumoto MM, Tsang E, Young WL, van Rooijen N, Kurihara C, et al. Roles of macrophages in flow-induced outward vascular remodeling. J Cereb Blood Flow Metab. 2009;29(3):495–503.
76. Brooks AL, Benjamin SA, McClellan RO. Toxicity of 90Sr-90Y in Chinese hamsters. Radiat Res. 1974;57:471–81.
77. Kaminski JM, Shinohara E, Summers JB, et al. The controversial abscopal effect. Cancer Treat Rev. 2005;31:159–72.
78. Lyng FM, Seymour CB, Mothersill C. Early events in the apoptotic cascade initiated in cells treated with medium from the progeny of irradiated cells. Radiat Prot Dosim. 2002;99:169–72.
79. Lyng FM, Seymour CB, Mothersill C. Initiation of apoptosis in cells exposed to medium from the progeny of irradiated cells: a possible mechanism for bystander-induced genomic instability? Radiat Res. 2002;157:365–70.
80. Hall EJ. The bystander effect. Health Phys. 2003;85:31–5.
81. Hall EJ, Hei TK. Genomic instability and bystander effects induced by high-LET radiation. Oncogene. 2003;22:7034–42.
82. Goh K, Sumner H. Breaks in normal human chromosomes: are they induced by a transferable substance in the plasma of persons exposed to total-body irradiation? Radiat Res. 1968;35:171–81.
83. Hollowell JG Jr, Littlefield LG. Chromosome damage induced by plasma of x-rayed patients: an indirect effect of x-ray. Proc Soc Exp Biol Med. 1968;129:240–4.
84. Sharpe HB, Scott D, Dolphin GW. Chromosome aberrations induced in human lymphocytes by x-irradiation in vitro: the effect of culture techniques and blood donors on aberration yield. Mutat Res. 1969;7:453–61.
85. Faguet GB, Reichard SM, Welter DA. Radiation-induced clastogenic plasma factors. Cancer Genet Cytogenet. 1984;12:73–83.
86. Ahmed MM, Sells SF, Venkatasubbarao K, et al. Ionizing radiation inducible apoptosis in the absence of p53 linked to transcription factor EGR-1. J Biol Chem. 1997;272:33056–61.
87. Hallahan DE, Spriggs DR, Beckett MA, et al. Increased tumor necrosis factor alpha mRNA after cellular exposure to ionizing radiation. Proc Natl Acad Sci U S A. 1989;86:10104–7.
88. Hallahan DE, Haimovitz-Friedman A, Kufe DW, et al. The role of cytokines in radiation oncology. Important Adv Oncol. 1993:71–80.
89. Hallahan DE, Virudachalam S, Sherman ML, et al. Tumor necrosis factor gene expression is mediated by protein kinase C following activation by ionizing radiation. Cancer Res. 1991;51:4565–9.
90. Unnithan J, Macklis RM. TRAIL induction by radiation in lymphoma patients. Cancer Investig. 2004;22:522–5.
91. Asur R, Butterworth KT, Penagaricano JA, et al. High dose bystander effects in spatially fractionated radiation therapy. Cancer Lett. 2015;356(1):52–7.
92. Peters ME, Shareef MM, Gupta S, et al. Potential utilization of bystander/Abscopal-mediated signal transduction events in the treatment of solid tumors. Curr Signal Transduct Ther. 2007;2:129–43.
93. Konoeda K. Therapeutic efficacy of pre-operative radiotherapy on breast carcinoma: in special reference to its abscopal effect on metastatic lymph nodes. Nihon Gan Chiryo Gakkai Shi. 1990;25:1204–14.
94. Gupta S, Zagurovskaya M, Wu X et al. Spatially fractionated Grid high dose radiation-induced tumor regression in A549 lung adenocarcinoma xenografts: cytokines and ceramide regulators balance in abscopal phenomena. Sylvester Comprehensive Cancer Center. 2014;20.
95. Camphausen K, Moses MA, Ménard C, et al. Radiation abscopal antitumor effect is mediated through p53. Cancer Res. 2003;63:1990–3.
96. Santana P, Peña LA, Haimovitz-Friedman A, et al. Acid sphingomyelinase deficient human lymphoblasts and mice are defective in radiation-induced apoptosis. Cell. 1996;86:189–99.

Organs at Risk (OAR) Tolerance in Hypofractionated Radiosurgery

13

Alfredo Conti

13.1 Introduction

The aim of any radiation treatment is based on the local control (LC) of the tumors while maintaining safety on the surrounding normal tissue, especially on the organs at risk (OAR), which are normal structures that can be specifically radiosensitive and/or in intimate relationship with the target. The physician aims to achieve the best therapeutic ratio, namely, the best compromise between the doses necessary to obtain a higher probability of tumor control (TCP) while maintaining the lowest normal probability of tissue complication (NTCP).

For CyberKnife radiosurgery, the precise interpretation of the TCP and NTCP ratio is exceptionally important due to the reverse planning algorithm and the non-isocentric irradiation geometry adopted by the system, which requires setting the dose constraints for any OAR together with the dose prescribed to the target. Unfortunately, after eight decades of radiotherapy practice, the current knowledge of both issues remains rather imprecise. The situation is even more complicated by the fact that OAR tolerance limits in self-treated schedules are mostly unexplored.

Last but not the least, the diversity of organs, the variety of complication endpoints for each organ, infinite variations in any combination of radiotherapy parameters such as fractionation, volume, overall time, etc. The physiological status of these organs before radiation, the severity of the disease, and the age of the patient are some of the factors that make this task enormously demanding.

Many of us, in the daily practice of radiotherapy, refer to the tolerance doses documented by Rubin and Cassarett, published about two decades ago. TD 5/5 (the probability of a 5% complication within 5 years of treatment) and TD 50/5 (the probability of a 50% complication within 5 years) which they introduced [1] are still the most prevalent and dominant in expressing the tolerance of normal tissues to radiation therapy.

In order to move from these data to an estimate of NTCP in hypofractionated treatments, it is fundamental to understand the basic radiobiological principles together with the basic radiobiological characteristics of the OAR found during central nervous system radiosurgery. Here, we summarize these principles of radiobiology and describe how we can use these principles to assist decision-making in hypofractionated treatments.

A. Conti (✉)
Alma Mater Studiorum University of Bologna, Bologna, Italy

IRCCS ISNB Istituto delle Scienze Neurologiche di Bologna, Bologna, Italy

Department of Neurosurgery, Charité Faculty of Medicine, Berlin, Germany
e-mail: alfredo.conti2@unibo.it

© Springer Nature Switzerland AG 2020
A. Conti et al. (eds.), *CyberKnife NeuroRadiosurgery*,
https://doi.org/10.1007/978-3-030-50668-1_13

13.2 Eye

In general, the eyeball is shaped as a single structure and, in most cases, beams should be forced to avoid crossing the eyeballs. However, this is not always possible, and dose loss is possible from the beams that come out of the targets lying close to the orbits. Therefore, it is very important to understand the dose limits of the structures included in the eyeball. The information provided does not refer to hypofraction, as there is currently no data available. Nevertheless, we can consider the following information a guide to interpret dose limits in hypofractionated regimens using the linear-quadratic (LQ) model and considering a very low α/β ratio (i.e., <1).

13.2.1 Cornea

Radiotherapy can injure the cornea by damaging the deeper layers of the stroma, but in most cases acute toxicity is due to the loss of the tear film. The maximum recommended dose (D_{max}) is <40 Gy. Corneal stroma edema appears at a dose of 40–50 Gy but is usually transient. With doses of 60 Gy, the possibility of corneal ulceration increases to 17–20%, which further increases if chemotherapy is added. [2].

13.2.2 Retina

This innermost layer of the globe is approximately 0.25 mm thick and is usually not displayed in MRI standard. Contoured with a 3 mm brush, the retina covers the posterior 5th/6th of the globe. Dose recommendations are 45 Gy maximum. Acute retinal toxicity is not reported. Usually there is a latent period of 6 months to 3 years before the onset of clinically significant retinopathy. The mean latent period is 19 months [2].

13.2.3 Lens

The structure is about 10 mm in diameter seen in the coronal plane. Dose recommendations (D_{max}): 5–10 Gy. Acute lens toxicity is not reported. A single dose of 2 Gy can cause cataract but is usually visually insignificant [2]. The time of onset is dose-related. For doses in the range of 2.5–6.5 Gy, the latency is 8 years with the possibility of 33% progressive cataract, whereas for doses of 6.5–11.5 Gy, the latency reduces to 4 years with the 66% risk of progressive cataract [2].

13.3 Anterior Optic Pathway

The optic nerves and chiasm are among the most radiosensitive structures encountered in the radiosurgery of brain lesions. The literature on tolerance limits for both the single fraction and the normofraction is widely available and provides sufficient evidence.

A maximum dose of ≤8 Gy has been traditionally used as a limit for the anterior optic pathways [3]. Such value came from a retrospective analysis of 62 cases of benign tumors of the cavernous sinus treated with Gamma Knife (GK). In the following studies, the incidence of radiation-induced optic neuropathy (RION) was ≤2% in patients receiving a maximal dose in a single fraction of 8–12 Gy, >10% in case of >12–15 Gy [4].

It should be noted, however, that these studies included patients who had received previous irradiation.

Papers published since 2010 report an increase in the maximum dose: Mayo et al. found in their review that in the case of SRS, the incidence of RION is rare for a D_{max} < 8 Gy, increases in the range 8–12 Gy, and becomes >10% in the range 12–15 Gy [5].

Hasegawa et al. have shown in their experience that a dose <14 Gy could be related to a low risk of RION, provided that the nerve is exposed only for a small tract and assuming that a dose of 14 Gy over a few mm could be better tolerated instead of dosage lower for the whole length of the nerve [6].

In 2013, Leavitt et al. suggested that 2–4 mm of the anterior optic tracts could tolerate doses ≥12 Gy, describing an incidence of RION of 0.5% in 222 patients which were not previously irradiated [7].

Owing to the abovementioned sensitivity to radiation of the optic apparatus, radiosurgery is conventionally precluded for "perioptic" meningiomas, namely, for lesions lying <3 mm of the anterior visual pathways [8, 9]. Nonetheless, it has been postulated that the use of hypofractionation can increase the potential of radiosurgery to treat perioptic tumors [10–12].

Here, we provide a radiobiological model to be used in hypofractionated schedules and can be used to adapt the dose that can be tolerated by the anterior optic structures.

13.3.1 The Isoeffect Model

In order to create this model, we used the linear-quadratic (LQ) model. The isoeffect model was used to estimate the α/β ratio of the optic nerve and chiasm and the threshold D_{max} using Eqs. 13.1 and 13.2.

The LQ formula $e^{-(\alpha D + \beta D\exp 2)}$ is often used to model biological response to radiation. For instance, when applied to single-fraction cell survival studies, the surviving fraction (SF) is generally expressed as:

$$SF = e^{-(\alpha D + \beta D \exp 2)} \quad (13.1)$$

where D is the dose in Gy, α is the cell kill per Gy of the initial linear component (on a log-linear plot), and β is the cell kill per Gy^2 of the quadratic component of the survival curve.Curves for the individual LQ components $e^{-\alpha D}$ and $e^{-\beta D\exp 2}$intersect at the dose where the αD and βD^2 components of cell killing are equal.

This intersection occurs at a dose equal to the ratio of α to β and is often referred to in the literature simply as α/β.

The biological effect (E) per fraction (n) of fractional dose (D) can be expressed as:

$$E_n = \left(\alpha D + \beta D^2\right) \quad (13.2)$$

The biologically effective dose (BED or E/α) is an approximate quantity with which it is possible to compare different radiotherapy fractionation regimens.

For instance, for an external beam radiotherapy (EBRT) regimen employing n equal fractions of conventional size, the BED will be:

$$\text{BED} = E/\alpha = nD\left(1 + D/\alpha/\beta\right) \quad (13.3)$$

where n = number of fractions, D = dose/fraction, and nD = total dose.

The LQ model is based on the assumption that the biological response to radiation can be described by an equation with two principal components, one proportional to the dose and another proportional to the square of the dose.

To relate the biological effect of a course of radiation to the total dose of radiation administered with the standard fractionation scheme (2 Gy/fraction), the concept of normalized total dose (NTD) or the biologically equivalent doses as calculated in 2 Gy daily fractions (EqD2) has been used.

Expression is derived from the linear-quadratic model:

$$\text{NTD} = \text{BED}/\left(1 + d/\alpha\beta\right) \quad (13.4)$$

To establish the equivalence with the standard fractionation scheme (2 Gy/fraction), in the second equation, d would take the value 2 Gy (EqD2).

We calculated a target BED to be used in modified fractionation schemes. To calculate this target BED, we analyzed the clinical data, currently available in the literature, on the irradiation of the optic nerve.

Sufficient clinical evidence could be found on the treatment in a single fraction and using conventional fractionation (2 Gy/day). We selected two different dose/fraction schemes that resulted in similar TCP/NTCP and were therefore biologically equivalent. Using the LQ model, it was possible to calculate the α/β of the optic nerves and the BED that can be considered safe when using modified fractionation schemes.

Actually, assuming that the BED_1 of the first treatment is equal to the BED_2 for a tissue of unknown a/b,

$$\text{D1}\left(1 + \frac{\text{d1}}{\frac{\alpha}{\beta}}\right) = \text{D2}\left(1 + \frac{\text{d2}}{\frac{\alpha}{\beta}}\right) \quad (13.5)$$

and

$$\frac{\alpha}{\beta} = \frac{\text{D1} \times \text{d1} - \text{D2} \times \text{d2}}{\text{D2} - \text{D1}} \quad (13.6)$$

Although BED calculation seems reliable for comparing conventional fractionation regimens, the standard LQ formulation upon which these calculations rests is suspect for fractions of very high dose.

To better address hypofractionation regimens (e.g., SBRT, SRS), which employ very high dose fractions, we used the linear quadratic-linear (LQ-L) model described by Astrahan [13] which describes the transition to a linear tail at high dose fractions.

To transition to a linear model for very high dose fractions, the standard LQ model can be extended to a bipartite $LQ_{\alpha/\beta} - L_{DT}$ form which can be expressed as:

$$\mathrm{SF} = \mathrm{e}^{-(\alpha D + \beta D \exp 2)} \text{ for fraction size } D < D_\mathrm{T}$$

and

$$\mathrm{SF} = \mathrm{e}^{-[\alpha D_\mathrm{T} + \beta D_\mathrm{T} \exp 2 + \gamma (D - D_\mathrm{T})]} \quad \text{for fraction size } D \geq D_\mathrm{T} \tag{13.7}$$

$$\mathrm{BED}_n = \left[D + D^2 / (\alpha / \beta) \right] \quad \text{for fraction size } D < D_\mathrm{T} \tag{13.8}$$

and

$$\mathrm{BED}_n = D_\mathrm{T} + D_\mathrm{T}^2 / (\alpha / \beta) + \left[(\gamma / \alpha)(D - D_\mathrm{T}) \right] \quad \text{for fraction size } D \geq D_\mathrm{T} \tag{13.9}$$

where D_T is the dose at which the survival curve becomes linear and γ is the cell kill per Gy in this high-dose linear portion of the survival curve. The term γ is thus reminiscent of D_0 (the dose that reduces survival by $1/e$ in the final linear portion of the survival curve) of the older multitarget survival model but is more in line with the Greek nomenclature of the LQ model. The γ/α ratio can be calculated from the tangent at D_T and the α/β term of the standard LQ model as:

$$\gamma / \alpha = 1 + \left[2D_\mathrm{T} / (\alpha / \beta) \right] \tag{13.10}$$

According to the abovementioned considerations, it is possible to consider two irradiation regimes that are similar in terms of complication rate to estimate the α/β of the anterior optic pathway. Considering isoeffectiveness in terms of NTCP, a D_{max} of 11.0 Gy in a single fraction [9] and 50.0 Gy in 25 fractions (5% complication probability at 5 years [14]), the α/β of the optic nerve and chiasm turned out to be 0.55 Gy. Using this α/β, the isoeffective BED and the D_{max} values to be used in hypofractionated treatments, having a NTCP similar to that of 50.0 Gy in 25 fractions are 15.5 Gy in 2 fractions, 19.0 Gy in 3 fractions, 22.0 Gy in 4 fractions, and 24.0 Gy in 5 fractions.

13.3.2 The Optic Ret Model

An alternative model has been provided in the literature. The "optic ret" model has been proposed by Goldsmith et al. [15]. This is a model predicting the total dose associated with a "low risk" of optic neuropathy when various doses per fraction are used.

The optic ret is based on the total dose and the number of fractions used:

$$\text{optic ret dose} = D(\mathrm{cGy}) / N^{0.53} \tag{13.11}$$

where N is the number of fractions used and D is the total physical dose in cGy. Goldsmith et al. [15] predicted, based on clinical data, that doses to the optic nerve or chiasm less than or equal to 890 optic ret would be safe. Therefore, for a particular number of fractions N, the optic tolerance would be predicted to be the total dose D, where

$$D = 890 \times N^{0.53} \tag{13.12}$$

This model predicts doses that are associated with a very low risk of optic neuropathy with single doses, 8.9 Gy, and up to at least 30 fractions, 54 Gy. This model can be used to predict the tolerance of the optic nerve in schedules using variable numbers of fractions.

13.3.3 Calculating the NTCP for Doses in Hypofractionated Schedules

Normal tissue complication probability (NTCP) has been calculated using the widely used NTCP model of Lyman. It is an empirical model employing four parameters and assumes that the probability of complications after uniform irradiation follows a sigmoid dose-response relationship; the formulation of the model has been discussed in detail elsewhere [16, 17].

Briefly, a four-parameter model was proposed by Lyman [16]. In this model, the complication probability $P(D, v)$ for a uniform irradiation of a normal tissue volume V with a dose D is given by [16–20]:

$$\mathrm{NTCP}\left(D,v\right)=\frac{1}{\sqrt{2\pi}}\int_{-\infty}^{t}\exp^{\frac{-x^2}{2}}dx \quad (13.13)$$

$$t=\frac{1}{m}\left(\frac{D}{\mathrm{TD}_{50}\left(v\right)}-1\right) \quad (13.14)$$

$$\mathrm{TD}_{50}\left(v\right)=\mathrm{TD}_{50}\left(1\right)\cdot v^{-n} \quad (13.15)$$

$$v=\frac{V}{V_{\mathrm{ref}}} \quad (13.16)$$

The four parameters of the model are given by TD_{50}, m, n, and V_{ref}, which have to be adjusted to clinical data for each tissue type using a specified biological end point. $TD_{50}(v)$ is the tolerance dose for the fractional volume v, m is related to the slope of the dose-response curve, n describes the volume effect, and V_{ref} is the reference volume to which the fractional volume refers to. V_{ref} may be chosen as the whole organ or as a part of it.

Equation (13.15) refers to the tolerance doses of the partial volume v to that of the reference volume ($v = 1$). Emami et al. [14] published tolerance doses for various tissues and fractional volumes that were derived from bibliographic research and clinical experience. The authors considered the uncertainty of these tolerance doses quite high. Next, the parameters of the Eq model. (13.10) were adjusted to fit these tolerance data [18].

In clinical practice, normal tissue will not be uniformly irradiated as assumed by the Lyman model; therefore, the model has been extended by introducing histogram reduction algorithms, which transform the multi-step dose volume histogram obtained for a specific treatment plan in a biologically isoeffective single-pass histogram, i.e., non-uniform irradiation is transformed into a biologically isoeffective one uniform irradiation.

Two different types of reduction algorithms have been proposed which lead to similar although not identical NTCP-values: the first one [21, 22] replaces the two rightmost bins (at doses D_n and D_{n-1} and volumes V_n and V_{n-1}) of the cumulative histogram by a single bin at dose D_{n-1} and Volume V_{n-1}. The dose D_{n-1} is calculated such that the new histogram has the same NTCP according to Eqs. 13.13–13.16. This procedure is iterated until a single-step histogram is achieved corresponding to a homogeneous irradiation of the reference volume ($v = 1$) with a dose D1 for which the Lyman model can directly be applied.

The second algorithm [17, 20] transforms the initial multi-step histogram (having the maximum dose $D_{\max}$) to a biologically isoeffective single-step histogram with an effective volume V_{eff} at the dose $D_{\max}$. For this approach, a volume effect according to Eq. 13.15 is assumed. The single-step histogram then corresponds to a homogeneous irradiation of the fractional volume $V_{\mathrm{eff}} = V_{\mathrm{eff}}/V_{\mathrm{ref}}$ and the NTCP is then calculated by Eqs. 13.13–13.16.

To calculate an NTCP using the Lyman formalism, single dose-volume numbers are calculated as an intermediate step with a dose-volume histogram (DVH) reduction method. For example, the DVH representing non-uniform irradiation of the organ must be transformed to a "single-step" DVH with an effective volume (V_{eff}) and a corresponding reference dose.

The resulting uniform single-step DVH gives an equivalent NTCP to the initial non-uniform DVH. This method has been discussed by Kutcher et al. [17]. Optic nerve, and optic chiasm tolerance values, summarized by Emami et al. [14], gave tolerance doses for 5% (TD_5) and 50% (TD_{50}) chances of a complication happening in 5 years for the whole organ irradiated to a uniform dose.

Burman et al. [18] fit the clinical data to provide parameters to be used in the Lyman model, and these have been used in our calculations (optic nerve and chiasma: volume parameter $n = 0.25$, adaptation parameter $m = 0.14$, $TD_{50} = 65$ Gy).

We calculated the NTCP of those doses on the optic apparatus according to the Lyman-Kutcher-Burman model [16–18]. Using the parameters provided in the Emami study [14], a Dmax of 24.0 Gy in 5 fractions to the optic nerve and chi-

Table 13.1 Estimates of normal tissue complication probability for some dose/fraction schemes using the Lyman-Kutcher-Burman model

Dose/fraction schedule	2-Gy dose equivalence	Normal tissue complication probability (%)
10 Gy/1f	EqD2 = 30 Gy	0
11 Gy/1f	EqD2 = 36 Gy	0
13 Gy/1f	EqD2 = 49 Gy	0.31
25 Gy/5f	EqD2 = 44 Gy	0.12
27.50 Gy/5f	EqD2 = 52 Gy	2.98

asm resulted in a NTCP of 0.02%, that is, 0.12 for 25.0 Gy in 5 fractions and 2.98% for 27.5 Gy in 5 fractions (Table 13.1).

13.4 Spinal Cord

Radiation-induced spinal cord injuries are rarely encountered in clinical practice and in the literature. However, when complications do occur, radiation-induced myelopathies can be very serious and range from pain, sensory deficits, Brown-Sequard syndrome, loss of bowel/bladder control, to complete paralysis. The pathogenesis of these lesions is generally attributed to vascular damage, glial cell injury, or both.

Clearly the question of dose limits is of primary importance for spinal radiosurgery. Unfortunately, precise limits and firm recommendations cannot simply be drawn from patients treated with fractional radiotherapy because of the difficulty in translating biologically equivalent dose effects between very different dosage schemes.

Compared to conventional radiotherapy, radiosurgery involves a much higher dose per fraction (whose biological equivalence to radiotherapy is not known precisely) but is directed to a very limited portion of the spinal cord (again, as this translates into recommendations for dose constraints it is not immediately clear).

Yet another question deserving attention is the tolerance of the spinal cord to re-irradiation, to which radiosurgery has been shown to be safely applied when larger-field conventional radiotherapy would likely cause radiation-induced complications. Below, we review some of the relevant data, noting both the complexity of the problem and the tentative recommendations that have been made to date.

Schultheiss [23] reviewed the literature on external beam radiotherapy and analyzed the incidence of radiation myelopathy in 335 and 1946 patients receiving radiotherapy to their cervical and thoracic spines, respectively. The risk of myelopathy as a function of dose was estimated using a probability distribution model. A good fit to the combined cervical and thoracic cord data was not possible, and separate analyses were performed. For the cervical cord data, at 2-Gy per fraction and assuming α/β of 0.87 Gy, the probability of myelopathy was estimated as 0.03% at 45 Gy, 0.2% at 50 Gy, and 50% at >69 Gy.

Schultheiss [23] considered that if there are n patients at risk who received dose D, the expected number of responders during the time interval t to t + dt after irradiation is:

$$n \cdot P(D) \cdot S(t) \cdot f(t)dt \qquad (13.17)$$

where $P(D)$ is the probability of exceeding tolerance with dose D, $S(t)$ is the survival function for the population at risk, and $f(t)$ is the distribution of latent periods for the complication.

The number of responders during the interval is the total number at risk, times the probability P that tolerance was exceeded, times the probability that the latency is in the time interval (given that tolerance was exceeded), times the probability that the patient survives to time t.

The total number of responders, r, is given by the integral of this expression:

$$r = n \cdot P(D) \cdot \int S(t) f(t) dt \qquad (13.18)$$

This expression can be used to estimate the value of $P(D)$, the probability of complication for dose D when r and n are known.

Generally, in reports of radiation myelopathy occurring after a group of patients has been treated with the same regimen, investigators limit the number of patient records that need to be examined by only examining the records of patients who survive for at least a minimum amount of time, typically 6–12 months.

The number of patients who survive this minimum time is given by "n" above. The investigators also give the number who survived for longer periods, typically in 6- or 12-month increments. Thus, the survival function $S(t)$ is identically equal to unity for patients surviving this minimum time.

Thereafter, the survival function should be used not of the entire initial population but of the n patients who survived in the minimum time. For the purposes of this study, Schultheiss [23] hypothesized that the form of the survival function was exponential and used linear interpolation (on a logarithmic scale) to estimate the values of $S(t)$ between the discrete points that were indicated in the original reports.

This study extends the use of the method to estimate probabilities from different dose regimens and applies the maximum likelihood estimate (MLE) to determine the parameters in a dose-response model and their confidence intervals (CI). If data are available for a variety of dosing regimens, the $P(D)$ estimates of each dosing regimen can be used in a likelihood function to estimate the parameter values in a dose-response function. In this case, the probability function is given by:

$$L = \Pi\left[P(D)\cdot g\right]^{r}\left[1 - P(D)\cdot g\right]^{n-r} \tag{13.19}$$

where $g = f(t)S(t)dt$. The parameter g is the probability that a patient whose tolerance has been exceeded will survive long enough to express the injury. For the present study, Schultheiss [23] used the logistic dose—response function for P with:

$$P(D) = \left\{1 + \exp\left[k\cdot\ln\left(D_{50}\right) - k\cdot\ln(D)\right]\right\}^{-1} = 1 \Big/ \left[1 + \left(\frac{D_{50}}{D}\right)^{k}\right] \tag{13.20}$$

where $P(D)$ is the probability of exceeding tolerance when the cord receives dose D, D_{50} is the median tolerance dose, k is the slope parameter, and D is the effective dose in 2-Gy fractions [21]. $f(t)$ has been shown to be a bimodal lognormal distribution [24], and $S(t)$ was estimated from data in the original reports. The effective dose, D, was determined from the fractionation scheme and converted to its 2-Gy-per-fraction equivalent using the LQ model. The dose-response parameters D_{50} and k and the α/β ratio from the LQ model were obtained by using MLE.

The value of g was calculated numerically by using the above function for $f(t)$ and the survival data in each report, interpolating between survival points, assuming an exponential function for $S(t)$. Data for the cervical cord come from reports listed in Table 13.2.

Results of the fit of the cervical cord data in Table 13.2 were D_{50} = 69.4 Gy, k = 18.8, and α/β = 0.87 Gy. Pearson's chi-square statistic was 2.1 with 5 df, indicating a good fit of the model to the data. The 95% CI for D_{50} was 66.4–72.6 Gy. The 95% CI for a/b was 0.54–1.19 Gy. The 95% CI for k was 13.3–27.4. Data and the dose-response function are shown in Fig. 13.1. Doses were converted into 2-Gy/fraction equivalents. Using these values, the probability of myelopathy is 0.03% at 45 Gy and 0.2% at 50 Gy.

Table 13.2 Data for cervical spinal cord response from published reports (from Shultheiss [23])

Authors	Dose (Gy)	Dose/fraction	r	n	Integral, g	2-Gy dose equivalent
McCunniff et al. [25]	60	2	1	12	0.929	60.0
McCunniff et al. [25]	65	1.63	0	24	0.929	56.6
Abbatucci et al. [26]	54	3	7	15	0.750	72.8
Atkins et al. [27]	19	9.5	4	13	0.704	68.6
Marcus et al. [28]	47.5	1.9	0	211	0.738	45.0
Marcus et al. [28]	52.5	1.9	0	22	0.738	49.8
Jeremic et al. [29]	60	2	2	19	0.891	60.0
Jeremic et al. [29]	65	1.63	0	19	0–891	56.6

r, n, and g are as used in Eqs. 13.17 and 13.18. $\alpha/\beta = 0.87$

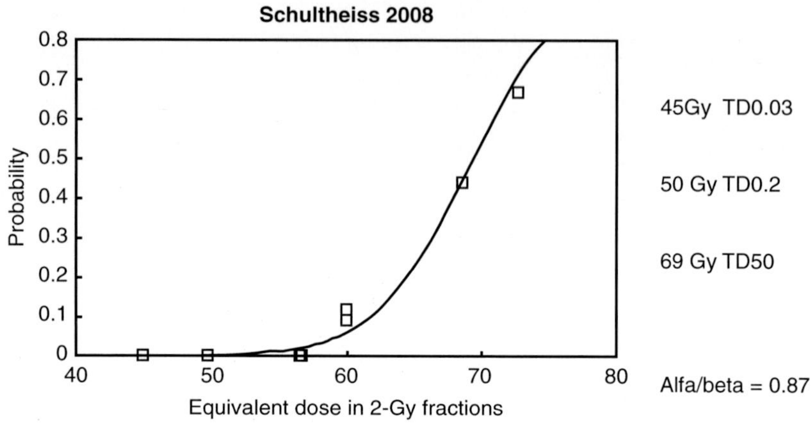

Fig. 13.1 The dose-response function for the cervical spinal cord as calculated by Schultheiss [23]

A second interesting study was published by Daly et al. [30] from Stanford University. Authors treated 24 spinal hemangioblastomas in 17 patients. Seventeen tumors received a single fraction with a median dose of 20 Gy (range, 18–30 Gy) and 7 received 20–25 Gy in 2 or 3 sessions, with cord maximum doses of 22.7 Gy (range, 17.8–30.9 Gy) and 22.0 Gy (range, 20.2–26.6 Gy), respectively. The Lyman-Kutcher-Burman model was used to calculate the biologically equivalent uniform dose and NTCP for each treatment by using conventional values for α/β (i.e., = 3), volume parameter n, 50% complication probability dose TD_{50}, and inverse slope parameter m [17]. In this study, one case (4%) of myelopathy occurred but the Lyman-Kutcher-Burman model using radiobiological parameters from Emami [$n = 0.05$, $m = 0.175$, $TD_{50} = 66.5$ Gy, and $\alpha/\beta = 3$ Gy] [14], and the logistic model with parameters from Schultheiss et al. [23] overestimated complication rates, predicting 13 complications (54%) and 18 complications (75%), respectively (Fig. 13.2a, b). Therefore, authors suggested that the spinal cord tolerance for doses common to SRS is higher than predicted by the Lyman-Kutcher-Burman model, and emphasized that radiobiological models traditionally used to estimate spinal cord NTCP may not apply to SRS.

The study is very interesting for implementing the understanding of radiobiological bases of irradiation of the spinal cord. So, we reviewed this study in greater detail. Applying the LKB model to the spinal SRS cohort, the authors performed NTCP calculations and compared the observed vs. predicted outcomes. First, the Emami parameters ($m = 0.175$, $TD_{50} = 66.5$, $n = 0.05$) were used to determine the NTCP, assuming $\alpha/\beta = 3$ Gy. With these parameters, considering the treatments as independent, 13 complications were predicted (Fig. 13.2a), greatly overestimating the actual number of complications observed (i.e., 1 case). Therefore, the authors adopted a maximum likelihood estimation (MLE) method to determine the optimum α/β ratio to best fit their clinical data by use of Emami parameters. Furthermore, they carried out a dual MLE LKB model optimization of α/β and n, keeping m and TD_{50} fixed at their Emami values, and *a* separate MLE logistic model optimization of n, keeping k and TD_{50} fixed at their Schultheiss values.

Because of the partial-volume irradiation typical of spinal radiosurgery, the radiation tolerance of the cord may exhibit aspects of parallel architecture. As such, optimizing the value of n to 0.31, to assume a greater degree of organization in parallel, as well as keeping the other Emami parameters constant, suggested a risk of less than 10% for the majority of treated lesions.

However, the model continues to generate an NTCP of greater than 90% for one unaffected patient. Furthermore, with an n of 0.31, the NTCP for the only patient in whom a complication developed approaches 0. Although the predicted NTCP with the optimized n of 0.31 (Fig. 13.2c) suggests a somewhat improved fit to the clinical data, this model with these parameters clearly does not fully explain the tolerance of the spinal

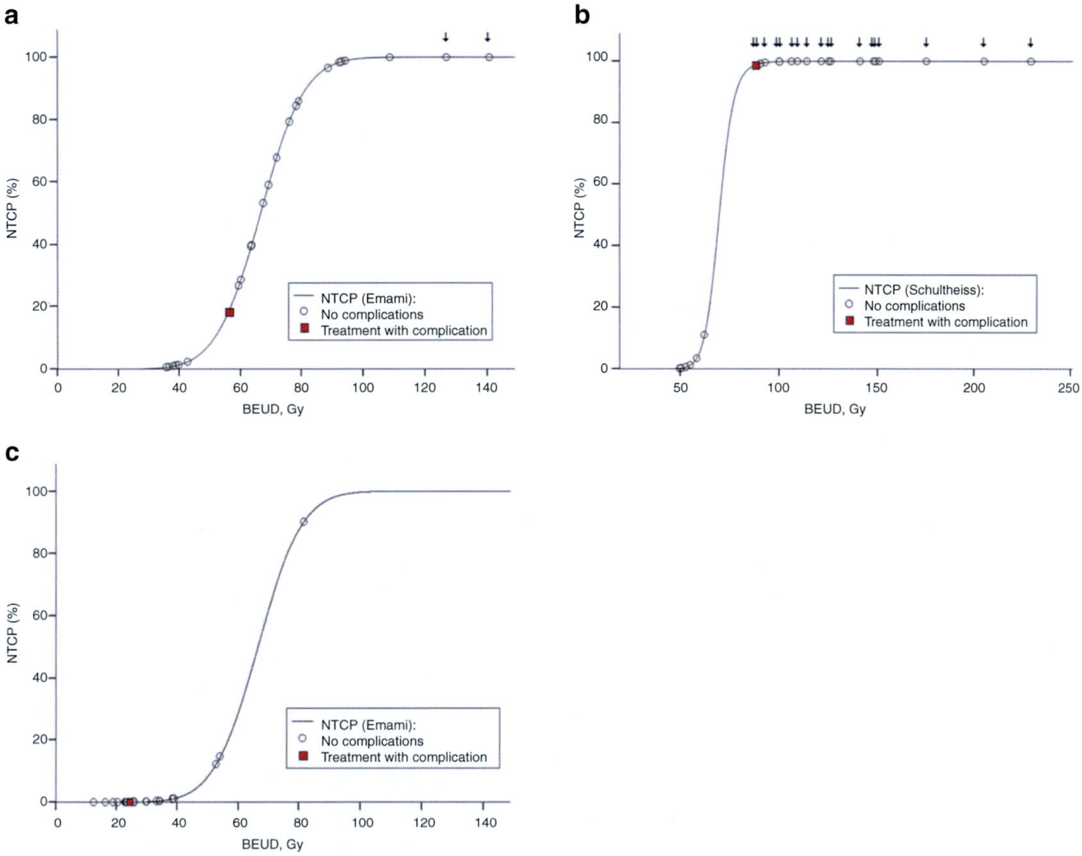

Fig. 13.2 (**a**) Predicted normal tissue complication probability (NTCP) for spinal cord with NTCP modeling of clinical outcomes based on parameters of Emami [14]. One cord toxicity was observed clinically (square) (with an NTCP of 19%). Seventeen other lesions with a higher NTCP, including two lesions with 100% NTCP (arrows), did not have toxicity. Therefore, these modeling parameters were not predictive of the clinical outcomes. (**b**) Predicted NTCP for spinal cord with NTCP modeling of clinical outcomes based on parameters of Schultheiss [5]. One cord toxicity was observed clinically (square) (with an NTCP of 99.8%), whereas 17 other lesions without toxicity had an NTCP of greater than 99% (arrows). These NTCP values overestimated the toxicity rates observed clinically. (**c**) With optimization of the volume parameter ($n = 0.31$), assuming a greater degree of parallel organization of the spinal cord. Although this optimization provided a better fit of the NTCP to the clinical data, these NTCP parameters continued to overestimate the risk of complications. From Daly et al. [30]

cord to this type of dosimetry. When the α/β ratio was adjusted to find a best fit to the data set (using the remainder of the Emami parameters as assumptions), an a/b ratio of 13.2 Gy was found that best fit the data (Fig. 13.3). However, there are no confirmatory data suggesting an α/β this high for the human spinal cord; rather, a number of studies suggest a much lower α/β (i.e., 0.87–3 Gy). Therefore, if this explanation were applicable, it would be restricted only to small-volume irradiation and is unlikely to be valid for conventional radiotherapy.

In conclusion, for spinal cord tolerance, the widely used LKB NTCP model does not accurately predict the risk of radiation myelopathy from low-volume, high-dose spinal SRS by use of any set of accepted parameters. With small-volume irradiation to the spinal cord, typical of SRS, we can assume a greater parallel organization (i.e., higher value for n) than usually considered. This improves the fit of the model but does not fully explain the observed data.

An α/β higher than the conventional one, corresponding to a higher repair efficiency, also improves

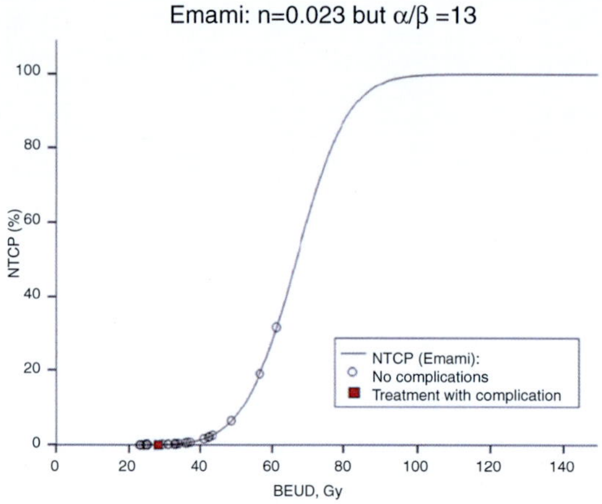

Fig. 13.3 Predicted normal tissue complication probability (NTCP) for spinal cord with a best-fit α/β and NTCP modeling with Emami parameters. An α/β ratio of 13.2 Gy provided the best fit of the curve to the clinical outcomes. This α/β, however, is much higher than and not consistent with any published value. *BEUD* biologically equivalent uniform dose. From Daly et al. [30]

the fitting of the model. However, this does not agree with the α/β values associated with a wider field radiation. Dose limits to spinal cord irradiation and reirradiation based on further clinical data are provided in dedicated chapters on spinal metastases.

13.5 Cochlea

It has been supposed that hSRT may result in lower toxicity and thus higher rates of hearing preservation, compared to either SRS or surgery [31, 32]. Similar hypofractionated schedules of three to five fractions showed slightly better hearing preservation rates that range from 50 to 93%, compared to 32–81% for SRS-treated lesions [31, 33, 34]. Andrews et al. noted a 2.5-fold higher hearing preservation rate (81%) for FSRT vs. (33%) for Gamma Knife SRS [35]. However, hearing preservation rates vary considerably with some studies demonstrating no difference or even poorer hearing outcomes for FSRT compared to SRS [36, 37].

The dose to the central cochlea is considered one major factor associated to hearing loss. In SRS treatments, patients who received a radiation dose of <4.2 Gy to the central cochlea had significantly better hearing preservation of the same Gardner-Robertson class. Twelve of 12 patients <60 years of age who had received a cochlear dose <4.2 Gy retained serviceable hearing at 2 years post-SRS [38]. Therefore, the maximal dose to the cochlea should be lower than 4 Gy in a single fraction to minimize the risk of hearing loss.

The equivalent dose for hSRT is more difficult to establish because of limited clinical data. Based on a dose- response model of clinical data sets, Rashid et al. [39] suggested that the 14 Gy in single fraction and the 27.5Gy in 5-fraction limit carry a 17.9% and 17.4% risk of hearing deterioration, respectively, whereas the 12 Gy in a single fraction and the 25 Gy in a 5-fraction limit had 11.8% and 13.8% risk, respectively [39].

An interesting study from the group of Stanford analyzed this topic [40]. A cochlea dose-volume histogram was generated for each of the 94 patients who were treated with 3-fraction hSRT and were qualified for the study. Gardner Robertson grade I–II hearing post-treatment was maintained in 74% of patients (70/94). Larger cochlear volume was associated with lower risk of hearing loss. Controlling for differences in cochlear volume among subjects, each additional mm^3 of cochlea receiving 10–16 Gy (single session equivalent doses of 6.6–10.1 Gy) significantly increased the odds of hearing loss by approximately 5% [40]. The role of cochlear volume, tumor volume, and prescribed cochlear dose in hearing preservation in hSRT is confirmed in a study by Tsai et al. [32].

13.6 Pituitary Gland

Dose recommendations: D_{max} 45 Gy (to avoid panhypopituitarism, lower for growth hormone (GH) deficiency). The anterior pituitary has five different types of cells, each with different radiosensitivities. The most sensitive is the GH axis, followed by the gonadotropin, ACTH, and TSH axis. GH deficiency has been noted in relatively lower doses and has been reported for whole brain irradiation and for doses as low as 10 Gy [23], but the incidence increases substantially after 30 Gy where the incidence can be as high as 50–100%.

13.7 Brain

Normal brain tolerance has been explored in relatively few studies. As for the maximum dose tolerated by the normal brain, Meyer and Sminia [41] analyzed the clinical data available on the re-irradiation of the glioma with respect to the tolerance dose of normal brain tissue. Clinical studies of brain re-irradiation from January 1996 to December 2006 on late adverse effects induced by radiation, i.e., brain tissue necrosis, were considered.

The studies were analyzed by using the LQ model to derive information on the cumulative BED (BED cumulative) and equivalent doses in 2-Gy fractions (EqD2) for the healthy human brain. Radiation-induced normal brain tissue necrosis is found to occur at EqD2 cumulative >100 Gy. Accordingly, it can be suggested that the above reported radiobiological models can be used for the brain considering 100 Gy as a threshold for radiation necrosis [41]. The rate of this risk is however difficult to be extrapolated, but it is, intuitively, associated with the volume of normal brain receiving this dose. Another point deserving attention is that to calculate the dose to normal brain using the LQ model, an $\alpha/\beta = 2$ should be used for the brain (whereas the α/β is = 10 for tumors).

The issue of maximal dose in a single fraction is another important issue in order to understand the real span of dose tolerated by the normal brain.

The literature from which we are able to extrapolate some data on this topic concerns the radiosurgery of arteriovenous malformations, due to the high doses used and the long life expectancy of patients which allows the development of late toxicity. Friedman et al. [42] analyzed dosimetric and clinical data for 269 patients undergoing AVM radiosurgery.

Of the 269 patients studied, 228 experienced no complication, 10 (3.7%) experienced a transient radiation-induced complication, 3 (1%) experienced a permanent radiation-induced complication, and 28 (10%) experienced posttreatment hemorrhage. The 12-Gy volume was predictive of permanent radiation-induced complications. Eloquent AVM location and 12-Gy volume were correlated with the occurrence of transient radiation-induced complications.

Similar results were obtained by Blonigen et al. [43] who reviewed data of 63 patients with a total of 173 brain metastases.

Table 13.3 Estimated percent risk of permanent side effects for AVMs measuring 1, 2, 3, and 4 cm according to location

AVM location	Risk of adverse radiation effects (%)			
	Nidus diameter			
	1 cm	2 cm	3 cm	4 cm
Low-risk regions				
Frontal	0.04	0.07	0.11	1.48
Temporal	0.59	0.94	1.45	16.95
Mid-risk regions				
Intraventricular	1.32	2.11	3.22	31.63
Cerebellum	1.65	2.62	4	36.68
Parietal	2.61	2.55	3.88	35.99
Moderate-risk regions				
Corpus callosum	3.73	5.88	8.8	57.32
Occipital lobe	3.87	6.09	9.11	58.2
High-risk regions				
Medulla	7.43	11.46	16.66	73.55
Thalamus	12.36	18.51	25.98	83
Basal ganglia	15.01	22.15	30.54	85.95
Pons/midbrain	44.02	55.89	66.19	96.46

From Lunsford et al. [44]

Most of the patients (63%) had received previous whole-brain irradiation. The mean prescribed SRS dose was 18 Gy. Symptomatic radionecrosis was observed in 10% and asymptomatic radionecrosis in 4% of the treated lesions. Multivariate regression analysis showed that V8 Gy–V16 Gy is predictive of symptomatic radionecrosis ($p < 0.0001$). Threshold volumes for significant rise in radionecrosis rates occurred with a volume of 10.45 cm^3 for V10 Gy and 7.85 cm^3 for V12 Gy. Accordingly, the authors propose that patients with V10 Gy >10.5 cm^3 or V12 Gy >7.9 cm^3 should be considered for hypofractionated rather than single-fraction treatment, to minimize the risk of symptomatic radionecrosis.

The normal brain site involved in irradiation is also relevant. Lunsford et al. [44] constructed a multivariate model of the effects of AVM location and tissue volume receiving at least 12 Gy (12 Gy volume) for the risk of developing permanent post-radiosurgery sequelae. AVM sites were associated with an increasing risk of significant post-surgical injury. The position score (between 0 and 10) was frontal, temporal, intraventricular, parietal, cerebellar, corpus callosum, occipital, medulla, thalamus, basal ganglia, and bridge/midbrain. The final statistical model predicted the risk of permanent symptomatic sequelae from the AVM position and from the 12-Gy volumes. Table 13.3 lists the risks of permanent symptomatic sequelae for AVMs that measure 1, 2, 3, and 4 cm in average diameter by location.

References

1. Rubin P, Casarett GW. Clinical radiation pathology as applied to curative radiotherapy. Cancer. 1968;22(4):767–78.
2. Jeganathan VS, Wirth A, MacManus MP. Ocular risks from orbital and periorbital radiation therapy: a critical review. Int J Radiat Oncol Biol Phys. 2011;79(3):650–9.
3. Tishler RB, Loeffler JS, Lunsford LD, Duma C, Alexander E, Kooy HM, et al. Tolerance of cranial nerves of the cavernous sinus to radiosurgery. Int J Radiat Oncol Biol Phys. 1993;27(2):215–21.
4. Leber KA, Berglöff J, Pendl G. Dose—response tolerance of the visual pathways and cranial nerves of the cavernous sinus to stereotactic radiosurgery. J Neurosurg. 1998;88(1):43–50.
5. Mayo C, Martel MK, Marks LB, Flickinger J, Nam J, Kirkpatrick J. Radiation dose–volume effects of optic nerves and chiasm. Int J Radiat Oncol Biol Phys. 2010;76(3):S28–35.
6. Hasegawa T, Kida Y, Yoshimoto M, Iizuka H, Ishii D, Yoshida K. Gamma knife surgery for convexity, parasagittal, and falcine meningiomas. J Neurosurg. 2011;114(5):1392–8.
7. Leavitt JA, Stafford SL, Link MJ, Pollock BE. Long-term evaluation of radiation-induced optic neuropathy after single-fraction stereotactic radiosurgery. Int J Radiat Oncol Biol Phys. 2013;87(3):524–7.
8. Leber KA, Bergloff J, Langmann G, Mokry M, Schrottner O, Pendl G. Radiation sensitivity of visual and oculomotor pathways. Stereotact Funct Neurosurg. 1995;64(Suppl 1):233–8.
9. Stafford SL, Pollock BE, Leavitt JA, Foote RL, Brown PD, Link MJ, et al. A study on the radiation tolerance of the optic nerves and chiasm after stereotactic radiosurgery. Int J Radiat Oncol Biol Phys. 2003;55(5):1177–81.
10. Conti A, Pontoriero A, Midili F, Iati G, Siragusa C, Tomasello C, et al. CyberKnife multisession stereotactic radiosurgery and hypofractionated stereotactic radiotherapy for perioptic meningiomas: intermediate-term results and radiobiological considerations. Springerplus. 2015;4:37.
11. Marchetti M, Conti A, Beltramo G, Pinzi V, Pontoriero A, Tramacere I, et al. Multisession radiosurgery for perioptic meningiomas: medium-to-long term results from a CyberKnife cooperative study. J Neuro-Oncol. 2019;143(3):597–604.
12. Puataweepong P, Dhanachai M, Hansasuta A, Dangprasert S, Sitathanee C, Ruangkanchanasetr R, et al. Clinical outcomes of perioptic tumors treated with hypofractionated stereotactic radiotherapy using CyberKnife(R) stereotactic radiosurgery. J Neuro-Oncol. 2018;139(3):679–88.
13. Astrahan M. Some implications of linear-quadratic-linear radiation dose-response with regard to hypofractionation. Med Phys. 2008;35(9):4161–72.
14. Emami B, Lyman J, Brown A, Coia L, Goitein M, Munzenrider JE, et al. Tolerance of normal tissue to therapeutic irradiation. Int J Radiat Oncol Biol Phys. 1991;21(1):109–22.
15. Goldsmith BJ, Rosenthal SA, Wara WM, Larson DA. Optic neuropathy after irradiation of meningioma. Radiology. 1992;185(1):71–6.
16. Lyman JT. Complication probability as assessed from dose-volume histograms. Radiat Res Suppl. 1985;8:S13–9.
17. Kutcher GJ, Burman C. Calculation of complication probability factors for non-uniform normal tissue irradiation: the effective volume method. Int J Radiat Oncol Biol Phys. 1989;16(6):1623–30.
18. Burman C, Kutcher GJ, Emami B, Goitein M. Fitting of normal tissue tolerance data to an analytic function. Int J Radiat Oncol Biol Phys. 1991;21(1):123–35.

19. Burman CM. Fitting of tissue tolerance data to analytic function: improving the therapeutic ratio. Front Radiat Ther Oncol. 2002;37:151–62.
20. Kutcher GJ. Quantitative plan evaluation: TCP/NTCP models. Front Radiat Ther Oncol. 1996;29:67–80.
21. Lyman JT, Wolbarst AB. Optimization of radiation therapy, III: a method of assessing complication probabilities from dose-volume histograms. Int J Radiat Oncol Biol Phys. 1987;13(1):103–9.
22. Lyman JT, Wolbarst AB. Optimization of radiation therapy, IV: a dose-volume histogram reduction algorithm. Int J Radiat Oncol Biol Phys. 1989;17(2):433–6.
23. Schultheiss TE. The radiation dose-response of the human spinal cord. Int J Radiat Oncol Biol Phys. 2008;71(5):1455–9.
24. Schultheiss TE, Thames HD, Peters LJ, Dixon DO. Effect of latency on calculated complication rates. Int J Radiat Oncol Biol Phys. 1986;12(10):1861–5.
25. McCunniff AJ, Liang MJ. Radiation tolerance of the cervical spinal cord. Int J Radiat Oncol Biol Phys. 1989;16(3):675–8.
26. Abbatucci JS, Delozier T, Quint R, Roussel A, Brune D. Radiation myelopathy of the cervical spinal cord: time, dose and volume factors. Int J Radiat Oncol Biol Phys. 1978;4(3–4):239–48.
27. Atkins HL, Tretter P. Time-dose considerations in radiation myelopathy. Acta Radiol Ther Phys Biol. 1966;5:79–94.
28. Marcus RB Jr, Million RR. The incidence of myelitis after irradiation of the cervical spinal cord. Int J Radiat Oncol Biol Phys. 1990;19(1):3–8.
29. Jeremic B, Djuric L, Mijatovic L. Incidence of radiation myelitis of the cervical spinal cord at doses of 5500 cGy or greater. Cancer. 1991;68(10):2138–41.
30. Daly ME, Luxton G, Choi CY, Gibbs IC, Chang SD, Adler JR, et al. Normal tissue complication probability estimation by the Lyman-Kutcher-Burman method does not accurately predict spinal cord tolerance to stereotactic radiosurgery. Int J Radiat Oncol Biol Phys. 2012;82(5):2025–32.
31. Morimoto M, Yoshioka Y, Kotsuma T, Adachi K, Shiomi H, Suzuki O, et al. Hypofractionated stereotactic radiation therapy in three to five fractions for vestibular schwannoma. Jpn J Clin Oncol. 2013;43(8):805–12.
32. Tsai JT, Lin JW, Lin CM, Chen YH, Ma HI, Jen YM, et al. Clinical evaluation of CyberKnife in the treatment of vestibular schwannomas. Biomed Res Int. 2013;2013:297093.
33. Meijer OW, Vandertop WP, Baayen JC, Slotman BJ. Single-fraction vs. fractionated linac-based stereotactic radiosurgery for vestibular schwannoma: a single-institution study. Int J Radiat Oncol Biol Phys. 2003;56(5):1390–6.
34. Poen JC, Golby AJ, Forster KM, Martin DP, Chinn DM, Hancock SL, et al. Fractionated stereotactic radiosurgery and preservation of hearing in patients with vestibular schwannoma: a preliminary report. Neurosurgery. 1999;45(6):1299–305. discussion 305–7.
35. Andrews DW, Suarez O, Goldman HW, Downes MB, Bednarz G, Corn BW, et al. Stereotactic radiosurgery and fractionated stereotactic radiotherapy for the treatment of acoustic schwannomas: comparative observations of 125 patients treated at one institution. Int J Radiat Oncol Biol Phys. 2001;50(5):1265–78.
36. Collen C, Ampe B, Gevaert T, Moens M, Linthout N, De Ridder M, et al. Single fraction versus fractionated linac-based stereotactic radiotherapy for vestibular schwannoma: a single-institution experience. Int J Radiat Oncol Biol Phys. 2011;81(4):e503–9.
37. McWilliams W, Trombetta M, Werts ED, Fuhrer R, Hillman T. Audiometric outcomes for acoustic neuroma patients after single versus multiple fraction stereotactic irradiation. Otol Neurotol. 2011;32(2):297–300.
38. Kano H, Kondziolka D, Khan A, Flickinger JC, Lunsford LD. Predictors of hearing preservation after stereotactic radiosurgery for acoustic neuroma: clinical article. J Neurosurg. 2013;119(Suppl):863–73.
39. Rashid A, Karam SD, Rashid B, Kim JH, Pang D, Jean W, et al. Multisession radiosurgery for hearing preservation. Semin Radiat Oncol. 2016;26(2):105–11.
40. Hayden Gephart MG, Hansasuta A, Balise RR, Choi C, Sakamoto GT, Venteicher AS, et al. Cochlea radiation dose correlates with hearing loss after stereotactic radiosurgery of vestibular schwannoma. World Neurosurg. 2013;80(3–4):359–63.
41. Mayer R, Sminia P. Reirradiation tolerance of the human brain. Int J Radiat Oncol Biol Phys. 2008;70(5):1350–60.
42. Friedman WA, Bova FJ, Bollampally S, Bradshaw P. Analysis of factors predictive of success or complications in arteriovenous malformation radiosurgery. Neurosurgery. 2003;52(2):296–307. discussion–8.
43. Blonigen BJ, Steinmetz RD, Levin L, Lamba MA, Warnick RE, Breneman JC. Irradiated volume as a predictor of brain radionecrosis after linear accelerator stereotactic radiosurgery. Int J Radiat Oncol Biol Phys. 2010;77(4):996–1001.
44. Lunsford LD, Niranjan A, Kondziolka D, Sirin S, Flickinger JC. Arteriovenous malformation radiosurgery: a twenty year perspective. Clin Neurosurg. 2008;55:108–19.

Part V

Oncology

14 Brain Metastases Surgical Management: Diagnostic, Therapeutic and Strategic Considerations

Philippe Metellus

14.1 Introduction

Brain metastases (BM) represent a major health problem in patients with cancer. It is estimated that approximately 20–40% of patients with malignant neoplasia develop brain metastases during their disease [1, 2]. These lesions, whose incidence is increasing due to the improvement of primary cancer management, represent the most frequent intra-axial brain tumors.

Whole-brain radiation therapy (WBRT) [3–5] has been the standard treatment of BM for a while. However, the advent of modern imaging techniques (CT and MRI), the improvement of surgical techniques, neuroanesthesia [6–9], and the positive impact of stereotactic radiotherapy [10] led to a reappraisal of local treatment modalities in BM management. Therapeutic decision depends on several factors related to tumor characteristics (number, radiological aspect, size, location), patient clinical status (neurological deficit, general condition, comorbidities, performance status), and primary disease status (controlled or uncontrolled, extracranial active metastatic disease) [11]. In this chapter, we will provide a review of available data on the impact of surgery in BM management and surgical indications in these patients.

P. Metellus (✉)
Département de Neurochirurgie, Ramsay Santé, Hôpital Privé Clairval, Marseille, France

14.2 Survival Impact of Surgery in BM

The effective impact, in terms of overall survival (OS), of surgery associated with WBRT in patients with single brain metastasis of solid cancers, in comparison with WBRT alone, has been demonstrated in several studies (Table 14.1).

In 1990, Patchell et al. [8] firstly showed that surgery associated with WBRT led to a significant increase of OS in patients with a unique brain metastasis compared to WBRT alone. In 1993, Vecht et al. [9] confirmed the positive impact on OS of the association of surgery and WBRT, in single brain metastasis. In 1996, Mintz et al. [12] did not find such a positive impact of surgery on OS. However, in this study only 21.4% of patients had a controlled extra-cerebral disease, and none of the patients had brain MRI assessment conversely to the two other studies, which suggests that these results should be interpreted with caution. Eventually, the survival impact of surgery associated with WBRT in comparison with surgery alone has then been evaluated [13, 14] (Table 14.1).

While adjuvant WBRT has led to a significant improvement in global intracranial control, it fails to improve the duration of functional and operating system independence [13, 14]. The conclusion was that in well-functioning patients with otherwise stable systemic disease and a limited number of BM [1–3] initially treated

© Springer Nature Switzerland AG 2020
A. Conti et al. (eds.), *CyberKnife NeuroRadiosurgery*,
https://doi.org/10.1007/978-3-030-50668-1_14

Table 14.1 Results of phase 3 randomized trials assessing the impact of local treatment on brain metastases

	Study design/level of evidence	Treatment	Population	Median survival	Patient with recurrence/progression Control rate	Median time to recurrence/progression
Patchell, 1990 [8]	Randomized trial Class 1	G1: WBRT ($n = 23$) G2: surgery + WBRT ($n = 25$)	Single metastasis	G1: 15 weeks G2: 40 weeks *Overall survival curves* *Log rank $p < 0.01$*	*Surgical site* G1: 12/23 (52%) G2: 5/25 (20%) *$p < 0.02$* *Remote site* G1: 3/23 (13%) G2: 5/25 (20%) *$p = NS$*	*Surgical site* G1: 21 weeks G2: >59 weeks *Local recurrence curves* *Log rank $p < 0.0001$*
Patchell, 1998 [13]	Randomized trial Class 1	G1: surgery ($n = 46$) G2: surgery + WBRT ($n = 49$)	Single metastasis	G1: 43 weeks G2: 48 weeks *Overall survival curves* *Log rank $p = NS$*	*Surgical site* G1: 21/46 (46%) G2: 5/49 (10%) *$p < 0.01$* *Remote site* G1: 17/46 (37%) G2: 7/49 (14%) *$p < 0.01$*	*Surgical site* G1: 27 weeks G2: >52 weeks *Local recurrence curves* *Log rank $p < 0.001$*
Kocher, 2011 [14]	Randomized trial Class 1	G1: surgery or SRS ($n = 179$) G2: surgery or SRS + WBRT ($n = 180$)	1–3 metastases	G1: 10.9 months G2: 10.7 months *Overall survival curves* *Log rank $p = 0.89$*	*Intracranial* G1: 139/179 (78%) G2: 57/180 (48%) *$p < 0.001$*	*Intracranial* G1: 3.4 months G2: 4.6 months *Local recurrence curves* *Log rank $p < 0.020$*
Mahajan, 2017 [18]	Randomized trial Class 1	G1: surgery + SRS ($n = 63$) G2: surgery ($n = 65$)	1–3 metastases	G1: 17 months G2: 18 months *Overall survival curves* *Log rank $p = 0.24$*	*Local control rate*[a] G1: 72% G2: 43% *$p = 0.015$* *Distant control rate*[a] G1: 42% G2: 33% *$p = 0.35$*	*Surgical site* G1: not reached G2: 7.6 months *Fine-Gray regression* *$p = 0.0097$*
Brown, 2017 [19]	Randomized trial Class 1	G1: surgery + SRS ($n = 98$) G2: surgery + WBRT ($n = 96$)	1–3 metastases	G1: 12.2 months G2: 11.6 months *Overall survival curves* *Log rank $p = 0.70$*	*Local control rate*[a] G1: 61.8% G2: 87.1% *$p = 0.00016$* *Distant control rate*[a] G1: 64.7% G2: 89.2% *$p = 0.00045$*	NR

Abbreviations: WBRT whole-brain radiation therapy, SRS stereotactic radiosurgery, NR no reported, NS not significant

[a]*Control rate at 12 months*

with just one surgery, WBRT can be withheld if neuroimaging monitoring is performed adequately.

However, although surgical techniques have improved significantly compared to the study reported by Patchell and colleagues [8], the local recurrence after surgery only reported in the available literature is still 50% at 6 months.

Furthermore, although the adjuvant WBRT allows better local control, the results of recent studies have shown an association with cognitive decline [14–16]. In order to avoid this toxicity, stereotactic radiosurgery (SRS) has been widely assessed in this population and has progressively replaced adjuvant WBRT although high-level evidence is still lacking [17].

Two phase 3 studies have recently been conducted to address the adjuvant strategy issue in patients with an oligometastatic disease treated with surgery. The first study by Mahajan and colleagues addressed, in a series of 132 patients, the value of postoperative SRS compared to observation in a surgical resection cavity.

They showed that SRS in the adjuvant tumor bed was associated with reduced local recurrence but failed to improve OS [18].

Twelve-month freedom from local recurrence was 43% in the observation group (68 patients) and 72% in the SRS group (64 patients) (Table 14.1). They also found the metastasis size to be inversely associated with better local control. Indeed, in patients harboring tumors up to 2.5 cm in maximal diameter, 12-month freedom from local recurrence was 91% versus 40% in patients with tumors of 2.5–3.5 cm in maximal diameter (HR 8.3 (95% CI 2.5–27.6), $p = 0.0005$) [18]. Another phase 3 study reported by Brown and colleagues addressed the value of WBRT compared to SRS in 194 patients with one resected brain metastasis and a resection cavity less than 5.0 cm in maximal extent. They failed to show any difference in terms of OS, but in the WBRT group, time to cognitive decline was significantly shorter. Actually, median time to cognitive deterioration was 3.7 months in the SRS group (98 patients) compared to 3.1 months in the WBRT group (96 patients) [19].

14.3 Surgical Indications

BM surgery goal is to improve brain tumor control, allow the patient's neurological symptoms relief, and provide an accurate tumor molecular characterization. Large tumors responsible for intracranial hypertension and symptomatic tumors located in eloquent areas represent a surgical indication.

In this situation, surgical resection is often indicated to relieve mass effect, offer seizure control, decrease intracranial pressure, and address neurological symptoms. Large masses can also result in significant cerebral edema, often requiring the use of corticosteroids; surgical resection of the offending lesion is the most effective way to reduce this edema and allow for the prompt cessation of steroid administration.

Additionally, if the lesion is in proximity to or involving the ventricular system, surgery can help prevent or address hydrocephalus from obstructed cerebrospinal fluid (CSF) flow. Furthermore, for larger lesions, surgery may offer superior local tumor control compared to radiation treatment modalities such as single-fraction stereotactic radiosurgery (SRS) [20–22].

Ebner et al. reported that lesions >3 cm in maximal diameter undergoing SRS had a lower 1-year local control rate (68%) than lesions less than 3 cm in size (86%) [23]. Specifically, in the treatment of large lesions, single-fraction SRS in particular becomes more limited, as larger volume masses require a reduced radiation dose to avoid toxicity, and treatment is thus more prone to failure [24].

A recent multi-institutional study evaluated the outcome of single-fraction SRS alone relative to surgical resection followed by postoperative SRS for brain metastases over 4 cm^3 in volume (diameter of ~2 cm) and reported a significantly lower local recurrence rate for patients treated with upfront surgery (36.7% and 20.5%, respectively; $p = 0.007$) [25]. As such, SRS is accepted as an ideal treatment for smaller lesions, particularly those less than 2 cm in maximal diameter. Surgery may also have a diagnostic role. In case of unknown primary, surgery is warranted to have a histological diagnosis.

Also, when a differential tumor diagnosis or pseudo-progression (radionecrosis) is suspected, a histological authentication may be necessary [6]. Finally, in some cases, it may be interesting to biologically document the cerebral metastatic disease. Indeed, molecular or gene expression changes may occur between the primary tumor and BM. This could impact surgical decision-making in patients with BM. Furthermore, for some patients whose initial tumor material is not available, biological metastatic disease documentation could identify patients eligible for a specific targeted therapy. Therefore, surgical resection of BM, in these cases, represents a pivotal step in the treatment strategy decision-making process that can lead to a change in the therapeutic management.

In summary, surgical excision, when possible, should be performed in the following situations:

- Therapeutic
  - Voluminous lesion >3 cm, symptomatic or not
  - Cystic or necrotic lesion with edema
  - Symptomatic lesion located in an eloquent area
  - Lesion located in the posterior fossa with a mass effect or associated hydrocephalus
- Diagnostic
  - Unknown primary cancer
  - Potential differential diagnosis
  - Suspected radionecrosis in previously irradiated patients
- Strategic
  - Biological documentation of brain metastatic disease in patients potentially eligible for new targeted therapy

Finally, surgical resection of brain metastatic lesions also contributes to the constitution of a BM tissue database that could allow for a better understanding of the molecular determinants underlying the brain metastatic disease and for identifying new potential molecular targets and its associated treatments.

14.4 Selection of Patients for Surgical Resection

In addition to radiographic and tumoral factors, multiple clinical issues should be considered when evaluating surgical candidacy.

A patient's systemic cancer status is a serious consideration, and generally patients with controlled or absent systemic disease are ideal surgical candidates. However, these decisions must be individualized and discussed in collaboration with the patient's medical oncology team to assess the patient's overall prognosis and the availability of additional therapy for systemic disease.

For instance, if a patient has concomitant brain and systemic metastases and is treatment naive, surgical resection may be a reasonable first step in the treatment plan. In addition to the oncological status, the patient's functional status is important. In particular a Karnofsky Performance Scale (KPS) score of at least 70 is desirable. In addition, patients with multiple severe comorbidities and coagulopathy or undergoing systemic chemotherapy may serve better with less invasive methods of treatment.

In order to assist in clinical decision-making, the Radiation Therapy Oncology Group (RTOG) developed the classification system of recursive partitioning analysis (RPA), which captures the salient factors that go into treatment planning. RPA is classified on the basis of KPS score, patient age, and extracranial disease status. The RPA I class is associated with the most favorable prognosis, while the RPA III patients have the worst expected result.

Tendulkar et al. analyzed the outcome of 271 patients undergoing resection of a single brain metastasis [26] and reported that patient survival significantly correlated with RPA class, with the mean survival times of RPA classes 1, 2, and 3 patients post-tumor resection being 21.4, 9, and 8.9 months, respectively, validating the prognostic significance of this scale.

The predictive impact of the RPA class has therefore been validated in several surgical series [27, 28]. Another prognostic score, the Graded Prognostic Assessment (GPA) is an updated prognostic index for patients with brain metastases.

This prognostic index is based on age, KPS score, number of intracranial lesions, and status of systemic disease. It was originally developed from a database of 1960 patients accrued to four RTOG protocols for patients with brain metastases [29]. The median overall survival times based on GPA score were 2.6 months for 0–1 point, 3.8 months for 1.5–2.5 points, 6.9 months for 3 points, and 11 months for 3.5–4 points. The GPA has been refined to include histology-specific prognostic indices based on multi-institutional analyses of 4259 patients with brain metastases from breast carcinoma, small cell and non-small cell lung carcinoma, GI cancers, melanoma, and renal cell carcinoma [30].

14.5 Impact of Extent of Resection (EOR) and Surgical Technique

When it is anatomically safe to perform it, gross total resection (GTR) is the goal of surgery for metastatic disease, as it improves the outcome, particularly in patients with single or solitary metastasis [26, 27]. A single institute study evaluated outcome predictors in 271 patients with single brain metastases. In this study, patients who received a GTR had a median survival time of 10.6 months compared to 8.7 months in patients who had a subtotal resection (STR) [26].

In another retrospective study analyzing the surgical outcome of 157 patients with brain metastases, 96 of whom (60%) had a single brain metastasis, the authors reported that extent of resection (EOR) significantly impacted patient survival. Patients who had a STR had a median survival time of 15.1 months compared to 20.4 months in patients where a GTR was achieved [27]. Furthermore, GTR strongly affected patients' functional status; KPS scores of the GTR group improved from 82 to 87, and those of the STR group changed from 79 to 77, and this difference was found to be statistically significant. It is important to note that even though patients with metastatic disease represent a higher-risk population, with diligent patient selection, maximal safe resection is often well tolerated and with an acceptable risk. A retrospective study examining the outcomes of 206 surgical patients with brain metastases reported mortality and morbidity rates of 0% and 10.3%, respectively [31]. This low perioperative morbidity was similar to that described in another retrospective study that included 208 surgical patients and reported an operative mortality of only 1.9% [28].

In addition to achieving maximum resection, there is extensive literature indicating that the surgical resection method also affects the outcome, in particular the value of en bloc resection. En bloc resection involves circumferential dissection of the metastatic lesion along the brain-tumor interface and avoiding the breech of the tumor capsule.

This technique has multiple practical benefits over fragmented resection (i.e., debulking and removal of the internal tumor), including avoidance of tumor cell spillage into the surrounding brain, reduction of intraoperative bleeding, and clearer visualization of tumor borders. In addition to its intraoperative benefits, en bloc resection also imparts clinical advantages.

A notable study analyzed the predictors of local recurrence in 570 patients with single brain metastasis who had undergone surgery where GTR was achieved. The authors demonstrated that patients who had a fragmented tumor resection were 1.7 times more likely to develop local recurrence than those whose tumors were removed en bloc [32].

In addition to its impact on local recurrence, resection technique also influences the risk of metastatic CSF dissemination, i.e., leptomeningeal disease (LMD), which carries a universally poor prognosis. In a surgical series of 242 patients with brain metastases (68% with a single lesion), 16% of patients subsequently developed LMD. Analysis of potential clinical predictors of LMD demonstrated that fragmented resection had a four times higher risk of developing LMD than fragmented resection [33].

In addition, another study focusing on surgically treated posterior fossa metastases (260 patients) also showed the benefit of en bloc resection [34].

Note that posterior fossa injuries are of particular interest in the development of LMD due to their proximity to CSF spaces. In this study, GTR was achieved in 96% of patients and 10% of patients developed LMD. Fragmented resection was significantly associated with an increased risk of LMD; in particular 13.9% of patients with fragmented resection eventually developed LMD compared to only 5.7% of patients with en bloc resection [34].

In addition to being effective, en bloc resection is safe, even for metastases found near functional (eloquent) cortex. Recent data indicate that en bloc resection technique is both feasible and safe in this setting.

Following an analysis of 1033 surgical patients, 62% of whom underwent en bloc resection, it was reported by the authors that en bloc resection was not associated with an increase in complication rates compared to fragmented resection, even for tumors localized in the eloquent cortex [35].

Intraoperative imaging and brain mapping technologies have emerged as powerful additions to maximize the extent of brain metastasis resection while minimizing morbidity. However, the clinical benefit of aggressive surgical resection is negated if it results in severe postoperative neurological deficits. New functional deficits can increase the risk of thromboembolic complications, remove eligibility for systemic treatment regimens or selected clinical trials, and seriously affect the quality of life of patients.

With this in mind, the use of surgical adjuncts has proven to be vital in making surgical resection effective and safe. Both preoperative and intraoperative imaging modalities can contribute to surgical success. Standard three-dimensional preoperative magnetic resonance (MR) imaging is routinely performed to facilitate accurate targeting of a specific lesion for surgical approach planning and tailoring of the craniotomy. In the situation where the lesion abuts or involves the eloquent cortex (e.g., speech or motor centers), additional functional imaging modalities are often required. Predicting the exact location in the eloquent cortex with standard anatomical imaging can be difficult at times due to distortion of the normal anatomy by the tumor or surrounding edema. Preoperative localization of eloquent brain regions using functional MR imaging, diffusion tensor (DT) imaging tractography, and/or transcranial magnetic stimulation (TMS) allows for more detailed surgical planning through visualization of the spatial relationship between the lesion and surrounding eloquent cortex [36–38]. These preoperative imaging data also assist the surgeon and treatment team when counselling the patient regarding preoperative risk and potential postoperative recovery time. Intraoperatively, ultrasound is a useful and cost-effective technological imaging adjunct that provides real-time data to confirm the extent of resection [39].

In addition, intraoperative MRI imaging is widely used as an adjunct to resection for infiltrative glial tumors, but it can also be used to assist resection of brain metastases, particularly deep lesions [40, 41].

Intraoperative neuro-monitoring is another key element for the resection of brain metastases in the eloquent regions. Even with notable advances in pre- and postoperative imaging techniques, the gold standard for identifying the eloquent cortex during surgery remains the use of intraoperative mapping. Brain mapping provides real-time information on the relationship between the lesion and the surrounding critical structures.

For metastases that adhere to or involve the precentral gyrus (posterior frontal lobe; motor cortex) or deep subcortical motor tracts, intraoperative localization of these motor fibers is critical to conduct safe resection. With the assistance of a neurophysiology team, the location of the motor cortex can be confirmed intraoperatively by placing a grid electrode on the cortical surface before the start of tumor resection.

The location of subcortical motor fibers (corticospinal tract) can be localized and continuously monitored during surgical resection using direct stimulation with a current-generating monopolar or bipolar electrode. The benefit of motor mapping in the management of brain metastases has been confirmed in the literature [42, 43]. For example, a retrospective study of 33 surgical patients with lesions in proximity to the motor cortex reported favorable outcomes using intraoperative mapping techniques. Specifically, GTR was achieved in 94% of patients. Six patients (18%) experienced worsening neurological symptoms, but all patients were neurologically recovered at their 3-month follow-up visit [43]. For lesions located in language areas (e.g., posterior temporal lobe, inferior frontal lobe, interior parietal lobule), intraoperative speech mapping is often needed, and this requires awake surgery. After the initial craniotomy and cortical exposure, a bipolar electrode is used to stimulate the cortical region of interest while language tasks are performed by the patient. Areas on the cortex that produce frank speech arrest or language disturbances (semantic errors, paraphasias, perseverations) are marked and subsequently avoided during the remainder of the surgical resection.

Overall, the advantage of intraoperative mapping is evident and has shown improved neurological outcomes in patients with brain metastases in difficult areas [44].

14.6 New Surgical Indications in the Era of Targeted Therapies

The interest in having a biological documentation of metastatic disease is to identify a potential molecular phenotype switch in metastatic tumors that could help the clinician to define the therapeutic strategy.

A recent study has shown that genetic and phenotypic heterogeneity in metastases of breast cancer explained the resistance to targeted therapies [45]. Indeed, it is well established that there can be a molecular phenotypic conversion between the primitive and metastatic disease, which is influenced by the time to onset of metastasis and by the metastatic site [46–53]. Thus, the possibility to obtain a molecular characterization of the cerebral metastatic disease may be warranted when the molecular status of the primary tumor is insufficiently documented or when the modern profiling tests used were not available at time of diagnosis. Indeed, molecular profiling of the metastatic disease not only can lead to a change of the local treatment but also could impact the systemic treatment strategy [46, 54–56].

This emphasizes the critical role of the surgeon who is not only a local treatment actor (large, symptomatic, and potentially life-threatening lesions) but mainly plays a key role in the overall decision-making therapeutic strategy [47, 48].

In 2012, a pioneering randomized phase 2 study compared the use of targeted therapies based on the molecular profile of tumors compared to conventional chemotherapy in all types of cancers in treatment failure. This study showed that this approach was well tolerated, feasible, and consistent with routine clinical practice [45]. However, if this study has shown that this approach is feasible, it remains to be shown that the choice of a target based on the molecular profile of the tumor improves the prognosis of the patients.

Thus, in this perspective, a French multicenter study led by the same group reported the interest of molecular screening by array CGH and high-throughput sequencing of metastatic breast cancer. This innovative approach consisted of identifying genomic alterations in metastatic tumors that could be targeted by new agents. However, the results of this study, while promising, were disappointing because to date, there are no effective molecular therapies available to target the identified genomic alterations.

Furthermore, this approach does not integrate other components of personalized medicine such as immunotherapy, DNA repair modulation, and intra-tumor heterogeneity [57]. More recently, several studies dedicated to brain metastases have reported genomic, post-genomic, and epigenomic profiling in primary and matched brain metastases [54–56, 58–61].

Brastianos and colleagues first reported the genomic characterization of brain metastases and their matched primary tumors. They showed that brain metastatic tumors shared genetic alterations that were frequently not detected in the primary tumor.

These data suggest that primary tumor sequencing may miss a considerable number of opportunities for targeted therapies [56].

Most of these studies were performed in breast cancer patients [46, 55, 60], and from this pioneering report, several studies have reported the comparison of genetic alterations between brain metastases and primary tumor.

Another study reported by Tyran et al. on DNA mutation and copy of numerical profiles of primary breast cancer and coupled brain metastases also provided strong evidence that BM tumor samples had more genetic alterations than their primary counterparts, emphasizing its potential interest in precision medicine [60].

In line with these published data, several recent works have provided strong evidence of the genetic heterogeneity between primary tumors and their brain metastases, pointing out the actual necessity to further characterize the biology of the metastasis. Gene expression profiling recently reported data uncovered recurrent gene expression acquisitions in brain metastases distinct from their matched primary tumors [54]. All these studies provided a growing body of evidence that there is a specific acquired molecular phenotype in brain metastases that is not present in the primary tumors and which warrants immediate clinical attention. The identification of these metastases-acquired aberrations in key oncogenic pathways could provide suitable therapeutic targets.

Therefore, genomic and post-genomic profiling of paired specimens represents a compelling and underutilized strategy to identify targeting dependencies in advanced cancer patients. All in all, these data underline the actual need to obtain tissues from patients with brain metastases because they offer immediate opportunities for a more informed decision-making process based on genetic analysis. From this perspective, the neurosurgeon does not act only as an actor of local treatment but rather as a key actor involved in all diagnostic, therapeutic, and strategic stages in patients' brain metastases management.

References

1. Mehta MP, Khuntia D. Current strategies in whole-brain radiation therapy for brain metastases. Neurosurgery. 2005;57(Suppl 5):S33–44. discusssion S1–4.
2. Mehta MP, Tsao MN, Whelan TJ, Morris DE, Hayman JA, Flickinger JC, et al. The American Society for Therapeutic Radiology and Oncology (ASTRO) evidence-based review of the role of radiosurgery for brain metastases. Int J Radiat Oncol Biol Phys. 2005;63(1):37–46.
3. Gaspar LE, Mehta MP, Patchell RA, Burri SH, Robinson PD, Morris RE, et al. The role of whole brain radiation therapy in the management of newly diagnosed brain metastases: a systematic review and evidence-based clinical practice guideline. J Neuro-Oncol. 2010;96(1):17–32.
4. Kalkanis SN, Kondziolka D, Gaspar LE, Burri SH, Asher AL, Cobbs CS, et al. The role of surgical resection in the management of newly diagnosed brain metastases: a systematic review and evidence-based clinical practice guideline. J Neuro-Oncol. 2010;96(1):33–43.
5. Linskey ME, Andrews DW, Asher AL, Burri SH, Kondziolka D, Robinson PD, et al. The role of stereotactic radiosurgery in the management of patients with newly diagnosed brain metastases: a systematic review and evidence-based clinical practice guideline. J Neuro-Oncol. 2010;96(1):45–68.
6. Al-Shamy G, Sawaya R. Management of brain metastases: the indispensable role of surgery. J Neuro-Oncol. 2009;92(3):275–82.
7. Lang FF, Sawaya R. Surgical management of cerebral metastases. Neurosurg Clin N Am. 1996;7(3):459–84.
8. Patchell RA, Tibbs PA, Walsh JW, Dempsey RJ, Maruyama Y, Kryscio RJ, et al. A randomized trial of surgery in the treatment of single metastases to the brain. N Engl J Med. 1990;322(8):494–500.
9. Vecht CJ, Haaxma-Reiche H, Noordijk EM, Padberg GW, Voormolen JH, Hoekstra FH, et al. Treatment of single brain metastasis: radiotherapy alone or combined with neurosurgery? Ann Neurol. 1993;33(6):583–90.
10. Andrews DW, Scott CB, Sperduto PW, Flanders AE, Gaspar LE, Schell MC, et al. Whole brain radiation therapy with or without stereotactic radiosurgery boost for patients with one to three brain metastases: phase III results of the RTOG 9508 randomised trial. Lancet. 2004;363(9422):1665–72.
11. Gaspar L, Scott C, Rotman M, Asbell S, Phillips T, Wasserman T, et al. Recursive partitioning analysis (RPA) of prognostic factors in three radiation therapy oncology group (RTOG) brain metastases trials. Int J Radiat Oncol Biol Phys. 1997;37(4):745–51.
12. Mintz AH, Kestle J, Rathbone MP, Gaspar L, Hugenholtz H, Fisher B, et al. A randomized trial to assess the efficacy of surgery in addition to

radiotherapy in patients with a single cerebral metastasis. Cancer. 1996;78(7):1470–6.
13. Patchell RA, Tibbs PA, Regine WF, Dempsey RJ, Mohiuddin M, Kryscio RJ, et al. Postoperative radiotherapy in the treatment of single metastases to the brain: a randomized trial. JAMA. 1998;280(17):1485–9.
14. Kocher M, Soffietti R, Abacioglu U, Villa S, Fauchon F, Baumert BG, et al. Adjuvant whole-brain radiotherapy versus observation after radiosurgery or surgical resection of one to three cerebral metastases: results of the EORTC 22952-26001 study. J Clin Oncol. 2011;29(2):134–41.
15. Soffietti R, Kocher M, Abacioglu UM, Villa S, Fauchon F, Baumert BG, et al. A European Organisation for Research and Treatment of Cancer phase III trial of adjuvant whole-brain radiotherapy versus observation in patients with one to three brain metastases from solid tumors after surgical resection or radiosurgery: quality-of-life results. J Clin Oncol. 2013;31(1):65–72.
16. Brown PD, Jaeckle K, Ballman KV, Farace E, Cerhan JH, Anderson SK, et al. Effect of radiosurgery alone vs radiosurgery with whole brain radiation therapy on cognitive function in patients with 1 to 3 brain metastases: a randomized clinical trial. JAMA. 2016;316(4):401–9.
17. Roberge D, Parney I, Brown PD. Radiosurgery to the postoperative surgical cavity: who needs evidence? Int J Radiat Oncol Biol Phys. 2012;83(2):486–93.
18. Mahajan A, Ahmed S, McAleer MF, Weinberg JS, Li J, Brown P, et al. Post-operative stereotactic radiosurgery versus observation for completely resected brain metastases: a single-centre, randomised, controlled, phase 3 trial. Lancet Oncol. 2017;18(8):1040–8.
19. Brown PD, Ballman KV, Cerhan JH, Anderson SK, Carrero XW, Whitton AC, et al. Postoperative stereotactic radiosurgery compared with whole brain radiotherapy for resected metastatic brain disease (NCCTG N107C/CEC.3): a multicentre, randomised, controlled, phase 3 trial. Lancet Oncol. 2017;18(8):1049–60.
20. Angelov L, Mohammadi AM, Bennett EE, Abbassy M, Elson P, Chao ST, et al. Impact of 2-staged stereotactic radiosurgery for treatment of brain metastases >/= 2 cm. J Neurosurg. 2018;129(2):366–82.
21. Petrovich Z, Yu C, Giannotta SL, O'Day S, Apuzzo ML. Survival and pattern of failure in brain metastasis treated with stereotactic gamma knife radiosurgery. J Neurosurg. 2002;97(5 Suppl):499–506.
22. Vogelbaum MA, Angelov L, Lee SY, Li L, Barnett GH, Suh JH. Local control of brain metastases by stereotactic radiosurgery in relation to dose to the tumor margin. J Neurosurg. 2006;104(6):907–12.
23. Ebner D, Rava P, Gorovets D, Cielo D, Hepel JT. Stereotactic radiosurgery for large brain metastases. J Clin Neurosci. 2015;22(10):1650–4.
24. Shaw E, Scott C, Souhami L, Dinapoli R, Kline R, Loeffler J, et al. Single dose radiosurgical treatment of recurrent previously irradiated primary brain tumors and brain metastases: final report of RTOG protocol 90-05. Int J Radiat Oncol Biol Phys. 2000;47(2):291–8.
25. Prabhu RS, Press RH, Patel KR, Boselli DM, Symanowski JT, Lankford SP, et al. Single-fraction stereotactic radiosurgery (SRS) alone versus surgical resection and SRS for large brain metastases: a multi-institutional analysis. Int J Radiat Oncol Biol Phys. 2017;99(2):459–67.
26. Tendulkar RD, Liu SW, Barnett GH, Vogelbaum MA, Toms SA, Jin T, et al. RPA classification has prognostic significance for surgically resected single brain metastasis. Int J Radiat Oncol Biol Phys. 2006;66(3):810–7.
27. Lee CH, Kim DG, Kim JW, Han JH, Kim YH, Park CK, et al. The role of surgical resection in the management of brain metastasis: a 17-year longitudinal study. Acta Neurochir. 2013;155(3):389–97.
28. Paek SH, Audu PB, Sperling MR, Cho J, Andrews DW. Reevaluation of surgery for the treatment of brain metastases: review of 208 patients with single or multiple brain metastases treated at one institution with modern neurosurgical techniques. Neurosurgery. 2005;56(5):1021–34.
29. Sperduto PW, Berkey B, Gaspar LE, Mehta M, Curran W. A new prognostic index and comparison to three other indices for patients with brain metastases: an analysis of 1,960 patients in the RTOG database. Int J Radiat Oncol Biol Phys. 2008;70(2):510–4.
30. Sperduto PW, Chao ST, Sneed PK, Luo X, Suh J, Roberge D, et al. Diagnosis-specific prognostic factors, indexes, and treatment outcomes for patients with newly diagnosed brain metastases: a multi-institutional analysis of 4,259 patients. Int J Radiat Oncol Biol Phys. 2010;77(3):655–61.
31. Schodel P, Schebesch KM, Brawanski A, Proescholdt MA. Surgical resection of brain metastases-impact on neurological outcome. Int J Mol Sci. 2013;14(5):8708–18.
32. Patel AJ, Suki D, Hatiboglu MA, Abouassi H, Shi W, Wildrick DM, et al. Factors influencing the risk of local recurrence after resection of a single brain metastasis. J Neurosurg. 2010;113(2):181–9.
33. Ahn JH, Lee SH, Kim S, Joo J, Yoo H, Lee SH, et al. Risk for leptomeningeal seeding after resection for brain metastases: implication of tumor location with mode of resection. J Neurosurg. 2012;116(5):984–93.
34. Suki D, Abouassi H, Patel AJ, Sawaya R, Weinberg JS, Groves MD. Comparative risk of leptomeningeal disease after resection or stereotactic radiosurgery for solid tumor metastasis to the posterior fossa. J Neurosurg. 2008;108(2):248–57.
35. Patel AJ, Suki D, Hatiboglu MA, Rao VY, Fox BD, Sawaya R. Impact of surgical methodology on the complication rate and functional outcome of patients with a single brain metastasis. J Neurosurg. 2015;122(5):1132–43.
36. Hendrix P, Senger S, Griessenauer CJ, Simgen A, Schwerdtfeger K, Oertel J. Preoperative navigated transcranial magnetic stimulation in patients with

motor eloquent lesions with emphasis on metastasis. Clin Anat. 2016;29(7):925–31.
37. Huberfeld G, Trebuchon A, Capelle L, Badier JM, Chen S, Lefaucheur JP, et al. Preoperative and intraoperative neurophysiological investigations for surgical resections in functional areas. Neurochirurgie. 2017;63(3):142–9.
38. Sollmann N, Wildschuetz N, Kelm A, Conway N, Moser T, Bulubas L, et al. Associations between clinical outcome and navigated transcranial magnetic stimulation characteristics in patients with motor-eloquent brain lesions: a combined navigated transcranial magnetic stimulation-diffusion tensor imaging fiber tracking approach. J Neurosurg. 2018;128(3):800–10.
39. Unsgaard G, Selbekk T, Brostrup Muller T, Ommedal S, Torp SH, Myhr G, et al. Ability of navigated 3D ultrasound to delineate gliomas and metastases—comparison of image interpretations with histopathology. Acta Neurochir. 2005;147(12):1259–69. discussion 69.
40. Senft C, Ulrich CT, Seifert V, Gasser T. Intraoperative magnetic resonance imaging in the surgical treatment of cerebral metastases. J Surg Oncol. 2010;101(5):436–41.
41. Tan TC, Black PM. Image-guided craniotomy for cerebral metastases: techniques and outcomes. Neurosurgery. 2007;61(1 Suppl):349–56. discussion 56–7.
42. Kellogg RG, Munoz LF. Selective excision of cerebral metastases from the precentral gyrus. Surg Neurol Int. 2013;4:66.
43. Sanmillan JL, Fernandez-Coello A, Fernandez-Conejero I, Plans G, Gabarros A. Functional approach using intraoperative brain mapping and neurophysiological monitoring for the surgical treatment of brain metastases in the central region. J Neurosurg. 2017;126(3):698–707.
44. Kamp MA, Dibue M, Niemann L, Reichelt DC, Felsberg J, Steiger HJ, et al. Proof of principle: supramarginal resection of cerebral metastases in eloquent brain areas. Acta Neurochir. 2012;154(11):1981–6.
45. Le Tourneau C, Kamal M, Tredan O, Delord JP, Campone M, Goncalves A, et al. Designs and challenges for personalized medicine studies in oncology: focus on the SHIVA trial. Target Oncol. 2012;7(4):253–65.
46. Priedigkeit N, Hartmaier RJ, Chen Y, Vareslija D, Basudan A, Watters RJ, et al. Intrinsic subtype switching and acquired ERBB2/HER2 amplifications and mutations in breast cancer brain metastases. JAMA Oncol. 2017;3(5):666–71.
47. Metellus P, Bialecki E, Le Rhun E, Dhermain F. Neurosurgical and radiosurgical decision making in brain metastasis patients in the area of targeted therapies? Chin Clin Oncol. 2015;4(2):19.
48. Metellus P, Tallet A, Dhermain F, Reyns N, Carpentier A, Spano JP, et al. Global brain metastases management strategy: a multidisciplinary-based approach. Cancer Radiother. 2015;19(1):61–5.
49. Houssami N, Macaskill P, Balleine RL, Bilous M, Pegram MD. HER2 discordance between primary breast cancer and its paired metastasis: tumor biology or test artefact? Insights through meta-analysis. Breast Cancer Res Treat. 2011;129(3):659–74.
50. Duchnowska R, Jassem J, Goswami CP, Dundar M, Gokmen-Polar Y, Li L, et al. Predicting early brain metastases based on clinicopathological factors and gene expression analysis in advanced HER2-positive breast cancer patients. J Neuro-Oncol. 2015;122(1):205–16.
51. Duchnowska R, Sperinde J, Chenna A, Huang W, Weidler JM, Winslow J, et al. Quantitative HER2 and p95HER2 levels in primary breast cancers and matched brain metastases. Neuro Oncol. 2015;17(9):1241–9.
52. Duchnowska R, Peksa R, Radecka B, Mandat T, Trojanowski T, Jarosz B, et al. Immune response in breast cancer brain metastases and their microenvironment: the role of the PD-1/PD-L axis. Breast Cancer Res. 2016;18(1):43.
53. Duchnowska R, Jarzab M, Zebracka-Gala J, Matkowski R, Kowalczyk A, Radecka B, et al. Brain metastasis prediction by transcriptomic profiling in triple-negative breast cancer. Clin Breast Cancer. 2017;17(2):e65–75.
54. Vareslija D, Priedigkeit N, Fagan A, Purcell S, Cosgrove N, O'Halloran PJ, et al. Transcriptome characterization of matched primary breast and brain metastatic tumors to detect novel actionable targets. J Natl Cancer Inst. 2019;111(4):388–98.
55. Lee JY, Park K, Lim SH, Kim HS, Yoo KH, Jung KS, et al. Mutational profiling of brain metastasis from breast cancer: matched pair analysis of targeted sequencing between brain metastasis and primary breast cancer. Oncotarget. 2015;6(41):43731–42.
56. Brastianos PK, Carter SL, Santagata S, Cahill DP, Taylor-Weiner A, Jones RT, et al. Genomic characterization of brain metastases reveals branched evolution and potential therapeutic targets. Cancer Discov. 2015;5(11):1164–77.
57. Le Tourneau C, Delord JP, Goncalves A, Gavoille C, Dubot C, Isambert N, et al. Molecularly targeted therapy based on tumour molecular profiling versus conventional therapy for advanced cancer (SHIVA): a multicentre, open-label, proof-of-concept, randomised, controlled phase 2 trial. Lancet Oncol. 2015;16(13):1324–34.
58. Orozco JIJ, Knijnenburg TA, Manughian-Peter AO, Salomon MP, Barkhoudarian G, Jalas JR, et al. Epigenetic profiling for the molecular classification of metastatic brain tumors. Nat Commun. 2018;9(1):4627.
59. De Mattos-Arruda L, Ng CKY, Piscuoglio S, Gonzalez-Cao M, Lim RS, De Filippo MR, et al. Genetic heterogeneity and actionable mutations in

HER2-positive primary breast cancers and their brain metastases. Oncotarget. 2018;9(29):20617–30.
60. Tyran M, Carbuccia N, Garnier S, Guille A, Adelaide J, Finetti P, et al. A comparison of DNA mutation and copy number profiles of primary breast cancers and paired brain metastases for identifying clinically relevant genetic alterations in brain metastases. Cancers. 2019;11(5):665.
61. Schulten HJ, Bangash M, Karim S, Dallol A, Hussein D, Merdad A, et al. Comprehensive molecular biomarker identification in breast cancer brain metastases. J Transl Med. 2017;15(1):269.

Brain Metastasis: The Experience of the Burdenko Institute of Neurosurgery

15

Elena R. Vetlova, Andrey V. Golanov,
Sergey M. Banov, Natalia A. Antipina,
Amayak A. Durgaryan, Elena N. Igoshina,
Mikhail V. Galkin, and Elena I. Butenko

15.1 Introduction

Brain metastases are characterized by:

- Presence of distinct borders with brain tissue (CTV = GTV)
- Opportunity to determine precisely the position of the target relative to the skull (allows the usage of minimal margin for PTV)
- Stability of the target position from fraction to fraction during the course of hypofractionated irradiation
- Immobility of the target during irradiation (ITV = CTV)

Thus, patients with metastatic brain damage need precise and conformal irradiation, which can be fully implemented with the CyberKnife LINAC.

E. R. Vetlova · N. A. Antipina · A. A. Durgaryan
E. N. Igoshina · M. V. Galkin · E. I. Butenko
National Medical Research Center of Neurosurgery named after Burdenko, Moscow, Russia

A. V. Golanov (✉) · S. M. Banov
National Medical Research Center of Neurosurgery named after Burdenko, Moscow, Russia

Moscow Gamma Knife Center, Moscow, Russia
e-mail: golanov@nsi.ru

15.2 Stereotactic Radiosurgery

Radiosurgical treatment is an effective and generally accepted alternative to whole brain irradiation in patients with a limited (up to four foci) metastatic brain disease due to good local tumor control and low frequency of complications [1, 2].

Radiosurgery has the following advantages for metastatic foci: the possibility of irradiating multiple targets located in the deep compartments or eloquent brain areas, preservation of cognitive functions of the brain, and the possibility of re-irradiation in case of local reccurence.

At the NMRC Burdenko, the results of the radiosurgical treatment of patients with BM were analyzed [3].

This retrospective analysis included 502 patients with 2782 BMs who received stereotactic radiosurgery. Local control of BM for a period of 6, 12, and 24 months was 96.2%, 90.9%, and 82.6%, respectively. The 12-month local growth control was lesions >4 cm^3 was 1.7 times worse than in the case of BM with a volume of less than 1 cm^3 (6.8% and 93.8%, respectively). The overall survival of patients with metastatic brain lesions at 12, 24, and 36 months was 42.5%, 24.6%, and 17.8%, respectively.

Therefore, considering our own experience as well as published data, the results of stereotactic radiosurgery for metastatic lesions with a volume

© Springer Nature Switzerland AG 2020
A. Conti et al. (eds.), *CyberKnife NeuroRadiosurgery*,
https://doi.org/10.1007/978-3-030-50668-1_15

of up to 4 cm^3 using the CyberKnife appear to be similar to those of other specialized radiosurgical devices.

Studies have shown the fundamental possibility of performing radiosurgery in patients with multiple (up to 10) BMs, which increments the possibilities of radiosurgical treatment of patients with BM [3].

There is a tendency to perform radiosurgery with the CyberKnife at the NMRC Burdenko in patients with 5–10 BMs in the case of extracranial stability of the extracranial disease and reserves of systemic antitumor treatment. The possibilities of radiosurgery are limited by the target volume, since the presence of a large-sized BM requires a dose reduction, which, in turn, reduces the indices of local control. In the study of dose escalation during radiosurgery, depending on the target volume, RTOG (Radiation Therapy Oncology Group) 90-05, the concept of large BMs was identified [4]. Large metastases were considered a poor target for radiosurgery.

The optimal dose of radiation for radiosurgery is 22–24 Gy in the presence of foci up to 2 cm in diameter (<4 cm^3), which allows local control in 85–96% of lesions with acceptable toxicity [5, 6].

BM from 2–3 cm and more (3–4 cm) in the maximum diameter can be irradiated with a dose of 18 Gy and 15 Gy, respectively, without exceeding the tolerance of normal brain tissue. Nevertheless, dose restrictions for BM larger than 3 cm in diameter led to a decrease in 12-month local control rate to less than 50% [7, 8].

Thus, performing radiosurgery is not always possible in the presence of large lesions, where the dose escalation per fraction is limited by the likelihood of early and late side effects.

Metastases to the brain can be characterized according to their diameter or volume. Lesioins larger than 2–3 cm in diameter or larger than 4 cm in volume will be considered as large BM.

The optimal treatment for such BM has not been established yet. For large BM, the preferred method is a combination of surgery with postoperative irradiation of the tumor bed. BM surgery can reduce the mass effect, alleviating neurological symptoms, while the addition of stereotactic radiation therapy can improve control of the tumor growth.

15.3 Surgical Resection of Brain Metastasis Followed by Stereotactic Radiotherapy

The retrospective analysis of the results of the surgical treatment of BM followed by stereotactic radiation therapy at NMRC Burdenko was also performed.

The aim of this study was to evaluate the effectiveness of adjuvant stereotactic radiation therapy of a tumor cavity using hypofractionated schedules.

Between 2013 and 2019, 261 patients received neurosurgical treatment for metastatic brain lesions. Of these, 140 patients underwent stereotactic radiation therapy of the tumor bed of the removed metastasis in the hypofractionated mode, and 121 patients only underwent surgical treatment.

The study included adult patients with BM who underwent surgical resection of at least one lesion and who had not previously received whole brain irradiation (WBRT) and did not require WBRT as adjuvant treatment. In the case of additional distant BM, stereotactic radiosurgery was performed simultaneously with irradiation of the cavity of the resected lesion. The characteristics of the patients are presented in Table 15.1.

According to preoperative magnetic resonance imaging (MRI), the median of the maximum diameter of the resected lesion in the group as a whole was 3.7 cm (1.5–8.5 cm).

In the group of surgical and combined treatment, the median of the maximum diameter of the resected lesion was 3.8 cm (2.5–6.0 cm) and 3.5 cm (2.2–8.5 cm), respectively. The median of the target volume of the irradiation was 21.29 cm^3 (2.8–114).

Twenty-five (21%) out of 121 patients in the combination treatment group had a residual tumor. The fractionation modes of stereotactic

Table 15.1 Features of patients with large BM, treated with different modes of stereotactic irradiation

	Surgery		Surgery → SRT		SRS → Surgery		SRT	
Patient characteristics	*n*	%	*n*	%	*n*	%	*n*	%
Patients/metastasis all	141/184	100	122/141		48/110	100	114/314	100
Large BM	164	100	220		62	100	131	100
Sex								
Female	83	59	81	66	31	65	73	64
Male	58	41	41	34	17	35	40	36
Age (years)								
Median (range)	55 (29–76)		55 (24–79)		53 (29–72)		53 (25–72)	
Primary disease								
NSCLC	27	19	27	22	12	25	25	22
Melanoma	30	21	24	20	8	17	15	13
Breast cancer	36	26	35	29	12	25	41	37
Colorectal cancer	12	8	12	10	5	10.5	19	17
Gynecologic cancer	9	6	7	6	4	8,5	4	3
Renal cell carcinoma	15	11	10	8	7	14	5	4
Other	12	9	7	5	–		5	4
Tumor max diameter (cm)								
Median (range)	3.7 (1.5–6.1)		3.8 (1.8–8.5)		–		2.8 (2.0–5.0)	
Tumor/target volume (cm^3)								
Median (range)	–		21.16[a] (0.77–114)		16.523 (2.1–59.1)		8.4 (3.8–46)	

[a]A target is defined as a postoperative cavity with a margin of 2 mm

radiation therapy were 3 fractions for 77 (64%) patients, 5 fractions for 35 (29%) patients, and 9 fractions for the others (7%).

Irradiation technique: A target is defined as a postoperative cavity and a contrast enhancement area with a margin of 2 mm (CTV = GTV + 2 mm). In the case of residual tumor, the radiation dose was increased to this area in the form of an integrated boost, and the treatment was carried out using a hypofractionated schedule.

Results: The 12-month overall survival in the observed group was 76.6%. The survival rate without local recurrence at 12 months was 40.4% in the surgical group compared to 80.7% in the combination treatment group ($p < 0.0001$). Radionecrosis was registered in 4.1% of patients, which was accompanied by neurotoxicity grades 2–3. Additional leptomeningeal progression was detected in 15.7% and 9.8% of cases in the groups of surgical resection alone and combined treatment, respectively.

Thus, the combination of surgery with further stereotactic radiation is a satisfactory treatment strategy for operable large brain metastases especially in patients with neurological symptoms and mass effect.

15.4 Preoperative (Neoadjuvant) Radiosurgery

Adjuvant stereotactic radiotherapy (radiosurgery or hypofractionated radiotherapy) is associated with the uncertainty of the contouring of the postoperative tumor cavity and the risk of spreading of micrometastases in tissue outside of the postoperative cavity, which requires additional margin to the target volume. This leads to an increase in final volume and, consequently, to an increase in the risk of postradiation complications [9, 10].

Given the high risk of leptomeningeal progression and postradiation complications after surgical resection of BM followed by adjuvant stereotactic radiotherapy, a new therapeutic approach has emerged in recent years: the use of preoperative stereotactic radiosurgery followed by surgical resection within 24–48 h [11].

The limited data from retrospective studies have shown the advantages of preoperative radiosurgery in terms of reducing the level of postradiation complications and leptomeningeal progression while maintaining the quality of life and a high level of local control that is comparable to adjuvant stereotactic radiation therapy [12, 13].

Preoperative radiosurgery has several potential advantages compared to postoperative radiotherapy. In the case of preoperative radiosurgery, an intact (previously untreated) BM is irradiated, which is well visualized on an MRI of the brain with a clear outline and does not require the margin for CTV.

In addition, in the case of preoperative radiosurgery, a metastatic focus is maintained with preserved blood supply and relatively good oxygenation, which determines the higher radiosensitivity of the irradiated focus. Thus, the biological effect with the same focal dose of radiation will be greater in the case of preoperative radiosurgery compared to the postoperative irradiation of the hypoxic bed of the tumor.

Preoperative radiosurgery decreases the time of combined treatment of BM, optimizing the time schedule. In the case of additional metastases that are not subject to resection, the delay of postoperative stereotactic radiation therapy can lead to an increase in their size (continued growth), which will negatively affect the treatment results. This does not occur in the case of preoperative irradiation of BM. Additionally preoperative radiosurgery reduces the risk of tumor cells seeding during surgery which in turn decreases the chance of leptomeningeal progression. Based on the available clinical data, it is necessary to conduct prospective studies of preoperative radiosurgery followed by surgical resection of BM, in comparison with stereotactic radiation therapy after surgical resection.

Two years ago, the application of the method of preoperative stereotactic radiosurgery with subsequent neurosurgical removal of the irradiated BMs after 24–48 h began at the NMRC Burdenko. The following irradiation technique was applied: a target was defined as contrast enchansement zone without additional margin (CTV = GTV). The average dose in the focus for preoperative radiosurgery was 15–20% higher than recommended by RTOG 90-05, based on dose loads of tolerance per 100 cm^3 of brain tissue. Surgical removal of the irradiated BM 24–48 h after radiosurgery was performed.

Currently, preoperative radiosurgery has been performed in 53 patients with 110 brain metastases, and the preliminary observation data are available in 45 patients.

The median volume of irradiated foci was 16.5 cm^3 (2.0–59.1 cm^3). The median of the total focal dose was 19.94 Gy (12–24).

The overall survival at 12 months was 62.5%. Survival rate without local recurrence at 12 months was 89.3%. Radionecrosis was registered in three (4.3%) patients with grades 2–3 neurotoxicity. Leptomeningeal progression was only detected in one patient.

Thus, this technique can be considered a method of choice in a number of patients with operable large brain metastases. Comparative characteristics of the different methods used by NMRC Burdenko are presented in Table 15.2.

15.5 Stereotactic Radiotherapy of the Large Brain Metastasis Using Hypofractionation Schedules

If the brain metastatic tumor has functionally significant localization, with high risk of increasing neurological symptoms after surgery, only whole brain irradiation (WBRT) or hypo-

Table 15.2 Results of various treatment methods for large BM (data from NMRC Burdenko)

Results	Surgery	Surgery → SRT	SRS → Surgery	SRT
Local recurrence (at 12 months) %	59.6	19.3	9.3	31
Radionecrosis %	–	4.1	4.3	17
Leptomeningeal progression %	15.7	9.8	2	5.2

fractionation stereotactic radiation therapy (dose-staged radiosurgery) may be performed [14–16].

Hypofractionation allows us to bring a relatively lower dose per fraction and increased total dose of irradiation while minimizing adverse events and providing good local control. However, the optimal schedule of fractionation (dose per fraction, as well as total dose) has not been determined yet [17].

Technical advances in stereotactic radiation therapy have increased hypofractionation (i.e., treatment of a lesion in a 2–7 fraction) acceptance.

According to published data, stereotactic irradiation in the hypofractionation regimen in the treatment of large BM provided a good balance between local control and postradiation complications in comparison to radiosurgery.

The optimal hypofractionation regimens for stereotactic radiation therapy for large (>2 cm in diameter) BM in an independent study were conducted at the Institute of Neurosurgery.

The aim of this study was to assess local control, postradiation complications, and overall survival during various dose-equivalent stereotactic radiation therapy using hypofractionated schedules.

Taking into account the radiobiological linear-quadratic model and the adapted formula of Pack and Anthon, the following dose-equivalent fractionation modes were calculated: three fractions of 8 Gy (total dose = 24 Gy; study group 1), five fractions of 6 Gy (total dose = 30 Gy; study group 2), and seven fractions of 5 Gy (total dose = 35 Gy; study group 3) [18].

Patients were recruited into groups using simple randomization. The study included patients with metastases of epithelial malignant tumors (excluding small cell lung cancer) or melanomas to the brain, with lesions from 20 to 40 mm in maximum diameter, who had not previously undergone radiation therapy/radiosurgery.

Other synchronous metastases in the brain (up to 10) with a diameter of up to 20 mm were additionally treated with radiosurgery.

The study was performed in NMRC Burdenko and was approved by the local ethics committee.

One hundred fourteen patients with 131 brain metastases were included in this study from March 2013 to June 2019. The median follow-up was 9.8 months (1–73 months). The median overall survival for the study group was 13 months (95% CI, 9.05–22.2). Indicators of overall survival at 12 and 24 months amounted to 51% and 36%, respectively.

The survival rate without local recurrence at 12 months in the entire observation group was 69%. At a period of 12 months in the first, second, and third group of the study, the survival rate without local recurrence was 73%, 78%, and 75.4%, respectively (p = 0.5). Survival without local recurrence, depending on the volume >7 cm^3 and <7 cm^3, was 63% and 83%, respectively (p = 0.03). Symptomatic radionecrosis (ARE) at the 12-month follow-up was detected in 17% of patients and was associated with neurotoxicity of the second or third degree. In groups 1, 2, and 3 of the study, radionecrosis was detected in 23%, 16.7%, and 9.3% of patients, respectively ($p > 0.05$).

Thus, the preliminary results of the study show that local control does not depend on the selected fractionation modes but on the volume of the brain metastases. On the other hand, radiation toxicity depends on the fractionation schedule, being less pronounced in modes with a higher number of fractions. Hypofractionated stereotactic radiation therapy is the optimal method for the treatment of large inoperable brain metastases. Table 15.2 presents the results of various treatment methods for large BM (data from NMRC Burdenko).

15.6 CyberKnife for Treatment of Intraocular (Choroidal) Metastases

The prevalence of intraocular metastases has been studied using complete ophthalmologic examination in cohorts of patients diagnosed with cancer [19, 20]. The prevalence of metastases ranged from 2 to 7%, with the exception of

Mewis et al. who reported 26.8% related to the large proportion of symptomatic patients referred for examination [21].

The most frequent primary tumor site of choroidal metastases is the breast, found in 40–53% of cases [19, 22].

The second most frequent site is the lung, found in 20–29% of cases [22, 23]. Less frequent primary tumors include carcinomas of the gastrointestinal tract (4%), prostate (2%), kidney (2%), and skin (2%) [22, 24, 25].

There is currently no consensus on the treatment strategy of intraocular metastases. In patients with a shorter life expectancy, systemic therapies such as those targeting oncogenic drivers or targeted immunotherapy can induce a regression of the choroidal metastases and may be sufficient to temporarily decrease visual symptoms. However, they often acquire resistance to systemic treatment and intraocular relapse which usually requires radiotherapy for additional tumor control.

Conventional RT with photons is the most common treatment for choroidal metastases as it is cost-effective and readily accessible worldwide. Complications of radiotherapy may occur months or years after irradiation, which is relevant to long-term survivors. Depending on the prognosis, the aims of radiotherapy are not only to control tumor growth but also to preserve vision in the long term by sparing the macula and the optic disc. When possible, the anterior chamber is excluded from the irradiation field. On the other hand, the macula is also sensitive to radiation. There are various forms of radiotherapy and various fractionation schemes depending on the local extent of the disease, its proximity to the macula, the number of lesions, the general prognosis, and the treatment goals. Thus, the radiotherapy technique and treatment scheme should be adapted carefully after evaluation of the risk-benefit ratio. More advanced forms of radiotherapy theoretically associated with fewer side effects can be proposed.

As demonstrated by research [26], stereotactic radiation therapy increases local control and reduces radiation complications compared to conventional radiation therapy.

It can be used as an alternative to brachytherapy, which makes therapy more accessible and minimally invasive. The optimistic results of stereotactic irradiation of choroidal metastasis (CM) are shown in a very small number of studies with a limited series of 7–10 patients, which requires additional study of the use of this technique [27, 28]. Stereotactic irradiation is feasible for choroidal metastasis treatment, but its value has to be evaluated in larger series, and it may require invasive eye fixation tools.

In 2010, the development and implementation of the method of stereotactic radiation therapy of choroidal metastases began in our center (Fig. 15.1).

Irradiation technique: The target is defined as the posterior pole of the eyeball, taking into account the creeping nature of metastases, the lack of visualization of the entire volume according to MRI, and eye mobility during irradiation. Additionally, a retinal attachment site (CTV = GTV) is included in the target volume. Considering the macular radiosensitivity, the treatment is carried out for seven fractions of 5 Gy to SOD = 35 Gy. The total dose (5% isodose) to the lens was 10.4 Gy.

At the Institute of Neurosurgery, we treated 27 patients and 36 eyes with metastases at the choroid. Of these, 23 were women and 4 were men. The median age was 50 (25–67). The most frequent primary tumor site of choroidal metastases was breast cancer, found in 19 cases, non-small cell lung cancer in 4, melanoma in 1, and gastrointestinal tract cancer in 2 cases. In 17 patients, metastases in the choroid were accompanied by a deterioration of visual function. Patients had concomitant brain metastases for which concomitant stereotactic radiation therapy was performed.

A retrospective analysis included 21 eyes. Local control at the follow-up period of 12 months was 100%, and complete regression, partial regression, and tumor stabilization were in nine, seven, and five patients' eyes, respectively. The dynamics of visual function corresponded to improvement, stabilization, and further deterioration in visual acuity in 5, 13, and 3 cases, respectively. Within 1 year of observation, the development of cataracts was seen in one patient.

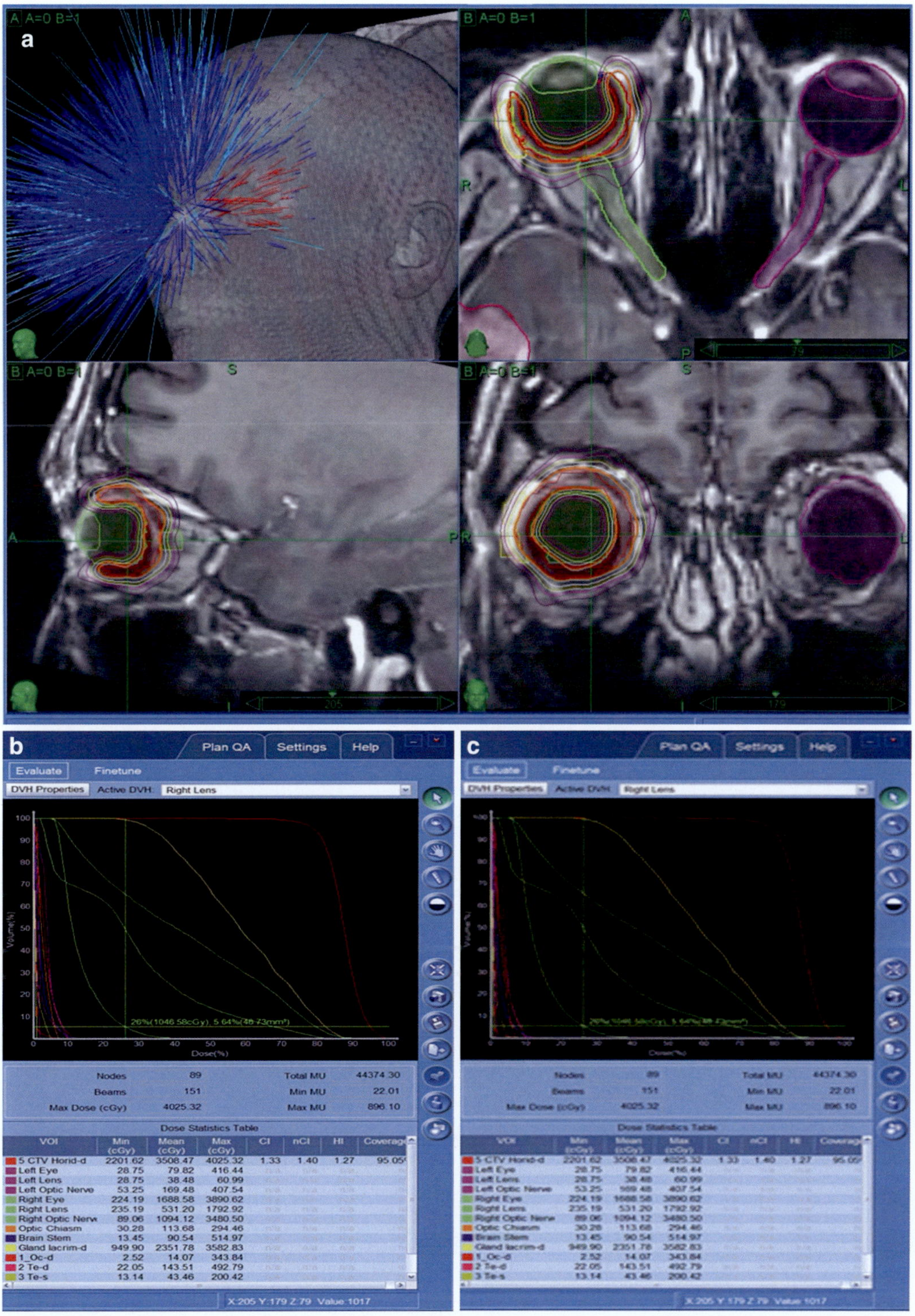

Fig. 15.1 Plan of irradistion of metastasis to the choroid, assessment of radiation load on critical structures. (**a**) Contouring of the target and critical organs. (**b**) 5% of the volume of the lens—10 Gy. (**c**) 50% of the lacrimal gland—23 Gy

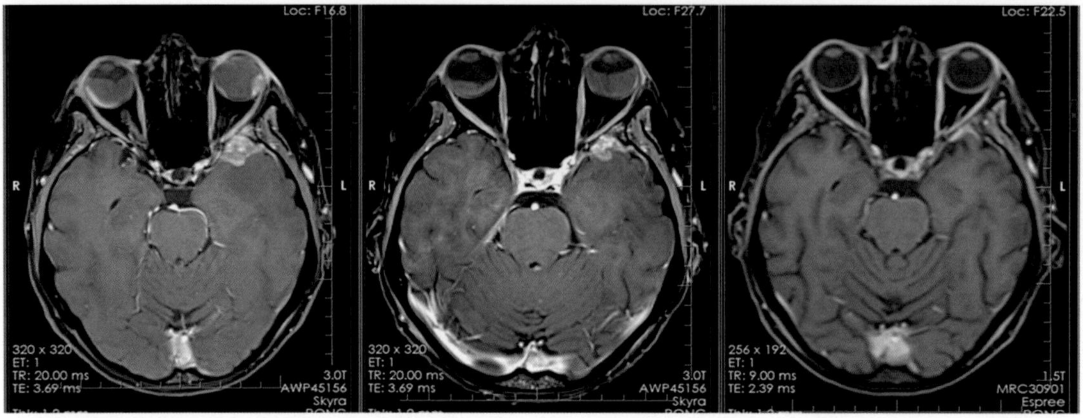

Fig. 15.2 Patient Z. 37 years old. Breast cancer and bilateral choroidal metastases. Conducted SRT in GF mode. Left: to STL, Vis OD/OS = 0.4/0.01. Middle: after 3 months, Vis OD/OS = 1.0/0.01. Right: after 10 months, Vis OD/OS = 1.0/0.4

The dynamics of visual function and treatment results according to MRI data illustrates a clinical example of patient Z, a 37-year-old female patient with breast cancer (triple-negative phenotype), presented in Fig. 15.2.

Thus, adequate treatment with the use of CyberKnife of intraocular metastases in the choroid improves the quality of life of cancer patients and may prevent the development of blindness.

Clinical Case Report

A 46-year-old non-smoking man, with a diagnosis of NSCLC of the lower lobe of the right lung T2N1M1, underwent observation at the Burdenko Institute of Neurosurgery. Examination revealed the presence of a metastatic lesion in the brain. On December 25, 2007, a metastasis of the right cerebellar hemisphere was removed. Histological examination of the surgical material revealed the presence of a moderately differentiated pulmonary adenocarcinoma (status EGFR/ALK/ROS1 neg.).

An extended lower lobe right lung lobectomy was performed on February 28, 2008 (2.1 months after diagnosis and the start of treatment). The morphological diagnosis was adenocarcinoma of moderate differentiation with metastases in the basal and paraesophageal lymph nodes.

During the postoperative period, six cycles of poly-chemotherapy were carried out according to the cisplatin + etoposide regimen. In October 2009, an intracranial progression in the form of solitary metastasis of the left frontal lobe was found. On October 28, 2009 (after 22.4 months), a stereotactic radiosurgical treatment was performed with median dose 20 Gy along the edge of the metastatic focus. MRI control on January 28, 2011 (after 37.7 months), showed stabilization of the previously irradiated lesion and the appearance of new metastatic lesions.

On February 8, 2011, stereotactic radiosurgery of three metastatic lesions in the left frontal lobe, the lenticular nucleus of the right hemisphere, and the right cerebellar hemisphere (22–24Gy) was performed. MRI on May 16, 2011 (after 41.3 months), showed a complete and partial regression of previously irradiated BMs and the appearance of new BMs.

Neuro-ophthalmological examination showed the appearance of intraocular metastasis in the region of the left eye disc. Extracranial disease progression and liver metastases were also seen. Chemotherapy was performed according to the paclitaxel regimen (175 mg / m^2 + cisplatin 75 m / m^2).

On May 23, 2011, a session of radiosurgical treatment was conducted with the CyberKnife system: BM radiosurgery in the left occipital region (CTV = 1.66 cm^3) and the caudate nucleus head region on the right (CTV = 0.287 cm^3). The average dose in the lesions was 22 Gy. The prescribed dose for 81% of the isodose line was 20 Gy. From May 26, 2011, to June 3, 2011, a course of stereotactic radiotherapy using CyberKnife was conducted on a metastatic lesion in the posterior pole of the left eye (CTV = 4.981 cm^3). The treatment was carried out in the hypofractionation mode: the average dose of radiation per fraction was 5 Gy to SOD 35 Gy. According to an examination by an ophthalmologist dated July 1, 2011 (after 42.8 months from the diagnosis and the start of treatment), a new metastatic lesion was revealed in the ciliary body of the left eye, next to the previously irradiated focus. Visual acuity in the left eye deteriorated (OS = 0.2). According to brain MRI, five new metastatic foci of the brain up to 5 mm in diameter were visualized. The patient categorically refused the proposed exposure to the entire brain. From July 7 to July 13, 2011, a cycle of stereotactic radiotherapy was carried out in hypofractionation mode (average radiation dose of 6 Gy daily to SOD of 30 Gy) on the metastatic lesion (CTV = 4.981 cm^3) in the ciliary body of the left eye.

On July 14, 2011, a radiosurgical session with CyberKnife was performed for treatment of the brain metastases: two foci in the left parietal region (CTV = 0.343 cm^3) and two foci in the right occipital region (CTV = 0.469 cm^3; CTV = 0.369 cm^3). The average radiation dose of 24 Gy was applied to the focus in the left occipital region (CTV = 0.474 cm^3) and to the focus of the right hemisphere of the cerebellum (CTV = 0.685 cm^3). The prescribed dose at the 82% isodose line was 22 Gy.

Ophthalmological examination on September 13, 2011 (45.3 months after diagnosis and treatment), showed a complete regression of intraocular metastatic foci, with the formation of foci of chorioretinal dystrophy. Positive dynamics of the visual functions of the left eye (visual acuity OS = 0.7) were also seen. According to brain MRI, radionecrosis was recorded in the left frontal lobe together with the appearance of two new metastatic lesions up to 0.5 cm in diameter and a partial and complete regression of previously irradiated lesions.

From September 12 to September 20, 2011, according to ultrasound stabilization of the extracranial process and an increase in single liver metastasis, a cycle of stereotactic radiation therapy was carried out with the CyberKnife using hypofractionation schedules (average radiation dose of 15 Gy per fraction up to a sum dose of 45 Gy) of the metastatic lesion in the liver (CTV = 54.485 cm^3 and PTV = 141.094 cm^3).

At the same time, on September 20, 2011, a stereotactic radiosurgery session was performed for metastases in the right thalamus (CTV = 0.216 cm^3) and left occipital lobe (CTV = 0.553 cm^3). The average radiation dose was 21 and 22 Gy, respectively.

Control MRI on December 22, 2011 (after 48.6 months), demonstrated stabilization of previously irradiated foci together with the detection of multiple metastatic (more than 20 new foci) brain lesions. Two radiosurgery sessions were performed (11 foci for each radiosurgery session). The total CTV was 4306 cm^3. The average dose of radiation for each session was 19 and 20 Gy. Given the subsequent irradiation of the entire brain, the dose of radiosurgical treatment was reduced. From January 10 to January 23, 2012, a conventional LINAC was used for WBRT in conventional regime with 3 Gy per fraction for ten fractions.

On the control MRI of the brain on May 23, 2012 (after 53.7 months), partial and complete regression of previously irradiated foci were observed. Complete regression of intraocular metastatic foci and visual functions without negative dynamics were noted. The patient notes a slight decrease in memory for current events. No neurocognitive impairment was detected (MMSE = 29 points). Extracranial stabilization and complete regression of a single metastasis in the liver were also noted.

According to brain MRI from August 29, 2012 (after 57.0 months), three new metastatic foci of the brain were revealed. The new radiosurgery session for three metastases was performed with CyberKnife: to the right occipital lobe (CTV = 2.625 cm^3), left parietal lobe (CTV = 0.429 cm^3), and left cerebellar hemisphere (CTV = 0.377 cm^3). The average dose of radiation per fraction was 22 Gy. The prescribed dose of 77% of the isodose line was 19.5 Gy.

On September 20, 2012 (after 57.7 months), the patient developed lower paraparesis with severe pain in the lower thoracic spine. The MRI of the spinal cord revealed metastatic lesions of the vertebral bodies Th9 and Th11 with intrathecal soft tissue component and spinal cord compression. From September 25 to September 27, 2012, a course of stereotactic radiotherapy using a hypofractionated schedule was performed using CyberKnife to the vertebral bodies Th9 (CTV = 28.967 cm^3) and Th11 (CTV = 32.823 cm^3). The average radiation dose was 7 Gy/fr to sum dose 21 Gy with integrated boost (average radiation dose of 8 Gy to sum dose of 24 Gy) to the soft tissue component in the regions of Th9 (GTV = 8.240 cm^3) and Th11 (GTV = 14.816 cm^3). The prescribed dose at the 74% isodose line was 20.7 Gy. Stabilization of the disease was achieved, with complete pain control.

In November 2012, at the latest evaluation of the patient, intracranial stabilization of metastatic lesions and extracranial disease progression in the form of multiple metastatic lesions to the lungs and liver were noted. Death from extracranial progression occurred on February 7, 2013 (62.4 months after the start of treatment). For most of the observation time, the patient maintained a good quality of life and continued to work and to maintain an active life's position.

He received a total of 21 radiosurgical and radiotherapeutic procedures for metastases with different localizations, including 17 for multiple BMs, 2 sessions of hypofractionation for choroidal metastasis, and also 1 hypofractionation for liver and 1 for lung metastases with good results and temporal remission of cancer disease.

Case Conclusion

This case demonstrates that in selected patients with multiple metastatic lesions of various locations, including the brain and spinal cord, constant observation and consistent stereotactic radiotherapy and radiosurgery can lead to long-term survival while maintaining a satisfactory quality of life.

References

1. Baschnagel AM, Meyer KD, Chen PY, Krauss DJ, Olson RE, Pieper DR, et al. Tumor volume as a predictor of survival and local control in patients with brain metastases treated with gamma knife surgery. J Neurosurg. 2013;119:1139–44. https://doi.org/10.3171/2013.7.JNS13431.
2. Sheehan JP, Sun M-H, Kondziolka D, Flickinger J, Lunsford LD. Radiosurgery for non-small cell lung carcinoma metastatic to the brain: long-term outcomes and prognostic factors influencing patient survival time and local tumor control. J Neurosurg. 2002;97:1276–81. https://doi.org/10.3171/jns.2002.97.6.1276.
3. Golanov AV, Banov SM, Il'yalov SR, Trunin YY, Maryashev SA, Vetlova ER, et al. Overall survival and intracranial relapse in patients with brain metastases after gamma knife radiosurgery alone. Voprosy Neirokhirurgii Buedenko. 2016;80:35–46. https://doi.org/10.17116/neiro201680235-46.
4. Shaw E, Scott C, Souhami L, Dinapoli R, Kline R, Loeffler J, Farnan N. Single dose radiosurgical treatment of recurrent previously irradiated primary brain tumors and brain metastases: final report of RTOG protocol 90-05. Int J Radiat Oncol Biol Phys. 2000;47:291–8. https://doi.org/10.1016/S0360-3016(99)00507-6.
5. Elliott RE, Rush SC, Morsi A, Mehta N, Spriet J, Narayana A, et al. Local control of newly diagnosed and distally recurrent, low-volume brain metastases with fixed-dose (20 gy) gamma knife radiosurgery. Neurosurgery. 2011;68:921–31;. ; discussion 931. https://doi.org/10.1227/NEU.0b013e318208f58e.
6. Schoeggl A, Kitz K, Ertl A, Reddy M, Bavinzski G, Schneider B. Prognostic factor analysis for multiple brain metastases after gamma knife radiosurgery: results in 97 patients. J Neurooncol. 1999;42:169–75. https://doi.org/10.1023/A:1006110631704.
7. Wiggenraad R, Verbeek-de Kanter A, Kal HB, Taphoorn M, Vissers T, Struikmans H. Dose-effect relation in stereotactic radiotherapy for brain metastases. A systematic review. Radiother Oncol. 2011;98:292–7. https://doi.org/10.1016/j.radonc.2011.01.011.
8. Vogelbaum MA, Angelov L, Lee S-Y, Li L, Barnett GH, Suh JH. Local control of brain metastases by stereotactic radiosurgery in relation to dose to the tumor margin. J Neurosurg. 2006;104:907–12. https://doi.org/10.3171/jns.2006.104.6.907.
9. Kirkpatrick JP, Wang Z, Sampson J, Kelsey C, Allen K, Duffy E, et al. Early results of a randomized trial to identify an optimal PTV in stereotactic radiosurgery of brain metastases. Int J Radiat Oncol Biol Phys. 2013;87:S50. https://doi.org/10.1016/j.ijrobp.2013.06.131.
10. Choi CYH, Chang SD, Gibbs IC, Adler JR, Harsh GR, Lieberson RE, Soltys SG. Stereotactic radiosurgery of the postoperative resection cavity for brain metasta-

ses: prospective evaluation of target margin on tumor control. Int J Radiat Oncol Biol Phys. 2012;84:336–42. https://doi.org/10.1016/j.ijrobp.2011.12.009.
11. Vetlova ER, Golbin AD, Golanov AV, Potapov AA, Banov SM, Antipina NA, et al. Preoperative stereotactic radiosurgery of brain metastases: preliminary results. Cureus. 2017;9:e1987.
12. Minniti G, D'Angelillo RM, Scaringi C, Trodella LE, Clarke E, Matteucci P, et al. Fractionated stereotactic radiosurgery for patients with brain metastases. J Neurooncol. 2014;117:295–301. https://doi.org/10.1007/s11060-014-1388-3.
13. Joshi R, Johnson MD, Maitz A, Marvin KS, Olson RE, Grills IS. Utility of graded prognostic assessment in evaluation of patients with brainstem metastases treated with radiosurgery. Clin Neurol Neurosurg. 2016;147:30–3. https://doi.org/10.1016/j.clineuro.2016.05.001.
14. Angelov L, Mohammadi AM, Bennett EE, Abbassy M, Elson P, Chao ST, et al. Impact of 2-staged stereotactic radiosurgery for treatment of brain metastases ≥ 2 cm. J Neurosurg. 2018;129:366–82. https://doi.org/10.3171/2017.3.JNS162532.
15. Murai T, Ogino H, Manabe Y, Iwabuchi M, Okumura T, Matsushita Y, et al. Fractionated stereotactic radiotherapy using CyberKnife for the treatment of large brain metastases: a dose escalation study. Clin Oncol (R Coll Radiol). 2014;26:151–8. https://doi.org/10.1016/j.clon.2013.11.027.
16. Jiang X-S, Xiao J-P, Zhang Y, Xu Y-J, Li X-P, Chen X-J, et al. Hypofractionated stereotactic radiotherapy for brain metastases larger than three centimeters. Radiat Oncol. 2012;7:36. https://doi.org/10.1186/1748-717X-7-36.
17. Kim KH, Kong D-S, Cho KR, Lee MH, Choi J-W, Seol HJ, et al. Outcome evaluation of patients treated with fractionated gamma knife radiosurgery for large (3 cm) brain metastases: a dose-escalation study. J Neurosurg. 2019:1–10. https://doi.org/10.3171/2019.5.JNS19222.
18. Inoue HK, Sato H, Suzuki Y, Saitoh J-I, Noda S-E, Seto K-I, et al. Optimal hypofractionated conformal radiotherapy for large brain metastases in patients with high risk factors: a single-institutional prospective study. Radiat Oncol. 2014;9:231. https://doi.org/10.1186/s13014-014-0231-5.
19. Demirci H, Shields CL, Chao A-N, Shields JA. Uveal metastasis from breast cancer in 264 patients. Am J Ophthalmol. 2003;136:264–71. https://doi.org/10.1016/S0002-9394(03)00192-2.
20. Ferry AP, Font RL. Carcinoma metastatic to the eye and orbit. I. A clinicopathologic study of 227 cases. Arch Ophthalmol. 1974;92:276–86. https://doi.org/10.1001/archopht.1974.01010010286003.
21. Mewis L, Young SE. Breast carcinoma metastatic to the choroid. Ophthalmology. 1982;89:147–51. https://doi.org/10.1016/S0161-6420(82)34838-1.
22. Shields CL, Shields JA, Gross NE, Schwartz GP, Lally SE. Survey of 520 eyes with uveal metastases. Ophthalmology. 1997;104:1265–76. https://doi.org/10.1016/S0161-6420(97)30148-1.
23. Shah SU, Mashayekhi A, Shields CL, Walia HS, Hubbard GB, Zhang J, Shields JA. Uveal metastasis from lung cancer: clinical features, treatment, and outcome in 194 patients. Ophthalmology. 2014;121:352–7. https://doi.org/10.1016/j.ophtha.2013.07.014.
24. de Potter P, Shields CL, Shields JA, Tardio DJ. Uveal metastasis from prostate carcinoma. Cancer. 1993;71:2791–6. https://doi.org/10.1002/1097-0142(19930501)71:9<2791::AID-CNCR2820710917>3.0.CO;2-Y.
25. Dieckert JP, Berger BB. Prostatic carcinoma metastatic to choroid. Br J Ophthalmol. 1982;66:234–9. https://doi.org/10.1136/bjo.66.4.234.
26. Mathis T, Jardel P, Loria O, Delaunay B, Nguyen A-M, Lanza F, et al. New concepts in the diagnosis and management of choroidal metastases. Prog Retin Eye Res. 2019;68:144–76. https://doi.org/10.1016/j.preteyeres.2018.09.003.
27. Bellmann C, Fuss M, Holz FG, Debus J, Rohrschneider K, Völcker HE, Wannenmacher M. Stereotactic radiation therapy for malignant choroidal tumors. Ophthalmology. 2000;107:358–65. https://doi.org/10.1016/S0161-6420(99)00081-0.
28. Cho KR, Lee KM, Han G, Kang SW, Lee J-I. Gamma knife radiosurgery for cancer metastasized to the ocular choroid. J Korean Neurosurg Soc. 2018;61:60–5. https://doi.org/10.3340/jkns.2016.0606.003.

16 Multiple Brain Metastases

Frederic Dhermain

16.1 Selecting Optimal Indications for Radiosurgery in a Rapidly Evolving Landscape

Since 2014, ASTRO contributed, by a "choosing wisely" publication policy, to identifying radiosurgery (RS) as the preferred option for patients presenting *a "limited number*" of brain metastases (BMs), namely, *up to four lesions* [1]. In contrast, for patients with multiple BMs, whole-brain radiotherapy (WBRT) continues to be a "first option" for most oncologists, even if this attitude is clearly decreasing [2, 3]. Furthermore, most neuro-oncologists suggest the role of many other parameters, such as the general and neurological status, extracranial disease control, size and/or volume of BMs, molecular profile of the primary and secondary tumors, and the expected outcome in the decision-making process.

Actually, over the past decade, a series of key events have occurred. Firstly, because of the large dissemination of "radiosurgery" systems, an increasing number of cancer centers had the possibility to propose an alternative to whole-brain radiotherapy (WBRT) for their patients presenting multiple BMs; not only Gamma Knife (GKN) or CyberKnife (CKN) devices, which were fully developed for RS, but also LINAC-based machines "dedicated" to stereotactic radiotherapy (SRT) are now available, even in hospitals of small-intermediate size. At the same time, more asymptomatic patients will present with multiple BMs, due to the increased access to MRI for neurological symptoms or simply as a "checkup," or before inclusion in clinical trials.

Secondly, since the "enrichment" of diagnostic and treatment opportunities of BM patients is now available with the routine use of molecular profiling and frontline immunotherapies, the outcome of an increasing part of them has been significantly improved, mostly in melanoma patients. Mainly immune checkpoint inhibitors (ICIs) have dramatically changed their prognosis, with durable intracranial overall response rates (ORR) almost comparable to extracranial results [4]. This positive trend is going to be translated, at a lower level, in "targetable" metastatic lung cancer patients with EGFR mutation [5] and ALK rearrangement [6], with impressive results. ICIs are also evaluated in retrospective and prospective studies as first-line therapy for metastatic lung cancer patients [7, 8]. As a consequence of the increased efficacy of these systemic treatments, more metastatic patients will be "long-term survivors" and consequently exposed to the risk of developing (new) and, mostly, multiple BMs for longer.

F. Dhermain (✉)
Department of Radiation Oncology, Gustave Roussy University Hospital, Villejuif Cedex, France
e-mail: Frederic.dhermain@gustaveroussy.fr

© Springer Nature Switzerland AG 2020
A. Conti et al. (eds.), *CyberKnife NeuroRadiosurgery*,
https://doi.org/10.1007/978-3-030-50668-1_16

Thirdly, beyond the basic calculation of the number of BMs and the RPA (Recursive Partitioning Analysis) index [9, 10], several new prognostic scores and tools are now available to better approach the outcome of metastatic patients, defining subgroups of different prognosis more precisely, from less than 3 months to more than 2 years of expected median survival. Indeed, this possibility to better predict the outcome for each type of primary, with a margin of error still recognized as too wide, is essential in the "choosing wisely" decision process [11]. Delivering WBRT to a patient with slowly evolving multiple BMs and an expected median survival of 18 months is as questionable as delivering an SRS to a patient who will present an explosion of new BMs or a leptomeningeal invasion in 3 months. Consequently, beginning with the RPA index and then refining the *Graded Prognostic Assessment index (GPA)* [12–15], new "Diagnosis-Specific" GPA indexes were published, dedicated for each histomolecular subgroup of patients, from melanoma, breast, colorectal, and lung cancers to renal cell carcinomas and sarcomas [16–20]. In parallel, the "*Velocity index*" could better predict the risk of an early indication of WBRT after an initial SRS delivery, making the latter questionable in some rapidly evolving cases [21–23]. Finally and for an optimal compromise between efficacy and toxicity, the recent concept of "*Cumulative intracranial tumor volume*" was proposed and evaluated [24, 25], not only for a prognostic evaluation but also to better exclude some RS indications: this category of patients with multiple and bulky BMs would possibly suffer more from neurological toxicities than "benefit" from RS.

16.2 Predicting Survival at "Individual" Level: Definitions, Thresholds, and Endpoints

Several prognostic tools were evaluated mainly based on RPA and then GPA scores, age, Karnofsky Performance Scale (KPS)/Performance Status (PS) score, number of brain metastases, and presence/absence of (active) extracranial metastases and either focused on expected survival (a basic "efficacy" marker) or the quality of life, the "toxicity" parameter being very heterogeneously evaluated [26]. For daily practice, the last DS-GPA classification for each histomolecular diagnosis could be proposed, since it evolves continuously over time, is user-friendly, integrates the advances in "personalized" systemic treatments, and clearly divides patients in four categories with different prognosis. Limitations include the retrospective aspect, the rapidly changing landscape of "personalized" treatments (second or third generation of targeted drugs/different anti-PD1, anti-PDL-1 molecules), and, importantly, the high spatial-temporal tumoral heterogeneity, with a possible clonal shift between primary and metastatic sites. BMs could have, in up to 50% of cases, distinctly different phylogenic origins to those of the dominant clones of the primary tumor [27], encouraging to resect operable BMs when it is functionally safe.

An interesting dynamic tool, both predictive and prognostic, was recently described: the "*Brain Metastasis Velocity*" (BMV) index, predicting clinical outcome after initial distant brain failure following upfront SRS alone. It was defined as "the cumulative number of new BMs since initial SRS/Total time between initial SRS and Time of new BMs." The subgroup with a BMV index of less than four new BMs per year presented the lowest risk of salvage WBRT, the best prognosis, and consequently the best indication for SRT [23].

Definition of the 'oligometastatic status' has evolved over time: in the initial RPA index, the "oligometastatic" status was defined by 1–3 BMs, even if the more recent DS-GPA scores consider that a patient presents "multiple" BMs from 5 to 10 BMs which are possibly "treatable" with RS up to 15 or even 20 BMs. Recently, Yamamoto and other authors strongly suggested that, for a highly selected population, patients treated with SRS presenting five to ten BMs seem to have the same prognostic as those with one to five lesions [28–31].

This highlights an important "new" parameter to consider: the "*Cumulative intracranial volume*" *(CIV)* of BMs (in mL or cm^3), which was introduced more than 10 years ago [32] and more recently suggested as a possibly better independent prognostic indicator than the number or the largest size of BMs (more than 3 cm) [24]. For example, *a threshold of 15 cm³* was an exclusion criterion in the Yamamoto study, and some ongoing trials exclude patients with a CIV superior to 20 cm^3.

Considering only studies including patients with multiple BMs (all but one retrospective) with a median follow-up of at least 6 months, it is interesting to note that older publications reported median overall survivals (mOS) of 4–8 months, in contrast to the more recent one which identified subgroups of patients with mOS as high as 11 months [28, 33]. This could be explained both by more stringent selection criteria with a larger part of asymptomatic patients and by the efficacy of new personalized systemic treatments, particularly for the melanoma group and an increasing proportion of lung cancer patients.

Consequently, with this important part of "long survivors" (more than 9–12 months of expected OS), the choice of primary endpoints is shifting from the local/intracranial control rate to the overall survival item and, furthermore, toward quality of life and neurocognitive evaluations [34]. The longer the expected survival, the more important the items assessing patient-reported outcomes (PRO), and, ideally, both clinical toxicity (as disabling radionecrosis/leukoencephalopathy) and OS should be evaluated as co-primary endpoints. It is the case in one of the most interesting ongoing trials testing RS versus WBRT, the NCT03550391 (Table 16.1).

16.3 Combining SRT with New "Precision Medicine," Is There Still a Place for "Modern" WBRT?

Most patients with multiple BMs are also extracranially metastatic patients and candidates for systemic frontline treatments. Consequently, the question of "do we have to" and "how to combine" targeted drugs and/or immunotherapies with SRT is increasing in our daily practice. Because there is no conclusive solid data based on results of already closed prospective randomized trials, we only have the ability to analyze published heterogeneous series mostly with a limited number of patients [35, 36]. However, available data are favoring the early introduction of SRT, "combined with" the systemic personalized treatments if the latter is necessary. Furthermore, the concurrent administration of immunotherapies with frontline SRS (and a minimal dose/no steroids) for these patients with multiple BMs could not only improve intracranial control (without a significant increase in clinical toxicity) but potentially improve overall survival [37]. Focusing on melanoma brain metastases, the question of introducing SRT frontline with or as salvage after introduction of targeted drugs/immunotherapy is the object of a randomized trial (the "Become-MB" trial NCT04074096).

In this context of early delivery of "precision medicine" therapies to most patients with multiple BMs, the place of WBRT seems more debatable, even for those with more than ten BMs. Due

Table 16.1 Clinical trials comparing SRS versus WBRT in patients with more than four brain metastases

Trial number	Group	Arms	Number of lesions/ Number of patients planned	HA	Opening/end expected	Primary endpoint(s)
NCT03550391	CCTG	SRS/WB*	5–15/206	*All* HA-WB	2018/2022	Survival and neurocognition
NCT01592968	MDACC	SRS/WB	4–15/100	No HA	2012/2020	Local control and neurocognition
NCT03075072	B & W	SRS**/WB	5–20/196	HA *if possible*	2017/2022	Quality of life at 6 months

SRS stereotactic radiosurgery, *SRS*** 1 to 5 fractions, *WB* whole-brain radiotherapy, *WB** with memantine, *HA* hippocampal avoidance, *CCTG* Canada Cancer Trials Group with Alliance and NRG groups, *MDACC* MD Anderson Cancer Center, *B & W* Birgham and Women's Hospital to the justified fear of unnecessary added neurotoxicity and the necessity of delivering WBRT during a period of 2 weeks, many oncologists are reluctant to stop or delay their systemic treatment, particularly if it is effective on extracranial metastatic disease. They will favor a shorter treatment such as SRT, with one to three fractions in a week, which will always spare more normal brain white matter than any hippocampal-avoiding (HA) modern WBRT, even if this technique seems to limit (marginally) its negative impact on some important neurocognitive functions [38]. Finally, and outside ongoing prospective trials, HA-WBRT could be proposed in some highly selected and more palliative indications (see Table 16.2 and Fig. 16.1), but clearly not as a "last option" for frail patients, in light of the QUARTZ study [39].

16.4 Ongoing Trials, Daily Practice, and Perspectives: A Case-by-Case Multidisciplinary Decision

Among the very few ongoing trials (see Table 16.1) still proposing WBRT as the "reference arm" for patients with multiple BMs (with or without memantine, with or without HA), the NCT03550391 trial seems to be the one that could best answer the two coupled questions that are still topical: What impact will a modern HA-WBRT choice have on survival and neurocognition? Other registered trials are either not yet recruiting or don't propose HA systematically in the WBRT arm or are slowly recruiting. Consequently, because there is no "level 1 evidence-based" data to definitively conclude pro or against HA-WBRT versus RS in patients with multiple BMs, a case-by-case interdisciplinary discussion will be the best option.

In daily practice, outside including patients in ongoing trials, the individual decision should integrate several key factors including clinical, radiological data (volumetric and dynamic) and also the histomolecular profile, if possible based on the more recent tissue available, as proposed in Table 16.2. For example, the "best candidate" for exclusive RS/hFSRT will meet both the following characteristics: a symptomatic patient with a favorable/intermediate expected survival and a low velocity index with "non-targetable" lesions of a "non-bulky" total cumulative volume (Fig. 16.2). Combination of RS/hFSRT and targeted drugs/immunotherapies could be preferably proposed for multiple BM patients who also present a favorable/intermediate prognostic, but needs to be controlled rapidly both intra- and extracranially with "targetable" lesions.

Table 16.2 Choosing between systemic treatments vs RS ± HFSRT vs a combination of both vs WBRT

	Systemic treatment (ST)	RS/HFSRT	Combination of ST and SRS/HFSRT*	Modern WBRT with HA
Molecular profile	Targetable	Non-targetable	Targetable	Non-targetable
Number of BMs	More than 10	4–10	4–10	More than 10
Total cumulative volume of BMs	More than 15–20 cc	Less than 15–20 cc	Less than 15 cc	More than 20 cc *(surgery if needed)*
BM velocity index	>13 new BMs/year	<4 new BMs/year	4–13 new BMs/year	>13 new BMs/ year
Survival^T	>3 months	>3 months	>6 months	3–12 months
Neurological status	No symptom	Symptomatic	+/– Symptomatic	Symptomatic

Survival^T expected median overall survival based on DS-GPA dedicated index, *DS-GPA* disease-specific Graded Prognostic Assessment, *ST* systemic treatment, essentially targeted drugs and/or immunotherapies as checkpoint inhibitors, *RS* radiosurgery (1 fraction), *HFSRT* hypofractionated stereotactic radiotherapy (3–5 fractions/1 week), *HSFRT** if possible, *before* ST or "*concomitant*" with ST (within 1 half-life of the drug)

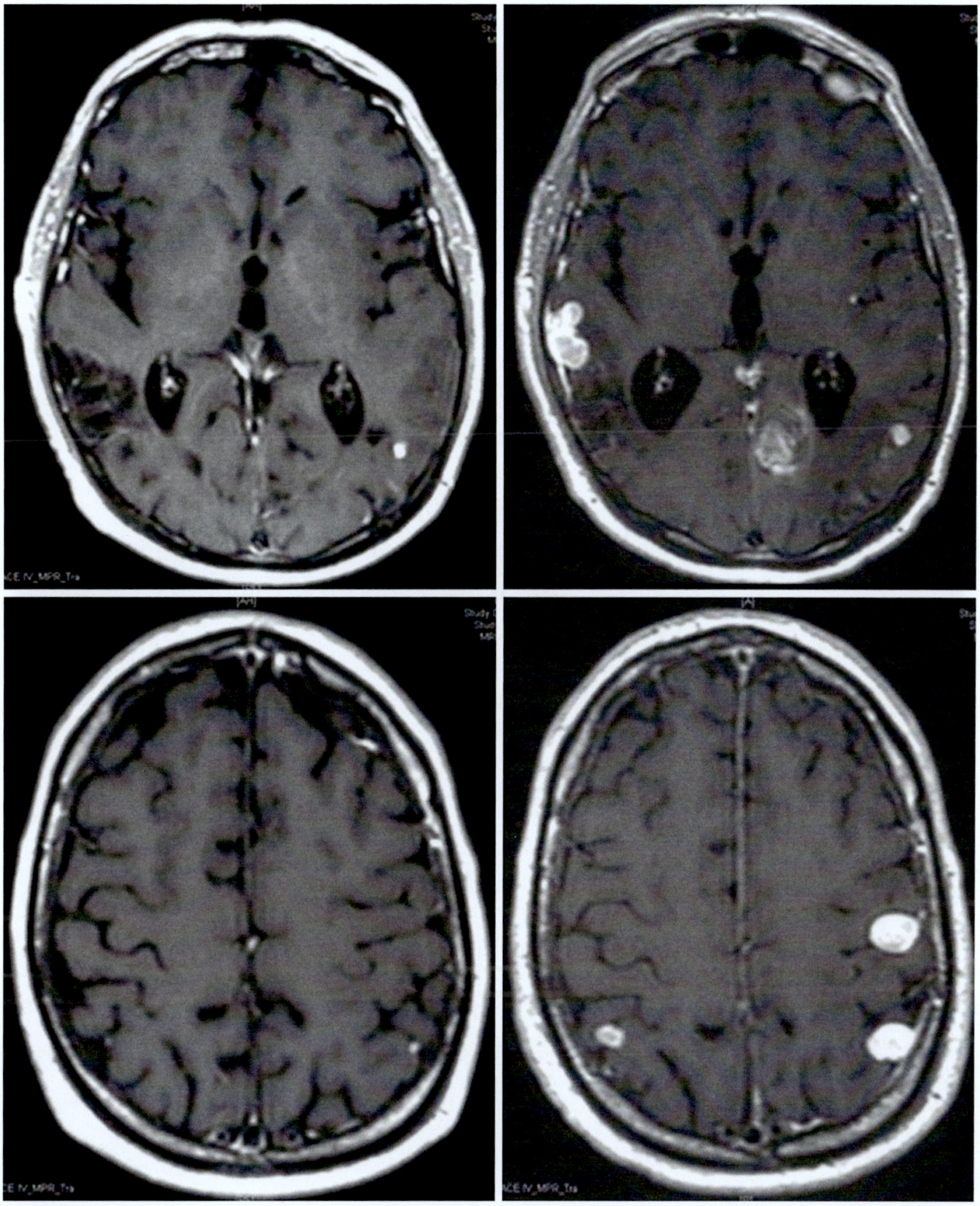

Fig. 16.1 A healthy 55-year-old patient, recently metastatic from a melanoma, progressing extracranially with an anti-BRAF anti-MEK treatment who switched to a checkpoint inhibitor (pembrolizumab). Contrary to a previously normal investigation obtained 3 months ago, the MRI showed rapid emergence of nine new lesions, some of which are not well defined. The cumulative volume was 20 cm^3 but without a definitive sign of leptomeningeal invasion. The patient underwent whole-brain radiotherapy, and ten fractions of 3 Gy were delivered in 2 weeks (between two cycles of pembrolizumab). However, the patient died quickly 3 months later due to intracranial progression

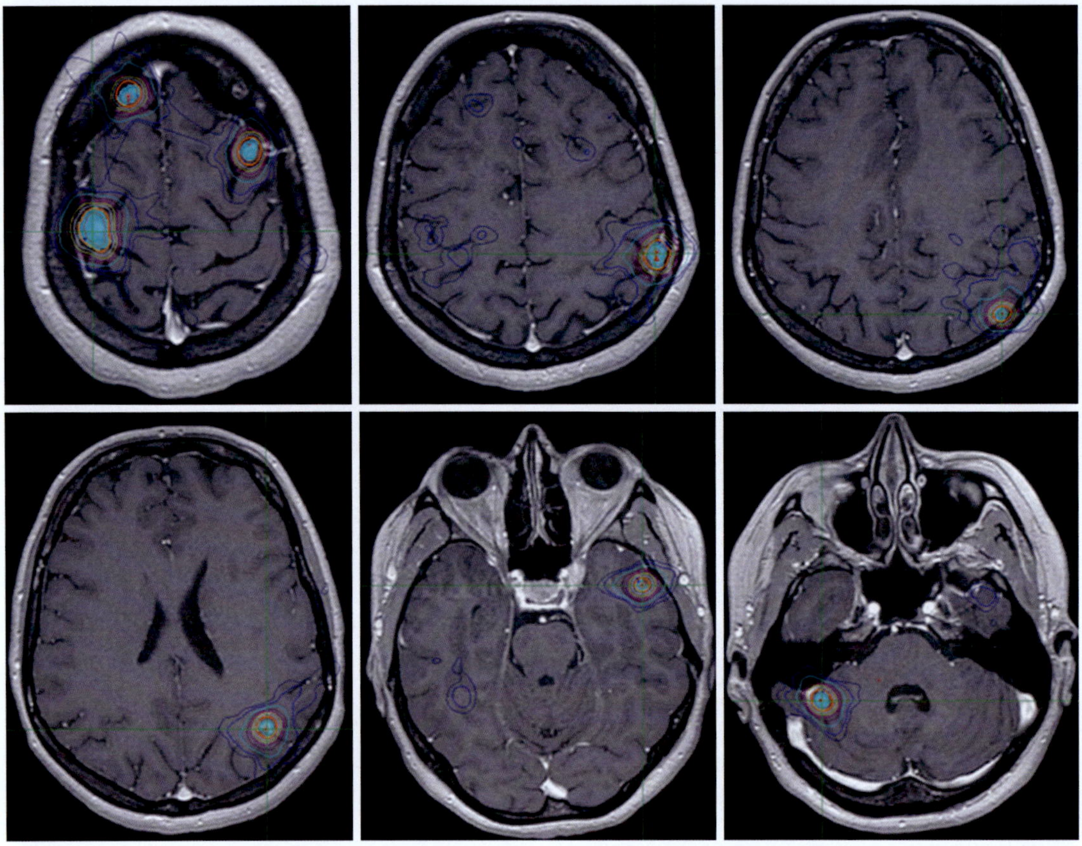

Fig. 16.2 A 35-year-old patient in good health, treated 3 years before for slowly evolving metastatic lesions of a primary carcinoid carcinoma of the lung. Multiple surgeries for "focal" secondary sites in the breast, ovary, and skin; five already known brain metastases (< 5 mm each) controlled with different lines of chemotherapy. Recently: headache, diplopia, and a new MRI showing a progression of BMs and three new BMs, the largest one near the brainstem. Total: eight different BMs but a cumulative volume of 10 cm^3, temozolomide ongoing. CyberKnife treatment was delivered: three fractions for the largest lesion and then one fraction for the other BMs. All lesions are controlled at 2 years with correction of symptoms and excellent general status

References

1. ASTRO releases second list of five radiation oncology treatments to question, as part of national Choosing Wisely® campaign. 2014. www.choosingwisely.org/astro-releases-second-list.
2. Soffietti R, Abacioglu U, Baumert B, et al. Diagnosis and treatment of brain metastases from solid tumors: guidelines from the European Association of Neuro-Oncology (EANO). Neuro Oncol. 2017;19(2):162–74.
3. Levy A, Faivre-Finn C, Hasan B, et al. Young Investigators EORTC Lung Cancer Group (YI EORTC LCG). Diversity of brain metastases screening and management in non-small cell lung cancer in Europe: results of the European Organisation for Research and Treatment of Cancer Lung Cancer Group survey. Eur J Cancer. 2018;93:37–46.
4. Kluger HM, Chiang V, Mahajan A, et al. Long-term survival of patients with melanoma with active brain metastases treated with Pembrolizumab on a phase II trial. J Clin Oncol. 2019;37(1):52–60.
5. Ramalingam SS, Vansteenkiste J, Planchard D, et al. FLAURA Investigators. Overall survival with Osimertinib in untreated, *EGFR*-mutated advanced NSCLC. N Engl J Med. 2020;382(1):41–50.
6. Camidge DR, Kim HR, Ahn MJ, et al. Brigatinib versus Crizotinib in ALK-positive non-small-cell lung cancer. N Engl J Med. 2018;379(21):2027–39.
7. Reck M, Rodríguez-Abreu D, Robinson AG, et al. KEYNOTE-024 Investigators. Pembrolizumab versus chemotherapy for PD-L1-positive non-small-cell lung cancer. N Engl J Med. 2016;375(19):1823–33.

8. Tamiya M, Tamiya A, Hosoya K, et al. Efficacy and safety of pembrolizumab as first-line therapy in advanced non-small cell lung cancer with at least 50% PD-L1 positivity: a multicenter retrospective cohort study (HOPE-001). Invest New Drugs. 2019;37(6):1266–73.
9. Gaspar LE, Scott C, Murray K, et al. Validation of the RTOG recursive partitioning analysis (RPA) classification for brain metastases. Int J Radiat Oncol Biol Phys. 2000;47(4):1001–6.
10. Sanghavi SN, Miranpuri SS, Chappell R, et al. Radiosurgery for patients with brain metastases: a multi-institutional analysis, stratified by the RTOG recursive partitioning analysis method. Int J Radiat Oncol Biol Phys. 2001;51(2):426–34.
11. Suh JH, Kotecha R, Chao ST, Ahluwalia MS, Sahgal A, Chang EL. Current approaches to the management of brain metastases. Nat Rev Clin Oncol. 2020;17(5):279–99.
12. Sperduto PW, Berkey B, Gaspar LE, et al. A new prognostic index and comparison to three other indices for patients with brain metastases: an analysis of 1,960 patients in the RTOG database. Int J Radiat Oncol Biol Phys. 2008;70(2):510–4.
13. Sperduto CM, Watanabe Y, Mullan J, et al. A validation study of a new prognostic index for patients with brain metastases: the graded prognostic assessment. J Neurosurg. 2008;109(Suppl):87–9.
14. Villà S, Weber DC, Moretones C, et al. Validation of the new graded prognostic assessment scale for brain metastases: a multicenter prospective study. Radiat Oncol. 2011;6:23.
15. Sperduto PW, Kased N, Roberge D, et al. Summary report on the graded prognostic assessment: an accurate and facile diagnosis-specific tool to estimate survival for patients with brain metastases. J Clin Oncol. 2012;30(4):419–25.
16. Sperduto PW, Yang TJ, Beal K, et al. Estimating survival in patients with lung cancer and brain metastases: an update of the graded prognostic assessment for lung cancer using molecular markers (Lung-molGPA). JAMA Oncol. 2017;3(6):827–31.
17. Sperduto PW, Jiang W, Brown PD, et al. Estimating survival in melanoma patients with brain metastases: an update of the graded prognostic assessment for melanoma using molecular markers (Melanoma-molGPA). Int J Radiat Oncol Biol Phys. 2017;99(4):812–6.
18. Sperduto PW, Deegan BJ, Li J, Jethwa KR, et al. Estimating survival for renal cell carcinoma patients with brain metastases: an update of the renal graded prognostic assessment tool. Neuro Oncol. 2018;20(12):1652–60.
19. Sperduto PW, Mesko S, Li J, et al. Beyond an updated graded prognostic assessment (Breast GPA): a prognostic index and trends in treatment and survival in breast cancer brain metastases from 1985 to today. Int J Radiat Oncol Biol Phys. 2020;107(2):334–43.
20. Patrikidou A, Chaigneau L, Isambert N, et al. Development of a disease-specific graded prognostic assessment index for the management of sarcoma patients with brain metastases (Sarcoma-GPA). BMC Cancer. 2020;20(1):117.
21. Farris M, McTyre ER, Cramer CK, et al. Brain metastasis velocity: a novel prognostic metric predictive of overall survival and freedom from whole-brain radiation therapy after distant brain failure following upfront radiosurgery alone. Int J Radiat Oncol Biol Phys. 2017;98(1):131–41.
22. Yamamoto M, Serizawa T, Nagano O, et al. Three-institution study on applicability of initial brain metastasis velocity for breast cancer brain metastasis patients undergoing stereotactic radiosurgery. J Neurooncol. 2020;147(1):177–84.
23. LeCompte MC, Hughes RT, Farris M, et al. Impact of brain metastasis velocity on neurologic death for brain metastasis patients experiencing distant brain failure after initial stereotactic radiosurgery. J Neurooncol. 2020;146(2):285–92.
24. Hirshman BR, Wilson B, Ali MA, et al. Superior prognostic value of cumulative intracranial tumor volume relative to largest intracranial tumor volume for stereotactic radiosurgery-treated brain metastasis patients. Neurosurgery. 2018;82(4):473–80.
25. Knoll MA, Oermann EK, Yang AI, et al. Survival of patients with multiple intracranial metastases treated with stereotactic radiosurgery: does the number of tumors matter? Am J Clin Oncol. 2018;41(5): 425–31.
26. Sheehan JP, Grills I, Chiang VL, et al. Quality of life outcomes for brain metastasis patients treated with stereotactic radiosurgery: pre-procedural predictive factors from a prospective national registry. J Neurosurg. 2018;131(6):1848–54.
27. Brastianos PK, Carter SL, Santagata S, et al. Genomic characterization of brain metastases reveals branched evolution and potential therapeutic targets. Cancer Discov. 2015;5(11):1164–77.
28. Yamamoto M, Serizawa T, Shuto T, et al. Stereotactic radiosurgery for patients with multiple brain metastases (JLGK0901): a multi-institutional prospective observational study. Lancet Oncol. 2014;15(4):387–95.
29. Soike MH, Hughes RT, Farris M, et al. Does stereotactic radiosurgery have a role in the management of patients presenting with 4 or more brain metastases? Neurosurgery. 2019;84(3):558–66.
30. Serizawa T, Yamamoto M, Higuchi Y, et al. Local tumor progression treated with gamma knife radiosurgery: differences between patients with 2-4 versus 5-10 brain metastases based on an update of a multi-institutional prospective observational study (JLGK0901). J Neurosurg. 2019;26:1–10.
31. Hughes RT, Masters AH, McTyre ER, et al. Initial SRS for patients with 5 to 15 brain metastases: results of a multi-institutional experience. Int J Radiat Oncol Biol Phys. 2019;104(5):1091–8.
32. Kim CH, Im YS, Nam DH, et al. Gamma knife radiosurgery for ten or more brain metastases. J Korean Neurosurg Soc. 2008;44(6):358–63.

33. Mohammadi AM, Recinos PF, Barnett GH, et al. Role of gamma knife surgery in patients with 5 or more brain metastases. J Neurosurg. 2012;117(Suppl):5–12.
34. Fogarty GB, Hong A, Gondi V, et al. Debate: adjuvant whole brain radiotherapy or not? More data is the wiser choice. BMC Cancer. 2016;16:372.
35. Tallet AV, Dhermain F, Le Rhun E, Noël G, Kirova YM. Combined irradiation and targeted therapy or immune checkpoint blockade in brain metastases: toxicities and efficacy. Ann Oncol. 2017;28(12):2962–76.
36. Nardin C, Mateus C, Texier M, et al. Tolerance and outcomes of stereotactic radiosurgery combined with anti-programmed cell death-1 (pembrolizumab) for melanoma brain metastases. Melanoma Res. 2018;28(2):111–9.
37. Kotecha R, Kim JM, Miller JA, et al. The impact of sequencing PD-1/PD-L1 inhibitors and stereotactic radiosurgery for patients with brain metastasis. Neuro Oncol. 2019;21(8):1060–8. https://doi.org/10.1093/neuonc/noz046.
38. Brown PD, Gondi V, Pugh S, et al; for NRG Oncology. Hippocampal avoidance during whole-brain radiotherapy plus Memantine for patients with brain metastases: phase III trial NRG oncology CC001. J Clin Oncol. 2020;38(10):1019–1029.
39. Mulvenna P, Nankivell M, Barton R, et al. Dexamethasone and supportive care with or without whole brain radiotherapy in treating patients with non-small cell lung cancer with brain metastases unsuitable for resection or stereotactic radiotherapy (QUARTZ): results from a phase 3, non-inferiority, randomised trial. Lancet. 2016;388(10055):2004–14.

17 CyberKnife Neuroradiosurgery for Large Brain Metastases and Tumor Bed

Maximilian I. Ruge

17.1 Introduction

Brain metastases are the most frequently diagnosed intracranial neoplasia with increasing incidence, which derives mostly from malignant melanoma, lung and breast cancer, and, less often, also solid uro-genital, gynecological, and gastrointestinal tumors. The continuous introduction of modern targeted or tumor-specific immune therapies beyond the classic chemotherapy schemes has substantially improved the prognosis for selected patients and tumor entities (i.e., malignant melanoma) [1, 2]. This improved prognosis with improved survival and understanding of the potential mid- and long-term negative impacts of external beam radiation therapy on cognitive functioning has now shifted the indications from whole brain radiation therapy to concepts of localized irradiation such as fractionated stereotactic radiation therapy (FSRT) and single- or hypo-fractionated stereotactic radiosurgery (SRS; hypo-fractionated SRS) [3, 4].

The principle of stereotactic radiosurgery (SRS) is the precise application of a highly focused single irradiation dose on a target while saving the surrounding tissue as much as possible. However, the greater the volume of treatment (i.e., large brain metastases or tumor bed after resection), the higher the risk of radiation-induced consequences such as edema or tissue necrosis.

For instance, if more than 10 cm^3 of normal brain receives a single stereotactically applied dose of 12 Gy, the risk of radiation-induced tissue necrosis rises up to 60%. Following this rationale, a maximum size of 3 cm in diameter (or 14 cm^3 in volume) for single-fraction SRS has been proposed in most guidelines by the results of RTOG Protocol 90-05 by Shaw et al. in 1996 [5–7].

With the introduction of frameless stereotactic systems like the CyberKnife, an image-guided robotic-assisted radiosurgery system from Accuray and other concepts, the option of repeatedly applying SRS with reduced dose to lower the radiation load on normal tissue, but still with a precision within a range of millimeters, larger target volumes beyond 3 cm in diameter can be treated using, for example, three or five sessions.

While first clinical data using the concept of treating large brain metastases with hypo-fractionated SRS (also referred as multi-fractionated SRS) are available, prospective (randomized) data systematically evaluating local control during follow-up, as well as adverse effects, like clinically relevant edema or necrosis, from large patient cohorts are missing. The same

M. I. Ruge (✉)
Department of Stereotaxy and Functional Neurosurgery, Centre of Neurosurgery, University Hospital of Cologne, Cologne, Germany
e-mail: maximilian.ruge@uk-koeln.de

© Springer Nature Switzerland AG 2020
A. Conti et al. (eds.), *CyberKnife NeuroRadiosurgery*,
https://doi.org/10.1007/978-3-030-50668-1_17

applies to the concept of applying single- or hypo-fractionated SRS to previously resected (large) brain metastases.

In the following relevant literature, the application of robotic (CyberKnife) SRS for the treatment of large brain metastases and the SRS application to the cavity after resection (tumor bed) of brain metastases will be systematically outlined.

A systematic review was performed in preparation for this chapter of the book using the PubMed database. Potential studies have been identified from a comprehensive bibliographic search using the following keywords (in various combinations): "CyberKnife," "robotic," "radiosurgery," "large brain metastasis," and "large brain metastases."

For the second part of the chapter, the keywords "cavity," "tumor bed," "resected," and "postoperative" have been added. Research was limited to full-text studies available and published in English between January 1980 and February 2020. Titles and abstracts were screened for the radiosurgical technique used. If the information was not included in the title or abstract, the full text was briefly reviewed.

17.2 CyberKnife SRS Treatment for Large Brain Metastases

Nine studies published between 2006 and 2015 including two review articles by Wowra et al. [8] and Masucci [9] met the research criteria reporting on the treatment of patients with large brain metastases by robotic (CyberKnife) radiosurgery:

Most recently, in 2019 Han et al. published a dosimetric comparison of fractionated radiosurgery plans using the frameless Gamma Knife ICON and CyberKnife systems with linear accelerator-based radiosurgery plans for multiple large brain metastases. The multi- fraction deliverable irradiation plans of ten patients with two or more large brain metastases (one of the target volumes had to be >10 cm^3) in treatment with LINAC radiosurgery were the basis of this comparison. In conclusion, all three modalities were capable of treating multiple large brain lesions with MF-SRS. Differences in flexibility in workflow and treatment time were found, while dosimetry between the Gamma Knife ICON and CyberKnife systems was comparable [10].

Murai et al. reported in 2014 on a dose escalation study using fractionated stereotactic radiotherapy using CyberKnife for the treatment of 61 large brain metastases (≥2.5 cm in maximum diameter) in 54 patients with a total of 102 brain metastases.

Toxicity and efficacy of the dosing regimens of (1) 18–30 Gy in three fractions for tumors with a diameter ≥2.5–4 cm and (2) 21–35 Gy in five fractions for tumors ≥4 cm were evaluated followed by a dose escalation in different levels starting at 18–22 Gy in three fractions to 31–35 Gy in five fractions. Overall survival rates were 52% and 31% at 6 months and 12 months, respectively.

Six- and 12-month local tumor control rates of the 61 large brain metastases were 77% and 69%, respectively, and no grade 3 or higher toxicity was observed. The authors concluded that dose levels of 27–30 Gy in three fractions and 31–35 Gy in five fractions seemed tolerable and effective in controlling large brain metastases and that these dose regimens can be used for further studies [11]. Even in 2014, Inoue et al. reported on a five-fraction CyberKnife radiosurgery regimen for large brain metastases in critical areas in relation to the impact of the surrounding brain volume to avoid radiation-induced tissue damage. Hereby, 78 patients with 85 brain metastases (including tumors >30 cm^3 or 4 cm in diameter) were treated with a median marginal dose of 31 Gy at a prescribed median isodose of 58% in five fractions, and parameters such as neurological changes, local tumor control, and adverse effects were evaluated.

Especially, the volume of the surrounding tissue which received 28.8 Gy (the dose equivalent to a single dose of 14 Gy = V14) was calculated. As a result, pre-existing neurological deficits (motor weakness, visual or speech disturbances) improved in 28 of 55 cases (50.9%), and local tumor control was achieved for 79 of 85 metastases (92.9%) during a median follow-up of 8 months. In ten patients, a symptomatic edema occurred which was, in two cases, surgically relevant (2.6%). Among these cases the calculated V14 ranged between 3.0 and 19.7 cm^3. While 16 tumors exceeded the V14 of ≥7.0 cm^3, which resulted in extensive brain edema in two cases, a V14 of less than 7.0 cm^3 did not induce surgically relevant brain edema. Thus, a five-fraction radiosurgery regimen and the consideration of the V14 to reduce the risk for radiation necrosis were concluded by the authors [12, 13].

For recurrent brain metastases after whole brain radiation therapy, Gwak et al. suggested in 2009 to consider the "radiation toxicity factor" (cumulative dose times tumor volume of <1000 Gy × cm^3) as a significant predictor of both acute and chronic CNS toxicities especially in patients with large tumors [14].

Nishizaki et al., the results of 71 patients with 148 brain metastases (mean/median tumor volume per lesion, 6.6/2.6 cm^3; range, 0.1–53.2) treated with CyberKnife were described in a retrospective study in 2006.

The median marginal dose was 20.7 Gy applied in a single-fraction for 108 lesions (median tumor volume of 4.3 cm^3) and with fractional dose schemes of 2 fractions (28 lesions, median volume 7.4 cm^3) and 3 fractions (12 lesions, median volume of 20.0 cm^3). Survival rates reported at 6 and 12 months were 74 and 47%, and local control was 83%. In conclusion, the role of fractional robotic radiosurgery treatment for the unfavorable cohort of patients with multiple and/or large brain metastases has been highlighted with comparable results in the literature [15].

Efficacy and safety using fractionated CyberKnife radiosurgery (median peripheral dose ranging from 30 to 41 Gy in 3 to 5 fractions) was also concluded by Jeong et al. in 2015 reporting on 37 patients with large brain metastases >3 cm.

The crude local tumor control after a median follow-up of 10 months (range 1–37 months) was 86.8% and estimated at 12 and 24 months as 87.0% and 65.2%, respectively. Median overall and progression-free survival (OS; PFS) were 16 and 11 months, respectively, and estimated at 6, 12, and 24 months as 81.1%, 65.5,% and 56.8% (OS) and 44.9%, 40.7%, and 25.7% (PFS), respectively. Improvement of performance status occurred in 20 of 35 patients (57.1%) and of pre-existing neurological deficits in 12 of 17 patients (70.6%). Treatment resulted in six lesions in radiation necrosis of toxicity grades 2 and 3 (15.8%) [16].

To investigate the differences between single- and multi-fractionated stereotactic radiosurgery for large brain metastases, Lehrer et al. performed an international meta-analysis of 24 trials with overall 1887 SRS-treated brain metastases. Using methods defined by the guidelines for research and data review, as well as for statistical calculations, outcome parameters such as local control and radionecrosis rates after either definitive or postoperative radiosurgery were calculated by forming groups of "large" brain metastases: Group A 4–14 cm^3 (or 2–3 cm in diameter) and Group B > 14 cm^3 (or about >3 cm in diameter). The authors concluded that the use of multi-fractionated SRS regimens to treat large brain metastases may offer a relative reduction of radionecrosis while maintaining or improving relative rates of 1-year local control as compared to single-fraction SRS [7].

In conclusion, the use of multiple fraction SRS (also referred to as HFRT (hypo-fractionated radiation therapy)) or using three fractions with doses ranging from 27 to 30 Gy or five fractions 31 to 35 Gy as suggested by Murai et al. in 2014 seems to provide a reasonable balance between the efficacy of local control and the onset of negative side effects induced by treatment for large brain metastases (≥ 2.5 cm) (Fig. 17.1).

The 2019 NCCN Guidelines recommend, for brain metastases >3 cm in diameter, 27 Gy in three fractions or 30 Gy in five fractions [17]. Furthermore, it seems useful to consider a V14 (equivalent dose of a single dose) of the surround-

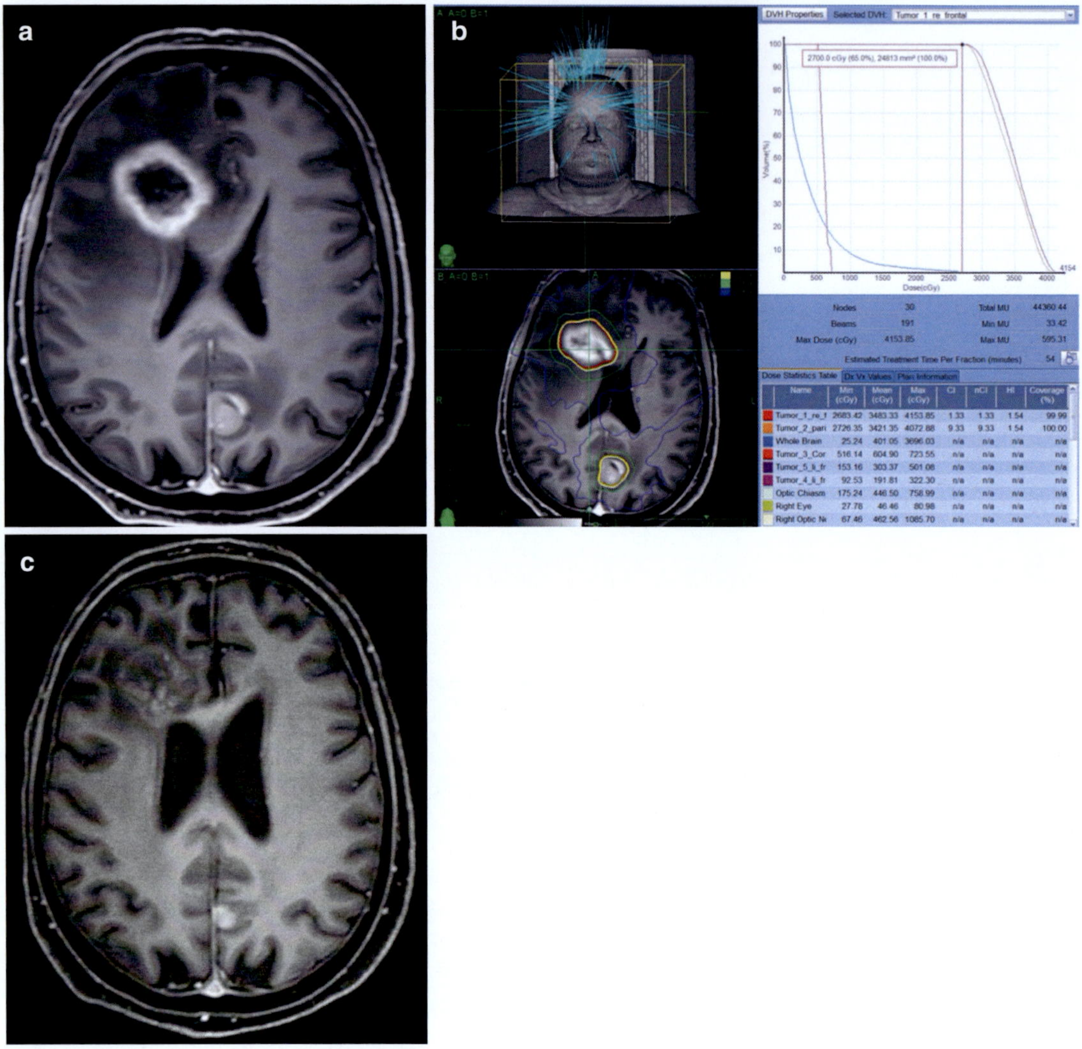

Fig. 17.1 (**a**) MRI scan from November 2016 from a 65-year-old female diagnosed with small cell lung cancer in March 2015. Besides multimodal systemic treatment, she received prophylactic whole brain radiation therapy (30 Gy; ten fractions) from August to September 2015. (**b**) As salvage treatment she received hypo-fractionated CyberKnife SRS with a dose of 27 Gy in three fractions for two brain metastases [right frontal (volume = 24.8 cm^3); left parietal (volume = 3.5 cm^3)] since a microsurgical resection was not indicated due to comorbidities and a—at that time—reduced Karnofsky performance status of <70 due to the general medical condition. (**c**) MRI follow-up 8 weeks later shows partial remission of both metastases [right frontal (volume = 8.9 cm^3); left parietal (volume = 0.4 cm^3)] and reduction of the initial focal edema. Neurologically she stayed stable but deceased in February 2017 due to massive progression of her systemic disease

ing tissue with a volume less than 7.0 cm^3 to avoid extensive brain edema or even radiation-induced necrosis as investigated by Inoue et al. [12].

Reviewing the literature, however, it becomes very clear that, besides the lack of a limited definition of "large" brain metastases, long-term data from large prospective trials are essentially necessary to evaluate the efficacy and safety of different SRS regimens for "large" brain metastases.

17.3 CyberKnife SRS Treatment for Cavity (Tumor Bed) After Resection of Brain Metastases

In 2008 Soltys et al. treated 72 patients with 76 cavities after resection of brain metastases from 1998 to 2006 delivering a median marginal dose of 18.6 Gy (15–30 Gy) to an average target volume of 9.8 cm^3 (0.1–66.8 cm^3) using CyberKnife radiosurgery. Single-session SRS was applied in 78% (6.8 cm^3 median target volume) of the treatments and multiple session (two in 9%; three in 12%; and four in 1%) for a median of 13.8 cm^3.

Follow-up imaging was available in 65 patients, and the median follow-up was 8.1 months (0.1–80.5 months). Actuarial local 6- and 12-month tumor control was 88% and 79%, respectively. Seven patients developed symptomatic post-treatment brain edema requiring steroids, and three of those underwent surgical resection due to radiation-induced necrosis.

None of the relevant treatment factors (treatment dose, target volume, or target number) were related. While neither the target volume, the dose, nor the number of sessions influenced local control in the univariate analysis, the high conformity indices were significantly correlated. Based on improved local control with less conformal plans, the authors recommended to include a 2 mm margin around the resection cavity [18].

Wang et al. treated 37 patients with large resection cavities of >3 cm in diameter with a hypo-fractionated stereotactic radiosurgery (HSRS) regimen of 800 cGy applied in 3 daily fractions.

Twelve-month follow-up was available from 35 patients with a median survival of 5.5 months and a 6-month actuarial local control rate of 80% and 3 patients with adverse events such as radiation necrosis (n = 1), prolonged use of steroids (n = 1), and new onset of seizures (n = 1). This treatment regimen was considered safe and effective as the authors concluded [19].

Also using a fractionated SRS regimen, Vogel et al. reported in 2015 the outcome of 30 patients with postoperative treatment of 33 large resection cavities >3 cm between 2011 and 2014. Median treatment volume was 25.1 cm^3 (range 4.7–90.9 cm^3), and median maximal postoperative cavity diameter was 3.8 cm (range 2.8–6.7). Twenty-six patients were treated with 30 Gy in five fractions and a median isodose level of 76%.

Five patients were treated with 24 Gy in three fractions and two patients with 18.5 Gy in one fraction. Estimated 6-month and 1-year local control rates available from 26 patients were 82.3% and 68.5%, respectively. The crude rate of local failure was 24% (seven patients), and the rate of the development of leptomeningeal tumor spreading was 34% (nine patients). In three patients (10%), a significant radionecrosis was detected resulting in steroid treatment [20].

Martinage et al. published in 2019 the, at that point, largest retrospective multicentric series with 160 patients using CyberKnife HFSRT (24 Gy/3 fractions in 52 patients, 30 Gy/5 fractions in 37 patients, 27–30 Gy/3 fractions in 34 patients, 30 Gy/6 fractions in 15 patients, others in 22 patients) for postoperative treatment of the cavity after resection of brain metastases. Estimated 1- and 2-year local control rates were 88% [95% CI, 81%–93%] and 81% [95% CI, 70%–88%], respectively, and no prognostic factor was associated with local progression.

With regard to the prognostic factors associated with the treatment, a higher planning target volume influenced the significantly negative overall survival in the uni- and multivariate analyses. Early and late adverse side effects according to the Common Terminology Criteria for Adverse Events (CTCAE) classification system version 4.03 of grades 2 and 3 occurred in five (3.4%) and ten (7.2%) patients, respectively. Radiation necrosis was observed in 13 (8.9%) patients. The authors also considered HFSRT for metastasectomy tumor beds to be effective for local control and well tolerated by patients [21]. A systematic review and meta-analysis for post-operative stereotactic radiosurgery following the excision of brain metastases was published in 2020 by Akanda et al. reporting results of 50 studies with 3458 patients treated overall with various radiosurgery techniques and various single and multi-fractionations. Parameters such as local control at 12 months, appearance of radiation necrosis, and leptomeningeal disease were investigated. Local control in all studies was 83.7%, and patients treated with fractionated SRS had better local control than those treated with single-fraction SRS (87.3% vs 80.0%, $p = 0.021$; univariate analysis). The addition of a margin did not improve local control (margin vs no margin = 84.3% vs 83.1%; $p = 0.71$).

Radiation necrosis was rare at 6.9% in all reported studies, and leptomeningeal disease was found to be 13% in all reported studies [22]. When consulting the 2019 NCCN Guidelines, 16–20 Gy for small resection cavities and 27 Gy in three fractions or 30 Gy in five fractions for large mounds are recommended [17]. In conclusion, fractionated SRS regimens should be considered for the treatment of cavities (tumor bed) after resection of brain metastases to omit whole brain radiation or local fractionated radiation therapy. Adding a safety margin seems not to improve local control. While the rate of severe early and late radiation-induced side effects seems to be reasonably low, the occurrence of leptomeningeal disease deserves attention. To potentially reduce the latter, the concept of pre-operative or also referred to as neoadjuvant SRS should be investigated. Also, prospective (randomized) data from lager cohorts are not yet available.

References

1. Moravan MJ, Fecci PE, Anders CK, Clarke JM, AKS S, Adamson JD, Floyd SR, Torok JA, Salama JK, Sampson JH, Sperduto PW, Kirkpatrick JP. Current multidisciplinary management of brain metastases. Cancer. 2020;126(7):1390–406. https://doi.org/10.1002/cncr.32714. Epub 2020 Jan 23.
2. Suh JH, Kotecha R, Chao ST, Ahluwalia MS, Sahgal A, Chang EL. Current approaches to the management of brain metastases. Nat Rev Clin Oncol. 2020;17:279. https://doi.org/10.1038/s41571-019-0320-3. [Epub ahead of print]. Review.
3. Brown PD, Ballman KV, Cerhan JH, Anderson SK, Carrero XW, Whitton AC, Greenspoon J, Parney IF, Laack NNI, Ashman JB, Bahary JP, Hadjipanayis CG, Urbanic JJ, Barker FG 2nd, Farace E, Khuntia D, Giannini C, Buckner JC, Galanis E, Roberge D. Postoperative stereotactic radiosurgery compared with whole brain radiotherapy for resected metastatic brain disease (NCCTG N107C/CEC·3): a multicentre, randomised, controlled, phase 3 trial. Lancet Oncol. 2017;18(8):1049–60. https://doi.org/10.1016/S1470-2045(17)30441-2. Epub 2017 Jul 4.
4. Soffietti R, Kocher M, Abacioglu UM, Villa S, Fauchon F, Baumert BG, Fariselli L, Tzuk-Shina T, Kortmann RD, Carrie C, Ben Hassel M, Kouri M, Valeinis E, van den Berge D, Mueller RP, Tridello G, Collette L, Bottomley A. A European Organisation for Research and Treatment of Cancer phase III trial of adjuvant whole-brain radiotherapy versus observation in patients with one to three brain metastases from solid tumors after surgical resection or radiosurgery: quality-of-life results. J Clin Oncol. 2013;31(1):65–72. https://doi.org/10.1200/JCO.2011.41.0639. Epub 2012 Dec 3.
5. Shaw E, Scott C, Souhami L, Dinapoli R, Bahary JP, Kline R, Wharam M, Schultz C, Davey P, Loeffler J, Del Rowe J, Marks L, Fisher B, Shin K. Radiosurgery for the treatment of previously irradiated recurrent primary brain tumors and brain metastases: initial report of radiation therapy oncology group protocol (90-05). Int J Radiat Oncol Biol Phys. 1996;34(3): 647–54.
6. Vogelbaum MA, Angelov L, Lee SY, Li L, Barnett GH, Suh JH. Local control of brain metastases by stereotactic radiosurgery in relation to dose to the tumor margin. J Neurosurg. 2006;104(6):907–12.
7. Lehrer EJ, Peterson JL, Zaorsky NG, Brown PD, Sahgal A, Chiang VL, Chao ST, Sheehan JP, Trifiletti DM. Single versus multifraction stereotactic radiosurgery for large brain metastases: an international meta-analysis of 24 trials. Int J Radiat Oncol Biol Phys. 2019;103(3):618–30. https://doi.org/10.1016/j.ijrobp.2018.10.038. Epub 2018 Nov 2.
8. Wowra B, Muacevic A, Tonn JC. CyberKnife radiosurgery for brain metastases. Prog Neurol Surg. 2012;25:201–9. https://doi.org/10.1159/000331193.
9. Masucci GL. Hypofractionated radiation therapy for large brain metastases. Front Oncol. 2018;8:379. https://doi.org/10.3389/fonc.2018.00379.
10. Han EY, Wang H, Luo D, Li J, Wang X. Dosimetric comparison of fractionated radiosurgery plans using frameless gamma knife ICON and CyberKnife systems with linear accelerator–based radiosurgery plans for multiple large brain metastases. J Neurosurg. 2019;132:1473. https://doi.org/10.3171/2019.1.JNS182769.
11. Murai T, Ogino H, Manabe Y, Iwabuchi M, Okumura T, Matsushita Y, et al. Fractionated stereotactic radiotherapy using CyberKnife for the treatment of large brain metastases: a dose escalation study. Clin Oncol. 2014;26:151–8. https://doi.org/10.1016/j.clon.2013.11.027.
12. Inoue HK, Sato H, Suzuki Y, Saitoh J, Noda SE, Seto K, et al. Optimal hypofractionated conformal radiotherapy for large brain metastases in patients with high risk factors: a single-institutional prospective study. Radiat Oncol. 2014a;9:231. https://doi.org/10.1186/s13014-014-0231-5.
13. Inoue HK, Sato H, Seto K, Torikai K, Suzuki Y, Saitoh J, Noda SE, Nakano T. Five-fraction CyberKnife

radiotherapy for large brain metastases in critical areas: impact on the surrounding brain volumes circumscribed with a single dose equivalent of 14 Gy (V14) to avoid radiation necrosis. J Radiat Res. 2014b;55(2):334–42. https://doi.org/10.1093/jrr/rrt127.
14. Gwak H-S, Yoo HJ, Youn S-M, Lee DH, Kim MS, Rhee CH. Radiosurgery for recurrent brain metastases after whole-brain radiotherapy: factors affecting radiation-induced neurological dysfunction. J Korean Neurosurg Soc. 2009;45(5):275–83.
15. Nishizaki T, Saito K, Jimi Y, Harada N, Kajiwara K, Nomura S, Ishihara H, Yoshikawa K, Yoneda H, Suzuki M, Gibbs IC. The role of CyberKnife radiosurgery/radiotherapy for brain metastases of multiple or large-size tumors. Minim Invasive Neurosurg. 2006;49(4):203–9.
16. Jeong WJ, Park JH, Lee EJ, Kim JH, Kim CJ, Cho YH. Efficacy and safety of fractionated stereotactic radiosurgery for large brain metastases. J Korean Neurosurg Soc. 2015;58:217–24. https://doi.org/10.3340/jkns.2015.58.3.217.
17. NCCN. Central Nervous System Cancers. NCCN Clinical Practice Guidelines in Oncology. 2019.
18. Soltys SG, Adler JR, Lipani JD, Jackson PS, Choi CY, Puataweepong P, White S, Gibbs IC, Chang SD. Stereotactic radiosurgery of the postoperative resection cavity for brain metastases. Int J Radiat Oncol Biol Phys. 2008;70(1):187–93. Epub 2007 Sep 19.
19. Wang CC, Floyd SR, Chang CH, Warnke PC, Chio CC, Kasper EM, Mahadevan A, Wong ET, Chen CC. Cyberknife hypofractionated stereotactic radiosurgery (HSRS) of resection cavity after excision of large cerebral metastasis: efficacy and safety of an 800 cGy × 3 daily fractions regimen. J Neurooncol. 2012;106(3):601–10. https://doi.org/10.1007/s11060-011-0697-z.
20. Vogel J, Ojerholm E, Hollander A, Briola C, Mooij R, Bieda M, Kolker J, Nagda S, Geiger G, Dorsey J, Lustig R, O'Rourke DM, Brem S, Lee J, Alonso-Basanta M. Intracranial control after Cyberknife radiosurgery to the resection bed for large brain metastases. Radiat Oncol. 2015;10:221. https://doi.org/10.1186/s13014-015-0523-4.
21. Martinage G, Geffrelot J, Stefan D, Bogart E, Rault E, Reyns N, Emery E, Makhloufi-Martinage S, Mouttet-Audouard R, Basson L, Mirabel X, Lartigau E, Pasquier D. Efficacy and tolerance of post-operative hypo-fractionated stereotactic radiotherapy in a large series of patients with brain metastases. Front Oncol. 2019;9:184. https://doi.org/10.3389/fonc.2019.00184. eCollection 2019.
22. Akanda ZZ, Hong W, Nahavandi S, Haghighi N, Phillips C, Kok DL. Post-operative stereotactic radiosurgery following excision of brain metastases: a systematic review and meta-analysis. Radiother Oncol. 2020;142:27–35. https://doi.org/10.1016/j.radonc.2019.08.024. Epub 2019 Sep 25.

18 Convexity and Parasagittal Meningiomas

Alfredo Conti

18.1 Introduction

Stereotactic radiosurgery (SRS) has progressively emerged as both an adjuvant treatment modality for residual tumors and an effective primary treatment of properly selected meningiomas. Ten-year local tumor control (LTC) rates range between 80 and 100%, depending on the size of the lesions, location, dose applied, and length of follow-up [1–6]. Radiosurgery is virtually noninvasive, but it does carry a risk of radiation-induced complications. For meningiomas, this risk ranges between 3 and 40% [7, 8]. Symptomatic post-treatment edema (PTE) causing seizures, focal deficits, and even intracranial hypertension is the most common complication in intracranial meningioma radiosurgery occurring in 6–35% of cases [9–11].

Such adverse effects induced by radiation appear more frequent in "nonbasal" meningiomas and, particularly, in parasagittal tumors [10–15]. Accordingly, the application of radiosurgery to these meningiomas is more controversial if compared to those located at the skull base.

A. Conti (✉)
Alma Mater Studiorum University of Bologna, Bologna, Italy

IRCCS ISNB Istituto delle Scienze Neurologiche di Bologna, Bologna, Italy

Department of Neurosurgery, Charité Faculty of Medicine, Berlin, Germany
e-mail: alfredo.conti2@unibo.it

© Springer Nature Switzerland AG 2020
A. Conti et al. (eds.), *CyberKnife NeuroRadiosurgery*,
https://doi.org/10.1007/978-3-030-50668-1_18

18.2 Tumor Control

Tumor control for convexity and parasagittal meningiomas appears to be slightly worse than for skull base tumors. In a multicenter study, involving 203 patients with parasagittal meningiomas treated by Gamma Knife radiosurgery (GKS), Kondziolka et al. [16] reported actuarial 5-year tumor control rates of 93 ± 4% and 60 ± 10% for patients receiving a primary and adjuvant treatment, respectively. The report also suggested that in most cases, failures resulted from out-of-field tumor progression. Similarly, in their more recent series of convexity meningiomas, the actuarial 3- and 5-year tumor control rates in patients with benign meningiomas and those who had not undergone prior surgery were 95% and 86%, respectively [8]. Hasegawa et al. reported an actuarial 5- and 10-year PFS rates of 78% and 55%, respectively, whereas the actuarial 5- and 10-year LTC rates were 87% and 71%, respectively [9].

Although the differences in tumor control among convexity and parasagittal meningiomas and skull base meningiomas may simply result from selection bias, it is possible that worse LTC rates result from a more difficult definition of target volumes. Sometimes, the differentiation between the tumor margin and normal dural tissue, especially for parasagittal lesions, cannot be trivial. In terms of prognostic factors, tumor control in patients who had previously undergone surgery was indeed significantly worse in the study by Hasegawa et al. [9]. The actuarial 5- and 10-year PFS rates were 93% and 84%, respectively, in patients who had GKS as the initial treatment, whereas the rates were 68% and 35%, respectively, in patients with a history of surgery. Again, this may result from suboptimal definition of the target volume. Prior surgery makes its definition volume more complicated, as it is difficult to distinguish residual tumor from postoperative changes on enhanced MR images. Furthermore, the residual tumor can often be spread out into multiple portions after surgery. These issues may eventually lead to out-of-field treatment failure.

18.3 Complications

It is fairly evident that "nonbasal" localization is associated with a significantly higher probability of peritumoral edema (PTE) after radiosurgery and related complications. There are sufficient data to quantify this risk. In the multicenter study reporting management data on 203 patients with parasagittal meningiomas, Kondziolka and colleagues [16] reported a 3- and 5-year actuarial rate of symptomatic edema of 16%. The risk of edema development was not related to tumor margin dose, sex, patient age, history of previous radiation therapy, lower isodoses (<50%), tumor volume, imaging finding of encephalomalacia, or maximum dose. In a following study, Kondziolka et al. [8] reported the results of a series of 972 patients who underwent GKS. The overall complication rate in this study was 7.7%, but the morbidity rate for meningiomas with parasagittal location was 9.7%. The association between a higher risk of complications and the parasagittal location of a meningioma has been strongly supported by other data. Chang et al. [17] reported their experience in 179 meningiomas treated by GKS. Magnetic resonance imaging showed complications after GKS in 35 lesions (25%) among the 140 lesions with follow-up data. Radiation-induced imaging changes were observed mostly in convexity, parasagittal, and falx meningiomas. About 60% of these imaging changes were asymptomatic; the overall rate of symptomatic imaging changes was 9.3%. The authors analyzed the factors related to peritumoral imaging changes on MR imaging after GKS. In the univariate analysis, tumor location ($p < 0.001$), maximum tumor dose ($p = 0.0002$), and tumor margin dose ($p = 0.037$) were significantly related to imaging changes. However, in the multivariate analysis, only tumor location was significant.

In the series of 76 meningiomas treated with GKS as reported by Singh and collaborators [18], the only factor related to edema development was the tumor site, but edema occurred most frequently in meningiomas of the parasagittal region. No correlation with tumor volume, tumor margin dose, mean or peak dose, or dose received by the surrounding brain tissue was found.

We have recently retrospectively analyzed our patients' data to identify factors associated with the development of symptomatic PTE in a series of 245 meningiomas in which 229 patients were treated by a single-fraction or multisession radiosurgery (2–5 fractions) or hypofractionated stereotactic radiotherapy (6–15 fractions) using the CyberKnife system (Accuray Inc., Sunnyvale, CA) at the University Hospital of Messina, Italy, between July 2007 and March 2014 [10]. Local tumor control was achieved in 200 out of 212 patients with World Health Organization (WHO) Grade I meningiomas (94%) at a mean follow-up of 62 months. Symptomatic PTE on MRI was diagnosed in 19 patients (8.3%) causing seizure (n = 17, 89%), aggravating headache (n = 12, 63%), or focal deficits (n = 13, 68%). Four variables were found to be associated with the likelihood of edema development, including tumor volume >4.5 mL, nonbasal tumor location, tight brain-tumor interface, and atypical histology. Nonetheless, when multivariate logistic regression analysis was performed, only tumor volume and brain-tumor interface turned out to be independent predictors of PTE development [10].

Our results suggest that larger tumors, an atypical histology, a convexity/parasagittal location, and tight brain-tumor interface were factors associated with the risk of developing symptomatic PTE. Among these factors, a larger tumor volume and a tight brain-tumor interface turned out to be independent predictors of symptomatic

PTE. Furthermore, patients with atypical parasagittal meningiomas without previous treatment had a 100% risk of developing symptomatic PTE. Also, larger convexity meningiomas, not previously operated and with an adherent brain-tumor interface, were also associated, in our series, with a risk of >90% of developing severe PTE. Noteworthily, no patient with a skull base meningioma developed symptomatic PTE, including those with large and very large lesions. The only cases in which irradiation of skull base meningiomas caused the development of brain edema were two unusual cases of meningiomas of the posterior fossa in patients with multiple sclerosis. In both cases, patients had PTE after treatment that was easily managed with steroids and resolved within 6 months. A remarkable shrinkage of the tumor was observed in both cases at 12 months.

Our results also suggested that dose staging, or hypofractionation, does not provide sufficient reassurance of the prevention of PTE. In fact, most patients with parasagittal meningiomas that developed PTE had previously received multisession radiosurgery. The development of PTE was independent of the invasion and consequential irradiation of peritumoral veins, including major sinuses. Indeed, parasagittal tumors were associated with high risk of PTE, whereas none of the patients with tumors involving the transverse sinuses developed PTE.

All patients with symptomatic PTE underwent high-dose steroid administration, but 11 out of 19 needed surgical resection with an almost immediate resolution of the PTE and associated symptoms, including drug-resistant seizures (Fig. 18.1). Our observation that the PTE almost immediately declines after meningioma resection unquestionably demonstrates that the factors responsible for the development of edema reside in the irradiated meningioma, therefore excluding a direct effect of radiotherapy on the peritumoral brain and vasculature, such as that responsible for PTE in arteriovenous malformations. This interpretation also justifies the lack of protective effect of hypofractionation that is, on the other hand, relevant in preventing radiosurgery-induced complications when dealing with large skull base meningiomas. Indeed, in these latter cases, hypofractionation may prevent direct effects of radiation on critically radiation-sensitive structures, including the brainstem, optic nerves and chiasm, and cranial nerves.

Peritumoral edema in meningiomas is vasogenic, not cytotoxic, and it is associated with increased intratumoral vascular permeability [7, 19–21]. Vasogenic edema is caused by an increased capillary permeability with extravasation of serum proteins and fluid into the extracellular spaces. It has been shown that irradiated meningiomas present high expression levels of markers of angiogenesis and hypoxia (vascular endothelial growth factor (VEGF) and hypoxia-inducible factor-1, respectively) that could be associated with the increased vascular permeability of the tumors [22, 23]. In fact, the VEGF pathway may participate in the formation of brain edema in meningiomas by inducing the formation of "leaky" capillaries, resulting in secretion of VEGF-A and plasma to the peritumoral brain tissue. Nonetheless, if the causes of PTE were all intrinsic to the meningioma and to its response to radiation, i.e., through an increased secretion of VEGF-A, there would be no reason to explain the prevalence of specific locations for the development of edema. Indeed, parasagittal and convexity meningiomas are associated with a significantly higher probability of edema after radiosurgery [6, 21, 24].

The somehow surprising fact that nonbasal meningiomas have a higher rate of complications than basal meningiomas is also confirmed by many other studies. Patil et al. [25] reported on 102 supratentorial meningiomas treated with CyberKnife SRS and fractionated radiosurgery. In this study, 9 (29%) of 31 patients with parasagittal meningiomas developed symptomatic edema. Hoe et al. [20] investigated the risks and patterns of evolution of PTE for asymptomatic intracranial meningiomas and found that the nonbasal location was an independent risk factor for the development of PTE. Therefore, PTE seems to be induced by the production of chemicals by the irradiated meningioma, but this occurs only in specific circumstances, such as a parasagittal or hemispheric location and, according to our

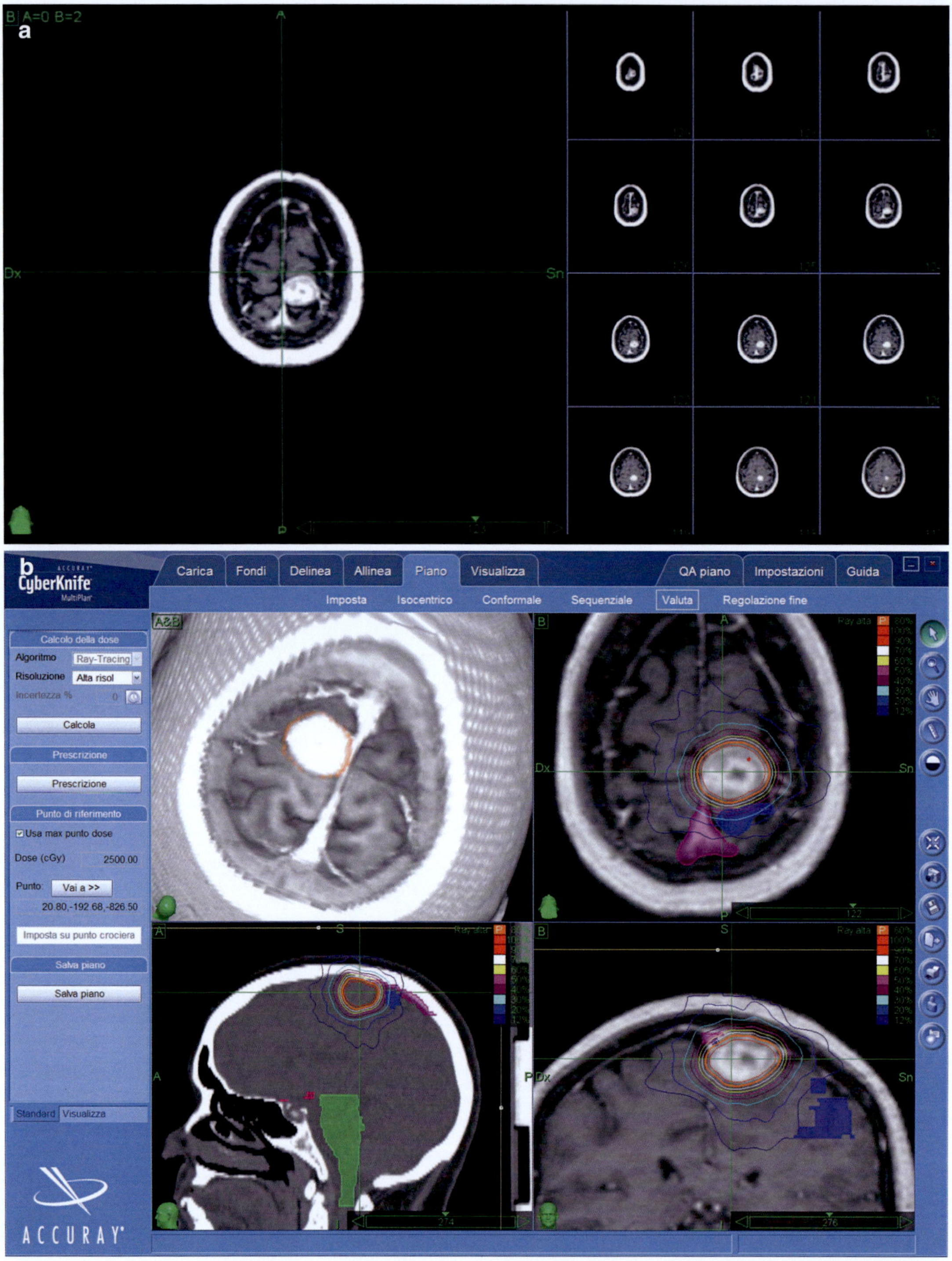

Fig. 18.1 Single-fraction CyberKnife radiosurgery treatment of a left parasagittal meningioma causing symptomatic post-treatment edema (PTE) with resolution after surgical resection of the tumor. (**a**) Premotor left parasagittal meningioma before treatment. (**b**) Radiosurgery treatment plan (single fraction; prescribed dose 13 Gy) for a parasagittal meningioma. Motor area and venous structures were contoured to reduce direct irradiation. (**c**) Left: 7 months after the treatment, the patient presented with confusion and seizures progressing to status epilepticus. The MRI showed severe perilesional edema. Right: The patient underwent resection of the meningioma with quick resolution of edema and symptoms

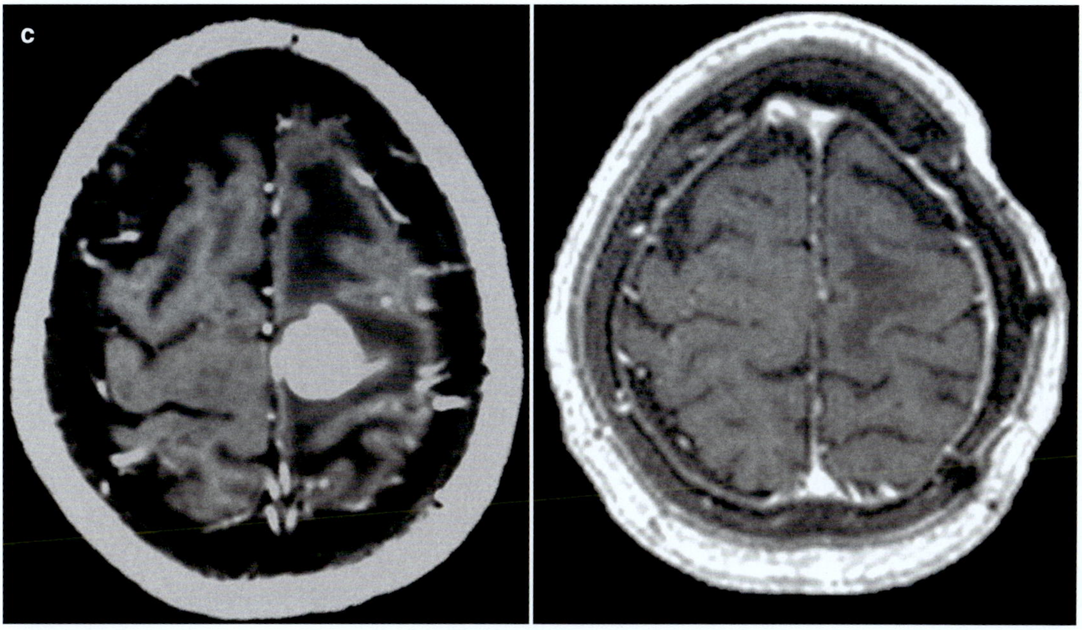

Fig. 18.1 (continued)

observations, the absence of a previous surgical manipulation. This suggests that the interface between the tumor and the brain may play a role. Cai et al. [15] suggested that the tumor-brain interface area is a strong predictor for the development of PTE and proposed a mechanistic relationship of the tumor-brain interface disruption by virtue of tumor growth (exerted by large tumors). Our results support the role of this factor showing that volume and tumor-brain interface are independent determinants of symptomatic PTE. Actually, skull base meningiomas are typically and almost invariably extraarachnoidal tumors. Convexity and, in particular, parasagittal meningiomas are often intraarachnoidal.

In the early 1990s, our group published the seminal paper on the brain-tumor interface in meningiomas describing three types of interface:

1. Smooth type, in which the tumor was well demarcated from the brain by a small preserved subarachnoid space; no peritumoral edema was usually present preoperatively in such cases.
2. Transitional type, in which vessels were often entrapped between the brain and tumor and the arachnoid membrane was very thin and extremely adherent to the tumor. The transitional type is associated with various degrees of halo-like peritumoral edema.
3. Invasive type is characterized by vessels crossing the brain-tumor interface [26].

The pial membrane is still present and extremely adherent to the tumor in some areas; however, a disruption of the cortical layer is systematically present in other areas in which the white matter is directly in contact with the tumor. This type of interface is associated with the presence of finger-like edema involving the white matter of the affected hemisphere.

Therefore, it appears that the arachnoid membrane can function as a mechanical and biochemical buffer against mediators released from a tumor. Tumor location where tumors are more likely to grow below the arachnoidal layer or to directly penetrate this layer (mostly nonbasal vs. skull base) is associated with a significantly increased risk of PTE. Another observation that we consider a proof of the role of brain-tumor interface as the major determinant of PTE development is the fact that parasagittal meningiomas are associated with a high risk of PTE only when SRS was the primary treatment. Remnants and small- to medium-sized recurrences were not associated to PTE in our series as a result of a looser interface in these tumors.

Peritumoral edema developed within months, reached its greatest extent at 11 months, and decreased thereafter over 2 years after SRS. Symptom onset and its duration approximated to this timeline. Permanent deficits from PTE after SRS treatment of meningiomas have been reported in less than 3% and were rarely disabling [6, 27, 28], although exceptional fatal cases have been also reported [29]. In our study, no patient harbored sustained neurological symptoms or reported permanent deficit attributable to PTE because patients with severe or worsening symptoms underwent surgical resection with a surprisingly quick recovery from symptoms, including drug-resistant epilepsy (Fig. 18.1).

18.4 Management Strategy

Despite a lower LTC, patients with tumor recurrence and residual tumor after open surgery are good candidates for radiosurgery. In particular, a tumor invading the superior sagittal sinus can be safely treated after removal of the extrasinusal portion of the tumor. Also, small- to medium-sized primary tumors can be relatively safely treated with a high rate of long-term tumor control.

Even though tumor size is relatively large (mean diameter ≥3 cm), low-dose radiosurgery is still effective at midterm. However, in patients harboring large symptomatic tumors or those with peritumoral edema demonstrated on preradiosurgical MR images, we recommend open surgery as the initial treatment, except in elderly patients or in those with medical comorbidities, in whom low-dose radiosurgery can be a treatment option. All patients with symptomatic PTE should receive high-dose steroid administration, but if symptoms persist, surgical resection provide the almost immediate resolution of the PTE and associated symptoms, including drug-resistant seizures.

In large parasagittal or falx meningiomas with peritumoral edema, surgical resection is recommended to avoid severe radiation-induced edema. Patients harboring asymptomatic tumors without peritumoral edema may be observed with follow-up serial imaging, or we may select open surgery or radiosurgery depending on age, tumor size, and location.

18.5 Conclusions

CyberKnife radiosurgery is effective for convexity and parasagittal meningiomas, and treatment results are not dissimilar from those obtained for skull base meningiomas in terms of LTC. Radiation doses to be delivered to these tumors are not very different from those used for skull base meningiomas. The application of CyberKnife radiosurgery to convexity and parasagittal meningiomas is, however, more controversial, because adverse effects induced by radiation appear to be more frequent in these meningiomas. Symptomatic post-treatment edema (PTE) causing seizures, focal deficits, and intracranial hypertension is a rather common complication when dealing with convexity and parasagittal meningiomas. Our results suggest that the factor associated with the risk of developing PTE is linked to the characteristics of meningioma rather than to the treatment modality used and that hypofractionation or other strategies to protect peritumoral veins are not protective. Thus, recommendations are mostly related to the indications rather than specific treatment modalities. An appropriate patient selection is the way to achieve safe treatment and long-term disease control. Variables associated with the likelihood of edema development include tumor volume, tight brain-tumor interface, and atypical histology.

References

1. Cohen-Inbar O, Lee CC, Sheehan JP. The contemporary role of stereotactic radiosurgery in the treatment of meningiomas. Neurosurg Clin N Am. 2016;27(2):215–28.
2. Cohen-Inbar O, Tata A, Moosa S, Lee CC, Sheehan JP. Stereotactic radiosurgery in the treatment of parasellar meningiomas: long-term volumetric evaluation. J Neurosurg. 2018;128(2):362–72.

3. Kim JW, Kim DG, Se YB, Kim SK, Chung HT, Paek SH, et al. Gamma knife radiosurgery for petroclival meningioma: long-term outcome and failure pattern. Stereotact Funct Neurosurg. 2017;95(4):209–15.
4. Kondziolka D, Patel AD, Kano H, Flickinger JC, Lunsford LD. Long-term outcomes after gamma knife radiosurgery for meningiomas. Am J Clin Oncol. 2016;39(5):453–7.
5. Seo Y, Kim DG, Kim JW, Han JH, Chung HT, Paek SH. Long-term outcomes after gamma knife radiosurgery for benign meningioma: a single institution's experience with 424 patients. Neurosurgery. 2018;83(5):1040–9.
6. Stafford SL, Pollock BE, Foote RL, Link MJ, Gorman DA, Schomberg PJ, et al. Meningioma radiosurgery: tumor control, outcomes, and complications among 190 consecutive patients. Neurosurgery. 2001;49(5):1029–37; discussion 37–8.
7. Bitzer M, Klose U, Geist-Barth B, Nägele T, Schick F, Morgalla M, et al. Alterations in diffusion and perfusion in the pathogenesis of peritumoral brain edema in meningiomas. Eur Radiol. 2002;12(8):2062–76.
8. Kondziolka D, Mathieu D, Lunsford LD, Martin JJ, Madhok R, Niranjan A, et al. Radiosurgery as definitive management of intracranial Meningiomas. Neurosurgery. 2008;62(1):53–60.
9. Hasegawa T, Kida Y, Yoshimoto M, Iizuka H, Ishii D, Yoshida K. Gamma knife surgery for convexity, parasagittal, and falcine meningiomas. J Neurosurg. 2011;114(5):1392–8.
10. Conti A, Pontoriero A, Siddi F, Iati G, Cardali S, Angileri FF, et al. Post-treatment edema after meningioma radiosurgery is a predictable complication. Cureus. 2016;8(5):e605.
11. Sheehan JP, Cohen-Inbar O, Ruangkanchanasetr R, Bulent Omay S, Hess J, Chiang V, et al. Post-radiosurgical edema associated with parasagittal and parafalcine meningiomas: a multicenter study. J Neurooncol. 2015;125(2):317–24.
12. Kondziolka D, Flickinger JC, Perez B. Judicious resection and/or radiosurgery for parasagittal meningiomas: outcomes from a multicenter review. Neurosurgery. 1998;43(3):405–13.
13. Lee JYK, Kondziolka D, Flickinger JC, Lunsford LD. Radiosurgery for intracranial meningiomas. In: Radiosurgery and pathological fundamentals. Basel: Karger; 2007. p. 142–9.
14. Lee SR, Yang KA, Kim SK, Kim S-H. Radiation-induced intratumoral necrosis and peritumoral edema after gamma knife radiosurgery for intracranial meningiomas. J Korean Neurosurg Soc. 2012;52(2):98–102.
15. Cai R, Barnett GH, Novak E, Chao ST, Suh JH. Principal risk of peritumoral edema after stereotactic radiosurgery for intracranial meningioma is tumor-brain contact interface area. Neurosurgery. 2010;66(3):513–22.
16. Flickinger JC, Kondziolka D, Maitz AH, Lunsford LD. Gamma knife radiosurgery of imaging-diagnosed intracranial meningioma. Int J Radiat Oncol Biol Phys. 2003;56(3):801–6.
17. Chang JH, Chang JW, Choi JY, Park YG, Chung SS. Complications after gamma knife radiosurgery for benign meningiomas. J Neurol Neurosurg Psychiatry. 2003;74(2):226–30.
18. Singh VP, Kansal S, Vaishya S, Julka PK, Mehta VS. Early complications following gamma knife radiosurgery for intracranial meningiomas. J Neurosurg. 2000;93(supplement_3):57–61.
19. Gilbert JJ, Paulseth JE, Coates RK, Malott D. Cerebral edema associated with meningiomas. Neurosurgery. 1983;12(6):599–605.
20. Hoe Y, Choi YJ, Kim JH, Kwon DH, Kim CJ, Cho YH. Peritumoral brain edema after stereotactic radiosurgery for asymptomatic intracranial meningiomas: risks and pattern of evolution. J Korean Neurosurg Soc. 2015;58(4):379–84.
21. Stevens JM, Ruiz JS, Kendall BE. Observations on peritumoural oedema in meningioma. Neuroradiology. 1983;25(3):125–31.
22. Kan P, Liu JK, Wendland MM, Shrieve D, Jensen RL. Peritumoral edema after stereotactic radiosurgery for intracranial meningiomas and molecular factors that predict its development. J Neurooncol. 2007;83(1):33–8.
23. Osawa T, Tosaka M, Nagaishi M, Yoshimoto Y. Factors affecting peritumoral brain edema in meningioma: special histological subtypes with prominently extensive edema. J Neurooncol. 2012;111(1):49–57.
24. Conti A, Pontoriero A, Salamone I, Siragusa C, Midili F, La Torre D, et al. Protecting venous structures during radiosurgery for parasagittal meningiomas. Neurosurg Focus. 2009;27(5):E11.
25. Patil CG, Hoang S, Borchers DJ, Sakamoto G, Soltys SG, Gibbs IC, et al. Predictors of peritumoral edema after stereotactic radiosurgery of supratentorial meningiomas. Neurosurgery. 2008;63(3):435–42.
26. Salpietro FM, Alafaci C, Lucerna S, Iacopino DG, Todaro C, Tomasello F. Peritumoral edema in meningiomas. Neurosurgery. 1994;35(4):638–42.
27. Chang SD, Adler JR. Treatment of cranial base meningiomas with linear accelerator radiosurgery. Neurosurgery. 1997;41(5):1019–27.
28. Kollová A, Liščák R, Novotný J, Vladyka V, Šimonová G, Janoušková L. Gamma knife surgery for benign meningioma. J Neurosurg. 2007;107(2):325–36.
29. Pollock BE, Stafford SL, Utter A, Giannini C, Schreiner SA. Stereotactic radiosurgery provides equivalent tumor control to Simpson grade 1 resection for patients with small- to medium-size meningiomas. Int J Radiat Oncol Biol Phys. 2003;55(4):1000–5.

Skull Base Meningiomas

19

Giuseppe Minniti, Alfredo Conti, and Antonio Pontoriero

19.1 Introduction

Skull base meningiomas are common primary brain tumors. According to the WHO classification, most meningiomas are benign lesions, whereas a minority are classified as atypical (Grade II) or malignant (Grade III). Surgical resection is the treatment of choice, and represents the definitive treatment for the majority of patients, especially those with benign tumors at favorable locations resulting in tumor growth control rates of about 75–90% at 10 years [1–3]. However, there are a group of complex tumors, including those strictly adjacent to the optic apparatus and encasing neurovascular structures, for which surgical resection is at higher risk for complications. Conventional external beam radiation therapy (RT) has traditionally been used to improve local tumor control after incomplete resection of a benign meningioma arising at unfavorable locations, or after surgical resection of atypical and malignant meningiomas, even following macroscopic removal. The reported control and survival following incomplete surgical resection and conventional RT are similar to those observed after complete resection and better than those achieved with incomplete resection alone [4–7].

In the last three decades, advances in radiological imaging and computer sciences, and their application to radiation planning and delivery techniques, have led to more accurate and focused treatment, rendering many commonly held views of the "old" RT obsolete. The application of conventional RT to skull base meningiomas has evolved with the development of conformal and stereotactic techniques which allow for a steeper dose gradient between the target and the surrounding normal tissue, thereby reducing the risk of long-term toxicity compared with conventional RT. Currently available advanced radiation techniques include fractionated stereotactic radiotherapy (FSRT), intensity-modulated radiotherapy (IMRT), and stereotactic radiosurgery (SRS) that allow for a steeper dose gradient, meaning a more favorable dose distribution on the target and surrounding normal tissue compared with conventional RT.

Specifically, SRS has progressively emerged as an accepted treatment option for both

G. Minniti
Radiation Oncology Unit, Department of Medicine, Surgery and Neurosciences, University of Siena, Policlinico Le Scotte, Siena, Italy
e-mail: giuseppe.minniti@unisi.it

A. Conti (✉)
Alma Mater Studiorum University of Bologna, Bologna, Italy

IRCCS ISNB Istituto delle Scienze Neurologiche di Bologna, Bologna, Italy

Department of Neurosurgery, Charité Faculty of Medicine, Berlin, Germany
e-mail: alfredo.conti2@unibo.it

A. Pontoriero
Department of Radiation Oncology, University of Messina, A.O.U. "G. Martino"—Cyberknife Centre of Messina, Messina, Italy
e-mail: apontoriero@unime.it

© Springer Nature Switzerland AG 2020
A. Conti et al. (eds.), *CyberKnife NeuroRadiosurgery*,
https://doi.org/10.1007/978-3-030-50668-1_19

incompletely resected and intact skull base meningiomas, i.e., cavernous sinus meningiomas. According to the American College of Radiology (ACR) and American Society for Radiation Oncology (ASTRO) Practice Guidelines for the Performance of Stereotactic Radiosurgery (SRS), SRS is defined as "radiation therapy delivered via stereotactic guidance with an approximately 1 mm targeting accuracy to intracranial targets in 1–5 fractions" [8]. Several terms have been used interchangeably for SRS delivered in 2–5 fractions, including "fractionated SRS," "multi-fraction SRS," "multi-dose SRS," "multi-session SRS," "hypofractionated SRS," and "hypofractionated stereotactic radiotherapy." Conversely, SRT refers to the treatment delivered in more than five fractions with the same level of accuracy.

Although SRS is virtually noninvasive, the treatment of tumors growing in the skull base region may carry a risk of radiation-induced toxicity ranging up to 15% [9]. Thus, the presence of structures that are highly sensitive to radiation, such as optic nerves and chiasm and the brainstem, represents the main concern regarding the use of SRS for skull base tumors. Nevertheless, adverse effects induced by radiation appear to be less frequent in skull base meningiomas as compared to convexity or parasagittal tumors. Here, we provide an overview of efficacy and toxicity of modern radiation techniques for those tumors, with special regard to the emerging role of fractionated SRS.

19.2 Fractionated Radiotherapy

19.2.1 Conventional External Beam Radiotherapy

Historically, conventional RT has been the only available technique for adjuvant treatment of benign skull base meningiomas. Using doses of 50–58 Gy in 30–33 daily fractions of 1.8–2.0 Gy, several published series report local tumor control rates of 70–90% at 10 and 20 years [3, 10–20] (Table 19.1). The reported control and survival after subtotal resection and RT are consistent with those observed after complete resection and better than those achieved with incomplete resection alone [30, 31]. With regard to the dose and fractionation, most of the published series showed no significant difference in tumor control with the use of doses ranging between 50 and 60 given in 1.8–2.0 Gy per fraction. However, a dose <50 Gy is associated with higher recurrence rates [3, 17, 20]. In clinical practice, most centers use total doses of 54–56 Gy, limiting the dose to 50–54 Gy for large meningiomas involving the optic pathway. Size and tumor site have been reported as predictors of tumor control. In a series of 54 patients with skull base meningiomas who received conventional RT, Connell et al. [32] observed 5-year control rates of 93% for skull base meningiomas less than 5 cm in size and 40% for those more than 5 cm in size, and similar findings have been reported by others [3, 33]. A worse outcome has been observed for meningioma of the convexity compared to those located at skull base [3, 32], as well for sphenoid ridge over other skull base meningiomas.

The reported late toxicity of conventional RT ranges from 0 to 24%, being less than 5% in most series, and includes the development of hypopituitarism, optic neuropathy, and deficits of cranial nerves of the cavernous sinus (III, IV, and VI), radiation brain necrosis, and cognitive deficits (Table 19.1). Radiation-induced optic neuropathy (RON) appearing as decreased visual acuity or visual field defects has been reported in 0–5% of patients receiving RT. However, RON is rarely observed for doses lower than 54 Gy delivered to the optic pathway at conventional fractionation of 1.8–2 Gy per fraction [17, 34, 35]. Cranial nerve deficits and brain necrosis have been reported in 0–3% of patients. Patients with parasellar meningiomas are at risk of developing late hypopituitarism; for those patients, the pituitary function should be assessed lifelong after the radiation treatment. Neurocognitive decline is a recognized consequence of large volume RT for brain tumors [36] and has occasionally been reported in irradiated patients with meningiomas, especially impairment of short-term memory [14, 37, 38]. High-dose radiation may be associated

Table 19.1 Summary of selected published studies on fractionated radiotherapy of intracranial meningiomas

Authors	Patients (*N*)	Technique	Volume (mL)	Dose (Gy)	Follow-up (months)	Local control (%)	Late toxicity (%)
Goldsmith et al. (1994) [3]	117	CRT	NA	54	40	89 at 5 and 77 at 10 years	3.6
Maire et al. (1995) [14]	91	CRT	NA	52	40	94	6.5
Nutting et al. (1999) [18]	82	CRT	NA	55–60	41	92 at 5 and 83 at 10 years	14
Vendrely et al. (1999) [20]	156	CRT	NA	50	40	79 at 5 years	11.5
Mendenhall et al. (2003) [21]	101	CRT	NA	54	64	92 at 10 and 15 years	8
Henzel et al. (2006) [22]	84	FSRT	11.1	56	30	100	NA
Tanzler et al. (2010) [23]	144	FSRT	NA	52.7	87	97 at 5 and 95 at 10 years	7
Minniti et al. (2011) [24]	52	FSRT	35.4	50	42	93 at 5 years	5.5
Slater et al. (2012) [25]	68	Proton beam	27.6	57	74	99 at 5 years	9
Weber et al. (2012) [26]	29	Proton beam	21.5	56	62	100 at 5 years	15.5
Solda et al. (2013) [27]	222	FSRT	12	50/55	43	100 at 5 and 10 years	4.5
Combs et al. (2013) [28]	507	FSRT/ IMRT	NA	57.6	107	91 at 10 years	1.8
Fokas et al. (2014) [29]	253	FSRT	14.4	55.8	50	93 at 5 and 87.5 at 10 years	3

CRT conventional radiation therapy, *FSRT* fractionated stereotactic radiation therapy, *IMRT* intensity-modulated radiation therapy, *NA* not assessed

with the development of a second brain tumor. In a large series of 426 patients with pituitary adenomas who received conventional EBRT at the Royal Marsden Hospital between 1962 and 1994, Minniti et al. [39] reported that the risk of second brain tumors was 2.0% at 10 years and 2.4% at 20 years, measured from the date of EBRT.

19.2.2 Advanced Fractionated Radiation Techniques

For patients with brain tumors, fractionated techniques have evolved from conventional RT to more sophisticated conformal and stereotactic techniques, including FSRT, IMRT, and VMAT. To deliver 3D conformal RT, IMRT combines two advanced concepts: inverse treatment planning with optimization by computer and computer-controlled intensity modulation of the beams during dose delivery treatment. VMAT is the delivery of IMRT while the gantry is in motion using dynamic leaf motion, thereby allowing for a reduction in treatment time. IMRT and VMAT result in better conformation of radiation to complex targets with concave regions, and reduction in radiation doses to surrounding sensitive structures, such as the optic pathway and the brainstem, as compared to conventional RT. The principal advances of stereotactic radiation techniques is improved immobilization using a precision mask system, with relocation accuracy in the region of 1–2 mm [40]. Treatment delivery is improved with the use of multiple (usually 4–8) fixed shaped beams employing a multileaf collimator (MLC) with smaller leaves (mini or micro MLC).

A summary of selected published series of FSRT and IMRT for skull base meningiomas is shown in Table 19.1. Using FSRT with doses of 50–58 Gy in 30–33 daily fractions, several large series have reported local control rates of 90–100% and overall survival times up to 100% at 10 years, including patients with large complex-shaped meningioma [41]. Toxicity is reported in up to 12% of patients and includes the development of RON, cranial nerve deficits, and hypopituitarism. Similar clinical outcomes have been observed in few series reporting on IMRT, with a reported local control of 93–97% at median follow-up of 19–36 months and low toxicity [34, 42, 43]. In a series of 506 patients with a skull base meningioma who received FSRT (n = 376) or IMRT (n = 131), Combs et al. [28] observed 10-year local control rates of 91% for benign meningiomas and 53% for high-risk meningiomas, with no significant differences between groups at a median follow-up of 107 months. The treatment was well tolerated. Quality of life was unchanged in 47.7% and improved in 37.5% of patients. In summary, published results support the efficacy and safety of both FSRT and IMRT for the treatment of skull base meningiomas of any size and/or involving neurovascular structures of the sellar and parasellar region.

19.2.3 Proton Beam Radiotherapy

Proton beam RT represents an important advance in the field of radiation, because of its ability to concentrate dose in the tumor while simultaneously sparing surrounding healthy tissue. The physical properties of proton irradiation can offer superior conformality in dose distribution when compared to IMRT, therefore offering the potential for better sparing of normal tissue particularly beyond the principal target. Data on proton treatment in meningiomas are available as either conventional fractionation or SRS [25, 26, 44, 45].

In a series of 51 patients with skull base meningioma who received combined photons and protons between December 1995 and December 1999 at the Centre de Protonthérapie d'Orsay (CPO), Noel et al. [46] observed 4-year local control and overall survival rates of 98% and 100%, respectively, using a median total dose of 60.6 cobalt gray equivalent (CGE). Neurological improvements were recorded in about 68% of cases. Grade III side effects occurred in two patients, one case of hearing loss and one case of complete pituitary deficiency. Wenkel et al. [47] reported the clinical outcomes of 46 patients with partially resected or recurrent benign meningiomas treated between 1981 and 1996 with combined photon and proton beam therapy at the Massachusetts General Hospital. At a median follow-up of 53 months, the observed overall survival rates were 93 and 77% at 5 and 10 years, respectively, and recurrence-free rate 100% and 88%, respectively. Eight (17%) patients developed severe long-term toxicity, including ophthalmologic (four patients), neurologic (four patients), and otologic (two patients) complications. In another series of 47 patients treated at Loma Linda University Medical Center with fractionated proton RT at doses of 57 Gy for a benign cavernous sinus meningioma, Slater et al. [25] reported a 5-year local control of 96% at a median follow-up of 74 months, with no severe long-term toxicity. Overall, the reported 5-year tumor control of about 90% and toxicity rates of 3–7% in published studies are consistent with those observed after fractionated photon RT (Table 19.1). Current data preclude to draw firm conclusions regarding the superiority of protons over photons radiation techniques in terms of efficacy and long-term toxicity in patients with skull base meningiomas.

19.2.4 Stereotactic Radiosurgery (SRS)

SRS, which typically refers to the use of single-fraction SRS, has been extensively employed in the treatment of benign skull base meningiomas as alternative treatment for lesions not amenable to surgical removal. The majority of published series report on the use of Gamma Knife (GK)

SRS, where patients are typically immobilized in a fixed frame and radiation dose is prescribed to the 50% isodose and delivered in a single session using multiple isocenter plans to optimize conformality and rapid dose fall-off. SRS technology has evolved with the development of frameless SRS, where patients are usually immobilized in a high precision mask fixation system and the treatment can be delivered as either single-fraction SRS or multi-fraction SRS (2–5 fractions). Commonly used frameless SRS techniques include the use of the image-guided robotic radiosurgery system CyberKnife (CK) or a modified linear accelerator (LINAC), e.g., Novalis, Truebeam Stx [30, 31, 48, 49]. The CyberKnife (Accuray, Sunnyvale, CA) combines a mobile linear accelerator mounted on a robotic arm with an image-guided robotic system [30]. A variable number of 80–200 overlapping beams are delivered to the target, and the number and direction of beams and analysis of dose distribution are chosen by a sophisticated computer optimization program through an inverse planning process. With LINAC-based SRS, the dose is delivered throughout multiple fixed fields or arcs shaped with a micro-multileaf collimator (2.5–3.0 mm leaf), and conformity improved by the use of IMRT and VMAT techniques. Further improvements of frameless CK and LINAC-based SRS techniques include improved accuracy of patient repositioning with the use of either orthogonal x-rays or cone beam computed tomography (CBCT) in-room imaging systems that are able to correct positioning errors by translating and rotating the treatment table in six directions with an accuracy <0.5–1 mm [50, 51]. A few studies have shown a comparable high degree of dose conformity for irregularly shaped brain tumors planned with GK, CK, and LINAC-based SRS [49, 52]. Despite several technical differences among GK, CK, and LINAC-based SRS, the reported results in terms of clinical stabilization, tumor growth control, and toxicity do not support the superiority of a technique over another, with equivalent 5-year tumor control rates of about 90–95%, with a low rate of treatment-related complication [53] (Table 19.2).

19.2.5 Results of Single-Fraction SRS

A summary of selected series of SRS for benign skull base meningiomas is shown in Table 19.2. In a large single-center series of 972 patients, mostly with skull base meningiomas, who underwent GK SRS at the University of Pittsburgh, Kondziolka et al. [61] reported tumor control rates of 93% at 5 years and 87% at 10 and 15 years using a median marginal dose of 13 Gy. Tumor volumes decreased in 34%, remained stable in 60%, and increased in 6% of patients. With respect to the clinical setting, there was no difference between the 384 patients who were treated with postoperative SRS and the 488 patients receiving upfront SRS. In another large retrospective multicentric study of 4565 consecutive patients harboring 5300 benign meningiomas who received GK SRS, 5-year and 10-year progression-free survival rates were 95.2% and 88.6%, respectively, at a median follow-up of 63 months. Tumor volumes decreased in 58%, remained unchanged in 34.5%, and increased in 7.5% of lesions, giving a control rate of 92.5%. In a meta-analysis of 2734 patients receiving GK and LINAC SRS in patients with brain meningiomas, Pannullo et al. [62] found an equivalent tumor control of 89% following GK and LINAC-based SRS, being similar for patients treated with upfront or postoperative SRS. In a few studies of CK SRS, the reported tumor control of 90–95% at 5 years is consistent with those observed following GK and LINAC-based SRS (Table 19.2).

SRS dose for skull base meningiomas is highly dependent upon the technique applied, the prescribed isodose, the proximity of organs at risk (OARs), as well as the size and configuration of the tumor. In most of published studies, doses range between 12 and 18 Gy, with a progressive dose reduction over the last years. Using doses of 12–14 Gy, the rates of tumor control at 5 years remain in the range of 90–95% as for higher doses [63–65]. The rate of tumor shrinkage varies in the different studies, ranging from 16 to 69%, and tends to increase in patients with longer follow-up. With regard to the factors predicting local tumor control, the majority of studies shows

Table 19.2 Summary of selected published studies on stereotactic radiosurgery (SRS) of intracranial meningiomas

Authors	Patients (*N*)	SRS technique	Volume (mL)	Dose (Gy)	Follow-up (months)	Local control rates (%)	Late toxicity (%)
Kreil et al. (2005) [54]	200	GK	6.5	12	95	98 at 5 and 97 at 10 years	4.5
Kollova et al. (2007) [64]	368	GK	4.4	12.5	60	98 at 5 years	15.9
Feigl et al. (2007) [55]	214	GK	6.5	13.6	24	86.3 at 4 years	6.7
Kondziolka et al. (2008) [61]	972	GK	7.4	14	48	87 at 10 and 15 years	7.7
Colombo et al. (2009) [53]	199	CK	7.5	16–25[a]	30	96	3.5
Skeie et al. (2010) [56]	100	GK	11.1	13	32	90.4 at 5 and 10 years	6
Halasz et al. (2011) [45]	50	Proton beam	27.4	13	36	94 at 3 years	5.9
Pollock et al. (2012) [57]	251	GK	7.7	15.8	62.9	99.4 at 10 years	11 at 5 years
Santacroce et al. (2012) [58]	3768	GK	4.8	14	63	95 at 5 and 88 at 10 years	6.6
Starke et al. (2014) [59]	254	GK	NA	13	71	93 at 5 and 84 at 10 years	6.4
Ding et al. (2014) [60]	177	GK	3.6	13	47	93 at 5 and 77 at 10 years	9
Sheehan et al. (2014) [84]	763	GK	4.1	13	66.7	95 at 5 and 82 at 10 years	9.6
Marchetti et al. (2016) [93]	143	CK	11	21–25[b]	44	93 at 5 years	5.1

GK Gamma Knife, *CK* CyberKnife
[a]16–25 Gy delivered in 2–5 fractions in 150 patients
[b]21–25 Gy delivered in 3–5 fractions

no significant differences between patients who received SRS as upfront treatment and those receiving SRS after incomplete resected or recurrent meningiomas [58, 61, 62]. Age, sex, site of meningioma, and neurological status did not affect significantly the outcome in most published series. However, larger meningiomas are associated with worse long-term local control [61, 66]. DiBiase et al. [66] reported a significant higher 5-year tumor control in patients with tumor volumes <10 mL than those with larger tumors (92% vs 68%, $p = 0.038$).

An important goal of SRS treatment is improving or maintaining neurological function. A variable improvement of neurological functions, including vision and ocular motility recovery, has been shown in 10–60% of patients. The rate of significant complications at doses of 13–14 Gy (as currently used in the majority of centers) is less than 8%, being represented by either transient or permanent complications, although a few series report a higher rate of long-term toxicity. Kondziolka et al. [61] reported permanent neurological deficits in 9% of patients at 10 and 15 years in 972 patients treated with GK SRS for intracranial meningiomas. The morbidity rate for cavernous sinus meningiomas was 6.3% and included optic neuropathy, sixth nerve palsy, and trigeminal neuropathy. In the series by Nicolato et al. [67], late complications occurred in 4.5% of patients, being transient in 80% of them, and similar complication rates have been reported in

Table 19.3 Summary of normal tissue constraints following conventional fractionation (2 Gy/fr) and stereotactic radiosurgery (SRS)

Organ	Type of radiation	Estimated toxicity rate and dose tolerance limits	Type of toxicity	References
Brain parenchyma	Conventional fractionation Single-fraction SRS Fractionated SRS	<3% for Dmax <60 Gy to whole brain <10% for Dmax 12 Gy to <10 mL brain volume <5% for Dmax 18 Gy/3fx to <26 mL brain volume	Symptomatic necrosis	[24, 70–73]
Brainstem	Standard fractionation Single-fraction SRS Fractionated SRS	<5% for Dmax <54 Gy to whole organ <5% for Dmax <12.5 Gy to whole organ <3% for Dmax of 18 Gy/3fx or 26 Gy/5fx to <1 mL	Permanent cranial deficit or necrosis	[70, 72, 74]
Optic nerve/ chiasm	Standard fractionation Single-fraction SRS Fractionated SRS	<3% for Dmax <55 Gy to whole organ <3% for Dmax <8 Gy and <10% for Dmax 8–12 Gy <3% for Dmax of 19.5 Gy/3fx and 25 Gy/5fx	Optic neuropathy	[70, 72, 74–76]
Cochlea	Standard fractionation Single-fraction SRS Fractionated SRS	<15% for mean doses ≤45 Gy to whole organ <25% for Dmax ≤14 Gy <3% for Dmax of 20/3fx and 27.5 Gy/5fx	Hearing loss	[70, 72, 74]
Pituitary gland	Standard fractionation Single-fraction SRS	20–40% at 5 years for Dmax ≤45 Gy to whole gland 10–30% at 5 years for Dmax <15 Gy	Hypopituitarism	[77–79]
Medulla oblongata	Standard fractionation Single-fraction SRS Fractionated SRS	1% for Dmax 54 Gy, 10% for Dmax of 61 Gy 1% for Dmax 13 Gy 1% for Dmax 22.5 Gy/3fx and 30 Gy/5fx	Myelopathy	[70, 72, 74, 80]

Dmax maximum dose

other large published series (Table 19.2). Other complications, such as epilepsy, internal carotid occlusion, and hypopituitarism have been rarely reported (less than 1–2%).

With respect to radiation-induced toxicity, i.e., cranial nerve deficits and risk of radionecrosis, a clear dose-volume relationship for side effects has been reported [57, 68, 69] after SRS (Table 19.3). The risk of clinically significant RON for patients receiving SRS is 1–2% following doses to optic chiasm below 8–10 Gy, although this percentage may significantly increase for higher doses [70, 71, 80, 81]. In contrast, cranial neuropathies and brain necroses are rarely reported using doses of less than 16 Gy [82, 83]. The risk to develop a new tumor after SRS appears to be significantly less than the risk seen following fractionated RT [39]. However longer follow-ups are needed to draw definitive conclusions. Factors related to higher risk of delayed onset of hypopituitarism include maximum doses of 15 Gy delivered to the pituitary gland and 7–10 Gy to the pituitary stalk [77, 78, 84]. Overall, the reported long-term toxicity of SRS at doses of 13–15 Gy is relatively low when radiation doses to organs at risk around the tumor are within the accepted maximum tolerance doses for normal brain structures (Table 19.3). Based on these data, limitations can be seen for complex volumes adjacent to organs at risk and with increasing size, while fractionated treatments are associated with a comparable dose profile independent of tumor volume or diameter [22, 24, 65, 85–88].

19.2.6 Results of Fractionated SRS

More recently, fractionated SRS (2–5 fractions by definition) has emerged as an effective treatment option for brain tumors with the aim of maintaining the precision and *accuracy* of *treatment delivery* while exploiting the potential *radiobiological advantage of fractionation in terms of tumor control and reduced toxicity* [89–91]. Thus, its use may represent an alternative treatment option to single-fraction SRS for large skull base meningiomas located in close proximity to critical anatomic structures such as the optic apparatus or the brainstem. Commonly used techniques to deliver fractionated SRS are the CK and modified LINAC. Recently, the last version of GK (Icon) has enabled the use of a mask system for frameless SRS. In a series of 199 benign intracranial meningiomas, 157 skull base tumors, Colombo et al. [53] reported a 5-year control of 93.5%. Tumors larger than 8 mL and/or located close to critical structures were treated with fractionated SRS, typically 21 Gy in three fractions or 25 Gy in five fractions. The tumor volume decreased in 36 patients, was unchanged in 148 patients, and increased in 7 patients. Clinical symptoms improved in 30 patients. Tumor control in 63 patients with tumor volume up to 65 mL treated with fractionated SRS was similar to that obtained in patients with smaller meningiomas receiving with single-fraction SRS. Neurological deterioration was observed in 4% of patients, represented mainly by visual deficits. In a series of 60 patients with a skull base meningioma treated at the University of Pittsburgh with CK with a median dose of 17.5 Gy (range 6–27 Gy) delivered in 2–5 fractions (mostly 3 fractions), Bria et al. [92] observed a local control of 96% at a median follow-up of 16.1 months. A subjective improvement in the existing, tumor-related symptoms occurred in 60% of the patients, with Grade III toxicity observed in one patient. In another large retrospective study of 143 patients treated with CK SRS at Besta Hospital in Milan for a perioptic meningioma, 15–25 Gy delivered in 3–5 fractions, Marchetti et al.[93] observed local control rates of 100%, 93%, and 90% after 3, 5, and 8 years, respectively. With respect to neurological outcome, vision improved in 42% and worsened in 3.7% of patients. Similar clinical outcomes and low toxicity have been reported by other authors [94–98].

The reported late neurological toxicity of fractionated SRS of doses 21–25 Gy in three to five fractions is low, with a reported incidence of RON and other cranial nerve deficits affecting visual motility in less than 2–3% of patients [53, 92, 94, 96]. In a large retrospective cooperative study of 167 patients with large skull base meningioma in close proximity to the anterior optic pathways who received fractionated SRS, 25 Gy in 5 fractions, at Besta Hospital in Milan and the University of Messina, Italy, Marchetti et al. [96] reported visual deterioration in 3.7% of patients, all with pretreatment visual deficits, at a median follow-up time of 51 months. In another series of 46 patients with a perioptic meningioma or a pituitary adenoma within 2 mm from the optic apparatus who received CK SRS, 18–25 Gy delivered in 2–5 sessions, Adler et al. [94] reported no visual impairments at a median follow-up of 49 months. Similar low toxicities have been reported in other few series [53, 92]. Although these results are of reassurance about the safety of hypofractionated schedules for skull base meningiomas, data on tolerance doses of the central nervous system (CNS) and organs at risk (OARs) to fractionated SRS, e.g., cranial nerve deficits, hypopituitarism, and neurocognitive function, are relatively limited. For three-fraction and five-fraction SRS, a summary of dose-volume data and clinical risk estimates for OARs is presented in Table 19.3.

19.3 Comparison of Radiation Techniques

Several retrospective published studies have suggested that SRT, given as either hypofractionation or conventional fractionation, may offer a better balance of efficacy and toxicity compared with single-fraction SRS in patients with large brain tumors and/or tumors located in close proximity to critical brain structures [99–101]. A recent systematic review has compared the safety and long-term efficacy of SRS and SRT, either fractionated SRS or FSRT, in patients with intracranial meningiomas [101]. Twelve retrospective studies including 1736 patients who received SRS, fractionated SRS, and FSRT were analyzed. The median tumor sizes at the time of treatment with SRS, fractionated SRS, and FSRT were 2.84 cm^3, 5.45 cm^3, and 12.75 cm^3, respectively. At a median follow-up of 36 months, SRS was associated with a significantly worse radiographic tumor control and higher risk of neurological toxicity compared with SRT. However, PFS at 4–10 years was not statistically significant between groups. A large Italian retrospective multicenter study has compared the clinical outcomes in 341 patients with skull base meningiomas receiving FSRT, 59.4 Gy in 33 fractions, or fractionated CK SRS, 25 Gy in 5 fractions [100]. At a median follow-up of 36 months, local control rates were 96.8% and 80.3% at 3 and 10 years in patients treated with fractionated SRS and 99% and 79.1% in those receiving FSRT, respectively. Grade III or more toxicity rates were 0.5% and 2.1%. In a German retrospective multicentric study of 927 patients treated with SRS or fractionated RT (FSRT or IMRT), at a median follow-up time of 79 months, Combs et al. [99] reported local control rates of 92% and 86% at 5 years and 10 years, respectively. There was no difference between fractionated RT and SRS groups. For patients receiving fractionated RT, local control was similar using doses of 54 Gy and 57.6 Gy. Side effects were below 5% in both groups without any severe treatment-related complications.

Although data indicate that fractionated SRS may offer optimal balance between efficacy and respect of dose-volume constraints for relatively large skull base meningiomas, published results need to be interpreted with caution. Prospective data need to evaluate the dose-volume constraints for all sellar and parasellar sensitive structures, including optic chiasm and cavernous sinus cranial nerves, brainstem, and pituitary gland and stalk, to limit potential long-term toxicity of SRS treatments. Moreover, prospective controlled studies need to evaluate the efficacy and toxicity of fractionated SRS over other radiation techniques.

19.4 Conclusions

The use of radiations to treat skull base benign meningiomas results in satisfactory results with local control rates that could rival those observed following complete surgical resection. The reported 5-year and 10-year local control is more than 80–90% following either SRS and FSRT, with comparable results among commonly used techniques, such as GK, CK, and LINAC-based SRS. Typical doses are 13–5 Gy for single-fraction SRS, 21–25 Gy in three or five fractions, and 50–58 Gy in 30–33 daily fractions. In respect of dose-volume constraints, the observed long-term toxicity, such as the development of RON and other cranial nerve deficits, is low. As for the European Association of Neuro-Oncology (EANO) guidelines [102], single-fraction SRS is the recommended treatment for small meningiomas, while fractionated SRS and conventionally fractionated RT should be preferred for larger lesions. In clinical practice, this means that fractionated SRS, usually three or five fractions, may represent a safer treatment option than single-fraction SRS for large benign skull base meningiomas larger than 2.5–3 cm or in close proximity to the optic chiasm, when single doses to the *optic apparatus* exceed 8–10 Gy. For patients with very large lesions involving the optic apparatus, FSRT, 54–56 Gy in 30–33 daily fractions, would be the recommended treatment option.

References

1. Cusimano M, Sekhar L, Sen C, et al. The results of surgery for benign tumors of the cavernous sinus. Neurosurgery. 1995;37:1–9.
2. Kallio M, Sankila R, Hakulinen T, Jaaskelainen J. Factors affecting operative and excess long-term mortality in 935 patients with intracranial meningioma. Neurosurgery. 1992;31:2–12.
3. Goldsmith B, Wara W, Wilson C, Larson D. Postoperative irradiation for subtotally resected meningiomas. J Neurosurg. 1994;80:195–201.
4. Mirimanoff RO, Dosoretz DE, Linggood RM, Ojemann RG, Martuza RL. Meningioma: analysis of recurrence and progression following neurosurgical resection. J Neurosurg. 1985;62(1):18–24.
5. Stafford S, Pollock B, Foote R, et al. Meningioma radiosurgery: tumor control, outcomes, and complications among 190 consecutive patients. Neurosurgery. 2001;49:1029–37.
6. Soyuer S, Chang EL, Selek U, Shi W, Maor MH, DeMonte F. Radiotherapy after surgery for benign cerebral meningioma. Radiother Oncol. 2004;71(1):85–90.
7. Taylor BW Jr, Marcus RB Jr, Friedman WA, Ballinger WE Jr, Million RR. The meningioma controversy: postoperative radiation therapy. Int J Radiat Oncol Biol Phys. 1988;15(2):299–304.
8. Seung SK, Larson DA, Galvin JM, et al. American College of Radiology (ACR) and American Society for Radiation Oncology (ASTRO) practice guideline for the performance of stereotactic radiosurgery (SRS). Am J Clin Oncol. 2013;36(3):310–5.
9. Albano L, Losa M, Flickinger J, Mortini P, Minniti G. Radiotherapy of parasellar tumours. Neuroendocrinology. 2020.
10. Barbaro N, Gutin P, Wilson C, Sheline G, Boldrey E, Wara W. Radiation therapy in the treatment of partially resected meningiomas. Neurosurgery. 1987;20:525–8.
11. Brell M, Villa S, Teixidor P, et al. Fractionated stereotactic radiotherapy in the treatment of exclusive cavernous sinus meningioma: functional outcome, local control, and tolerance. Surg Neurol. 2006;65:28–33.
12. Carella R, Ransohoff J, Newall J. Role of radiation therapy in the management of meningioma. Neurosurgery. 1982;10:332–9.
13. Forbes A, Goldberg I. Radiation therapy in the treatment of meningioma: the Joint Center for Radiation Therapy experience 1970 to 1982. J Clin Oncol. 1984;2:1139–43.
14. Maire J, Caudry M, Guerin J, et al. Fractionated radiation therapy in the treatment of intracranial meningiomas: local control, functional efficacy, and tolerance in 91 patients. Int J Radiat Oncol Biol Phys. 1995;33:315–21.
15. Metellus P, Regis J, Muracciole X, et al. Evaluation of fractionated radiotherapy and gamma knife radiosurgery in cavernous sinus meningiomas: treatment strategy. Neurosurgery. 2005;57:873–86.
16. Milker-Zabel S, Zabel A, Schulz-Ertner D, Schlegel W, Wannenmacher M, Debus J. Fractionated stereotactic radiotherapy in patients with benign or atypical intracranial meningioma: long-term experience and prognostic factors. Int J Radiat Oncol Biol Phys. 2005;61:809–16.
17. Miralbell R, Linggood R, de la Monte S, Convery K, Munzenrider J, Mirimanoff R. The role of radiotherapy in the treatment of subtotally resected benign meningiomas. J Neurooncol. 1992;13:157–64.
18. Nutting C, Brada M, Brazil L, et al. Radiotherapy in the treatment of benign meningioma of the skull base. J Neurosurg. 1999;90:823–7.
19. Peele K, Kennerdell J, Maroon J, et al. The role of postoperative irradiation in the management of sphenoid wing meningiomas. Ophthalmology. 1996;103:1761–6.
20. Vendrely V, Maire J, Darrouzet V, et al. [Fractionated radiotherapy of intracranial meningiomas: 15 years' experience at the Bordeaux University Hospital Center]. Cancer Radiother. 1999;3:311–317.
21. Mendenhall W, Morris C, Amdur R, Foote K, Friedman W. Radiotherapy alone or after subtotal resection for benign skull base meningiomas. Cancer. 2003;98:1473–82.
22. Henzel M, Gross M, Hamm K, Surber G, Kleinert G, Failing T. Stereotactic radiotherapy of meningiomas: symptomatology, acute and late toxicity. Strahlenther Onkol. 2006;182:382–8.
23. Tanzler E, Morris CG, Kirwan JM, Amdur RJ, Mendenhall WM. Outcomes of WHO grade I meningiomas receiving definitive or postoperative radiotherapy. Int J Radiat Oncol Biol Phys. 2011;79(2):508–13.
24. Minniti G, Clarke E, Cavallo L, Osti M, Esposito V, Cantore G. Fractionated stereotactic conformal radiotherapy for large benign skull base meningiomas. Radiat Oncol. 2011;6:36.
25. Slater JD, Loredo LN, Chung A, et al. Fractionated proton radiotherapy for benign cavernous sinus meningiomas. Int J Radiat Oncol Biol Phys. 2012;83(5):e633–7.
26. Weber DC, Schneider R, Goitein G, et al. Spot scanning-based proton therapy for intracranial meningioma: long-term results from the Paul Scherrer Institute. Int J Radiat Oncol Biol Phys. 2012;83(3):865–71.
27. Solda F, Wharram B, De Ieso PB, Bonner J, Ashley S, Brada M. Long-term efficacy of fractionated radiotherapy for benign meningiomas. Radiother Oncol. 2013;109(2):330–4.
28. Combs SE, Adeberg S, Dittmar JO, et al. Skull base meningiomas: long-term results and patient self-reported outcome in 507 patients treated with fractionated stereotactic radiotherapy (FSRT) or intensity modulated radiotherapy (IMRT). Radiother Oncol. 2013;106(2):186–91.

29. Fokas E, Henzel M, Surber G, Hamm K, Engenhart-Cabillic R. Stereotactic radiation therapy for benign meningioma: long-term outcome in 318 patients. Int J Radiat Oncol Biol Phys. 2014;89(3):569–75.
30. Kuo JS, Yu C, Petrovich Z, Apuzzo ML. The CyberKnife stereotactic radiosurgery system: description, installation, and an initial evaluation of use and functionality. Neurosurgery. 2003;53(5):1235–9; discussion 1239.
31. Keeling V, Algan O, Ahmad S, Hossain S. Dosimetric comparison of intracranial metastasis treatment using two radiosurgery systems: TrueBeam STx with VMAT and gamma knife model 4C. J Radiosurg SBRT. 2016;4(3):235–43.
32. Connell P, Macdonald R, Mansur D, Nicholas M, Mundt A. Tumor size predicts control of benign meningiomas treated with radiotherapy. Neurosurgery. 1999;44:1194–9.
33. Condra KS, Buatti JM, Mendenhall WM, Friedman WA, Marcus RB Jr, Rhoton AL. Benign meningiomas: primary treatment selection affects survival. Int J Radiat Oncol Biol Phys. 1997;39(2):427–36.
34. Pirzkall A, Debus J, Haering P, et al. Intensity modulated radiotherapy (IMRT) for recurrent, residual, or untreated skull-base meningiomas: preliminary clinical experience. Int J Radiat Oncol Biol Phys. 2003;55:362–72.
35. Glaholm J, Bloom HJ, Crow JH. The role of radiotherapy in the management of intracranial meningiomas: the Royal Marsden Hospital experience with 186 patients. Int J Radiat Oncol Biol Phys. 1990;18(4):755–61.
36. Crossen J, Garwood D, Glatstein E, Neuwelt E. Neurobehavioral sequelae of cranial irradiation in adults: a review of radiation-induced encephalopathy. J Clin Oncol. 1994;12:627–42.
37. Dufour H, Muracciole X, Metellus P, Regis J, Chinot O, Grisoli F. Long-term tumor control and functional outcome in patients with cavernous sinus meningiomas treated by radiotherapy with or without previous surgery: is there an alternative to aggressive tumor removal? Neurosurgery. 2001;48:285–94.
38. Maguire P, Clough R, Friedman A, Halperin E. Fractionated external-beam radiation therapy for meningiomas of the cavernous sinus. Int J Radiat Oncol Biol Phys. 1999;44:75–9.
39. Minniti G, Traish D, Ashley S, Gonsalves A, Brada M. Risk of second brain tumor after conservative surgery and radiotherapy for pituitary adenoma: update after an additional 10 years. J Clin Endocrinol Metab. 2005;90:800–4.
40. Minniti G, Valeriani M, Clarke E, et al. Fractionated stereotactic radiotherapy for skull base tumors: analysis of treatment accuracy using a stereotactic mask fixation system. Radiat Oncol. 2010;5:1.
41. Minniti G, Clarke E, Lanzetta G, et al. Stereotactic radiosurgery for brain metastases: analysis of outcome and risk of brain radionecrosis. Radiat Oncol. 2011;6:48.
42. Sajja R, Barnett G, Lee S, et al. Intensity-modulated radiation therapy (IMRT) for newly diagnosed and recurrent intracranial meningiomas: preliminary results. Technol Cancer Res Treat. 2005;4:675–82.
43. Uy N, Woo S, Teh B, et al. Intensity-modulated radiation therapy (IMRT) for meningioma. Int J Radiat Oncol Biol Phys. 2002;53:1265–70.
44. Amichetti M, Amelio D, Minniti G. Radiosurgery with photons or protons for benign and malignant tumours of the skull base: a review. Radiat Oncol. 2012;7:210.
45. Halasz LM, Bussiere MR, Dennis ER, et al. Proton stereotactic radiosurgery for the treatment of benign meningiomas. Int J Radiat Oncol Biol Phys. 2011;81(5):1428–35.
46. Noel G, Bollet MA, Calugaru V, et al. Functional outcome of patients with benign meningioma treated by 3D conformal irradiation with a combination of photons and protons. Int J Radiat Oncol Biol Phys. 2005;62(5):1412–22.
47. Wenkel E, Thornton A, Finkelstein D, et al. Benign meningioma: partially resected, biopsied, and recurrent intracranial tumors treated with combined proton and photon radiotherapy. Int J Radiat Oncol Biol Phys. 2000;48:1363–70.
48. Chang SD, Adler JR. Treatment of cranial base meningiomas with linear accelerator radiosurgery. Neurosurgery. 1997;41(5):1019–27.
49. Gevaert T, Verellen D, Tournel K, et al. Setup accuracy of the Novalis ExacTrac 6DOF system for frameless radiosurgery. Int J Radiat Oncol Biol Phys. 2012;82(5):1627–35.
50. Lamba M, Breneman JC, Warnick RE. Evaluation of image-guided positioning for frameless intracranial radiosurgery. Int J Radiat Oncol Biol Phys. 2009;74(3):913–9.
51. Wurm RE, Erbel S, Schwenkert I, et al. Novalis frameless image-guided noninvasive radiosurgery: initial experience. Neurosurgery. 2008;62(5 Suppl):A11–7; discussion A17–18.
52. Kim H, Potrebko P, Rivera A, et al. Tumor volume threshold for achieving improved conformity in VMAT and gamma knife stereotactic radiosurgery for vestibular schwannoma. Radiother Oncol. 2015;115(2):229–34.
53. Colombo F, Casentini L, Cavedon C, Scalchi P, Cora S, Francescon P. Cyberknife radiosurgery for benign meningiomas: short-term results in 199 patients. Neurosurgery. 2009;64(2 Suppl):A7–13.
54. Kreil W, Luggin J, Fuchs I, Weigl V, Eustacchio S, Papaefthymiou G. Long term experience of gamma knife radiosurgery for benign skull base meningiomas. J Neurol Neurosurg Psychiatry. 2005;76:1425–30.
55. Feigl GC, Samii M, Horstmann GA. Volumetric follow-up of meningiomas: a quantitative method to evaluate treatment outcome of gamma knife radiosurgery. Neurosurgery. 2007;61(2):281–6; discussion 286–287.

56. Skeie BS, Enger PO, Skeie GO, Thorsen F, Pedersen PH. Gamma knife surgery of meningiomas involving the cavernous sinus: long-term follow-up of 100 patients. Neurosurgery. 2010;66(4):661–8; discussion 668–669.
57. Pollock B, Stafford S, Link M, Garces Y, Foote R. Single-fraction radiosurgery for presumed intracranial meningiomas: efficacy and complications from a 22-year experience. Int J Radiat Oncol Biol Phys. 2012;83:1414–8.
58. Santacroce A, Walier M, Regis J, et al. Long-term tumor control of benign intracranial meningiomas after radiosurgery in a series of 4565 patients. Neurosurgery. 2012;70(1):32–9; discussion 39.
59. Starke R, Kano H, Ding D, et al. Stereotactic radiosurgery of petroclival meningiomas: a multicenter study. J Neurooncol. 2014;119(1):169–76.
60. Ding D, Starke RM, Kano H, et al. Gamma knife radiosurgery for cerebellopontine angle meningiomas: a multicenter study. Neurosurgery. 2014;75(4):398–408; quiz 408.
61. Kondziolka D, Mathieu D, Lunsford L, et al. Radiosurgery as definitive management of intracranial meningiomas. Neurosurgery. 2008;62:53–8.
62. Pannullo SC, Fraser JF, Moliterno J, Cobb W, Stieg PE. Stereotactic radiosurgery: a meta-analysis of current therapeutic applications in neuro-oncologic disease. J Neurooncol. 2011;103(1):1–17.
63. Ganz J, Reda W, Abdelkarim K. Gamma knife surgery of large meningiomas: early response to treatment. Acta Neurochir. 2009;151:1–8.
64. Kollova A, Liscak R, Novotny J, Vladyka V, Simonova G, Janouskova L. Gamma Knife surgery for benign meningioma. J Neurosurg. 2007;107:325–36.
65. Minniti G, Amichetti M, Enrici RM. Radiotherapy and radiosurgery for benign skull base meningiomas. Radiat Oncol. 2009;4:42.
66. DiBiase S, Kwok Y, Yovino S, et al. Factors predicting local tumor control after gamma knife stereotactic radiosurgery for benign intracranial meningiomas. Int J Radiat Oncol Biol Phys. 2004;60:1515–9.
67. Nicolato A, Foroni R, Alessandrini F, Maluta S, Bricolo A, Gerosa M. The role of gamma knife radiosurgery in the management of cavernous sinus meningiomas. Int J Radiat Oncol Biol Phys. 2002;53:992–1000.
68. Morita A, Coffey R, Foote R, Schiff D, Gorman D. Risk of injury to cranial nerves after gamma knife radiosurgery for skull base meningiomas: experience in 88 patients. J Neurosurg. 1999;90:42–9.
69. Novotny J, Kollova A, Liscak R. Prediction of intracranial edema after radiosurgery of meningiomas. J Neurosurg. 2006;105(Suppl):120–6.
70. Emami B, Lyman J, Brown A, et al. Tolerance of normal tissue to therapeutic irradiation. Int J Radiat Oncol Biol Phys. 1991;21(1):109–22.
71. Lawrence YR, Li XA, el Naqa I, et al. Radiation dose-volume effects in the brain. Int J Radiat Oncol Biol Phys. 2010;76(3 Suppl):S20–7.
72. Marks LB, Yorke ED, Jackson A, et al. Use of normal tissue complication probability models in the clinic. Int J Radiat Oncol Biol Phys. 2010;76(3 Suppl):S10–9.
73. Minniti G, Osti MF, Niyazi M. Target delineation and optimal radiosurgical dose for pituitary tumors. Radiat Oncol. 2016;11(1):135.
74. Timmerman RD. An overview of hypofractionation and introduction to this issue of seminars in radiation oncology. Semin Radiat Oncol. 2008;18(4):215–22.
75. Mayo C, Martel MK, Marks LB, Flickinger J, Nam J, Kirkpatrick J. Radiation dose–volume effects of optic nerves and chiasm. Int J Radiat Oncol Biol Phys. 2010;76(3):S28–35.
76. Pollock BE, Link MJ, Leavitt JA, Stafford SL. Dose-volume analysis of radiation-induced optic neuropathy after single-fraction stereotactic radiosurgery. Neurosurgery. 2014;75(4):456–60.
77. Sicignano G, Losa M, del Vecchio A, et al. Dosimetric factors associated with pituitary function after gamma knife surgery (GKS) of pituitary adenomas. Radiother Oncol. 2012;104(1):119–24.
78. Leenstra JL, Tanaka S, Kline RW, et al. Factors associated with endocrine deficits after stereotactic radiosurgery of pituitary adenomas. Neurosurgery. 2010;67(1):27–32; discussion 32–3.
79. Marek J, Ježková J, Hána V, et al. Is it possible to avoid hypopituitarism after irradiation of pituitary adenomas by the Leksell gamma knife? Eur J Endocrinol. 2011;164(2):169–78.
80. Kirkpatrick JP, Marks LB, Mayo CS, Lawrence YR, Bhandare N, Ryu S. Estimating normal tissue toxicity in radiosurgery of the CNS: application and limitations of QUANTEC. J Radiosurg SBRT. 2011;1(2):95–107.
81. Leavitt JA, Stafford SL, Link MJ, Pollock BE. Long-term evaluation of radiation-induced optic neuropathy after single-fraction stereotactic radiosurgery. Int J Radiat Oncol Biol Phys. 2013;87(3):524–7.
82. Leber KA, Bergloff J, Langmann G, Mokry M, Schrottner O, Pendl G. Radiation sensitivity of visual and oculomotor pathways. Stereotact Funct Neurosurg. 1995;64(Suppl 1):233–8.
83. Tishler RB, Loeffler JS, Lunsford LD, et al. Tolerance of cranial nerves of the cavernous sinus to radiosurgery. Int J Radiat Oncol Biol Phys. 1993;27(2):215–21.
84. Sheehan JP, Starke RM, Kano H, et al. Gamma knife radiosurgery for sellar and parasellar meningiomas: a multicenter study. J Neurosurg. 2014;120(6):1268–77.
85. Combs S, Adeberg S, Dittmar J, Welzel T, Rieken S, Habermehl D. Skull base meningiomas: long-term results and patient self-reported outcome in 507 patients treated with fractionated stereotactic radiotherapy (FSRT) or intensity modulated radiotherapy (IMRT). Radiother Oncol. 2013;106(2):186–91.

86. Debus J, Wuendrich M, Pirzkall A, et al. High efficacy of fractionated stereotactic radiotherapy of large base-of-skull meningiomas: long-term results. J Clin Oncol. 2001;19:3547–53.
87. Hamm K, Henzel M, Gross M, Surber G, Kleinert G, Engenhart-Cabillic R. Radiosurgery/stereotactic radiotherapy in the therapeutical concept for skull base meningiomas. Zentralbl Neurochir. 2008;69:14–21.
88. Henzel M, Gross M, Hamm K, et al. Significant tumor volume reduction of meningiomas after stereotactic radiotherapy: results of a prospective multicenter study. Neurosurgery. 2006;59:1188–94.
89. Kirkpatrick JP, Soltys SG, Lo SS, Beal K, Shrieve DC, Brown PD. The radiosurgery fractionation quandary: single fraction or hypofractionation? Neuro Oncol. 2017;19(suppl_2):ii38–49.
90. Chang SD, Adler JR. Robotics and radiosurgery—the cyberknife. Stereotact Funct Neurosurg. 2001;76(3–4):204–8.
91. Shrieve DC. Basic principles of radiobiology applied to radiotherapy of benign intracranial tumors. Neurosurg Clin N Am. 2006;17(2):67–78, v.
92. Bria C, Wegner RE, Clump DA, et al. Fractionated stereotactic radiosurgery for the treatment of meningiomas. J Cancer Res Ther. 2011;7(1):52–7.
93. Marchetti M, Bianchi S, Pinzi V, et al. Multisession radiosurgery for sellar and parasellar benign meningiomas: long-term tumor growth control and visual outcome. Neurosurgery. 2016;78(5):638–46.
94. Adler JR Jr, Gibbs IC, Puataweepong P, Chang SD. Visual field preservation after multisession cyberknife radiosurgery for perioptic lesions. Neurosurgery. 2006;59(2):244–54; discussion 244–254.
95. Conti A, Pontoriero A, Midili F, et al. CyberKnife multisession stereotactic radiosurgery and hypofractionated stereotactic radiotherapy for perioptic meningiomas: intermediate-term results and radiobiological considerations. Springerplus. 2015; 4:37.
96. Marchetti M, Conti A, Beltramo G, et al. Multisession radiosurgery for perioptic meningiomas: medium-to-long term results from a CyberKnife cooperative study. J Neurooncol. 2019;143(3):597–604.
97. Tuniz F, Soltys SG, Choi CY, et al. Multisession cyberknife stereotactic radiosurgery of large, benign cranial base tumors: preliminary study. Neurosurgery. 2009;65(5):898–907; discussion 907.
98. Han J, Girvigian MR, Chen JC, et al. A comparative study of stereotactic radiosurgery, hypofractionated, and fractionated stereotactic radiotherapy in the treatment of skull base meningioma. Am J Clin Oncol. 2014;37(3):255–60.
99. Combs SE, Farzin M, Boehmer J, et al. Clinical outcome after high-precision radiotherapy for skull base meningiomas: pooled data from three large German centers for radiation oncology. Radiother Oncol. 2018;127(2):274–9.
100. Alfredo C, Carolin S, Guliz A, et al. Normofractionated stereotactic radiotherapy versus CyberKnife-based hypofractionation in skull base meningioma: a German and Italian pooled cohort analysis. Radiat Oncol. 2019;14(1):201.
101. Fatima N, Meola A, Pollom EL, Soltys SG, Chang SD. Stereotactic radiosurgery versus stereotactic radiotherapy in the management of intracranial meningiomas: a systematic review and meta-analysis. Neurosurg Focus. 2019;46(6):E2.
102. Goldbrunner R, Minniti G, Preusser M, et al. EANO guidelines for the diagnosis and treatment of meningiomas. Lancet Oncol. 2016;17(9):e383–91.

High-Grade Meningiomas and Hemangiopericytomas

20

Güliz Acker

20.1 Atypical and Anaplastic Meningiomas

20.1.1 Epidemiology

Meningiomas are the most common primary intracranial tumor [1, 2]. Malignant meningiomas include atypical and anaplastic meningiomas (classified by the WHO as grade II and III, respectively). The 2007 WHO classification edited the definition of the grade II and III meningiomas making the diagnosis more objective and less debatable [3].

The classification of meningiomas has not undergone major revisions since the WHO classification of 2016, which implies the introduction of brain invasion as a criterion for the diagnosis of atypical meningioma [4]. Ostrom et al. have recently published a statistical report on primary brain cancer and other central nervous system tumors diagnosed in the United States in 2012–2016, where meningiomas were the most frequent brain tumor, representing 37.6% of cancers overall [2]. In this recent report, 17.7% of documented meningiomas were WHO grade II, while only 1.7% were WHO grade III meningiomas [2].

20.1.2 Treatment

High-grade meningiomas are challenging to treat due to high recurrence rates. These tumors require frequent re-treatments including repeated surgeries and radiotherapy still with often disappointing survival rates [5–7]. Surgical resection is the treatment of choice for high-grade meningiomas when the tumor is in an accessible location [6–9].

For anaplastic meningiomas, adjuvant radiotherapy is recommended regardless of the extent of surgical resection [9]. However, for atypical meningiomas the postoperative management remains controversial. Retrospective studies regarding adjuvant radiotherapy in atypical meningiomas have demonstrated inconsistent results. While various retrospective studies demonstrated lower recurrence rates and improved overall survival (OS) for adjuvantly irradiated WHO grade II meningiomas, [10–12] numerous other studies have found no definite advantage of adjuvant radiotherapy [13–16].

Importantly, the first report for intermediate-risk meningiomas from the randomized trial RTOG 0539 supported the postoperative radiotherapy for newly diagnosed gross totally resected WHO grade II meningiomas based on

G. Acker (✉)
Department of Neurosurgery and Center for Stroke research Berlin (CSB), Charité Universitätsmedizin Berlin, Berlin, Germany

Berlin Institute of Health (BIH), Berlin, Germany

Charité CyberKnife Center, Charité Universitätsmedizin Berlin, Berlin, Germany
e-mail: gueliz.acker@charite.de

© Springer Nature Switzerland AG 2020
A. Conti et al. (eds.), *CyberKnife NeuroRadiosurgery*,
https://doi.org/10.1007/978-3-030-50668-1_20

excellent 3-year progression-free survival results [17].

Rogers et al. have also recently published the initial outcomes for high-risk meningiomas (defined as a new or recurrent WHO grade III meningioma of any resection extent, recurrent WHO grade II of any resection extent, or new WHO grade II after subtotal resection (STR)) [18]. In this report, patients with high-risk meningiomas treated by radiotherapy experienced 3-year progression-free survival of 58.8%. The combined acute and late adverse events were mainly limited to grades I–III. The authors concluded that these results supported postoperative radiotherapy for high-risk meningiomas [18]. Further results from RTOG 0539 and another ongoing randomized-controlled trial ROAM/EORTC-1308 will contribute to establish high-level evidence-based guidelines [17, 19]. Until then, the decision-making, especially for the adjuvant treatment of grade II meningiomas, seems to remain case-based.

20.1.3 Radiotherapy

If radiotherapy is indicated, for example, after subtotal resection or as a salvage treatment in local or distant progression for grade II meningiomas, conventionally fractionated external beam radiotherapy (EBRT) is widely accepted as adjuvant treatment [8]. In recent years, stereotactic radiosurgery (SRS) has been increasingly used for high-grade meningiomas as an alternative to EBRT with comparable local control rates [6, 7, 20–24].

A very recent meta-analysis reported a comparable progression-free survival at 4–10 years for both techniques, a 6-week conventionally fractionated course of radiotherapy versus SRS (89% vs. 88.8%) [25]. SRS enables high patient comfort due to short treatment course.

Most of the series published on SRS for high-quality meningiomas are a Gamma Knife radiosurgery series with reasonable local control rates (Table 20.1) [22, 23, 26–29]. A few SRS CyberKnife series (CK; Accuray Inc., Sunnivale, USA) have also been reported [24, 30–32]. In the next section, we will focus on the application of CK-SRS in high-grade meningiomas compared to other SRS series reported.

20.1.4 Image-Guided CyberKnife Stereotactic Radiosurgery

So far, four reports have been published using CK-SRS on high-grade meningiomas to our knowledge [24, 30–32]. Here, two studies belong to the same group; the first includes only atypical meningiomas with shorter follow-up, [30] while the most recent study included both pathologies of high-risk meningioma [24]. Di Franco et al. did not specify the tumor grades in their analysis, so it is not possible to make an adequate comparison of our results [31]; therefore we have only listed the study by Zhang et al. and ours in Table 20.1 as representative studies using CyberKnife [24, 32].

Our group published the most recent CK-SRS series with 127 treated lesions (105 atypical and 22 anaplastic) in 35 patients (Table 20.1) [32]. More importantly, our results confirmed that SRS could achieve a reasonable local control rate with mild radiation-related morbidity. In our series, the 12- and 24-month local control rates for atypical meningiomas were excellent with 97% and 89%, respectively. The results for anaplastic meningiomas were sobering with 66% local control rates for both time points [32]. For instance, the comparison of our results with the study of Pasquier et al. with 82 patients with grade II meningioma treated with fractional radiotherapy highlights comparable local control rates (5-year disease-free survival rate, Pasquier et al. 58%; our series, 67%) [24, 30–33].

The local control of our patients appeared to be slightly better than those reported by Zhang et al. also using CK-SRS (atypical meningiomas, 12, 36, and 60 mos., 90%, 71%, and 49% vs. our series 97%, 77%, and 67%; anaplastic meningiomas, 12 and 24 mos., 57% and 50% vs. our series 66% each; Table 20.1) [24]. In the abovementioned previous study from the same group, the 36-month local control was 74% for atypical meningiomas [30].

Table 20.1 Summary of previous literature in comparison to our results depending on the dose

References	Number of patients	Prescription dose (Gy)	Volume (cm³)	Local tumor control	Follow-up (median months)
Representative studies using CyberKnife					
Our series [32]	II: 27 III: 8	II: median 16 Gy III: median 18 Gy Overall: median 16 Gy	II: 1.55 cm³ (range, 0.06–16.3) III: 2.38 cm3 (range, 0.29–22.5)	II: 97, 77, and 67% at 1, 3, and 5 years III: 66% at 1 and 2 years	23 months (range, 2.1–60.3)
Zhang et al. 2016 [24]	II:44 III:9	II: Median 20 (15–35) III: Median 20 (12–40)	II: Median 3.33 (0.33–26.0) III: Median 3.36 (0.13–35.3)	II: 90, 71, and 49% at 1, 3, and 5 years, LC III: 93, 57, and 50%, at 1, 3, and 5 years, LC	II: 29 III:17
Representative studies using gamma knife					
Kuhn et al. 2013 [29]	II: 41 III: 48	Median (with I) 12 (8.8–20)	Median (with I) 3.25 (0.0367–414.7)	II, III: 72.3, 57.7, and 52.9% at 1, 3, and 5 years, LC II, III: 62.5, 37.1, and 29.7% at 1, 3, and 5 years, PFS	34.2
Attia et al. 2012 [26]	II: 24	Median 14 (10.5–18)	Median 6.2 (0.168–44.08)	II: 75, 51, and 44% at 1, 2, and 5 years, LC	42.5
Pollock et al. 2012 [23]	II: 37 III: 13	Median 15 (9–20)	Median 14.6 (1.8–97.7)	II, III: 85 and 45% at 1 and 5 years, LC II, III: 76 and 40% at 1 and 5 years, PFS	38
Aboukais et al. 2015 [34]	II: 27	Mean 15.2 (12–21)	Mean 5.4 (0.192–14.2)	II: 75, 52, and 40% at 1, 2, and 3 years, LC II: 75, 48, and 33% at 1, 2, and 3 years, RC	56.4
Kim et al. (2012) [35]	II: 25 III: 10	Mean 16 (12–21)	Mean 3.5 (0.3–25.3)	II, III: 78, 53, and 36% at 1, 2, and 3 years, LC II, III: 35 and 10% at 1 and 2 years, LC	33
Refaat et al. 2017 [36]	II: 75	Mean marginal 16 (12–21)	Mean 3.5 (0.3–25.3)	II: 68.9 and 55.7% at 3 and 5 years, LC	41
Ferraro et al. [37]	II: 31 III: 4	Median 18 (14–24)	Median 3.90 (0.19–33.1)	II: 95.7 and 70.1% at 1 and 3 years, PFS III: 0 and 0% at 1 and 3 years, PFS	34.5

local control (LC), progression-free survival (PFS); the studies are listed from lowest to highest prescription dose applied (adapted partly from Acker, G., et al., *Image-Guided Robotic Radiosurgery for Treatment of Recurrent Grade II and III Meningiomas. A Single-Center Study.* World Neurosurg, 2019. 131: p. e96–e107)

We applied a median prescribed dose of 16 Gy for grade II and 18 Gy for grade III meningiomas (Table 20.2). This dose regime was comparable to doses that were used in GK-SRS series within a range of 14–18 Gy (see Table 20.1).

The local response rates vary between the published studies with a tendency for dose dependency.

For instance, Refaat et al. published the largest series for atypical meningiomas ($n = 97$) with a

Table 20.2 A summary of the treatment regimens

Atypical (n = 105)			*Anaplastic (n = 22)*		
	n	%		n	%
1 fraction	88	83.8	*1 fraction*	12	54.5%
15 Gy	34		15 Gy	7	
16 Gy	49		16 Gy	2	
17 Gy	5		17 Gy	1	
			18 Gy	2	
3 fractions	7	6.7	*3 fractions*		45.5%
21 Gy	3		24 Gy	10	
22,5 Gy	2				
24 Gy	2				
4 fractions	1	1.0			
20 Gy	1				
5 fractions	5	4.8			
25 Gy	5				

(from Acker, G., et al., *Image-Guided Robotic Radiosurgery for Treatment of Recurrent Grade II and III Meningiomas. A Single-Center Study.* World Neurosurg, 2019. 131: p. e96–e107)

relatively low median dose of 14.5 Gy and reported 68.9 and 57.5% local control rates at 3 and 5 years and determined doses below and above 13.5 Gy and tumor size as prognostic factors for local control [36]. The second debatable prognostic factor for the local response is the target volume [23, 38, 39]. In the series of El-Khatib et al., [40] a correlation between volume and local response could not be confirmed for LINAC SRS similar to our series [32]. Another possible prognostic factor might be the conformality index as suggested by Attia et al. [26]. However, the conformality index has not been constantly reported in the majority of the published series; thus, a sufficient comparison is not possible.

The complication rate in our cohort was low with no severe adverse events (CTCAE ≥III) [32]. Zhang et al. [24] reported 7.5% severe neurological deficits in their series where a higher dose and volume range was applied in comparison to our study [24, 32].

The role of re-irradiation as a possible cofounder for toxicity is also a topic of discussion [24, 30, 32, 41]. Overall, a prospective randomized trial is still required in order to identify the relevant predictors for local failure and to establish the best dose and treatment planning algorithms to achieve the best local control with minimal toxicity.

Nevertheless, regional and distant failures do frequently occur which increases the challenge for the treatment of these tumors.

For instance, the 5-year recurrence rate for anaplastic meningiomas was reported up to 70% [42]. Since SRS is only a precise local treatment with high ablative potential within the target field with almost none out-of-field efficacy, SRS cannot prevent the need for repeated treatments in patients with grade II and III meningiomas with frequent distant recurrences. In conclusion, high-grade meningiomas require a multimodality treatment, and the data so far support the application of SRS on recurring or residual meningiomas with reasonable local control rates and low toxicity.

20.1.4.1 CyberKnife SRS Treatment Guide

In this section, we summarize the published CK-SRS treatment protocol for high-grade meningiomas in our institution as a practical guidance [32].

Preparation: A thermoplastic mask is needed individually for each patient for the immobilization during the treatment. A high-resolution thin-slice (0.75 mm) computed tomography (CT) by a 16-slice CT scanner after contrast agent injection is performed at site. The patients also need a recent MRI scan (T1-weighted MPRAGE using 7 mL Gadovist, 1.0 mm slice thickness).

Planning: The treatment planning is carried out on this contrast-enhanced CT fused with MRI images using MultiPlan v. 4.5 (Accuray Inc., Sunnyvale, CA). In our department, we prefer to use DOTATOC-PET MRI based on the data published so far to better visualize meningiomas [43].

The decision on the marginal and maximal doses and the number of fractions are dependent on various factors, such as histology (grade II or III), tumor volume, adjacent organs at risk (optic nerve, chiasm, brainstem), and, if applicable, the previously irradiated tumor volume.

An inverse treatment planning algorithm is used to generate steep dose gradients by delivering non-isocentric rays of up to 1600 incident beams, thereby allowing optimal tumor coverage and a minimum dosage for organs and tissues at risk of late radiation damage [32, 44–48].

The ray-tracing algorithm has been routinely used for this purpose. The physical treatment planning process included (1) selection of the adequate size and number of collimators, avoiding beams through the eyeballs; (2) the addition of help structures to reduce dose in specific brain regions; (3) definition of dose constraints and their weight for the target volume and critical structures; and (4) maximization of dose resolution using a calculation grid fully covering the CT scan to evaluate distant scattering radiation at distant sites of the body.

The GTV is defined as the absolute tumor volume based on fused CT and MRI images. PTV is created by expanding the GTV by 0–1 mm for grade II and III, respectively.

Dose regimes: So far, we have applied four different dose regimens overall depending on the site and size of the lesion including single and multisession radiosurgery as defined by Barnett et al. [49] These included either single fraction SRS in the range of 15–18 Gy; four fractions of 20 Gy; three fractions of 7–8 Gy up to 21–24 Gy, respectively; or five fractions summing up to 25 Gy, always prescribed at the 70% isodose of the PTV (Table 20.2). For single fraction 16–17 Gy for atypical and 18 Gy for anaplastic meningiomas should be aimed. The biological equivalent dose with 2 Gy per fraction (EQD_2) was calculated according to the LQ model assuming an α/β ratio of 10 Gy for high-grade meningiomas.

After therapy: Patients routinely received 4 mg dexamethasone after SRS for the prevention of side effects like headache, nausea, vomiting, or neurologic deficits due to post-radiosurgical tumor or normal tissue swelling.

Follow-up: Clinical and radiological follow-up with contrast-enhanced MRI was carried out every 3–6 months after CK-SRS for the first 2 years for grade III and II, respectively, and then every 6–12 months each year.

20.2 Solitary Fibrous Tumor/ Hemangiopericytomas

Hemangiopericytomas (HPC) are rare tumors that represent about 0.4% of primary CNS tumors and 2.4% of meningiomas [50–52]. The 2016 WHO classification introduced the combined term solitary fibrous tumor/hemangiopericytoma to describe these lesions in the future [4].

In this chapter, we will use the abbreviation "HPC" to simplify the nomenclature.

The main treatment modality for this aggressive tumor pathology is surgery. A larger extent of resection was associated with an increase in overall survival in a retrospective study [53]. However, since healthcare professionals are known to have high recurring rates, standard treatment includes gross total resection combined with adjuvant radiation therapy [50, 53, 54].

Radiosurgery had already been proposed as an alternative option to fractionated radiotherapy for these tumors in the 1990s. However, due to the rarity of this pathology, the experience so far is still based on small patient series (Table 20.3) [52, 54–65]. In most series, Gamma Knife was used, and the local control rate at the last follow-up varied between 46.4 and 100% depending on the follow-up period (Table 20.3).

Cohen-Inbar et al. analyzed the results of the multicenter Gamma Knife of 90 patients with 133 lesions reviewing management and outcome following stereotactic radiosurgery (SRS) for recurrent or newly discovered HPCs [52].

This represents the largest series of radiosurgery for this pathology. This study showed progression-free actuarial rates from the first Gamma Knife therapy at 2, 4, 6, 8, and 10 years as 81.7%, 66.3%, 54.5%, 37.2%, and 25.5%, respectively [52]. In this series 32 patients underwent 48 repeated SRS procedures for 76 lesions demonstrating the safety and efficacy of repeated SRS treatments. The authors assessed the adverse effects using the RTOG scale. RTOG II–IV grades were detected in 6.7% [52].

So far, the experience with CyberKnife is less than with Gamma Knife. The most recent and

Table 20.3 Published studies on stereotactic radiosurgery for hemangiopericytoma (adapted partly from Veeravagu et al., 2010) [64]

Series	Treatment modality	No. of patients/ lesions	Mean marginal dose (Gy)	Mean follow-up (months)	Tumor control at last FU (%)
Coffey (1993) [56]	Gamma Knife	5/11	15.5	14.8	81.8
Galanis (1998) [58]	Gamma Knife	10/20	12–18	6–36	100[a]
Payne (2000) [62]	Gamma Knife	10/12	14	24.8	75
Sheehan (2002) [54]	Gamma Knife	14/15	15	31.3	80
Chang (2003) [55]	LINAC, CyberKnife	8/8	20.5	44	75
Ecker (2003) [57]	Gamma Knife	15/45	16	45.6	93[b]
Kano (2008) [59]	Gamma Knife	20/29	15	37.9	72.4
Sun (2009) [63]	Gamma Knife	22/58	13.5	26	89.7
Iwai (2009) [65]	Gamma Knife	8/13	15.1	61	100
Olson (2010) [61]	Gamma Knife	21/28	17	69	46.4
Kim (2010) [60]	Gamma Knife	9/17	18.1	49	82.4
Veeravagu (2010) [64]	CyberKnife	14/22	21.2	37	81.8
Kim (2017) [5]	Gamma Knife	18/40	16(median)	76.6	80
Cohen-Inbar (2017) [52]	Gamma Knife	90/133[c]	15(median)	59(median)	54.9

[a]Tumors responded to GKS with decrease or stability in volume, but effect lasted less than 1 year in the majority of patients. Study also includes the five patients from Coffey et al. [56] manuscript
[b]Also includes five patients from Coffey et al. [56] manuscript
[c]Includes partly the other Gamma Knife series listed above

largest patient series treated by CyberKnife included only 14 patients with 24 tumors [64]. This study supplied promising results for this treatment modality. Over half of the treated tumors decreased in size (54.5%), six remained unchanged (27.3%), and only four showed recurrence (18.2%) after CK-RS. Progression-free survival rate was 95, 71.5, and 71.5% at 1, 3, and 5 years after multiple CK treatments. None of the patients described worsening of initial clinical presentation. The 5-year survival rate after CK was 81% [64].

In this series, the mean tumor volume was 9.16 cm^3, and the mean marginal dose to the tumors was 21.2 Gy (16–30 Gy) with varying treatment regimens in regard to the size and location and history of prior radiation. The mean isodose line was 77.5%.

In conclusion, reports on patients treated so far by radiosurgery are encouraging for these treatment modalities, in particular regarding the necessary repeated treatments. Given the rarity of this tumor, an international prospective registry should be initiated to be able to establish treatment algorithms.

References

1. Buerki RA, et al. An overview of meningiomas. Future Oncol. 2018;14(21):2161–77.
2. Ostrom QT, et al. CBTRUS statistical report: primary brain and other central nervous system tumors diagnosed in the United States in 2012–2016. Neuro Oncol. 2019;21(Suppl. 5):v1–v100.
3. Louis DN, et al. The 2007 WHO classification of tumours of the central nervous system. Acta Neuropathol. 2007;114(2):97–109.
4. Louis DN, et al. The 2016 World Health Organization classification of tumors of the central nervous system: a summary. Acta Neuropathol. 2016;131(6):803–20.
5. Kim M, et al. Analysis of the results of recurrent intracranial meningiomas treated with re-radiosurgery. Clin Neurol Neurosurg. 2017;153:93–101.
6. Rogers L, et al. Meningiomas: knowledge base, treatment outcomes, and uncertainties. A RANO review. J Neurosurg. 2015;122(1):4–23.
7. Sun SQ, et al. An evidence-based treatment algorithm for the management of WHO Grade II and III meningiomas. Neurosurg Focus. 2015;38(3):E3.
8. Hwang KL, et al. The role of radiotherapy in the management of high-grade meningiomas. Chin Clin Oncol. 2017;6(Suppl. 1):S5.
9. Goldbrunner R, et al. EANO guidelines for the diagnosis and treatment of meningiomas. Lancet Oncol. 2016;17(9):e383–91.

10. Aghi MK, et al. Long-term recurrence rates of atypical meningiomas after gross total resection with or without postoperative adjuvant radiation. Neurosurgery. 2009;64(1):56–60.. discussion 60
11. Komotar RJ, et al. The role of radiotherapy following gross-total resection of atypical meningiomas. J Neurosurg. 2012;117(4):679–86.
12. Park HJ, et al. The role of adjuvant radiotherapy in atypical meningioma. J Neurooncol. 2013;115(2):241–7.
13. Hammouche S, et al. Long-term survival analysis of atypical meningiomas: survival rates, prognostic factors, operative and radiotherapy treatment. Acta Neurochir. 2014;156(8):1475–81.
14. Graffeo CS, et al. Revisiting adjuvant radiotherapy after gross total resection of World Health Organization Grade II meningioma. World Neurosurg. 2017;103:655–63.
15. Mair R, et al. Radiotherapy for atypical meningiomas. J Neurosurg. 2011;115(4):811–9.
16. Champeaux C, Houston D, Dunn L. Atypical meningioma. A study on recurrence and disease-specific survival. Neurochirurgie. 2017;63(4):273–81.
17. Rogers L, et al. Intermediate-risk meningioma: initial outcomes from NRG Oncology RTOG 0539. J Neurosurg. 2018;129(1):35–47.
18. Rogers CL, et al. High-risk meningioma: initial outcomes from NRG Oncology/RTOG 0539. Int J Radiat Oncol Biol Phys. 2020;106(4):790–9.
19. Jenkinson MD, et al. The ROAM/EORTC-1308 trial: radiation versus observation following surgical resection of atypical meningioma: study protocol for a randomised controlled trial. Trials. 2015;16:519.
20. Albert A, et al. Adjuvant treatment of meningioma with stereotactic radiation surgery and hypofractionated stereotactic radiation surgery: patterns of care and survival in a large, hospital database. Adv Radiat Oncol. 2018;3(3):280–7.
21. Mattozo CA, et al. Stereotactic radiation treatment for recurrent nonbenign meningiomas. J Neurosurg. 2007;106(5):846–54.
22. Mori Y, et al. Gamma knife stereotactic radiosurgery for atypical and malignant meningiomas. Acta Neurochir Suppl. 2013;116:85–9.
23. Pollock BE, et al. Stereotactic radiosurgery of World Health Organization grade II and III intracranial meningiomas: treatment results on the basis of a 22-year experience. Cancer. 2012;118(4):1048–54.
24. Zhang M, et al. CyberKnife stereotactic radiosurgery for atypical and malignant meningiomas. World Neurosurg. 2016;91:574–81. e1
25. Fatima N, et al. Stereotactic radiosurgery versus stereotactic radiotherapy in the management of intracranial meningiomas: a systematic review and meta-analysis. Neurosurg Focus. 2019;46(6):E2.
26. Attia A, et al. Patterns of failure after treatment of atypical meningioma with gamma knife radiosurgery. J Neurooncol. 2012;108(1):179–85.
27. Hardesty DA, et al. The impact of adjuvant stereotactic radiosurgery on atypical meningioma recurrence following aggressive microsurgical resection. J Neurosurg. 2013;119(2):475–81.
28. Harrison G, et al. Quantitative tumor volumetric responses after Gamma Knife radiosurgery for meningiomas. J Neurosurg. 2016;124(1):146–54.
29. Kuhn EN, et al. Patterns of recurrence after stereotactic radiosurgery for treatment of meningiomas. Neurosurg Focus. 2013;35(6):E14.
30. Choi CY, et al. Cyberknife stereotactic radiosurgery for treatment of atypical (WHO grade II) cranial meningiomas. Neurosurgery. 2010;67(5):1180–8.
31. Di Franco R, et al. Radiosurgery and stereotactic radiotherapy with cyberknife system for meningioma treatment. Neuroradiol J. 2018;31(1):18–26.
32. Acker G, et al. Image-guided robotic radiosurgery for treatment of recurrent Grade II and III meningiomas. A single-center study. World Neurosurg. 2019;131:e96–e107.
33. Pasquier D, et al. Atypical and malignant meningioma: outcome and prognostic factors in 119 irradiated patients. A multicenter, retrospective study of the Rare Cancer Network. Int J Radiat Oncol Biol Phys. 2008;71(5):1388–93.
34. Aboukais R, Zairi F, Lejeune JP, et al. Grade 2 meningioma and radiosurgery. J Neurosurg. 2015;122:1157–62.
35. Kim JW, Kim DG, Paek SH, et al. Radiosurgery for atypical and anaplastic meningiomas: histopathological predictors of local tumor control. Stereotact Funct Neurosurg. 2012;90:316–24.
36. Refaat T, et al. Gamma knife stereotactic radiosurgery for Grade 2 meningiomas. J Neurol Surg B Skull Base. 2017;78(4):288–94.
37. Ferraro DJ, Funk RK, Blackett JW, et al. A retrospective analysis of survival and prognostic factors after stereotactic radiosurgery for aggressive meningiomas. Radiation Oncol. 2014;9:38.
38. Harris AE, et al. The effect of radiosurgery during management of aggressive meningiomas. Surg Neurol. 2003;60(4):298–305.. discussion 305
39. Ojemann SG, et al. Radiosurgery for malignant meningioma: results in 22 patients. J Neurosurg. 2000;93(Suppl. 3):62–7.
40. El-Khatib M, et al. Stereotactic LINAC radiosurgery for incompletely resected or recurrent atypical and anaplastic meningiomas. Acta Neurochir. 2011;153(9):1761–7.
41. Stafford SL, et al. Meningioma radiosurgery: tumor control, outcomes, and complications among 190 consecutive patients. Neurosurgery. 2001;49(5):1029–37.. discussion 1037–8
42. Yang SY, et al. Atypical and anaplastic meningiomas: prognostic implications of clinicopathological features. J Neurol Neurosurg Psychiatry. 2008;79(5):574–80.
43. Acker G, et al. Impact of 68Ga-DOTATOC PET/MRI on robotic radiosurgery treatment planning in meningioma patients: first experiences in a single institution. Neurosurg Focus. 2019;46(6):E9.

44. Conti A, et al. Frameless stereotactic radiosurgery for treatment of multiple sclerosis-related trigeminal neuralgia. World Neurosurg. 2017;103:702–12.
45. Conti A, et al. CyberKnife multisession stereotactic radiosurgery and hypofractionated stereotactic radiotherapy for perioptic meningiomas: intermediate-term results and radiobiological considerations. Springerplus. 2015;4:37.
46. Conti A, et al. Integration of functional neuroimaging in CyberKnife radiosurgery: feasibility and dosimetric results. Neurosurg Focus. 2013;34(4):E5.
47. Conti A, et al. Protecting venous structures during radiosurgery for parasagittal meningiomas. Neurosurg Focus. 2009;27(5):E11.
48. Conti A, et al. Post-treatment edema after meningioma radiosurgery is a predictable complication. Cureus. 2016;8(5):e605.
49. Barnett GH, et al. Stereotactic radiosurgery—an organized neurosurgery-sanctioned definition. J Neurosurg. 2007;106(1):1–5.
50. Guthrie BL, et al. Meningeal hemangiopericytoma: histopathological features, treatment, and long-term follow-up of 44 cases. Neurosurgery. 1989;25(4):514–22.
51. Kleihues P, et al. The WHO classification of tumors of the nervous system. J Neuropathol Exp Neurol. 2002;61(3):215–25.. discussion 226–9
52. Cohen-Inbar O, et al. Stereotactic radiosurgery for intracranial hemangiopericytomas: a multicenter study. J Neurosurg. 2017;126(3):744–54.
53. Melone AG, et al. Intracranial hemangiopericytoma—our experience in 30 years: a series of 43 cases and review of the literature. World Neurosurg. 2014;81(3–4):556–62.
54. Sheehan J, et al. Radiosurgery for treatment of recurrent intracranial hemangiopericytomas. Neurosurgery. 2002;51(4):905–10.. discussion 910–1
55. Chang SD, Sakamoto GT. The role of radiosurgery for hemangiopericytomas. Neurosurg Focus. 2003;14(5):e14.
56. Coffey RJ, Cascino TL, Shaw EG. Radiosurgical treatment of recurrent hemangiopericytomas of the meninges: preliminary results. J Neurosurg. 1993;78(6):903–8.
57. Ecker RD, et al. Hemangiopericytoma in the central nervous system: treatment, pathological features, and long-term follow up in 38 patients. J Neurosurg. 2003;98(6):1182–7.
58. Galanis E, et al. Management of recurrent meningeal hemangiopericytoma. Cancer. 1998;82(10):1915–20.
59. Kano H, et al. Adjuvant stereotactic radiosurgery after resection of intracranial hemangiopericytomas. Int J Radiat Oncol Biol Phys. 2008;72(5):1333–9.
60. Kim JW, et al. Gamma knife stereotactic radiosurgery for intracranial hemangiopericytomas. J Neurooncol. 2010;99(1):115–22.
61. Olson C, et al. Radiosurgery for intracranial hemangiopericytomas: outcomes after initial and repeat Gamma knife surgery. J Neurosurg. 2010;112(1):133–9.
62. Payne BR, et al. Gamma surgery for hemangiopericytomas. Acta Neurochir. 2000;142(5):527–36.. discussion 536–7
63. Sun S, Liu A, Wang C. Gamma knife radiosurgery for recurrent and residual meningeal hemangiopericytomas. Stereotact Funct Neurosurg. 2009;87(2):114–9.
64. Veeravagu A, et al. CyberKnife stereotactic radiosurgery for recurrent, metastatic, and residual hemangiopericytomas. J Hematol Oncol. 2011;4:26.
65. Iwai Y, Yamanaka K. Gamma knife radiosurgery for other primary intra-axial tumors. Prog Neurol Surg. 2009;22:129–41.

Perioptic Meningiomas

21

Laura Fariselli, Valentina Pinzi, Sara Morlino, and Marcello Marchetti

21.1 Introduction

Meningiomas represent 24–33% of primary brain tumors, and the incidence is 6 per 100,000, increasing with age [1]. According to the 2016 WHO classification, meningiomas are divided into three major groups: most meningiomas (80%) are grade I, benign and slow-growing tumors, whereas a criterion for grade II meningioma diagnosis is brain invasion presence. However, this feature does not represent a certain predictive factor for outcome, recurrence, or response to treatment [2]. Lastly, grade III malignant meningiomas show more aggressive progress, being highly recurrent neoplasms with dismal prognosis [3, 4].

A variety of chromosomal alterations have been identified such as deletion of chromosomes 1 and 14, having independent prognostic value for relapse-free survival. Also, the tumor suppressor gene neurofibromin 2 (NF2), located on chromosome 22, is mutated or lost in several cases.

That seems to confirm that neurofibromatosis type 2 patients are likely to be diagnosed with meningiomas. Furthermore, the PIK3CA (phosphatidylinositol-4,5-bisphosphate 3-kinase catalytic subunit alpha) gene seems to correlate with the mutational landscape and grade of meningiomas: in the future, a specific target could be selected for targeted therapies [5–8].

L. Fariselli (✉) · V. Pinzi · S. Morlino · M. Marchetti
Radiotherapy Unit, Fondazione IRCCS Istituto Neurologico Carlo Besta, Milan, Italy
e-mail: laura.fariselli@istituto-besta.it; valentina.pinzi@istituto-besta.it; sara.morlino@istituto-besta.it; marcello.marchetti@istituto-besta.it

21.2 Radiosurgery Treatment

The treatment of perioptic meningiomas aims to obtain local control, in terms of tumor growth, while stabilizing or improving the patients' neurological status. In this scenario radiosurgery (RS) can be performed as first-line treatment when surgery is excluded (upfront RS), following partial resection (adjuvant RS) or in case of recurrence/progression after surgery (salvage RS).

There is no common consensus on the need to irradiate after partial resection, even though evidence suggests that previous surgery may affect tumor control [9, 10]. This correlation may be due to the postoperative artifacts, which can make clear visualization and interpretation of target images difficult.

There is also no evidence that adjuvant RS may produce better results than salvage RS. The choice may be based on the clinician's evaluation, taking into account a variety of factors such as tumor volume and location, previous surgery, patient age, symptoms, and clinical conditions [11].

© Springer Nature Switzerland AG 2020
A. Conti et al. (eds.), *CyberKnife NeuroRadiosurgery*,
https://doi.org/10.1007/978-3-030-50668-1_21

Literature data have shown the effectiveness of single-session radiosurgery (sRS) [12–14]. Reported local control rates range from 85 to 95% at 3–5 years [14, 18, 19] with low toxicity risk. The prescription doses range from 12 to 16 Gy, mostly depending on location. In fact, a peripheral dose of 12–14 Gy is accepted for meningiomas of the cavernous sinus, since it is considered safe in terms of local neurotoxicity (cranial nerves IV, VI).

On the contrary, the proximity to the anterior optic pathway (AOP) can lead to dose reduction, in order to keep the maximum dose to the AOP structures lower than 10 Gy [15–17].

According to the literature, a maximum dose of 8–12 Gy can be considered safe for the optic pathway [18–25]. Unfortunately, this dose may be insufficient to effectively treat meningiomas.

Beyond proximity to critical structures, large volumes also require a decrease in the treatment dose and thus the efficacy of sRS. Recent published studies reported poorer local control for large-volume lesions when sRS was performed [26–31]. In this scenario, multisession radiosurgery (mRS) may represent a valid option. Therefore it is increasingly used for meningioma treatment when lesions are greater than 3 cm or lying near critical structures, i.e., lesions located close to optic apparatus.

A 25–30 Gy dose, administered in up to five fractions, is usually prescribed for mRS treatments.

The preliminary results from early studies are encouraging. Local disease control rate from 94 to 100% and a radiation-induced optic neuropathy rate from 0 to 6% have been reported. The visual improvement rate in the same groups ranged from 16 to 100% [12, 15, 16, 32].

The number of studies that indicate the potential of mRS is now on the rise [19]. In their 2017 study, Han et al. observed that, in a series of relatively large lesions, the tumor control rate did not differ between sRS and mRS, while complications were significantly higher in the sRS group [33].

Manabe et al. [34] treated 9 patients with sRS (mean volume 4.6 cm^3) and 32 patients with mRS (mean volume 11.3 cm^3). Among the patients with a tumor volume <13.5 mL, the authors did not observe a significant difference in terms of progression-free survival (PFS) rate between the patients treated with mRS (n = 14, median dose 25 Gy) and those treated with sRS (n = 8, median dose 16.5 Gy). Nevertheless, the authors noticed a 3-year worse PFS rate for the 14 patients with a tumor volume greater than 13.5 mL compared to the 27 patients with a tumor volume smaller than 13.5 mL ($P = 0.031$) [34].

A recent study by Conti et al. [14] appears to confirm the efficacy and safety of mRS in meningiomas that develop near the AOP. Sixty-four patients were treated (39 of which were prospectively assessed) with a marginal dose between 18 and 40 Gy. Local control was 100% for both retrospectively and prospectively evaluated patients, with a median follow-up of 57.5 and 15 months, respectively.

The absence of a significant visual deterioration and a good improvement rate (up to 20%) also supported the potential value of the mRS.

Similarly, a retrospective study of a larger series by Marchetti et al. [35] focused on mRS for perioptic benign meningiomas. The prescription doses ranged from 15 to 25 Gy delivered in three to five fractions, with a median follow-up of 32 months (range 12–113 months). PFS rate at 3, 5, and 8 years was 100%, 93%, and 90%, respectively. The overall visual deterioration rate was 7.4%. Excluding patients with progression disease (PD), the visual deterioration rate was 5.1%. The improvement rate was 36%.

A recent multicentric study by Marchetti et al. seems to confirm these encouraging results [36]. One-hundred sixty-seven patients were included in the analysis, and 101 underwent mRS as primary treatment, while 66 received it after previous surgery. The median follow-up was 51 months (range 36–129 months). Progression-free survival rate at 3, 5, and 8 years was 98%, 94%, and 90%, respectively. Excluding the progressive disease patients, the visual worsening rate was 3.7%. Forty-two percent of the patients with a pre-treatment visual deficit experienced improvement in vision.

Another study [1] analyzed 52 patients with intracranial meningioma treated with RS. The authors concluded that single and fractionated approaches with CyberKnife could offer particular benefits to patients with tumors that were large or located in critical site, allowing good disease control. Moreover, the authors reported a visual deficit improvement in 50% of patient with perioptic lesions.

The main limit of the described studies is their retrospective nature. However, the good results on local control and the low toxicity on optic pathway are promising, even though they should be confirmed by prospective trials. The 25 Gy in five fractions schedule could be the new possible paradigm for the treatment of meningiomas of the optic tract.

21.3 Radiological Features to Draw the Target Volume and Organs at Risk

For both fractional and single-session approaches, the planning target volume (PTV) is generally defined as the T1-weighted contrast-enhancing lesion on post-contrast MRI images [37]. However, identifying the target volume often remains difficult. Since low-grade meningiomas are generally associated with long-term survival and good overall prognosis, the key to maintaining local control is the correct identification of the optimal PTV while preserving very radiosensitive structures, such as the AOP.

Therefore, improvement of target volume delineation is a first essential step in the radiation oncology treatment planning process. Additional PET imaging with DOTATOC or DOTANOC tracers could improve target volume definition [38, 39]. In meningiomas located in proximity of the bony structures, which are difficult to distinguish on MRI and CT, the volumes generally increase by adding PET [40–44].

Additionally, it is currently under debate whether the dural tail should be included in the gross tumor volume definition (GTV): although 75% of recurrent tumors involve the dura outside the treatment field, a precise determination of its extension is not an easy task.

Indeed, the evaluation of the extent of infiltration is currently only based on the experience of the radiotherapy oncologist.

The nodular areas have to be taken into consideration when drawing the PTV. Future trials focusing on RS treatments should carefully assess the importance of this issue. The same goes for the hyperostotic bone, although there is little evidence that hyperostotic bone irradiation yields better results in terms of local control.

Organs at risk (OARs) have to be defined through MRI (chiasma, pituitary, brainstem, cranial nerves, and carotid artery) and CT scan (eyes, lens, cochlea) in all plans. Before contouring skull base meningiomas/perioptic meningiomas, the correct imaging has to be carefully chosen. Axial T1 is preferably obtained by means of 3D-GE sequences (MPRAGE/TFE/SPGR), owing to their high spatial definition and excellent gray-white contrast.

In addition, they allow data reformatting in any plan and guarantee a good elaboration in all three plans. Gadolinium compounds are used to appraise the degree of permeability of the blood-tumor barrier and extension of the lesion.

Fat suppression applied to T1-weighted images is commonly used to suppress the signal from adipose tissue and better distinguish the contours of the meningioma.

The latter can also be assessed via FSE T2 sequences (preferably 2D acquisition with thin slices and minimal gap, to better determine the relationship of the tumor to the surrounding structures) [41]. CT scans provide useful insights into the bone features (hyperostosis, re-arrangement) and calcification [37, 41].

Therefore, PET can be a useful tool to integrate CT and MRI during the planning process, to improve the accuracy of the treatment, even for meningiomas located in proximity of the bony structures [40–44].

21.4 Conclusion

Multisession radiosurgery can represent the answer to the dilemma on whether to treat perioptic meningiomas or not. Taking into account the likely very low risk of developing late visual toxicities, mRS can be offered to our patients as exclusive treatment. Prospective studies need to be drawn to confirm these preliminary results.

References

1. Di Franco R, Borzillo V, Ravo V, Falivene S, Romano FJ, Muto M, et al. Radiosurgery and stereotactic radiotherapy with CyberKnife system for meningioma treatment. Neuroradiol J. 2018;31(1):18–26.
2. Louis DN, Perry A, Reifenberger G, Von Deimling A, Figarella-Branger D, Cavenee WK, et al. The 2016 World Health Organization classification of tumors of the central nervous system: a summary. Acta Neuropathol. 2016;131(6):803–20.
3. Kshettry VR, Ostrom QY, Kruchko C, Al-Mefty O, Barnett GH, Barnholtz-Sloan JS. Descriptive epidemiology of world health organization grades II and III intracranial meningiomas in the United States. Neuro Oncol. 2015;17(8):1166–73.
4. Aghi MK, Carter BS, Cosgrove GR, Ojemann RG, Amin-Hanjani S, Martuza RL, et al. Long-term recurrence rates of atypical meningiomas after gross total resection with or without postoperative adjuvant radiation. Neurosurgery. 2009;64(1):56–60.
5. Juratli TA, McCabe D, Nayyar N, Williams EA, Silverman IM, Tummala SS, et al. DMD genomic deletions characterize a subset of progressive/higher-grade meningiomas with poor outcome. Acta Neuropathol. 2018;136(5):779.
6. Sahm F, Schrimpf D, Olar A, Koelsche C, Reuss D, Bissel J, et al. TERT promoter mutations and risk of recurrence in meningioma. J Natl Cancer Inst. 2016;108(5):djc377.
7. Suppiah S, Nassiri F, Bi WL, Dunn IF, Hanemann CO, Horbinski CM, et al. Molecular and translational advances in meningiomas. Neuro-Oncology. 2019;21(Suppl1):i4–i17.
8. Patel AJ, Wan Y-W, Al-Ouran R, Revelli J-P, Cardenas MF, Oneissi M, et al. Molecular profiling predicts meningioma recurrence and reveals loss of DREAM complex repression in aggressive tumors. Proc Natl Acad Sci U S A. 2019;116(43):21,715–26.
9. Islim AI, Mohan M, Moon RDC, Srikandarajah N, Mills SJ, Brodbelt AR, et al. Incidental intracranial meningiomas: a systematic review and meta-analysis of prognostic factors and outcomes. J Neuro-Oncol. 2019;142(2):211.
10. Sanford NN, Yeap BY, Larvie M, Daartz J, Munzenrider JE, Liebsch NJ, et al. Prospective, randomized study of radiation dose escalation with combined proton-photon therapy for benign meningiomas. Int J Radiat Oncol Biol Phys. 2017;99(4): 787–96.
11. Ganau M, Foroni RI, Gerosa M, Zivelonghi E, Longhi M, Nicolato A. Radiosurgical options in neuro-oncology: a review on current tenets and future opportunities. Part I: therapeutic strategies. Tumori. 2014;100(4):459–65.
12. Adler JR, Gibbs IC, Puataweepong P, Chang SD. Visual field preservation after multisession CyberKnife radiosurgery for perioptic lesions. Neurosurgery. 2006;59(2):244–54.
13. Pinzi V, Bisogno I, Prada F, Ciusani E, Fariselli L. Radiotherapy of meningioma: a treatment in need of radiobiological research. Int J Radiat Biol. 2018;94(7):621–7.
14. Pinzi V, Biagioli E, Roberto A, Galli F, Rizzi M, Chiappa F, et al. Radiosurgery for intracranial meningiomas: a systematic review and meta-analysis. Crit Rev Oncol Hematol. 2017;113:122–34.
15. Conti A, Pontoriero A, Midili F, Iatì G, Siragusa C, Tomasello C, et al. CyberKnife multisession stereotactic radiosurgery and hypofractionated stereotactic radiotherapy for perioptic meningiomas: intermediate-term results and radiobiological considerations. Springer Plus. 2015;4:37.
16. Marchetti M, Bianchi S, Milanesi I, Bergantin A, Bianchi L, Broggi G, et al. Multisession radiosurgery for optic nerve sheath meningiomas—an effective option: preliminary results. Neurosurgery. 2011;69(5):1116–23.
17. Milano MT, Grimm J, Soltys SG, Yorke E, Moiseenko V, Tomé WA, et al. Single- and multi-fraction stereotactic radiosurgery dose tolerances of the optic pathways. Int J Radiat Oncol Biol Phys. 2018:1–13. https://doi.org/10.1016/j.ijrobp.2018.01.053.
18. Biau J, Khalil T, Verrelle P, Lemaire JJ. Fractionated radiotherapy and radiosurgery of intracranial meningiomas. Neurochirurgie. 2018;64(1):29–36.
19. Lee CC, Trifiletti DM, Sahgal A, DeSalles A, Fariselli L, Hayashi M, et al. Stereotactic radiosurgery for benign (World Health Organization Grade I) cavernous sinus meningiomas-International Stereotactic Radiosurgery Society (ISRS) practice guideline: a systematic review. Neurosurgery. 2018;83(6):1128–42.
20. Sheehan JP, Starke RM, Kano H, Kaufmann AM, Mathieu D, Zeiler FA, et al. Gamma knife radiosurgery for sellar and parasellar meningiomas: a multicenter study. J Neurosurg. 2014;120(6):1268–77.
21. Stafford SL, Pollock BE, Leavitt JA, Foote RL, Brown PD, Link MJ, et al. A study on the radiation tolerance of the optic nerves and chiasm after stereotactic radiosurgery. Int J Radiat Oncol Biol Phys. 2003;55:1177–81.
22. Leavitt JA, Stafford SL, Link MJ, Pollock BE. Long-term evaluation of radiation-induced optic neuropathy after single fraction stereotactic radiosurgery. Int J Radiat Oncol Biol Phys. 2013;87:524–7.
23. Pollock BE, Link MJ, Leavitt JA, Stafford SL. Dose–volume analysis of radiation-induced optic neuropathy after single fraction stereotactic radiosurgery. Neurosurgery. 2014;75:456–60.
24. Mayo C, Martel MK, Marks LB, Flickinger J, Nam J, Kirkpatrick J. Radiation dose–volume effects of optic nerves and chiasm. Int J Radiat Oncol Biol Phys. 2010;76(3):S28–35.

25. Hasegawa T, Kobayashi T, Kida Y. Tolerance of the optic apparatus in single-fraction irradiation using stereotactic radiosurgery: evaluation in 100 patients with craniopharyngioma. Neurosurgery. 2010;66:688–95.
26. Azar M, Kazemi F, Chanideh I, Amirjamshidi A, Amini E, Ghanavati P. Gamma knife radiosurgery in sphenopetroclival meningiomas: preliminary experience at the Iran Gamma Knife Center. World Neurosurg. 2016;93:39–43.
27. El-Khatib M, El Majdoub F, Hunsche S, Hoevels M, Kocher M, Sturm V, et al. Stereotactic LINAC radiosurgery for the treatment of typical intracranial meningiomas. Efficacy and safety after a follow-up of over 12 years. Strahlenther Onkol. 2015;191(12):921–7.
28. Kondziolka D, Mathieu D, Lunsford LD, Martin JJ, Madhok R, Niranjan A, et al. Radiosurgery as definitive management of intracranial meningiomas. Neurosurgery. 2008;62(1):53–8.
29. Pollock BE, Stafford SL, Link MJ, Brown PD, Garces YI, Foote RL. Single-fraction radiosurgery of benign intracranial meningiomas. Neurosurgery. 2012;71(3):604–12.
30. Williams BJ, Yen CP, Starke RM, Basina B, Nguyen J, Rainey J, et al. Gamma knife surgery for parasellar meningiomas: long-term results including complications, predictive factors, and progression-free survival: clinical article. J Neurosurg. 2011;114(6):1571–7.
31. Di Biase SJ, Kwok Y, Yovino S, Arena C, Naqvi S, Temple R, et al. Factors predicting local tumor control after gamma knife stereotactic radiosurgery for benign intracranial meningiomas. Int J Radiat Oncol Biol Phys. 2004;60:1515–9.
32. Romanelli P, Wowra B, Muacevic A. Multisession CyberKnife radiosurgery for optic nerve sheath meningiomas. Neurosurg Focus. 2007;23(6):E11.
33. Han MS, Jang WY, Moon KS, Lim SH, Kim IY, Jung TY, et al. Is fractionated gamma knife radiosurgery a safe and effective treatment approach for large-volume (>10 cm3) intracranial meningiomas? World Neurosurg. 2017;99:477–83.
34. Manabe Y, Murai T, Ogino H, Tamura T, Iwabuchi M, Mori Y, et al. CyberKnife stereotactic radiosurgery and hypofractionated stereotactic radiotherapy as first-line treatments for imaging-diagnosed intracranial meningiomas. Neurol Medico-Chir. 2017;57(12):627–33.
35. Marchetti M, Bianchi S, Pinzi V, Tramacere I, Fumagalli ML, Milanesi IM, et al. Multisession radiosurgery for sellar and parasellar benign meningiomas: longterm tumor growth control and visual outcome. Neurosurgery. 2016;78(5):638–46.
36. Marchetti M, Conti A, Beltramo G, Pinzi V, Pontoriero A, Tramacere I, et al. Multisession radiosurgery for perioptic meningiomas: medium-to-long term results from a CyberKnife cooperative study. J Neuro-Oncol. 2019;143:597–604.
37. Milker-Zabel S, Zabel-du Bois A, Henze M, Huber P, Schulz-Ertner D, Hoess A. Improved target volume definition for fractionated stereotactic radiotherapy in patients with intracranial meningiomas by correlation of CT, MRI, and [68Ga]-DOTATOC-PET. Int J Radiat Oncol Biol Phys. 2006;65(1):222–7.
38. Acker G, Kluge A, Lukas M, Conti A, Pasemann D, Meinert F, et al. Impact of 68Ga-DOTATOC PET/MRI on robotic radiosurgery treatment planning in meningioma patients: first experiences in a single institution. Neurosurg Focus. 2019;46(6):E9.
39. Nyuyki F, Plotkin M, Graf R, Michel R, Steffen I, Denecke T, et al. Potential impact of (68) Ga-DOTATOC PET/CT on stereotactic radiotherapy planning of meningiomas. Eur J Nucl Med Mol Imaging. 2010;37(2):310–8.
40. Kunz WG, Jungblut LM, Kazmierczak PM, Vettermann FJ, Bollenbacher A, Tonn JC, et al. Improved detection of transosseous meningiomas using (68)Ga-DOTATATE PET/CT compared with contrast-enhanced MRI. J Nucl Med. 2017;58(10):1580–7.
41. Ranabhat K, Bishokarma S, Agrawal P, Shrestha P, Panth R, Ghimire RK. Role of MR morphology and diffusion-weighted imaging in the evaluation of meningiomas: radio-pathologic correlation. JNMA J Nepal Med Assoc. 2019;57(215):37–44.
42. Kessel KA, Weber W, Yakushev I, Fischer H, Voglhuber T, Diehl C, et al. Integration of PET-imaging into radiotherapy treatment planning for low-grade meningiomas improves outcome. Eur J Nucl Med Mol Imaging. 2020;47(6):1391–9.
43. Dittmar JO, Kratochwil C, Dittmar A, Welzel T, Habermehl D, Rieken S, et al. First intraindividual comparison of contrast-enhanced MRI, FET- and DOTATOC-PET in patients with intracranial meningiomas. Radiat Oncol. 2017;12(1):169.
44. Galldiks N, Albert NL, Sommerauer M, Grosu AL, Ganswindt U, Law I, et al. PET imaging in patients with meningioma-report of the RANO/PET Group. Neuro-Oncology. 2017;19(12):1576–87.

22 Optic Nerve Sheath Meningiomas

Carolin Senger

22.1 Introduction

Primary optic nerve sheath meningiomas (ONSM) account for about 1–2% of all intracranial meningiomas [1, 2]. ONSM grow from the meninges that surround the optic nerve, located on the orbital or canalicular part of the nerve. In contrast, tumors that are located within <2–3 mm of the optical structures are called "perioptic meningiomas" [3]. ONSM are more common in middle-aged women. Compared to perioptic or intracranial meningiomas, ONSM occur at an earlier age. In 1992, Dutton et al. [2] published a large review of ONSM with a mean age of 40.8 years, of which 61% were female. In several series, the proportion of women is even up to 80% [4–6].

Anatomically, the optic nerve is a part of the central nervous system. For this reason, the fibers of the optic nerve are covered with myelin, which is produced by oligodendrocytes instead of Schwann cells of the peripheral nervous system, and are additionally enveloped by the meninges. The growth pattern of ONSM is typically circular along the optic nerve, resulting in nerve atrophy due to compression on the one hand and failure of the pial vascular supply on the other.

A lesion that displaces the nerve peripherally is usually an early stage of ONSM, followed by encapsulation of the nerve over time. As the lesion progresses, ONSM can spread intracranially through the optic nerve canal, involving the optic chiasm with bilateral visual disturbances through direct extension or by tension and distortion of the chiasm. Nevertheless, the spread of initially unilateral ONSM via the optic chiasma to the contralateral side is rare. The actual incidence of bilateral isolated ONSM is difficult to estimate, some are new independent tumors, and others are associated with type 2 neurofibromatosis [7, 8]. Overall, bilateral ONSM occur in up to 5–10% of patients [1, 2]. In childhood, ONSM are very rare and only described in a few cases [9].

22.2 Imaging and Ophthalmological Diagnostics

A Danish retrospective analysis showed an increased incidence of ONSM over the past 25 years, most likely due to the more frequent use and improved magnetic resonance imaging (MRI) techniques in recent years [10].

C. Senger (✉)
Department of Radiation Oncology, Charité Universitätsmedizin Berlin, Berlin, Germany

Charité CyberKnife Center, Charité Universitätsmedizin Berlin, Berlin, Germany
e-mail: carolin.senger@charite.de

© Springer Nature Switzerland AG 2020
A. Conti et al. (eds.), *CyberKnife NeuroRadiosurgery*,
https://doi.org/10.1007/978-3-030-50668-1_22

The MRI has become the gold standard for the diagnosis and has obviated the need for tissue biopsy, especially gadolinium-enhanced fat-suppressed MRI sequences (Fig. 22.1) that show the typical thickening of the optic nerve, sometimes associated with calcification within the tumor [11]. The axial MRI shows the characteristic "tram track" sign, which corresponds to the enhancing outer ONSM surrounding the inner, non-enhancing nerve. On the coronal images, ONSM are typically represented with a "dot" sign [12]. Occasionally ONSM cannot be distinguished from optic nerve gliomas by imaging. Another diagnostic tool is the ^{68}Ga-DOTATOC PET/MRI, which can additionally be used to confirm the diagnosis of ONSM [13].

Compared to intracranial meningiomas, ONSM are usually diagnosed at the onset of symptoms, mainly due to visual impairment or even complete blindness. ONSM lead to chronic compression of the optic nerve, which typically causes painless progressive loss of visual function. The most common symptoms at initial diagnosis are reduced visual acuity, visual field defect, color vision disturbance, and proptosis [2]. The triad of progressive painless loss of visual function, optic nerve atrophy, and optociliary shunt vessels together, mentioned by Spencer et al. (1972) [14] and Friesen et al. (1973) [15], is pathognomonic for ONSM.

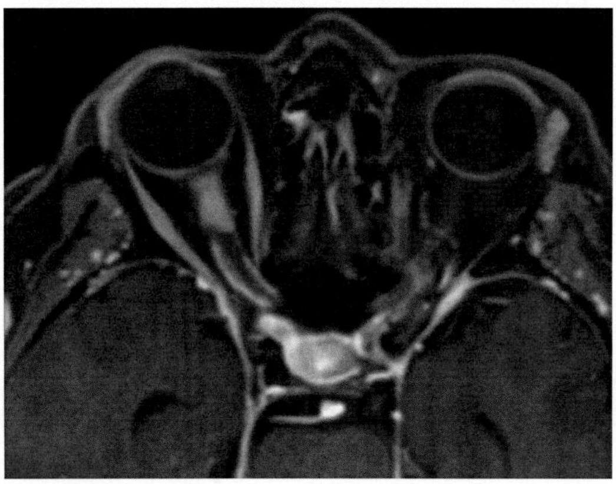

Fig. 22.1 Gadolinium-enhanced fat-suppressed axial T1-weighted MRI showing a right retrobulbar optic nerve sheath meningioma

22.3 Treatment of ONSM

Primary ONSM are challenging to treat because of their intimidate relation to the optic nerve and their sharing of common pial blood supply. The decision on treatment depends on treatment-related factors, e.g., possible neurological consequences of surgery and/or radiation, the likelihood of complete resection and/or complete irradiation with SRS, treatability in case of tumor progression, available surgical or radiation oncology expertise, and resources [16]. In general, ONSM are best managed by a multidisciplinary team of neurosurgeons, neuro-ophthalmologists, radiation oncologists, and neurologists.

Microsurgical resection is usually reserved for patients with complete vision loss of the affected eye or clinically relevant exophthalmos. Encasement of the optic nerve almost always makes it impossible to perform a radial resection without seriously damaging the optic nerve or vascular supply. A recent systemic review showed a decline in visual outcome in 56% of patients who underwent surgery for the intraorbital segment of the optic nerve [17]. The high risk of vision loss after microsurgical resection encouraged a non-surgical approach.

Stereotactic fractionated radiation therapy (SFRT) with a total dose of 50.4–54.0 Gy (1.8 Gy/fraction), either as adjuvant or primary treatment, has been shown to be effective and reasonably well tolerated [1, 5, 6, 18–20]. SFRT can maintain visual function in the majority of patients treated, despite low risks of radiation-induced retinopathy or optic neuropathy. A radiation retinopathy was detected in the retrospective work of Arvold et al. [5] in 3 of 25 clinically asymptomatic patients by an ophthalmological examination. Paulsen et al. [1] published the evaluation of 113 patients with primary and secondary ONSM. Visual acuity was preserved in 94.8% of cases after 3 years and 90.9% of cases after 5 years. Radiographic tumor control was 100% at 3 years and 98% at 5 years. Primary SFRT treatment is recommended to maintain vision in patients with

ONSM, rather than an observation strategy alone. While SFRT relies on a combination of tissue sparing (which still allows dose leakage via the retina and other nearby structures) and daily fractionation to protect the optical structures, SRS offers extremely narrow dose distributions and enhanced dose conformality, focusing the irradiation of the tumor and providing maximum protection for the organs at risk [21].

22.4 CyberKnife Stereotactic Radiosurgery

In recent years, radiosurgery has been increasingly used. The frame-based Gamma Knife SRS was rarely used because the single-fraction SRS dose far exceeds the tolerance dose of the optic nerve [22]. Frameless SRS devices, such as the CyberKnife, can provide sub-millimeter accuracy and extremely tight conformality, combined with the ability to split the radiosurgical dose into multiple fractions, known as multisession, staged, or hypo-fractionated SRS [23–25]. By treating in several sessions, it is possible to take advantage of the different speeds of recovery of normal and pathological tissue, thus limiting or preventing damage to the visual pathway while controlling tumor growth. Finally, higher doses per session allow a more significant shortening of treatment compared to standard fractionated radiotherapy, which is beneficial for most patients when everything else is equal.

In 2010 Marchetti et al. [4] published the largest series of primary ONSM treated by CyberKnife radiosurgery. The authors included 21 patients using multisession radiosurgery with five fractions to a total dose of 25 Gy, prescribed to the 75–85% isodose line. The median pretreatment ONSM volume was 2.8 mL. In this cohort, all patients tolerated SRS well, with only one patient developing mild optic neuropathy (which resolved after systemic steroid therapy). There were no other acute or late radiation-induced toxicities observed. No patient showed ONSM progression (10% responded partially) on subsequent MRI with a mean follow-up period of 30 months. None of the patients had a deterioration in visual function, which was stable in 65% of the patients and improved in 35% of the patients.

22.4.1 Radiosurgery Treatment Planning

For planning purposes, a contrast-enhanced thin-slice (0.75–1.0 mm) CT should be performed. For immobilization during treatment, a thermoplastic mask is required for each patient individually. Contouring of the organs at risk (OAR) is performed on a co-registered T1-weighted MPRAGE. The gross tumor volume (GTV) is defined as the ONSM volume based on the planning CT and a second co-registered T1-weighted fat-suppressed MRI dataset, a safety margin of 0–1 mm could be added, and overlap with the optic nerve should be avoided. Additional ^{68}Ga-DOTATOC PET/MRI (Fig. 22.2) could improve target volume contouring [26]. Multidisciplinary input for treatment planning is recommended.

An inverse optimization algorithm is used to generate steep dose gradients by non-isocentric beam guidance of about 100–400 incident beams, allowing optimal tumor coverage and minimal dose to the OAR (i.e., minimize dose to the optic pathway). The ray-tracing dose calculation algorithm is routinely used. Dose volume histograms are useful in the evaluation process to select the optimal treatment plan as a trade-off between target coverage and OAR dose.

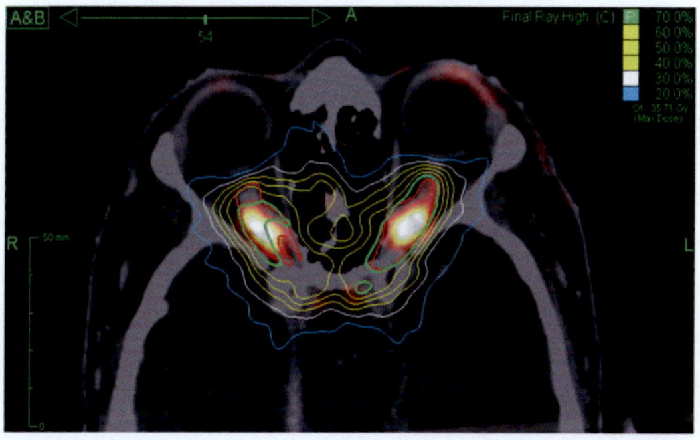

Fig. 22.2 Treatment planning with PET/MRI to identify the full extent of the tumor, in particular bilateral optic nerve sheath meningiomas

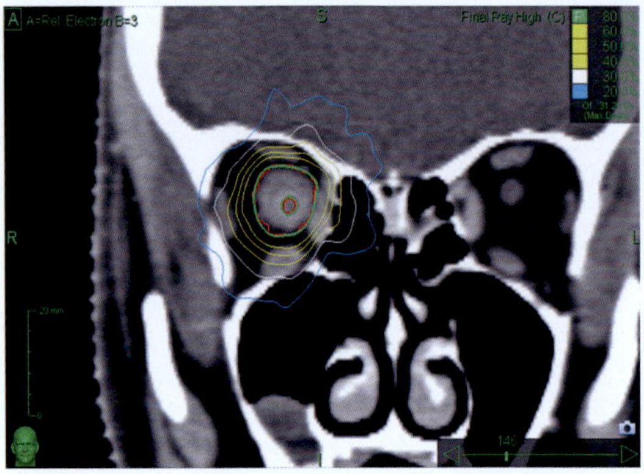

Fig. 22.3 Treatment planning showing a large optic nerve sheath meningioma, in addition of the green line for the 70% isodose and clear blue line outside for 20% isodose

22.4.2 Radiosurgery Dose Concepts

Two different dose concepts as a function of the extent of the tumor and vision are widely used with respect to the circumference of the optic nerve surrounded by the tumor. Dose regimens depending on extension of the ONSM include either single-fraction SRS in the range of 12–14 Gy prescribed at the 70% isodose for a lesion displacing the nerve peripherally or hypo-fractionated SRS (25 Gy in 5 fractions) usually prescribed at the 80–85% isodose for circumferential growing ONSM (Fig. 22.3) [16]. Patients with ONSM are often not candidates for single-fraction SRS because of concerns about radiation-induced optic neuropathy. However, these patients have been successfully treated with hypo-fractionated SRS (usually 5 fractions).

Two publications on multisession radiosurgery by Romanelli et al. [21, 27] show very promisingly that the CyberKnife offers the highest degree of conformity and accuracy and in addition the ability to perform the treatment in multiple sessions, thus improving the tolerance of the optic nerve to radiation.

22.4.3 Organs at Risk and Optic System Tolerance Dose

Critical structures that are important to preserve are the eyes, ipsilateral and contralateral optic nerves, chiasma, retinae, lacrimal glands, pituitary gland, and the brainstem.

Hiniker et al. [28] summarized different data on the visual pathway constraints and showed that the Dmax limits of 12.0 Gy in one fraction from QUANTEC [29], 19.5–20 Gy in three fractions from Timmerman et al. [30], and 25.0 Gy in five fractions from AAPM Task Group 101 [31] all had less than 1% risk of optic complications. Equal dose constraints were concluded in a recent work by Milano et al. [32] which summarized 34 studies with a total of 1578 patients. Recommended dose constraints to organs at risk (OAR) for one fraction/three fractions/five fractions of SRS are as follows: $<0.20\ cm^3$ of the optic pathway was allowed to receive 8.0/15.3/23.0 Gy with a maximum dose of 10.0/17.4/25.0 Gy in $\leq 0.035\ cm^3$ [31]. If the single-fraction maximum dose for the optic system exceeds 10 Gy in $\leq 0.035\ cm^3$, a hypo-fractionated SRS concept (e.g., 25 Gy in 5 fractions) should be used. The treatment dose, fractionation, or isodose should be adjusted if the OAR restrictions were not applicable due to individual tumor extension.

22.4.4 Clinical and Imaging Follow-Up

Patients routinely received 4 mg of dexamethasone after each fraction of the SRS to prevent side effects such as headache and transient visual disturbances due to edema. In a study by Adler et al. [24], mild headaches and occasional temporary diplopia responding to short course dexamethasone were rarely reported. No other acute or subacute side effects were observed in this study.

The response to the treatment is evaluated by a contrast-enhanced MRI control and an ophthalmological examination. Vision is assessed by visual acuity and visual field tests, which should be carried out by an ophthalmologist after 3, 6, and 12 months and then annually. Post-radiation MRI is recommended after 6 and 12 months, then at intervals of 12 months for 5 years, and afterward every 1–3 years, as clinically indicated. A less frequent follow-up may be considered after 5–10 years [16].

22.5 Conclusion

Their origin and location make ONSM one of the most demanding treatment challenges in radiosurgery. ONSM can be safely treated with multisession SRS and in some cases with a single-fraction SRS with high rates of tumor control and preservation of visual function. Hypofractionated radiosurgery is an especially effective alternative to surgery or conventional SFRT in the management of ONSM. CyberKnife SRS enables a high level of patient comfort due to the short course of treatment.

References

1. Paulsen F, et al. Fractionated stereotactic radiotherapy in patients with optic nerve sheath meningioma. Int J Radiat Oncol Biol Phys. 2012;82:773–8.
2. Dutton JJ. Optic nerve sheath meningiomas. Surv Ophthalmol. 1992;37:167–83.
3. Conti A, et al. CyberKnife multisession stereotactic radiosurgery and hypofractionated stereotactic radiotherapy for perioptic meningiomas: intermediate-term results and radiobiological considerations. Springerplus. 2015;4:37.
4. Marchetti M, et al. Multisession radiosurgery for optic nerve sheath meningiomas—an effective option: preliminary results of a single-center experience. Neurosurgery. 2011;69(5):1116–22.
5. Arvold ND, et al. Visual outcome and tumor control after conformal radiotherapy for patients with optic nerve sheath meningioma. Int J Radiat Oncol Biol Phys. 2009;75(4):1166–72.
6. Eckert F, et al. Retrospective analysis of fractionated intensity-modulated radiotherapy (IMRT) in the interdisciplinary management of primary optic nerve sheath meningiomas. Radiat Oncol. 2019;14:1–9.
7. Bosch MM, et al. Ophthalmologic findings and long-term course in patients with neurofibromatosis type 2. Am J Ophthalmol. 2006;141(6):1068–77.
8. Lekovic GP, et al. Intra-orbital meningioma causing loss of vision in neurofibromatosis type 2: case series and management considerations. Front Surg. 2018;5:60.
9. Narayan DA-O, et al. Natural history of primary paediatric optic nerve sheath meningioma: case series and review. Br J Ophthalmol. 2018;102(8):1147–53.
10. Lindegaard J, Heegaard S, Prause JU. Histopathologically verified non-vascular optic nerve lesions in Denmark 1940-99. Acta Ophthalmol Scand. 2002;80(1):32–7.
11. Miller NR. New concepts in the diagnosis and management of optic nerve sheath meningioma. J Neuroophthalmol. 2006;26(3):200–8.
12. Radiopaedia. Optic nerve sheath meningioma. 29 Mar 2020. https://radiopaedia.org/articles/optic-nerve-sheath-meningioma.
13. Al Feghali KA, et al. The use of (68)Ga-DOTATATE PET/CT in the non-invasive diagnosis of optic nerve sheath meningioma: a case report. Front Oncol. 2018;8:454.
14. Spencer WH. Primary neoplasms of the optic nerve and its sheaths: clinical features and current concepts of pathogenetic mechanisms. Trans Am Ophthalmol Soc. 1972;70:490–528.
15. Frisen L, Royt WF, Tengroth BM. Optociliary veins, disc pallor and visual loss. A triad of signs indicating spheno-orbital meningioma. Acta Ophthalmol (Copenh). 1973;51(2):241–9.
16. National Comprehensive Cancer Network. Central nervous system cancers. 2019 Mar 30, 2020. https://www.nccn.org/professionals/physician_gls/pdf/cns_blocks.pdf.
17. Henaux PL, et al. Modern management of meningiomas compressing the optic nerve: a systematic review. World Neurosurg. 2018;118:e677–86.
18. Adeberg S, et al. Prior surgical intervention and tumor size impact clinical outcome after precision radiotherapy for the treatment of optic nerve sheath meningiomas (ONSM). Radiat Oncol. 2011;6:117.
19. Metellus P, et al. Fractionated conformal radiotherapy for management of optic nerve sheath meningiomas: long-term outcomes of tumor control and visual function at a single institution. Int J Radiat Oncol Biol Phys. 2011;80(1):185–92.
20. Inoue TA-O, et al. Early intervention using high-precision radiotherapy preserved visual function for five consecutive patients with optic nerve sheath meningioma. Int J Clin Oncol. 2018;23(5):826–34.
21. Romanelli P, et al. Staged image guided robotic radiosurgery for optic nerve sheath meningiomas. Comput Aided Surg. 2011;16(6):257–66.
22. Jin J, et al. Optic nerve sheath meningioma: preliminary analysis of the role of radiation therapy. Brain Tumor Res Treat. 2018;6(1):8–12.
23. Marchetti M, et al. Multisession radiosurgery for perioptic meningiomas: medium-to-long term results from a CyberKnife cooperative study. J Neuro-Oncol. 2019;143:597–604.
24. Adler JR Jr, et al. Visual field preservation after multisession cyberknife radiosurgery for perioptic lesions. Neurosurgery. 2006;62(Suppl 2):733–43.
25. Pham CJ, et al. Preliminary visual field preservation after staged CyberKnife radiosurgery for perioptic lesions. Neurosurgery. 2004;54(4):799–810.
26. Acker G, et al. Impact of 68Ga-DOTATOC PET/MRI on robotic radiosurgery treatment planning in meningioma patients: first experiences in a single institution. Neurosurg Focus. 2019;46(6):E9.
27. Romanelli P, Wowra A, Fau B, Muacevic A. Multisession CyberKnife radiosurgery for optic nerve sheath meningiomas. Neurosurg Focus. 2007;23(6):E11.

28. Hiniker SM, et al. Dose-response modeling of the visual pathway tolerance to single-fraction and hypofractionated stereotactic radiosurgery. Semin Radiat Oncol. 2016;26(2):97–104.
29. Bentzen SM, et al. Quantitative analyses of normal tissue effects in the clinic (QUANTEC): an introduction to the scientific issues. Int J Radiat Oncol Biol Phys. 2010;76:S3–9.
30. Timmerman RD. An overview of hypofractionation and introduction to this issue of seminars in radiation oncology. Semin Radiat Oncol. 2008;18(4):215–22.
31. Benedict SH, et al. Stereotactic body radiation therapy: the report of AAPM task group 101. Med Phys. 2010;37(8):4078–101.
32. Milano MT, et al. Single- and multi-fraction stereotactic radiosurgery dose tolerances of the optic pathways. Int J Radiat Oncol Biol Phys. 2018; https://doi.org/10.1016/j.ijrobp.2018.01.053.

23 Vestibular Schwannomas

Alexander Muacevic, Paul Windisch, Joerg-Christian Tonn, Christoph Fürweger, Berndt Wowra, Markus Kufeld, and Christian Schichor

23.1 Introduction

Stereotactic radiosurgery (SRS) is becoming increasingly popular, mostly for small vestibular schwannomas (VS), due to its treatment efficiency and ease of use compared to surgical tumor resection. However, long-term data with reasonably large patient cohorts are lacking, and few quality of life assessments after SRS for VS has been presented so far [1, 2]. Moreover, most published series comprise smaller heterogeneous patient groups treated with inconsistent radiosurgical techniques and doses [3–5].

While there has been a tendency to treat ever smaller and sometimes asymptomatic tumors, concerns have been raised about long-term auditory toxicity [6] and cases of suspected malignant transformation [7]. Here we report the functional outcome, local tumor control, and quality of life after SRS for VS of a large group of patients for up to 10 years and take into consideration the published literature.

A. Muacevic (✉) · P. Windisch · B. Wowra
M. Kufeld
European CyberKnife Center, Munich, Germany
e-mail: alexander.muacevic@cyber-knife.net; info@cyber-knife.net

J.-C. Tonn · C. Schichor
Department of Neurosurgery, Ludwig Maximilian University, Munich, Germany

C. Fürweger
European CyberKnife Center, Munich, Germany

Department of Stereotaxy and Functional Neurosurgery, University of Cologne, Faculty of Medicine and University Hospital Cologne, Cologne, Germany

23.2 Materials and Methods

Treatment records of 1378 patients with 1384 VS treated with CyberKnife-based SRS (Accuray Inc., Sunnyvale, CA, USA), at the European CyberKnife Center in Munich between 2005 and 2018, were collected in a database for SRS [8]. CyberKnife is a frameless, image-guided robotic SRS system [9]. The therapeutic radiation is generated by a 6-MV compact linear accelerator mounted on a six-axis robotic manipulator. In a typical VS treatment, 100–200 non-isocentric, non-coplanar beams are directed at the tumor. Intra-fraction patient motion is compensated by the automatic adaptation of beam directions based on stereoscopic X-ray images of the patient's skull acquired periodically during treatment. Patients who received SRS as a treatment for recurrence after previous radiotherapy were excluded. Two cases where the tumor was considered a surgery-induced metastasis and four cases where patients were treated in more than a single fraction were excluded as well. Nine hundred ninety-six patients with 1002 tumors had at least 1 year of follow-up after SRS and were included for analysis. Follow-ups consisted of a

© Springer Nature Switzerland AG 2020
A. Conti et al. (eds.), *CyberKnife NeuroRadiosurgery*,
https://doi.org/10.1007/978-3-030-50668-1_23

clinical examination and magnetic resonance imaging (MRI). Audiograms were recorded by otorhinolaryngologists elsewhere and added to our database during follow-up.

Follow-ups were performed after 6 months, every year for 2 years, and every 2 years thereafter. Tumor response was assessed by MRI. Shrinkage and no change in size were scored as locally controlled disease. Increased size in two consecutive follow-ups was interpreted as a local recurrence.

Facial nerve palsy was assessed using the House-Brackmann (HB) score.

Hearing function and ototoxicity were assessed using bilateral serial pure tone audiometry as described previously [10]. Only patients with testable hearing prior to SRS (defined as Gardner-Robertson class 1–4) were included in the analysis.

First, to determine the overall hearing loss, bilateral serial pure tone audiometry was performed including the frequencies 0.5, 1, 2, 4, and 8 kHz. Then, the net hearing loss was calculated at each frequency as the difference between the hearing thresholds of the healthy ear and the affected ear. The mean of the net hearing loss values at the frequencies of 0.5, 1, 2, 4, and 8 kHz defined the overall hearing loss in decibels (dB). Hearing loss attributable to radiosurgery was calculated, as the difference between overall hearing loss at the time of radiosurgery and during follow-up. Worsening of hearing loss attributable to radiosurgery between SRS and follow-up by more than 20 dB was defined as ototoxicity.

Patient-reported quality of life (PR-QoL) was assessed via the 4-week recall SF-12v2® (German Version 2.0, Optum, Inc., Eden Prairie, MN, USA) which has been used and evaluated in numerous different populations [11, 12] and covers eight concepts: physical functioning (PF), role physical (RP), bodily pain (BP), general health (GH), vitality (VT), social functioning (SF), role emotional (RE), and mental health (MH). SF-12 health surveys were evaluated with SF Health Outcomes Scoring Software (QualiMetric Inc., Lincoln, RI, USA). Higher scores on the 0–100 scale represent improved quality of life.

Statistical analysis was performed with the Statistical Package for Social Sciences (SPSS) v. 23.0 (IBM SPSS Statistics, Armonk, NY, USA) and Prism v. 8.0 (GraphPad, San Diego, CA, USA). The significance of time to event data was assessed using the Cox proportional hazards model and the log-rank test. Variables tested for predictive significance concerning local recurrence were age, sex, side of the tumor, NF2 status, prior surgery, tumor volume, and radiosurgical prescription dose. In the case of toxicity analysis, tumor recurrence was also included as a variable in the models, and multivariate analysis was performed accordingly. Local control was plotted as a Kaplan-Meier survival curve for each variable of interest. Continuous variables were split into two groups at their respective median. All data was gathered in accordance with the World Medical Association Declaration of Helsinki.

23.3 Results

23.3.1 Patient Characteristics

Patient characteristics are depicted in Table 23.1. The median age at SRS was 55.1 years (range, 15.1–85.2 years), and median follow-up was 3.6 years (1–12.5 years). All tumors were treated in a single fraction, with a median prescription dose of 13 Gy (11.5–15 Gy). The median prescription isodose line was 65% (55–80%). Tumors that had undergone surgical resection prior to SRS received a median dose of 13.5 Gy. While 827 tumors (82.5%) had not been treated previously, 175 tumors (17.5%) had undergone surgical resection. Of those 175 tumors, 39 received SRS due to subtotal resection, while the remaining 136 schwannomas had recurred. Median tumor volume was 0.61 cm^3 (0.03–13.5 cm^3). Thirty-one tumors (3.1%) were NF2-associated.

23.3.2 Tumor Control

Three-, 5-, and 10-year follow-up data was available for 609, 321, and 48 tumors, respectively, showing Kaplan-Meier estimates for local con-

Table 23.1 Patient characteristics

Number of patients	996	
Number of tumors	1002	
Localization		
Left	530	52.7%
Right	472	46.9%
Sex		
Male	459	46.1%
Female	537	53.7%
Median age (yr)	55.1	(15.1–85.2)
Pre-treatment		
None	827	82.2%
Surgery (residual tumor)	39	3.9%
Surgery (local recurrence)	136	13.5%
Follow-up		
Median (years)	3.6	(1.0–12.5)
≥1 year	1002	
≥3 years	609	
≥5 years	321	
≥10 years	48	
NF2-associated tumors	31	3.1%
Median tumor volume (cm^3)	0.61	(0.03–13.5)
Median dose (Gy)	13	(11.5–15)
Median isodose (%)	65	(55–80)

Numbers in parentheses denote ranges if not specified otherwise

trol of 96.6% (95% CI: 94.9%–97.7%), 92.3% (95% CI: 89.8%–94.3%), and 90.8% (95% CI: 87.2%–93.9%) (Fig. 23.1).

Tumor volume was a significant predictor of local control with larger volumes being associated with worse control in both the Cox proportional hazards model and the log-rank test (Table 23.2). When splitting the tumors in two groups at the median (0.61 cm^3), Kaplan-Meier-estimated local control at 3, 5, and 10 years was 97.4%, 94.4%, and 94.4% for smaller and 95.7%, 90.3%, and 87.7% for larger tumors (Fig. 23.1). Age, sex, side, NF2 status, and dose were not predictive of local control (Fig. 23.2). Surgery prior to SRxfS was only significant in the univariate analysis.

Creating three subgroups of tumor volumes (<0.5 cm^3, 0.5–2 cm^3, >2 cm^3) revealed that the smallest tumors showed significantly improved local control compared to both the middle-sized ($p = 0.0153$) and larger schwannomas ($p = 0.038$) who, in turn, did not differ significantly from each other.

Of the 49 patients who experienced tumor recurrence, 13 received an additional CyberKnife treatment at a median of 3.2 years (2–8.5 years) after initial SRS, while 16 underwent surgery. The remaining 20 cases were either very recent so that the additional therapy had not been documented at the time of this study or the patients were lost to follow-up. The median follow-up for SRS retreatment was 4.6 years (range, 0.5–8.1 years). Local control was achieved in all cases, while no grave toxicity was observed. The median volume of the re-irradiated tumors was 1.94 cm^3 (range, 0.59–5.34 cm^3).

23.3.3 Toxicity

The HB score prior to treatment was available for 997 tumors. Of the 943 cases with good facial function (HB grade I–II) before SRS, 14 (1.5%) experienced worsening to HB grade III–V which was transient in five cases.

In six of the nine cases where worsening of facial nerve function was permanent, the tumor recurred, and three of these recurrences had already been treated with surgical resection. No patient experienced total facial nerve palsy (HB VI) following SRS.

Of the 54 patients with HB grade III–VI prior to SRS, four patients (7.4%) experienced an improvement of facial nerve function to HB I–II, which was permanent in all cases.

Valid audiograms prior to treatment and at 1 year (6–18 months) post-treatment were available for 210 patients. Fifty-five patients had valid pre-treatment and 5-year (48–72 months) post-treatment audiograms. Median hearing loss prior to SRS was 17 dB and increased to 23 dB at 1 year and 29 dB at 5 years post-treatment.

At 1 year, 63 ears (30%) experienced an improvement in hearing compared to the healthy ear, while five patients (2.4%) had no change in hearing deficit and 142 patients (67.6%) experienced worsening. However, only 23 of these patients (10.9%) experienced ototoxicity as defined by an increase of hearing loss ≥20 dB. Results of audiograms at 1-year post-SRS are depicted in Table 23.3.

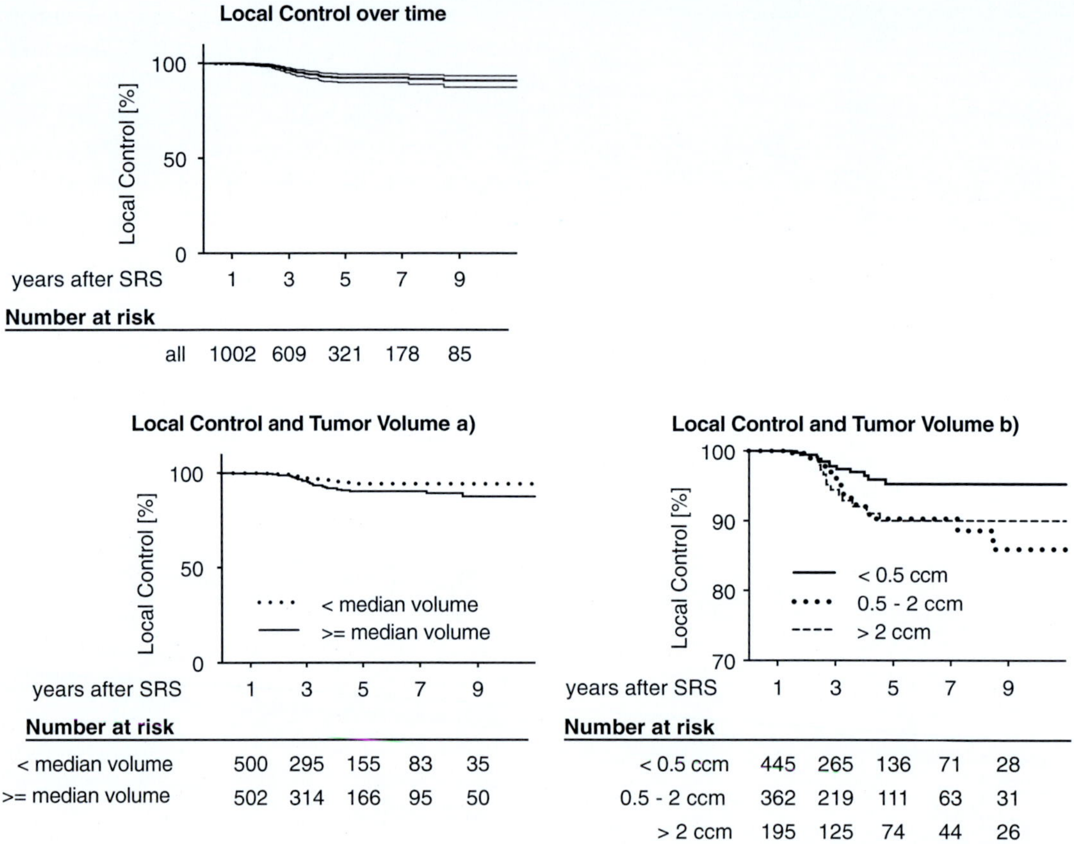

Fig. 23.1 Kaplan-Meier estimates of local control over time. The 3-, 5-, and 10-year local tumor control was 96.6%, 92.3%, and 90.8%, respectively. Thin lines in "local control over time" plot indicate 95% CI intervals. Number of tumors at risk for each subgroup is depicted below each graph. Larger tumors are associated with significantly reduced local control

Table 23.2 Cox proportional hazards model predicting local control reveals tumor volume as the only significant predictive variable

	Hazard ratio	95% CI	*p*	Log-rank
Age (years)	0.99	0.97–1.01	0.169	0.414
Sex (f/m)	1.36	0.76–2.42	0.299	0.217
Side (r/l)	1.23	0.70–2.17	0.464	0.309
NF2	0.25	0.03–1.97	0.19	0.486
Surgery	1.72	0.93–3.16	0.086	**0.035**
Tumor vol (cm^3)	1.16	1.02–1.33	**0.033**	**0.026**
Dmin (Gy)	0.82	0.65–1.02	0.072	0.262

Numbers in parentheses denote ranges if not specified otherwise. **p<0.05**

At 5 years, 12 patients (22.8%) experienced an improvement in hearing compared to the healthy ear, while the remaining 43 patients (78.2%) experienced worsening. However, only 13 (23.6%) of these patients experienced ototoxicity as defined by an increase of hearing loss ≥20 dB post-SRS. Results of audiograms at 5 years post-SRS are depicted in Table 23.4.

Two patients had seizures during their follow-up period without any hints suggesting an association with the tumor or the treatment.

Treatment-associated hydrocephalus requiring shunt implantation could be observed in five

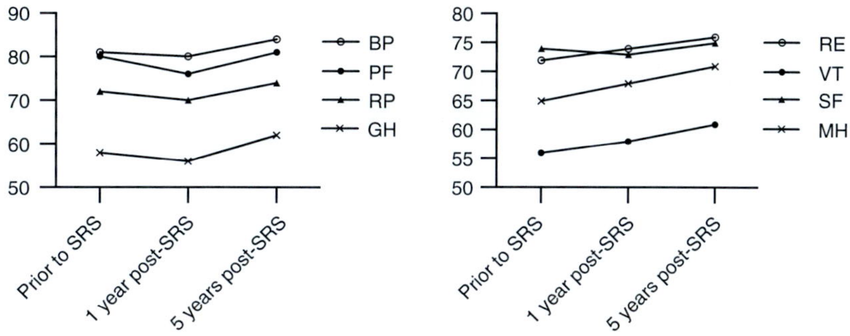

Fig. 23.2 Kaplan-Meier estimates of local control over time. Number of tumors at risk for each subgroup is depicted below each graph

Table 23.3 Hearing toxicity at year one post-SRS

Number of patients	210	
Median HL pre-treatment (dB)	17	(-45 to 72)
Median HL at 1 year (dB)	23	(-40 to 90)
Patients with ototoxicity (HL ≥ 20 dB)	23	10.9%
Patients with improved hearing	63	30.0%
Patients with worsened hearing	142	67.7%
Patients with unchanged hearing	5	2.4%

Numbers in parentheses denote ranges if not specified otherwise

Table 23.4 Hearing toxicity at year five post-SRS

Number of patients	55	
Median HL pre-treatment (dB)	17	(-64 to 53)
Median HL at 1 year (dB)	29	(-7 to 91)
Patients with ototoxicity (HL ≥ 20 dB)	13	23.6%
Patients with improved hearing	12	21.5%
Patients with worsened hearing	43	78.2%
Patients with unchanged hearing	0	0.0%

Numbers in parentheses denote ranges if not specified otherwise

patients (0.5%). One patient developed hydrocephalus due to local recurrence and received shunt implantation combined with microsurgical resection. Median tumor volume of patients with treatment-associated hydrocephalus requiring shunt implantation was 3.38 cm^3 (0.27–7.88 cm^3) with five of the six tumors being larger than 2.4 cm^3.

Thirty-one patients (3.1%) who reported no trigeminal sensory dysfunction at treatment started to present symptoms during follow-up examinations. However, these were permanent in only five patients (0.5%) as defined by having trigeminal sensory dysfunction at each patient's most recent respective follow-up.

No case of malignant tumor transformation was observed.

23.3.4 Patient-Reported Quality of Life

PR-QoL was available for 801, 542, and 226 patients before and at 1 and 5 years after treatment, respectively. Results are depicted in Table 23.5 and Fig. 23.3. At 1 year, PF showed significant reduction and MH significant improvement, while none of the other categories reached significance. GH, VT, RE, and MH showed significant improvement after 5 years. PF, RP, BP, and SF improved slightly without reaching significance at the 5-year follow-up.

23.4 Discussion

23.4.1 Local Tumor Control

As most of the studies on long-term safety and efficacy of SRS for VS are based on different treatment technologies (such as the Gamma Knife), this study, to the best of our knowledge, analyzes the largest patient collectives treated exclusively with CK, as well as the largest collective treated with SRS in general.

Local control results of over 90% even 10 years after treatment are consistent with other studies on the long-term efficacy of SRS for VS [13] using Gamma Knife radiosurgery. Given

Table 23.5 Mean SF-12 scores before and 1 year and 5 years after treatment

	Mean prior to SRS	Standard deviation	1 year post-SRS	Standard deviation	*D* 1 year	*p* (1 year vs prior)	5 years post-SRS	Standard deviation	*D* 5 years	*p* (5 years vs prior)
PF	80.0	26.86	76.0	28.55	−4.1	**0.008**	81.1	27.32	1.1	0.469
RP	71.8	26.23	69.5	25.65	−2.3	0.064	74.3	23.71	2.4	0.350
BP	80.9	26.00	80.0	25.10	−0.9	0.228	83.7	22.67	2.8	0.316
GH	58.2	22.24	56.3	22.36	−1.9	0.116	62.0	22.06	3.8	**0.021**
VT	56.5	24.74	57.9	23.37	1.4	0.363	60.7	24.06	4.3	**0.022**
SF	73.6	27.57	73.2	26.20	−0.4	0.541	75.4	26.75	1.8	0.409
RE	72.0	26.58	73.6	25.40	1.6	0.410	76.4	24.08	4.5	**0.045**
MH	65.0	20.76	67.9	19.28	2.9	**0.019**	70.6	20.61	5.6	**<0.001**

BP Bodily pain, *PF* Physical functioning, *RP* Role physical, *GH* General health, *RE* Role emotional, *VT* Vitality, *SF* Social functioning, *MH* Mental health. Significance was assessed using Mann–Whitney *U*-test to compare pre-treatment and 1-year/5-year follow-up scores

that in some cases local recurrence is diagnosed within less than 2 years following treatment, the actual tumor control might be even higher, as pseudoprogression is a frequent cause of volume change after SRS for VS, especially in the first 24 months post-treatment [14].

The association of reduced local control with increased tumor volumes has been the subject of ongoing discussion [15]. Analyzing local control for tumors smaller and larger than the median volume (0.61 cm^3) resulted in local control of 97.4%, 94.4%, and 94.4% for the smaller and 95.7%, 90.3%, and 87.7% for the larger group at 3, 5, and 10 years which corresponds to a recent study by Ruess et al. where for a group of 335 patients with a median tumor volume of 1.1 cm^3, local control was 89% and 87% at 5 and 10 years, respectively [16].

These differences, especially in long-term control, should be considered when deciding to place smaller tumors under surveillance.

The finding that previous surgery is a significant factor predicting local control in the univariate but not in the multivariate analysis might be due to the fact that tumors that had already undergone surgery before the SRS have significantly larger volumes (median volume 1.31 cm^3, $p < 0.0001$).

The missing dose effect on tumor control that has been reported by previous publications including one from this institution [10, 17] could be explained by the narrow dose range as 963 of 1002 tumors were irradiated with 12.5–13.5 Gy.

Contrary to most research on SRS for VS, NF2 was not associated with reduced local control. Given the limited number of NF2-associated tumors ($n = 31$) in this study (but also in many other publications), drawing definite conclusions from this finding is difficult [18–20]. However, the median tumor volume for NF2-associated tumors was 0.82 cm^3 which is considerably smaller than in many existing publications [20]. Mathieu et al. report local control rates of 85% and 81% at 5 and 10 years following Gamma Knife radiosurgery of NF2-associated VS for 74 tumors with a mean volume of 5.7 cm^3 [19].

It could therefore be hypothesized that the effect of reduced local control associated with NF2 loses predictive significance once tumors have been irradiated at a sufficiently small volume. Since many studies on schwannomas associated with NF2 already report tumor volume as a predictor of local control [18, 19], subgroup analyses of these collectives could answer the question on whether treatment of small NF2-associated tumors may determine local control rates and is equally good as the treatment of sporadic VS.

23.4.1.1 Retreatment

In several cases where the initial SRS treatment could not stop tumor growth, patients were suitable to receive SRS retreatment. Although there is still very little data on retreatment with SRS for VS (which makes it difficult to assess the risk associated with radiation dose accumulation),

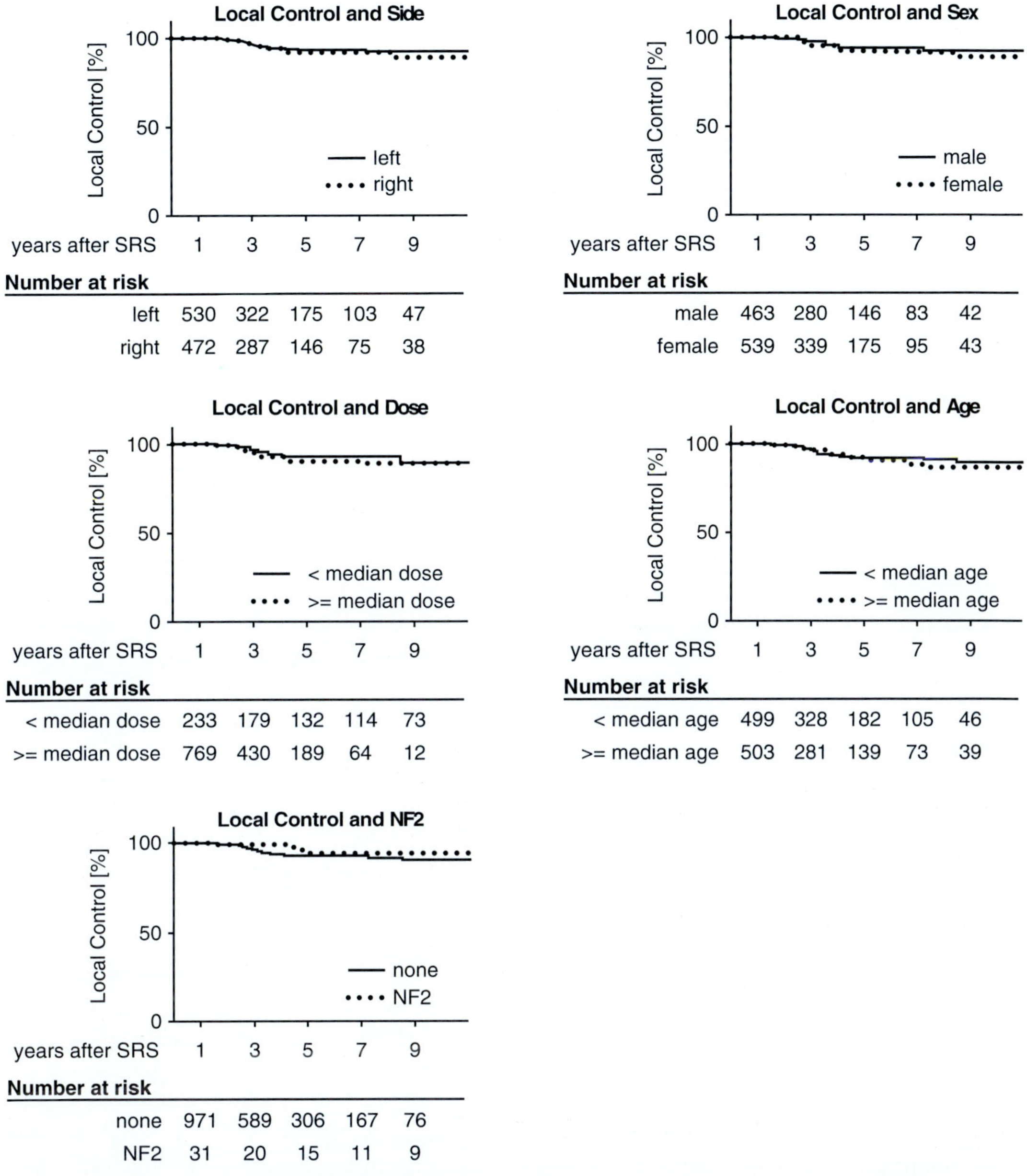

Fig. 23.3 Mean SF-12 scores before and 1 year and 5 years after treatment. Data is split into two graphs for visibility. Left: *BP* Bodily pain, *PF* Physical functioning, *RP* Role physical, *GH* General health. Right: *RE* Role emotional, *VT* Vitality, *SF* Social functioning, *MH* Mental health

retreatment seems to be a safe and effective option. Others also reported good tumor control on Gamma Knife withdrawal but noted a slightly higher rate of facial nerve toxicity, at least compared to the group of patients who received SRS as an initial treatment in this study [21].

23.4.2 Hearing

A meta-analysis of hearing outcomes following SRS for VS and a study by Santa Maria et al. comprising 344 patients with audiograms and more than 3 years of follow-up reported hearing preservation of 51% and 50%, respectively, at 3 years post-treatment [6, 22].

Even though ototoxicity as defined in this study only occurred in 23.2% of cases, there is a risk of underestimating the extent of ototoxicity, as patients who have no residual hearing on the treated (or both) ears might stop doing audiograms as part of their follow-up and as a reduction in hearing capacity of the healthy ear, which reduces the difference between the healthy and the affected ear. This should be considered when irradiating very small tumors that have not shown significant growth in an attempt to save the patient's hearing.

However, if tumor growth is present, hearing function has been reported to decline fairly quickly with patients often losing serviceable hearing within the first 5 years if the tumor has not been treated [23].

Therefore, in the case of small tumors and good hearing, we suggest to monitor the hearing function closely at 6-month intervals and suggest treatment only when a hearing decline can be documented and/or the tumor is growing.

23.4.3 Facial Nerve Toxicity

While facial nerve toxicity is rare compared to hearing toxicity, it can severely impact the patient's quality of life. A meta-analysis covering 1908 patients by Yang et al. described an association between lower marginal doses of 13 Gy or less and reduced facial nerve toxicity for Gamma Knife radiosurgery [24].

However, reducing the dose may result in tumor recurrence which was observed in the majority of cases with facial nerve toxicity in this study. Additionally, increased tumor volume was associated with higher rates of facial nerve toxicity. Overall, the authors reported a facial nerve preservation rate of 96.2%.

In a study of facial nerve function after translabyrinthine vestibular schwannoma surgery on 392 patients, 81% had HB grade I–II 1 year after surgery, while 12 patients experienced total facial nerve palsy (HB grade VI) [25].

Falcioni et al. reported anatomical interruption of the facial nerve in 48 out of 1151 cases. Thirty-five percent of the remaining cases where the facial nerve could be preserved had a postoperative HB grade III or worse. Smaller tumors had better facial nerve outcome with postoperative HB grade III or worse occurring in 14% of 444 patients with tumor diameters of less than 1 cm [26].

23.4.4 Hydrocephalus

A study by Lee et al. reported a hydrocephalus incidence of 4.1% for 702 patients treated with Gamma Knife radiosurgery and found age, tumor origin, and tumor volume to be significant predictors [27]. The higher incidence could be due to a higher mean tumor volume of 3.6 cm^3 compared to 1.25 cm^3 in this study. Median age of the patients who developed hydrocephalus in this study was 55.9 years (48.4–74.6 years), only marginally higher than the median age of the whole collective. Hydrocephalus is also a rare complication when treating VS surgically.

23.4.5 Patient-Reported Quality of Life

PR-QoL data following SRS for VS is lacking, especially as baseline measurements are frequently missing which is why existing studies report good quality of life comparable to the general population [1]. In contrast, the data in this study suggests the ability of CK-based SRS to significantly improve the long-term quality of life of VS patients in several categories. A possible explanation for the reduction in quality of life in PF at the 1-year follow-up could be the development of transient treatment-related toxicity such as trigeminal sensory dysfunction which spontaneously regressed in more than 83% (26/31) of cases in this study.

The good PR-QoL outcome is in accordance with other studies on the functional outcome of SRS for VS. Régis et al. report no functional deterioration after SRS for VS in 91% of 97 patients treated with Gamma Knife radiosurgery [28].

Limitations of quality of life assessment in this study include the use of a general health instead of a VS-specific survey and the fact that patients who suffered from recurrence and thereby reduced quality of life might have stopped attending their follow-up.

23.5 Conclusion

SRS is a safe and effective treatment option for treating VS with tolerable toxicities and a favorable PR-QoL over time. Though additional and particularly prospective studies are desirable, SRS should be considered as a primary treatment option for small- and medium-sized vestibular schwannomas.

Acknowledgments Parts of this manuscript have been presented at the annual meeting of German CyberKnife centers in Cologne 2018 and have been published in the *Cureus Journal of Medical Science* (https://www.cureus.com/articles/25873).

References

1. Berkowitz O, Han Y-Y, Talbott EO, et al. Gamma knife radiosurgery for vestibular schwannomas and quality of life evaluation. Stereotact Funct Neurosurg. 2017;95(3):166–73.
2. Myrseth E, Pedersen P-H, Møller P, Lund-Johansen M. Treatment of vestibular schwannomas. Why, when and how? Acta Neurochir. 2007;149(7):647–60; discussion 660.
3. Watanabe S, Yamamoto M, Kawabe T, et al. Stereotactic radiosurgery for vestibular schwannomas: average 10-year follow-up results focusing on long-term hearing preservation. J Neurosurg. 2016;125(Suppl 1):64–72.
4. Rutten I, Baumert BG, Seidel L, et al. Long-term follow-up reveals low toxicity of radiosurgery for vestibular schwannoma. Radiother Oncol. 2007;82(1):83–9.
5. Mahboubi H, Sahyouni R, Moshtaghi O, et al. CyberKnife for treatment of vestibular schwannoma: a meta-analysis. Otolaryngol Head Neck Surg. 2017;157(1):7–15.
6. Santa Maria PL, Shi Y, Gurgel RK, et al. Long-term hearing outcomes following stereotactic radiosurgery in vestibular schwannoma patients-a retrospective cohort study. Neurosurgery. 2019;85(4):550–9. https://doi.org/10.1093/neuros/nyy407.
7. Maducdoc MM, Ghavami Y, Linskey ME, Djalilian HR. Evaluation of reported malignant transformation of vestibular schwannoma: de novo and after stereotactic radiosurgery or surgery. Otol Neurotol. 2015;36(8):1301–8.
8. Kufeld M, Fürweger C, Drexler CG, Wowra B, Muacevic A. Implementation of a medical database system for a radiosurgery center. Cureus. 2009;1(12):e4. https://assets.cureus.com/uploads/review_article/pdf/2/1511801903-20171127-2074-tvu3b1.pdf.
9. Adler JR Jr, Chang SD, Murphy MJ, Doty J, Geis P, Hancock SL. The Cyberknife: a frameless robotic system for radiosurgery. Stereotact Funct Neurosurg. 1997;69(1–4 Pt 2):124–8.
10. Wowra B, Muacevic A, Fürweger C, Schichor C, Tonn J-C. Therapeutic profile of single-fraction radiosurgery of vestibular schwannoma: unrelated malignancy predicts tumor control. Neuro Oncol. 2012;14(7):902–9.
11. Resnick B, Nahm ES. Reliability and validity testing of the revised 12-item short-form health survey in older adults. J Nurs Meas. 2001;9(2):151–61.
12. Lim LL, Fisher JD. Use of the 12-item short-form (SF-12) health survey in an Australian heart and stroke population. Qual Life Res. 1999;8(1–2):1–8.
13. Hasegawa T, Kida Y, Kato T, Iizuka H, Kuramitsu S, Yamamoto T. Long-term safety and efficacy of stereotactic radiosurgery for vestibular schwannomas: evaluation of 440 patients more than 10 years after treatment with gamma knife surgery. J Neurosurg. 2013;118(3):557–65.
14. Hayhurst C, Zadeh G. Tumor pseudoprogression following radiosurgery for vestibular schwannoma. Neuro Oncol. 2012;14(1):87–92.
15. Huang C-W, Tu H-T, Chuang C-Y, et al. Gamma knife radiosurgery for large vestibular schwannomas greater than 3 cm in diameter. J Neurosurg. 2018;128(5):1380–7.
16. Rueß D, Pöhlmann L, Hellerbach A, et al. Acoustic neuroma treated with stereotactic radiosurgery: follow-up of 335 patients. World Neurosurg. 2018;116:e194–202.
17. Anniko M, Arndt J, Norén G. The human acoustic neurinoma in organ culture. II. Tissue changes after gamma irradiation. Acta Otolaryngol. 1981;91(3–4):223–35.
18. Kruyt IJ, Verheul JB, Hanssens PEJ, Kunst HPM. Gamma knife radiosurgery for treatment of growing vestibular schwannomas in patients with neurofibromatosis type 2: a matched cohort study with sporadic vestibular schwannomas. J Neurosurg. 2018;128(1):49–59.
19. Mathieu D, Kondziolka D, Flickinger JC, et al. Stereotactic radiosurgery for vestibular schwannomas in patients with neurofibromatosis type 2: an analysis of tumor control, complications, and hearing preser-

vation rates. Neurosurgery. 2007;60(3):460–8; discussion 468–470.
20. Sharma MS, Singh R, Kale SS, Agrawal D, Sharma BS, Mahapatra AK. Tumor control and hearing preservation after gamma knife radiosurgery for vestibular schwannomas in neurofibromatosis type 2. J Neurooncol. 2010;98(2):265–70.
21. Fu VX, Verheul JB, Beute GN, et al. Retreatment of vestibular schwannoma with gamma knife radiosurgery: clinical outcome, tumor control, and review of literature. J Neurosurg. 2018;129(1):137–45.
22. Yang I, Sughrue ME, Han SJ, et al. A comprehensive analysis of hearing preservation after radiosurgery for vestibular schwannoma. J Neurosurg. 2010;112(4):851–9.
23. Kondziolka D, Mousavi SH, Kano H, Flickinger JC, Lunsford LD. The newly diagnosed vestibular schwannoma: radiosurgery, resection, or observation? Neurosurg Focus. 2012;33(3):E8.
24. Yang I, Sughrue ME, Han SJ, et al. Facial nerve preservation after vestibular schwannoma gamma knife radiosurgery. J Neurooncol. 2009;93(1):41–8.
25. Brackmann DE, Cullen RD, Fisher LM. Facial nerve function after translabyrinthine vestibular schwannoma surgery. Otolaryngol Head Neck Surg. 2007;136(5):773–7.
26. Falcioni M, Fois P, Taibah A, Sanna M. Facial nerve function after vestibular schwannoma surgery. J Neurosurg. 2011;115(4):820–6.
27. Lee S, Seo S-W, Hwang J, et al. Analysis of risk factors to predict communicating hydrocephalus following gamma knife radiosurgery for intracranial schwannoma. Cancer Med. 2016;5(12):3615–21.
28. Régis J, Pellet W, Delsanti C, et al. Functional outcome after gamma knife surgery or microsurgery for vestibular schwannomas. J Neurosurg. 2002;97(5):1091–100.

Large Vestibular Schwannomas

24

Alfredo Conti

24.1 Theoretical Bases

The single-fraction, high-dose radiosurgery, according to radiobiological criteria, represents the best option for slow-growing lesions that respond late to radiation therapy like vestibular schwannomas (VS) [1, 2]. These lesions respond to a single, high-dose radiation better than to a conventionally fractionated radiotherapy (50/55 Gy in 25/30 fractions). Late responders have a low α/β ratio, where the coefficient α represents a lethal single-impact injury and coefficient β a lethal injury due to accumulation of sublethal doses.

Single-fraction radiosurgery has demonstrated its efficacy in thousands of patients with VS. Nonetheless, some results are not fully satisfactory, such as the 10-year actuarial hearing preservation rate that has been measured as low as 44.5%, with hearing loss that may develop as late as 6 years after treatment. This has not been significantly improved by modern radiosurgical techniques and lower marginal doses that are delivered today as compared to the past [3]. Furthermore, single-fraction radiosurgery is reserved to small- to medium-sized lesions.

The current standard therapeutic dose (12–14 Gy) may be too high to be tolerated by healthy surrounding nervous structures, such as the brainstem, that are in direct contact with large schwannomas. Furthermore, there is a direct correlation between tumor size and facial nerve damage.

The lower dose per fraction used in fractionated stereotactic radiotherapy (FSRT) is, in theory, less harmful for the surrounding healthy structures allowing for higher rates of hearing, facial, and trigeminal nerve preservation, especially in such large lesions.

The FSRT is usually delivered using a linear accelerator (LINAC)-based system with a relocatable (not invasively fixed to the patient head) stereotactic frame for the patient. Multileaf or micro-multileaf collimators (micro-MLC) have replaced traditional cylindrically shaped collimators. This allows the conformal shaping of the radiation beam while it is delivered with dose distributions that are much more favorable. Modern LINACs such as Novalis (BrainLAB AG, Heimstetten, Germany), Versa HD (Elekta AB, Stockholm, Sweden), and TrueBeam STx and Edge (Varian, Medical Systems, Palo Alto, CA) have image guidance and dosimetric characteristics that make them suitable for FSRT applications. Image guidance and sophisticated dose delivery techniques offer the opportunity to perform "hypofractionated" stereotactic radiotherapy (hSRT) (usually delivered in 2–5 fractions).

A. Conti (✉)
Alma Mater Studiorum University of Bologna, Bologna, Italy

IRCCS ISNB Istituto delle Scienze Neurologiche di Bologna, Bologna, Italy

Department of Neurosurgery, Charité Faculty of Medicine, Berlin, Germany
e-mail: alfredo.conti2@unibo.it

© Springer Nature Switzerland AG 2020
A. Conti et al. (eds.), *CyberKnife NeuroRadiosurgery*,
https://doi.org/10.1007/978-3-030-50668-1_24

The CyberKnife (Accuray, Sunnyvale, CA) is a frameless system for stereotactic radiosurgery system [4–8]. It uses a lightweight high-energy radiation source and a robotic delivery system to deliver SRS in single or multiple sessions.

Such extremely hypofractionated treatments allow the delivery of radiosurgical doses still ablative to the lesion in association with advanced protection of adjacent tissues that receive lower doses and lower dose rates but also have precious time to recover between the fractions (Fig. 24.1).

24.2 Clinical Outcome

24.2.1 Tumor Control

With the limitation that very long-term results are not currently available, the tumor control rates of FSRT are similar to those of SRS. In many different series, the reported tumor control rate at medium term is >90% [9–16].

The first report in the literature of hSRT is from Lederman and colleagues [17] in 1997 with four or five weekly fractions to a total dose of 20 Gy. They described 100% tumor control with no permanent cranial neuropathies. Similarly, good results have been reported by the Stanford group [14, 18].

The Stanford group's initial experience with hypofractionation for vestibular schwannoma was with the rigid fixation of an SRS frame with three fractions spaced 8 h apart to a total dose of 21 Gy [14]. With this technique, they reported a 97% tumor control with a 77% hearing reservation rate at a median follow-up interval of 24 months.

Later, the Stanford group used the frameless CyberKnife system to deliver hSRT [19]. With this technique, they treated patients with three 6 Gy fractions delivered on consecutive days to a total dose of 18 Gy. In the latest report [20], the group led by Adler reported the results of 383 patients with median follow-up duration of 3.6 years (ranging 1–10 years) and median tumor volume of 1.1 cm^3 (range 0.02–19.8 cm^3). Ninety percent of patients were treated to 18 Gy in three fractions, while 9.6% were treated to 21 Gy in three fractions. The 3- and 5-year Kaplan-Meier resection/repeat SRS-free tumor control rates were 99% and 96%, respectively [20].

In a systematic review of 11 studies reporting data on treatment outcomes with CyberKnife for VS including 800 patients studied between 1998 and 2012, the reported collective mean tumor control rate was 96.3% (95% CI: 94.0%–98.5%) [21].

Hypofractionated treatments can be adopted in large lesions when the patient is not a candidate for surgical treatment. In one study on large vestibular schwannomas (mean 3.3 cm; range 2.5–5.0 cm), 85% tumor control at 5 years and 80% at 10 years were reported [22]. Casentini et al. [23] reported results of 33 patients with large VS (median volume 9.4 cm^3; range 8–24 cm^3) who were treated by multi-session CyberKnife. The treatment was delivered in two to five fractions with 14–19.5 Gy to 70–85%. Actuarial progression-free survival rates at 1 year and 5 years were 97% and 83%, respectively. Hearing was also preserved in seven of the eight patients with serviceable baseline hearing. Vertigo, tongue paresthesia, and trigeminal neuralgia were recorded in one case each. Similar results were reported by Teo et al. [24] on 30 Koos grade IV tumors, of which 19 had undergone primary CyberKnife and 11 were >3 cm after previous resection. Patients were treated by a median of three fractions at 18 Gy. Overall, 80% of large VSs were adequately controlled by CK with 97 months of median follow-up [24].

According to available data and radiobiological calculations [23], a dose of 6 Gy × 3 fractions may correspond to 11.5 Gy delivered in single fraction. This dose can therefore be considered the lower limit for the treatment of medium-sized lesions, while higher doses would be suitable for small tumors.

24.2.2 Hearing Preservation

It has been supposed that hSRT may result in lower toxicity and thus higher rates of hearing preservation, compared to either SRS or surgery [25, 26]. Similar hypofractionated schedules of three to five fractions showed slightly better hearing preservation rates that range from 50% to

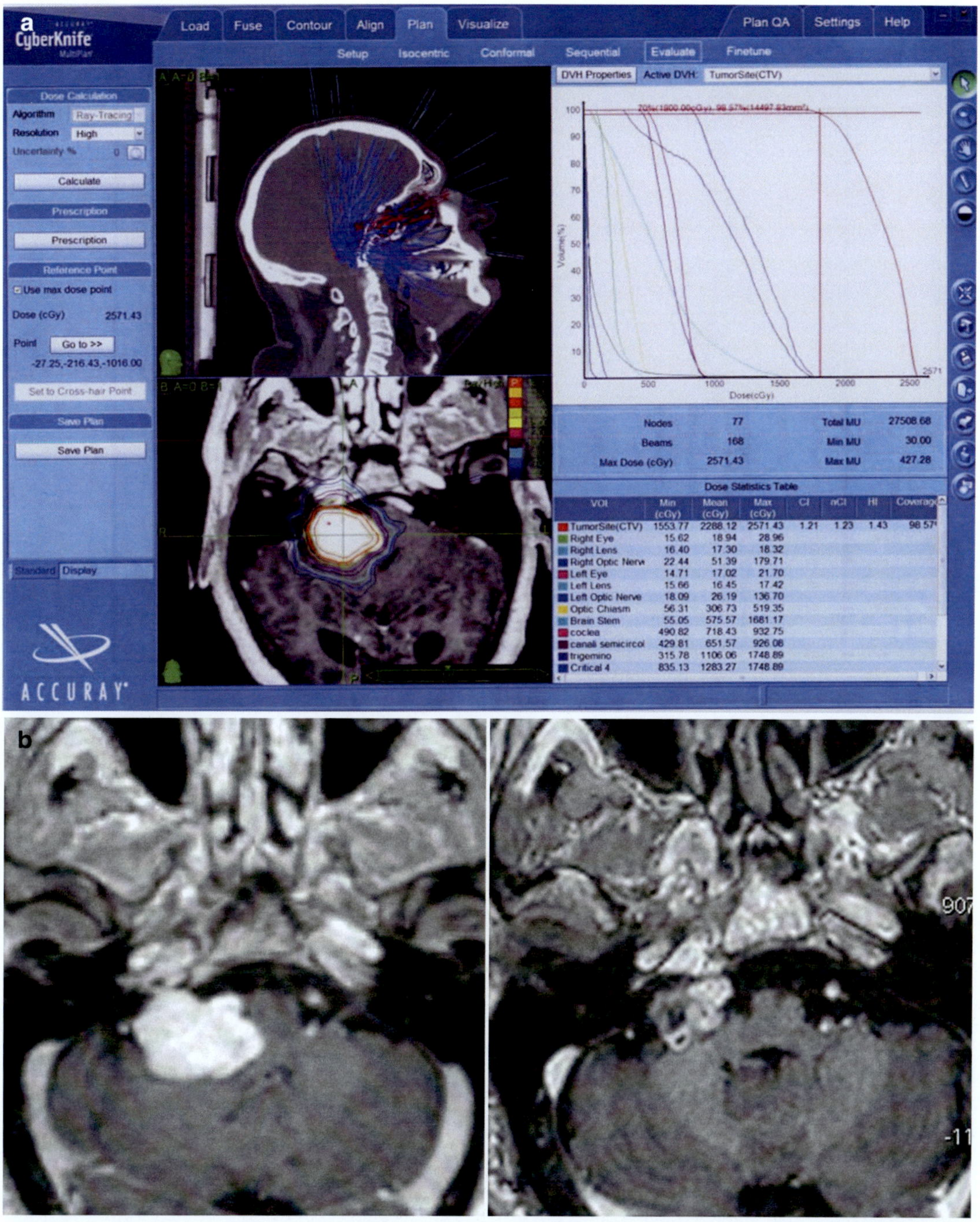

Fig. 24.1 Large vestibular schwannoma treated in three fractions. (**a**) Treatment plan showing that the dose prescribed was 18 Gy at the 70% isodose line. (**b**) Pre- and post-treatment MR showing a significant shrinkage of the tumor and the decompression of the brainstem

93%, compared to 32–81% for SRS-treated lesions [13, 14, 25]. Andrews et al. noted a 2.5-fold higher hearing preservation rate (81%) for FSRT vs. (33%) Gamma Knife SRS [27].

However, hearing preservation rates vary considerably with some studies demonstrating no difference or even poorer hearing outcomes for FSRT compared to SRS [12, 28]. Combs and coll. Evaluated long-term clinical outcome and determined prognostic factors for local control, hearing preservation, and cranial nerve toxicity in a large German multicentric study. 449 patients treated for 451 (VS) with radiosurgery (n = 169; 38%) or fractionated stereotactic radiotherapy (FSRT; n = 291; 62%). After treatment, "useful hearing" was preserved in 85% of the patients. Loss of useful hearing was observed in the FSRT group in 14% and in the SRS group in 16% of the patients. For patients treated with SRS ≤13 Gy, useful hearing deterioration was 13% [29].

A recent review showed an aggregate estimate of 37% for hearing preservation with high variability (95% CI: [19–59%]) in patients treated by hSRT [30], whereas the collective hearing preservation rate was 79.1% (95% CI: 71.0–87.3%) in 427 patients with measurable hearing in a systematic review of CyberKnife hSRT [21].

This disparity in hearing outcomes across hSRT studies may be attributed to the fact that larger lesions are preferentially treated with hSRT, which may result in lower hearing preservation rates.

Many possible prognostic factors of hearing preservation have been proposed, including the Gardner-Robertson grade, radiation dose to the cochlea, transient volume expansion after SRS, length of irradiated cochlear nerve, marginal dose to the tumor, and age. However, we still do not clearly understand why patients lose their hearing after SRS for VS.

Flickinger et al. [31] noted that in SRS-treated VS, increased marginal dose was associated with decreased testable speech discrimination. Age may also play a role in hearing preservation. Fong et al. suggested that LINAC-based SRS may be more beneficial to younger patients, while LINAC-based FSRT may be more beneficial to older patients [31]. Better hearing preservation rates have also been shown to associate with younger age in both FSRT and SRS [32].

The dose to the central cochlea is considered one major factor associated to hearing loss. In SRS treatments, patients who received a radiation dose <4.2 Gy to the central cochlea had significantly better hearing preservation of the same Gardner-Robertson class. Twelve of 12 patients <60 years of age who had received a cochlear dose <4.2 Gy retained serviceable hearing at 2 years post-SRS [33]. Therefore, the maximal dose to the cochlea should be lower than 4 Gy in single fraction to minimize the risk of hearing loss.

The equivalent dose for hSRT is more difficult to establish because of limited clinical data. Based on a dose-response model of clinical data sets, Rashid et al. [34] suggested that the 14 Gy in single-fraction and the 27.5 Gy in a five-fraction limit carry a 17.9% and 17.4% risk of hearing deterioration, respectively, whereas the 12 Gy in single-fraction and the 25 Gy in a five-fraction limit had 11.8% and 13.8% risk, respectively [34].

An interesting study from the Stanford group analyzed this topic [35]. A cochlea dose-volume histogram was generated for each of the 94 patients who were treated with three-fraction hSRT and were qualified for the study. Gardner Robertson grade I–II hearing post-treatment was maintained in 74% of patients (70/94). Larger cochlear volume was associated with lower risk of hearing loss. Controlling for differences in cochlear volume among subjects, each additional mm^3 of cochlea receiving 10–16 Gy (single session equivalent doses of 6.6–10.1 Gy) significantly increased the odds of hearing loss by approximately 5% [35]. The role of cochlear volume, tumor volume, and prescribed cochlear dose in hearing preservation in hSRT is confirmed in a study by Tsai et al [26]

24.2.3 Facial and Trigeminal Nerve Deficits

Many studies using hSRT to treat VS report excellent facial and trigeminal nerve preservation rates of 92–100% [13, 16, 18, 36]. For the single-fraction SRS series with a marginal dose of 12–14 Gy, the 5-year facial nerve preservation rate ranged from 95 to 100%. In the six series [9, 13, 27–29, 37] which directly compare the facial preservation rate of single-fraction SRS (9.7–

16 Gy) versus FSRT, no study reported statistically significant differences in facial preservation rate at 5 years [9, 13, 27–29, 37]. Like hearing preservation, facial nerve preservation is dose-dependent with a higher marginal dose correlated with increased facial nerve palsy development [38]. Furthermore, hSRT has been shown to have marginally better trigeminal nerve preservation [9, 33]. However, these studies are limited by their follow-up time and sample size. One study on CyberKnife hSRT has shown that the degree of tinnitus will not be significantly worsened by the radiation treatment [39].

24.3 Other Complications

Besides the cranial nerve toxicity, two significant complications of radiation therapy of VS are hydrocephalus and radionecrosis. One case of brainstem necrosis after FSRT was reported [40]. The authors analyzed dose distributions and concluded that the dose delivered was within the tolerance limits and therefore cause remains unclear. However, we have to consider that in FSRT the volume of the brainstem receiving high cumulative doses of radiations is rather relevant and therefore a minimal risk does exist. In hSRT the dose distribution is similar to that of single-fraction radiosurgery, and usually the dose to the brainstem is negligible, unless the treated tumor has a large or very large size.

The development of chronic hydrocephalus is a relatively frequent complication. In a recent report on 235 patients treated by single-fraction SRS, 15 (6.38%) exhibited hydrocephalus [41].

A recent paper looking specifically at the risk of hydrocephalus in patients treated with FSRT for VS found an incidence of 11% within 19 months after radiotherapy [42]. On univariate analysis, pretreatment factors predictive of hydrocephalus were maximum tumor diameter, proximity to midline, displacement of the fourth ventricle, partial effacement of the fourth ventricle, contact with the medulla, and more brainstem structures. Age is another important factor that has been associated with the development of hydrocephalus in other series [23, 43].

Therefore, hydrocephalus should be concerned when FSRT and hSRT are used to treat large tumors, with fourth ventricle dislocation in elderly patients.

24.4 Malignant Transformation

As with any treatment involving radiation, malignant transformation is a chief concern. Thirteen cases of radiation-induced brain malignancies have been reported in patients treated with SRS for VS [44]. There are nine cases of radiation-induced malignant peripheral nerve sheath tumor, which appears to be the most common tumor type in this category. The true rate of malignant transformation in VS is unknown. Rowe et al. [45] retrospectively assessed the safety of radiosurgery in 137 patients with NF2 and von Hippel-Lindau disease. A total of 146 VS were treated with radiosurgery. Two patients experienced suspected malignant transformation.

Pollock et al. [46] examined over 1800 patients who underwent initial single-fraction SRS treatment for benign intracranial lesions or arteriovenous malformations at follow-up longer than 5 years. The study noted a 15-year risk of radiation-induced tumor of 0.0% and risk of malignant transformation of 0.3%, occurring in 1 out of 358 treated VS patients.

Of note, the study excluded patients who had an underlying susceptibility to tumor development, which may have contributed to a larger number of malignant transformations. Although no cases of malignant transformation were recorded following hSRT, studies with longer follow-up times may be necessary to provide an accurate evaluation.

Table 24.1 provides some practical suggestions and rates of possible complications for hypofractionated treatment of large vestibular schwannomas.

Table 24.1 Practical suggestions and rates of possible complications for hypofractionated treatment of large vestibular schwannomas

Parameter	Suggestion
Imaging	Contrast-enhanced CT Contrast-enhanced T1 MPRAGE FIESTA/CISS
Maximal volume	4.5 cm Large cystic lesions represent a relative contraindication
Number of fractions	3
Dose to the target	18/19.5 Gy
Dose to the brainstem	16 Gy
Dose to the trigeminal nerve	18 Gy
Dose to the cochlea	10 Gy
Dose to the labyrinth	10 Gy
5-Year expected tumor control	>80%
Expected hearing preservation	70%
Expected tinnitus	Stable
Expected facial nerve deficit	<1%
Expected chronic hydrocephalus	5–10%
Malignant transformation	0.3%

References

1. Liu L, Bassano DA, Prasad SC, Hahn SS, Chung CT. The linear-quadratic model and fractionated stereotactic radiotherapy. Int J Radiat Oncol Biol Phys. 2003;57(3):827–32.
2. Vernimmen FJ, Slabbert JP. Assessment of the alpha/beta ratios for arteriovenous malformations, meningiomas, acoustic neuromas, and the optic chiasma. Int J Radiat Biol. 2010;86(6):486–98.
3. Chopra R, Kondziolka D, Niranjan A, Lunsford LD, Flickinger JC. Long-term follow-up of acoustic schwannoma radiosurgery with marginal tumor doses of 12 to 13 Gy. Int J Radiat Oncol Biol Phys. 2007;68(3):845–51.
4. Adler JR Jr, Chang SD. Cyberknife image-guided radiosurgery. Neurosurgery. 2009;64(2 Suppl):A1.
5. Adler JR Jr, Chang SD, Murphy MJ, Doty J, Geis P, Hancock SL. The Cyberknife: a frameless robotic system for radiosurgery. Stereotact Funct Neurosurg. 1997;69(1–4 Pt 2):124–8.
6. Conti A, Pontoriero A, Siddi F, et al. Post-treatment edema after meningioma radiosurgery is a predictable complication. Cureus. 2016;8(5):e605.
7. Romanelli P, Conti A, Bianchi L, Bergantin A, Martinotti A, Beltramo G. Image-guided robotic radiosurgery for trigeminal neuralgia. Neurosurgery. 2017;83:1023–30.
8. Yu C, Main W, Taylor D, Kuduvalli G, Apuzzo ML, Adler JR Jr. An anthropomorphic phantom study of the accuracy of Cyberknife spinal radiosurgery. Neurosurgery. 2004;55(5):1138–49.
9. Anderson BM, Khuntia D, Bentzen SM, et al. Single institution experience treating 104 vestibular schwannomas with fractionated stereotactic radiation therapy or stereotactic radiosurgery. J Neuro-Oncol. 2014;116(1):187–93.
10. Fuss M, Debus J, Lohr F, et al. Conventionally fractionated stereotactic radiotherapy (FSRT) for acoustic neuromas. Int J Radiat Oncol Biol Phys. 2000;48(5):1381–7.
11. Maire JP, Huchet A, Milbeo Y, et al. Twenty years' experience in the treatment of acoustic neuromas with fractionated radiotherapy: a review of 45 cases. Int J Radiat Oncol Biol Phys. 2006;66(1):170–8.
12. McWilliams W, Trombetta M, Werts ED, Fuhrer R, Hillman T. Audiometric outcomes for acoustic neuroma patients after single versus multiple fraction stereotactic irradiation. Otol Neurotol. 2011;32(2):297–300.
13. Meijer OW, Vandertop WP, Baayen JC, Slotman BJ. Single-fraction vs. fractionated linac-based stereotactic radiosurgery for vestibular schwannoma: a single-institution study. Int J Radiat Oncol Biol Phys. 2003;56(5):1390–6.
14. Poen JC, Golby AJ, Forster KM, et al. Fractionated stereotactic radiosurgery and preservation of hearing in patients with vestibular schwannoma: a preliminary report. Neurosurgery. 1999;45(6):1299–305; discussion 305–7.
15. Shirato H, Sakamoto T, Takeichi N, et al. Fractionated stereotactic radiotherapy for vestibular schwannoma (VS): comparison between cystic-type and solid-type VS. Int J Radiat Oncol Biol Phys. 2000;48(5):1395–401.
16. van de Langenberg R, Dohmen AJ, de Bondt BJ, Nelemans PJ, Baumert BG, Stokroos RJ. Volume changes after stereotactic LINAC radiotherapy in vestibular schwannoma: control rate and growth patterns. Int J Radiat Oncol Biol Phys. 2012;84(2):343–9.
17. Lederman G, Lowry J, Wertheim S, et al. Acoustic neuroma: potential benefits of fractionated stereotactic radiosurgery. Stereotact Funct Neurosurg. 1997;69(1–4 Pt 2):175–82.
18. Chang SD, Gibbs IC, Sakamoto GT, Lee E, Oyelese A, Adler JR Jr. Staged stereotactic irradiation for acoustic neuroma. Neurosurgery. 2005;56(6):1254–61; discussion 61–3.
19. Sakamoto GT, Blevins N, Gibbs IC. Cyberknife radiotherapy for vestibular schwannoma. Otolaryngol Clin N Am. 2009;42(4):665–75.
20. Hansasuta A, Choi CY, Gibbs IC, et al. Multisession stereotactic radiosurgery for vestibular schwannomas:

single-institution experience with 383 cases. Neurosurgery. 2011;69(6):1200–9.
21. Mahboubi H, Sahyouni R, Moshtaghi O, et al. CyberKnife for treatment of vestibular schwannoma: a meta-analysis. Otolaryngol Head Neck Surg. 2017;157(1):7–15.
22. Ishihara H, Saito K, Nishizaki T, et al. CyberKnife radiosurgery for vestibular schwannoma. Minim Invasive Neurosurg. 2004;47(5):290–3.
23. Casentini L, Fornezza U, Perini Z, Perissinotto E, Colombo F. Multisession stereotactic radiosurgery for large vestibular schwannomas. J Neurosurg. 2015;122(4):818–24.
24. Teo M, Zhang M, Li A, et al. The outcome of hypofractionated stereotactic radiosurgery for large vestibular schwannomas. World Neurosurg. 2016;93:398–409.
25. Morimoto M, Yoshioka Y, Kotsuma T, et al. Hypofractionated stereotactic radiation therapy in three to five fractions for vestibular schwannoma. Jpn J Clin Oncol. 2013;43(8):805–12.
26. Tsai JT, Lin JW, Lin CM, et al. Clinical evaluation of CyberKnife in the treatment of vestibular schwannomas. Biomed Res Int. 2013;2013:297093.
27. Andrews DW, Suarez O, Goldman HW, et al. Stereotactic radiosurgery and fractionated stereotactic radiotherapy for the treatment of acoustic schwannomas: comparative observations of 125 patients treated at one institution. Int J Radiat Oncol Biol Phys. 2001;50(5):1265–78.
28. Collen C, Ampe B, Gevaert T, et al. Single fraction versus fractionated linac-based stereotactic radiotherapy for vestibular schwannoma: a single-institution experience. Int J Radiat Oncol Biol Phys. 2011;81(4):e503–9.
29. Combs SE, Engelhard C, Kopp C, et al. Long-term outcome after highly advanced single-dose or fractionated radiotherapy in patients with vestibular schwannomas—pooled results from 3 large German centers. Radiother Oncol. 2015;114(3):378–83.
30. Nguyen T, Duong C, Sheppard JP, et al. Hypofractionated stereotactic radiotherapy of five fractions with linear accelerator for vestibular schwannomas: a systematic review and meta-analysis. Clin Neurol Neurosurg. 2018;166:116–23.
31. Flickinger JC, Kondziolka D, Niranjan A, Lunsford LD. Results of acoustic neuroma radiosurgery: an analysis of 5 years' experience using current methods. J Neurosurg. 2001;94(1):1–6.
32. Ito K, Kurita H, Sugasawa K, Mizuno M, Sasaki T. Analyses of neuro-otological complications after radiosurgery for acoustic neurinomas. Int J Radiat Oncol Biol Phys. 1997;39(5):983–8.
33. Kano H, Kondziolka D, Khan A, Flickinger JC, Lunsford LD. Predictors of hearing preservation after stereotactic radiosurgery for acoustic neuroma: clinical article. J Neurosurg. 2013;119(Suppl):863–73.
34. Rashid A, Karam SD, Rashid B, et al. Multisession radiosurgery for hearing preservation. Semin Radiat Oncol. 2016;26(2):105–11.
35. Hayden Gephart MG, Hansasuta A, Balise RR, et al. Cochlea radiation dose correlates with hearing loss after stereotactic radiosurgery of vestibular schwannoma. World Neurosurg. 2013;80(3–4):359–63.
36. Sakanaka K, Mizowaki T, Arakawa Y, et al. Hypofractionated stereotactic radiotherapy for acoustic neuromas: safety and effectiveness over 8 years of experience. Int J Clin Oncol. 2011;16(1):27–32.
37. Kopp C, Fauser C, Muller A, et al. Stereotactic fractionated radiotherapy and LINAC radiosurgery in the treatment of vestibular schwannoma-report about both stereotactic methods from a single institution. Int J Radiat Oncol Biol Phys. 2011;80(5):1485–91.
38. Hasegawa T, Kida Y, Kato T, Iizuka H, Kuramitsu S, Yamamoto T. Long-term safety and efficacy of stereotactic radiosurgery for vestibular schwannomas: evaluation of 440 patients more than 10 years after treatment with gamma knife surgery. J Neurosurg. 2013;118(3):557–65.
39. Vivas EX, Wegner R, Conley G, et al. Treatment outcomes in patients treated with CyberKnife radiosurgery for vestibular schwannoma. Otol Neurotol. 2014;35(1):162–70.
40. Woolf DK, Williams M, Goh CL, et al. Fractionated stereotactic radiotherapy for acoustic neuromas: long-term outcomes. Clin Oncol (R Coll Radiol). 2013;25(12):734–8.
41. Kim JH, Jung HH, Chang JH, Chang JW, Park YG, Chang WS. Predictive factors of unfavorable events after gamma knife radiosurgery for vestibular schwannoma. World Neurosurg. 2017;107:175–84.
42. Powell C, Micallef C, Gonsalves A, Wharram B, Ashley S, Brada M. Fractionated stereotactic radiotherapy in the treatment of vestibular schwannoma (acoustic neuroma): predicting the risk of hydrocephalus. Int J Radiat Oncol Biol Phys. 2011;80(4):1143–50.
43. Roche PH, Khalil M, Soumare O, Regis J. Hydrocephalus and vestibular schwannomas: considerations about the impact of gamma knife radiosurgery. Prog Neurol Surg. 2008;21:200–6.
44. Germano IM, Sheehan J, Parish J, et al. Congress of neurological surgeons systematic review and evidence-based guidelines on the role of radiosurgery and radiation therapy in the management of patients with vestibular schwannomas. Neurosurgery. 2018;82(2):E49–51.
45. Rowe JG, Radatz MW, Walton L, Soanes T, Rodgers J, Kemeny AA. Clinical experience with gamma knife stereotactic radiosurgery in the management of vestibular schwannomas secondary to type 2 neurofibromatosis. J Neurol Neurosurg Psychiatry. 2003;74(9):1288–93.
46. Pollock BE, Link MJ, Stafford SL, Parney IF, Garces YI, Foote RL. The risk of radiation-induced tumors or malignant transformation after single-fraction intracranial radiosurgery: results based on a 25-year experience. Int J Radiat Oncol Biol Phys. 2017;97(5):919–23.

Pituitary Adenomas

25

Hiromitsu Iwata and Yuta Shibamoto

25.1 Introduction

Pituitary adenoma (PA) is a benign tumor that mainly occurs in adults between 20 and 50 years of age and constitutes approximately 10–20% of all intracranial tumors [1–3]. PA is classified into functioning and non-functioning adenomas. Functioning PA have been divided into growth hormone (GH)-secreting adenoma, prolactin-secreting adenoma, adrenocorticotropic hormone (ACTH)-secreting adenoma, thyroid-stimulating hormone-secreting adenoma, and gonadotropin (luteinizing hormone and follicle-stimulating hormone)-secreting adenoma. The purpose and method of treatment differ between the two entities. Treatment for functioning PA aims to prevent the excessive secretion of anterior pituitary lobe hormones, whereas that for non-functioning PA is typically intended to control tumor volume and prevent or reverse visual disorders and endocrinopathies. Tumors that cause visual symptoms are primarily treated with transsphenoidal surgery or craniotomy, and if patients are asymptomatic, a wait-and-see approach may be taken. Non-functioning PA is not necessarily treated by immediate radiotherapy (RT) after resection. However, previous studies reported recurrence in approximately 20–50% of cases treated with surgery alone [4–6].

A basic approach to the treatment of functioning PA is surgery (transcranial and transsphenoidal) combined with drug therapy. Since decreases in the secretion of hormones typically take years after RT, functioning PA is also not necessarily treated by immediate adjuvant RT. Nevertheless, it is mandatory to obtain hormone secretion control. Regarding GH-PA, for example, if excess GH persists for a long time, the incidence of cardiovascular events increases, and failure to achieve a GH level within the biochemical criterion range has been reported to lead to a death rate that is two- to fourfold higher than that in healthy individuals and also decreases life expectancy by 10–15 years [7, 8].

25.2 Rationale for RT

RT is a treatment option considered for residual or recurrent tumors invading the cavernous sinus or in cases in which repeated surgeries have resulted in fibrosis and inoperable non-functioning PA. Conventional RT was previously used to treat these cases [9, 10]. External beam

H. Iwata (✉)
Department of Radiation Oncology, Nagoya Proton Therapy Center, Nagoya City West Medical Center, Nagoya, Japan

Department of Radiology, Nagoya City University Graduate School of Medical Sciences, Nagoya, Japan
e-mail: h-iwa-ncu@nifty.com

Y. Shibamoto
Department of Radiology, Nagoya City University Graduate School of Medical Sciences, Nagoya, Japan

© Springer Nature Switzerland AG 2020
A. Conti et al. (eds.), *CyberKnife NeuroRadiosurgery*,
https://doi.org/10.1007/978-3-030-50668-1_25

RT is a relatively safe option for tumors close to organs at risk (OAR), such as the brainstem and optic apparatus [11, 12]. However, due to its convenience and precision as well as the reduced risk of hypopituitarism, the use of stereotactic irradiation is increasingly adopted. Regarding functioning PA, RT is also considered for poor responders to surgery and pharmacotherapy or when visual disorders must be reverted. While the targeting accuracy and dose falloff of Gamma Knife stereotactic radiosurgery (GKS) are excellent, it may not be suitable for large tumors or for those adjacent to the anterior optic pathways, because the dose limitation for these structures is considered to be 8–10 Gy in a single session [13–15]. The sparing of normal tissues, particularly late-responding tissues presumably with a low α/β ratio ($\leq$3 Gy), such as the optic pathways and brain stem, may be more efficient by using lower daily doses with hypofractionated radiation than with SRS [16–18]. Figure 25.1 shows an example of treatment planning and dose volume histograms of CyberKnife hypofractionated stereotactic radiotherapy (HSRT) for the Planning Target Volume (PTV) and OAR, including the chiasm, optic nerves, pituitary stalk, and brain stem.

25.3 Single Fraction Radiosurgery

25.3.1 Non-Functioning PA

Table 25.1 shows representative results of GKS for non-functioning PA [19–24]. Representative reports of Gamma Knife SRS indicate 5-year local control rates of 92–97% with a median follow-up period of 5 years or longer. Post-radiation visual disorder and hypopituitarism developed in 2–30% of patients. Complication rates were acceptable. However, tumor volumes that are safely treatable with Gamma Knife SRS may be less than 10 cm^3.

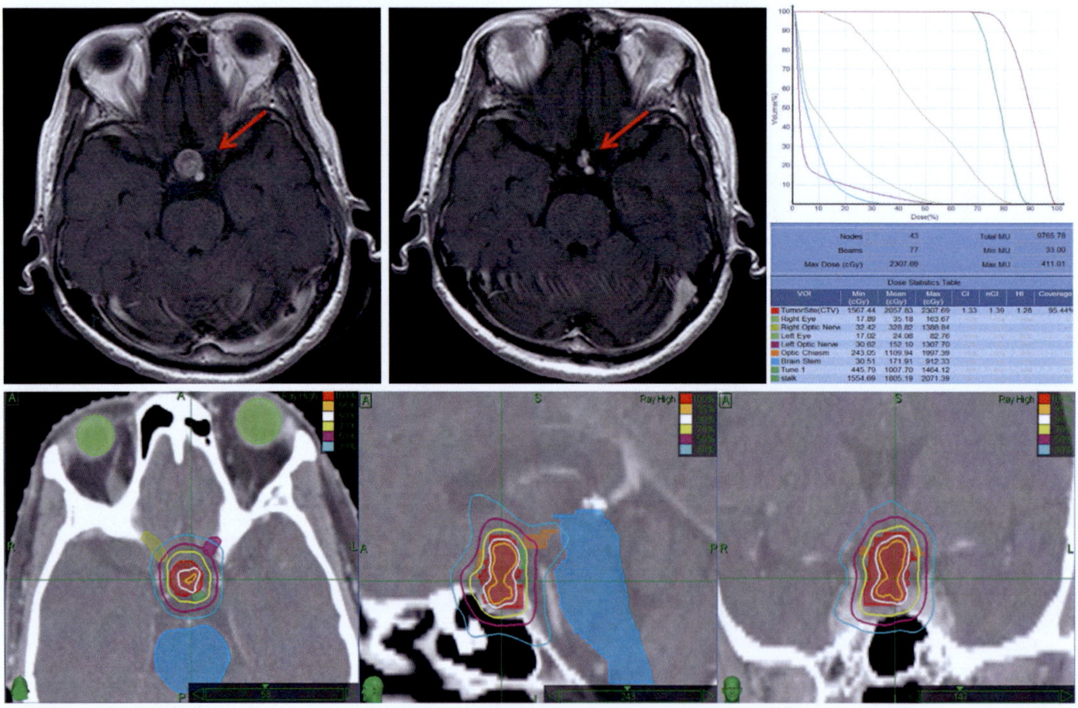

Fig. 25.1 CyberKnife treatment of non-functioning pituitary adenoma. Pre- (upper left) and post-treatment (upper middle) imaging and treatment plan. The dose delivered to the tumor was 18 Gy/3 fractions at the 78% isodose line

Table 25.1 Representative reported results of GKSRS, FSRT, and CKHSRT for non-functioning pituitary adenoma

Author (year)	Treatment modality	Patient number	Marginal or total dose	Treatment outcome (local control rate, visual disorders, hypopituitarism)	Median follow-up (months)
Mingione (2006) [19]	GKS	90	18.5 Gy/1 Fr	92% (4y), 0%, 25%	45
Pollock (2008) [20]	GKS	62	16 Gy/1 Fr	95% (3y, 7y), 0%, 32%	64
Kobayashi (2009) [21]	GKS	60	14.1 Gy/1 Fr	97% (≥3y), 4%, 8%	63
Sheehan (2013) [22]	GKS	512	16 Gy/1 Fr	95% (5y) and 85% (10y), 7.9%, 21%	36
Lee (2014) [23]	GKS	41	12 Gy/1 Fr	94% (5y) and 85% (10y), 2.4%, 24.4%	48
Bir (2015) [24]	GKS	57	15 Gy/1 Fr	98% (5y) and 90% (10y), 8.8%, 8.8%	45.5
Iwata (2011) [17]	CKhFSRT	100	21.0 Gy/3 Fr or 25.0 Gy/5 Fr	98% (3y), 1%, 2%	33
Puataweepong (2016) [30]	CKhFSRT	40	25.0 Gy/5 Fr	98% (3y), 0%, 0%	38.5

Abbreviations: GKS Gamma Knife stereotactic radiosurgery, *FSRT* Fractionated stereotactic radiotherapy, *CKhFSRT* CyberKnife hypofractionated stereotactic radiotherapy, *Fr* Fractions, *y* Years, *PA* Pituitary adenoma

25.3.2 Functioning PA (GH-PA)

Table 25.2 shows representative results of GKS for GH-PA [25–29]. These studies indicated biochemical hormone control rates of 30–82% with a median follow-up period of 5 years or longer. However, most evaluations in these studies were based on lax criteria, such as GH <2.5 (ng/mL), rather than the Cortina consensus criteria. Many studies reported that 5–10 years or longer is required before hormone levels decrease. Post-radiation visual disorder and hypopituitarism developed in 3–30% of patients. A higher radiation dose appeared to be needed in order to obtain better results.

25.4 Image-Guided Hypofractionated Radiotherapy

25.4.1 Non-Functioning PA

More recent studies have indicated promising outcomes with hypofractionated stereotactic radiotherapy hFSRT. Table 25.1 also shows the representative outcomes of CyberKnife fSRT (CKhS) for non-functioning PA [17, 30] with 3-year local control rates of 98% in a median follow-up period of 3 years or longer. Post-radiation visual disorder and hypopituitarism developed in less than 5% of patients. Acceptable rates of complications were observed. Approximately 10–20% of patients had large tumors (>15 cm^3) that are not normally treated with Gamma Knife SRS or tumors that were adjacent to the optical pathways. Despite these potential limitations to radiation treatment, the outcomes of CyberKnife hFSRT compare favorably with those reported previously.

25.4.2 Functioning PA (GH-PA)

Table 25.2 includes representative results of CyberKnife hFSRT for GH-PA [31], showing biochemical hormone control rates of only 17% with a median follow-up period of 5 years or longer. Post-radiation visual disorder and hypopituitarism developed in less than 5% of patients.

The authors reported many cases where tumor shrinkage or complete response was observed after CK hFSRT when evaluated by imaging. However, the effects observed on these images were not consistent with the biochemical data. Although hFSRT significantly reduced hormone levels, complete endocrinological remission was achieved only in 17% of all cases and in 9% of

Table 25.2 Representative reported results of SRS for growth hormone-secreting pituitary adenoma

Author (year)	Treatment modality	Patient number	Marginal dose (mean or median)	Treatment outcome: biochemical remission for GH and IGF-1, toxicity	Follow-up (months)
Ikeda (2001) [25]	GKS	17[a]	20.0 Gy/1 Fr	82% (not CC), no toxicity	–
Attanasio (2003) [26]	GKS	30	20.0 Gy/1 Fr	30% (not CC), hormone deficiency: 7%, visual deterioration: 0%	46 (median)
Wang (2003) [27]	GKS	149	20.9 Gy/1 Fr	64.9% (not CC), no toxicity	72.5 (mean)
Kobayashi (2009) [21]	GKS	67	18.9 Gy/1 Fr	4.8% (CC), hormone deficiency: 14.5%, visual deterioration: 11.1%	63.3 (mean)
Lee (2014) [28]	GKS	136	25 Gy/1 Fr	31.7%, 64.5%, 73.4% 82.6% (2, 4, 6, 8y) (not CC), hormone deficiency: 31.6%, visual deterioration: 3%	61.5 (median)
Gupta (2018) [29]	GKS	46[b]	25 Gy/1 Fr	28% (not CC), hormone deficiency: 19.6%, visual deterioration: Not reported	69.5
Iwata (2016) [31]	CKhFSRT	52	21.0 Gy/3 Fr or 25.0 Gy/5 Fr	17% (CC), hormone deficiency: 2%, visual deterioration: 0%	60 (median)

Abbreviations: GKS Gamma Knife stereotactic radiosurgery, *CKhS* CyberKnife hypofractionated stereotactic radiotherapy, *Fr* Fractions, *CC* Cortina consensus, *y* Years
[a]Invading the cavernous sinus
[b]Including Cushing's disease (21 patients)

symptomatic cases when evaluated with the Cortina consensus criteria. Therefore, a higher radiation dose or other fractionation seemed necessary for better results.

25.5 Toxicity

The use of poorly conformal irradiation increases the risk of substantial irradiation of the surrounding normal structures, resulting mainly in visual and pituitary deficits.

25.5.1 Visual Deficits

In GKS, the dose delivered to the anterior optic pathways is generally limited to 8–10 Gy. Hypofractionated irradiation may also reduce the incidence of complications in normal late-responding tissues.

Although linear-quadratic formalism applies with limitation to these fractionation schedules [32, 33], 21 Gy in 3 fractions corresponds to 13.1 Gy in 1 fraction, assuming an α/β of 3 Gy, while 25 Gy in 5 fractions corresponds to 12.7 Gy. The actual efficacy of hypofractionation is considered to be approximately 15% higher [32]. Therefore, 21 Gy in 3 fractions may correspond to approximately 15.1 Gy in 1 fraction and 25 Gy in 5 fractions to approximately 14.6 Gy. Available data showed an incidence of visual disorders of 0–10% with GKS and CyberKnife hFSRT (Tables 25.1 and 25.2).

Fractionation was considered to be desirable from the viewpoint of adverse events [17, 30, 31]. Milano et al. [34] reported the dose tolerance of the optic pathways from pooled data that were extracted from previous studies (PubMed indexed between 1990 and June 2015).

In patients with no prior radiation therapy receiving SRS/HSRT in 1–5 fractions, optic apparatus maximum point doses resulting in <1% radiation-induced optic nerve/chiasm neuropathy (RION) risks may be 12 Gy in 1 fraction (which is greater than our recommendation of 10 Gy in 1 fraction), 20 Gy in 3 fractions, and 25 Gy in 5 fractions. Moreover, Hiniker et al. [35] reported dose-response modeling of the visual pathway tolerance. Their model suggested <1% RION risk in their group of patients treated with an anterior optic pathway maximum point dose of 12 Gy in 1 fraction, 19.5 Gy in 3 fractions, and 25 Gy in 5 fractions.

25.5.2 Pituitary Deficits

Literature data showed an occurrence of new pituitary deficits of 0–10% with GKS and CyberKnife hFSRT (Tables 25.1 and 25.2). Post-radiation hypopituitarism developed less often in CyberKnife hFSRT than in GKS. The reasons for this are unclear. However, in addition to the use of hypofractionation, possible explanations include differences in GTV contouring, margin setting, apparatus, and dose specifications. In SRS and hFSRT, a correlation was observed between the dose to the pituitary stalk and the incidence of anterior pituitary hypofunction [36]. By controlling the dose to the pituitary stalk, the incidence of hypopituitarism may be reduced. Conti et al. [37] reported that the retrospective and prospective data of 64 patients with "perioptic" meningiomas treated by CyberKnife multisession radiosurgery or hFSRT (18–40 Gy/2–5 Fr) showed no new pituitary function deficits.

25.5.3 Other Complications

There have been few cases of radiation-induced brain necrosis or paralysis of the oculomotor or abducens nerve. Iwata et al. [17] reported that transient cyst enlargement occurred in three cases at 3, 6, and 9 months, respectively, and they developed transient slight visual field disturbance. However, the attenuation of their symptoms was noted as their cyst diminished at 6, 9, and 12 months, respectively. Internal carotid artery stenosis and aneurysms have also been reported [38].

25.6 Conclusion

CyberKnife hFSRT using schedules of 21 Gy in 3 fractions or 25 Gy in 5 fractions is safe and effective for non-functioning PA and also for GH-PA and has been rated as effective based on diagnostic imaging findings. However, when SRT is applied to patients with symptomatic GH-PA without any other therapy, it may be difficult in most cases to satisfy the Cortina consensus criteria. To protect the visual nerve and neuroendocrine function, hFSRT appears to be preferable, particularly for tumors located near the optic pathways and for large tumors. Further investigations of CyberKnife hFSRT for PA with longer follow-ups are warranted to confirm clinical outcomes and define its role in the treatment of GH-PA.

Practical suggestions for the hypofractionated treatment of pituitary adenoma

Parameter	Suggestion
Imaging	Contrast-enhanced CT T1-weighted contrast-enhanced images
Maximal tumor volume	Large cystic lesion may be contraindicated
Number of fractions	3 or 5
Dose to the target	21 or 25 Gy (non-functioning PA) More than 21 or 25 Gy (functioning PA)
Dose to the brainstem	<21 or 25 Gy
Dose to the optic nerves	<21 or 25 Gy
Dose to the optic chiasm	<21 or 25 Gy
Dose to the hypothalamus	<21 or 25 Gy
5-year expected tumor control	>90% (imaging criteria) <50% (hormonal criteria)
Expected visual deficits	<5%
Expected hypothalamic dysfunction	<1%
Expected pituitary deficits	<10%
Internal carotid artery stenosis or aneurysms	Unknown
Secondary carcinogenesis	<1%

Acknowledgments The authors are grateful to Drs. Seiji Ohta, Koshi Tatewaki, Naoki Yokota, Kengo Sato, and Mitsuhiro Inoue, and the staffs at the Yokohama CyberKnife Center for their valuable help.

References

1. Ezzat S, Asa SL, Couldwell WT, et al. The prevalence of pituitary adenomas: a systematic review. Cancer. 2004;101:613–9.
2. Molitch ME. Diagnosis and treatment of pituitary adenomas: a review. JAMA. 2017;317:516–24. https://doi.org/10.1001/jama.2016.19699.
3. Hoffman S, Propp JM, McCarthy BJ. Temporal trends in incidence of primary brain tumors in the United States, 1985-1999. Neuro-Oncology. 2006;8:27–37.
4. Brochier S, Galland F, Kujas M, et al. Factors predicting relapse of nonfunctioning pituitary macroadenomas after neurosurgery: a study of 142 patients. Eur J Endocrinol. 2010;163:193–200.
5. Park P, Chandler WF, Barkan AL, et al. The role of radiation therapy after surgical resection of nonfunctional pituitary macroadenomas. Neurosurgery. 2004;55:100–6.
6. Losa M, Mortini P, Barzaghi R, et al. Early results of surgery in patients with nonfunctioning pituitary adenoma and analysis of the risk of tumor recurrence. J Neurosurg. 2008;108:525–32.
7. Bihan H, Espinosa C, Valdes-Socin H, et al. Long-term outcome of patients with acromegaly and congestive heart failure. J Clin Endocrinol Metab. 2004;89:5308–13.
8. Damjanovic SS, Neskovic AN, Petakov MS, et al. High output heart failure in patients with newly diagnosed acromegaly. Am J Med. 2002;112:610–6.
9. Sasaki R, Murakami M, Okamoto Y, et al. The efficacy of conventional radiation therapy in the management of pituitary adenoma. Int J Radiat Oncol Biol Phys. 2000;47:1337–45.
10. Erridge SC, Conkey DS, Stockton D, et al. Radiotherapy for pituitary adenomas: long-term efficacy and toxicity. Radiother Oncol. 2009;93:597–601.
11. Jenkins PJ, Bates P, Carson MN, et al. Conventional pituitary irradiation is effective in lowering serum growth hormone and insulin-like growth factor-I in patients with acromegaly. J Clin Endocrinol Metab. 2006;91:1239–45.
12. Barrande G, Pittino-Lungo M, Coste J, et al. Hormonal and metabolic effects of radiotherapy in acromegaly: long-term results in 128 patients followed in a single center. J Clin Endocrinol Metab. 2000;85:3779–85.
13. Leber KA, Berglöff J, Pendl G. Dose-response tolerance of the visual pathways and cranial nerves of the cavernous sinus to stereotactic radiosurgery. J Neurosurg. 1998;88:43–50.
14. Stafford SL, Pollock BE, Leavitt JA, et al. A study on the radiation tolerance of the optic nerves and chiasm after stereotactic radiosurgery. Int J Radiat Oncol Biol Phys. 2003;55:1177–81.
15. Tishler RB, Loeffler JS, Lunsford LD, et al. Tolerance of cranial nerves of the cavernous sinus to radiosurgery. Int J Radiat Oncol Biol Phys. 1993;27:215–21.
16. Shigematsu N, Kunieda E, Kawaguchi O, et al. Indications of stereotactic irradiation for brain lesions. Acta Oncol. 2000;39:597–603.
17. Iwata H, Sato K, Tatewaki K, Yokota N, et al. Hypofractionated stereotactic radiotherapy with CyberKnife for nonfunctioning pituitary adenoma: high local control with low toxicity. Neuro-Oncology. 2011;13:916–22. https://doi.org/10.1093/neuonc/nor055.
18. Hoban PW, Jones LC, Clark BG. Modeling late effects in hypofractionated stereotactic radiotherapy. Int J Radiat Oncol Biol Phys. 1999;43:199–210.
19. Mingione V, Yen CP, Vance ML, et al. Gamma surgery in the treatment of nonsecretory pituitary macroadenoma. J Neurosurg. 2006;104:876–83.
20. Pollock BE, Cochran J, Natt N, et al. Gamma knife radiosurgery for patients with nonfunctioning pituitary adenomas: results from a 15-year experience. Int J Radiat Oncol Biol Phys. 2008;70:1325–9.
21. Kobayashi T. Long-term results of stereotactic gamma knife radiosurgery for pituitary adenomas. Specific strategies for different types of adenoma. Prog Neurol Surg. 2009;22:77–95. https://doi.org/10.1159/000163384.
22. Sheehan JP, Starke RM, Mathieu D, et al. Gamma Knife radiosurgery for the management of nonfunctioning pituitary adenomas: a multicenter study. J Neurosurg. 2013;119:446–56. https://doi.org/10.3171/2013.3.JNS12766.
23. Lee CC, Kano H, Yang HC, et al. Initial Gamma Knife radiosurgery for nonfunctioning pituitary adenomas. J Neurosurg. 2014;120:647–564. https://doi.org/10.3171/2013.11.JNS131757.
24. Bir SC, Murray RD, Ambekar S, et al. Clinical and radiologic outcome of Gamma Knife radiosurgery on nonfunctioning pituitary adenomas. J Neurol Surg B Skull Base. 2015;76:351–7. https://doi.org/10.1055/s-0035-1549309.
25. Ikeda H, Jokura H, Yoshimoto T. Transsphenoidal surgery and adjuvant gamma knife treatment for growth hormone-secreting pituitary adenoma. J Neurosurg. 2001;95:285–91.
26. Attanasio R, Epaminonda P, Motti E, et al. Gamma-knife radiosurgery in acromegaly: a 4-year follow-up study. J Clin Endocrinol Metab. 2003;88:3105–12.
27. Wang MH, Liu P, Liu AL, et al. Efficacy of gamma knife radiosurgery in treatment of growth hormone-secreting pituitary adenoma. Zhonghua Yi Xue Za Zhi. 2003;83:2045–8.

28. Lee CC, Vance ML, Xu Z, et al. Stereotactic radiosurgery for acromegaly. J Clin Endocrinol Metab. 2014;99:1273–81. https://doi.org/10.1210/jc.2013-3743.
29. Gupta A, Xu Z, Kano H, et al. Upfront gamma knife radiosurgery for Cushing's disease and acromegaly: a multicenter, international study. J Neurosurg. 2018;131:1–7. https://doi.org/10.3171/2018.3.JNS18110.
30. Puataweepong P, Dhanachai M, Hansasuta A, et al. The clinical outcome of hypofractionated stereotactic radiotherapy with CyberKnife robotic radiosurgery for perioptic pituitary adenoma. Technol Cancer Res Treat. 2016;15:NP10–5.
31. Iwata H, Sato K, Nomura R, et al. Long-term results of hypofractionated stereotactic radiotherapy with CyberKnife for growth hormone-secreting pituitary adenoma: evaluation by the cortina consensus. J Neuro-Oncol. 2016;128:267–75. https://doi.org/10.1007/s11060-016-2105-1.
32. Iwata H, Shibamoto Y, Murata R, et al. Estimation of errors associated with use of linear-quadratic formalism for evaluation of biologic equivalence between single and hypofractionated radiation doses: an in vitro study. Int J Radiat Oncol Biol Phys. 2009;75:482–8.
33. Park C, Papiez L, Zhang S, et al. Universal survival curve and single fraction equivalent dose: useful tools in understanding potency of ablative radiotherapy. Int J Radiat Oncol Biol Phys. 2008;70:847–52.
34. Milano MT, Grimm J, Soltys SG, et al. Single- and multi-fraction stereotactic radiosurgery dose tolerances of the optic pathways. Int J Radiat Oncol Biol Phys. 2018; https://doi.org/10.1016/j.ijrobp.2018.01.053.
35. Hiniker SM, Modlin LA, Choi CY, et al. Dose-response modeling of the visual pathway tolerance to single-fraction and hypofractionated stereotactic radiosurgery. Semin Radiat Oncol. 2016;26:97–104. https://doi.org/10.1016/j.semradonc.2015.11.008.
36. Sheehan JP, Niranjan A, Sheehan JM, et al. Stereotactic radiosurgery for pituitary adenomas: an intermediate review of its safety, efficacy, and role in the neurosurgical treatment armamentarium. J Neurosurg. 2005;102:678–91.
37. Conti A, Pontoriero A, Midili F, et al. CyberKnife multisession stereotactic radiosurgery and hypofractionated stereotactic radiotherapy for perioptic meningiomas: intermediate-term results and radiobiological considerations. Springerplus. 2015;4:37. https://doi.org/10.1186/s40064-015-0804-2.
38. Inoue H, Kawano T, Ohmori Y, et al. Internal carotid artery aneurysms diagnosed after stereotactic radiosurgery for a growth hormone-secreting pituitary adenoma: a case report and literature review. Acta Neurochir. 2019;161:1191–5. https://doi.org/10.1007/s00701-019-03840-5.

Craniopharyngiomas

26

Andrey V. Golanov, Alexander N. Savateev, Yury Y. Trunin, Natalia A. Antipina, Konstantin V. Nikitin, and Alexander N. Konovalov

26.1 Introduction

The incidence of craniopharyngioma is 0.5–2.5 cases per 1 million, that is, 2–5% of all primary brain tumors in adults [1] and 5.6–13% in children [2]. Most often craniopharyngiomas appear in the age groups of 5–14 years and 50–74 years [3]. These tumors form the largest group of non-glial tumors in children. CP contribute to up to 56% all tumors of the chiasmo-sellar region in children [2]. In 2011, Nielsen et al. performed a meta-analysis of 15 epidemiological studies of CP (with total of 1232 patients). According to these data, the incidence was 1.34 (1.24–1.46) per 1 million people and 1.44 (1.33–1.56) per 1 million children [4].

Taking into account the benign nature of craniopharyngiomas, the main method of treatment is the removal of the tumor. However, the tendency of these tumors to invade critical structures (such as optic pathways, the hypothalamic-pituitary system, the Willis circle vessels) often limits the possibility of a radical surgery [5, 6].

Craniopharyngiomas of the third ventricle represent the greatest challenge for surgery [7, 8]. After radical operations, hypothalamic disorders often occur, including not only obesity but also cognitive, emotional, mental, and metabolic disturbances. Metabolic disorders associated with damage to the hypothalamus progress after surgery and lead to impaired functions of the internal organs. This process is irreversible and, in many cases, becomes the direct cause of the patient's death. The life expectancy of patients with the surgically affected hypothalamus is significantly shorter than without them. The incidence of hypothalamic disorders after surgery can reach 40% [9].

Even with macroscopically total resection, craniopharyngiomas can recur in 10–30% of cases [10–13], and in the presence of tumor remnants and with no further radiation treatment, the risk of recurrence significantly increases [7, 13, 14] to up to 50–70% according to various studies [15, 16] and even up to 85% [11, 17, 18]. For this

A. V. Golanov (✉)
National Medical Research Center of Neurosurgery named after Burdenko, Moscow, Russia

Moscow Gamma Knife Center, Moscow, Russia
e-mail: golanov@nsi.ru

A. N. Savateev
Moscow Gamma Knife Center, Moscow, Russia
e-mail: asavateev@nsi.ru

Y. Y. Trunin · N. A. Antipina · K. V. Nikitin
A. N. Konovalov
National Medical Research Center of Neurosurgery named after Burdenko, Moscow, Russia
e-mail: ytrunin@nsi.ru; nantipina@nsi.ru; knitin@si.ru; akonovalov@nsi.ru

© Springer Nature Switzerland AG 2020
A. Conti et al. (eds.), *CyberKnife NeuroRadiosurgery*,
https://doi.org/10.1007/978-3-030-50668-1_26

reason, the observation of patients with residual tumors after surgery is an incorrect strategy.

Radiation therapy significantly improves progression free survival (PFS), and the use of stereotactic irradiation techniques ensures conformity of irradiation of tumor remnants with a complicated shape and location [19–22], which potentially reduces the risk of undesirable post-radiation effects. Therefore, the quality of life in patients with craniopharyngiomas infiltrating the anterior third ventricle is significantly higher after non-radical operations with subsequent stereotactic radiation than after a total or subtotal removal [22, 23].

26.2 Rationale for Radiation Therapy

Attempts to use radiation therapy (RT) for the treatment of CP have been undertaken since the beginning of the twentieth century. The first results of irradiation of patients with CP, published in 1937, turned out to be unsatisfactory and the authors concluded that "tumors of the pituitary stalk can be resistant to radiation exposure" [24]. However, in 1950 Love et al. obtained good radiation results after partial removal of the tumor [25]. Subsequently, conventional fractionated RT became a routine treatment for patients with CP.

In 1961, Kramer was one of the first to publish the results of "limited" and safe surgical removal of CP followed by radiation therapy: a 15-year PFS was observed in all six patients who underwent this treatment [26]. In 1993, a publication appeared from Royal Marsden Hospital (London): 77 patients after non-radical surgical removal underwent a course of radiation therapy to a total dose of about 56 Gy. In 1950–1986, 5- and 10-year PFS was 83% and 79%, respectively [27].

Later, with the development of technology, stereotactic irradiation techniques appeared, including stereotactic radiosurgery (SRS) and hypofractionated radiotherapy. The use of radiation therapy with incomplete removal of CP allowed to the increase of progression-free survival up to 75–90% [21, 27–30] (Table 26.1). Currently, only stereotactic irradiation techniques should be used in CP, including standard fractionated radiation therapy, radiosurgery and hypofractionated RT.

26.3 Single Fraction Radiosurgery

For the treatment of residual tumor or relapses of craniopharyngioma, a stereotactic radiosurgery (SRS) technique can be used with a relatively high dose of ionizing radiation during a single session. Through this radiation technique, the dose outside the target decreases sharply without causing damage to healthy brain tissue [19, 31].

According to literature, PFS after radiosurgery with CP is comparable to survival after fractionated stereotactic RT. The 5-year PFS of patients who received radiosurgical treatment immediately after surgery or for relapse of CP was 56.7–91.6%, while the 5- and 10-year OS were 86–97% and 88–91%, respectively [32–35] (Table 26.2). The weighted average value of the 5-year PFS, calculated on the basis of data from four studies (231 patients) [32, 34–36] (Table 26.2) was 67%. And most relapses occur outside the target volume of radiosurgical irradiation.

Jeon et al. after an analysis of 50 observations did not disclose significant differences in the efficacy of SRS and fractionated RT [36]. Xu et al. in 2011 found out that prognostic factors for better tumor control were the volume of the solid component of the tumor being smaller than 1.6 cm^3 and the dose being more than 14.5 Gy [32]. The authors attributed the absence of a cystic component of the tumor and the minimum number of surgical operations before radiation treatment as additional factors associated with a good response of CP to SRS [32].

26.4 Image-Guided Radiosurgery and Hypofractionated Radiotherapy

Stereotactic navigation during radiosurgery can be performed with a frame (Gamma Knife, Novalis, etc.) or using frameless navigation (CyberKnife).

Table 26.1 Effects of irradiation in various combinations with surgical treatment (Clark, 2013 [61] with additions)

Authors	Total number of patients	Age (years)	Median follow-up (years)	Mean dose (Gy)	Treatments	Patients groups	5-year PFS (%)	10-year PFS (%)	20-year PFS (%)	Good funtional outcome
Wen (1989) [62]	52	–	–	50–70	Total resection	20	TC = 50%	–	–	–
					Non-total resection + RT	8	TC = 100%			
					Non-total resection	20	TC = 15%			
Hetelekidis (1993) [28]	61	7.5 (10 months–21 year)	10 years (2–20.5 years)	54.6	Only surgery	15 (24%)	–	31%	–	–
					Surgery + RT/SRS	37 (61%)		86%		
					Only RT	9 (15%)		100%		
Regine (1993) [49]	58	<16 years = 19 >16 years = 39	17 years	55.9–62.4	Surgery + RT/SRS	56	–	79%	–	42
					Only RT	2				
Rajan (1993) [27]	173	19 (3–68)	12 years	50	Total resection + RT	4 (2.3%)	–	100%[a]	–	94
					Non-total resection + RT	99 (57.3%)		88%[a]	84%[a]	
					Biopsy + RT	14 (8%)		77%[a]	74%[a]	
					Cyst aspiration + RT	34 (19.7%)				
					Only RT	22 (12.7%)		77%[a]	71%[a]	
Habrand (1999) [29]	37	7.4 (1–15)	>5 years	50	–	–	78%	56.5%	–	66
Varlotto (2002) [48]	24	29	12 years	60	Surgery + RT/SRS	22	–	89%	–	–
					Only RT	2				
Stripp (2004) [63]	75	8.5 (1.5–24.8)	7.6 years	54 (44–55.8)	Only surgery	57	–	TC = 42%	–	–
					Surgery + adjuvand RT	18		TC = 84%		
					Surgery + salvage RT	22		TC = 83%		
Pemberton (2005) [60]	87	<15 years = 28 >15 years = 59	8 years	42.5	–	–	–	78%	–	75
Karavitaki (2005) [64]	121	<16 years = 42 >16 years = 79 (2.5–83)	8.6 years (0.3 months –39 years)	–	Total resection	16	–	100%	–	–
					Total resection + RT	3		100%		
					Partial resection	51		38%		
					Partial resection + RT	33		77%		
					Cyst aspiration	6	1 year PFS—58%	0%		
					Cyst aspiration + RT	3	–	100%		

(continued)

Table 26.1 (continued)

Authors	Total number of patients	Age (years)	Median follow-up (years)	Mean dose (Gy)	Treatments	Patients groups	5-year PFS (%)	10-year PFS (%)	20-year PFS (%)	Good funtional outcome
Merchant (2006) [59]	28	7.3	3	54/55.8	–	–	90% (3-year PFS)	–	–	–
Combs (2007) [43]	40	41 (6–70) <18 years = 6	8.2 years (3 months –27 years)	52.2 (50.4–56)	Total resection + RT	7	100%	100%	4—Compl. Resp. 25—Part. Resp. 11—Witout progression	–
					Subtotal resection + RT	23				
					Partial resection or Ommaya + RT	10				
Lee (2008) [40]	11	34.5 (13–71)	1.3 years (4 months –5 years)	Marg. 21.6 Max. 29.9	Surgery + RT (hypofractionated)	11	TC = 91%	–	–	–
Lin (2008) [65]	31	8.1 (1.1–21)	6.5 years	52.2 (50.4–56)	Surgery + RT/SRS	10	100%	100%	–	91
					Cyst aspiration + isotope	1				
					Total surgery + observ.	20				
Smee (2011) [66]	41	<16 years = 12 >16 years = 29	23 years	50	Conventional RT	11	–	–	95.12%	–
				14	SRS	3				
				14.28	SRS with LINAC	1				
				50.4	FSRT	18				
				51.09	IMRT	7				
				18	Brachytherapy	1				
Schoenfeld (2012) [9]	122	30 years (11–52) <18 years = 46	4.7 years (1.6–12)	–	Total resection	30	≈60%	≈42%	–	–
					Total resection + RT	3				
					Subtotal resection	41	≈26%	≈10%		
					Subtotal resection + RT	48	≈71%	≈56%		
Greenfield (2015) [67]	24	Children	8.9 years	50.4 (49.8–54)	IMRT	24	PFS—65.8%	PFS – 60.7%	–	–
							CFS—70.2%	CFS – 65.2%	–	
							NFS—90.7%	NFS – 90.7%	–	

CyberKnife										
Iwata (2012) [20]	43	44 (3–85)	3.3 years (1–7.5 years)	21 (13.3–25)	Hypofractionated RT	40	3-year PFS–78% 5-year PFS–60%	–	–	–
				14.3 (13–16.3)	SRS (CyberKnife)	3				
Meta-analysis—109 articles										
Clark (2013) [61]	531	–	–	–	Total resection	–	77%	–	–	–
					Subtotal resection + RT		73%			
					Subtotal resection		43%			
Weighted average PFS depending on the radicality of surgery and RT/SRS					**Total resection**	**70**	**68.6%**	**65.2%**	**–**	**–**
					Surgery + RT	**622**	**77.4% (*n* = 210)**	**79.4% (*n* = 622)**	**–**	
					Non-total resection	**149**	**25.6% (*n* = 67)**	**31.8% (*n* = 149)**	**–**	

≈ Graph estimate (the exact numbers are not given in the original article)
TC Tumor control
CFS Cystic free survival
NFS Nodular free survival
[a]Differences are not statistically significant

Table 26.2 Effectiveness of SRS in craniopharyngioma

Authors	Number of patients	Mediam age (years)	Median tumor volume (sm^3)	Treatment method	Nubmer of procedures	Prescribed dose (Gy)	Maximum dose (Gy)	Follow-up period (years)	3-years PFS (%)	5-years PFS (%)	10-years PFS (%)
Kobayashi (2005) [35]	98	–	3.5	SRS	–	11.5	21.8	5.5 (6–12.3)	–	60.8	–
Niranjan (2010) [34]	46	23.5 (4–77)	1.0 (0.07–8.0)	SRS	51	13 (9–20)	26 (20–50)	5.2 (1–19.3)	91.6	91.6	–
Xu (2011) [32]	37	36 (4–78)	1.6 (0.1–18.6)	SRS[a]	39	14.5 (6–25)	30 (15.6–60)	4.2 (0.7–17.7)	84.8	67	–
Jeon (2011) [36]	50	33.6 (3–70)	–	SRS	13	11 (10–12 Gy 50% isodose)		5.9 (0.8–13.9)	–	56.7	45.3
				SRT	37	ср. СОД = 50,2 (50–60)					
Mediana	**231 (total)**	**31 years**	**2.5 sm^3**	**–**	**–**	**12.2 Gy**	**24 Gy**	**5.3 years**	**88.6%**	**67%**	**45.3%**

[a]Only a solid component was irradiated

Fixing the frame can be difficult in patients after surgery, especially after craniectomy or large bifrontal approaches, and CyberKnife has no such limitation. In addition, frameless navigation provides the possibility of multiple uniform positioning of the patient, which allows for hypofractional irradiation. The geometric accuracy of the CyberKnife system is higher than 0.5 mm [37–39], and the doses used are similar to those using the Gamma Knife [20, 40].

Since the 2000s, the use of hypofractionation mode in the treatment of intracranial tumors has actively entered into practice. This was facilitated by the widespread use of a robotic linear accelerator for the frameless stereotaxis of CyberKnife, in which radiation treatment is carried out in modes of radiosurgery or hypofractionated radiotherapy.

Hypofractionated radiation therapy (2–10 fractions) has a number of advantages in comparison with the standard course of radiotherapy and radiosurgery. First of all, unlike radiosurgical treatment, it is possible to irradiate patients with sufficiently large tumors located close to or inside critical structures (visual pathways, brainstem, pituitary gland, hypothalamus). At the same time, according to the biological response of the tumor, the hypofractionation mode is similar to radiosurgery. Secondly, in the treatment of patients with cystic CP, the use of a fast fractionation mode (3–5 days) allows the avoidance of an increase in the cystic component of the tumor during the cycle of radiotherapy and, accordingly, the tumor borders beyond the limits of the radiation volume, which can happen when using standard 6-week course of RT. Finally, from the radiobiological point of view, the use of hypofractionation allows us to target hypoxic cells.

The hypofractionation regimen implies a single dose (SD) <3 Gy. Irradiation in this mode can be carried out at many linear electron accelerators. However, in most published studies concerning hypofractionated RT, the system used was the CyberKnife.

Lee et al. reported a 90% tumor control with preservation of visual functions in 11 patients with CP after hypofraction with CyberKnife (total dose was 20–25 Gy in 3–5 fractions) in the setting of compression of the optic chiasm by the tumor; no complications have been reported [40].

Iwata et al. analyzed results of CyberKnife hypofractionated radiotherapy (2–5 fractions with marginal dose of 13–25 Gy) in 40 patients with CP with a median follow-up of 3 years; PFS was 85%. The following hypofractionation regimens were used: 8 Gy × 2 fractions, 7 Gy × 3 fractions, and 5 Gy × 5 fractions. Tumor volume was 0.09–20.8 cm^3. The author noted a temporary increase of cystic component after irradiation in nine patients, but no serious complications have been reported [20].

Currently, with hypofractionated irradiation of craniopharyngiomas, the dose of 25–27.5 Gy is considered to be optimal, as tolerant doses to critical structures are observed.

26.5 Toxicity

The vast majority of studies of side effects of radiation treatment reflect the risks associated with the use of conventional RT—a method which is currently no longer used in CP. Stereotactic irradiation methods provide much lower doses on critical structures minimizing complications. Thus, with stereotactic radiation therapy, only visual, endocrine, and cognitive complications should be monitored.

26.5.1 Visual Deficits

It is believed that the maximal tolerance single dose to the visual pathways is 10–12 Gy [21, 41]. The risk of radiation damage to the visual pathways is associated with fraction dose and total dose. According to the published data, radiation injury to the optic nerves is observed in 1–2% of patients who received a dose of 50 Gy or more with conventional irradiation and more often is observed in patients who had previously shown visual disturbances RT [27, 42].

Among patients who received 50–55 Gy with 1.8 Gy per fraction, the risk of visual deficit is less than 2.5% [43–47]. However, the frequency of this complication significantly increases at doses of 55–60 Gy [48, 49]. After stereotactic radiosurgery and hypofractionated RT, damage to the optic nerves was noted only in a few cases [40, 50].

26.5.2 Endocrinological Dysfunctions

Most patients with CP have hypopituitarism of varying severity after surgery and before radiation treatment [16, 51]. According to Mazerkina N.A. after surgical treatment, panhypopituitarism develops in 75% of patients [16]. Therefore, RT as a risk factor for endocrine deficiency is important only in 25% of patients with partially or fully preserved endocrine functions. New hormonal deficiency develops on average in 30–50% of patients within 6–12 months after SRS and 5–10 years after conventional RT. Most of these patients require hormone *replacement* therapy [27–29]. Development of diabetes insipidus after stereotactic RT/SRS occurs in the vast majority of cases due to tumor regrowth, rather than to radiation damage [52]. According to Vladyka (2003), based on the analysis of radiosurgery of pituitary adenomas, the average single dose of 15 Gy or less to adenohypophysis does not cause hypothyroidism and hypogonadism, and a single dose of 18 Gy or less does not cause hypocorticism [53].

The frequency of radio-induced endocrinopathies is a dose-dependent parameter. With conventional irradiation with a dose of superior to 60 Gy, the new endocrine deficiency was observed in more than 80% of cases and with use a dose of 54–60 Gy—in 36% [54]. Currently, doses greater than 60 Gy are not used in RT.

According to Xu et al., SRS in patients with pituitary adenoma of endosellar localization causes pituitary disorders to 30% of irradiated patients within 3 years, developing most often somatotropic insufficiency and hypothyroidism, less often hypogonadism, and even less often hypocorticism.

The high dose and suprasellar growth of the tumor are independent predictors of post-radiation hypopituitarism [55].

26.5.3 Cognitive and Neuropsychological Dysfunction

Cognitive deficiency was observed in the 1990s after radiation treatment of tumors of sellar and parasellar localization [56, 57]. The correlation between cognitive impairment and exposure of large volume of the brain in conventional RT is well known. Radiation-related cognitive deficit became less frequent and less pronounced with the invention of stereotactic RT and SRS [27, 49, 58–60] (Table 26.1).

In 2006, Merchant analyzed IQ in 27 patients with CP before and after stereotactic RT. Observation lasted 48 months after irradiation and a significant difference in the IQ was revealed in patients younger and older than the age of 7.4 at the time of RT. Furthermore, the intelligence level of children younger than 7.4 years after RT decreased linearly over time, while in older patients it remained almost unchanged [59].

In 2010, Kiehna together with Merchant analyzed 32 articles about RT in children with CP and noted that IQ remains stable for 5 years after conformal RT and then it may decline [52]. Based on their own data, the authors found negative prognostic factors for cognitive functions: an age less than 7 years old at the time of radiotherapy, female gender, presence of hydrocephalus, large cystic component in CP, traumatic surgery, and diabetes insipidus before surgery [52].

26.6 Our Experience

Since March 2005, more than 200 patients with craniopharyngiomas have been treated in the Department of Radiotherapy and Radiosurgery. From April 2009 to January 2015, 68 patients (38 men and 30 women) were irradiated with the CyberKnife. The median age was 26 (ranged 2–81). All patients had previous surgery, and 85% of patients ($n = 57$) had more than 1 surgery. Fifteen percent of patients ($n = 10$) had the Ommaya reservoir or a puncture biopsy of the tumor cyst before the RT.

Nine patients with a small tumor volume (<5 cm^3) and a sufficient distance to the visual pathways were selected for radiosurgery. The average dose was 14.3 Gy (12–16 Gy), and the average tumor volume was 1.8 cm^3 (0.07–4.1 cm^3). Fifty-nine patients with larger tumors (>10 cm^3) or with a tumor adjacent to critical structures underwent hypofractionated RT with the following regimens: 7 Gy × 3 fractions, 5 Gy × 5 fractions, 5.5 Gy × 5 fractions. The average tumor volume was 3.1 cm^3 (0.25–15.3 cm^3).

The median follow-up was 48.4 months (1.4–95.1 months). Progression of cystic component occurred in three patients (4.4%), which required emptying of the cyst or its removal. Progression of the solid component of the tumor was not noted. So, overall tumor control was 95.6%.

26.7 Clinical Cases

Case 1

Patient M., 53 years old with endosuprasellar CP was subjected to transcranial removal of the tumor. Immediately after the surgery panhypopituitarism and a bilateral amaurosis developed (though before the surgery, the visual acuity of the right eye was 0.8, and that of the left eye was 0.02). Three months after the surgery, a large remnant tumor in the third ventricle was revealed on MRI (Fig. 26.1a).

At ophthalmological examination, an atrophy of the optic nerve discs was confirmed. Five months after surgery, the patient underwent stereotactic radiation with CyberKnife in the hypofractionation mode: 25 Gy (median dose) was delivered to the tumor (PTV = 2.3 cm^3) in 5 fractions (Fig. 26.1b). Follow-up MRI at 1 year after irradiation (Fig. 26.1c) showed a minimal residual tumor in the third ventricle; symptoms were stable.

Case 2

Patient B., 18 years old. Admitted at the age of 12 because of growth delay; 5 years later an intra-suprasellar tumor was detected on MRI and endocrine disorders identified (growth hormone deficiency, secondary hypothyroidism, hypocortisolism). Patient underwent endoscopic endonasal intracapsular tumor removal in Burdenko National Center of Neurosurgery. During surgery the tumor capsule could not be removed.

Eight months after surgery an MRI showed small tumor regrowth (Fig. 26.2a). Given the small size of the tumor and the 4 mm distance between the optic pathways and the tumor, the decision to perform radiosurgery with 13.5 Gy (median dose) with CyberKnife (Fig. 26.2b) was made. The pituitary gland cannot be identified on MRI. The dose to 5% of chiasm did not exceed 6 Gy (Fig. 26.2c).

A marked shrinkage of the tumor was noted on MR images at 3 (Fig. 26.3a) and 6 months after SRS (Fig. 26.3b). At 1.5 years after irradiation, a follow-up MRI showed no residual tumor (Fig. 26.3c). No side effects of radiation appeared.

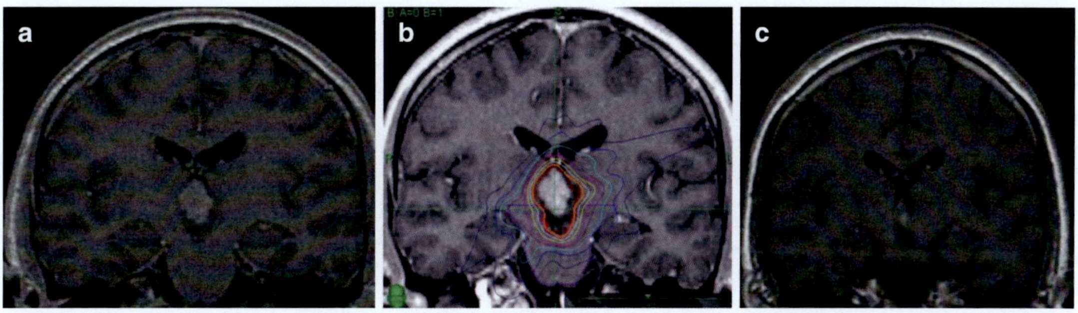

Fig. 26.1 (Illustrative Case 1). (**a**) Pretreatment MRI. (**b**) Dose planning. (**c**) 1 year after irradiation

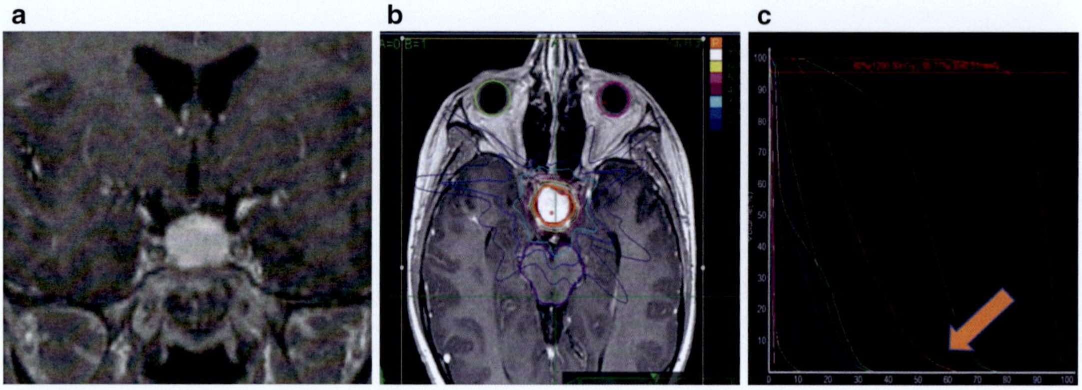

Fig. 26.2 (Illustrative Case 2). (**a**) MRI at the day of SRS. (**b**) Dose planning. (**c**) Dose–volume histogram, dose to 5% of volume of optic chiasm does not exceed 6 Gy

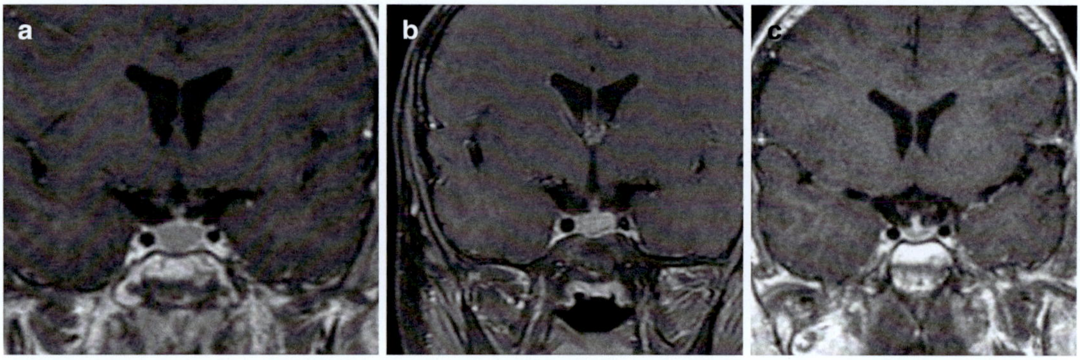

Fig. 26.3 (Illustrative Case 2 continued). (**a**) MRI 3 months after SRS. (**b**) MRI 6 months after SRS. (**c**) MRI 1.5 year after SRS

26.8 Conclusions

The use of stereotactic radiotherapy or radiosurgery in a setting of presence of a residual tumor, tumor relapse, or progression of craniopharyngioma significantly increases the disease-free survival after non-radical surgery to a level similar to that obtained after total resection of the tumor. SRT and SRS after non-radical surgery are safer for visual function preservation than total removal of the craniopharyngioma.

Stereotactic irradiation in patients with CP rarely exacerbates hormonal deficiency (6.7% of cases). Hypofractionated SRT and SRS do not lead to a worsening of the diencephalic disorders. In patients with CP that infiltrates the third ventricle, the quality of life is higher after non-radical operations followed by stereotactic radiation than after total or subtotal removal of the tumor.

The use of CyberKnife for radiosurgery and stereotactic irradiation in the hypofractionation mode allows an effective and maximally safe radiation treatment for craniopharyngiomas, despite the difficulty in localizing these tumors both after incomplete removal of the tumor and as an independent treatment.

26.9 Practical Guide

It is important to include both solid and cystic components of the tumor in the GTV when planning radiation therapy. Any cyst wall fragments in the tumor bed should be incorporated in

Table 26.3 Parameters for stereotactic irradiation in different modes

	Suggestion	
Parameter	Hypo-SRT	SRS
Imaging	Topometric non-contrast CT Topometric contrast enhanced T1 (slices 1 mm) Volumetric T2 (slices 1 mm)	
PTV definition	CTV + 1–2 mm	
Maximal tumor volume	Large cystic lesions can be contraindicated	About 10 sm^3
Number of fractions	5	1
Dose to the target (mean)	25 or 27.5 Gy (pediatric and adults)	12–16 Gy
Dose to the brainstem	<25 Gy	<100 mm^3 receiving 12 Gy
Dose to the optic nerves	22–24 Gy	<12 Gy
Dose to the optic chiasm	22–24 Gy	<12 Gy
Dose to the hypothalamus	<25 Gy	<16 Gy
5-year expected tumor control	>95%	>82%
Expected visual deficits	5%	
Expected hypothalamic dysfunction	None	
Expected new pituitary deficits	6.5%	
Malignant transformation	None	

GTV. As a result, GTV should also include cyst walls and all calcifications, visible on high resolution MRI and CT scans.

The tumor often grows around the vessels of the Willis circle, chiasm, and optic nerves. In addition to MRI and CT studies at the time of radiation therapy, it is helpful to use a preoperative MR and CT scan, taking into account the data of the operation protocol. It is recommended to use CTV equal to GTV, and PTV is formed as CTV plus 1–2 mm margin for SRS or hypofractionated SRT.

For cases with a small GTV and when the tumor allows us to exclude optic chiasm and the bottom of the third ventricle from the PTV, single-fraction SRS is chosen. Hypofractionated SRT can be used even in cases where critical structures are inside the tumor (Table 26.3).

References

1. Samii M, Tatagiba M. Craniopharyngioma. In: Kaye AH, Laws Jr ER, editors. Brain tumors: an encyclopedic approach. New York: Churcill Livingstone; 1995. p. 873–94.
2. Rickert CH, Paulus W. Epidemiology of central nervous system tumors in childhood and adolescence based on the new WHO classification. Childs Nerv Syst. 2001;17(9):503–11.
3. Bunin GR, et al. The descriptive epidemiology of craniopharyngioma. J Neurosurg. 1998;89(4):547–51.
4. Nielsen EH, et al. Incidence of craniopharyngioma in Denmark (n = 189) and estimated world incidence of craniopharyngioma in children and adults. J Neuro-Oncol. 2011;104(3):755–63.
5. Elliott RE, Wisoff JH. Surgical management of giant pediatric craniopharyngiomas. J Neurosurg Pediatr. 2010;6(5):403–16.
6. Mortini P, et al. Neurosurgical treatment of craniopharyngioma in adults and children: early and long-term results in a large case series. J Neurosurg. 2011;114(5):1350–9.
7. Gorelyshev SK. Surgical treatment of craniopharyngiomas of the III ventricle in children. The dissertation for the degree of candidate of Medicine Doctor. USSR Academy of Medical Sciences, N. N. Burdenko Research Institute of Neurosurgery. Moscow, 1989.
8. Gonzalez LF, et al. Working area and angle of attack in three cranial base approaches: pterional, orbitozygomatic, and maxillary extension of the orbitozygomatic approach. Neurosurgery. 2002;50(3):550–5; discussion 555–7.
9. Schoenfeld A, et al. The superiority of conservative resection and adjuvant radiation for craniopharyngiomas. J Neuro-Oncol. 2012;108(1):133–9.
10. Hoffman HJ, et al. Aggressive surgical management of craniopharyngiomas in children. J Neurosurg. 1992;76(1):47–52.
11. Fahlbusch R, et al. Surgical treatment of craniopharyngiomas: experience with 168 patients. J Neurosurg. 1999;90(2):237–50.
12. Duff J, et al. Long-term outcomes for surgically resected craniopharyngiomas. Neurosurgery. 2000;46(2):291–302; discussion 302–5.
13. Semenova, ZhB. Histobiology of craniopharyngiomas and features of the disease course. The dissertation for the degree of Doctor of Medicine. Moscow, 2000.
14. Konovalov AN, Vihert TM, Korshunov AG, Gorelyshev SK. Assessment of the removal radicality of craniopharyngiomas from third ventricle in children and possible sources of continued growth

and relapse. Zh Vopr Neirokhir Im N N Burdenko. 1988;6:7–12.
15. Elliott RE, Jane JA Jr, Wisoff JH. Surgical management of craniopharyngiomas in children: meta-analysis and comparison of transcranial and transsphenoidal approaches. Neurosurgery. 2011;69(3):630–43; discussion 643
16. Mazerkina NA, et al. Endocrine disorders in craniopharyngiomas in children: dependence on the site of a tumor. Zh Vopr Neirokhir Im N N Burdenko. 2008;1:23–9; discussion 29.
17. Zuccaro G. Radical resection of craniopharyngioma. Childs Nerv Syst. 2005;21(8–9):679–90.
18. Shi XE, et al. Craniopharyngioma: surgical experience of 309 cases in China. Clin Neurol Neurosurg. 2008;110(2):151–9.
19. Golanov AV, et al. The first experience of using the Gamma Knife unit for radiosurgical treatment of intracranial tumors. Zh Vopr Neirokhir Im N N Burdenko. 2007;1:3–11.
20. Iwata H, et al. Single and hypofractionated stereotactic radiotherapy with CyberKnife for craniopharyngioma. J Neuro-Oncol. 2012;106(3): 571–7.
21. Aggarwal A, Fersht N, Brada M. Radiotherapy for craniopharyngioma. Pituitary. 2013;16(1):26–33.
22. Savateev AN, Trunin YY, Mazerkina NA. Radiotherapy and radiosurgery in treatment of craniopharyngiomas. Zh Vopr Neirokhir Im N N Burdenko. 2017;81(3):94–106.
23. Muller HL. Paediatrics: surgical strategy and quality of life in craniopharyngioma. Nat Rev Endocrinol. 2013;9(8):447–9.
24. Carpenter RC, Chamberlin GW, Frazier CH. The treatment of hypophyseal stalk tumors by evacuation and irradiation. Am J Roentgenol. 1937;38:162–77.
25. Love JG, Marshall TM. Craniopharyngiomas. Surg Gynecol Obstet. 1950;90(5):591–601.
26. Kramer S, McKissock W, Concannon JP. Craniopharyngiomas. Treatment by combined surgery and radiation therapy. J Neurosurg. 1961;18:217–26.
27. Rajan B, et al. Craniopharyngioma—a long-term results following limited surgery and radiotherapy. Radiother Oncol. 1993;26(1):1–10.
28. Hetelekidis S, et al. 20-year experience in childhood craniopharyngioma. Int J Radiat Oncol Biol Phys. 1993;27(2):189–95.
29. Habrand JL, et al. The role of radiation therapy in the management of craniopharyngioma: a 25-year experience and review of the literature. Int J Radiat Oncol Biol Phys. 1999;44(2):255–63.
30. Van Effenterre R, Boch AL. Craniopharyngioma in adults and children: a study of 122 surgical cases. J Neurosurg. 2002;97(1):3–11.
31. Leksell L. The stereotaxic method and radiosurgery of the brain. Acta Chir Scand. 1951;102(4):316–9.
32. Xu Z, et al. Outcomes of Gamma Knife surgery for craniopharyngiomas. J Neuro-Oncol. 2011;104(1):305–13.
33. Hasegawa T, Kobayashi T, Kida Y. Tolerance of the optic apparatus in single-fraction irradiation using stereotactic radiosurgery: evaluation in 100 patients with craniopharyngioma. Neurosurgery. 2010;66(4):688–94; discussion 694–5.
34. Niranjan A, et al. Radiosurgery for craniopharyngioma. Int J Radiat Oncol Biol Phys. 2010;78(1):64–71.
35. Kobayashi T, et al. Long-term results of gamma knife surgery for the treatment of craniopharyngioma in 98 consecutive cases. J Neurosurg. 2005;103(6 Suppl):482–8.
36. Jeon C, et al. The therapeutic efficacy of fractionated radiotherapy and gamma-knife radiosurgery for craniopharyngiomas. J Clin Neurosci. 2011;18(12):1621–5.
37. Iwata H, et al. Hypofractionated stereotactic radiotherapy with CyberKnife for nonfunctioning pituitary adenoma: high local control with low toxicity. Neuro-Oncology. 2011;13(8):916–22.
38. Chang SD, et al. An analysis of the accuracy of the CyberKnife: a robotic frameless stereotactic radiosurgical system. Neurosurgery. 2003;52(1):140–6; discussion 146–7.
39. Antypas C, Pantelis E. Performance evaluation of a CyberKnife G4 image-guided robotic stereotactic radiosurgery system. Phys Med Biol. 2008;53(17):4697–718.
40. Lee M, et al. Radiation therapy and CyberKnife radiosurgery in the management of craniopharyngiomas. Neurosurg Focus. 2008;24(5):E4.
41. Leber KA, Bergloff J, Pendl G. Dose-response tolerance of the visual pathways and cranial nerves of the cavernous sinus to stereotactic radiosurgery. J Neurosurg. 1998;88(1):43–50.
42. Brada M, Thomas DG. Craniopharyngioma revisited. Int J Radiat Oncol Biol Phys. 1993;27(2):471–5.
43. Combs SE, et al. Achievement of long-term local control in patients with craniopharyngiomas using high precision stereotactic radiotherapy. Cancer. 2007;109(11):2308–14.
44. Kalapurakal JA, et al. Clinical outcome in children with recurrent craniopharyngioma after primary surgery. Cancer J. 2000;6(6):388–93.
45. Merchant TE. Craniopharyngioma radiotherapy: endocrine and cognitive effects. J Pediatr Endocrinol Metab. 2006;19(Suppl 1):439–46.
46. Minniti G, et al. Fractionated stereotactic conformal radiotherapy following conservative surgery in the control of craniopharyngiomas. Radiother Oncol. 2007;82(1):90–5.
47. Selch MT, et al. Initial clinical results of stereotactic radiotherapy for the treatment of craniopharyngiomas. Technol Cancer Res Treat. 2002;1(1):51–9.
48. Varlotto JM, et al. External beam irradiation of craniopharyngiomas: long-term analysis of tumor control and morbidity. Int J Radiat Oncol Biol Phys. 2002;54(2):492–9.
49. Regine WF, Mohiuddin M, Kramer S. Long-term results of pediatric and adult craniopharyngio-

mas treated with combined surgery and radiation. Radiother Oncol. 1993;27(1):13–21.
50. Chiou SM, et al. Stereotactic radiosurgery of residual or recurrent craniopharyngioma, after surgery, with or without radiation therapy. Neuro-Oncology. 2001;3(3):159–66.
51. Tiulpakov AN, et al. Growth in children with craniopharyngioma following surgery. Clin Endocrinol. 1998;49(6):733–8.
52. Kiehna EN, Merchant TE. Radiation therapy for pediatric craniopharyngioma. Neurosurg Focus. 2010;28(4):E10.
53. Vladyka V, et al. Radiation tolerance of functioning pituitary tissue in gamma knife surgery for pituitary adenomas. Neurosurgery. 2003;52(2):309–16; discussion 316–7.
54. Regine WF, Kramer S. Pediatric craniopharyngiomas: long term results of combined treatment with surgery and radiation. Int J Radiat Oncol Biol Phys. 1992;24(4):611–7.
55. Xu Z, et al. Hypopituitarism after stereotactic radiosurgery for pituitary adenomas. Neurosurgery. 2013;72(4):630–7; 636–7.
56. Grattan-Smith PJ, et al. Neuropsychological abnormalities in patients with pituitary tumours. Acta Neurol Scand. 1992;86(6):626–31.
57. Peace KA, et al. The effect of treatment variables on mood and social adjustment in adult patients with pituitary disease. Clin Endocrinol. 1997;46(4):445–50.
58. Merchant TE, et al. Craniopharyngioma: The St. Jude Children's Research Hospital experience 1984-2001. Int J Radiat Oncol Biol Phys. 2002;53(3):533–42.
59. Merchant TE, et al. Phase II trial of conformal radiation therapy for pediatric patients with craniopharyngioma and correlation of surgical factors and radiation dosimetry with change in cognitive function. J Neurosurg. 2006;104(2 Suppl):94–102.
60. Pemberton LS, et al. Experience of external beam radiotherapy given adjuvantly or at relapse following surgery for craniopharyngioma. Radiother Oncol. 2005;77(1):99–104.
61. Clark AJ, et al. A systematic review of the results of surgery and radiotherapy on tumor control for pediatric craniopharyngioma. Childs Nerv Syst. 2013;29(2):231–8.
62. Wen BC, et al. A comparison of the roles of surgery and radiation therapy in the management of craniopharyngiomas. Int J Radiat Oncol Biol Phys. 1989;16(1):17–24.
63. Stripp DC, et al. Surgery with or without radiation therapy in the management of craniopharyngiomas in children and young adults. Int J Radiat Oncol Biol Phys. 2004;58(3):714–20.
64. Karavitaki N, et al. Craniopharyngiomas in children and adults: systematic analysis of 121 cases with long-term follow-up. Clin Endocrinol. 2005;62(4):397–409.
65. Lin LL, et al. Long-term outcome in children treated for craniopharyngioma with and without radiotherapy. J Neurosurg Pediatr. 2008;1(2):126–30.
66. Smee RI, et al. Modern radiotherapy approaches in the management of craniopharyngiomas. J Clin Neurosci. 2011;18(5):613–7.
67. Greenfield BJ, et al. Long-term disease control and toxicity outcomes following surgery and intensity modulated radiation therapy (IMRT) in pediatric craniopharyngioma. Radiother Oncol. 2015;114(2):224–9.

Malignant Gliomas

27

Franziska Loebel

27.1 Introduction

Malignant glial tumors include a variety of tumor entities, including astrocytomas, oligodendrogliomas, ependymomas, or neurocytomas. In general, "anaplastic astroctytomas" and "anaplastic oligodendrogliomas" (WHO grade III) as well as "glioblastomas" (WHO grade IV) are summarized as malignant gliomas due to their aggressive natural course of disease.

Of these, glioblastoma multiforme (GBM, WHO IV) is the most prevalent and lethal type of all malignant gliomas and primary malignant brain tumors in adults. It accounts for 46.1% of primary malignant brain and CNS tumors and has an annual incidence of 3.1 per 100,000 people. According to the Central Brain Registry of the United States, the relative 5-year survival rate of 5.1% foreshadows a median life expectancy of about 15–18 months from initial diagnosis. GBMs are mostly found in patients in their sixth and seventh decade, with a higher incidence in the male population [1].

The current standard of treatment for malignant gliomas consists of radical surgical tumor resection, followed by concurrent chemoradiation to a total dose of 60 Gy (30 x 2 Gy) with daily temozolomide (TMZ) chemotherapy, followed by up to six cycles of TMZ with a monthly dose of 150–200 mg/m^2 [2, 3]. Despite this protocol, disease recurrence is certain after an average of 6 to 9 months, and subsequent treatment options are limited to re-excision, re-irradiation, or systemic chemotherapy, all of which have limited activity and might compromise the patient's quality of life [4]. Stereotactic radiosurgery (SRS, e.g., using the CyberKnife or Gamma Knife systems) is a safe and effective non-invasive treatment alternative for malignant glioma patients that can be performed as an outpatient procedure and is typically used as a salvage treatment in recurrent cases or difficult-to-resect lesions [5]. This chapter illuminates the technical aspects, in vitro experiments, and results of in vivo studies applying SRS as treatment for malignant gliomas.

27.2 Technical Aspects of Radiosurgery for Malignant Gliomas

Re-irradiation of tissue that previously received conventional radiotherapy is often restricted due to increased risk of radiation toxicity.

Tightly focused external photon beams used in radiosurgery allow for high accuracy in delivering high doses to small target volumes, thus reducing therapy-related side effects to surrounding brain tissue. Radiosurgery for malignant gliomas is

F. Loebel (✉)
Department of Neurosurgery & CyberKnife Center, Charité University Medicine, Berlin, Germany
e-mail: franziska.loebel@charite.de

© Springer Nature Switzerland AG 2020
A. Conti et al. (eds.), *CyberKnife NeuroRadiosurgery*,
https://doi.org/10.1007/978-3-030-50668-1_27

mostly administered in a single fraction but can also be delivered in a hypofractionated fashion for larger target volumes [5]. Most commonly, single fraction doses around 20 Gy are applied. However, a variety of doses ranging from 14 to 32 Gy have been described in the literature. Hypofractionated concepts usually have a median prescription dose of 23 Gy (range 12–28 Gy), with a median three fractions applied. Median isodoses used range from 70% to 80% [6].

The conformality and precise targeting with SRS may be especially useful in the delivery of radiation treatment in malignant gliomas in close proximity to critical anatomical structures (such as eyes, optic nerves, optic chiasm, or brainstem), in order to prevent late radiation-induced complications. When used in adjunct to conventional radiation therapy, CyberKnife radiosurgery has been shown to improve the targeting accuracy in the treatment of malignant gliomas [7].

27.3 In Vitro Studies of SRS in Malignant Gliomas

Several in vitro studies have tried to elucidate the effects of CyberKnife irradiation on glioma cell lines. In these experimental studies, a significant decrease of surviving fraction was observed after intermittent irradiation using a CyberKnife system when compared to continuous irradiation using a conventional linear accelerator (LINAC) [8]. However, further studies suggested that CyberKnife-driven irradiation significantly increased the invasion potential of human glioma cell lines as well as resulted in elevation of levels of TGF-β and β1-integrin [9]. The exact effects of CyberKnife irradiation on a cellular level still need further investigation.

27.4 Newly Diagnosed Malignant Gliomas

The earliest results of radiosurgical treatment of malignant gliomas were reported by Yoshikawa et al. in 2006. Their 25 patients with malignant gliomas were treated with the CyberKnife system applying a median dose of 20.3 Gy (range 13.9–26.4 Gy) to 44 lesions. Median survival in the GBM patients was 20.7 months after diagnosis. Only one patient suffered from symptoms due to delayed radiation necrosis. No other acute or delayed neurological morbidity was seen, suggesting SRS to be a well-tolerated treatment strategy in malignant gliomas [10].

Following the results of a large prospective randomized study, the "STUPP-protocol" (conventional radiation to a total dose of 60 Gy and up to six cycles of TMZ) has become the standard postoperative neuro-oncological care for newly diagnosed GBM [2]. To elucidate whether radiosurgical therapy is more beneficial in newly diagnosed or recurrent GBM, a multicenter study in 2009 was conducted by Villavicienco et al. The outcome of 20 patients who received CyberKnife radiation upon initial diagnosis of GBM were compared with 26 patients who underwent CyberKnife treatment for tumor recurrence. Mean survival of the patients treated with CyberKnife initially was 11.5 months vs. 21 months in the patients with recurrent tumor. Median survival following CyberKnife radiation treatment was 9.5 months vs. 7 months. In fact, there seemed to be no apparent survival advantage to using CyberKnife for newly diagnosed patients, and it should be reserved for tumor recurrence [11].

27.5 Recurrent Malignant Gliomas

Typically, radiosurgery is used as a salvage treatment in recurrent GBM after initial concurrent chemoradiation and other therapies and has been shown to be effective and achieve a more favorable outcome in the treatment of recurrent GBM (Fig. 27.1).

The first report to investigate this strategy was published in 2008. After undergoing gross total resection, subtotal resection, or biopsy for recurrent GBM, 20 patients were treated with CyberKnife radiosurgery. In some patients, additional adjuvant ACNU or Vincristine chemotherapy was prescribed. The median overall survival rate was 16 months, which is favorable

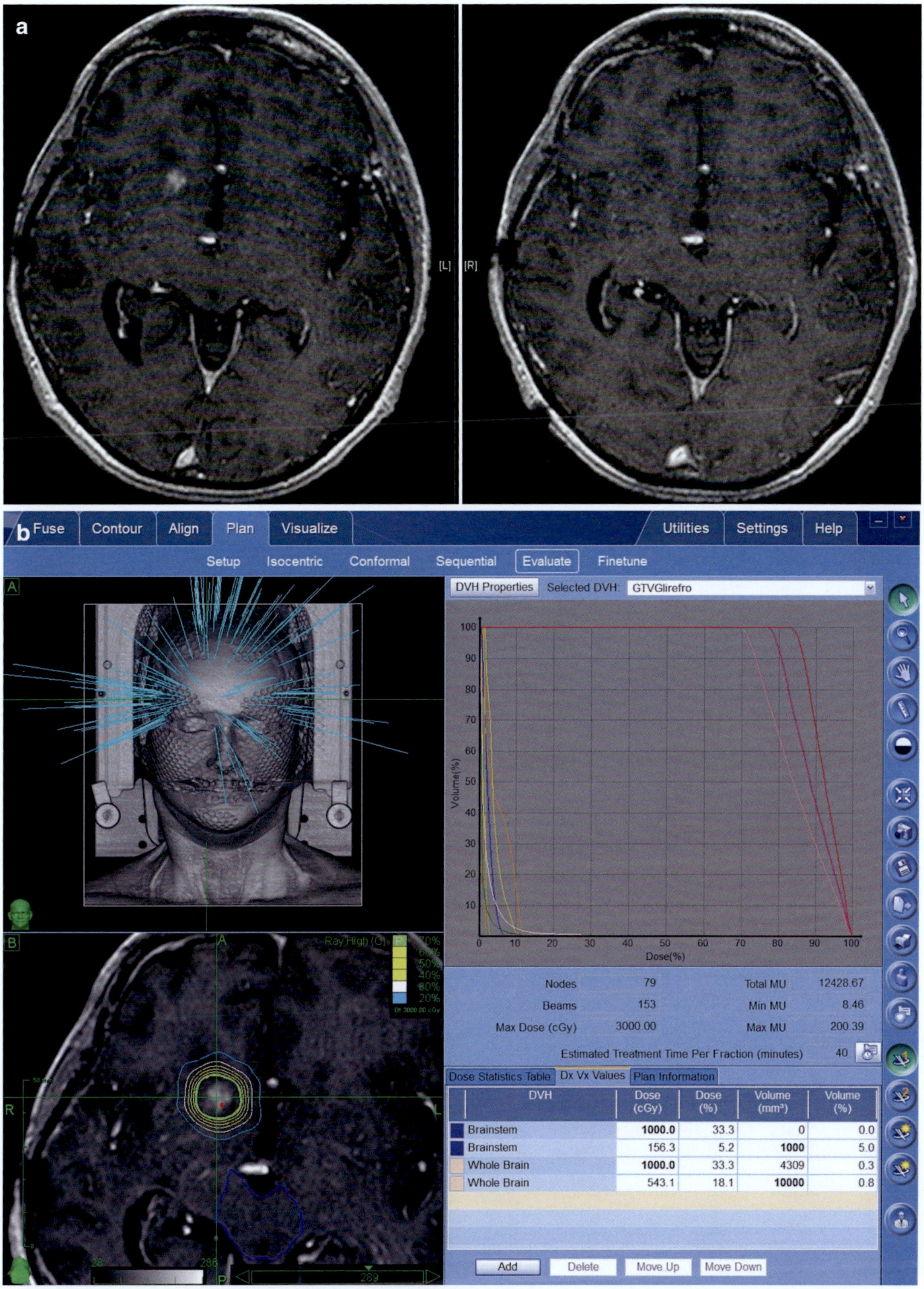

Fig. 27.1 (**a**) *Left:* preoperative MRI of 67-year-old female diagnosed with glioblastoma multiforme (GBM) IDHwt and MGMT+ diagnosed and resected. The patient received STUPP chemoradiation until May 2017. The patient presented the small recurrence showed in the MRI scan in May 2018. *Right:* the patient was treated with CyberKnife radiosurgery and the treatment consisted of 21 Gy in single fraction. The contrast enhancing lesion disappeared in August 2018. (**b**) The CyberKnife treatment plan showing dose distribution and dose-volume histograms (DVH)

when compared to historical controls [12]. In another large cohort of 128 recurrent high-grade glioma patients with 161 lesions, median survival from initial diagnosis was 32 months with median survival following SRS of 11.5 months [6]. But even if the survival benefit is only marginal, a single-fraction or hypofractionated therapy is preferable to a multi-week therapy from a patients' point of view and increases quality of life in these patients with very limited life expectancy.

Levy et al. investigated factors associated with progression-free survival (PFS) following stereotactic irradiation for recurrent malignant gliomas. In 13 patients, they found the median survival after SRS to be 14 months, with relapse-free survival of 3.7 months. Factors associated with prolonged PFS were patient age, total dose, dose per fraction, and number of fractions. The authors concluded that stereotactic re-irradiation for recurrent malignant glioma is well tolerated and a dose of more than 30 Gy delivered in 5 or more doses seems to prolong relapse-free survival [13]. Other studies suggested that in patients with recurrent malignant gliomas, performance status (as graded on the Karnofsky scale) at time of re-irradiation is a significant predictor of PFS. Median OS and PFS periods following SRS for recurrent malignant gliomas were reported to be 9.0 and 3.0 months, respectively [14].

Concerning overall survival (OS) in patients treated with radiosurgery, factors that are significantly improving OS have been found to be age <40 years, salvage surgery before SRS, and additional use of other post-SRS therapies [6].

Recurrence of malignant gliomas after bevacizumab failure represents a major clinical problem with very low overall survival rates (median OS <4 months). In these cases, SRS might serve as well-tolerated salvage therapy with low toxicity rates, good local control, and potential prolongation of OS, as has been shown by some authors. The median OS after salvage SRS after bevacizumab failure has been reported around 4.8 months [15].

27.6 Rare Entities: Radiosurgery of Malignant Brainstem Glioma and Optic Nerve Glioma

Brainstem glioma is one of the most challenging entities of malignant gliomas with a very poor prognosis. Surgical options are limited due to neurological eloquence and close proximity to critical structures. Conventional radiation is problematic, as diffuse radiation might impact surrounding tissue. Radiosurgical therapy presents a feasible treatment alternative because of the high precision and steep dose gradient. A recent study of 21 brainstem gliomas treated with CyberKnife radiosurgery showed efficacy with only mild toxicity. Median OS was 19 months [16].

Cases of optic pathway glioma which predominantly occur in children are equally challenging. Surgery is technically difficult with a high risk of visual deficits and almost always results in residual tumor. Accurately targeting these lesions with SRS has the potential to preserve vision in these young patients. Given the scarcity of these lesions, scientific reports on treatment parameters and outcome are rare. Case reports suggest potential for regression after fractionated SRS treatment with low toxicity [17]. However, more extended studies need to be performed.

27.7 Combination with Other Treatments

A combined approach that uses SRS in combination with chemotherapy produces a potential radiosensitization effect and an increase in cytotoxicity on tumor cells in distant areas.

"Dose-dense" administration of SRS and TMZ for recurrent GBM even increases median survival from 7 to 12 months ($p < 0.01$) as well as the 6-month progression-free survival (6-PFS) (66.7% vs. 18% [18]). The rates of radiation necrosis or hematological toxicity have been shown to be acceptable [19].

Also, concurrent application of angiogenesis inhibitors or immunotherapy (e.g., anti-epidermal growth factor receptor (125)l-m-antibody 425) is a feasible and safe treatment strategy and might prolong survival [20]. Further studies are needed to verify these findings.

27.8 Therapeutic Side Effects

Besides the observed survival benefits with radiosurgical treatment for malignant gliomas, especially in the setting of disease recurrence, treatment-associated toxicity remains a major point of discussion. Radiation necrosis is a possible complication after CyberKnife treatment. However, the incidence reported is small. In an early study, radiation necrosis was found in 4 of 61 patients with GBM treated with 3–6 fractions [21]. In a large cohort of 128 recurrent high-grade glioma patients with 161 lesions, the incidence of radiation necrosis was found to be around 6%, while other high-grade toxicity was not reported in any of the patients [6]. Similar rates of radiation necrosis were found elsewhere (11.4%) [14]. Tumor volumes and dose delivery are not associated with the appearance of unwanted treatment-induced necrosis. Careful planning and management of doses have been shown to eliminate the risk of radiation toxicity.

Total doses >40 Gy are associated with radiation necrosis and the use of hypofractionation reduces the onset and severity of the necrosis in patients with malignant glioma [22].

Interestingly, the appearance of radiation necrosis or pseudoprogression does not always implicate a worse outcome. As seen in a study of 37 patients treated with SRS for recurrent GBM with a median OS of 10.6 months after SRS, the patients showing pseudoprogression on their initial MRI scan experienced a significantly longer survival when compared to those with regression, stable disease, or progressive disease [23].

27.9 Conclusions

Radiosurgery can be applied as a salvage therapy in patients with malignant gliomas, especially in the setting of disease recurrence after initial concurrent chemoradiation or in cases that are considered technically difficult-to-resect due to eloquent anatomical location and high risk of perioperative deficits or complications. High target accuracy may be achieved through the application of tightly focused beams and reduces treatment-induced toxicity to surrounding brain tissue or critical structures in close proximity. While there is no apparent survival benefit found with SRS treatment in newly diagnosed malignant gliomas, favorable rates of overall survival as well as prolonged progression-free survival were reported with the use of radiosurgery in recurrent malignant glioma cases. Incidence of treatment-related toxicity, such as radiation necrosis, remains low in cases treated with SRS, and the risk of radiation toxicity can be minimized by careful treatment planning and management of doses. Hypofractionation may reduce the occurrence and severity of necrosis in malignant glioma patients even further, making SRS a safe and well-tolerated treatment in malignant glioma patients [6].

In conclusion, despite occasional incidence of treatment-induced toxicity, SRS represents a safe and feasible option to treat patients with recurrent malignant glioma or gliomas in proximity to critical structures with low complication rates and potential survival benefits.

References

1. Ostrom QT, et al. CBTRUS statistical report: primary brain and central nervous system tumors diagnosed in the United States in 2008–2012. Neuro Oncol. 2015;17(Suppl 4):iv1–iv62.
2. Stupp R, et al. Radiotherapy plus concomitant and adjuvant temozolomide for glioblastoma. N Engl J Med. 2005;352(10):987–96.
3. Wirsching H-G, Galanis E. Glioblastoma. In: Handbook of Clinical Neurology. Berger MS, Weller M, eds. 2016. San Diego: Elsevier. pp. 381–97
4. Mallick S, et al. Management of glioblastoma after recurrence: a changing paradigm. J Egypt Natl Canc Inst. 2016;28(4):199–210.
5. Bucholz RD, Laycock KA, Cuff LE. CyberKnife stereotactic radiosurgery for intracranial neoplasms, with a focus on malignant tumors. Technol Cancer Res Treat. 2010;9(6):541–50.
6. Pinzi V, et al. Radiosurgery reirradiation for high-grade glioma recurrence: a retrospective analysis. Neurol Sci. 2015;36(8):1431–40.

7. Oermann E, et al. CyberKnife enhanced conventionally fractionated chemoradiation for high grade glioma in close proximity to critical structures. J Hematol Oncol. 2010;3:22.
8. Canazza A, et al. Increased migration of a human glioma cell line after in vitro CyberKnife irradiation. Cancer Biol Ther. 2011;12(7):629–33.
9. Canazza A, et al. In vitro effects of Cyberknife-driven intermittent irradiation on glioblastoma cell lines. Neurol Sci. 2011;32(4):579–88.
10. Yoshikawa K, et al. CyberKnife stereotactic radiotherapy for patients with malignant glioma. Minim Invasive Neurosurg. 2006;49(2):110–5.
11. Villavicencio AT, et al. Survival following stereotactic radiosurgery for newly diagnosed and recurrent glioblastoma multiforme: a multicenter experience. Neurosurg Rev. 2009;32(4):417–24.
12. Lipani JD, et al. Survival following CyberKnife radiosurgery and hypofractionated radiotherapy for newly diagnosed glioblastoma multiforme. Technol Cancer Res Treat. 2008;7(3):249–55.
13. Levy S, et al. Reirradiation of gliomas under stereotactic conditions: prognostic factors for survival without relapse or side effects, a retrospective study at Tours Regional University Hospital (France). Cancer Radiother. 2017;21(8):759–65.
14. Adachi K, et al. Feasibility of salvage re-irradiation with stereotactic radiotherapy for recurrent Glioma using CyberKnife. Anticancer Res. 2019;39(6):2935–40.
15. Shi W, et al. Salvage fractionated stereotactic re-irradiation (FSRT) for patients with recurrent high grade gliomas progressed after bevacizumab treatment. J Neuro-Oncol. 2018;137(1):171–7.
16. Zhang J, et al. Clinical efficacy of CyberKnife radiosurgery for adult brainstem glioma: 10 years experience at Tianjin CyberKnife center and review of the literature. Front Oncol. 2019;9:257.
17. Uslu N, et al. Optic nerve glioma treatment with fractionated stereotactic radiotherapy. J Neurosurg Pediatr. 2013;11(5):596–9.
18. Conti A, et al. Efficacy and toxicity of CyberKnife re-irradiation and "dose dense" temozolomide for recurrent gliomas. Acta Neurochir. 2012;154(2):203–9.
19. Ekici K, et al. Efficacy of stereotactic radiotherapy as salvage treatment for recurrent malignant gliomas. J BUON. 2014;19(4):1029–34.
20. Hasan S, et al. Salvage fractionated stereotactic radiotherapy with or without chemotherapy and immunotherapy for recurrent glioblastoma multiforme: a single institution experience. Front Oncol. 2015;5:106.
21. Sato K, et al. Radiation necrosis and brain edema association with CyberKnife treatment. Acta Neurochir Suppl. 2003;86:513–7.
22. Vordermark D, et al. Hypofractionated stereotactic re-irradiation: treatment option in recurrent malignant glioma. BMC Cancer. 2005;5:55.
23. Yazici G, et al. Hypofractionated stereotactic reirradiation for recurrent glioblastoma. J Neuro-Oncol. 2014;120(1):117–23.

Pilocytic Astrocytomas

28

Yury Y. Trunin, Andrey V. Golanov, Alexander N. Konovalov, Natalia A. Antipina, and Konstantin V. Nikitin

28.1 Introduction

Pilocytic astrocytoma (PA) is a low-grade glial tumor that accounts for 25–30% of all central nervous system (CNS) tumors in children and 2–5% in adult patients [1, 2]. Surgical removal of the tumor is the main treatment for newly diagnosed and relapsing PA [3–5]. In the case of radical surgery, 5-year progression-free survival (PFS) ranges 90–100%.

With incomplete removal of the tumor, however, 10-year PFS does not exceed 15–50% with a high risk of recurrence in the first 2–3 years after surgery [3]. After total removal of the tumor, dynamic observation is performed; with incomplete removal, adjuvant treatment is required.

Chemotherapy (ChT) is the treatment of choice in newborn children with anterior optic gliomas, which may delay or replace radiation treatment in some patients [6, 7]. An increased risk of developing late toxicity (endocrine and neurocognitive disorders) in younger patients with tumors located in the chiasm-sellar region is the main reason to postpone radiation treatment at a later date, until the onset of progression of the disease [7–9].

Relapse-free survival rates with chemotherapy alone are on average two times lower than after radiation treatment (40–45% versus 75–90%) [6, 7, 9–15]. Radiation therapy has been the standard of care for patients with low grade glioma (PA) for many years and can achieve high survival rates comparable to those of radical tumor resection [9, 13, 15].

28.2 Approaches to Radiation Treatment

In radiation treatment of patients with PA, the standard regimen is usually adopted (1.8–2.0 Gy/fraction, up to a total of 50–54 Gy). This is due to the frequent localization of the tumor in proximity of critical structures (chiasma, brainstem, subcortical eloquent areas), the presence of a diffuse component in the tumor, and the need to irradiate the tumor bed after surgical removal. Furthermore, conventional irradiation, previously used, is only possible in the standard fractionation mode.

This radiation treatment regimen has been the most reported in the literature [10, 12, 13, 16]. Nonetheless, the development of novel modalities of irradiation—stereotactic radiosurgery and stereotactic fractionated radiotherapy—providing an improved profile of radiation exposure of critical structures, has led to inclusion of these

Y. Y. Trunin · A. V. Golanov (✉) · A. N. Konovalov · N. A. Antipina · K. V. Nikitin
National Medical Research Center of Neurosurgery named after Burdenko, Moscow, Russia
e-mail: ytrunin@nsi.ru; golanov@nsi.ru; akonovalov@nsi.ru; nantipina@nsi.ru; knitin@si.ru

© Springer Nature Switzerland AG 2020
A. Conti et al. (eds.), *CyberKnife NeuroRadiosurgery*,
https://doi.org/10.1007/978-3-030-50668-1_28

techniques in the standard radiation treatment modalities of patients with PA [14, 17].

28.3 Radiosurgery and Hypofractionation

The hypofractionated stereotactic radiation therapy (hFSRT) is close in its radiobiological effects to radiosurgery. Both have similar indications for radiation treatment: a limited tumor volume and clear tumor boundaries, especially in patients with relapses.

Kida et al. observed that the use of the hypofractionation regimen (2–7 fractions) allows to bring a higher biologically effective dose (BED), with risks of complications comparable to radiosurgery. This is particularly true for tumor locations close to critical structures (brain stem, visual pathways) [18].

To date, there are no clear indications for stereotactic irradiation in primary patients with low-grade gliomas (LGG), because in most cases, primary treatment involves the use of marginal expansion in the area of healthy tissue in the definition of the CTV. The inclusion of healthy tissues involves the use of a standard fractionation modality, in which the effectiveness of the treatment is based on the different radiosensitivity of normal versus tumor cells.

During the treatment planning of radiosurgery and hFSRT, the CTV does not exceed 1 and 2 mm, respectively. This is however associated with a high risk of functional and structural damages to the normal tissue that fall into the treatment volume. Successful use of stereotactic radiation in radiosurgery and hFSRT modalities in a number of studies confirms the fact that not all primary PA require marginal expansion. Its value needs to be revised. There are only some studies where this approach was used and the number of observations in these papers is not relevant [19–21].

Hadjipanayis et al. [20] evaluated the results using of radiosurgery for patients with PA of various localization in 37 children (54% with relapses): 18 tumors were in the brainstem, 5 in the thalamus, 3 in the corpus callosum, optic tract, and hypothalamic region; the remaining PAs were located in the cerebral hemispheres. The average dose to the border of the tumor was 15 Gy (9.6–22.5 Gy). A series of control MRI studies revealed a complete tumor response in ten patients, a partial tumor response in eight patients, and stabilization in seven patients. Tumor progression was observed in 12 children.

The authors noted that tumor growth control indicators were slightly better (71%) in patients who underwent primary radiation therapy than patients who underwent treatment after a tumor relapse (64%). Simonova et al. [19] evaluated the results of radiotherapy treatment in patients with I–II grade astrocytomas (PA 47%) using hypofractionation technique. Dose per fraction was 5 Gy, to a total dose of 25 Gy. Dynamic observation was carried out for 5 years. Partial or complete tumor response was observed in 83%, stabilization of the disease in 11%, and progression in 6% of cases. The average tumor response time was 18 months after the treatment 5-year PFS was 88% complications were noted in 6%. The results of radiosurgical and hFSRT in patients with PA are summarized in Table 28.1.

The total number of observations in the literature analyzed was 309 patients, of which 18 patients were treated by hypofractionation. The median follow-up for all studies is 72 months. The mean value of tumor growth control after radiosurgical treatment is 80.9%. The weighted average of 5-year OS is 94.4%. The average dose of radiosurgical treatment is 14.9 Gy (9.6–22.5 Gy).

28.4 NMRC Burdenko's Experience

From April 2005 to December 2018 at the Department of Radiotherapy and Radiosurgery of NMRC named after N.N. Burdenko, 431 patients with a diagnosis of intracranial pilocytic astrocytoma (PA) underwent stereotactic radiation treatment. Of these, 118 patients underwent hFSRT, whereas 57 patients received radiosurgery treatment. The median age was 12.9 years (2–52 years). The tumor was most often localized in the cerebellum, 23 patients (40.4%); the brain-

Table 28.1 Results of radiosurgical and hypofractionated stereotactic radiotherapy in patients with pilocytic astrocytoma

Author (year)	*N* pts	Median of FU time (mo)	Mode of SRT	Total dose Gy	Survival%	
					Tumor control/ PFS	Overall
Kida et al. (2000) [22]	12	24	SRS	12	TC 92%	–
Boethius et al. (2002) [23]	17	72	SRS	10–20	TC100%	5 OS 100%
Hadjipanayis et al. (2002) [20]	37	–	SRS	15 (9.6–22.5)	TC 68%	7 OS 76%
Kano et al. (2009) [21]	50	54	SRS	11–22	TC 54%	10 OS 97.4%
Lizarraga et al. (2012) [24]	3	144	SRS	18.75 (16.7–20)	TC 33%	OS 91.7%
Hallemeler et al. (2012) [25]	18	96	SRS	15 (12–20)	61.2% 5 PFC 41%	10 OS 71%
Simonova et al. (2005) [19]	5	181	SRS	16	TC 96% 10 PFS 80%	10 OS 96%
Simonova et al. (2016) [26]	18		SRT hypo	25/5fr		
Trifiletti et al. (2017) [27]	149	62	SRS	17 (4–20)	TC 93% 12 PFS 80%	10 OS 100%
Total (9 papers)	*N* 309	72	7SRS 1SRT	SRS weighted average dose (Gy) 15 9.6–22.5 Gy	TC 80.9% (33–100%)	5–10 OS weighted average OS 94.4% (76–100%)
NMRC Burdenko experience	57	45	SRS	18 Gy (12–30)	5 PFS 97.5%	5 OS 99%
NMRC Burdenko experience	61		SRT hypo	30 Gy (21–30)		

stem, 13 patients (22.8%); in the area of basal ganglia and thalamus, 11 patients (19.3%); and in the cerebral hemispheres in 10 patients (17.5%).

Fifty-one patients (89.5%) had a history of surgery and histological verification of the pathology. In six patients (10.5%), the diagnosis was made on the basis of clinical and radiological data. The majority of patients underwent treatment due to recurrence: 33 patients (57.9%). 18 patients (31.6%) underwent adjuvant treatment due to the presence of a residual tumor after surgery. The median tumor volume (GTV) was 1.9 cm^3 (0.14–19.3 cm^3). The average radiosurgical dose was 18 Gy (12–30 Gy).

Sixty-one patients underwent hFSRT. The median age was 8 years (1–66). The hypofractionation regimen was often performed in patients with brain stem tumors, 25 patients (41%); located in the optic tract, 23 patients (37.7%); with cerebellar location, 8 patients (14.1%); in subcortical structures, 4 patients (6.6%); and cerebral hemispheres, 1 patient (1.6%). The ratio between primary and relapsing patients was 32/29. A total of 43 patients (70.5%) were operated on, 10 of them 2 or more times. The median tumor volume (GTV) was 6.24 cm^3 (0.1–36.9 cm^3). The dose per fraction was 5–5.5 Gy, average total dose –28 Gy (24–35 Gy).

28.5 The Results of the Treatment

The median follow-up from the time of diagnosis was 82 months (13–313 months); the median follow-up after treatment was 48 months (3–151 months). At the time of the follow-up, all patients were alive. In 23 patients (19.5%), some events occurred: in the majority—21 patients (17.8%)—a phenomenon of "pseudoprogression" was noted, which we defined as a temporary increase in volume of the tumor by 10% or more with its subsequent spontaneous regression or stabilization without additional treatment (Fig. 28.1).

In two patients, there was a progression of the disease outside the irradiation area: the appearance of two intracerebral foci in one patient and multiple diffusions along the subarachnoid spaces in the second. Therefore, the 5-year PFS indicator amounted to 97.5% and 5-year OS – 100%.

An indicator of 5-year survival without pseudoprogression (5-y PFS) was evaluated depending on the irradiation technique. There was no statistically significant difference in patients who underwent radiosurgery or hFSRT (Fig. 28.2a).

There was also no statistical difference in patients PFS treated with hypofractionation, depending on the total dose greater than or less than 30 Gy (Fig. 28.2b). Acute toxicity headaches were observed in 10% of patients during radiation therapy and in the early stages after treatment. Long-term complications were not recorded.

28.6 Clinical Case

Child A., 5 years old, the disease appeared 3 months before admission with the development of unsteadiness when walking and then nausea and periodic vomiting. MRI revealed a tumor of the pons and medulla oblongata. On 25/02/2010, the child was operated on at the NMRC Burdenko to remove a brainstem glioma. Histological diagnosis was pilocytic astrocytoma with polymorphism of nuclei and areas of dense cell arrangement.

On the control MRI of the brain before and after contrast enhancement performed 3 months after the operation, a residual tumor was disclosed, and therefore, radiation treatment was planned (Fig. 28.3a). At the time of the radiation

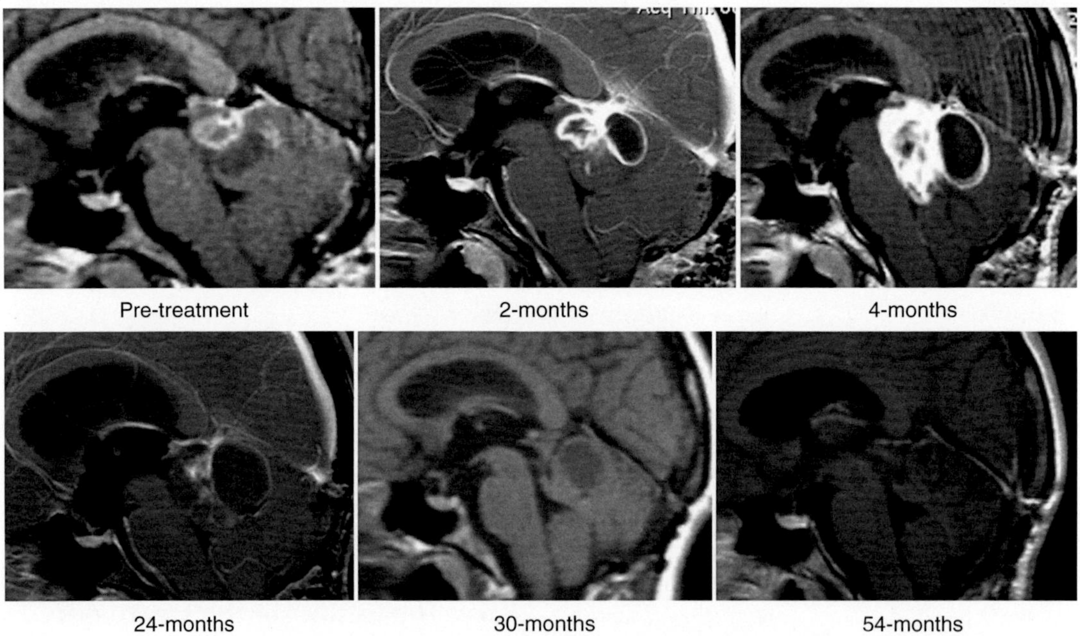

Fig. 28.1 Patient C, 12 years with remaining part after incomplete surgical removal of PA, was treated in the conventional fractionation regimen with a median total dose of 56 Gy. At 2 and 4 months, the contrast area and cyst increased. Taking into account the patient's stable condition, it was decided to delay the intervention. 24 and 30 months after treatment with RT, the growth of the cyst and contrast is gradually decreasing. After 4.5 years of treatment, the residual tumor was not noticed

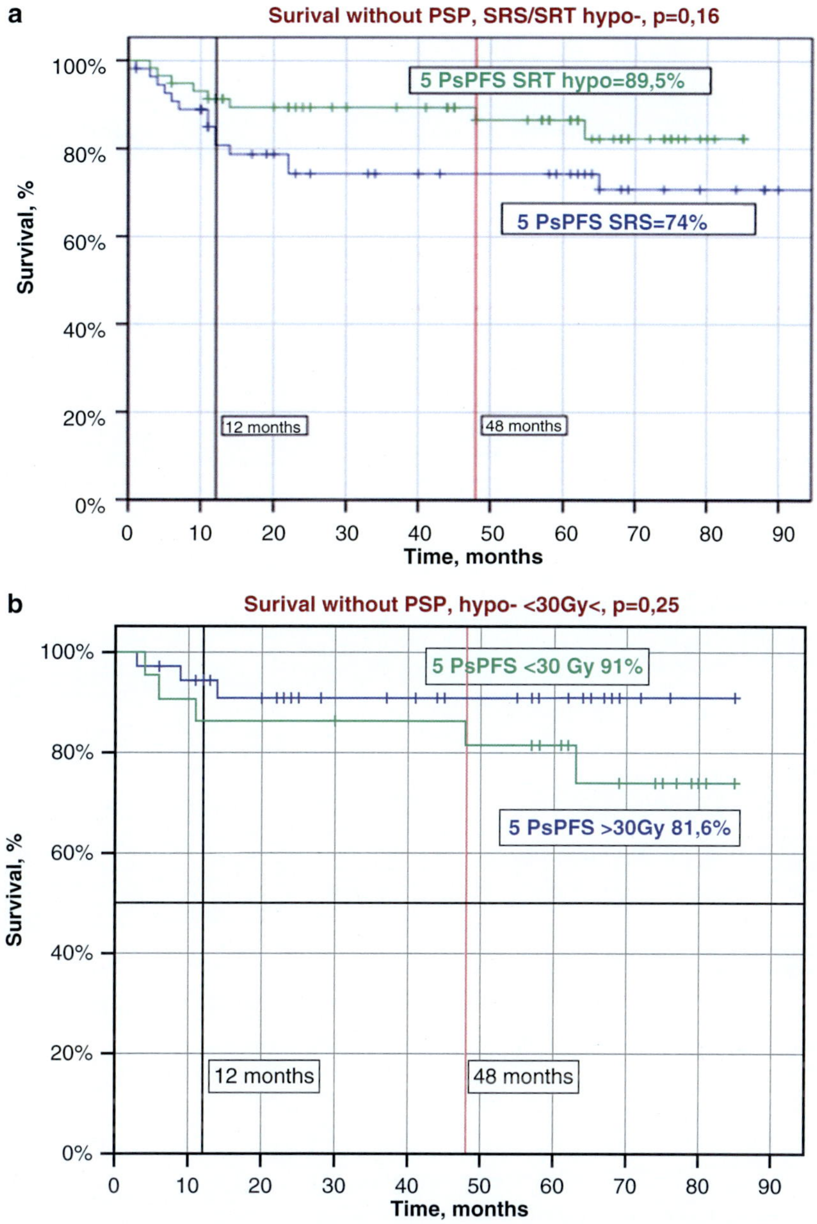

Fig. 28.2 5-year survival without pseudoprogression (5-y PFS) in patients undergoing CyberKnife treatment of pilocytic astrocytoma. (**a**) There was no statistically significant difference in patients who underwent radiosurgery or hFSRT. (**b**) No statistical difference in patients treated with hFSRT was evidenced for a total dose greater than or less than 30 Gy

treatment, the neurological examination revealed minimal neurological symptoms as a horizontal nystagmus with XII and central type VII cranial nerves palsy. Given the presence of a clear border of the tumor in all MRI sequences, its small volume, we decided to deliver an hFSRT adjuvant treatment (Fig. 28.3b).

An hFSRT course was conducted with the CyberKnife linear accelerator from 29/06/2019 to 06/07/2019: to the target area (CTV = PTV = 12.32 cm^3) using the multiple-beam technique, 6 fractions × 5 Gy were to a total dose of 30 Gy were delivered. The patient tolerated the treatment satisfactorily. During the

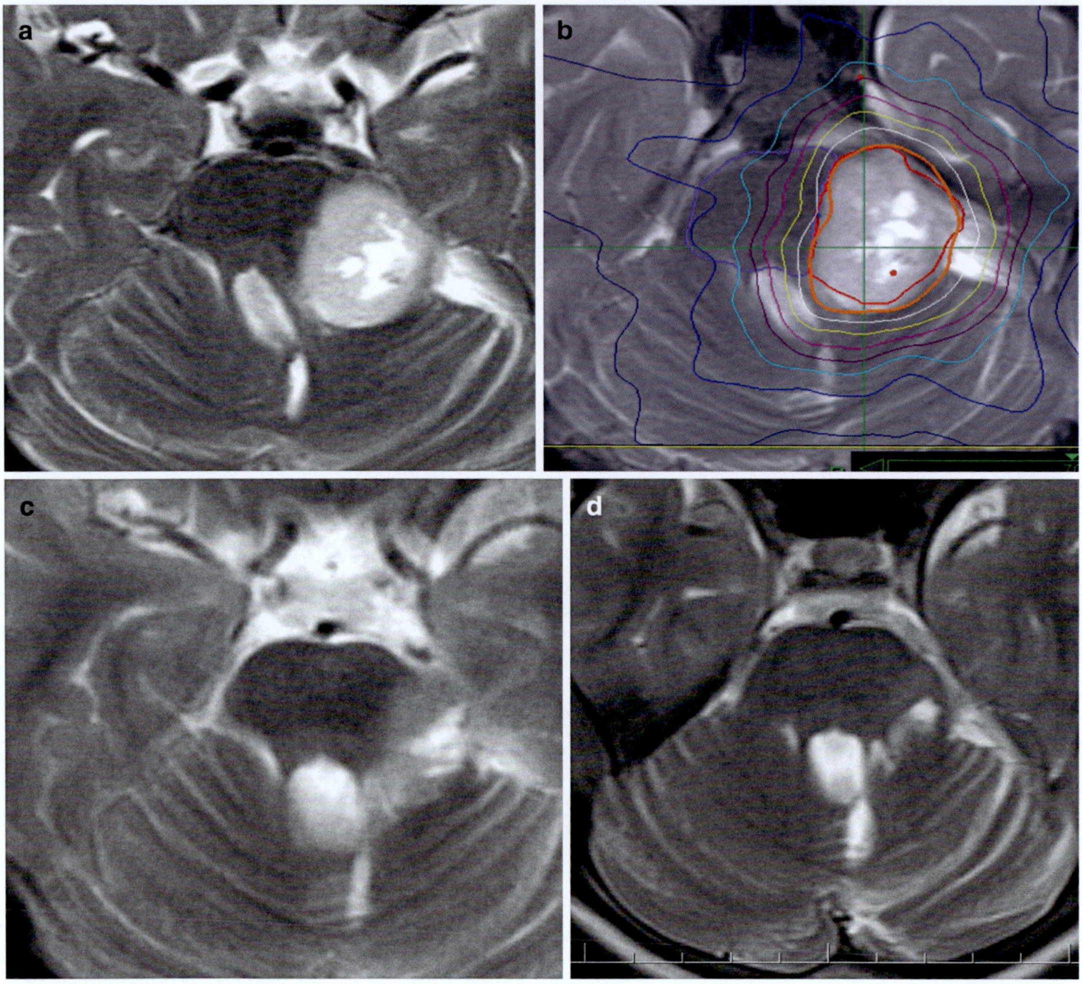

Fig. 28.3 (**a**) 3 months post-surgical MRI of a 5-year-old patient with residual pilocytic astrocytoma (**b**) A hFSRT course was planned using the CyberKnife. Target volume (CTV = PTV = 12.32 cm^3). Six fractions × 5 Gy were delivered to a total dose of 30 Gy. (**c**) 7 months after the RT course, a fast tumor response to the treatment was observed. (**d**) 9 years after irradiation, no apparent residual tumor was detected

follow-up examination 7 months after the RT course, a fast tumor response to the treatment was observed (Fig. 28.3c). Nine years after irradiation, no apparent residual tumor was detected (Fig. 28.3d). No clinical deterioration was reported.

28.7 Conclusion

The use of CyberKnife for radiosurgery and hypofractionation provides high levels of growth control (98.8%) of pilocytic astrocytomas, with low risk of complications.

The high rate of tumor growth control in our clinical series after radiosurgical treatment, which exceeds the weighted average values in the literature, is most likely associated with careful patient selection for radiosurgical treatment and a relatively small median observation time.

There only are a few studies dedicated to hypofractionated irradiation in the literature. Rates of tumor growth after hypofractionation are comparable to those of radiosurgery. A low risk of complications can be associated with the absence of margins during hypofractionated irradiation, which is confirmed by literature data on the standard fractionation.

Prospective studies on hFSRT in patients with pilocytic astrocytomas and a comparison of this therapeutic regimen with other methods are necessary to definitively establish its efficacy and safety.

References

1. Lee KJ, Marchan E, Peterson J, Harrell AC, Quinones-Hinojosa A, Brown PD, Trifiletti DM. Management and survival of adult patients with pilocytic astrocytoma in the National Cancer Database. World Neurosurg. 2018;112:e881–7. https://doi.org/10.1016/j.wneu.2018.01.208.
2. Ohgaki H, Kleihues P. Population-based studies on incidence, survival rates, and genetic alterations in astrocytic and oligodendroglial gliomas. J Neuropathol Exp Neurol. 2005;64(6):479–89.
3. Wisoff JH, Sanford RA, Heier LA, Sposto R, Burger PC, Yates AJ, et al. Primary neurosurgery for pediatric low-grade gliomas. A prospective multi-institutional study from the Children's Oncology Group. Neurosurgery. 2011;68(6):1548–54; discussion 1554–5. https://doi.org/10.1227/NEU.0b013e318214a66e.
4. Serova NK. Clinical neuroophtalmology: neurosurgical aspects. Tver': Triada; 2011. (in russian).
5. Berger MS, Deliganis AV, Dobbins J, Keles GE. The effect of extent of resection on recurrence in patients with low grade cerebral hemisphere gliomas. Cancer. 1994;74(6):1784–91.
6. Gnekow AK, Walker DA, Kandels D, Picton S, Giorgio P, Grill J, et al. A European randomised controlled trial of the addition of etoposide to standard vincristine and carboplatin induction as part of an 18-month treatment programme for childhood (≤16 years) low grade glioma—a final report. Eur J Cancer (Oxford, England: 1990). 2017;81:206–25. https://doi.org/10.1016/j.ejca.2017.04.019.
7. Ater JL, Zhou T, Holmes E, Mazewski CM, Booth TN, Freyer DR, et al. Randomized study of two chemotherapy regimens for treatment of low-grade glioma in young children. A report from the Children's oncology group. J Clin Oncol. 2012;30(21):2641–7. https://doi.org/10.1200/JCO.2011.36.6054.
8. Grill J, Couanet D, Cappelli C, Habrand JL, Rodriguez D, Sainte-Rose C, Kalifa C. Radiation-induced cerebral vasculopathy in children with neurofibromatosis and optic pathway glioma. Ann Neurol. 1999;45(3):393–6.
9. Merchant TE, Conklin HM, Wu S, Lustig RH, Xiong X. Late effects of conformal radiation therapy for pediatric patients with low-grade glioma. Prospective evaluation of cognitive, endocrine, and hearing deficits. J Clin Oncol. 2009;27(22):3691–7. https://doi.org/10.1200/JCO.2008.21.2738.
10. Gnekow AK, Falkenstein F, von Hornstein S, Zwiener I, Berkefeld S, Bison B, et al. Long-term follow-up of the multicenter, multidisciplinary treatment study HIT-LGG-1996 for low-grade glioma in children and adolescents of the German Speaking Society of Pediatric Oncology and Hematology. Neuro-Oncology. 2012;14(10):1265–84. https://doi.org/10.1093/neuonc/nos202.
11. Laithier V, Le Grill J, Deley M-C, Ruchoux M-M, Couanet D, Doz F, et al. Progression-free survival in children with optic pathway tumors. Dependence on age and the quality of the response to chemotherapy—results of the first French prospective study for the French Society of Pediatric Oncology. J Clin Oncol. 2003;21(24):4572–8. https://doi.org/10.1200/JCO.2003.03.043.
12. Saran FH, Baumert BG, Khoo VS, Adams EJ, Garré ML, Warrington AP, Brada M. Stereotactically guided conformal radiotherapy for progressive low-grade gliomas of childhood. Int J Radiat Oncol Biol Phys. 2002;53(1):43–51. https://doi.org/10.1016/S0360-3016(02)02734-7.
13. Oh KS, Hung J, Robertson PL, Garton HJ, Muraszko KM, Sandler HM, Hamstra DA. Outcomes of multidisciplinary management in pediatric low-grade gliomas. Int J Radiat Oncol Biol Phys. 2011;81(4):e481–8. https://doi.org/10.1016/j.ijrobp.2011.01.019.
14. Paulino AC, Mazloom A, Terashima K, Su J, Adesina AM, Okcu MF, et al. Intensity-modulated radiotherapy (IMRT) in pediatric low-grade glioma. Cancer. 2013;119(14):2654–9. https://doi.org/10.1002/cncr.28118.
15. Greenberger BA, Pulsifer MB, Ebb DH, MacDonald SM, Jones RM, Butler WE, et al. Clinical outcomes and late endocrine, neurocognitive, and visual profiles of proton radiation for pediatric low-grade gliomas. Int J Radiat Oncol Biol Phys. 2014;89(5):1060–8. https://doi.org/10.1016/j.ijrobp.2014.04.053.
16. Müller K, Gnekow A, Falkenstein F, Scheiderbauer J, Zwiener I, Pietsch T, et al. Radiotherapy in pediatric pilocytic astrocytomas. A subgroup analysis within the prospective multicenter study HIT-LGG 1996 by the German Society of Pediatric Oncology and Hematology (GPOH). Strahlenther Onkol. 2013;189(8):647–55. https://doi.org/10.1007/s00066-013-0357-7.
17. Cherlow JM, Shaw DWW, Margraf LR, Bowers DC, Huang J, Fouladi M, et al. Conformal radiation therapy for pediatric patients with low-grade glioma. Results from the children's oncology group phase 2 study ACNS0221. Int J Radiat Oncol Biol Phys. 2019;103(4):861–8. https://doi.org/10.1016/j.ijrobp.2018.11.004.
18. Kida Y, Mori Y. Stereotactic radiotherapy with fractionation for the lesions in and around the brainstem and optic nerve. Cureus. 2019;11:e6087. https://doi.org/10.7759/cureus.6087.
19. Simonova G, Novotny J Jr, Liscak R. Low-grade gliomas treated by fractionated gamma knife surgery. J Neurosurg. 2005;102(Suppl):19–24.

20. Hadjipanayis CG, Kondziolka D, Gardner P, Niranjan A, Dagam S, Flickinger JC, Lunsford LD. Stereotactic radiosurgery for pilocytic astrocytomas when multimodal therapy is necessary. J Neurosurg. 2002;97(1):56–64. https://doi.org/10.3171/jns.2002.97.1.0056.
21. Kano H, Niranjan A, Kondziolka D, Flickinger JC, Pollack IF, Jakacki RI, Lunsford LD. Stereotactic radiosurgery for pilocytic astrocytomas part 2. Outcomes in pediatric patients. J Neuro-Oncol. 2009;95(2):219–29. https://doi.org/10.1007/s11060-009-9912-6.
22. Kida Y, Kobayashi T, Mori Y. Gamma knife radiosurgery for low-grade astrocytomas. Results of long-term follow up. J Neurosurg. 2000;93:42–6. https://doi.org/10.3171/jns.2000.93.supplement_3.0042.
23. Boëthius J, Ulfarsson E, Ráhn T, Lippitz B. Gamma knife radiosurgery for pilocytic astrocytomas. J Neurosurg. 2002;97:677–80. https://doi.org/10.3171/jns.2002.97.supplement_5.0677.
24. Lizarraga KJ, Gorgulho A, Lee SP, Rauscher G, Selch MT, DeSalles AAF. Stereotactic radiation therapy for progressive residual pilocytic astrocytomas. J Neuro-Oncol. 2012;109(1):129–35. https://doi.org/10.1007/s11060-012-0877-5.
25. Hallemeier CL, Pollock BE, Schomberg PJ, Link MJ, Brown PD, Stafford SL. Stereotactic radiosurgery for recurrent or unresectable pilocytic astrocytoma. Int J Radiat Oncol Biol Phys. 2012;83(1):107–12. https://doi.org/10.1016/j.ijrobp.2011.05.038.
26. Simonova G, Kozubikova P, Liscak R, Novotny J Jr. Leksell Gamma Knife treatment for pilocytic astrocytomas: long-term results. J Neurosurg Pediatr. 2016;18(1):58–64. https://doi.org/10.3171/2015.10.PEDS14443.
27. Trifiletti DM, Peach MS, Xu Z, Kersh R, Showalter TN, Sheehan JP. Evaluation of outcomes after stereotactic radiosurgery for pilocytic astrocytoma. J Neuro-Oncol. 2017;134(2):297–302. https://doi.org/10.1007/s11060-017-2521-x.

Pineal Tumors

29

Eun Suk Park and Young Hyun Cho

29.1 Introduction

Pineal region tumors are rare, accounting for 2.6–3.2% of primary brain tumors in children and adolescents, or 0.4–1.2% overall and in young adults [1]. These tumors represent a heterogeneous group of diverse histological entities, which originate from different cell types that form the pineal gland. Pinealocytes, arranged in lobules to form the pineal parenchyma, contribute about 95% of cells of the pineal gland, with the remainders mainly consisting of interstitial cells such as astrocytes and microglia which are embedded in a network of blood vessels and nerve fibers [2].

The pineal gland contains nerve endings from sympathetic nervous innervation to the pinealocytes [3]. The ependymal cells of the third ventricle adjoin the gland along its anterior border [4]. Tumors in the pineal region arise from these histological origins. Meanwhile, germ cell tumors (GCTs) are the most common type of pineal region tumors, which arise from pluripotent germ cells usually not inhabiting the pineal gland.

Theoretically, these germ cells mistakenly migrate to the pineal gland during embryogenesis and fail to undergo apoptosis [5]. GCTs account for more than 50% of tumors in this region [6, 7]. Pineal parenchymal tumors (PPTs) are the next most common entity, which are classified into pineocytomas, that are pineal parenchymal tumors of intermediate differentiation (PPTIDs) and pineoblastomas based on their cell maturity and aggressiveness in behavior. A newer entity named papillary tumors of the pineal region (PTPRs), which presumably originate from specialized ependymocytes of the sub-commissural organ located in the lining of the posterior commissure, has been included in the World Health Organization (WHO) classification of tumors of the central nervous system (CNS) in 2007 [8, 9]. Based on the review of the French Register of primary CNS tumors with 25,756 cases [10], pineal region tumors consist of 27% GCTs, 27% PPTs, 17% gliomas, 8% PTPRs, 7% pineal cysts, and 1% primitive neuroectodermal tumors. PPTs are represented by 13% pineocytomas, 66% PPTIDs, and 21% pineoblastomas.

Practically, tumors arising in the pineal gland region can be classified into five main categories: GCTs, PPTs, PTPRs, glial tumors, and other miscellaneous tumors such as meningioma, choroid plexus papilloma, and lymphoma [4].

E. S. Park
Department of Neurosurgery, Ulsan University Hospital, University of Ulsan College of Medicine, Ulsan, Republic of Korea
e-mail: espark@uuh.ulsan.kr

Y. H. Cho (✉)
Department of Neurosurgery, Asan Medical Center, University of Ulsan College of Medicine, Seoul, Republic of Korea
e-mail: yhyunc@amc.seoul.kr

© Springer Nature Switzerland AG 2020
A. Conti et al. (eds.), *CyberKnife NeuroRadiosurgery*,
https://doi.org/10.1007/978-3-030-50668-1_29

Given their heterogeneity, appropriate management for pineal region tumors can be complex. The procedure of obtaining tumor tissue for diagnosis is crucial because of varied biological behaviors according to different histological entities, as well as the lack of diagnostic specificity of imaging alone [4, 7].

Biopsy in any form of stereotactic, endoscopic, or craniotomy should be an initial step for optimal management of tumors in this region. Following histological verification, specific therapy including surgical resection, radiation therapy (RT), and chemotherapy is administered.

In the recent era, stereotactic radiosurgery (SRS) has emerged as a useful or added alternative to surgery or fractionated RT in a variety of intracranial tumors. Here, we have critically reviewed current knowledge on the use of SRS in the treatment of pineal region tumors.

Table 29.1 The 2016 WHO classification of tumors of the pineal region and germ cell tumors [35]

Tumor class	Grade	ICD-O code
Tumors of the pineal region		
Pineocytoma	I	9361/1
Pineal parenchymal tumor of intermediate differentiation	II or III	9362/3
Papillary tumor of the pineal region	II or III	9395/3
Pineoblastoma	IV	9362/3
Germ cell tumors		
Germinoma		9064/3
Embryonal carcinoma		9070/3
Yolk sac tumor		9071/3
Choriocarcinoma		9100/3
Teratoma		9080/1
Mature teratoma		9080/0
Immature teratoma		9080/3
Teratoma with malignant transformation		9084/3
Mixed germ cell tumor		9085/3

WHO World Health Organization, *ICD-O* The International Classification of Diseases for Oncology

29.2 Germ Cell Tumors

GCTs are the most frequently encountered tumor type in the pineal region and comprise germinomas and non-germinomatous germ cell tumors (NGGCTs). NGGCTs include choriocarcinomas, endodermal sinus tumors, embryonal carcinomas, teratomas, and mixed tumors (Table 29.1). GCTs, both germinomas and NGGCTs, are sensitive to radiation and chemotherapy. Therefore, the role of surgical resection beyond diagnostic biopsy is controversial except in case of teratomas.

Owing to the high sensitivity to radiation, all germinoma patients were treated with craniospinal irradiation (CSI) alone until the early 1990s, which yielded a cure rate of over 90% [11, 12]. However, to reduce long-term toxicities such as neurocognitive insufficiency and endocrinopathy, combined treatment with primary chemotherapy and RT was used.

Since then, multicenter or international trials, such as the French SFOP (Société Française d'Oncologie Pédiatrique/French Pediatric Oncology Society) in 1990 [13], the Japanese Pediatric Brain Tumor Study Group trial in 1995 [14], the European SIOP (International Society of Paediatric Oncology) study in 1996 [15], and the North American COG (Children's Oncology Group) ACNS 0232/1123 [16], were conducted. Based on these trials, radiation dose and volume were reduced. Moreover, CSI is no longer prescribed for the treatment of localized germinoma, and thus the whole ventricular system provides the reference for RT target volume [16].

Currently, the European SIOP GCT CNS II protocol (NCT01424839) states that patients with localized germinoma are primarily treated with chemotherapy (two cycles of carboplatin and etoposide alternating with two cycles of ifosfamide and etoposide). And then, if there is a complete response at reassessment, whole ventricular irradiation (WVI) alone (24 Gy in 15 fractions) is added. If there is a partial response, WVI plus focal boost (16 Gy in 10 fractions) is administered. Or, if there is a stable disease, surgery followed by RT is recommended. As per the current ACNS 1123 protocol (NCT01602666), germinoma patients who present with a complete response after chemotherapy (four cycles of carboplatin and etoposide) receive WVI (18 Gy) and a boost to the primary tumor (12 Gy). In patients with a partial response with residual tumor less than 1.5 cm, WVI (24 Gy) followed by a focal boost (12 Gy) is added without second-look surgery. According to the Japanese protocol [14], patients in the good prognosis group should be treated with 3 cycles of chemotherapy (carboplatin 450 mg/m^2 on day 1 and etoposide 150 mg/m^2 on day 1–3), followed by 23.4 Gy WVI. The response-based adjuvant chemotherapy (3 cycles of ICE, viz., ifosfamide 900 mg/m^2, cisplatin 20 mg/m^2 and etoposide 60 mg/m^2 on day 1–5) can be used for non-complete response patient group. Patients in the intermediate group should be treated with 3 cycles of chemotherapy (carboplatin 450 mg/m^2 on day 1 and etoposide 150 mg/m^2 on day 1–3), followed by 50.4 Gy WVI. In case of disseminated germinomas, patients are treated with the localized germinoma protocol, along with CSI and/or local boost RT.

NGGCTs are less radiosensitive than germinomas, and RT-alone treatment has provided both 5- and 10-year survival rates of 36% [17], whereas chemotherapy-alone treatment conferred poor outcome in these patients [18, 19]. Hence, NGGCTs are treated in combination with surgical resection, chemotherapy, and RT to obtain the best outcome. Especially for management of teratomas or tumors harboring teratoma components, surgery is preferred because of the resistant nature of these tumors to RT and chemotherapy [20]. In short, except for teratomas, localized NGGCTs are managed by multimodality therapy with chemotherapy followed by local RT [16].

The outcomes of management for GCTs varies across histological subtypes, where 10 year overall survival (OS) for germinomas is more than 90% [21–23], but that for NGGCTs remains 60–80% [23–25]. Considering the outcomes obtained from current management strategy, SRS would be the best option for treatment of resistant or recurrent tumors. Accordingly, most of the previous studies reported the utility of SRS as an adjuvant or salvage therapy rather than primary modality (Table 29.2) [26–34]. Better outcomes have been obtained in germinoma subtype as well as in the case of residual or recurrent tumors. In eight patients with germinomas, Kobayashi et al. observed 100% tumor control during 26-month follow-up, where patients were treated with conventional chemotherapy and fractionated RT followed by adjuvant SRS therapy [26]. In a study by Mori et al., 16 patients with germinoma who underwent SRS as a part of their management [30] showed local control in 82% of cases and progression-free survival (PFS) in 63% of cases, both at 5 years. Recently, Iorio-Morin et al. reported that 80% local control was obtained at 20 years for four patients with germinoma treated with SRS as an adjuvant boost following initial fractionated RT, but one patient with recurrence could not survive the disease [34].

Collectively, SRS can be considered a safe and effective adjuvant treatment for germinomas.

However, studies on NGGCTs suggest a relatively poor outcome compared with germinomas. Kobayashi et al. reported that in 13 patients with malignant GCTs, 50% local tumor control was obtained during 21-month mean follow-up, where patients were treated with SRS as an adjuvant therapy after the conventional treatment [26]. Hasegawa et al. observed 75% local control during 25-month mean follow-up in four patients with NGGCT, with death of one patient due to disease progression [27]. In the study by Mori et al. [30], 22 patients with NGGCT showed 5-year local control of 62% and 5-year PFS of 37%. In short, despite relatively poor outcomes in NGGCTs, SRS may offer a reasonable option in adjuvant or salvage settings, considering the aggressive nature of these tumors.

29.3 Pineal Parenchymal Tumors

The WHO classification in 2016 has categorized PPTs into three subtypes with up to four different grade categories: pineocytomas (grade I), PPTIDs (grade II or III), and pineoblastomas (grade IV) (Table 29.1) [35]. This classification was not changed from the 2007 WHO classification. Management and prognosis of patients are highly dependent on histological subtype and grade.

Pineocytomas are slowly growing tumors with favorable prognosis with 5-year survival of 64–91% [36]. Tumors cause symptoms by local

Table 29.2 Stereotactic radiosurgery for germ cell tumors

Author	Year	No. of cases treated with SRS (FU cases)	Median age in years (range)	Histology	SRS modality	Mean tumor vol. in cm^3 (range)	Median marginal dose in Gy (range)	Median maximal dose in Gy (range)	FU duration (mo)	Local tumor control (%)	Survival outcome	Complication
Kobayashi et al. [26]	2001	8 (7)	17 (13–22)	Germinoma	GK	ND	16.8	34.5	26.7 (10–36)	LTC 100% (CR2, PR5)	No death	ND
Kobayashi et al. [26]	2001	4 (3)	23.3 (16–41)	G + STGC	GK	ND	13.5	30.0	11 (10–12)	LTC 67% (PR2, PD1)	1 Death during FU	ND
Kobayashi et al. [26]	2001	13 (12)	24.8 (12–66)	Malignant GCT	GK	ND	15.7	30.8	21.4 (6–81)	LTC 50% (CR3, PR3, PD6)	5 Death during FU	ND
Hasegawa et al. [27]	2003	4	16.5 (9–21)	NGGCT	GK	10.5 (2.2–23.4)	14 (12–16)	28 (24–32)	25 (4–58)	LTC 75% (CR1, PR1, SD1, PD1)	1 Death during FU	None
Amendola et al. [28]	2005	13	21.8 (8–65)	GCT	GK	15.8 (0.52–52.1)	11.2 (8–13)		ND	ND	2 Death during FU	None
Amendola et al. [28]	2005	1	5	Teratoma	GK	0.1	10		ND	ND	No death	None
Lekovic et al. [29]	2007	1	13	NGGCT	GK	1.2	15	30	73	LTC 100% (CR1)	No death	None
Lekovic et al. [29]	2007	1	18	Malignant teratoma	GK	2.1	13	26	15	LTC 100% (PR1)	No death	None
Mori et al. [30]	2009	16	21 (8–66)	Germinoma	GK	3.3 (0.1–22)	15.5 (9.9–25.7)	27.4 (14.5–40)	49 (3–192)	LTC (3 yr) 82%, LTC (5 yr) 82%	PFS (3 yr) 79%, PFS (5 yr) 63%	ND

Mori et al. [30]	2009	22	21 (8–66)	NGGCT	GK	3.3 (0.1–22)	15.5 (9.9–25.7)	27.4 (14.5–40)	49 (3–192)	LTC (3 yr) 72%, LTC (5 yr) 62%)	PFS (3 yr) 43%, PFS (5 yr) 37%	ND
Yianni et al. [31]	2012	2	ND	Teratoma	GK	ND	15	ND	62.5 (6–240)	LTC 100%	PFS (1 yr) 100%	ND
Li et al. [33]	2015	40	14.4 (6–20)	GCT	GK	ND	ND	ND	ND	LTC (1 yr) 96.7%, LTC (3 yr) 88%, LTC (5 yr) 77.27%	No death	ND
Balossier A et al. [32]	2015	1	7	NGGCT	GK	1.9	18	ND	10	PD	No death	None
Balossier A et al. [32]	2015	1	18	Teratoma	GK	1.3	17	ND	ND	ND	No death	None
Iorio-Morin C et al. [34]	2017	5	36 (3.5–83)	Germinoma	GK	1.5 (1.3–9.8)	14 (13–18)	28 (26–36)	47 (4.6–260)	LTC (1 yr) 80%, LTC (3 yr) 80%, LTC (20 yr) 80%	OS (1 yr) 100%, OS (3 yr) 80%, OS (20 yr) 80%	ND
Iorio-Morin C et al. [34]	2017	1	9	NGGCT	GK	9.68	16	32	4	PD	1 Death during FU	ND
Iorio-Morin C et al. [34]	2017	2	36 (3.5–83)	Teratoma	GK	16.4 (9.3–23.4)	13 (12–14)	26 (24–28)	105.5 (58–153)	LTC (3 yr) 100%, LTC (5 yr) 50%	PFS (5 yr) 50%, OS (5 yr) 100%	ND

No Number, *SRS* Stereotactic radiosurgery, *FU* Follow-up, *Gy* Gray, *mo* Month, *GK* Gamma Knife radiosurgery, *ND* Data was not determined, *LTC* Local tumor control, *CR* Complete response, *PR* Partial response, *G + STGC* Germinoma with syncytiotrophoblastic giant cells, *PD* Progressive disease, *GCT* Germ cell tumors, *NGGCT* Non-germinomatous germ cell tumors, *SD* Stable disease, *yr* Year, *PFS* Progression-free survival, *OS* Overall survival compression. Although complete surgical resection can be considered in a curative intention, the risk of operation-related complications is not negligible. Hence, SRS can be adopted either as primary or adjuvant treatment for residual or recurrent tumors (Table 29.3) [26, 27, 29–32, 34, 37–41]. In a retrospective study by Hasegawa et al., tumor control was observed during 69-month follow-up in all ten patients who underwent SRS as primary or adjuvant treatment for pineocytomas, except for one patient who succumbed to secondary leptomeningeal tumor spread [27]. Reyns et al. reported that of eight patients with pineocytoma, one showed complete and four showed partial regression, and two showed stable disease, following primary or adjuvant SRS treatment [38]. Tumor control was achieved in all patients without death during the mean follow-up of 32 months. Kano et al. also reported 100% tumor control in 13 patients, with complete tumor regression in 3, partial regression in 8, and stable status in 2 [39]. In addition, 5-year overall survival rate was 92.3%. In our own experience, all three patients showed sustained tumor control (one complete and two partial regression) during 99-month follow-up after SRS treatment (Fig. 29.1) [41]. Collectively, available data in the literature uniformly support high tumor control and patient survival rates, both up to 100% following SRS, indicating the role of SRS as an effective alternative or adjunct to surgical resection for management of pineocytomas.

PPTIDs share some features with both pineocytomas and pineoblastomas. Five-year survival rates for grade II and III tumors are estimated at 74% and 39%, respectively [32]. Pineoblastomas are considered malignant tumors with mean survival of around 2 years [42] and have a high rate of recurrence and metastasis. PPTIDs have histological features associated with an increased risk of recurrence and are commonly managed with surgical resection. However, the role of fractionated RT or SRS is not clearly elucidated to date. Most studies in the literature (Table 29.4) [31, 32, 34, 39, 41] did not opt for histological stratification, and the results for PPTIDs were pooled with those of pineoblastomas or pineocytomas, complicating a sound interpretation of the data. One recent study by Iorio-Morin et al. showed that patients with PPTID received SRS upfront or at recurrence and represented 5-year tumor control and survival rates of 50% and 56%, respectively [34]. In our own series of five patients with biopsy-confirmed PPTID, 100% local tumor control (two complete and three partial responses) and 100% survival were observed during 103-month follow-up following SRS (Fig. 29.2). With limited data available currently, the therapeutic role of SRS in PPTIDs needs to be further investigated, despite some promising outcomes in select cases.

Malignant pineoblastomas are managed with maximal surgical resection followed by fractionated RT and chemotherapy. SRS is usually reserved for treatment of recurrent tumors or as a local boost after primary therapy (Table 29.4) [31, 34, 39]. Reyns et al. reported 75% local control with complete or partial regression in three patients, although two patients succumbed to disease progression and death during 40-month follow-up [38]. Iorio-Morin et al. reported worse outcomes of 5-year tumor control and survival rates of 27% and 48%, respectively, where SRS was applied as a local boost or salvage. The utility of SRS appears to be limited in pineoblastomas, and it is usually indicated for recurrent tumors.

29.4 Papillary Tumors of the Pineal Region

The WHO introduced PTPRs, a rare grade II–III pineal lesion with specific histological and immunohistochemical features as a newer entity in 2007 [43]. These tumors present an immunohistochemical profile similar to that of choroid plexus tumors [43, 44].

However, they are morphologically less differentiated than choroid plexus papillomas and more differentiated than choroid plexus carcinomas. As a result, earlier PTPRs were frequently misdiagnosed as either ependymomas or choroid plexus tumors [45].

Table 29.3 Stereotactic radiosurgery for pineal parenchymal tumors (pineocytoma)

Author	Year	No. of cases treated with SRS (FU cases)	Median age in years (range)	Histology	SRS modality	Mean tumor vol. in cm^3 (range)	Median marginal dose in Gy (range)	Median maximal dose in Gy (range)	FU duration (mo)	Local tumor control (%)	Survival outcome	Complication
Kobayashi et al. [26]	2001	3	34.3 (11–69)	Pineocytoma	GK	ND	16.6	ND	21.7 (12–38)	LTC 100% (CR2, PR1)	No death	ND
Hasegawa et al. [27]	2003	10	38.3 (8–68)	Pineocytoma	GK	5.0 (0.9–14.2)	15.7 (12–20)	31.4 (24–40)	69.1 (14–108)	LTC 100% (CR2, PR7, SD1)	1 Death during FU	1 temporally upward gaze palsy
Deshmukh et al. [37]	2004	5	45 (24–63)	Pineocytoma	GK	6.4 (1.9–9.6)	14.8 (14–16)	ND	14.6 (6–31)	LTC 100% (PR5)	No death	None
Amendola et al. [28]	2005	1	37	Pineocytoma	GK	8.1	11	ND	ND	ND	No death	None
Reyns et al. [38]	2006	8 (7)	35.9 (13–74)	Pineocytoma	GK	ND	ND	28.8 (24–36)	32 (6–72)	LTC 100% (CR1, PR4, SD2)	No death	None
Lekovic et al. [29]	2007	8	49 (24–70)	Pineocytoma	GK	6.0 (1.9–12.4)	14.4 (13–16)	28.8 (26–32)	17 (1–57)	LTC 100% (PR4, SD4)	1 Death during FU	None
Mori et al. [30]	2009	5	21 (8–66)	Pineocytoma	GK	3.7 (0.3–23)	16.7 (12.5–20.3)	27.9 (17–40)	49 (3–192)	LTC (3 yr) 85%, LTC (5 yr) 85%	PFS (3 yr) 80%, PFS (5 yr) 80%	ND
Kano et al. [39]	2009	13	34 (3.5–68.4)	Pineocytoma	GK	4.4 (0.9–14.2)	15.2 (12–20)	30.4 (24–40)	51.6	LTC 100% (CR3, PR8, SD2)	OS (1 yr) 100%, OS (3 yr) 92.3%, OS (5 yr) 92.3%	4 ARE (1 ptosis, 2 upward gaze palsy, 1 asymptomatic edema)
Yianni et al. [31]	2012	6	ND	Pineocytoma	GK	3.8	20	ND	62.5 (6–240)	LTC 83% (PD1)	2 Deaths during FU, PFS (1 yr) 81%, PFS (5 yr) 77%	ND

Table 29.3 (continued)

Author	Year	No. of cases treated with SRS (FU cases)	Median age in years (range)	Histology	SRS modality	Mean tumor vol. in cm^3 (range)	Median marginal dose in Gy (range)	Median maximal dose in Gy (range)	FU duration (mo)	Local tumor control (%)	Survival outcome	Complication
Author	Year	No. of cases treated with SRS (FU cases)	Median age in years (range)	Histology	SRS modality	Mean tumor vol. in cm^3 (range)	Median marginal dose in Gy (range)	Median maximal dose in Gy (range)	FU duration (mo)	Local tumor control (%)	Survival outcome	Complication
Wilson et al. [40]	2012	5	43.6 (24–57)	Pineocytoma	GK	2.6 (1.9–9.7)	14.6 (14–16)	29.2 (28–32)	64.8	LTC 100% (CR2, PR1 SD2)	No death	None
Park et al. [41]	2015	3 (GK:1, CK:2)	49 (44–53)	Pineocytoma	GK/CK	8.2 (3.6–13.4)	GK:12/ CK:36	ND	99 (32–223)	LTC 100% (CR1, PR2)	No death	1 Temporary memory impairment
Balossier A et al. [32]	2015	3	61 (57–64)	Pineocytoma	GK	4.3 (1.8–8.7)	15 (12–17)	30 (24–34)	42 (15–72)	LTC 100% (CR1, PR1, SD1)	No death	None
Iorio-Morin et al. [34]	2017	26	36 (3.5–83)	Pineocytoma	GK	3.3 (0.89–14.2)	16 (12–20)	32 (21–50)	47 (4.6–260)	LTC (1 yr) 100%, LTC (3 yr), 100%, LTC (20 yr) 80%	OS (1 yr) 100%, OS (20 yr) 80%	ND

No Number, SRS Stereotactic radiosurgery, *FU* Follow-up, *Gy* Gray, *mo* Month, *GK* Gamma Knife radiosurgery, *ND* Data was not determined, *LTC* Local tumor control, *CR* Complete response, *PR* Partial response, *SD* Stable disease, *yr* Year, *PFS* Progression-free survival, *OS* Overall survival, *PD* Progressive disease, *CK* CyberKnife radiosurgery

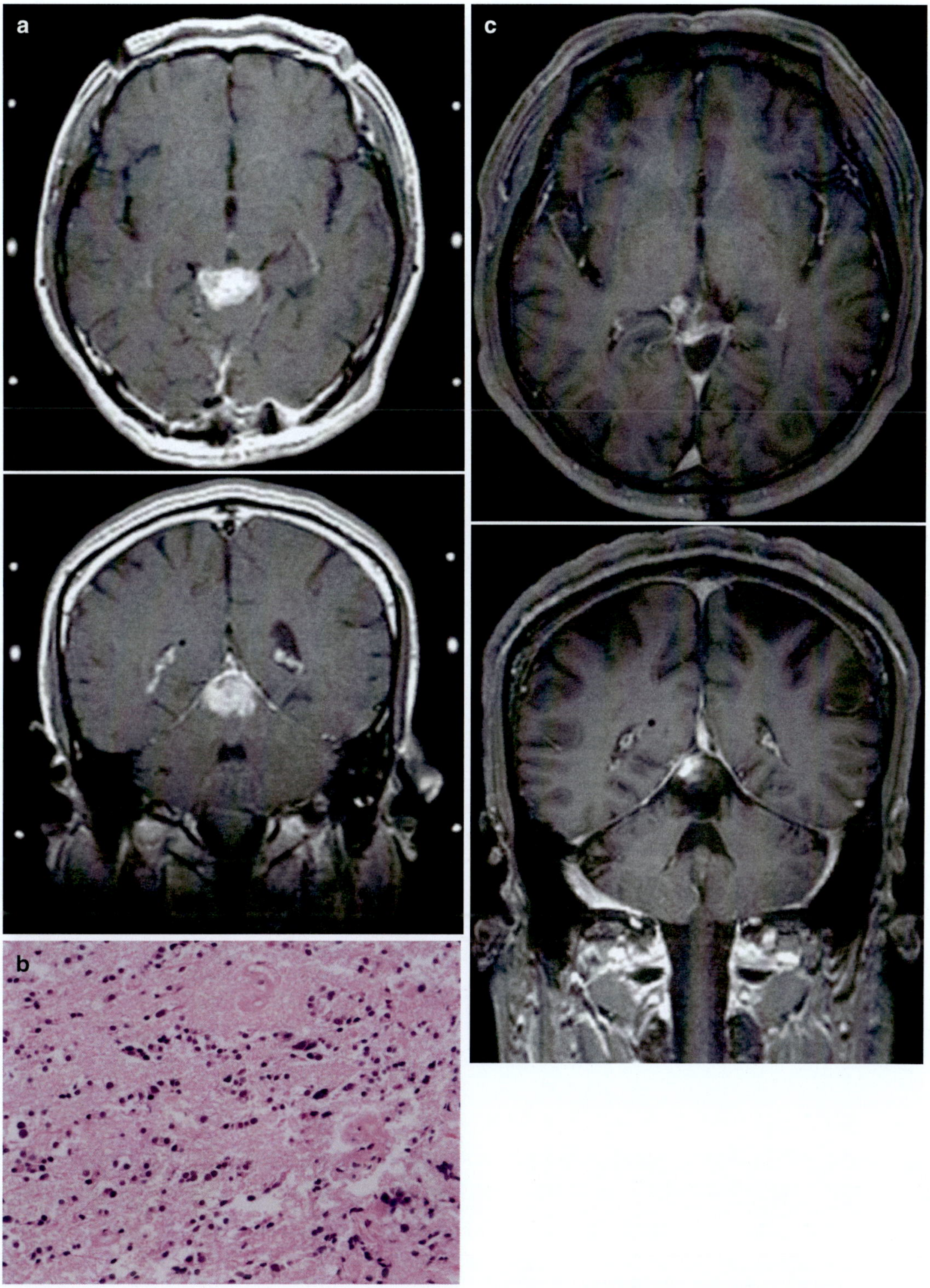

Fig. 29.1 A 44-year-old man with a pineocytoma. Gadolinium-enhanced magnetic resonance images at the time of Gamma Knife (GK) treatment (**a**). Tumor consists of relatively small, uniform, mature cells resembling normal pineocytes, and cell-free spaces filled with cell processes are forming vague rosettes (**b**). After GK treatment with marginal dose of 12 Gy, he achieved durable tumor response with more than 18 years of follow-up (**c**)

Table 29.4 Stereotactic radiosurgery for pineal parenchymal tumors of intermediate differentiation (PPTIDs) and pineoblastoma

Author	Year	No. of cases treated with SRS (FU cases)	Median age in years (range)	Histology	SRS modality	Mean tumor vol. in cm^3 (range)	Median marginal dose in Gy (range)	Median maximal dose in Gy (range)	FU duration (mo)	Local tumor control (%)	Survival outcome	Complication
Kobayashi et al. [26]	2001	3	22.5 (3–42)	Pineoblastoma	GK	ND	16.6	ND	21.7 (12–38)	LTC 100% (CR2, PR1)	No death	ND
Hasegawa et al. [27]	2003	6	34.8 (3–66)	2PPTID + 4Pineoblastoma	GK	5.0 (0.9–14.2)	14.5 (13–16)	29 (26–32)	23 (7–47)	LTC 100% (CR2, PR1, SD1, 2 cases of no image)	4 Death during FU	1 upward And downward gaze palsies and ataxia
Reyns et al. [38]	2006	5 (8 lesions)	28 (10–45)	Pineoblastoma	GK	ND	ND	33.6 (28–50)	40.6 (10–88)	LTC 75% (CR4 PR2, PD2)	2 Death during FU	None
Lekovic et al. [29]	2007	1	22	Pineoblastoma	GK	23	14	28	45	SD1	No death	None
Mori et al. [30]	2009	3	21 (8–66)	1PPTID + 2Pineoblastoma	GK	3.7 (0.3–23)	16.7 (12.5–20.3)	27.9 (17–40)	49 (3–192)	LTC (2 yr) 30%	PFS (2 yr) 33%	ND
Kano et al. [39]	2009	7	34 (3.5–68.4)	2 Mixed PPTs + 5pineoblastoma	GK	4.4 (0.9–14.2)	15.2 (12–20)	30.4 (24–40)	54.1	LTC 66.7% (CR2, PR1, NC1, PD2, f/u loss 1)	4 Death during FU, PFS (1 yr) 100% PFS (3 yr) 66.7%	None
Yianni et al. [31]	2012	5	ND	3PPTID + 2Pineoblastoma	GK	ND	20	ND	62.5 (6–240)	LTC 80% (PD1)	1 Death during FU, PFS (1 yr) 81%, PFS (5 yr) 77%	ND
Park et al. [41]	2015	5 (GK2, CK3)	37.4 (31–50)	PPTID	GK/CK	7.18 (3.8–11.9)	GK14 (12–16) /CK30	ND	77.6 (14–213)	LTC 100% (CR2, PR3)	No death	None
Balossier A et al. [32]	2015	6	59.3 (45–73)	PPTID	GK	3.1 (1.16–5.2)	15.5 (14–18)	31 (28–36)	24.2 (5–76)	LTC 100% (SD5, f/u loss 1)	No death	None
Iorio-Morin et al. [34]	2017	20	36 (3.5–83)	7PPTID + 13Pineoblastoma	GK	PPTID 2.6 (0.61–38.3) /PB 3.3 (0.78–10.5	PPTID 17 (10–20) /PB 15 (12–18)	PPTID 34 (20–40) /PB 30 (21.54–36)	47 (4.6–260)	PPTID LTC (5 yr) 50% /PB LTC (5 yr) 27%	PPTID OS (5 yr) 56% /PB OS (5 yr) 48%	ND

No Number, *SRS* Stereotactic radiosurgery, *FU* Follow-up, *Gy* Gray, *mo* Month, *GK* Gamma Knife radiosurgery, *ND* Data was not determined, *LTC* Local tumor control, *CR* Complete response, *PR* Partial response, *PPTID* Pineal parenchymal tumors of intermediate differentiation, *SD* Stable disease, *PD* Progressive disease, *yr* Year, *PFS* Progression-free survival, *PPTs* Pineal parenchymal tumors, *CK* CyberKnife radiosurgery, *PB* Pineoblastoma, *OS* Overall survival

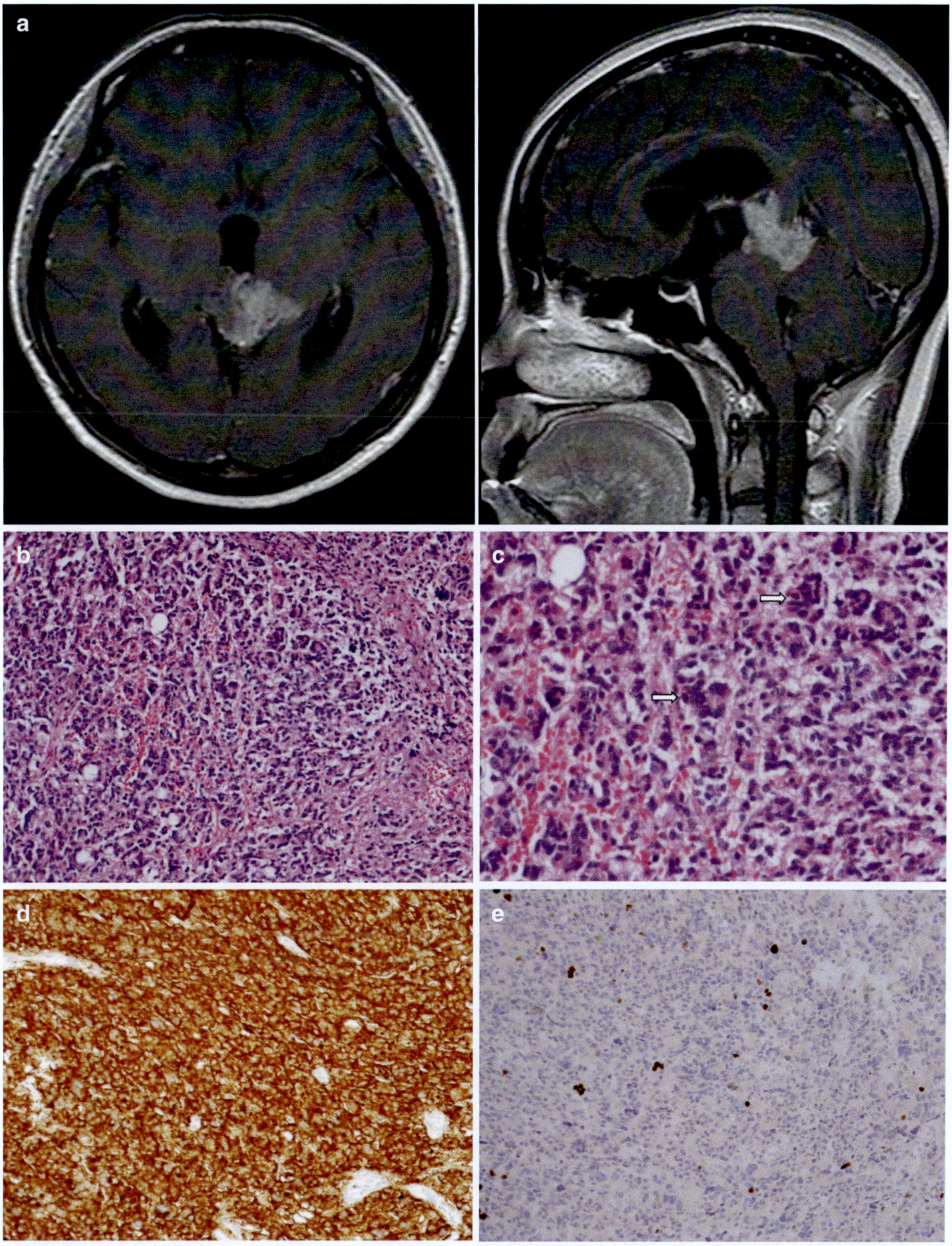

Fig. 29.2 A 34-year-old woman with pineal parenchymal tumor of intermediate differentiation. Gadolinium-enhanced magnetic resonance images showing a 3.2 × 2.4 × 2.0 cm sized tumor in the pineal region with obstructive hydrocephalus (**a**). Tumor shows diffusely high cellularity and tumor cell nuclei are pleomorphic (**b**). Multinucleated giant tumor cells are frequently seen (**c**, arrows). Tumor cells are diffusely immunoreactive for synaptophysin (**d**). Proliferation activity assessed by MIB-1 is low (**e**). She received 5-fraction CyberKnife (CK) treatment with marginal dose of 30 Gy (**f**). Tumor completely disappeared in 9 months after CK, and no evidence of disease was seen at the last follow-up of 5 years (**g**)

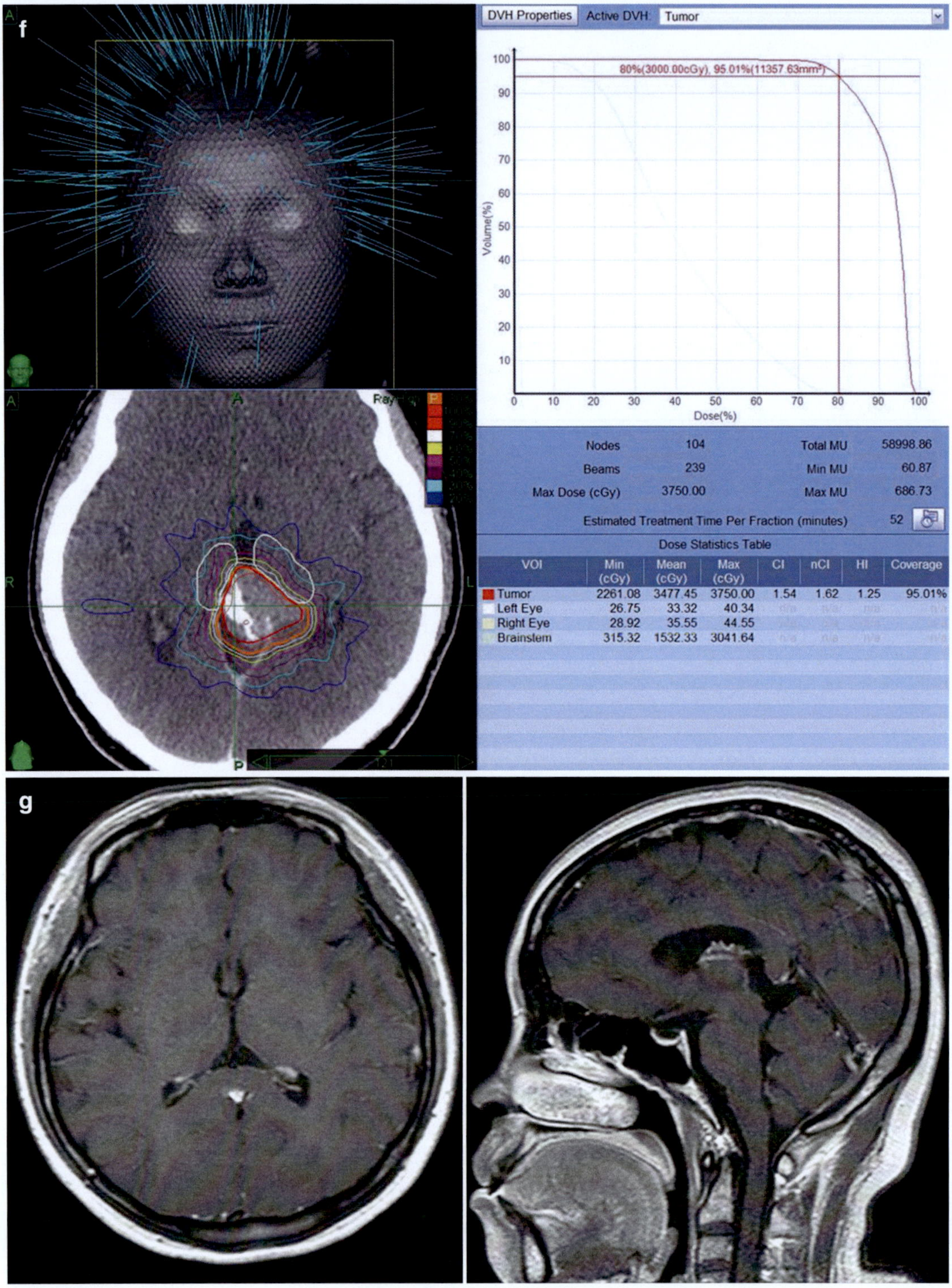

Fig. 29.2 (continued)

Optimal management remains debatable, and upfront RT or chemotherapy have not led to reduction in the risk of recurrence [46].

Focusing on the high potential for local recurrence [47], several groups are investigating the role of SRS in the treatment of PTPRs (Table 29.5) [32, 34, 46, 48–50]. In a retrospective study by Fauchon et al., out of 43 patients with PTPR, only 2 patients opted for SRS following partial resection, but both showed tumor recurrence and 1 succumbed to death [46].

Iorio-Morin et al. reported that five patients with histologically confirmed PTPR opted for SRS as an initial management and another one patient was treated for recurrence after gross total resection (GTR). Among them, five patients experienced local recurrence yielding 5-year tumor control and survival rates of 33% and 100%, respectively. All patients with recurrent tumors underwent repeat SRS and prolonged local control was achieved in four patients [34]. Fernandez-Mateos et al. reported that treatment of two patients with PTPR using SRS following biopsy showed excellent outcomes without recurrence during 15-year follow-up [50]. SRS therefore appears to be a viable option as primary or adjuvant treatment for residual or recurrent PTPRs.

29.5 Pineal Glioma and Miscellaneous Tumors

Pineal region gliomas arise either from the pineal region itself or from the adjacent structures such as the thalamus or the midbrain. Various glial histologies including pilocytic astrocytomas, fibrillary astrocytomas, anaplastic astrocytomas, glioblastomas, oligodendrogliomas, and ependymomas have been reported [4].

In general, maximal safe resection is applied for the management of pineal gliomas. But, the success rate to achieve GTR varies from 21 to 88% depending on the surgeon's skill and experience [51]. Adjuvant RT and chemotherapy are used for treatment of malignant gliomas based on the histology. Although scarce data are available (Table 29.6) [28, 29, 31], SRS appears to be useful for local tumor control with less radiation toxicity than conventional RT. However, detailed analyses on more clinical data are needed to define the role of SRS in the treatment of pineal gliomas.

Various other tumor types such as meningiomas, choroid plexus papillomas, and metastatic tumors may arise in the pineal region. Although the number of cases are limited, these tumors have been treated using SRS, representing similarly fair outcomes (Table 29.6) [29, 31].

In certain clinical situations, obtaining patient tumor tissue is not possible due to various reasons such as patient comorbidities, refusal for surgery, or limited available tissue despite surgery. Li et al. reported a large cohort consisting of 147 patients who underwent SRS for pineal lesions based on imaging and clinical diagnosis alone [33]. They observed regression of the initial tumor in 69% of the cases, with local control rates of 97%, 94%, and 91% after 1, 3, and 5 years, respectively, following SRS. In addition, patient survival rates were 80%, 72%, and 67% after 1, 3, and 5 years of follow-up respectively. Iorio-Morin et al. opted for SRS in 10 patients based on imaging diagnosis without histological confirmation and obtained 5-year tumor control and survival rates of 61% and 67%, respectively, which were similar to the aggregate results of their entire series [34].

These observations support the utility of SRS in selected patients even in the absence of histopathological confirmation [26, 30, 31, 33, 34].

29.6 The Role of SRS for Pineal Region Tumors

Given the limited number of reports, it is difficult to draw a clear conclusion on the utility of SRS for the treatment of pineal region tumors. The authors had to combine various histological subgroups of tumors due to insufficient cohort size, complicating the analyses on tumor control and survival outcomes. Recently, Iorio-Morin et al. tried to overcome this hurdle by performing histology-stratified analyses to provide better quality data to guide patient management.

Table 29.5 Stereotactic radiosurgery for papillary tumors of the pineal region (PTPR)

Author	Year	No. of cases treated with SRS (FU cases)	Median age in years (range)	Histology	SRS modality	Mean tumor vol. in cm^3 (range)	Median marginal dose in Gy (range)	Median maximal dose in Gy (range)	FU duration (mo)	Local tumor control (%)	Survival outcome	Complication
Fauchon et al. [46]	2013	2	52.5 (52–53)	PTPR	GK	ND	12	ND	63.5 (36–91)	ND	No death	None
Riis et al. [48]	2013	1	20	PTPR	GK	12.3	12	24	50	LTC 100% (PR1)	No death	None
Shakir et al. [49]	2015	1	31	PTPR	GK	4.2	18	36	9 yr	LTC 100% (PR1)	No death	None
Balossier et al. [32]	2015	1	27	PTPR	GK	1.1	15	30	6	LTC 100% (SD1)	No death	None
Iorio-Morin et al. [34]	2017	6	36 (3.5–83)	PTPR	GK	2.2 (0.42–4.97)	16 (14–18)	32 (26.67–36)	47 (4.6–260)	LTC 16% (5PD) LTC (5 yr) 33%	No death	ND
Fernandez-Mateos et al. [50]	2018	2	36.5 (27–46)	PTPR	GK	8.5(4.1–13)	11(10–12)	20	17.5 yr. (15–20 yr)	LTC 100% (PR2)	No death	None

No Number, *SRS* Stereotactic radiosurgery, *FU* Follow-up, *Gy* Gray, *mo* Month, *PTPR* Papillary tumors of the pineal region, *GK* Gamma Knife radiosurgery, *ND* Data was not determined, *LTC* Local tumor control, *PR* Partial response, *SD* Stable disease, *PD* Progressive disease, *yr* Year

Table 29.6 Stereotactic radiosurgery for miscellaneous tumors of the pineal region

Author	Year	No. of cases treated with SRS (FU cases)	Median age in years (range)	Histology	SRS modality	Mean tumor vol. in cm^3 (range)	Median marginal dose in Gy (range)	Median maximal dose in Gy (range)	FU duration (mo)	Local tumor control (%)	Survival outcome	Complication
Lekovic et al. [29]	2007	1	55	Choroid plexus adenoma	GK	3.9	16	32	96	LTC 100% (PR1)	No death	None
Lekovic et al. [29]	2007	1	54	Neurocytoma	GK	1.4	18	36	54	LTC 100% (PR1)	No death	None
Lekovic et al. [29]	2007	1	29	Anaplastic astrocytoma	GK	3.9	12	24	8	PD1	1 Death during FU	None
Lekovic et al. [29]	2007	1	55	PNET	GK	8.3	12	24	44	LTC 100% (PR1)	No death	None
Lekovic et al. [29]	2007	1	27	Low-grade astrocytoma	GK	1.6	14	28	21	LTC 100% (CR1)	No death	None
Mori et al. [30]	2009	2	21 (8–66)	Unknown	GK	ND	ND	ND	26 (22–30)	LTC 100%	No death	ND
Yianni et al. [31]	2012	11		Astrocytoma	GK		17.3		62.5 (6–240)		PFS (1 yr) 100%, PFS (5 yr) 100%	None
Yianni et al. [31]	2012	6		Ependymoma	GK		16.5		62.5 (6–240)		PFS(1 yr) 67%, PFS (5 yr) 67%	None
Yianni et al. [31]	2012	1		Choroid plexus papilloma	GK		?18		62.5 (6–240)		PFS(1 yr) 95%, PFS (5 yr) 81%	None
Yianni et al. [31]	2012	1		Meningioma	GK		?18		62.5 (6–240)		PFS(1 yr) 95%, PFS (5 yr) 81%	None
Yianni et al. [31]	2012	5		Unknown	GK		?18		62.5 (6–240)		PFS(1 yr) 95%, PFS (5 yr) 81%	None
Iorio-Morin et al. [34]	2017	10	36 (3.5–83)	Unknown	GK	2.2 (0.58–7.14)	15 (13–17)	30 (26–35)	47 (4.6–260)	LTC (2 yr) 61%	OS (2 yr) 67%	ND

No Number, *SRS* Stereotactic radiosurgery, *FU* Follow-up, *Gy* Gray, *mo* Month, *GK* Gamma Knife radiosurgery, *LTC* Local tumor control, *PR*: Partial response, *PD* Progressive disease, *CR* Complete response, *ND* Data was not determined, *PFS* Progression-free survival, *yr* Year

Overall, currently available data in the literature supports that SRS can be a useful treatment option for select patients with pineal region tumors. In germinomas, SRS can be used as a focal radiation boost to the tumor bed serving as an alternative to fractionated RT, or as salvage for recurrence. For NGGCTs, SRS can be used as a focal boost to residual tumor following surgical resection or as a salvage after recurrence, itself alone or in combination with fractionated RT and/or chemotherapy. Similar strategy could be an option for pineoblastomas.

In pineocytomas, SRS appears to provide long-term tumor control and patient survival, suggesting upfront SRS as a viable alternative to surgical resection. Although this idea may be applied to PPTIDs as well, cautions are required to interpret the results of studies with these tumors merged with pineocytomas or pineoblastomas. Finally, SRS may serve as a reasonable primary or adjuvant option for patients with PTPRs given their high propensity for local recurrence even in case of GTR.

In most studies, SRS dose to tumor margin varied from 10 to 20 Gy, and the optimal dosage could not be formulated given the rarity of available data. Since SRS is frequently used as an adjunct or as salvage after previous fractionated RT, careful dose adjustment is recommended accordingly, with application of dose constraints to critical structures such as the diencephalon and the brainstem.

29.7 Conclusions

Evidence on the role of SRS to guide the management of pineal region tumors is still insufficient. However, SRS may be useful as an effective and safe modality in different tumor types based on histological verification. In pineocytomas and PTPRs, it can be used as an alternative to surgery for primary treatment or as an adjunct/salvage for residual/recurrent disease. In case of GCTs and pineoblastomas, SRS helps as a part of multimodality management or as a salvage option for recurrence.

Further clinical studies are needed to elucidate more clearly the role of SRS in the treatment of tumors in this particular region.

References

1. Ostrom QT, et al. CBTRUS statistical report: primary brain and other central nervous system tumors diagnosed in the United States in 2011-2015. Neuro-Oncology. 2018;20:iv1–iv86. https://doi.org/10.1093/neuonc/noy131.
2. Moller M, Baeres FM. The anatomy and innervation of the mammalian pineal gland. Cell Tissue Res. 2002;309:139–50. https://doi.org/10.1007/s00441-002-0580-5.
3. Fevre-Montange M, Vasiljevic A, Champier J, Jouvet A. Histopathology of tumors of the pineal region. Future Oncol. 2010;6:791–809. https://doi.org/10.2217/fon.10.28.
4. Gaillard F, Jones J. Masses of the pineal region: clinical presentation and radiographic features. Postgrad Med J. 2010;86:597–607. https://doi.org/10.1136/pgmj.2009.087460.
5. Hoei-Hansen CE, et al. New evidence for the origin of intracranial germ cell tumours from primordial germ cells: expression of pluripotency and cell differentiation markers. J Pathol. 2006;209:25–33. https://doi.org/10.1002/path.1948.
6. Drummond KJ, Rosenfeld JV. Pineal region tumours in childhood. A 30-year experience. Childs Nerv Syst. 1999;15:119–26; discussion 127. https://doi.org/10.1007/s003810050347.
7. Al-Hussaini M, Sultan I, Abuirmileh N, Jaradat I, Qaddoumi I. Pineal gland tumors: experience from the SEER database. J Neuro-Oncol. 2009;94:351–8. https://doi.org/10.1007/s11060-009-9881-9.
8. Louis DN, et al. The 2007 WHO classification of tumours of the central nervous system. Acta Neuropathol. 2007;114:97–109. https://doi.org/10.1007/s00401-007-0243-4.
9. Chang AH, et al. MR imaging of papillary tumor of the pineal region. AJNR Am J Neuroradiol. 2008;29:187–9. https://doi.org/10.3174/ajnr.A0784.
10. Mottolese C, Szathmari A, Beuriat PA. Incidence of pineal tumours. A review of the literature. Neurochirurgie. 2015;61:65–9. https://doi.org/10.1016/j.neuchi.2014.01.005.
11. Maity A, et al. Craniospinal radiation in the treatment of biopsy-proven intracranial germinomas: twenty-five years' experience in a single center. Int J Radiat Oncol Biol Phys. 2004;58:1165–70. https://doi.org/10.1016/j.ijrobp.2003.08.028.
12. Bamberg M, et al. Radiation therapy for intracranial germinoma: results of the German cooperative prospective trials MAKEI 83/86/89. J Clin Oncol. 1999;17:2585–92. https://doi.org/10.1200/JCO.1999.17.8.2585.

13. Bouffet E, et al. Combined treatment modality for intracranial germinomas: results of a multicentre SFOP experience. Societe Francaise d'Oncologie Pediatrique. Br J Cancer. 1999;79:1199–204. https://doi.org/10.1038/sj.bjc.6690192.
14. Matsutani M, Japanese Pediatric Brain Tumor Study Group. Combined chemotherapy and radiation therapy for CNS germ cell tumors—the Japanese experience. J Neuro-Oncol. 2001;54:311–6. https://doi.org/10.1023/a:1012743707883.
15. Calaminus G, et al. SIOP CNS GCT 96: final report of outcome of a prospective, multinational nonrandomized trial for children and adults with intracranial germinoma, comparing craniospinal irradiation alone with chemotherapy followed by focal primary site irradiation for patients with localized disease. Neuro-Oncology. 2013;15:788–96. https://doi.org/10.1093/neuonc/not019.
16. Bowzyk Al-Naeeb A, et al. Current management of intracranial germ cell tumours. Clin Oncol (R Coll Radiol). 2018;30:204–14. https://doi.org/10.1016/j.clon.2018.01.009.
17. Fuller BG, Kapp DS, Cox R. Radiation therapy of pineal region tumors: 25 new cases and a review of 208 previously reported cases. Int J Radiat Oncol Biol Phys. 1994;28:229–45. https://doi.org/10.1016/0360-3016(94)90162-7.
18. Baranzelli MC, et al. An attempt to treat pediatric intracranial alphaFP and betaHCG secreting germ cell tumors with chemotherapy alone. SFOP experience with 18 cases. Societe Francaise d'Oncologie Pediatrique. J Neuro-Oncol. 1998;37:229–39. https://doi.org/10.1023/a:1005863601481.
19. Balmaceda C, et al. Chemotherapy without irradiation—a novel approach for newly diagnosed CNS germ cell tumors: results of an international cooperative trial. The First International Central Nervous System Germ Cell Tumor Study. J Clin Oncol. 1996;14:2908–15. https://doi.org/10.1200/JCO.1996.14.11.2908.
20. Lee YH, et al. Treatment and outcomes of primary intracranial teratoma. Childs Nerv Syst. 2009;25:1581–7. https://doi.org/10.1007/s00381-009-0974-8.
21. Alapetite C, et al. Pattern of relapse and outcome of non-metastatic germinoma patients treated with chemotherapy and limited field radiation: the SFOP experience. Neuro-Oncology. 2010;12:1318–25. https://doi.org/10.1093/neuonc/noq093.
22. Ogawa K, et al. Long-term results of radiotherapy for intracranial germinoma: a multi-institutional retrospective review of 126 patients. Int J Radiat Oncol Biol Phys. 2004;58:705–13. https://doi.org/10.1016/j.ijrobp.2003.07.001.
23. Acharya S, DeWees T, Shinohara ET, Perkins SM. Long-term outcomes and late effects for childhood and young adulthood intracranial germinomas. Neuro-Oncology. 2015;17:741–6. https://doi.org/10.1093/neuonc/nou311.
24. Kim JW, et al. A multimodal approach including craniospinal irradiation improves the treatment outcome of high-risk intracranial nongerminomatous germ cell tumors. Int J Radiat Oncol Biol Phys. 2012;84:625–31. https://doi.org/10.1016/j.ijrobp.2011.12.077.
25. Jinguji S, et al. Long-term outcomes in patients with pineal nongerminomatous malignant germ cell tumors treated by radical resection during initial treatment combined with adjuvant therapy. Acta Neurochir. 2015;157:2175–83. https://doi.org/10.1007/s00701-015-2614-2.
26. Kobayashi T, Kida Y, Mori Y. Stereotactic gamma radiosurgery for pineal and related tumors. J Neuro-Oncol. 2001;54:301–9. https://doi.org/10.1023/a:1012727306066.
27. Hasegawa T, Kondziolka D, Hadjipanayis CG, Flickinger JC, Lunsford LD. Stereotactic radiosurgery for CNS nongerminomatous germ cell tumors. Report of four cases. Pediatr Neurosurg. 2003;38:329–33. https://doi.org/10.1159/000070417.
28. Amendola BE, Wolf A, Coy SR, Amendola MA, Eber D. Pineal tumors: analysis of treatment results in 20 patients. J Neurosurg. 2005;102(Suppl):175–9. https://doi.org/10.3171/jns.2005.102.s_supplement.0175.
29. Lekovic GP, et al. Role of Gamma Knife surgery in the management of pineal region tumors. Neurosurg Focus. 2007;23:E12. https://doi.org/10.3171/FOC-07/12/E12.
30. Mori Y, Kobayashi T, Hasegawa T, Yoshida K, Kida Y. Stereotactic radiosurgery for pineal and related tumors. Prog Neurol Surg. 2009;23:106–18. https://doi.org/10.1159/000210057.
31. Yianni J, et al. Stereotactic radiosurgery for pineal tumours. Br J Neurosurg. 2012;26:361–6. https://doi.org/10.3109/02688697.2011.635818.
32. Balossier A, et al. Role of radiosurgery in the management of pineal region tumours: indications, method, outcome. Neurochirurgie. 2015;61:216–22. https://doi.org/10.1016/j.neuchi.2013.11.007.
33. Li W, et al. Gamma knife radiosurgery (GKRS) for pineal region tumors: a study of 147 cases. World J Surg Oncol. 2015;13:304. https://doi.org/10.1186/s12957-015-0720-5.
34. Iorio-Morin C, et al. Histology-stratified tumor control and patient survival after stereotactic radiosurgery for pineal region tumors: a report from the international gamma knife research foundation. World Neurosurg. 2017;107:974–82. https://doi.org/10.1016/j.wneu.2017.07.097.
35. Louis DN, et al. The 2016 World Health Organization classification of tumors of the central nervous system: a summary. Acta Neuropathol. 2016;131:803–20. https://doi.org/10.1007/s00401-016-1545-1.
36. Han SJ, Clark AJ, Ivan ME, Parsa AT, Perry A. Pathology of pineal parenchymal tumors. Neurosurg Clin N Am. 2011;22:335–40, vii. https://doi.org/10.1016/j.nec.2011.05.006.
37. Deshmukh VR, Smith KA, Rekate HL, Coons S, Spetzler RF. Diagnosis and management of pineocytomas. Neurosurgery. 2004;55:349–55; discussion 355-347. https://doi.org/10.1227/01.neu.0000129479.70696.d2.

38. Reyns N, et al. The role of gamma knife radiosurgery in the treatment of pineal parenchymal tumours. Acta Neurochir. 2006;148:5–11; discussion 11. https://doi.org/10.1007/s00701-005-0626-z.
39. Kano H, Niranjan A, Kondziolka D, Flickinger JC, Lunsford D. Role of stereotactic radiosurgery in the management of pineal parenchymal tumors. Prog Neurol Surg. 2009;23:44–58. https://doi.org/10.1159/000210052.
40. Wilson DA, et al. Long-term radiosurgical control of subtotally resected adult pineocytomas. J Neurosurg. 2012;117:212–7. https://doi.org/10.3171/2012.5.JNS1251.
41. Park JH, et al. Upfront stereotactic radiosurgery for pineal parenchymal tumors in adults. J Korean Neurosurg Soc. 2015;58:334–40. https://doi.org/10.3340/jkns.2015.58.4.334.
42. Hanft SJ, Isaacson SR, Bruce JN. Stereotactic radiosurgery for pineal region tumors. Neurosurg Clin N Am. 2011;22:413–20, ix. https://doi.org/10.1016/j.nec.2011.05.002.
43. Gutenberg A, et al. Common molecular cytogenetic pathway in papillary tumors of the pineal region (PTPR). Brain Pathol. 2011;21:672–7. https://doi.org/10.1111/j.1750-3639.2011.00493.x.
44. Hasselblatt M, et al. Immunohistochemical profile and chromosomal imbalances in papillary tumours of the pineal region. Neuropathol Appl Neurobiol. 2006;32:278–83. https://doi.org/10.1111/j.1365-2990.2006.00723.x.
45. Cardenas R, et al. Papillary tumor of pineal region: prolonged control rate after gamma knife radiosurgery—a case report and review of literature. Neurol India. 2010;58:471–6. https://doi.org/10.4103/0028-3886.66051.
46. Fauchon F, et al. Role of surgery, radiotherapy and chemotherapy in papillary tumors of the pineal region: a multicenter study. J Neuro-Oncol. 2013;112:223–31. https://doi.org/10.1007/s11060-013-1050-5.
47. Choque-Velasquez J, et al. Papillary tumor of the pineal region in children: presentation of a case and comprehensive literature review. World Neurosurg. 2018;117:144–52. https://doi.org/10.1016/j.wneu.2018.06.020.
48. Riis P, van Eck AT, Dunker H, Bergmann M, Borm W. Stereotactic radiosurgery of a papillary tumor of the pineal region: case report and review of the literature. Stereotact Funct Neurosurg. 2013;91:186–9. https://doi.org/10.1159/000344023.
49. Shakir HJ, Qiu J, Prasad D, Mechtler LL, Fenstermaker RA. Papillary tumor of the pineal region with extended clinical and radiologic follow-up. Surg Neurol Int. 2015;6:S451–4. https://doi.org/10.4103/2152-7806.166782.
50. Fernandez-Mateos C, Martinez R, Vaquero J. Long-term follow-up after radiosurgery of papillary tumor of pineal region: 2 case reports and review of literature. World Neurosurg. 2018;116:190–3. https://doi.org/10.1016/j.wneu.2018.05.080.
51. Li D, et al. Pineal region gliomas: a single-center experience with 25 cases. World Neurosurg. 2019;133:e6. https://doi.org/10.1016/j.wneu.2019.06.189.

30 Reirradiation of Skull Base Tumors

Sławomir Blamek

30.1 Malignant Skull Base Tumors

30.1.1 Introduction

In this chapter, a wide spectrum of challenging uses of radiosurgery is described. Radiosurgery is usually required as a salvage treatment for recurring tumors that already received surgery and conventionally fractionated radiotherapy. Nevertheless, previous irradiation increases the risk of adverse effects and is associated with reduced probability of response since the recurring tumor demonstrates a high resistance to radiation due to clonal selection. Stereotactic techniques in these patients are used to deliver a single dose of radiation (stereotactic radiosurgery - SRS), hypofractionated irradiation schedule (HSRT hypofractioned stereotactic radiotherapy) or even to complete conventionally fractionated treatment (FSRT fractionated stereotactic radiotherapy).

S. Blamek (✉)
Department of Radiotherapy, Maria Skłodowska-Curie National Research Institute of Oncology, Gliwice Branch, Gliwice, Poland
e-mail: slawomir.blamek@io.gliwice.pl

© Springer Nature Switzerland AG 2020
A. Conti et al. (eds.), *CyberKnife NeuroRadiosurgery*,
https://doi.org/10.1007/978-3-030-50668-1_30

30.1.2 Primary and Salvage Radiosurgery for Head and Neck Carcinomas

Most radiosurgery studies on head and neck carcinomas involving the skull base report on patients previously treated with conventional techniques. Cengiz et al. published one of the largest series consisting of 46 patients treated for recurrent head and neck tumors [1]. Most patients (29 constituting 63% of the group) were treated with 30 Gy in five fractions. A five-fraction regimen was used also to deliver doses of 25 and 35 Gy in two and nine other patients, respectively (see Table 30.1). No patient in this series received chemotherapy during irradiation. The response to treatment could not be evaluated in nine patients. Complete response was achieved in 27% of the remainder, partial response in 29.8%, and stabilization of the tumor in 27%. No relationship between the result and the dose or pathology has been reported. Adverse treatment effects were reported inconsistently and eight carotid ruptures were not counted as early or late side effects. In total, nine bleedings were recorded, but the additional one was attributed to tumor progression and infiltration of the trachea by esophageal cancer. Of the remaining eight patients, seven died due to carotid rupture and one was saved by intra-arterial embolization. Interestingly, the carotid blow-out syndrome occurred only in patients with carotid artery encased for at minimum 180°

Table 30.1 Summary of the treatment parameters and outcome of patients after radiosurgical salvage for malignant head and neck tumors

Study	Patients evaluated	Pathology		Primary treatment dose (Gy) range (median)	SRS dose (Gy)/ number of fractions	Median interval between treatments (months)	LC (%)	Grade >3 late toxicities (%)	OS
Cengiz [1]	44	Squamous carcinoma	30	30–70 (51)	30 (18–35)/1–5	38	83.8	28.6 including 8 (17.8%) carotid ruptures	47% @1 y, median 11.9
		Adenoid cystic carcinoma	7						
		Mucoepidermoid cancer	1						
		Medullary thyroid cancer	1						
		Papillary thyroid cancer	1						
		Adenocarcinoma	1						
		Acinic cell cancer	1						
		Melanoma	1						
		Chordoma	1						
		Hemangiopericytoma	1						
		Malignant mesenchymal tumor	1						
Özyiğit [4]	24	Primary nasal carcinoma	24	48–70 (70)	30/5	38	82@2 y	29.16	64@2 y
Roh [8]	36	Squamous carcinoma	24	39.6–134.4 (70.2)	18–40/3–5	24	52.2	8.3	30.9@2 y
		Primary nasal carcinoma	8						
		Adenoid cystic carcinoma	7						
		Chondrosarcoma	4						
		Melanoma	1						
Kaplan [9]	7	Squamous carcinoma	4	n/a	17.5–35/1	n/a	100[a]	n/a[b]	n/a
		Adenoid cystic carcinoma	2						
		Mucoepidermoid carcinoma	1						
Xiao [5]	50	Primary nasal carcinoma	50	70–80 +/– 50[c]	14–35/1–5	1–4[d]	DFS 64.9@2 y[f]	16[e]	65@2 y
Orecchia [6]	13	Primary nasal carcinoma	13	63–72 (66)	24/2–4	23	100[g]	0	31%@3 y
Dhanachai [7]	32	Primary nasal carcinoma	32	63–82 (68.6)	30–50/6–25	15	LPFS 67.8%@1 y	3.12	71.2@3 y

SRS Stereotactic radiosurgery, *DFS* Disease-free survival, *LPFS* Local progression-free survival
[a]Follow-up ranged between 1 and 11 months
[b]Two cranial nerve palsies were reported, one resolved, no data on the severity were given
[c]Patients after two courses of radiotherapy were allowed
[d]In the groups with residual disease or after second course of radiotherapy
[e]Eight fatal hemorrhages, no other severe late toxicity was reported
[f]73.97% in patients with residual tumor and 46.53% in patients with a recurrent tumor
[g]Evaluated after radiosurgery, five had complete remission, two partial response (>50%), four minor response, and two stabilization of the tumor

of its circumference. They also observed that no carotid ruptures occurred in patients with carotid dose less than 100% of the prescribed dose. Conversely, Voynov et al. reported no bleeding or grade 4 or 5 acute or late grade adverse effects [2]. They used 1–8 daily fractions of 3–16 Gy resulting in total doses of 10–36 Gy (median 24 Gy). The most common scheme was 25 Gy in 5 fractions. Note that only one lesion was located in the skull base and four in nasopharynx among 22 patients enrolled into the study. Most of the lesions involved cervical lymph nodes and/or the oropharynx. The favorable toxicity profile resembled that from Phase I study on SBRT for recurrent head and neck cancers conducted in Pittsburgh. Five dose levels of 25–44 Gy delivered in five fractions in five patients per group were tested. No grade 3/4 or dose-limiting toxicities were recorded in 25 patients enrolled in the study [3].

A group of researchers from University of Hacettepe in Ankara published a study on the comparison between conventional radiation therapy and CyberKnife radiosurgery for salvage treatment in patients with nasopharyngeal carcinoma [4]. They demonstrated similar effectiveness of both techniques in terms of local control and overall survival with a better safety profile of CK radiosurgery. Grade 3 or higher late toxicity was observed in 21% of patients in the CK group, whereas severe late toxicity was observed in 48% of patients in the three-dimensional conformal radiotherapy group ($p = 0.04$). The doses used were 30 Gy in five fractions administered daily in a group of 24 patients and 57 Gy (median) with 2 Gy per fraction in a group of 27 patients, respectively [4]. Of note, they had also treated four patients with a third course of radiotherapy using CyberKnife, but the results were not included in the analysis. A series of 50 patients treated for primary nasal carcinoma was presented by Xiao et al. [5]. Patients with residual tumors after primary radiation therapy and patients with relapses after one or two cycles of radiation therapy were treated. The destruction of the bone structures of the base of the skull was documented in 31 patients. Treatment schedules were complicated and different for almost each patient. Doses of 6–8 to 15 Gy per fraction were delivered with 4–6 day intervals. Tumor response was observed in almost all cases, with a complete response of 76% and a partial response of 18%.

The number of fatal hemorrhages was similar to that reported by Cengiz et al. [1]. The analysis of factors possibly associated with the risk of fatal bleeding indicated that tumors involving the Rosenmueller fossa and invading deeply into the foramen lacerum are more likely to cause bleeding, as this part of the carotid is more vulnerable to damage [5]. The other factors that could contribute to the risk of bleeding were high doses of radiation from previous radiation therapy cycles, overly high doses used in HSRT, and diabetes [5]. On the other hand, no serious late sequelae were reported by Orecchia et al. [6], and in the Dhanachai et al. [7] study, only one patient had a grade 3 hearing impairment requiring a hearing aid but no carotid ruptures were reported [6, 7]. For details, please refer to Table 30.1. A series of 36 patients treated for 44 sites of recurrence with CyberKnife hypofractionated radiotherapy was presented by Roh et al. [8]. The doses applied ranged between 18 and 40 Gy (median 30 Gy), delivered in 3–5 fractions. All patients had been treated with conventionally fractionated radiotherapy before salvage radiosurgery, and the doses delivered ranged between 39.6 and 134.4 Gy. Grade 3 acute toxicity of the treatment was reported in 13 patients, late in 3. All three had bone necrosis, one in the mandible and two at the base of the skull, associated with soft tissue necrosis in one and chronic ulceration and trismus in the last two. All three patients with late toxicity were previously irradiated with doses of 66–70 Gy and the salvage dose of 30–33 Gy in all of them was delivered in 3 fractions. The authors state that currently a 5-fraction scheme is preferable in their institutional protocol, and doses of 25–40 Gy are prescribed without serious complications since the introduction of this protocol [8]. One of the older, but worthy of attention, studies on salvage Gamma Knife radiosurgery (GKRS) for recurrent head and neck tumors was presented by Kaplan et al. [9]. A series of seven patients with ten lesions treated with single doses of 17.5–35 Gy was analyzed. All patients were treated due to symptoms caused by recurrent tumors, like diplopia, facial numbness, cranial nerve palsy, etc. In all patients the symptoms improved or remained stable; in all also radiographic evaluation demonstrated stabilization or regression of the tumor. Three patients were treated for out-of-field recurrence with further radiosurgery.

Karam et al. [10] presented a small series of patients with salivary gland cancer treated with intensity modulated radiotherapy (IMRT) and radiosurgical boost [10]. Three of ten patients presented had skull base invasion. All were treated with IMRT with a median dose of 64.8 Gy (range 50–75.6 Gy), mainly with 1.8 Gy per fraction. The median boost dose was 17.5 Gy and ranged between 10 and 30 Gy administered in 3–6 fractions (median 5). The median interval between IMRT and radiosurgery was 1 week (range 0–2).

At a median follow-up of 29 months, progression-free survival (PFS) and overall survival (OS) were 68, and 79%, respectively. Median values for any survival measure were not reached. Treatment toxicity was acceptable according to the authors, acute grade 4 and 5 toxicity were not reported. Long-term toxicity included osteoradionecrosis of the mandible in one and graft ulceration in another patient.

Four patients suffered from sensorineural hearing loss, three from xerostomia and six from fibrosis.

The same team reported on the re-radiation of recurrent salivary gland tumors [11]. The median dose delivered with HSRT was 30 Gy in 5 fractions. The risk of late soft tissue necrosis in this group was associated with cumulative doses of at least 90 Gy. This late effect occurred in 4 of 18 patients in this series, and the authors stated that the numbers were too small to draw general conclusions [11]. To sum up, the most popular CyberKnife treatment regimen in case of recurrent head and neck tumors appears to be 25–35 Gy delivered in 5 fractions. The dose is chosen individually with respect to the previous treatment doses and proximity of critical structures. The most worrying and dangerous side effects are carotid blow-out syndrome, bone and soft tissue necrosis, and ulceration.

30.1.3 Plasma Cell Tumors

There are clinical cases in the literature reporting high effectiveness of CyberKnife stereotactic radiosurgery in patients with cranial plasma cell tumors. The first patient was reported by Wong et al. [12]. They irradiated a clivus plasmacytoma extending to foramen magnum with excellent local control evaluated at 12 months after radiosurgery [12]. In another patient with no evidence of systemic disease, a tumor also localized in clivus was treated with 21 Gy in three fractions prescribed on 80% isodose line with 97% coverage [13]. MRI performed 6 months after the treatment showed complete response, but the disease eventually evolved into a widespread form requiring systemic treatment. CyberKnife radiosurgery cannot be considered a standard treatment in case of plasma cell tumors, but localized high-dose irradiation in case of limited disease can be an interesting option for selected patients. Additional data is needed to draw more general conclusions on the value of radiosurgery in this group of patients.

30.1.4 Pituitary Carcinoma

The diagnosis of pituitary carcinoma can be made only after discovery of metastases, or the presence of tumor is not contiguous to the primary sellar tumor. The presence of a high Ki-67 labeling index, mitoses, or nuclear atypia indicate aggressive behavior but is not sufficient to diagnose pituitary carcinoma [14]. Radiosurgery in patients with pituitary carcinoma is usually employed to treat metastatic tumors. Experiences with SRS in this setting are very limited and the outcome is poor. Tuleasca et al. [15] described a case of a patient with a non-functioning pituitary adenoma subject to surgery and conventional radiotherapy (50.4 Gy). During radiotherapy, the patient was operated on due to sudden visual disturbances—bitemporal hemianopia and partial third cranial nerve paralysis. Microscopic examination revealed a cyst and radiotherapy was continued. Eighteen years after diagnosis, the patient was subjected to a second surgery and 18 Gy gamma knife radiosurgery for tumor progression. Three years later, another out-of-field relapse was treated with 18 Gy GKRS. Additionally, a resection of a (separate) left frontal lesion was performed, and, due to its atypical features, adjuvant radiotherapy was prescribed (60 Gy in 30 fractions). Postoperative MRI revealed another lesion in the left orbital gyrus which was also treated with radiosurgery (18 Gy). Spinal MRI also revealed metastatic lesions at the S1 and S2 levels which were treated with radiotherapy and temozolomide. Control imaging performed 19 months after radiosurgery for the first metastasis showed that all lesions treated with radiosurgery were stable [15].

Park et al. also described a case of pituitary carcinoma treated with GKRS 3 years after conventional radiotherapy (50.4 Gy) for residual pituitary prolactinoma. The metastatic tumor was controlled for 3 years after radiosurgery with marginal dose of 16 Gy. In this case, radiosurgery was also used as a salvage treatment after progression of the disease in a patient subjected to prior conventional irradiation [16]. The other malignancy sometimes occurring in the pituitary region is metastatic cancer. Due to the rarity of such lesions, no specific recommendations are available for pituitary metastases. On the other hand, metastases represent about 1% to even 2.8% of surgically treated pituitary tumors as reported by Chon et al., which in fact makes it a more frequent pituitary malignancy than primary pituitary carcinoma observed in 0.1–0.2% of pituitary tumors [17, 18]. In general, the proposed management is similar to the management of cerebral metastases, described in detail in specific chapters of this book (see Chaps. 14–17). Interestingly, there is a study on fractionated CyberKnife radiosurgery for pituitary metastases showing excellent results in terms of local control (LC in all seven patients with complete remission in four) [17]. All patients were irradiated with 31 Gy delivered in 5 fractions. The authors calculated the biologically equivalent dose (BED) assuming an alpha/beta = 10 Gy to obtain a scheme radiobiologically equivalent to 18 Gy in a single fraction. Three patients in this series died from cancer progression outside the pituitary, the rest were alive at the time of evaluation. The median overall survival was 14 months. No adverse effects were reported, and in all patients, the pre-existing symptoms of diabetes insipidus improved. An improvement was observed in all three patients with visual function impairment. Similar results in terms of improvement of endocrine and visual function and local control were also noted after GKRS [19, 20]. Adverse effects of radiosurgery for pituitary tumors are not uncommon and are related in most cases to the function of the anterior pituitary, but they develop as late effects of the treatment and can probably be considered of secondary importance in the case of malignant tumors. The details of pituitary radiosurgery are discussed elsewhere. It is worth noting here that in addition to the doses for the pituitary gland itself, the role of the dose in the pituitary stalk and hypothalamus has also been discussed in the context of hypopituitarism associated with radiosurgery for perisellar tumors. According to Feigl et al., in patients with pituitary stalk dose less than 4.1 ± 2.8 Gy and pituitary dose less than 9.5 ± 7.1 Gy, the risk of radiation-induced deficiency is low [21].

30.1.5 Malignant Peripheral Nerve Sheath Tumor

Malignant peripheral nerve sheath tumors (MPNSTs) are rare and are usually discussed in the context of secondary malignancies after radiotherapy or radiosurgery. The treatment is usually surgical, often followed by adjuvant radiotherapy. Nevertheless, there are some data in the literature on the use of radiosurgery for MPNST treatment, both in patients subjected to prior irradiation and in patients with MPNSTs arising without previous radiotherapy. Yang et al. reported a case of MPNST originating from the hypoglossal nerve. The tumor was presumed to be a schwannoma and was treated with CyberKnife with a single dose of 13.6 Gy at the

85% isodose encompassing the tumor. Nine years after CyberKnife radiosurgery the tumor progressed. The patient had microsurgical resection of the tumor with adjuvant conventional radiotherapy (50 Gy) and a radiosurgical boost of 10 Gy at 50% isodose [22]. In this case, CyberKnife radiosurgery might be first involved in malignant transformation of the tumor, but eventually it was used also to boost the radiation dose to the malignant lesion. Bashir et al. reported a case of MPNST arising 42 months after excision of pathologically proven vestibular schwannoma, without prior irradiation. After surgery, the patient was referred to radiotherapy and received 54 Gy delivered with 1.8 Gy per fraction, as a fractionated stereotactic radiotherapy. The patient had no evidence of tumor 9 months after irradiation [23]. Conventionally fractionated stereotactic radiotherapy as an adjuvant treatment for a resected, spontaneous MPNST was also reported by Belyaev et al. [24]. Radiotherapy was delivered to a progressing residual tumor after surgery, but after 10 months, both local and distant progression were diagnosed, and the patient was treated with another surgery and GKRS for metastases.

Another similar case was presented by Karami et al., but here, the MPNST diagnosis was made after the first surgery. The patient had undergone surgical treatments and adjuvant intensity-modulated radiotherapy (70 Gy). After radiotherapy, GKRS for three residual foci was performed and doses of 14–16 Gy were delivered with good local effect, but it did not prevent further dissemination of the tumor [25]. Raper et al. described a case of primary MPNST treated with surgery and radiotherapy (56 Gy in 28 fractions) with a GKRS boost of 15 Gy prescribed at 50% isodose line to the residual tumor. The tumor was stable at the time of the last follow-up [26]. The authors pointed out that reports on radiosurgery for MPNST were scarce and listed the available papers showing the results of GKRS with doses of 12–20 Gy. The reported responses varied from "slight tumor shrinkage" to 97% decrease of the tumor volume [25, 27, 28]. In one of these cases, in fact, a CNS metastasis from a peripheral tumor was treated [28]. All the anecdotal reports on the use of radiosurgery for MPNST treatment do not allow to make general recommendations either concerning the dose or fractionation schedules. Nevertheless, irradiation with stereotactic techniques can be considered a treatment option that apparently allows for lasting local control.

30.1.6 Skull Base Sarcomas

The use of CyberKnife radiosurgery for chondrosarcomas as most often malignant mesenchymal tumors of the base of the skull is described in detail in a dedicated chapter. To supplement its content, it is worth noting that there are anecdotal descriptions of radiosurgery for other types of skull-based sarcomas and repeated radiosurgery for chondrosarcomas in this location.

Jiang et al. reported on a series of patients with chondrosarcoma, including three with CyberKnife radiosurgery for recurrent tumor after conventional external beam radiation therapy, IMRT and GKRS. Three patients with chondrosarcoma metastases also underwent radiation therapy before CK radiosurgery. Two of them had intracranial metastases, one treated with a single fraction of 24 Gy and the other with 24 Gy delivered in 3 fractions. Both had radiographic progression.

Radiographic progression was also diagnosed in the one of the three patients treated for primary tumor who received 30 Gy in 5 fractions. The other two patients had radiological stabilization and improvement after a single fraction of 27 and 18 Gy delivered in 3 fractions, respectively. No complications related to CyberKnife radiosurgery were reported in any of them [29]. No stratification according to pathological grade was performed. The authors noted better local control in patients treated with doses above 24 Gy, but this trend did not reach the statistical significance threshold ($p = 0.09$) [29]. The use of radiosurgery for skull base malignancies is also mentioned by Wilson et al. in their paper on CyberKnife radiosurgery for skull base tumors. Most of the patients in this series had benign tumors and the outcome was presented only for this subgroup. No specific outcome data were available for patients with malignant lesions [30].

An unusual case of fibro-odontosarcoma primarily located in the oral cavity and recurring in the sphenoid was described in a small cohort of pediatric patients treated with fractionated radiosurgery for recurrent or oligometastatic intracranial tumors. The patient was treated with CyberKnife fractionated radiotherapy with 42 Gy delivered in 14 fractions. The treatment was performed 15.5 months after the first radiotherapy, and the cumulative BED amounted to 456 Gy. The patient experienced late grade 3 toxicity manifesting as brain necrosis and osteonecrosis [31].

Two cases of rhabdomyosarcoma, two Ewing sarcomas and one leiomyosarcoma were included in the series published by Coppa et al. [32]. One patient in this series, treated for melanoma metastasis, underwent stereotactic radiosurgery before CyberKnife treatment, four had conventional radiotherapy, and one with renal cell carcinoma metastasis was salvaged with second cycle of CK radiosurgery. Doses of 12.6–35 Gy were delivered in a median of 5 fractions (range: 2–7) for the whole series. In two patients with Ewing sarcoma treated after EBRT, the tumor was stable, as was the leiomyosarcoma and both rhabdomyosarcomas, one treated after EBRT and one without prior irradiation, but only the last patient survived. The other four died due to progression of the disease outside the treated lesion. An interesting clinical case on a skull base osteosarcoma was presented by Yamada et al. A 78-year-old woman was irradiated with CyberKnife hypofractionated stereotactic radiotherapy with five fractions and a marginal dose of 39.38 Gy after subtotal transnasal transsphenoidal surgery. Six months after the treatment, the patient was admitted to hospital due to coma and high fever. A CSF leakage was diagnosed and attributed to tumor shrinkage after radiosurgery. The leakage was surgically repaired and the patient fully recovered [33].

All the papers referenced here are clinical cases or small series of cases that do not allow general recommendations, but it appears that CyberKnife radiosurgery is a reasonable option for patients with unresectable sarcomas at the base of the skull, in particular recurring after previous treatment or not amenable to surgery. In addition to the risk of bone or brain necrosis, the risk of cerebrospinal fluid leakage after regression should also be considered, especially in patients who have undergone previous surgery.

30.1.7 Esthesioneuroblastoma

Surgical resection is the primary treatment of esthesioneuroblastoma. After surgery, conventional radiotherapy and/or chemotherapy are used as adjuvant treatment. Stereotactic radiosurgery can be used as a salvage treatment. Data on inclusion of radiosurgery as a part of primary therapy are extremely scarce. Unger et al. described a series of 14 patients treated with endoscopic sinus surgery and adjuvant radiosurgery with marginal doses of 15–34 Gy delivered to residual tumor. Twelve patients were treated de novo, 2 received radiosurgery after surgical resection performed 24 and 39 months earlier [34]. Tumor progression occurred in five patients, in every case outside the irradiated lesion. One patient with recurrence was salvaged with surgical resection. The remaining four had another radiosurgery. Tumor progression or recurrence occurred after a median of 34 months. All patients were alive at the time of publication (follow-up ranged between 13 and 128 months). Most of the reported side effect were surgery-related. Cephalea and dizziness attributed to radiosurgery resolved within 48 h. The largest series of patients treated with recurrent esthesioneuroblastoma radiosurgery has been published by Van Gompel et al., but the number of patients included was lower than in the Unger study [35]. A total of 31 recurrences in 13 patients were analyzed. Ten patients had fractionated radiotherapy before radiosurgery and the dose delivered ranged between 27 and 69 Gy (median 52.2 Gy). The median marginal dose during radiosurgery was 15 Gy and ranged between 7 and 18 Gy. All patients were treated by GKRS. Failure (tumor growth greater than 2 mm in any plane) was diagnosed in 3 of 27 tumors. In this study, no procedure-related adverse effects were reported either.

30.1.8 Hemangiopericytoma

The largest series of patients treated with CyberKnife for recurrent or residual hemangiopericytomas was published by Veeravagu et al. [36]. Fourteen patients (3 with spinal and 11 with intracranial lesions) with 24 tumors were treated with mean marginal doses of 21.2 Gy delivered in 1–5 fractions. In most cases, a single fraction of 16–24 Gy was delivered. Some tumors were irradiated with two fractions of 8, 10 or 11 Gy. Three-fraction regimens were used for 3 lesions, and the prescribed dose varied between 22 and 27 Gy. One patient received 30 Gy in 5 fractions. Progression-free survival was 71.5% at 5 years, and overall survival was 81%. No significant procedure-related adverse effects were reported and the authors noted no dependence on tumor volume or dose and outcome. An interesting series of patients treated with radiosurgery for intracranial hemangiopericytomas was presented by Kim et al. They treated nine patients with 17 tumors with GKRS. All patients had previously been treated with surgery and conventional radiotherapy with mean doses ranging from 50.4 to 61.2 Gy (mean 56.9 Gy). At the time of radiosurgery, median marginal doses of 20 Gy (range 11–22 Gy) were employed. Contrary to the previous study, they found that better local control can be expected in case of tumors irradiated with doses equal or above 17 Gy [37]. This is slightly higher dose than the most commonly prescribed 15 Gy in single fraction [38–40]. A group of eight skull base hemangiopericytomas was reported by Pan et al. among a larger series of 43 patients with primary and secondary malignant skull base tumors treated with radiosurgery [41]. In two patients with hemangiopericytoma, progression occurred but no further details were given as no specific subgroups of tumors were analyzed separately. Of 43 patients treated, 9 experienced adverse effects including 6 with deterioration of cranial nerve functions, but again, no relation to histology was presented.

30.1.9 Skull Base Metastases

Numerous studies deal with radiosurgery for skull base metastases [1, 32]. Very often they are metastases of peripheral carcinomas, and usually the treatment is similar to that prescribed for skeletal metastases in other locations, but some also concern metastases of unusual intracranial tumors like pineoblastoma [42]. The described patient received craniospinal irradiation (36 Gy) with 20 Gy local boost but experienced dissemination of the tumor. A sphenoid lesion was resected, and the metastatic tumor bed irradiated with CyberKnife with 30 Gy delivered in 5 fractions. Two months later, multiple leptomeningeal metastases were treated with WBRT with CK boost delivered in single fractions of 13.4–13.6 Gy. After 4 months, another tumor was irradiated with a single dose of 24 Gy and six others subsequently received 20 Gy in a single fraction each. The patient was also treated for spinal metastases with CK radiosurgery. Most of the irradiated lesions regressed, but the spread continued and the patient died during the salvage therapy with temozolomide. This example shows that the flexibility of frameless radiosurgery that allows the delivery of a single fraction or a fractionated treatment, according to current needs, allows effective local control of irradiated lesions and previous therapies almost never affect the possibility of salvage radiosurgery.

30.2 Benign Skull Base Tumors

30.2.1 Vestibular Schwannomas

Probably most of the informative literature on repeated stereotaxic irradiation for skull base tumors concerns re-radiation of vestibular schwannomas (Table 30.2) [43–48]. These tumors have a specific biology with initial tumor volume increase, and real progression should not be diagnosed earlier than two years after treatment, unless rapidly progressing. Available sources provide information on the results of retreatment with GKRS that can be tailored to the needs of CyberKnife treatment planning. These observations concern single-fraction treatment; there are no reliable data on repeat HSRT for vestibular schwannomas.

Lonneville et al. showed feasibility of partial irradiation of the vestibular schwannoma which was performed in patients who experi-

Table 30.2 Summary of the results of studies on radiosurgical retreatment of vestibular schwannomas

Study	Patients evaluated	Dose SRS1	Dose SRS2	Dose SRS3	Interval between treatments (months)	LC (%)
Kano et al. [44]	6	13	11	–	63	100
Dewan et al. [48]	11	12	12	–	51	81.8
Roman Liščák, Gokhan Özyiğit [43]	24	13	13	–	43	91.6
Fu et al. [45]	38	11	11.5	–	49	100
Yomo et al. [46]	8	12	12	–	46	100
Lonneville et al. [47]	25	12	12	14	45	85
Hasegawa et al. [49]	4	–[a]	–	–	30.25	75

[a]Mean margin dose in the whole group of 111 initially treated patients was 14.6 Gy

Table 30.3 Summary of the outcome reported in studies on radiosurgical retreatment of functioning pituitary adenomas

	No. of patients evaluated	Secreted hormone	SRS 1 dose	SRS 2 dose	Interval between treatments	Local control (%)	Hormonal control (%)	New deficits (%)
Alonso et al. [52]	18 imaging, 21 endocrine	GH	17 Gy	23 Gy	Median 5 years	83.3	42.9	19
Mehta et al. [51]	20	ACTH	21.8 Gy	20 Gy	1.3–9.7 years	85%	53% @ 10 y	10

enced progression of a certain part of the tumor. This option deserves to be considered in heavily pretreated patients with organs at risk already irradiated to the tolerance limits or in patients with preserved hearing with a growing part of the tumor located outside the internal auditory meatus. In four of the six patients treated with partial tumor irradiation, planning was guided with PET study. PET was also performed after the treatment showing metabolic regression in all patients [47]. In a series described by Hasegawa et al., 4 patients of the 80 analyzed were treated with radiosurgery for the second time due to progression. One patient even had three procedures, but neither detailed outcome nor treatment parameters were provided [49].

30.2.2 Pituitary Adenomas

The secretory function of pituitary adenomas is more difficult to control with irradiation than their growth potential, and the need for retreatment after initial management in case of secreting adenomas is not a rare issue. After initial normalization of cortisol levels after radiosurgery, for example, hormonal recurrence can be expected in 18% of patients according to Mehta et al. [50]. Nevertheless, the reports on stereotactic reirradiation of pituitary tumors are limited (Table 30.3) [51, 52]. Only a few studies focused on the results after CyberKnife radiosurgery for functioning pituitary adenomas. The reports on the use of CK for reirradiation are anecdotal. A patient with acromegaly was reported by Sala et al. [53]. After initial success, the disease recurred and the patient was irradiated again with CK, but no details on dose or fractionation in these patients were available. The patient had normal pituitary function after the first procedure but experienced hypopituitarism after the second. Another patient was reported in a series published by Roberts et al. [54]. The patient was irradiated 3.5 years after prior CK radiosurgery offered as upfront treatment. The tumor received 20 Gy in one fraction and complete hormonal response with no deficit in pituitary function at last follow-up was reported. It is worth noting that a possibility of malignant transformation after repeated radiosurgery for recurrent pituitary adenoma has been suggested, but metastatic spread and diagnosis of pituitary carcinoma can also occur in patients without previous radiotherapy [55, 56].

The use of radiotherapy in these patients associated with aggressive clinical behavior of the adenomas which ultimately progresses to the metastatic form of pituitary carcinoma can be an unusual but occasionally seen as an undesirable outcome of radiosurgery. An interesting review of the literature on pituitary carcinomas in patients undergoing radiotherapy has been published by Lall et al. They presented 45 cases of pituitary carcinoma associated with prior radiotherapy, including their own case subject to 5 cycles of irradiation including fractionated CyberKnife treatment [57]. No details on dose and fraction were given but the patient developed a carotid pseudoaneurysm in the left cavernous sinus which was an intractable source of nosebleeds and discussed as a possible radiation injury.

The literature data on retreatment after primary radiosurgery failure for nonfunctioning pituitary adenomas are scarce. Radiotherapy for inoperable, recurrent, or residual nonfunctioning adenomas produces excellent results, and the need for retreatment is infrequent. Consequently, it was not possible to find reliable bibliographic data on repeated radiosurgical treatment for nonfunctioning pituitary adenomas. The available reports are anecdotal and do not allow general conclusions to be drawn. Pollock et al. described a case of repeated radiosurgery for a progressive nonfunctioning pituitary adenoma with good results and no visual impairment reported [58].

The toxicity of repeated treatment in case of perisellar tumors is always a cause for concern. In the Cushing's disease group, two patients experienced cranial nerve deficiency after a second radiosurgery without evidence of radiological progression of the tumor [51].

One patient suffered from a third cranial nerve palsy that did not resolve after treatment, and the other had a visual field deficit 1 week after SRS which resolved 8 weeks later.

The maximum point dose delivered to the optic apparatus was 7 Gy during the initial treatment and 3.6 Gy at the time of the retreatment. It is not clear whether the maximum doses overlapped but even if so, the resulting total dose was 10.6 Gy, which is close to a single dose of 10 Gy, deemed safe according to the AAPM guidelines [59].

Due to the not negligible morbidity of repeated radiosurgery, great caution is advised. In any case, it is necessary to make an individual decision taking into account the doses administered to the critical organs during the initial treatment and the geometric characteristics of the target lesion such as its volume, size, and relationship with neighboring structures. A more precise account of the details of the repeated treatments, in particular of the cumulative dose-volume histograms of all treatments, is certainly necessary. This could allow a reliable estimate of the dose constraints for previously irradiated critical structures known for their limited tolerance, such as elements of the optic pathway, pituitary gland, infundibulum, and hypothalamus in case of reradiation of sellar and perisellar tumors.

Even if maximum doses are reported in critical structures, no information is usually given regarding the location of the maximum dose at the time of retreatment in relation to the location of the maximum dose point at the initial treatment, which precludes drawing conclusions on the tolerance of previously irradiated structures. It should also be considered that tolerance can depend not only on the maximum dose but also on the volume of the structure irradiated with doses lower than the maximum dose for the structure and, of course, on previous damage associated, for example, with surgery or due to the tumor itself.

30.2.3 Meningiomas

Repeated radiosurgery for meningioma is discussed in this section. No report specifically addressing the role reirradiation of skull base meningiomas could be found. Kim et al. described 33 patients with repeated radiosurgery for recurrent meningiomas. Twelve of them had skull base tumors but the outcome was not reported separately for convexity or parafalcine and skull base tumors. Low grade meningiomas and tumors without pathological confirmation had better reirradiation results in terms of progression-free survival than the WHO grade II and grade III meningiomas [60]. The results of the retreatment were not satisfactory: less than half of the group

Table 30.4 Summary of the outcome reported in studies on radiosurgical re-treatment of meningiomas

	No. of patients evaluated	SRS 1 dose	SRS 2 dose	Interval between treatments	Local control (%)	New deficits (%)
Kim et al. [60]	33[a] (12 skull base)	12.58 Gy	14.21 Gy	60.91 months	48.48	19
Wojcieszynski et al. [61]	19 (11 skull base)	15/54 for fSRT	15/50.4 for fSRT	40	42	0% grade 3 or higher

[a]36 patients had second radiosurgery but data on 3 treated with fractionated schedule with CyberKnife were not given

(16 patients) maintained local control, while 17 had progression (Table 30.4).

Similar outcome was reported by Wojcieszynski et al. [61]. This suggests that meningiomas that resist the primary treatment with radiotherapy are composed of highly radioresistant cells, and the results of repeated radiosurgery are markedly different from those observed, e.g., in the case of acoustic neuromas. The aggressive behavior of meningiomas progressing after radiosurgery was also pointed out in other studies [62].

30.2.4 Skull Base Chordomas

The details of CyberKnife radiosurgery for chordomas have been described elsewhere. Due to high radioresistance of these tumors, recurrences after irradiation are not unusual. The largest series treated with stereotactic radiosurgery was reported by Kano et al. They found that previous radiation therapy was a factor that adversely affects survival after radiosurgery in both univariate and multivariate analysis. However, patients treated after prior radiotherapy had applied significantly lower doses than patients who had not been irradiated before SRS (mean margin doses respectively 13.4 Gy versus 16.3 Gy) [63]. In another study, 3 patients out of 15 treated with GK for skull base chordomas were retreated due to out-of-field progression. Two of them demonstrated tumor stabilization at the last follow-up. In one of the two with local control, facial pain in the area innervated by the ophthalmic branch of the trigeminal nerve appeared after the second treatment, but symptoms of imbalance and diplopia improved. In the second patient, the complaints about balance, facial numbness, headache, and diplopia remained unchanged. The mean prescription dose was 15.3 at the initial treatment and 13.7 Gy at the second treatment; the mean maximum dose was 36.6 Gy and 32.8 Gy, respectively [64]. It is also worth noting that Raza et al. reported that in patients with progression after radiotherapy, salvage resection was not associated with benefits for the patient whereas there was a trend toward improvement after stereotactic radiosurgery [65]. The doses prescribed ranged between 16 and 20 Gy.

30.2.5 Craniopharyngiomas

In a series of 11 patients treated with CyberKnife at the Stanford University School of Medicine only one had previously undergone EBRT receiving 54 Gy and received 19.5 Gy in three fractions but no detailed outcome information was given [66]. This also refers to a patient subject to GKRS after EBRT in a series from Milano [67]. Five of the seven patients who progressed after the first GKRS were treated with another GKRS, four with single-fraction and one with fractionated treatment. In this series, multisession GKRS was also used in 21 of 50 patients included into the study and the mean dose per fraction was 6.9 ± 0.6 Gy for the fractionated treatment and 14.9 ± 0.6 Gy for single-fraction. There is, however, no direct statement that second GKRS procedures were performed with the same doses. No serious side effects were reported after the second procedure.

In another GKRS series, there were two patients after GKRS, three after EBRT, and one after both types of radiotherapy included, but again, no specific information on outcome in this subset of patients was provided [68]. In this series, multisession GKRS was also used in some patients (median marginal dose 6 Gy, range 5–7.5 Gy, three fractions) apart from single-session radiosurgery of 15 Gy. In the univariate analysis, there was a difference between outcomes after single- and multisession radiosurgery which was not confirmed in the multivariate analysis. The total doses used in the fractionated treatment, however, can be considered insufficient in terms of biological efficiency as compared to a single fraction of 15 Gy.

Despite many drawbacks, the only valuable data on repeated radiosurgery for craniopharyngioma actually come from Gamma Knife centers. Repeating radiosurgery for craniopharyngiomas may improve local control rates, as reported by Lee et al. [69]. Among 137 patients initially treated with GKRS, 21 were re-treated with radiosurgery due to in-field progression, out-of-field recurrence, or progression of the cystic component of the tumor. Thanks to the repeated radiosurgery, the tumor control rates were 82% at 3 years, 77% at 5 years, 64% at 8 years, and 61% at 10 years as compared to 75% at 3 years, 70% at 5 years, 54% at 8 years, and 44% at 10 years after one treatment. This difference was statistically significant. Note that 10 out of the 137 patients had already been treated with EBRT before radiosurgery, but no specific data on outcome in this subset of patients was given and no information on whether any of them had multiple GKRS procedures.

30.3 Final Remarks

There is no reliable evidence in the literature to specify dose constraints for critical structures in the setting of repeated radiosurgical treatment. It should be assumed that the tolerable doses will be lower at the second treatment than those specified for non-irradiated structures. Due to the lack of any standard for reporting the cumulative doses in critical structures after repeated treatments, no specific recommendations for certain structures can be made. There is definitely a need for new reports concerning retreatment with detailed information on doses delivered to organs at risk.

Contemporary treatment planning systems allow us to add treatment plans, some allow for non-rigid registration of images and dose distributions. These functionalities should be routinely used and the results reported in order to create any evidence-based recommendations related to repeated radiosurgery. Unfortunately, in the currently available literature, in most cases the essential data that allow the formulation of detailed guidelines are lacking. The analysis of the reference studies only allows us to highlight some issues to consider before planning a treatment for reirradiation.

- Patients with carcinomas entrapping the carotid should be irradiated with caution, especially those with more than 180° of its circumference involved. Hotspots in this area should be strictly avoided and special attention should be paid to tumors located in the region of Rosenmueller fossa and foramen lacerum.
- Five-fraction irradiation schedules and total doses of 25–35 Gy appear to be most frequently used in patients with carcinomas involving the skull base due to their efficiency and relatively low toxicity.
- Pituitary carcinomas treated with radiosurgery are usually tumors in patients previously irradiated for residual pituitary adenoma. Consequently, the (small) risk of malignant progression should be discussed with patients irradiated for pituitary adenoma. The causative role of radiation is disputable, since malignant progression can occur without irradiation as in the case of vestibular schwannomas progressing to MPNSTs. Local control after radiosurgery is satisfactory given the severe prognosis associated with the diagnosis, but eventually the disease progresses anyway.
- Repeated treatment with hypofractionated schedules may prove more advantageous than single fraction treatment, at least in terms of radiation-induced toxicity, but this hypothesis requires confirmation in specially designed studies.
- Local control after repeated radiosurgery for meningiomas tends to be lower than other benign tumors such as vestibular schwannomas or pituitary adenomas treated for the second time.

References

1. Cengiz M, Özyiit G, Yazici G, et al. Salvage reirradiaton with stereotactic body radiotherapy for locally recurrent head-and-neck tumors. Int J Radiat Oncol Biol Phys. 2011;81(1):104–9. https://doi.org/10.1016/j.ijrobp.2010.04.027.
2. Voynov G, Heron DE, Burton S, et al. Frameless stereotactic radiosurgery for recurrent head and neck carcinoma. Technol Cancer Res Treat. 2006;5(5):529–35. https://doi.org/10.1177/153303460600500510.
3. Heron DE, Ferris RL, Karamouzis M, et al. Stereotactic body radiotherapy for recurrent squamous cell carcinoma of the head and neck: results of a phase I dose-escalation trial. Int J Radiat Oncol Biol Phys. 2009;75(5):1493–500. https://doi.org/10.1016/j.ijrobp.2008.12.075.
4. Ozyigit G, Cengiz M, Yazici G, et al. A retrospective comparison of robotic stereotactic body radiotherapy and three-dimensional conformal radiotherapy for the reirradiation of locally recurrent nasopharyngeal carcinoma. Int J Radiat Oncol Biol Phys. 2011;81(4):263–8. https://doi.org/10.1016/j.ijrobp.2011.02.054.
5. Xiao J, Xu G, Miao Y. Fractionated stereotactic radiosurgery for 50 patients with recurrent or residual nasopharyngeal carcinoma. Int J Radiat Oncol Biol Phys. 2001;51(1):164–70. https://doi.org/10.1016/s0360-3016(01)01623-6.
6. Orecchia R, Grazia Ruo Redda M, Ragona R, et al. Results of hypofractionated stereotactic re-irradiation on 13 locally recurrent nasopharyngeal carcinomas. Radiother Oncol. 1999;53(1):23–8. https://doi.org/10.1016/S0167-8140(99)00130-9.
7. Dhanachai M, Kraiphibul P, Dangprasert S, et al. Fractionated stereotactic radiotherapy in residual or recurrent nasopharyngeal carcinoma. Acta Oncol (Madr). 2007;46(6):828–33. https://doi.org/10.1080/02841860601103050.
8. Roh KW, Jang JS, Kim MS, et al. Fractionated stereotactic radiotherapy as reirradiation for locally recurrent head and neck cancer. Int J Radiat Oncol Biol Phys. 2009;74(5):1348–55. https://doi.org/10.1016/j.ijrobp.2008.10.013.
9. Kaplan ID, Adler JR, Hicks WL, Fee WE, Goffinet DR. Radiosurgery for palliation of base of skull recurrences from head and neck cancers. Cancer. 1992;70(7):1980–4. https://doi.org/10.1002/1097-0142(19921001)70:7<1980::aid-cncr2820700728>3.0.co;2-l.
10. Karam SD, Rashid A, Snider JW, et al. IMRT with stereotactic body radiotherapy boost for high risk malignant salivary gland malignancies: a case series. Front Oncol. 2014;4:268. https://doi.org/10.3389/fonc.2014.00268.
11. Karam SD, Snider JW, Wang H, et al. Reirradiation of recurrent salivary gland malignancies with fractionated stereotactic body radiation therapy. J Radiat Oncol. 2012;1(2):147–53. https://doi.org/10.1007/s13566-012-0010-6.
12. Wong ET, Lu XQ, Devulapalli J, Mahadevan A. Cyberknife radiosurgery for basal skull plasmacytoma. J Neuroimaging. 2006;16(4):361–3. https://doi.org/10.1111/j.1552-6569.2006.00062.x.
13. Alafaci C, Grasso G, Conti A, Caffo M, Salpietro FM, Tomasello F. Cyberknife radiosurgery for cranial plasma cell tumor. Turk Neurosurg. 2014;24(2):272–5. https://doi.org/10.5137/1019-5149.JTN.7049-12.0.
14. Heaney AP. Pituitary carcinoma: difficult diagnosis and treatment. J Clin Endocrinol Metab. 2011;96(12):3649–60. https://doi.org/10.1210/jc.2011-2031.
15. Tuleasca C, Messerer M, Levivier M, Daniel RT. Combined modalities of surgery, radiotherapy, radiosurgery and chemotherapy for invasive pituitary carcinoma. Ann Endocrinol (Paris). 2018;79(2):82–5. https://doi.org/10.1016/j.ando.2017.01.003.
16. Park KS, Hwang JH, Hwang SK, Kim S, Park SH. Pituitary carcinoma with fourth ventricle metastasis: treatment by excision and gamma-knife radiosurgery. Pituitary. 2014;17(6):514–8. https://doi.org/10.1007/s11102-013-0537-6.
17. Chon H, Yoon K, Kwon DH, Kim CJ, Kim M-S, Cho YH. Hypofractionated stereotactic radiosurgery for pituitary metastases. J Neuro-Oncol. 2017;132(1):127–33. https://doi.org/10.1007/s11060-016-2346-z.
18. Fassett DR, Couldwell WT. Metastases to the pituitary gland. Neurosurg Focus. 2004;16(4):1. https://doi.org/10.3171/foc.2004.16.4.9.
19. Kano H, Niranjan A, Kondziolka D, Flickinger JC, Lunsford LD. Stereotactic radiosurgery for pituitary metastases. Surg Neurol. 2009;72(3):248–55. https://doi.org/10.1016/j.surneu.2008.06.003.
20. Mori Y, Kobayashi T, Shibamoto Y. Stereotactic radiosurgery for metastatic tumors in the pituitary gland and the cavernous sinus. J Neurosurg. 2006;105(Supplement):37–42. https://doi.org/10.3171/sup.2006.105.7.37.
21. Feigl GC, Pistracher K, Berghold A, Mokry M. Pituitary insufficiency as a side effect after radiosurgery for pituitary adenomas: the role of the hypothalamus. J Neurosurg. 2010;113(Suppl):153–9.
22. Yang T, Juric-Sekhar G, Born D, Sekhar L. A case of malignant peripheral nerve sheath tumor of the hypoglossal nerve after stereotactic radiosurgery treatment. J Neurol Surg Rep. 2014;75(01):e42–6. https://doi.org/10.1055/s-0033-1358797.
23. Bashir A, Poulsgaard L, Broholm H, Fugleholm K. Late malignant transformation of vestibular schwannoma in the absence of irradiation: case report. J Neurosurg. 2016;125(2):372–7. https://doi.org/10.3171/2015.6.JNS1544.
24. Belyaev A, Usachev D, Shimansky V, et al. Spontaneous transformation of vestibular schwannoma into malignant peripheral nerve sheath tumor. Asian J Neurosurg. 2018;13(3):810–3. https://doi.org/10.4103/ajns.AJNS_251_16.
25. Karami KJ, Kelkar PS, Verdon MP, Grills IS, Bojrab DI, Pieper DR. Malignant peripheral nerve sheath

tumor of the vestibulocochlear nerve and brainstem: multimodality treatment with survival of 27 months. A case report and review of the literature. Neurosurgery. 2011;69(5):E1152–65; discussion E1165. https://doi.org/10.1227/NEU.0b013e318223bc2a.
26. Raper DMS, Sweiss F, Almira-Suarez MI, Helm G, Sheehan JP. Malignant peripheral nerve sheath tumor at the cerebellopontine angle treated with gamma knife radiosurgery: case report and review of the literature. J Radiosurg SBRT. 2013;2(2):147–53.
27. Scheithauer BW, Erdogan S, Rodriguez FJ, et al. Malignant peripheral nerve sheath tumors of cranial nerves and intracranial contents: a clinicopathologic study of 17 cases. Am J Surg Pathol. 2009;33(3):325–38. https://doi.org/10.1097/PAS.0b013e31818d6470.
28. van Eck ATCJ, Horstmann GA. Gamma Knife surgery for multiple brain metastases from a malignant schwannoma of the penis. J Neurosurg. 2006;105(Suppl):238–40. https://doi.org/10.3171/sup.2006.105.7.238.
29. Jiang B, Veeravagu A, Feroze AH, et al. CyberKnife radiosurgery for the management of skull base and spinal chondrosarcomas. J Neuro-Oncol. 2013;114(2):209–18. https://doi.org/10.1007/s11060-013-1172-9.
30. Wilson HP, Price PM, Ashkan K, et al. CyberKnife radiosurgery of skull-base tumors: a UK Center experience. Cureus. 2018;10(3):e2380. https://doi.org/10.7759/cureus.2380.
31. Chandy E, Taylor H, Gaito S, et al. Hypofractionated stereotactic ablative radiotherapy for recurrent or oligometastatic tumours in children and young adults. Clin Oncol. 2019;32:316. https://doi.org/10.1016/j.clon.2019.11.005.
32. Coppa ND, Raper DMS, Zhang Y, et al. Treatment of malignant tumors of the skull base with multi-session radiosurgery. J Hematol Oncol. 2009;2:16. https://doi.org/10.1186/1756-8722-2-16.
33. Yamada SM, Ishii Y, Yamada S, et al. Skull base osteosarcoma presenting with cerebrospinal fluid leakage after CyberKnife® treatment: a case report. J Med Case Rep. 2013;7:116. https://doi.org/10.1186/1752-1947-7-116.
34. Unger F, Haselsberger K, Walch C, Stammberger H, Papaefthymiou G. Combined endoscopic surgery and radiosurgery as treatment modality for olfactory neuroblastoma (esthesioneuroblastoma). Acta Neurochir. 2005;147(3):595–601. https://doi.org/10.1007/s00701-005-0521-7.
35. Van Gompel JJ, Link MJ, Sheehan JP, et al. Radiosurgery is an effective treatment for recurrent esthesioneuroblastoma: a multicenter study. J Neurol Surg B Skull Base. 2014;75(6):409–14. https://doi.org/10.1055/s-0034-1378151.
36. Veeravagu A, Jiang B, Patil CG, et al. CyberKnife stereotactic radiosurgery for recurrent, metastatic, and residual hemangiopericytomas. J Hematol Oncol. 2011;4(1):26. https://doi.org/10.1186/1756-8722-4-26.
37. Kim JW, Kim DG, Chung HT, et al. Gamma knife stereotactic radiosurgery for intracranial hemangiopericytomas. J Neuro-Oncol. 2010;99(1):115–22. https://doi.org/10.1007/s11060-010-0114-z.
38. Kano H, Niranjan A, Kondziolka D, Flickinger JC, Lunsford LD. Adjuvant stereotactic radiosurgery after resection of intracranial hemangiopericytomas. Int J Radiat Oncol Biol Phys. 2008;72(5):1333–9. https://doi.org/10.1016/j.ijrobp.2008.03.024.
39. Sheehan J, Kondziolka D, Flickinger J, et al. Radiosurgery for treatment of recurrent intracranial hemangiopericytomas. Neurosurgery. 2002;51(4):905–11. https://doi.org/10.1097/00006123-200210000-00008.
40. Coffey RJ, Cascino TL, Shaw EG. Radiosurgical treatment of recurrent hemangiopericytomas of the meninges: preliminary results. J Neurosurg. 1993;78(6):903–8. https://doi.org/10.3171/jns.1993.78.6.0903.
41. Pan J, Liu AL, Wang ZC. Gamma knife radiosurgery for skull base malignancies. Clin Neurol Neurosurg. 2013;115(1):44–8. https://doi.org/10.1016/j.clineuro.2012.04.013.
42. Golbin D, Nikitin KV, Konovalov AN, et al. Intraosseous metastasizing of pineoblastoma into the anterior skull base, calvarial bones, and vertebrae. Cureus. 2015;7(12):e437. https://doi.org/10.7759/cureus.437.
43. Liscak R, Vladyka V, Urgosik D, Simonova G, Vymazal J. Repeated treatment of vestibular schwannomas after gamma knife radiosurgery. Acta Neurochir. 2009;151(4):317–24; discussion 324. https://doi.org/10.1007/s00701-009-0254-0.
44. Kano H, Kondziolka D, Niranjan A, Flannery TJ, Flickinger JC, Lunsford LD. Repeat stereotactic radiosurgery for acoustic neuromas. Int J Radiat Oncol Biol Phys. 2010;76(2):520–7. https://doi.org/10.1016/j.ijrobp.2009.01.076.
45. Fu VX, Verheul JB, Beute GN, et al. Retreatment of vestibular schwannoma with gamma knife radiosurgery: clinical outcome, tumor control, and review of literature. J Neurosurg. 2018;129(1):137–45. https://doi.org/10.3171/2017.3.JNS162033.
46. Yomo S, Arkha Y, Delsanti C, Roche P-H, Thomassin J-M, Régis J. Repeat Gamma Knife surgery for regrowth of vestibular schwannomas. Neurosurgery. 2009;64(1):48–55. https://doi.org/10.1227/01.NEU.0000327692.74477.D5.
47. Lonneville S, Delbrouck C, Renier C, Devriendt D, Massager N. Repeat Gamma Knife surgery for vestibular schwannomas. Surg Neurol Int. 2015;6(1):153. https://doi.org/10.4103/2152-7806.166173.
48. Dewan S, Norén G. Retreatment of vestibular schwannomas with Gamma Knife surgery. J Neurosurg. 2008;109(Suppl):144–8. https://doi.org/10.3171/JNS/2008/109/12/S22.
49. Hasegawa T, Kida Y, Kobayashi T, Yoshimoto M, Mori Y, Yoshida J. Long-term outcomes in patients with vestibular schwannomas treated using gamma

knife surgery: 10-year follow up. J Neurosurg. 2005;102(1):10–6. https://doi.org/10.3171/jns.2005.102.1.0010.
50. Mehta GU, Ding D, Patibandla MR, et al. Stereotactic radiosurgery for Cushing disease: results of an international, multicenter study. J Clin Endocrinol Metab. 2017;102(11):4284–91. https://doi.org/10.1210/jc.2017-01385.
51. Mehta GU, Ding D, Gupta A, et al. Repeat stereotactic radiosurgery for Cushing's disease: outcomes of an international, multicenter study. J Neuro-Oncol. 2018;138(3):519–25. https://doi.org/10.1007/s11060-018-2817-5.
52. Alonso CE, Bunevicius A, Trifiletti DM, et al. Safety and efficacy of repeat radiosurgery for acromegaly: an international multi-institutional study. J Neuro-Oncol. 2019;145(2):301–7. https://doi.org/10.1007/s11060-019-03296-8.
53. Sala E, Moore JM, Amorin A, et al. CyberKnife robotic radiosurgery in the multimodal management of acromegaly patients with invasive macroadenoma: a single center's experience. J Neuro-Oncol. 2018;138(2):291–8. https://doi.org/10.1007/s11060-018-2793-9.
54. Roberts BK, Ouyang DL, Lad SP, et al. Efficacy and safety of CyberKnife radiosurgery for acromegaly. Pituitary. 2007;10(1):19–25. https://doi.org/10.1007/s11102-007-0004-3.
55. Tanaka T, Kato N, Aoki K, et al. Long-term follow-up of growth hormone-producing pituitary carcinoma with multiple spinal metastases following multiple surgeries: case report. Neurol Med Chir (Tokyo). 2013;53(10):707–11. https://doi.org/10.2176/nmc.cr2012-0152.
56. Phillips J, East HE, French SE, et al. What causes a prolactinoma to be aggressive or to become a pituitary carcinoma? Hormones. 2012;11(4):477–82. https://doi.org/10.14310/horm.2002.1380.
57. Lall RR, Shafizadeh SF, Lee KH, et al. Orbital metastasis of pituitary growth hormone secreting carcinoma causing lateral gaze palsy. Surg Neurol Int. 2013;4(1):59. https://doi.org/10.4103/2152-7806.110658.
58. Pollock BE, Cochran J, Natt N, et al. Gamma knife radiosurgery for patients with nonfunctioning pituitary adenomas: results from a 15-year experience. Int J Radiat Oncol Biol Phys. 2008;70(5):1325–9. https://doi.org/10.1016/j.ijrobp.2007.08.018.
59. Benedict SH, Yenice KM, Followill D, et al. Stereotactic body radiation therapy: the report of AAPM Task Group 101. Med Phys. 2010;37(8):4078–101. https://doi.org/10.1118/1.3438081.
60. Kim M, Lee DH, Kim RNHJ, Cho YH, Kim JH, Kwon DH. Analysis of the results of recurrent intracranial meningiomas treated with re-radiosurgery. Clin Neurol Neurosurg. 2017;153:93–101. https://doi.org/10.1016/j.clineuro.2016.12.014.
61. Wojcieszynski AP, Ohri N, Andrews DW, Evans JJ, Dicker AP, Werner-Wasik M. Reirradiation of recurrent meningioma. J Clin Neurosci. 2012;19(9):1261–4. https://doi.org/10.1016/j.jocn.2012.01.023.
62. Mattozo CA, De Salles AAF, Klement IA, et al. Stereotactic radiation treatment for recurrent nonbenign meningiomas. J Neurosurg. 2007;106(5):846–54. https://doi.org/10.3171/jns.2007.106.5.846.
63. Kano H, Iqbal FO, Sheehan J, et al. Stereotactic radiosurgery for chordoma: a report from the North American Gamma Knife Consortium. Neurosurgery. 2011;68(2):379–89. https://doi.org/10.1227/NEU.0b013e3181ffa12c.
64. Dassoulas K, Schlesinger D, Yen CP, Sheehan J. The role of Gamma Knife surgery in the treatment of skull base chordomas. J Neuro-Oncol. 2009;94(2):243–8. https://doi.org/10.1007/s11060-009-9846-z.
65. Raza SM, Bell D, Freeman JL, Grosshans DR, Fuller GN, DeMonte F. Multimodality management of recurrent skull base chordomas: factors impacting tumor control and disease-specific survival. Oper Neurosurg. 2018;15(2):131–43. https://doi.org/10.1093/ons/opx201.
66. Lee M, Kalani MYS, Cheshier S, Gibbs IC, Adler JR, Chang SD. Radiation therapy and CyberKnife radiosurgery in the management of craniopharyngiomas. Neurosurg Focus. 2008;24(5):E4. https://doi.org/10.3171/FOC/2008/24/5/E4.
67. Losa M, Pieri V, Bailo M, et al. Single fraction and multisession Gamma Knife radiosurgery for craniopharyngioma. Pituitary. 2018;21(5):499–506. https://doi.org/10.1007/s11102-018-0903-5.
68. Dho Y-S, Kim YH, Kim JW, et al. Optimal strategy of gamma knife radiosurgery for craniopharyngiomas. J Neuro-Oncol. 2018;140(1):135–43. https://doi.org/10.1007/s11060-018-2943-0.
69. Lee C-C, Yang H-C, Chen C-J, et al. Gamma Knife surgery for craniopharyngioma: report on a 20-year experience. J Neurosurg. 2014;121(Suppl_2):167–78. https://doi.org/10.3171/2014.8.GKS141411.

31 Chordoma and Chondrosarcoma

Steve Braunstein

31.1 Introduction

Chordomas are uncommon tumors in adults arising from remnants of embryonic notochord found within the skull base, vertebral spine, and sacral-coccygeal anatomic regions [1, 2]. Chondrosarcomas are likewise rare tumors which occur in the skull base and spine [3]. Clinical presentations vary but are often associated with regional cranial or sacral nerve dysfunction. Chordomas can be distinguished from chondrosarcomas by the tissue expression of the brachyury transcription factor [4]. Both tumor types are generally hypo- or isointense on T1-weighted MRI with hyperintensity following contrast administration [5, 6]. Tumors appear hyperintense on T2-weighted MRI sequences. Both tumors are characteristically locally aggressive and managed primarily with maximal safe surgical resection and adjuvant radiotherapy. The extent of resection is the strongest prognostic factor for local control and survival [7, 8]. Unfortunately, the anatomic location of these tumors often prohibits gross total resection given the regional critical neurovascular structures. Adjuvant high-dose radiotherapy improves local control. However, these tumors have a high rate of local recurrence. Recurrences are likewise managed with combinations of surgery and radiotherapy, with the potential for increased morbidity. There is limited evidence supporting the role of systemic therapy in management of chordoma and chondrosarcoma. Recently, some efficacy has been demonstrated with molecularly targeted therapy, including imatinib and erlotinib for PDGFRB expressing chordoma [9, 10]. Metastases, while uncommon, are observed with increasing frequency in the setting of a multiply recurrent tumor [11]. Five-year overall survival for chordoma is in the range of 50–80% [7, 12, 13], whereas the overall survival rates vary highly as a function of grade in chondrosarcoma from greater than 90% down to 30% [3].

31.2 Rationale for Radiation Therapy

Due to the anatomic locations and typical surgical approaches, including endoscopic skull base surgery, en bloc gross total resection is rarely achievable without significant morbidity. Adjuvant radiotherapy thus may augment local control. However, as with other mesenchymal tumors, chordomas and chondrosarcomas are not radiation-sensitive compared with epithelial and glandular tumors. The intrinsic radioresistance may be attributed to hypoxic regions within the tumor [14]. Some early series of photon-based radiotherapy reporting outcomes as a function of dose

S. Braunstein (✉)
University of California, San Francisco, USA
e-mail: steve.braunstein@ucsf.edu

© Springer Nature Switzerland AG 2020
A. Conti et al. (eds.), *CyberKnife NeuroRadiosurgery*,
https://doi.org/10.1007/978-3-030-50668-1_31

demonstrated improved local control at doses greater than 60 Gy, administered with conventional fractionation [15, 16]. Thus, higher-dose conventionally fractionated radiotherapy has been the most common treatment approach, generally employing doses in excess of 70 Gy [17].

A challenge with the application of high-dose radiotherapy is the limitation of the therapeutic ratio (Bloomer Hellman NEJM 1975). Radiotherapy delivery to gross tumor and "high-risk" regions of microscopic disease yields a dose deposition into regional normal tissues. The utility of radiotherapy is predicated upon greater radiation sensitivity of the target tumor tissue than that of regional normal tissue, largely attributed to a greater ability of normal cells to repair radiation-induced damage. This favorable therapeutic ratio allows for the delivery of tumor-killing radiotherapy at doses below the threshold of severe toxicity for normal tissue. Radiation sensitivity is influenced by a number of intrinsic and extrinsic tissue factors, and for most normal tissue, the therapeutic ratio can be further augmented by fractionation. Thus, most radiotherapy protocols have employed fractionated approaches to allow for tumor dose-escalation with the preservation of a favorable therapeutic ratio.

Advances in radiotherapy delivery have allowed for more conformal delivery, enabling the sparing of regional normal tissue. Historically, the physical characteristics of particle therapy, including proton-based and carbon ion-based techniques, have enabled dose escalation to radioresistant tumors while mitigating toxicity to surrounding tissue due to high dose gradients outside of the target, sparing normal, uninvolved structures [18]. Reports of proton-based radiotherapy techniques in chordomas and chondrosarcomas indicate 5-year local control generally at 75–94% [19–22], with better rates of control for primary rather than recurrent tumors [21]. Subsequently, technological developments in photon-based radiotherapy techniques, including the implementation of intensity-modulated radiation therapy and stereotactic radiotherapy, have allowed for similar gains in the therapeutic ratio of radiotherapy. Radiosurgery, in particular, may engender enhanced sensitivity within radioresistant chordoma and chondrosarcomas by yielding dense tracts of DNA damage in target tissue, refractory to repair, which may otherwise occur at doses employed during conventional fractionation [23]. Estimations of an alpha/beta ratio of 2.45 Gy of chordomas further support the greater sensitivity of these tumors to hypofractionation [24].

Radiotherapy for chordoma and chondrosarcoma can be delivered in a variety of settings, including preoperatively, postoperatively, as well as in the absence of surgery, a definite therapy, in cases where patients may be considered non-operable due to comorbidities or the location and extent of the tumor. Early adjuvant radiotherapy is associated with superior outcomes [25, 26]. Radiosurgery, in particular, can offer expedited treatment in such settings, which may be of value for interdisciplinary coordinated care, to quickly and effectively ablate microscopic disease at the tumor periphery in the preoperative field, thus potentially yielding reduced rates of recurrence post-resection.

31.3 Single-Fraction Radiosurgery

Radiosurgery emerged as a treatment modality for chordoma and chondrosarcoma in recent decades. A summary of recently reported single-fraction radiosurgery series is presented in Table 31.1. Among the largest modern series, a report from the North American Gamma Knife Consortium on single-session stereotactic radiosurgery for patients with chordoma [27] reported outcomes in 71 patients treated across six centers for both primary radiosurgery and salvage radiosurgery following prior external beam therapy. The median margin dose was 15 Gy (range 9–25 Gy) with a median target volume of 7.1 cm^3 (range 0.9–109 cm^3). The target volume delineation was non-standardized but generally included the T1 post contrast and the T2 infiltrative disease without margin. Five-year local control among patients who did vs those who did not receive prior radiotherapy was 62% vs 69%. A margin dose of ≥15 Gy was associated with improved local control. In contrast, larger tumor volume

Table 31.1 Single session radiosurgery for chordoma and chondrosarcoma

Study	Tumor	No. patients	Median dose (range)	Vol	LC	OS	Median FU
Liu et al. (2008) [29]	Chordoma	31	12.7 Gy (10–16)	0.5–28 cm^3	21%@5 yr	76%@5 yr	28 mo
Kano et al. (2011) [27]	Chordoma	71	15 Gy (9–25)	0.9–109 cm^3	66%@5 yr	80%@5 yr	60 mo
Iyer et al. (2012) [30]	Chondrosarcoma	22	15 Gy (10.5–20)	0.9–28 cm^3	72%@5 yr	75%@5 yr	75 mo
Yamada et al. (2013) [28]	Chordoma	24	24 Gy (18–24)	20–859 cm^3	95%@2 yr	67%@4.5 yr	24 mo
Kano et al. (2015) [31]	Chondrosarcoma	36	15 Gy (10.5–20)	0.9–28 cm^3	85%@5 yr	86%@5 yr	75 mo

Table 31.2 Fractionated radiosurgery for chordoma and chondrosarcoma

Study	Tumor	No. patients	Median dose (range)	Vol	LC	OS	Median FU
Henderson et al. (2009) [24]	Chordoma	18	35 Gy (24–40)	12–457 cm^3	59%@5.5 yr	74%@5.5 yr	46 mo
Jiang et al. (2013) [33]	Chondrosarcoma	16	22–30 Gy in 1–5 fx	0.8–391 cm^3	41%@5 yr	55%@5 yr	41 mo
Vasudevan et al. (2017) [32]	Chordoma Chondrosarcoma	20	37.5 (25–40)	7.1–314 cm^3	90%@3 yr	90%@3 yr	28 mo

(>7 cm^3) was associated with poorer local control. Improvement in cranial nerve deficits was up to 50% following radiosurgery. An adverse radiation effect was noted in 9% of patients, three of which were grade 3, and all occurred in patients who had prior radiotherapy.

Another large early series of single-fraction radiosurgery at MSKCC demonstrated a 24-month local control of 95% in 24 patients with spinal chordoma [28]. The median prescription dose to target volume was 24 Gy (range 18–24 Gy). Target volume included gross tumor plus a CTV margin to include potential microscopic spread plus a 2–3 mm PTV expansion. Several patients treated in the preoperative setting demonstrated high rates of necrosis in the tumor specimen at resection. Grade ≥3 toxicity was limited and included vocal cord paralysis (in a previously irradiated patient) and vertebral fractures.

31.4 Image-Guided Radiosurgery and Hypofractionated Radiotherapy

Fractionated radiosurgery may increase the therapeutic ratio for larger tumors or those in close apposition to radiosensitive organs at risk. Recent series of fractionated radiosurgery are summarized in Table 31.2. Of note, these series all represent very heterogeneous patient cohorts who received fractionated radiosurgery in both the upfront and salvage settings. Henderson et al. reported one of the first large institutional series of fractionated CyberKnife radiosurgery in management of chordomas [24]. Outcomes on 18 patients, largely treated in the postoperative setting, were reported. Median target volume was 128 cm^3 (range 12–457 cm^3) with median dose of 35 Gy (24–40 Gy) delivered in five daily fractions. Target volumes included a 1 cm margin on gross tumor for the clinical target volume. Five-year local control rate was 59% with overall survival of 74%. A study by Vasudevan et al. [32] reported outcomes of 20 patients with chordoma or chondrosarcoma who received fractionated radiosurgery. Treatment consisted of five-fraction radiosurgery to medial total dose of 37.5 Gy (range 25–40 Gy). Local control was reported at 90% over a median follow-up period of 28 months. High-grade toxicity (grade 4 and 5) was reported in two patients who had recurrent disease and multiple prior courses of radiotherapy.

A representative fractionated CyberKnife radiosurgery plan is shown in Fig. 31.1. This patient had gross total resection of a clival chordoma and received 40 Gy in five fractions to the surgical bed, prescribed to the 75% isodose. The target volume included the surgical bed (GTV) plus a 10 mm anisotropic expansion (CTV) and 2 mm setup margin (PTV).

31.5 Toxicity

The use of high-dose radiotherapy can translate to higher toxicity for regional normal tissue in proximity to targets. Moreover, the anatomic location of chordoma and chondrosarcoma within the craniospinal axis encompasses several radiosensitive organs at risk, including the spinal cord, brainstem, and the cranial and sacral nerves. Thus, the therapeutic ratio is rendered very narrow in the management of chordoma and chondrosarcoma. Fortunately, conformal techniques such as proton therapy and radiosurgery can provide steep dose gradients to minimize high-dose toxicity to regional tissue.

Modern series of proton-based high-dose radiotherapy demonstrate acute and late grade ≥3 toxicity of less than 10% [22, 34, 35]. Toxicities are a function of anatomic location. Infrequent toxicities (<15%) of spine chordoma/chondrosarcoma include wound infection and dehiscence,

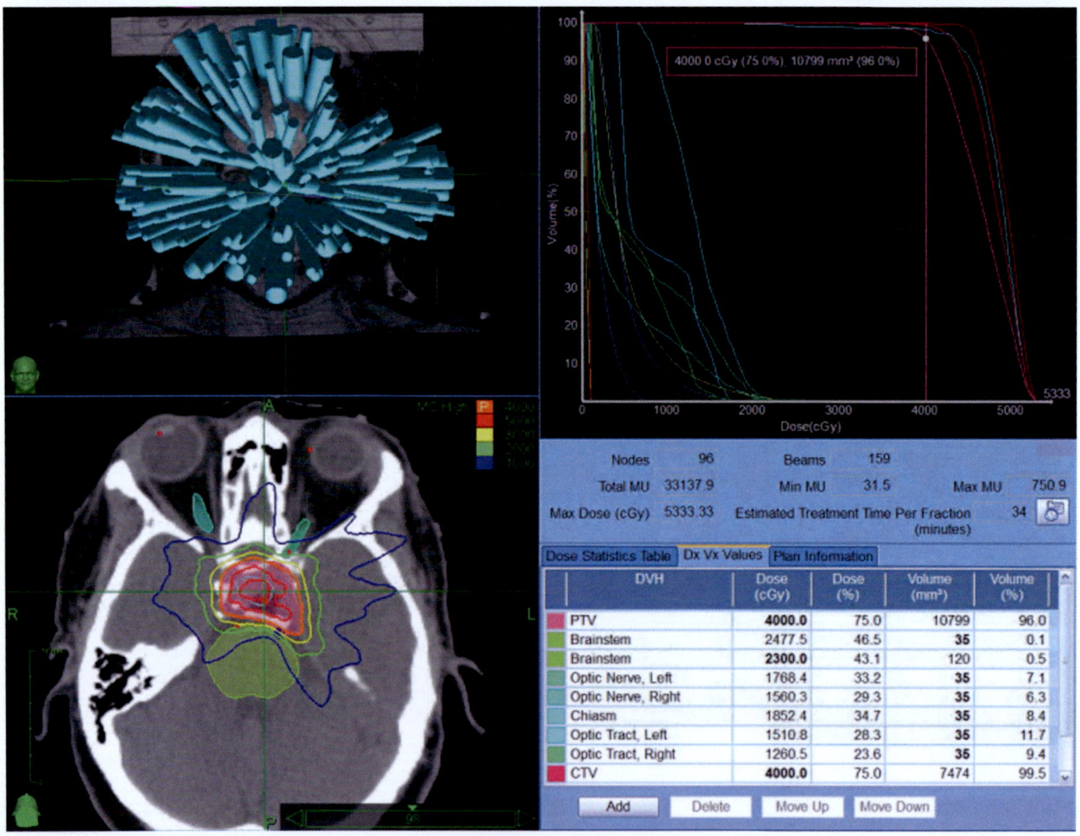

Fig. 31.1 Example CyberKnife fractionated radiosurgery plan for chordoma

proctitis, and late secondary malignancies though sacral insufficiency fractions may exceed 20% [21, 36]. Reported rates of toxicity are likewise limited (<20% grade ≥3) in modern radiosurgical series. Toxicities may include cranial and peripheral neuropathies, mucositis, and surgical wound infections. Of note, high-grade toxicities such as radiation-related vasculopathy can be observed, especially in the setting of re-irradiation with high-dose techniques [24, 32].

31.6 Conclusion

Chordomas and chondrosarcomas, while generally considered of relatively indolent malignant behavior, are ultimately highly challenging to effectively treat by virtue of their anatomic location near critical neurovascular structures that limits their resectability. Moreover, their relative lack of chemotherapy and targeted therapy sensitivity and intrinsic radioresistance largely limits the efficacy of adjuvant therapy. More advanced radiotherapy techniques, including charged particle therapy, have enabled dose escalation and better rates of local control. The application of newer photon-based radiosurgery technology has likewise allowed for dose escalation to yield improved rates of local control with limited high-grade toxicity in initial reports of heterogeneous patient cohorts. Long-term outcomes may further validate the efficacy of radiosurgery for local control which is expected to translate into a benefit in overall survival for these types of locally aggressive tumors.

Practical guide to treatment planning

Fraction no.[a]	Total dose[b]	Target volume[c]	Organs at risk[d]
1 Fx	18–24 Gy	GTV: Gross disease CTV: GTV + ≤10 mm PTV: CTV + ≤2 mm	Optic pathway $V_{8Gy} < 0.2$ cm^3, Max 10 Gy Brainstem $V_{10Gy} < 0.5$ cm^3, Max 15 Gy Spinal cord $V_{10Gy} < 0.35$ cm^3, Max 14 Gy Brachial plexus $V_{14Gy} < 5$ cm^3, Max 17.5 Gy Cauda Equina $V_{14Gy} < 5$ cm^3, Max 16 Gy Sacral plexus $V_{14.4Gy} < 5$ cm^3, Max 16 Gy
5 Fx	35–40 Gy	GTV: Gross disease CTV: GTV + ≤10 mm PTV: CTV + ≤2 mm	Optic pathway $V_{23Gy} < 0.2$ cm^3, Max 25 Gy Brainstem $V_{23Gy} < 0.5$ cm^3, Max 31 Gy Spinal cord $V_{23Gy} < 0.35$ cm^3, Max 30 Gy Brachial plexus $V_{27Gy} < 5$ cm^3, Max 30.5 Gy Cauda Equina $V_{30Gy} < 5$ cm^3, Max 32 Gy Sacral plexus $V_{30Gy} < 5$ cm^3, Max 32 Gy

[a]Consider fractionated radiosurgery for larger target volumes (>7 cm^3). Three fraction regimens can also be considered

[b]Doses at the higher end of the range should be considered for gross disease and/or chordoma histology

[c]MRI should be obtained for treatment planning. T1 post-contrast sequences (SPGR/CUBE) to assess gross tumor volume as well as T2-weighted sequences (FSE/FLAIR) to assess infiltrative tumor should be used for target delineation. Metal suppression sequences and CT/MR myelogram may be considered to better delineate the target and OARs in the postoperative setting. Target volumes should include gross disease and up to a 10 mm margin on the regional osseous structures. Fat graft in resected and reconstructed skull base tumors should not be included in GTV. Fat suppression MRI sequences may be of utility in these instances. CTV margins on the higher end of the range should be considered in the setting of recurrence, as limited by regional OAR tolerance. PTV margin is determined by setup and image guidance. More limited margins may be appropriate with frame-fixed procedures and/or real time fiducial or skull base tracking

[d]Organ at risk constraints per TG101

References

1. McMaster ML, et al. Chordoma: incidence and survival patterns in the United States, 1973-1995. Cancer Causes Control. 2001;12(1):1–11.
2. Walcott BP, et al. Chordoma: current concepts, management, and future directions. Lancet Oncol. 2012;13(2):e69–76.
3. van Praag Veroniek VM, et al. Incidence, outcomes and prognostic factors during 25 years of treatment of chondrosarcomas. Surg Oncol. 2018;27(3):402–8.
4. Vujovic S, et al. Brachyury, a crucial regulator of notochordal development, is a novel biomarker for chordomas. J Pathol. 2006;209(2):157–65.
5. Pamir MN, Ozduman K. Analysis of radiological features relative to histopathology in 42 skull-base chordomas and chondrosarcomas. Eur J Radiol. 2006;58(3):461–70.
6. Meyers SP, et al. Chondrosarcomas of the skull base: MR imaging features. Radiology. 1992;184(1):103–8.
7. Bakker SH, et al. Chordoma: a systematic review of the epidemiology and clinical prognostic factors predicting progression-free and overall survival. Eur Spine J. 2018;27(12):3043–58.
8. Osaka S, et al. Clinical significance of a wide excision policy for sacrococcygeal chordoma. J Cancer Res Clin Oncol. 2006;132(4):213–8.
9. Stacchiotti S, et al. Phase II study of imatinib in advanced chordoma. J Clin Oncol. 2012;30(9):914–20.
10. Launay SG, et al. Efficacy of epidermal growth factor receptor targeting in advanced chordoma: case report and literature review. BMC Cancer. 2011;11:423.
11. Bergh P, et al. Prognostic factors in chordoma of the sacrum and mobile spine: a study of 39 patients. Cancer. 2000;88(9):2122–34.
12. Sun HH, et al. Survival analysis of patients with spinal chordomas. Neurosurg Rev. 2019;42(2):455–62.
13. Debus J, et al. Stereotactic fractionated radiotherapy for chordomas and chondrosarcomas of the skull base. Int J Radiat Oncol Biol Phys. 2000;47(3):591–6.
14. Austin JP, et al. Probable causes of recurrence in patients with chordoma and chondrosarcoma of the base of skull and cervical spine. Int J Radiat Oncol Biol Phys. 1993;25(3):439–44.
15. Pearlman AW, Friedman M. Radical radiation therapy of chordoma. Am J Roentgenol Radium Therapy, Nucl Med. 1970;108(2):332–41.
16. Rich TA, et al. Clinical and pathologic review of 48 cases of chordoma. Cancer. 1985;56(1):182–7.
17. Palm RF, et al. The role of dose escalation and proton therapy in perioperative or definitive treatment of chondrosarcoma and chordoma: an analysis of the National Cancer Data Base. Cancer. 2019;125(4):642–51.
18. Imai R, et al. Effect of carbon ion radiotherapy for sacral chordoma: results of Phase I-II and Phase II clinical trials. Int J Radiat Oncol Biol Phys. 2010;77(5):1470–6.
19. Hug EB, et al. Proton radiation therapy for chordomas and chondrosarcomas of the skull base. J Neurosurg. 1999;91(3):432–9.
20. DeLaney TF, et al. Long-term results of Phase II study of high dose photon/proton radiotherapy in the management of spine chordomas, chondrosarcomas, and other sarcomas. J Surg Oncol. 2014;110(2):115–22.
21. Rotondo RL, et al. High-dose proton-based radiation therapy in the management of spine chordomas: outcomes and clinicopathological prognostic factors. J Neurosurg Spine. 2015;23(6):788–97.
22. Indelicato DJ, et al. A prospective outcomes study of proton therapy for Chordomas and chondrosarcomas of the spine. Int J Radiat Oncol Biol Phys. 2016;95(1):297–303.
23. Brown JM, Koong AC. High-dose single-fraction radiotherapy: exploiting a new biology? Int J Radiat Oncol Biol Phys. 2008;71(2):324–5.
24. Henderson FC, et al. Treatment of chordomas with CyberKnife: Georgetown university experience and treatment recommendations. Neurosurgery. 2009;64(2 Suppl):A44–53.
25. Holliday EB, et al. Postoperative proton therapy for chordomas and chondrosarcomas of the spine: adjuvant versus salvage radiation therapy. Spine (Phila PA 1976). 2015;40(8):544–9.
26. Hug EB, et al. Locally challenging osteo- and chondrogenic tumors of the axial skeleton: results of combined proton and photon radiation therapy using three-dimensional treatment planning. Int J Radiat Oncol Biol Phys. 1995;31(3):467–76.
27. Kano H, et al. Stereotactic radiosurgery for chordoma: a report from the North American Gamma Knife Consortium. Neurosurgery. 2011;68(2):379–89.
28. Yamada Y, et al. Preliminary results of high-dose single-fraction radiotherapy for the management of chordomas of the spine and sacrum. Neurosurgery. 2013;73(4):673–80; discussion 680.
29. Liu AL, et al. Gamma knife radiosurgery for residual skull base chordomas. Neurol Res. 2008;30(6):557–61.
30. Iyer A, et al. Stereotactic radiosurgery for intracranial chondrosarcoma. J Neuro-Oncol. 2012;108(3):535–42.
31. Kano H, et al. Skull base chondrosarcoma radiosurgery: report of the North American Gamma Knife Consortium. J Neurosurg. 2015;123(5):1268–75.
32. Vasudevan HN, et al. Management of chordoma and chondrosarcoma with fractionated stereotactic radiotherapy. Front Surg. 2017;4:35.
33. Jiang B, et al. Management of intracranial and extracranial chordomas with CyberKnife stereotactic radiosurgery. J Clin Neurosci. 2012;19(8):1101–6.
34. Baumann BC, et al. A prospective clinical trial of proton therapy for chordoma and chondrosarcoma: feasibility assessment. J Surg Oncol. 2019;120(2):200–5.
35. Holtzman AL, et al. Proton therapy for skull-base chondrosarcoma, a single-institution outcomes study. J Neuro-Oncol. 2019;142(3):557–63.
36. Osler P, et al. Sacral insufficiency fractures are common after high-dose radiation for sacral chordomas treated with or without surgery. Clin Orthop Relat Res. 2016;474(3):766–72.

Paragangliomas of the Head and Neck

32

Kaan Oysul and Sait Sirin

32.1 Introduction

Paragangliomas that are also called glomus tumors, chemodectomas, or nonchromaffin tumors are highly vascular neoplasms and that embryologically originate from the extra-adrenal paraganglia of the neural crest [1–3]. Paragangliomas originate from the sympathetic paraganglia (85% arise below the diaphragm) or from the parasympathetic paraganglia (head-and-neck paragangliomas). All paragangliomas arising from the parasympathetic ganglia are denoted as head-and-neck paragangliomas, although they may also arise from the anterior and middle mediastinum along the vagus nerve. Over 95% of head-and-neck paragangliomas are non-functioning and do not overproduce noradrenalin, although a small number (1–3%) of head-and-neck paragangliomas can overproduce it [4]. Paragangliomas may occur sporadically or may even be inherited familial tumors. Classic tumor syndromes associated with paragangliomas include multiple endocrine neoplasia of type II (MEN II), von Hippel-Lindau disease, and neurofibromatosis of type I (NF I) [5]. Multicentric paragangliomas occur in 10–20% of sporadic cases and up to 80% of hereditary cases [6].

Paragangliomas commonly develop in four locations in the head and neck: (1) the carotid bifurcation (glomus caroticum tumor), (2) the jugular bulb region (glomus jugulare tumor), (3) the middle ear cavity (glomus tympanicum tumor), and (4) the inferior ganglion region (ganglion nodosum) and cervical portion of the vagus nerve (glomus vagale or vagal body tumor) [7].

The carotid body, which is located within the carotid bifurcation, is a discrete, oval structure that directly receives its blood supply from the carotid bifurcation via the glomic arteries. Carotid body tumors are the most common, accounting for approximately 60% of paragangliomas [8] and typically present as a painless, mobile, slow-growing neck mass that may be pulsating and transmit bruits.

If these tumors extend into the parapharyngeal space, they can be associated with cranial nerve palsies (typically X and XII). Glomus jugulare tumors may be associated with bone destruction. Patients with these tumors can present with cranial nerve deficits, typically IX through XII. These tumors can originate and spread along

K. Oysul
Department of Radiation Oncology, CyberKnife Radiosurgery Center, Medicana International Ankara Hospital, Ankara, Turkey
e-mail: kaan@oysul.com

S. Sirin (✉)
Department of Neurosurgery, CyberKnife Radiosurgery Center, Medicana International Ankara Hospital, Ankara, Turkey
e-mail: ssirin@medicana.com.tr

© Springer Nature Switzerland AG 2020
A. Conti et al. (eds.), *CyberKnife NeuroRadiosurgery*,
https://doi.org/10.1007/978-3-030-50668-1_32

the tympanic canaliculus, invading superiorly into the middle ear and inferiorly toward the jugular fossa. Large glomus jugulare tumors can also cause cranial nerves VII and VIII deficits from mass effect [1, 2].

Distributed within the temporal bone in close association with the Jacobson's nerve, which is the tympanic branch of the glossopharyngeal nerve, is the jugulotympanic paraganglia. Typically, temporal bone paraganglia are located in the jugular fossa, and symptoms may involve early functional impairment of cranial nerves IX, X, XI within the jugular foramen, and XII as it exits the hypoglossal canal. Glomus tympanicum is likely to present with hearing loss, pulsatile tinnitus, and disequilibrium and on physical examination may appear as a reddish mass behind the eardrum. Unlike the glomus jugulare, ossicular chain destruction is unusual, but it may spread to the mastoid air cells, Eustachian tube, and nasopharynx. Vagal paraganglia are distinctly separate from jugulotympanic paraganglia because they do not form discrete bodies. They may be interspersed within the vagal nerve fibers in the pars nervosa of the jugular foramen (which transmits lower cranial nerves IX, X, and XI) or located within the vagus nerve beneath the perineurium. The superior vagal ganglion is visible at the level of the jugular foramen.

The origin of most vagal paragangliomas is the nodose vagal ganglion, which is located approximately 1–2 cm below the jugular foramen. Both the superior and nodose vagal ganglia are proximal to the pars venosa of the jugular foramen, cranial nerves IX to XII, and the ascending portion of the petrous internal carotid artery. Glomus vagale tumors can present as an intraoral parapharyngeal mass that anteriorly displaces the tonsil, or as a painless insidious lateral neck mass behind the angle of the mandible.

They can derive from any three of the vagal ganglia but usually come from the largest and most caudal: the ganglion nodosum.

Somewhat similar to glomus jugulare tumors, deficits in cranial nerves X through XII can be seen as these tumors progress, and Horner syndrome can develop. Therefore, vagal paragangliomas have distinct therapeutic sequela based on their close anatomical association with the superior portion of the vagal nerve and other adjacent neurovascular structures [1, 2].

Imaging is the primary investigative modality for paragangliomas of the head and neck. A combination of contrast-enhanced computed tomography (CT) scanning, magnetic resonance imaging (MRI), and angiography is ideal for proper diagnosis and localization of the tumors [9–11]. Lesions show a characteristic signature on the images, which is based on its location. CT imaging is excellent at demonstrating cervical masses along the course of the carotid artery, but findings of skull-base soft-tissue details can be limited. However, CT imaging is superb for explaining characteristic bony destructive skull-base changes. CT scanning is also best in the diagnosis of paragangliomas when a satisfactory bolus of contrast material is administered.

If maximum opacification of the tumor is not achieved during the CT scan, the mass may be misinterpreted as a non-injurious schwannoma or a nodal lesion. Magnetic resonance imaging can well demonstrate soft-tissue masses and their relationships with adjacent structures in multiple imaging planes. This capability is particularly helpful in skull-base imaging, in which both extracranial and intracranial components can be evaluated. MRIs can fail to depict enhancement if the contrast agent bolus is inadequate. MRI is inherently limited in its ability to show subtle areas of bony destruction, which may be necessary for proper diagnosis.

Angiography is generally reserved for patients undergoing preoperative evaluation or for cases where the presence of neovascularization can help in focusing the differential diagnosis.

Angiography is a minimally invasive test and, therefore, not the imaging study of choice. Diagnostic angiography can rarely show soft-tissue neovascularity and other types of abnormalities, such as those encountered with hypervascular lymphadenopathy or nodular fasciitis.

These tumors can have a profound neovascularization that mimics that of the paragangliomas.

Typically, paragangliomas demonstrate a median growth rate of 1.0 mm/year with a median

Table 32.1 Glasscock-Jackson classification of paragangliomas

Type 1	Involves jugular bulb, middle ear, mastoid process
Type 2	Extends under internal auditory canal
Type 3	Extends into petrous apex
Type 4	Extends beyond petrous apex into clivus or infratemporal fossa

Note: Types 2–4 may have intracranial extension

Table 32.2 Fisch classification of paragangliomas

A	Limited to middle ear cleft
B	Limited to the tympanomastoid area
C	Involving the infralabyrinthine compartment and petrous apex of the temporal bone
C1	Tumor with limited involvement of the vertical portion of the carotid canal
C2	Tumor invading the vertical portion of the carotid canal
C3	Tumor invasion of the horizontal portion of the carotid canal
D1	Intracranial extension <2 cm in greatest dimension
D2	Intracranial extension >2 cm in greatest dimension

tumor doubling time of 4 years [12]. Nearly all paragangliomas that occur in the head and neck region are benign tumors. The likelihood of malignancy depends on location, with 5% of temporal bone paragangliomas and 15% of carotid body/glomus vagale tumors [13]. The Glasscock-Jackson (Table 32.1) [14] and Fisch (Table 32.2) [15] classifications of paragangliomas are widely used. Both are based on the extension of the tumor to surrounding anatomic structures and are closely related to mortality and morbidity.

32.2 Treatment Options

The current treatment options for head-and-neck paragangliomas include surgical resection facilitated by endovascular embolization, conventional radiation therapy, stereotactic radiosurgery (SRS), or a combination of these modalities [16].

The first reported surgical resection was performed by Seiffert in 1934 [17]. The goal of surgery is complete tumor removal. Despite this, the incidence of tumor recurrence ranges from 3 to 20% [18–20]. The postoperative complication rate is considerable. Although many of these are transient, a low cranial nerve palsy has been reported in 22–59% of patients after surgery. Other postoperative complications include CSF leak, aspiration, wound infections, ischemia, pneumonia, and meningitis.

Surgical resection is now a safer and more reasonable choice than in previous years. However, after 70 years of advancement in microsurgical techniques, tumor removal is still challenging due to the high risk of hemorrhage and cranial nerve damage.

Surgery still involves high morbidity (a stroke rate of 8–20%, a cranial nerve rate between 22 and 44%, and a mortality rate of 5–13%). Advances in microsurgical techniques, imaging technologies for intraoperative navigation, preoperative embolization, and perioperative monitoring, have significantly improved the surgical results for paragangliomas [21].

External beam radiation was first utilized as adjuvant therapy for the treatment of recurrent paragangliomas or in subtotal resections in the 1950s [22, 23]. Fractionated external beam irradiation has been used to treat paragangliomas in elderly patients, for residual or recurrent lesions with tumor control rates ranging from 74 to 97% [24–26] and complication rates of 4–20% [27]. Nevertheless, the potential long-term risks involved in wide-field radiation treatment are still major concerns, namely, carcinogenesis and long-term neurovascular damage.

Tumor control rates are high with external radiation. However, it requires a 5-week course of treatment, which can be logistically unattractive, and patients are also exposed to radiation risks such as skin changes, xerostomia, and possible induction of secondary malignancies.

The use of stereotactic radiosurgery to treat paragangliomas was introduced in the 1990s with lower radiation risks and high control rates [28]. Over the past two decades, radiosurgery has emerged as a promising approach to the management of paragangliomas.

Today, the Gamma Knife, LINAC, and CyberKnife systems are those used for radiosurgery of paragangliomas [29]. Radiosurgery offers a high degree of accuracy, exquisite precision, and a rapid reduction in the dose of radiation to the periphery of the target lesions, allowing the clinician to deliver a high dose of radiation to the neoplastic tissue and save healthy brain tissue.

This is especially important in benign and indolent tumors, such as paragangliomas. Although microsurgical resection requires prolonged hospital stay and carries the risk of perioperative complications, radiosurgery is a relatively non-invasive treatment that can be performed as an outpatient procedure.

Despite the theoretical advantages of radiosurgery, a limited number of studies have reported on the use of this treatment modality to deal with paragangliomas.

Those that have been published are compromised by their small sample size, limited follow-up, and lack of control groups. No randomized controlled trials comparing the use of radiosurgery and other treatment modalities for the management of paragangliomas have been published, and all the published studies have been retrospective reviews.

Most of these studies have been conducted at a single institution. In addition, since surgery is considered the initial standard reference treatment of these tumors, most patients treated with radiosurgery in these studies were either poor surgical candidates or patients with recurrent or residual disease after microsurgical resection, thus obscuring the comparison between surgery and radiosurgery [30].

Radiosurgery, particularly for smaller lesions, leads to very high local control rates (77–98%) protecting the surrounding tissue from ablative radiation doses [16].

In several reports, vast majority of patients show neurological improvement or stability, and only a minority worsens clinically, most often experiencing hearing deterioration [31, 32].

Subtotal resection, while preserving neurological function and followed by postoperative radiosurgery or definitive radiosurgery, has been adopted as the standard treatment for many large tumors. Today, stereotactic radiosurgery has an increasing role in the management of these tumors either as a definitive or as an adjuvant approach after subtotal resection. However, a subset of patients will present with cranial nerve deficits. Death from paraganglioma is infrequent, and treatment of these tumors aims at minimizing morbidity rather than improving survival. Radiosurgery could be the most appropriate treatment option for paragangliomas.

32.3 CyberKnife Radiosurgery Technique

32.3.1 Patient Selection

After the patient's medical history, clinical examinations, and neuroimaging studies have provided a precise diagnosis, treatment decisions should be made by the multidisciplinary team. The age of the patient, the size of the tumor, the rate of growth (as determined by neuroimaging), symptoms, neurological deficits, and the patient's overall health should be recorded. Young patients with large symptomatic tumors might undergo resection with or without preoperative embolization. Smaller to moderately sized paragangliomas can be treated safely using radiosurgery as the primary intervention. Larger tumors can occasionally be treated with multisession radiosurgery when resection is not feasible. Secreting tumors are also an indication for radiosurgery, because resection enables a rapid reduction in catecholamine levels, which makes medical management more successful. Furthermore, biochemical remission does not usually occur until several years after radiosurgery.

32.3.2 Imaging

The target of stereotactic radiosurgery must be clearly and accurately imaged. The patient undergoes a 1.5 T planning MRI with two sequences consisting of a 2D T2 or a 3D 1 mm isotropic T2 series and a gadolinium-enhanced 3D 1 mm isotropic T1 series. The patient is then immobilized supine in a thermoplastic mask. Lastly, a simulation CT is acquired with 1 mm thickness while the patient is immobilized with a thermoplastic mask.

32.3.3 Treatment Planning

In the planning system, the first step is the fusion of the CT and MR images. The registration of all CT and MR image sets could be done automatically or manually co-registered using anatomical structures as principal landmarks. Peer review of registered image series is highly recommended. The second step is the contouring of the target and critical structures. The gross tumor volume (GTV) is contoured on the T1 contrast-enhanced MR images while checking on the T2 images. Optic nerves, cochlea, parotid, carotid, optic chiasm, area postrema, and brainstem could be contoured as critical structures. No margin is added to create a clinical target volume (CTV) and a planning target volume (PTV). The third step is planning. A shell structure is created 2 mm beyond the PTV. Collimator size is selected due to the size of the tumor. The plan is then optimized so that the PTV is covered by at least 95% of the prescription dose with the 80–85% isodose volume. The last step is evaluating and approving the plan. Conformity index, gradient index, dose uniformity, and the isodose lines are used for the evaluation.

32.3.4 Dose Selection

As with the treatment of other benign intracranial tumors, single-session radiosurgical margin doses for paragangliomas vary from 12 to 18 Gy. Care is required to avoid excessive high doses or "hot spots" in order to protect critical neurovascular structures, such as cranial nerves or the internal jugular vein around the jugular foramen. Most cranial nerves are more resistant to the effects of radiation than the optic nerve. However, numerous reports of cranial neuropathy have been reported following repeated radiosurgery. The tolerable limit of lower cranial nerves to radiation has yet to be fully determined. Some researchers have reported that single-session radiosurgical doses of between 19 and 30 Gy afford good outcomes without appreciable side effects.

Where doubts about the irradiation of adjacent healthy tissues can arise, radiosurgery can be performed in 2–5 sessions in order to adapt the dosage plan to the specifics of a particular case.

32.3.5 Treatment Delivery

The patient lies down the treatment couch, as is in the planning CT, and the treatment is delivered. Because treatment delivery takes time, the first most comfortable positioning and immobilization are extremely important.

32.3.6 Follow-up

After regular radiosurgery, follow-up is suggested, which typically includes interval neurological examinations and neuroimaging after 3 months and then every 6–12 months. Regular MRI examinations should consist of sequential measurements of tumor volume and changes in tumor characteristics.

32.4 The Clinical and Imaging Outcomes

Most of the published reports on radiosurgical treatment of paragangliomas have involved Gamma Knife technology.

In 2012, the International Gamma Knife Research Foundation reported a large-scale series in which tumor control was achieved in 93% of glomus tumors [33]. That series included patients from various medical centers that employ the same radiosurgical modalities and adhere to the same radiobiological principles. However, the study type can be hindered by differences in patient selection, radiosurgical techniques, and follow-up protocols. There is also substantial evidence that stereotactic radiosurgery CyberKnife or LINAC- based approaches have excellent outcomes with LC rates exceeding 95% and a favorable toxicity profile [29].

As far as we know, there are 13 publications in the literature that use CyberKnife in the treatment of paragangliomas; one is French and the other is Russian. Two are reviews, three are from the same center which includes the same group of patients, and three are for all benign tumors, including paragangliomas.

The first study is retrospective analyses of patients treated with CyberKnife radiosurgery for paragangliomas at Stanford University that were performed and published in 2003, 2004, and 2007. The last one consists of the other analyses dated from 1991 to 2006. Sixteen tumors in 13 patients were treated with the CyberKnife. All but five patients were treated with one fraction. Four patients received three fractions, and one patient underwent two fractions. Prescribed doses (typically to the 80% isodose line) to the periphery of the tumor ranged from 1400 to 2700 cGy. Follow-up of patients ranged from 6 to 162 months on radiographs (mean 46 months, median 30 months) and 6 to 213 months with clinical follow-up (mean 66 months, median 35 months). Patients ranged in age from 24 to 85 with a mean of 58 years old.

Four patients had prior open surgeries for their tumors, with two patients having undergone multiple attempts at resection. Two patients had multiple tumors.

Tumor sizes ranged from 1.2 to 6.2 cm at the largest measurable diameter, with an average of 3.04 cm. Isodose lines ranged from 72 to 90%, with an average of 79%. Posttreatment, three patients experienced transient worsening of preprocedural cranial nerve deficits: the first complained of temporary ipsilateral tongue atrophy and hearing loss, and the second reported worsened post-procedure voice hoarseness (confirmed through laryngoscopy) that resolved over 8 months.

The third patient experienced transient hearing loss. The remaining patients experienced no side effects. All 16 tumors were stable at the time of follow-up, which ranged from 6 to 162 months per radiograph. Six tumors, all of whom were treated with stereotactic radiosurgery alone, had a regression of their tumor size. Of the 16 tumors, 3 were followed out on an average of greater than 10 years. All tumors remained unchanged in size. No patients experienced permanent side effects [34].

In the second study, nine consecutive patients with paragangliomas were referred to the "Fondazione Istituto Neurologico C. Besta" in Milan, Italy, from August 2004 to December 2007. All patients presented with radiological diagnoses of paraganglioma of the head and neck, eight with glomus jugulare tumors and one with a carotid body tumor. The mean age at the time of treatment was 52.4 years. The most common presenting symptom was conductive hearing loss (six patients), but two patients experienced pulsatile tinnitus. Two had a facial weakness, three had dysphasia, one suffered headaches, and one patient reported disequilibrium. One patient presented with a hypertensive crisis due to the secretion of catecholamines. Five patients had undergone previous surgery.

No patient received prior radiotherapy treatment. The median dose to the tumor was 12.5 Gy (range 11–13 Gy) to the 72–83% isodose line to limit the cranial nerves' dose. Eight patients underwent a single-fraction treatment, and one was treated with three fractions (total dose of 24 Gy). The fractionated treatment was chosen because the maximum diameter of the lesion exceeded 6 cm. The median tumor volume was 5848 mm^3 (range 1602–12,782 mm^3). The largest diameter ranged from 2.4 to 6.4 cm. Follow-up evaluations were performed 2 months after radiosurgery, every 6 months during the first 2 years, and then annually. The mean clinical and radiological follow-up period was 20 months. All patients tolerated the treatment well. One patient died during follow-up due to unrelated causes. Two patients (25% of this series) improved after treatment. The rest showed a stable neurological status. None of the patients experienced worsening of the pre-existing neurological deficits. No new cranial nerve deficits developed. Two patients reported a transient headache immediately after treatment. One patient developed a non-specific difficulty walking that resolved after steroid administration. Post-radiosurgical imaging with MRI was available for eight of the nine patients. During the follow-up period, no local tumor progression was observed.

The third study is a retrospective analysis of 14 patients with glomus jugulare tumors between June 2007 and September 2010 in Turkey. One patient was male, and 13 were female. All patients had a radiological diagnosis of glomus jugulare. The median age was 68 years (range, 31–76 years). One patient had undergone previous surgery. An extra 1.0 mm was added to GTV for planning target volume (PTV).

The median dose to the tumor was 25 Gy (range 18–30 Gy) in median five fractions (range 1–5 fractions). Only one patient was treated with a single fraction. The GTV was median 15.8 cm^3 (range 2–64 cm^3). The dose was normalized to 80% isodose line (range 70–88%). The median homogeneity and conformity indices were 1.24 (1.0–1.43) and 1.61 (1.29–3.02), respectively. The median follow-up was 39 months (range 7–60 months). In eight of the patients, the lesions were stable based on the last available MRI. They did not observe any disease progression at the time of reporting. Over the full length of follow-up, 6 of 14 lesions (43%) demonstrated tumor regression.

Based on the last follow-up, the rate of local control was 100%. Before radiosurgery, two patients had a headache, seven patients had pulsatile tinnitus, and one patient had a loss of hearing functions at the lesion side. Complete clinical improvement was observed in eight patients. No treatment-related toxicity in this group of patients was observed [35].

The fourth study is the University of Texas at Southwestern Medical Center, reporting their experience of 31 consecutive patients treated from 2007 to 2013. The median age of patients in this series was 58.5 years. Follow-up time ranged from 4 to 78 months, with a median follow-up of 24 months. The majority of the tumors (58%) in this series involved both the middle ear and jugular foramen (glomus jugulotympanicum). Twelve tumors involved the jugular foramen exclusively (glomus jugulare), and one tumor involved the carotid body. Before radiation treatment, 58.1% of patients had pulsatile tinnitus. Eight patients (25.8%) had undergone previous surgical resection of their tumor and were referred either for residual tumor or tumor progression. The remaining patients' tumors manifested with vocal paralysis, headache, hearing loss, and epiphora (excessive tearing). A dose of 25 Gy in five fractions was prescribed to ensure adequate coverage of the planning target volume with a median prescription isodose line of 65%. The percent of tumors covered by the 50% isodose line ranged from 95 to 99.25%. The mean conformity index for these plans was 1.41, which is similar to conformity indices for stereotactic Gamma Knife plans [36]. Following CyberKnife radiation therapy, LC, progression-free survival, and overall survival were 100%. Posttreatment MRIs revealed no tumor progression in any patient, with some patients demonstrating tumor volume reduction on follow-up MRIs of the head. The most common neurologic symptom was pulsatile tinnitus, which was reported by 65% (n = 20) of patients before radiation therapy. Of the 20 patients with tinnitus before radiation therapy, 60% (n = 12) had subjective improvement of the tinnitus, of whom 50% (n = 6) had complete resolution of tinnitus. There was no change in subjective tinnitus in 40% of patients (n = 8). There were no reports of subjective worsening of tinnitus or onset of new cranial nerve deficits. Only two patients (6.4%) reported subjective deterioration of hearing on follow-up. Thus, overall symptom control was 94% in our series. Radiosurgery was well tolerated with no patient experiencing serious acute toxicity defined as grade 3 or higher. Acute grade 1–2 toxicity occurred in 19% (n = 6) of patients. The most common side effect was a headache, which occurred in three patients. The single grade 2 toxicity in this series was headache requiring short-term steroid administration in a patient whose tumor was 42.53 cm^3 (largest tumor in this series). Grade 1 xerostomia occurred in two patients, and grade 1 nausea occurred in one patient. Although follow-up was relatively short, no late radiation toxicities were reported, and no secondary tumors occurred that could be attributed to radiation therapy at the time of this analysis. There was a nonsignificant reduction in tumor volume size of 37.3% for all patients (p = 0.16). However, among patients with 24 months of follow-up or longer, there was a significant 49% reduction in tumor volume (p = 0.01). For patients who reported improvement of their tinnitus, there was an average of a 47.2% reduction in tumor volume. In patients who reported resolution of their tinnitus, there was an average of a 43.5% reduction in tumor volume [37].

The fifth study is also from Turkey, with a total of 12 patients with head-and-neck paragangliomas, treated with radiosurgery between March 2009 and June 2014. Six male (50%) and six female (50%) patients were analyzed in this report. The median age was 42 years. Three patients had jugular PGLs, five patients had carotid body PGLs, and three patients had tympanic PGLs. All patients presented either with Fisch Type C (50%) or D (50%) disease.

The most common symptoms were pulsatile tinnitus (five patients) and pain (five patients); other symptoms exhibited were pulsatile mass (three patients), hearing loss (three patients), neck mass (two patients), hoarseness (two patients), vertigo (one patient), and hypoglossal paralysis (one patient). One patient had bilateral neck PGLs (right neck; carotid body PGL, left neck; jugular PGL).

It was learned that his father and sister had PGLs as well, and genetic studies have revealed a mutation in the succinate dehydrogenase D gene. Due to the presentation of bilateral disease and family history, this patient was considered a hereditary PGLs.

Seven patients had prior surgery and recurrent tumor. Five patients had no previous treatment and, in addition, no histopathological diagnosis. GTV was defined according to the radiological findings. PTV was created by setting a 2–5 mm margin to GTV (median 3 mm). The median number of fractions was 3 (range, 3–5), and the total dose ranged between 21 and 30 Gy (median 24 Gy), and prescription isodose lines were selected between 67 and 90% (median 75%). The median maximum tumor diameter was 56 mm (range 26–92 mm), and the median tumor volume was 35.5 cm^3 (range 5.3–113.8 cm^3).

In the case of bilateral disease, both lesions at the neck were treated with radiosurgery using the CyberKnife system 1 month apart. The median follow-up was 30 months (range 0–66 months). Two patients had no follow-up. No local tumor progression was observed. Seven of 13 tumors (54%) had a partial response, and 5 tumors (46%) were considered a stable disease.

The decrease in tumor volume was not related to symptomatic improvement.

An overall local control rate of 100% was obtained. No acute or late toxicity related to radiosurgery was seen after treatment. None of the patients developed new cranial nerve deficits. All of our patients' clinical status was stable, and one patient had symptomatic relief [38].

The last two studies are from Stanford University for large paragangliomas. They reported their four cases treated between 1999 and 2008 and six cases between 2007 and 2018. Actually, these two studies reported radiosurgery for large benign intracranial tumors and paragangliomas just as a subgroup of the cohort. Therefore, tumor characteristics, tumor treatment details, and toxicities were not well understood, but they reported a local control rate of 100% [39, 40]. Eighty-seven paragangliomas (42 glomus jugulare, 29 glomus tympanicum, and 16 glomus caroticum) were treated in 86 patients between July 2013 and December 2019 at the Medicana International Ankara Hospital, Ankara, Turkey.

The prescribed dose ranged 13–30 Gy in 1–5 fractions. The median follow-up time was 38 months (range 1–66 months) (Fig. 32.1). At last follow-up, 81 of 83 evaluable tumors either demonstrated stable disease in tumor volume (59 tumors) or partial response in tumor volume (20 tumors) (Fig. 32.2). The overall tumor control rate was calculated at 98%. One patient experienced radiographic progression at 26 months, and the other patient experienced at 32 months

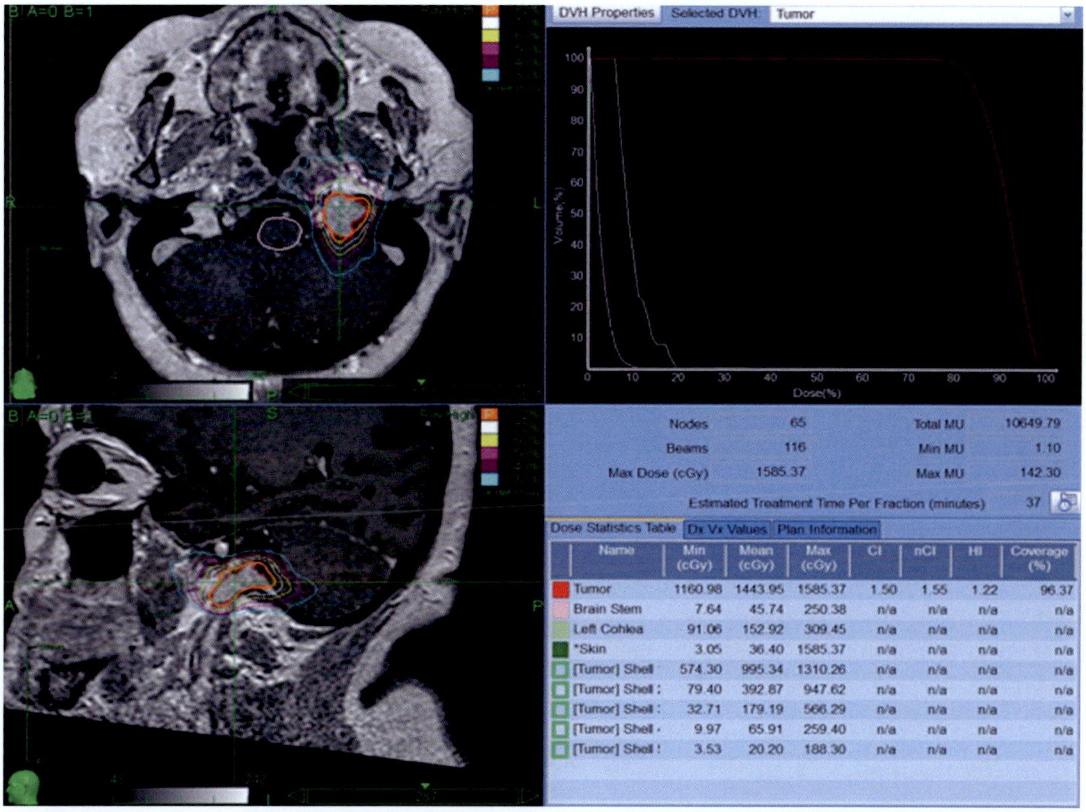

Fig. 32.1 Dose distributions in multiplan for left glomus jugulare tumor

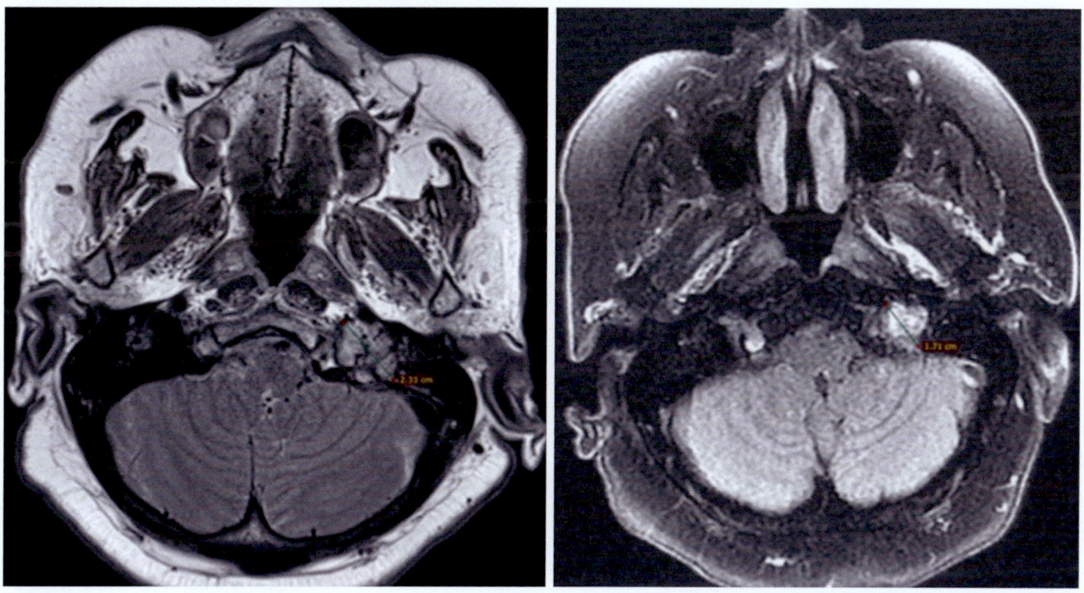

Fig. 32.2 Radiological response in the glomus jugulare patients. Pre-radiosurgery and post-radiosurgery (15 months)

following radiosurgery. No evidence of worsening cranial nerve function was noted.

32.5 Conclusion

Paragangliomas are rare neoplasms that arise from the paraganglia of the chemoreceptor system. The ideal treatment for patients with these cancers still remains a matter of debate. Treatment options include resection, endovascular embolization, fractionated radiation therapy, and SRS, alone or in combination. The difficulties associated with the surgical treatment of paragangliomas are primarily related to excessive intraoperative bleeding and the adhesion of tumors to critical neurovascular structures. In the modern era, craniotomy is generally not necessary for tissue biopsy in cases involving a paraganglioma due to the specific imaging characteristics and location of the neoplasm.

Surgery is suitable for patients in good medical condition, provided the tumor size and location make the risk of associated morbidity low. Fractionated radiation therapy for head and neck paragangliomas is a safe and efficacious treatment that is associated with a high probability of cure and a low incidence of morbidity. Stereotactic radiosurgery with CyberKnife has proven to be highly effective in the treatment of glomus tumors and conveys a relatively low risk of complications. Further extended-duration multi-institutional analyses are required to evaluate the long-term effects of this treatment modality for single-session radiosurgery and the emerging role of multisession radiosurgery for the rare neuropathology of glomus tumors.

References

1. Hu K, Persky MS. Multidisciplinary management of paragangliomas of the head and neck, part 1. Oncology (Williston Park). 2003a;17(7):983–93.
2. Hu K, Persky MS. The multidisciplinary management of paragangliomas of the head and neck, part 2. Oncology (Williston Park). 2003b;17(8):1143–61.
3. Persky MS, Hu KS. Paragangliomas of the head and neck. In: Harrison LB, Hong WK, Sessions RB, editors. Head and neck Cancer: a multidisciplinary approach. 3rd ed. Philadelphia, PA: Lippincott Williams & Wilkins; 2009. p. 655–87.
4. Nölting S, Ullrich M, Pietzsch J, et al. Current management of pheochromocytoma/paraganglioma: a guide for the practicing clinician in the era of precision medicine. Cancers (Basel). 2019;11(10):1505.
5. Neumann HP, Bausch B, McWhinney SR, Bender BU, Gimm O, Franke G, Schipper J, Klisch J, Altehoefer C, Zerres K, Januszewicz A, Eng C, Smith WM, Munk R, Manz T, Glaesker S, Apel TW, Treier M, Reineke M, Walz MK, Hoang-Vu C, Brauckhoff M, Klein-Franke A, Klose P, Schmidt H, Maier-Woelfle M, Peçzkowska M, Szmigielski C, Eng C, Freiburg-Warsaw-Columbus pheochromocytoma Study Group. Germ-line mutations in nonsyndromic pheochromocytoma. N Engl J Med. 2002;346(19):1459–566.
6. Lee JH, Barich F, Karnell LH, Robinson RA, Zhen WK, Gantz BJ, Hoffman HT, American College of Surgeons Commission on Cancer, American Cancer Society. National Cancer Data Base report on malignant paragangliomas of the head and neck. Cancer. 2002;94(3):730–7.
7. Mafee MF, Raofi B, Kumar A, Muscato C. Glomus faciale, glomus jugulare, glomus tympanicum, glomus vagale, carotid body tumors, and simulating lesions: role of MR imaging. Radiol Clin N Am. 2000;38:1059–76.
8. Lack EE. Tumors of the adrenal gland and extra-adrenal paraganglia, Atlas of tumor pathology, third series, fascicle 19. Washington, DC: Armed Forces Institute of Pathology; 1997.
9. Som PM, Curtin HD. Head and neck imaging. 3rd ed. St. Louis: Mosby-Year Book; 1996; 932–6, 1484–96.
10. van den Berg R, van Gils AP, Wasser MN. Imaging of head and neck paragangliomas with three-dimensional time-of-flight MR angiography. AJR Am J Roentgenol. 1999;172(6):1667–73.
11. Vogl TJ, Juergens M, Balzer JO, et al. Glomus tumors of the skull base: combined use of MR angiography and spin-echo imaging. Radiology. 1994;192(1):103–10.
12. Jansen JC, van den Berg R, Kuiper A, et al. Estimation of growth rate in patients with head and neck paragangliomas influences the treatment proposal. Cancer. 2000;88:2811–6.
13. Manolidis S, Shohet JA, Jackson CG, Glasscock ME 3rd. Malignant glomus tumors. Laryngoscope. 1999;109:30–4.
14. Jackson CG, Glasscock ME, Harris PF. Glomus tumors: diagnosis, classification, and management of large lesions. Arch Otolaryngol. 1982;108(7):401–6.
15. Oldring D, Fisch U. Glomus tumors of the temporal region: surgical therapy. Am J Otol. 1979;1:7–18.
16. Foote RL, Pollock BE, Gorman DA, Schomberg PJ, Stafford SL, Link MJ, Kline RW, Strome SE, Kasperbauer JL, Olsen KD. Glomus jugulare tumor: tumor control and complications after stereotactic radiosurgery. Head Neck. 2002;24:332–8.

17. Lundgren N. Tympanic body tumors in the middle ear: tumors of carotid body type. Acta Otolaryngol. 1949;37:366.
18. Chretien PB, Engelman K, Hoye RC, Geelhoed GW. Surgical management of intravascular glomus jugulare tumor. Am J Surg. 1971;122:740–3.
19. Gottfried ON, Liu JK, Couldwell WT. Comparison of radiosurgery and conventional surgery for the treatment of glomus jugulare tumors. Neurosurg Focus. 2004;17(2):E4.
20. Springate SC, Weichselbaum RR. Radiation or surgery for chemodectoma of the temporal bone: a review of local control and complications. Head Neck. 1990;12(4):303–7.
21. Lee CC, Trifiletti DM, Sheehan JP. Radiosurgery for glomus tumors. Prog Neurol Surg. 2019;34:215–22.
22. Jordan JA, Roland PS, McManus C, Weiner RL, Giller CA. Stereotastic radiosurgery for glomus jugulare tumors. Laryngoscope. 2000;110:35–8.
23. Mukherji SK, Kasper ME, Tart RP, Mancuso AA. Irradiated paragangliomas of the head and neck: CT and MR appearance. AJNR Am J Neuroradiol. 1994;15:357–63.
24. Cole JM, Beiler D. Long-term results of treatment for glomus jugulare and glomus vagale tumors with radiotherapy. Laryngoscope. 1994;104:1461–5.
25. de Jong AL, Coker NJ, Jenkins HA, Goepfert H, Alford BR, et al. Radiation therapy in the management of paragangliomas of the temporal bone. Am J Otol. 1995;16:283–9.
26. Springate SC, Haraf D, Weichselbaum RR. Temporal bone chemodectomas—comparing surgery and radiation therapy. Oncology (Williston Park). 1991;5:131–7.
27. Cummings BJ, Beale FA, Garrett PG, Harwood AR, Keane TJ, Payne DG, Rider WD. The treatment of glomus tumors in the temporal bone by megavoltage radiation. Cancer. 1984;53:2635–40.
28. Kida Y, Kobayashi T, Tanaka T, Oyama H, Niwa M. A new strategy for the treatment of jugular foramen tumors using radiosurgery. No Shinkei Geka. 1995;23:671–5.
29. Sager O, Beyzadeoglu M, Dincoglan F, Oysul K, et al. Evaluation of linear accelerator-based stereotactic radiosurgery in the management of glomus jugulare tumors. Tumori. 2014;100(2):184–8.
30. Guss ZD, Batra S, Limb CJ, et al. Radiosurgery of glomus jugulare tumors: a meta-analysis. Int J Radiat Oncol Biol Phys. 2011;81(4):e497–502.
31. Eustacchio S, Trummer M, Unger F, Schröttner O, Sutter B, Pendl G, et al. The role of gamma knife radiosurgery in the management of glomus jugular tumours. Acta Neurochir Suppl. 2002;84:91–7.
32. Saringer W, Khayal H, Ertl A, Schoeggl A, Kitz K. Efficiency of gamma knife radiosurgery in the treatment of glomus jugulare tumors. Minim Invasive Neurosurg. 2001;44:141–6.
33. Sheehan JP, Tanaka S, Link MJ, Pollock BE, Kondziolka D, Mathieu D, Duma C, Young AB, Kaufmann AM, McBride H, Weisskopf PA, Xu Z, Kano H, Yang HC, Lunsford LD. Gamma knife surgery for the management of glomus tumors: a multicenter study. J Neurosurg. 2012;117:246–54.
34. Lim M, Bower R, Nangiana JS, Adler JR, Chang SD. Radiosurgery for glomus jugulare tumors. Technol Cancer Res Treat. 2007;6(5):419–23.
35. Hurmuz P, Cengiz M, Ozyigit G, et al. Robotic stereotactic radiosurgery in patients with unresectable glomus jugulare tumors. Technol Cancer Res Treat. 2013;12(2):109–13.
36. Nakamura JL, Verhey LJ, Smith V, et al. Dose conformity of gamma knife radiosurgery and risk factors for complications. Int J Radiat Oncol Biol Phys. 2001;51:1313–9.
37. Chun SG, Nedzi LA, Choe KS, et al. A retrospective analysis of tumor volumetric responses to five-fraction stereotactic radiotherapy for paragangliomas of the head and neck (glomus tumors). Stereotact Funct Neurosurg. 2014;92(3):153–9.
38. Tosun I, Atalar B, Sahin B, et al. Robotic radiosurgery of head and neck paragangliomas: a single institution experience. Asia Pac J Clin Oncol. 2018;14(2):e3–7.
39. Fatima N, Meola A, Pollom E, Chang SD, Soltys S. Stereotactic radiosurgery for large benign intracranial tumors. World Neurosurg. 2020;134:e172–80.
40. Tuniz F, Soltys SG, Choi CY, et al. Multisession cyberknife stereotactic radiosurgery of large, benign cranial base tumors: preliminary study. Neurosurgery. 2009;65(5):898–907.

33 Paragangliomas: A Case Series from Burdenko Center of Neurosurgery

Svetlana V. Zolotova, Andrey V. Golanov, Natalia A. Antipina, Anastasia V. Barabanova, and Konstantin V. Nikitin

33.1 Introduction

The incidence of paragangliomas (PGLs) has been reported as 1 in 1.5 million people per year [1–3]. Women are affected more frequently than men, with reported ratios of 6:1 to 6:4 [4]. PGL is classified by the WHO as a tumor of indeterminate biology M code XXXX1 [5]. Malignant paragangliomas are uncommon, and their diagnosis can only be confirmed by the presence of metastatic disease. Although all mutated SDHx genes carry a risk for metastasis, SDHB mutations confer the highest risk at ~30%, SDHD carriers have smaller risk of about 3–4%. SDHB mutations are also associated with the poorest survival (11–36% at 5 years) [6]. Multicentric tumors occur in 10–20% of all head and neck paragangliomas [7]. However, reports of much higher incidence of multiple tumors, like 40% for sporadic form and 80% for familial variety, can be found in the literature [8].

In the head and neck region, the most common location of PGL is at the carotid body (60%), followed by the temporal bone (glomus tympanicum, 18%), arising from paraganglia associated with Arnold and Jacobson nerves within the middle ear, or glomus jugulare (12%), arising from paraganglia in the adventitia of the jugular vein, and upper pharyngeal space (glomus vagale, 5%), arising from the inferior vagal ganglion [9]. The most effective treatment modality for PGLs remains undetermined. Their involvement of major vessels, proximity to cranial nerves, and their propensity for intracranial extension can result in significant morbidity from surgical resection. Complications from resection include stroke (8–20%), cranial nerve injury (7–49%) [10], meningitis (6–9%), and cerebrospinal fluid leak (8.3%). In addition, the overall mortality rate was 1–5% [11]. Radiation therapy has the advantage of avoiding the morbidity of surgery while offering an equal possibility of cure. Among the 804 patients included in 34 different radiotherapy series between 1962 and 2009, the median local control rate is generally in excess of 90% [12]. Patients achieved symptomatic improvement in more than 70% of cases across 34 published series [13], and partial/complete resolution of symptoms can be estimated to be achieved in more than 60% of cases [14, 15].

33.2 Rationale for Radiation Therapy

The first review of the literature concerning the place of radiotherapy in the management of the head and neck paragangliomas was made by

S. V. Zolotova · A. V. Golanov (✉) · N. A. Antipina · A. V. Barabanova · K. V. Nikitin
National Medical Research Center of Neurosurgery named after Burdenko, Moscow, Russia
e-mail: szolotova@nsi.ru; golanov@nsi.ru; nantipina@nsi.ru; abrabanova@nsi.ru; knitin@si.ru

© Springer Nature Switzerland AG 2020
A. Conti et al. (eds.), *CyberKnife NeuroRadiosurgery*,
https://doi.org/10.1007/978-3-030-50668-1_33

Springate and Weichselbaum [16]. Nineteen series reporting 379 patients treated between 1932 and 1983 were reviewed. These patients received radiotherapy as the primary, preoperative, postoperative, or salvage treatment. Of 405 patients treated by surgery, 349 (86%) were reported as locally controlled. Of 379 patients who received radiotherapy, 344 (90%) were locally controlled, whereas the control rate for radiotherapy without surgery was even higher: 182 of 195 (93%) patients. Despite high control rates, external beam radiation requires large field sizes, resulting in high rate of complications: radionecrosis of the bone, brain necrosis/abscess, and xerostomia [17]. The advent of intensity-modulated radiation therapy has reduced the extent of normal tissue exposed to radiation, which in the future will likely be accompanied by a decrease in side effects.

With the development of technology, intensity-modulated radiotherapy (IMRT) delivers a highly conformal, three-dimensional (3D) distribution of radiation doses that is not possible with conventional methods. Henzel et al. [18] reported no severe (grade 3 or 4) acute or late toxicity in 16 patients, and 100% freedom from progression, with a median dose of 57 Gy. Mendenhall W.M. et al. [19] summarized the long-term outcomes of 149 patients treated with RT between May 1968 and September 2016 at the University of Florida College of Medicine. IMRT has been used to treat essentially all patients since 2001. No patient developed a new CN palsy after RT. No patient experienced a severe complication after RT or developed a radiation-induced malignancy. Table 33.1 shows the main series on conventionally fractionated radiotherapy treatment of PGLs.

33.3 Single-Fraction Radiosurgery

Compared with conventional radiotherapy, stereotactic radiosurgery involves a shorter treatment time (it usually takes 1 day, compared with 4–6 weeks for conventionally fractionated external beam radiation and with several weeks of postoperative recovery for resection), precise stereotactic localization, and a small volume of irradiated normal tissue. In 1997, Foote et al. [20] published the first report as a preliminary study. The goal of their study was to evaluate the immediate, acute, and chronic toxicity and the efficacy of stereotactic radiosurgery in patients with unresectable or subtotally resected glomus tumors. No acute or chronic toxicity was demonstrated, and eight of nine tumors remained stable in size at a median clinical follow-up duration of 20 months.

Recently, Shapiro et al. carried out very interesting meta-analyses on tumor control, symptomatic control, and complication rates of stereotactic radiosurgery as the primary treatment of glomus jugulare tumors [21]. The inclusion criteria were (1) no previous treatment of any kind, (2) follow-up with magnetic resonance imaging for at least 12 months, and (3) reported pre- and post-treatment symptoms, tumor control, or complications. Fifteen studies on 91 patients met the criteria. Tumor control was achieved in 92% of patients, symptoms control - in 93%, and complications occurred in 8%. There was one major complication. The recommended marginal tumor dose (prescribed most commonly to the 50% isodose line when the Gamma Knife is used) is 15–18 Gy, resulting in a maximum dose of 30–36 Gy. Table 33.2 shows the main series on single-fraction SRS treatment of PGLs.

33.4 Hypofractionated Radiotherapy

The efficacy and feasibility of CyberKnife radiotherapy was initially reported by investigators at Stanford University where patients were treated to a dose of 14–25 Gy in a single fraction or 18–25 Gy in three fractions [22, 23]. An Italian series of nine patients treated with CyberKnife radiotherapy for skull base paragangliomas reported doses ranging from 11 to 13 Gy in a single fraction and 24 Gy in three fractions [24]. In this series, local control was 100%, and 25% of patients had improvement in their symptoms. The results of these clinical series suggest that

Table 33.1 Summary of clinical series of conventionally fractionated radiotherapy for treatment of head and neck paragangliomas

Author	Place	Year	Treatment modality	No. of patients	Follow-up (years)	Tumor control	Toxicity
Nguyen [27]	Grenoble, France	1973–1996	EBRT	18	5.5	75% (5 years)	Radionecrosis
Lightowlers [28]	Cambridge, Great Britain	1998–2008	EBRT/IMRT	21	4.5	95% (5 years)	Xerostomia
Chino [29]	Durham, NC, USA	1963–2005	EBRT	31	9	96% (5 years) 90% (10 years)	Xerostomia, cataract
de Jong [30]	Houston, TX, USA	1956–1991	EBRT	38	11.5	79% primary RT, 100% combined treatment, 91% salvage RT	?
Combs [31]	Heidelberg, Germany		SRS/SRT/IMRT	39	10.5	97% (10 years)	Maxillary bone abscess, middle ear effusion
Breen [32]	Rochester, MN, USA	1973–2015	EBRT/IMRT	41 with malignant PG	9.7	81% (5 years)	Neuropathy
Huy [33]	Paris, France	1988–2003	EBRT	41	5.5	96%	Xerostomia, serous otitis media
Pemberton [34]	Manchester, UK	1965–1987	EBRT	49	7.4	92% (5 years) 92% (10 years)	None
Powell [35]	Boston, MA, USA	1949–1985	EBRT	64	9	73% (25 years) 100% combined treatment	Facial nerve palsy
Dupin [25]	Paris, France	1990–2009	EBRT	66	4.1	100% (5 years) 98.7% (10 years)	Xerostomia, stroke, radionecrosis, aphthous ulcer
Mendenhall [19]	Gainesville, FL, USA	1968–2016	SRS/SRT/EBRT	149	11.1	99% (5 years) 96% (10 years)	None
Total				557		93.6 (5 years) 95.6 (10 years)	

Table 33.2 Summary of clinical series of stereotactic radiosurgery for treatment of head and neck paragangliomas

Author	Place	Year	Treatment modality	No. of Patients	Follow-up (months)	Tumor response				CN morbidity (pts)
						Regression	Stable	Progression	Unknown	
Hurmuz [36]	Ankara, Turkey	2007–2010	CK	14	39 (7–60)	6	8	0	0	0
Schuster [37]	Nashville, Tennessee, USA	1998–2014	LINAC	14	31.7 (18.6–56.1)	6	7	1	0	1 (7%)
de Andrade [38]	São Paulo (SP), Brazil	2006–2011	LINAC	15	35.4 (3–61)	0	15	0	0	0
Bitaraf [39]	Tehran, Iran	2004–2006	GK	16	18.5 (4–28)	6	8	0	2	1 (6%)
Genç [40]	Istanbul, Turkey	1999–2008	GK	18	52.7 (12–116)	17	0	1	0	1 (5.5%)
Gerosa [41]	Verona, Italy	1996–2005	GK	20	51 (12–110)	9	11	0	0	1 (5%)
Marchetti [42]	Milan, Italy	2004–2014	CK	20/21	46.3 (12–111)	7	11	0	3	3 (15%)
El Majdoub [43]	Cologne, Germany	1991–2011	LINAC	27	9.6 years (5–19)	12	15	0	0	1 (3.7%)
Sallabanda [44]	Madrid, Spain	1993–2014	29 LINAC, 2 CK	30/31	4.6 years (1.5–12)	9	21	1	0	1 (3%)
Lieberson [13]	Stanford, CA, USA	1991–2009	6 LINAC, 30 CK	36/41	3.9 years (0.32–15.45)	18	19	0	4	4 (11%)
Winford [45]	Winston-Salem, NC, USA	2000–2015	GK	38	39.1 (5.5–141)	29		4	5	10 (26%)
Martín [46]	Granada, Spain	1997–2012	LINAC	39	71 (3–200)	22	14	3	0	7 (17.9%)
Hafez [47]	Cairo, Egypt	2005–2014	GK	40	84 (36–156)	12	27	1	0	3 (7.5%)
Sharma [48]	Cleveland, OH, USA	1997–2016	GK	42/43	62.3 (3–219)	31	1	6	5	8 (19%)
Liscak [49]	Prague, Czech Republic	1992–2003	GK	46	118 (12–217)	34	9	1	2	2 (4%)
Ibrahim [50]	Newport, United Kingdom	1994–2010	GK	75/76	51.5 (12–230)	43	28	5	0	12 (16%)
Patel [51]	Rochester, MN, USA	1990–2017	GK	85	66 (7–202)	41	14[a]	5	25	2 (3%) 7 NSH (20%)
Sheehan [52]	8 USA GK centers	1988–2010	GK	132/134	50.5 (5–220)	49	65	0	20	15 (11%)
Total				707						4 (0.6%)

[a]Tumor growth <15%; *NSH* Non-serviceable hearing

Table 33.3 Summary of clinical series of conventionally hypofractionated stereotactic radiotherapy for treatment of head and neck paragangliomas

Author	Place	Year	Fractions # × doses (Gy)	No. of patients	Follow-up (months)	Tumor control	CN morbidity (pts)
Lim [53]	Stanford, CA, USA	1991–2006	3 × 6–8.5	6	60 (6–162)	100%	3
Tosun [54]	Istanbul, Turkey	2009–2014	3 × 7–10 5 × 5	12	30 (0–66)	100%	0
Tse [55]	San Francisco, CA, USA	2010–2012	3 × 7–8 5 × 5–6	12	52 (31–74)	92.3%	8
Hurmuz [36]	Ankara, Turkey	2007–2010	5 × 5–6	13	39 (7–60)	100%	0
Lieberson [13]	Stanford, CA, USA	1991–2009	2 × 9–10 3 × 6.5–8 5 × 5	14	3.9 years (0.32–15.45)	100%	4
Marchetti [42]	Milan, Italy	2004–2014	3 × 8 5 × 5–6	14	46.3 (12–111)	100%	3
Chun [56]	Dallas, TX, USA	2007–2013	5 × 5	31	24 (4–78)	100%	2
Total		1991–2014		102		98.9%	3

CyberKnife may be used to treat PGLs with hypofractionated stereotactic radiotherapy that results in equivalent treatment outcomes as conventionally fractionated radiotherapy with improved patient convenience.

The experience of CyberKnife radiotherapy for treatment of the head and neck paragangliomas gained to date is presented in Table 33.3. The most frequently used regimens are three fractions per 8 Gy and five fractions per 5–6 Gy.

33.5 Toxicity

33.5.1 Acute Toxicity

Acute toxicity (during RT and within 3 months of its completion) can present with nausea, dermatitis with severe desquamation and fragility of the external auditory canal skin, headache, xerostomia, weight loss, mucositis, or ophthalmic zoster. After radiation treatment of PGLs, adverse events are rare and usually mild or moderate (grades 1–2). Only in a study Dupin et al. [25] described grade 3–4 acute toxicity: 9 out of 66 patients were hospitalized for weight loss, nausea, grade 3 mucositis, or ophthalmic zoster.

33.5.2 Late Toxicity

Springate and Weichselbaum [16] in the first systematic literature review of treatment modalities for paragangliomas of the temporal bone showed that complications after radiotherapy are very rare: bone necrosis (1.7%), brain necrosis/abscess (0.84%), and second malignant transformation (fibrosarcoma, 1 of 356 or 0.28%). These severe complications were observed in the dose range of 54–70 Gy. Complications became less frequent and less pronounced with the introduction of stereotactic RT and SRS.

33.6 Cranial Nerve Morbidity

Following SRS of jugular paragangliomas, 9.7% of patients had a post-treatment cranial nerve (CN) IX deficit, 9.7% had a post-treatment deficit of CN X, 12% had a post-treatment deficit of CN XI, and 8.7% of patients had a post-treatment CN

XII deficit. Importantly, patients suffered from lower cranial nerve neuropathy when treated with SRS alone [10]. Patients undergoing gross total resection reported worse rates of CN IX–XI deficit compared to those undergoing SRS. However, the CN XII deficit rates were comparable.

The auditory results after stereotactic radiosurgery for jugular paraganglioma are only described in detail in a study by Patel et al. [26] at the Mayo Clinic (Rochester, MN, USA): 7 out of 35 patients developed non-serviceable hearing. The estimated hearing preservation rates according to Kaplan-Meier at 1, 3, and 5 years after SRS were 91%, 80%, and 80%, respectively.

33.7 Own Experience

Three hundred and sixty-six patients with 381 PGLs were treated with SRS and SRT at our center between March 2005 and December 2018. From April 2009 to December 2018, 158 patients with 162 paragangliomas (127 women and 31 men) were treated with CyberKnife G4 system. The median age was 52 years (range 12–84). Forty-four patients have undergone microsurgery (28%), 12 of them repeatedly, 3 patients were operated 3 times, and 1 was operated 9 times. Eleven of the patients (7%) had undergone embolization alone before irradiation. CyberKnife radiosurgery was the primary treatment modality in the remaining 103 patients (65%). When surgery was not performed and thus the histological diagnosis was not confirmed, the diagnosis was based on CT and CT perfusion, MR imaging, angiography, and clinical findings. There were 85 jugulotympanic paragangliomas, 35 tumors of the glomus jugulare, 23 of glomus tympanicum, 8 carotid body paragangliomas, 5 tumors of the glomus vagale, 1 paraganglioma of the glomus ciliare, 1 spinal paraganglioma of the filum terminale, and 4 metastases of malignant paragangliomas.

Twenty-three paragangliomas with a mean of volume 3 cm^3 (range 0.5–7.7) were irradiated using a single fraction. Mean radiosurgical dose was 17.5 Gy (range 15–24). The higher mean doses of 22 and 24 Gy were used for metastasis of malignant paragangliomas. One hundred thirty-nine lesions with mean volume 17.6 cm^3 (range 0.2–73) underwent multisession CK treatment with the following regimes: 3 fractions per 7 Gy (74 cases–53%), 5 fractions per 5.5–6 Gy (54 cases–39%), and 7 fractions per 4.5–5 Gy (11 tumors—8%).

The median follow-up was 36 months (range 5–105). Follow-up time was calculated from the last day of the CK procedure. Seventeen patients were lost to follow-up. Forty-six percent of the patients had noticeable tumor shrinkage. In 72 patients (50%), the tumor size remained unchanged.

According to MRI control, the progression of tumor growth was observed in six cases (4%) after treatment: two of them were metastases of malignant paragangliomas (there were CK reradiation treatments that were used after the conventional fractionation of RT). In two cases with follow-up less than 1 year, minimal tumor enlargement was observed. In these cases, verification of the real continuous growth was required, and we continue to monitor these patients. As a result, actuarial local control was 96% at 3 years.

33.8 Clinical Case

A 45-year-old female patient was to the center to pulsatile tinnitus and reduced right-side hearing, which had occurred over the previous 7 years. The patient was also experiencing dizziness with nausea and vomiting, which occurred occasionally. Partial surgical excision with previous endovascular embolization was performed via a trans-canal approach 7 months before irradiation. After surgery and embolization, the patient had VII, IX, X, and XII CN dysfunction. Control MRI (Fig. 33.1a) showed a jugulotympanic paraganglioma with a pronounced extracranial extension (tumor volume was 17.8 cm^3). We treated her with hypofractionated radiotherapy with the CyberKnife (Fig. 33.1b) with a mean dose of 30 Gy in five fractions (prescribed dose 27 Gy to the 79% isodose line). A marked shrinkage of the tumor was noted on MR images at 4 years after SRT (Fig. 33.1c). There was no acute or chronic toxicity after procedure. The patient had improvement of IX and X cranial nerves function after CK treatment.

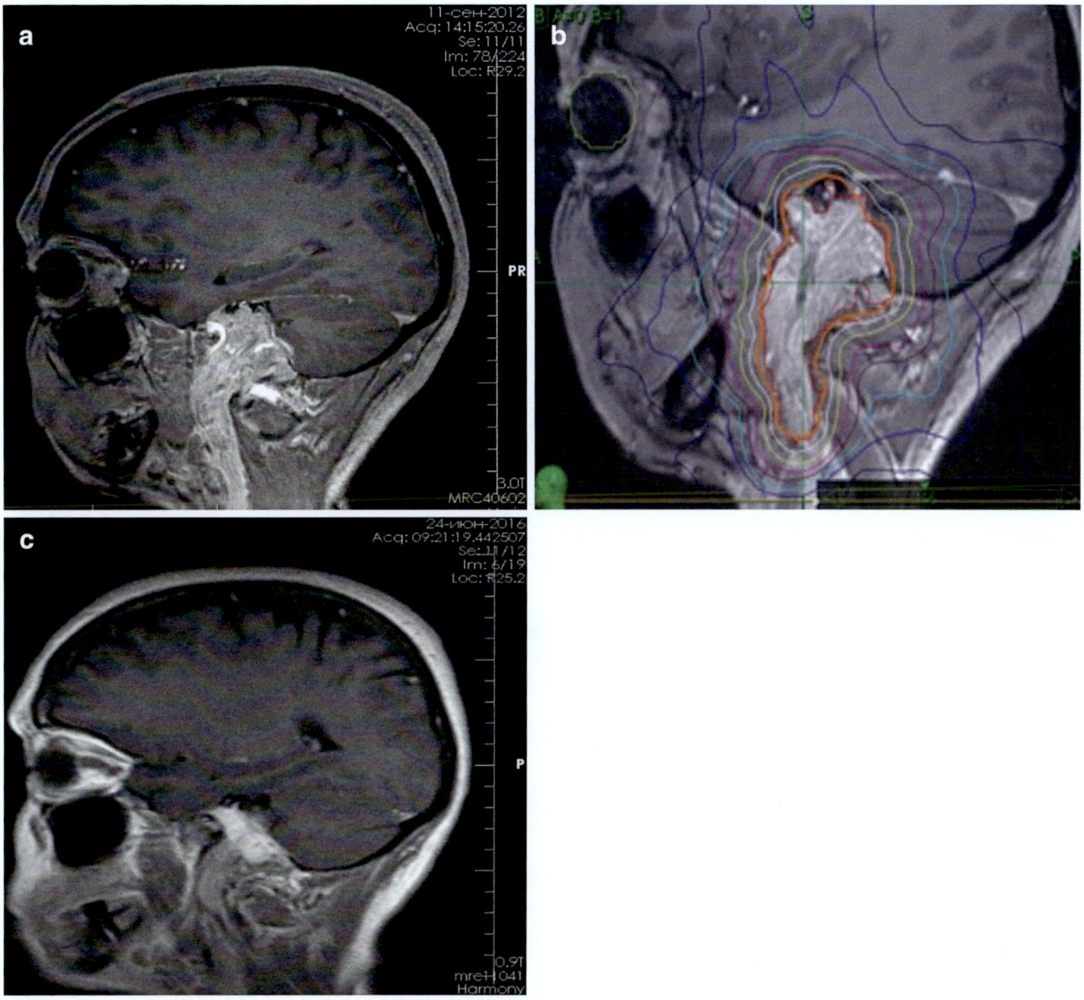

Fig. 33.1 (**a**) Pre-treatment MRI of a jugulotympanic paragangliomas with a pronounced extracranial extension (tumor volume was 17.8 cm^3). (**b**) CyberKnife treatment plan (five fractions per 6 Gy). Prescribed dose was 27 Gy to the 79% isodose line. (**c**) Post-treatment MRI obtained 4 years after SRT showing a marked shrinkage of the tumor

33.9 Conclusion

We have presented a brief history and overview of radiation treatment for glomus jugulare tumors, focusing on recent radiosurgical results. Due to the complex anatomy surrounding the tumors, resection often carries high rates of morbidity and mortality. Since 2000, multiple GKS-, LINAC-, and CyberKnife-based series have been reported. Collectively, they show excellent tumor control and relatively low complication rates. Table 33.4 summarizes practical suggestions for hypofractionated treatment of head and neck paragangliomas. Although longer-term follow-up studies are still in progress, the results in outcome studies published to date as well as our own data are good enough to justify the use of stereotactic irradiation as a method of choice and first-line treatment strategy for glomus jugulare tumors. The main problem in cases without histological conformation is differential diagnosis of paragangliomas with other tumors as benign (schwannomas, meningiomas, capillary hemangiomas, etc.) and malignant tumors (cancers, sarcomas, endolymphatic sac tumors, etc.) of the temporal bone and neck.

Table 33.4 Practical guide for hypofractionated treatment of head and neck paragangliomas

Parameter	Suggestion
Imaging	CT or MRI perfusion for differential diagnosis 3D T1 $_{TCE}$ MPRAGE, VIBE, or SPGR 3D T1w/o CE MPRAGE, VIBE, or SPGR Volumetric T2
Maximal tumor volume	75 cm^3 (may be less with substantial intracranial extension and brain stem compression)
Number of fractions	3 5
Mean dose per fraction (Gy)	7.5 6
Dose to the brain stem	<18 Gy for 0.15 ccm for 3 fractions <22.5 Gy for 0.15 ccm for 5 fractions
Dose to the cochlea	<15 Gy for 5% for 3 fractions <22 Gy for 5% for 5 fractions
10-year expected tumor control	95%
Serviceable hearing preservation at 5 years	>80%
Malignant transformation	<0.28%

References

1. van der Mey AGL, Frijns JH, Cornelisse CJ, et al. Does intervention improve the natural course of glomus tumors? A series of 108 patients seen in a 32-year period. Ann Otol Rhinol Laryngol. 1992;101(8):635–42. https://doi.org/10.1177/000348949210100802.
2. Guss ZD, Batra S, Limb CJ, et al. Radiosurgery of glomus jugulare tumors: a meta-analysis. Int J Radiat Oncol Biol Phys. 2011;81(4):e497–502.
3. Erickson D, Kudva YC, Ebersold MJ, et al. Benign paragangliomas: clinical presentation and treatment outcomes in 236 patients. J Clin Endocrinol Metabol. 2001;86(11):5210–6. https://doi.org/10.1210/jcem.86.11.8034.
4. Laviv Y, Thomas A, Kasper EM. Hypervascular lesions of the cerebellopontine angle: the relevance of angiography as a diagnostic and therapeutic tool and the role of stereotactic radiosurgery in management. A comprehensive review. World Neurosurg. 2017;100:100–17. https://doi.org/10.1016/j.wneu.2016.12.091.
5. Williams MD, Tischler AS. Update from the 4th edition of the World Health Organization classification of head and neck tumours: paragangliomas. Head Neck Pathol. 2017;11(1):88–95. https://doi.org/10.1007/s12105-017-0786-1.
6. Sethi RV, Sethi RKV, Herr MW, et al. Malignant head and neck paragangliomas: treatment efficacy and prognostic indicators. Am J Otolaryngol. 2013;34(5):431–8. https://doi.org/10.1016/j.amjoto.2013.03.010.
7. Papaspyrou K, Mewes T, Rossmann H, et al. Head and neck paragangliomas. Report of 175 patients (1989-2010). Head Neck. 2012;34(5):632–7. https://doi.org/10.1002/hed.21790.
8. Szymańska A, Szymański M, Czekajska-Chehab E, et al. Diagnosis and management of multiple paragangliomas of the head and neck. Eur Arch Otorhinolaryngol. 2015;272:1991–9.
9. Pellitteri PK, Rinaldo A, Myssiorek D, et al. Paragangliomas of the head and neck. Oral Oncol. 2004;40(6):563–75.
10. Ivan ME, Sughrue ME, Clark AJ, et al. A meta-analysis of tumor control rates and treatment-related morbidity for patients with glomus jugulare tumors. J Neurosurg. 2011;114(5):1299–305.
11. Gottfried ON, Liu JK, Couldwell WT. Comparison of radiosurgery and conventional surgery for the treatment of glomus jugulare tumors. Neurosurg Focus. 2004;17(2):E4.
12. Suárez C, Rodrigo JP, Boedeker CC, et al. Jugular and vagal paragangliomas: systematic study of management with surgery and radiotherapy. Head Neck. 2013;35(8):1195–204. https://doi.org/10.1002/hed.22976.
13. Lieberson RE, Adler JR, Soltys SG, et al. Stereotactic radiosurgery as the primary treatment for new and recurrent paragangliomas: is open surgical resection still the treatment of choice? World Neurosurg. 2012;77(5–6):745–61. https://doi.org/10.1016/j.wneu.2011.03.026.
14. Capatina C, Ntali G, Karavitaki N, et al. The management of head-and-neck paragangliomas. Endocr Relat Cancer. 2013;20(5):305. https://doi.org/10.1530/ERC-13-0223.
15. Suárez C, Rodrigo JP, Mendenhall WM, et al. Carotid body paragangliomas: a systematic study on management with surgery and radiotherapy. Eur Arch Otorhinolaryngol. 2014;271(1):23–34. https://doi.org/10.1007/s00405-013-2384-5.
16. Springate SC, Weichselbaum RR. Radiation or surgery for chemodectoma of the temporal bone: a

review of local control and complication. Head Neck. 1990;12(4):303–7.
17. Mendenhall WM, Hinerman RW, Amdur RJ, et al. Treatment of paragangliomas with radiation therapy. Otolaryngol Clin N Am. 2001;34(5):1007–20.
18. Henzel M, Hamm K, Gross MW, et al. Fractionated stereotactic radiotherapy of glomus jugulare tumors. Strahlenther Onkol. 2007;183(10):557–63.
19. Mendenhall WM, Morris CG, Amdur RJ, et al. Radiotherapy for benign head and neck paragangliomas. Head Neck. 2019;41(7):2107–10. https://doi.org/10.1002/hed.25664.
20. Foote RL, Coffey RJ, Gorman DA, et al. Stereotactic radiosurgery for glomus jugulare tumors. A preliminary report. Int J Radiat Oncol Biol Phys. 1997;38(3):491–5. https://doi.org/10.1016/s0360-3016(97)89482-5.
21. Shapiro S, Kellermeyer B, Ramadan J, Jones G, et al. Outcomes of primary radiosurgery treatment of glomus jugulare tumors. Systematic review with meta-analysis. Otol Neurotol. 2018;39(9):1079–87. https://doi.org/10.1097/MAO.0000000000001957.
22. Lim M, Gibbs IC, Adler JR, et al. The efficacy of linear accelerator stereotactic radiosurgery in treating glomus jugulare tumors. Technol Cancer Res Treat. 2003;2(3):261–5.
23. Lim M, Gibbs IC, Adler JR, et al. Efficacy and safety of stereotactic radiosurgery for glomus jugulare tumors. Neurosurg Focus. 2004;17(2):E11.
24. Bianchi LC, Marchetti M, Brait L, et al. Paragangliomas of head and neck: a treatment option with CyberKnife radiosurgery. Neurol Sci. 2009;30(6):479–85.
25. Dupin C, Lang P, Dessard-Diana B, et al. Treatment of head and neck paragangliomas with external beam radiation therapy. Int J Radiat Oncol Biol Phys. 2014;89(2):353–9.
26. Patel NS, Link MJ, Driscoll CLW, et al. Hearing outcomes after stereotactic radiosurgery for jugular paraganglioma. Otol Neurotol. 2018;39(1):99–105. https://doi.org/10.1097/MAO.0000000000001636.
27. Nguyen DQ, Boulat E, Troussier J, et al. Les paragangliomes tympano-jugulaires. A propos de 41 cas (the jugulotympanic paragangliomas: 41 cases report). Rev Laryngol Otol Rhinol (Bord). 2005;126(1):7–13.
28. Lightowlers S, Benedict S, Jefferies SJ, et al. Excellent local control of paraganglioma in the head and neck with fractionated radiotherapy. Clin Oncol. 2010;22(5):382–9. https://doi.org/10.1016/j.clon.2010.02.006.
29. Chino JP, Sampson JH, Tucci DL, et al. Paraganglioma of the head and neck. Long-term local control with radiotherapy. Am J Clin Oncol. 2009;32(3):304–7. https://doi.org/10.1097/COC.0b013e318187dd94.
30. de Jong AL, Coker NJ, Jenkins HA, et al. Radiation therapy in the management of paragangliomas of the temporal bone. Am J Otol. 1995;16(3):283–9.
31. Combs SE, Salehi-Allameh B, Habermehl D, et al. Clinical response and tumor control based on long-term follow-up and patient-reported outcomes in patients with chemodectomas of the skull base and head and neck region treated with highly conformal radiation therapy. Head Neck. 2014;36(1):22–7. https://doi.org/10.1002/hed.23274.
32. Breen W, Bancos I, Young WF, et al. External beam radiation therapy for advanced/unresectable malignant paraganglioma and pheochromocytoma. Adv Radiat Oncol. 2018;3(1):25–9. https://doi.org/10.1016/j.adro.2017.11.002.
33. Huy PTB, Kania R, Duet M, et al. Evolving concepts in the management of jugular paraganglioma. A comparison of radiotherapy and surgery in 88 cases. Skull Base. 2009;19(1):83–91. https://doi.org/10.1055/s-0028-1103125.
34. Pemberton LS, Swindell R, Sykes AJ. Radical radiotherapy alone for glomus jugulare and tympanicum tumours. Oncol Rep. 2005;14(6):1631–3.
35. Powell S, Peters N, Harmer C. Chemodectoma of the head and neck. Results of treatment in 84 patients. Int J Radiat Oncol Biol Phys. 1992;22(5):919–24. https://doi.org/10.1016/0360-3016(92)90788-J.
36. Hurmuz P, Cengiz M, Ozyigit G, et al. Robotic stereotactic radiosurgery in patients with unresectable glomus jugulare tumors. Technol Cancer Res Treat. 2013;12(2):109–13. https://doi.org/10.7785/tcrt.2012.500303.
37. Schuster D, Sweeney AD, Stavas MJ, et al. Initial radiographic tumor control is similar following single or multi-fractionated stereotactic radiosurgery for jugular paragangliomas. Am J Otolaryngol. 2016;37(3):255–8. https://doi.org/10.1016/j.amjoto.2016.01.002.
38. de Andrade EM, Brito JR, Mario SD, et al. Stereotactic radiosurgery for the treatment of glomus jugulare tumors. Surg Neurol Int. 2013;4(7):S429–35.
39. Bitaraf MA, Alikhani M, Tashili-Fahadan P, et al. Radiosurgery for glomus jugulare tumors: experience treating 16 patients in Iran. J Neurosurg. 2006;105(Suppl):168–74.
40. Genç A, Bicer A, Abacioglu U, et al. Gamma knife radiosurgery for the treatment of glomus jugulare tumors. J Neuro-Oncol. 2010;97(1):101–8. https://doi.org/10.1007/s11060-009-0002-6.
41. Gerosa M, Visca A, Rizzo P, et al. Glomus jugulare tumors: the option of gamma knife radiosurgery. Neurosurgery. 2006;59(3):561–9. https://doi.org/10.1227/01.NEU.0000228682.92552.CA.
42. Marchetti M, Pinzi V, Tramacere I, et al. Radiosurgery for paragangliomas of the head and neck: another step for the validation of a treatment paradigm. World Neurosurg. 2017;98:281–7. https://doi.org/10.1016/j.wneu.2016.10.132.
43. El Majdoub F, Hunsche S, Igressa A, et al. Stereotactic LINAC-radiosurgery for glomus jugulare tumors. A long-term follow-up of 27 patients. PLoS One. 2015;10(6):e0129057. https://doi.org/10.1371/journal.pone.0129057.
44. Sallabanda K, Barrientos H, Romero DAI, et al. Long-term outcomes after radiosurgery for glomus jugulare tumors. Tumori. 2018;104(4):300–6. https://doi.org/10.1177/0300891618765576.

45. Winford TW, Dorton LH, Browne JD, et al. Stereotactic radiosurgical treatment of glomus jugulare tumors. Otol Neurotol. 2017;38(4):555–62. https://doi.org/10.1097/MAO.0000000000001336.
46. Martín IT, Ávila RDM, Herrera MZ, et al. Role of radiosurgery in the management of glomus tumors. Head Neck. 2016;38(S1):E798–804. https://doi.org/10.1002/hed.24103.
47. Hafez RFA, Morgan MS, Fahmy OM, et al. Long-term effectiveness and safety of stereotactic gamma knife surgery as a primary sole treatment in the management of glomus jagulare tumor. Clin Neurol Neurosurg. 2018;168:34–7. https://doi.org/10.1016/j.clineuro.2018.02.037.
48. Sharma M, Meola A, Bellamkonda S, et al. Long-term outcome following stereotactic radiosurgery for glomus jugulare tumors. A single institution experience of 20 years. Neurosurgery. 2018;83(5):1007–14. https://doi.org/10.1093/neuros/nyx566.
49. Liscak R, Urgosik D, Chytka T, et al. Leksell Gamma Knife radiosurgery of the jugulotympanic glomus tumor: long-term results. J Neurosurg. 2014;121(Suppl 2):198–202.
50. Ibrahim R, Ammori MB, Yianni J, et al. Gamma Knife radiosurgery for glomus jugulare tumors: a single-center series of 75 cases. J Neurosurg. 2017;126(5):1488–97. https://doi.org/10.3171/2016.4.JNS152667.
51. Patel NS, Carlson ML, Pollock BE, et al. Long-term tumor control following stereotactic radiosurgery for jugular paraganglioma using 3D volumetric segmentation. J Neurosurg. 2018:1–9. https://doi.org/10.3171/2017.10.JNS17764.
52. Sheehan JP, Tanaka S, Link MJ, et al. Gamma Knife surgery for the management of glomus tumors: a multicenter study. J Neurosurg. 2012;117(2):246–54.
53. Lim M, Bower R, Nangiana JS, et al. Radiosurgery for glomus jugulare tumors. Technol Cancer Res Treat. 2007;6(5):419–23.
54. Tosun İ, Atalar B, Şahin B, et al. Robotic radiosurgery of head and neck paragangliomas. A single institution experience. Asia Pac J Clin Oncol. 2018;14(2):e3–7. https://doi.org/10.1111/ajco.12695.
55. Tse V, Sillanpaa J, Minn AY, et al. Glomus tumors treated with stereotactic radiosurgery. A retrospective study. J Radiosurg SBRT. 2017;5(1):73–81.
56. Chun SG, Nedzi LA, Choe KS, et al. A retrospective analysis of tumor volumetric responses to five-fraction stereotactic radiotherapy for paragangliomas of the head and neck (glomus tumors). Stereotact Funct Neurosurg. 2014;92(3):153–9. https://doi.org/10.1159/000360864.

Brainstem Tumors

34

Gokhan Ozyigit and Sezin Yuce Sari

34.1 Introduction

Even though they are rare, metastatic and primary tumors and benign conditions can arise from the brainstem. The brainstem has vital functions including breathing, blood pressure, heart rate, swallowing, awareness of hunger and thirst, as well as consciousness, learning, and memory. It is a very sensitive organ to irradiation. Therefore, the treatment of brainstem lesions must be given special attention. Patients with brainstem lesions can show a wide spectrum of symptoms such as headache, nausea and/or vomiting, and seizures, on one hand, and ataxia, vertigo, syncope, tinnitus, confusion, pyramidal motor symptoms, and nuclear palsies, on the other hand, specific of brainstem lesions. It is extremely difficult to perform surgery for the lesions in and around the brainstem without complication. Therefore, RT comes forward for the treatment of brainstem lesions. With the advantage of a homogeneous dose distribution and better preservation of critical organs, SRS or FSRT is the main treatment of choice for primary and recurrent tumors as well as metastases. However, the complication risk is still high, and extra caution must be given by avoiding moderately high doses to the brainstem.

G. Ozyigit (✉) · S. Y. Sari
Department of Radiation Oncology, Hacettepe University, Faculty of Medicine, Ankara, Turkey
e-mail: gozyigit@hacettepe.edu.tr

34.2 Rationale for Radiation Therapy

The indication for surgery is very limited in the treatment of lesions in and around the brainstem. Most tumors, either metastatic or primary, are not suitable for surgery except for dorsally exophytic gliomas and selected tumors abutting the brainstem surface [1]. Adjuvant RT is indicated for all high-grade tumors and incompletely resected low-grade tumors. However, since complete surgical resection is not safely possible in many of the brainstem tumors, RT is the definitive treatment of choice. Brainstem gliomas are rare tumors with a poor prognosis, and RT is a safe and effective modality for these tumors.

More than 60% of the patients will have their symptoms relieved with 50–55 Gy prescribed in conventional fractions [2]. Many dose escalation studies with total doses ranging from 62 to 78 Gy utilizing hyperfractionated RT have been undertaken [3–8]. However, a survival benefit could not be confirmed with either the use of hyperfractionated RT or higher doses. The recommended standard dose for conventional RT is 54–60 Gy in 30 fractions over 6 weeks. On the other hand, Janssens et al. [9] have recently compared hypofractionated (39 Gy/13 fractions or 44.8 Gy/16 fractions) and conventional (54 Gy/30 fractions) RT regimens, finding similar survival rates with similar toxicity.

© Springer Nature Switzerland AG 2020
A. Conti et al. (eds.), *CyberKnife NeuroRadiosurgery*,
https://doi.org/10.1007/978-3-030-50668-1_34

Malignant brainstem gliomas have a poorer prognosis with a clinical and radiographical response rate of 13% [10–13]. Poor prognostic factors other than higher tumor grade are older age (>40 years), a duration of symptoms of less than 3 months, poor performance status, and presence of contrast enhancement or necrosis on imaging [12]. Besides, the prognosis of children is apparently worse than that of adults.

Brainstem metastases have an extremely poor prognosis with survival limited to 1–6 months in most cases [14, 15]. In order to allow for an optimal sparing of normal tissues and structures surrounding the brainstem, RT should ideally be performed with high-precision techniques such as intensity-modulated RT, SRS, or FSRT. Due to the rarity of brainstem lesions, data on these treatment techniques are mainly based on retrospective series. In a recent trial comparing 268 patients with brainstem metastasis treated with SRS or surgical resection, the rate of LC was found to be similar [16]. However, they also reported that patients that underwent surgical resection had a much higher risk of early local recurrence (LR), whereas they had a lower risk at 9 months or longer. Brainstem tumors have generally been excluded from prospective SRS trials and often treated more conservatively with whole brain RT (WBRT).

However, there are limited prospective studies on brainstem metastases with SRS or FSRT either alone or sequential to WBRT in literature. In general, small lesions in the vicinity of or inside the brainstem can be safely treated with SRS. Nevertheless, a significant higher risk of complications exists for large lesions for which FSRT is a much safer option [17]. Brown et al. [18] compared postoperative SRS with WBRT after surgical resection and reported that the rate of overall survival (OS) was similar, but cognitive-deterioration-free survival was significantly higher in patients treated with postoperative SRS. Moreover, postoperative SRS comes with the difficulty in target delineation and the risk of leptomeningeal dissemination and radionecrosis [19–21].

SRS and FSRT induce hyalinization and thickening of blood vessel walls and lead to thrombo-obliterative response [22]. The main advantages of hypofractionated stereotactic regimens are the irreparable sublethal damage, negligible repopulation, early apoptosis, and prevention of re-oxygenation due to ablative doses. A single-fraction stereotactic RT is completed in a long treatment time in which a fair amount of sublethal damage repair occurs [23, 24]. It was reported that this repair has a biphasic component; the half-life of the fast component is a median 0.3 h, while the slow component's is approximately 4 h [23]. Therefore, when irradiation lasts more than half an hour such as in SRS and FSRT, 10% of the biological efficacy is lost due to damage repair. The loss is expected to be much larger for normal tissue compared to tumors because of their lower α/β ratio [25]. Furthermore, vascular damage and following chaotic intratumoral environment (e.g., hypoxia, acidic pH, and malnutrition) prevent the repair of radiation damage due to high fraction doses [26]. Compensatory repopulation of tumor cells starts at weeks 3–4 of conventional RT. Therefore, repopulation is negligible in the case of stereotactic RT as the total treatment time is too short. Moreover, the cells pause at the phase they are irradiated without continuing in the cell cycle suffering from apoptosis after an ablative radiation dose. This increases the probability of tumor death when apoptotic death is added to the mitotic death which is dominant in radiation-dependent cell killing. In addition, the vascular damage caused by ablative doses prevents the re-oxygenation of the tumor cells. Along with the oxygenated cells, hypoxic cells are also killed with these doses. Based on all these factors, the damage of an ablative radiation dose cannot be repaired.

34.3 Single-Fraction Radiosurgery

Surgery is the mainstay of treatment in primary brain tumors as it prolongs survival. However, it comes with a high rate of morbidity and mortality for brainstem tumors. If possible, subtotal resection is still recommended followed by adjuvant RT or observation. Survival is also based on the tumor grade. Median survival was reported to be 77, 21, and 15 months in patients with grade II, III, and IV tumors, respectively, according to the World Health Organization (WHO) [27]. For low-grade gliomas which are well-circumscribed and not locally invasive, SRS can be an option either as an adjuvant or definitive treatment method. Although follow-up durations are short, satisfactory results were reported with SRS doses of 10–22 Gy [28–31].

Radiation doses over 20 Gy are generally associated with higher complication rates. This was reported by Kihlström et al. who suggested 14 Gy as the threshold dose [28, 30, 32].

In an interesting case report, Liu et al. [33] treated a 13-year-old patient with a brainstem juvenile pilocytic astrocytoma with a 5 Gy prescription dose and 14 Gy maximum dose.

Receiving a brainstem dose ≤5 Gy, the patient was reported to be alive with no serious complications after 20 years of follow-up. A prospective study from Vietnam reported the results of 37 children and adult patients with low-grade glioma treated with a median of 12 Gy (range: 8–16 Gy) SRS [34]. After 36 months of follow-up, the median survival was 30 months, and they achieved a 3-year response rate of 87.5% with the best outcome observed in patients that received >13 Gy without increased toxicity. Fuchs et al. [35] reported 21 children and adult patients with low- and high-grade tumors treated with 10–18 Gy SRS. The response rate was 87% with a mean follow-up of 78 months. The authors observed transitory extrapyramidal symptoms and fluctuating impairment of consciousness in one patient at 6 months and a stroke at 8 years. Therapeutic results of SRS for metastatic brainstem tumors in selected trials are summarized in Table 34.1.

Table 34.1 Selected trials of SRS for metastatic brainstem tumors

Trial (year)	SRS modality	No. of patients/ lesions	Median SRS dose (range) (Gy)	Median FU (months)	Median OS (months)	Overall LC (%)
Winograd et al. (2019) [36]	GK	41/45	16 (12–20)	NA	11.6	85
Patel et al. (2018) [37]	GK	14/19	17.5 (14–22)	15.3	17.2	87.5
Murray et al. (2017) [38]	GK	44/48	15 (10–22)	16.1	5.4	76.9
Joshi et al. (2016) [39]	GK	48/51	15 (10–18)	4.8	7.3	1 y: 89
Trifiletti et al. (2016) [40]	GK	547/596	16 (8–25)	5.6	5.6	1 y: 82
Voong et al. (2015) [41]	GK	74/77	16 (10–20)	5.5	8.5	94
Trifiletti et al. (2015) [42]	GK	161/189	18 (8–25)	5.4	5.5	87.3
Kilburn et al. (2014) [43]	GK	44/52	18 (10–22)	6	6	1 y: 88
Peterson et al. (2014) [44]	GK	41/>41	17 (10–22.5)	NA	4.4	91
Jung et al. (2013) [45]	GK	32/32	13 (8–20)	12.5	5.2	87.5
Sengoz et al. (2013) [46]	GK	44/46	16 (10–20)	NA	8	96
Kawabe et al. (2012) [47]	GK	200/222	18 (12–25)	5.8	6	2 y: 82
Li et al. (2012) [48]	GK	28/32	16 (12–20)	NA	9	90.6

(continued)

Table 34.1 (continued)

Trial (year)	SRS modality	No. of patients/ lesions	Median SRS dose (range) (Gy)	Median FU (months)	Median OS (months)	Overall LC (%)
Lin et al. (2012) [49]	LINAC	45/48	14 (10–17)	NA	11.6	88
Yoo et al. (2011) [50]	GK	32/32	15.9 (6–23)	12	7.7	87.5
Kelly et al. (2011) [51]	LINAC	24/24	13 (8–16)	6.6	5.3	78.6
Valery et al. (2011) [52]	LINAC	30/30	13.4 (8.2–15)	10.4	10	90
Hatiboglu et al. (2011) [53]	LINAC	60/60	15 (8–18 Gy)	5.3	4	76
Koyfman et al. (2010) [54]	GK	43/43	15 (9.6–24)	5.3	5.8	1 y: 85
Lorenzoni et al. (2009) [55]	GK	25/27	20 (15–24)	10.5	11.1	95
Samblas et al. (2009) [56]	LINAC	28/30	11.1 (5–20)	NA	16.8	NA
Kased et al. (2008) [57]	GK	42/44	16 (10–19.8)	NA	9	1 y: 77
Hussain et al. (2007) [58]	GK	22/22	16 (14–23)	NA	8.5	100
Yen et al. (2006) [59]	GK	53/53	18 (9–25)	9.8	16	81
Fuentes et al. (2006) [60]	GK	28/28	19.6 (11–30)	NA	12	92
Shuto et al. (2003) [61]	GK	25/31	13 (8–18)	5.2	4.9	77.4
Huang et al. (1999) [62]	GK	26/27	16 (12–20)	9.5	11	95

Abbreviations: SRS Stereotactic radiosurgery, *N* Number, *FU* Follow-up, *OS* Overall survival, *LC* Local control, *GK* Gamma Knife®, *LINAC* Linear accelerator, *NA* Not applicable

Figure 34.1a, b shows the CyberKnife® plan of a patient with a pontine metastasis treated in our department. We prescribed a total dose of 12 Gy SRS. The target lesion is delineated in pink and the brainstem is in cyan. In the upper left corners of Fig. 34.1a, b, the light blue rays are the active and the dark blue rays are the inactive beams during the treatment. The isodose lines are shown in transverse, coronal, and sagittal images in the right upper, left lower, and right lower corners, respectively. In Fig. 34.1b on the right, critical organ doses are shown. In this plan, the maximum dose to the brainstem is 13.79 Gy.

34.4 Image-Guided Radiosurgery and Hypofractionated Radiotherapy

Fractionation can be used for several reasons. It is well established that higher fraction doses lead to higher late toxicity rates due to decreased recovery from sublethal damage to critical structures. Fractionated regimens also allow re-oxygenation at the target tissue and increase radiosensitivity. The α/β ratios for normal brain and benign and malignant brain tumors are assumed to be 2, 4, and 10, respectively [25]. Brainstem tumors can be treated with a single

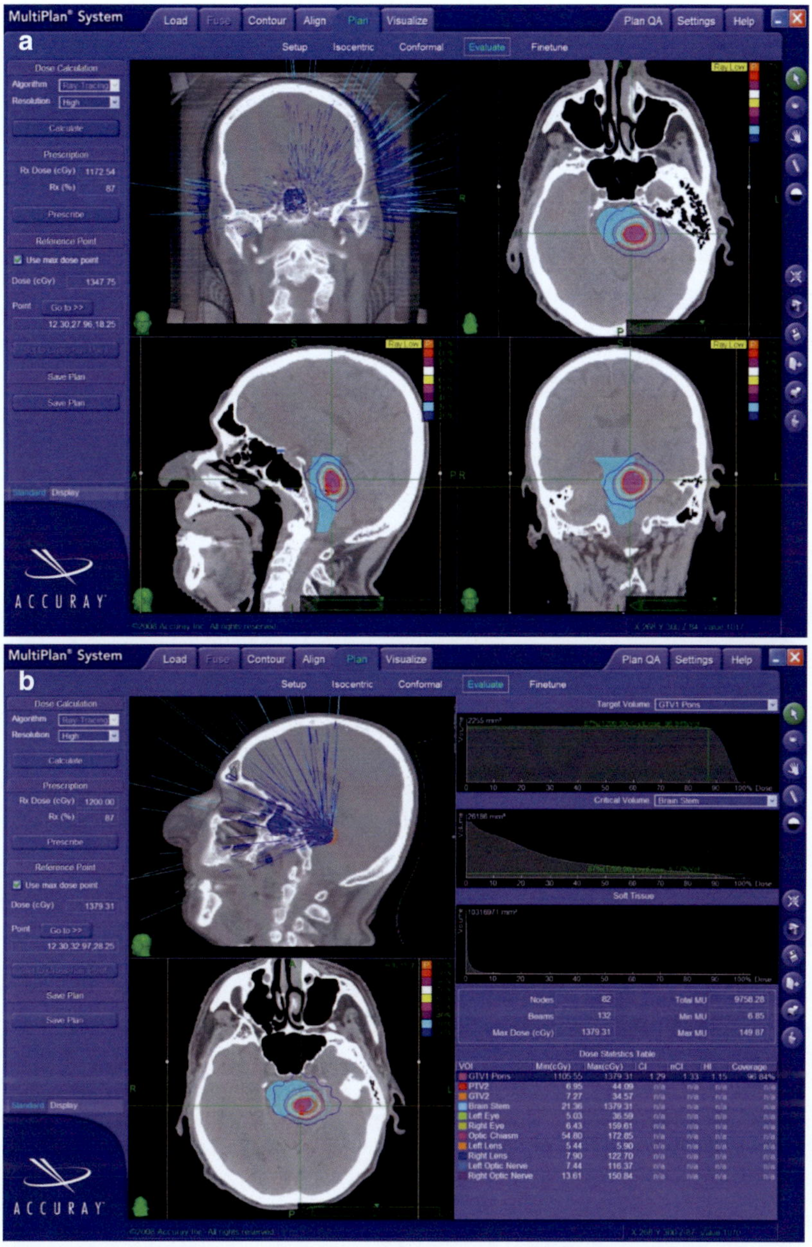

Fig. 34.1 (**a**) The CyberKnife® plan of a patient with a pontine metastasis prescribed with 12 Gy SRS. (**b**) The critical organ doses for the same patient

fraction of 15 Gy by SRS. To maintain the similar biologically effective dose (BED), i.e., 127.5 Gy, 2–5 fractions of 10.35 to 6.2 Gy are required. On the contrary, if fractionation is performed to maintain the same BED for the tumor, BED in the normal brain tissue can be significantly reduced. Therefore, when FSRT is used, BED can be increased by approximately 10.9% for benign tumors and 33.9% for malignant tumors compared to a single fraction [25].

SRS is the most preferred technique due to shorter overall treatment duration. On the other hand, limited studies on FSRT for benign and malignant brainstem tumors exist. Hall and Brenner [63] recommended 5–6 fractions of stereotactic RT instead of SRS in order to decrease the late complications in normal tissues and take a chance on hypoxic cells for re-oxygenation. Similarly, Fowler [64] claimed FRST should be preferred over SRS for the hypoxic cells to re-oxygenate and become radiosensitive. A Chinese study reported the results of 21 adult brainstem glioma patients treated with 14–33 Gy (median 26 Gy) CyberKnife® in 2–6 fractions (median 5 fractions) on consecutive days [65]. The isodose line of the planning target volume (PTV) ranged between 72% and 85% (median 80%). The BED ranged between 33.6 Gy and 76.56 Gy (median 59.8 Gy) when an α/β of 5 was calculated for tumor and 2 for brainstem. The maximum dose for brainstem was 31 Gy in five fractions. Six patients received concurrent temozolomide. The median OS was 19 months with a 1- and 2-year rate of 88% and 52%, respectively. Median progression-free survival (PFS) was 15 months with a 1- and 2-year rate of 69% and 41%, respectively. The authors stated that contrast enhancement on magnetic resonance imaging (MRI) is a poor prognostic factor for brainstem gliomas as demonstrated in previous studies [2, 66]. During a median of 54.5-month follow-up, three patients developed grade II (headache, $n = 1$; persistent dizziness, $n = 2$) and one patient developed grade III (vomiting and dizziness) radiation-related complications, which were controlled with oral medications. The median time to development of these symptoms was 3.4 months in this cycle.

Data on FSRT of brainstem metastases are limited. Leeman et al. [67] treated 38 tumors in 36 patients with a median 17 Gy (range 12–24 Gy) in 1–5 fractions. They reported the 6-month LC and OS rate of 93% and 27%, respectively. No serious treatment-related toxicity was observed. Nakamura et al. [68] reported the results of 20 patients with 26 brainstem metastases treated with 18–30 Gy in 3–5 fractions via CyberKnife®. After a median follow-up of 6.5 months, the 1-year OS, LC, and symptomatic control rate was 53%, 90%, and 76%, respectively. Liu et al. [69] reported the results of 54 patients with brainstem metastases treated with a median dose of 18 Gy in 1–4 fractions after a median follow-up of 5 months. They found the LC rate to be 80% and median OS to be 5 months. The most important finding in this study was a better outcome in patients with a better performance status.

In addition, stereotactic RT techniques give the opportunity to re-irradiate brainstem progressing tumors after the initial RT. Fontanilla et al. [70] initially demonstrated that selected patients with symptomatic or radiographic progression could safely undergo re-irradiation and experience symptomatic improvement. Although not specific to brainstem tumors, a recent review has justified the use of SRS for re-irradiation of progressive brain metastases after initial WBRT and initial SRS [71]. Noel et al. [72] recommended that the dose representing the 70% isodose level should be 14 Gy at most with SRS after initial WBRT because higher doses do not increase the LC rate but toxicity. In case of a progressive brain metastasis after initial SRS, Kim et al. [73] recommend to repeat SRS but not WBRT in order to avoid the neurocognitive toxicities caused by WBRT. A recent study also reported that a second cycle of SRS in three daily fractions is a feasible treatment option for selected patients with recurrent or progressive brain metastases [74]. Although SRS avoids some aspects of the severity of the neurocognitive toxicities of WBRT, it carries a significant risk of radiation necrosis. In 2008, Meyer et al. [75] reviewed the clinical brain re-irradiation held between 1996 and 2006 and reported that the cumulative normal tissue dose (NTD) in conventional re-irradiation series was lower than in the FSRT or SRS series. The reported time interval between the initial and repeat RT was not correlated with the incidence of radionecrosis. However, radiation-induced normal brain tissue necrosis was found to occur at cumulative NTD >100 Gy.

Figure 34.2a, b shows the CyberKnife® plan and critical organ doses of a patient with a metastasis in the pons treated in our department. She was prescribed 2 × 9 Gy because of the large volume of the lesion. The target lesion is delineated in pink and the brainstem is in cyan. In the upper left corners of Fig. 34.2a, b, the light blue rays are the active and the dark blue rays are the inactive beams during the treatment. The isodose lines are

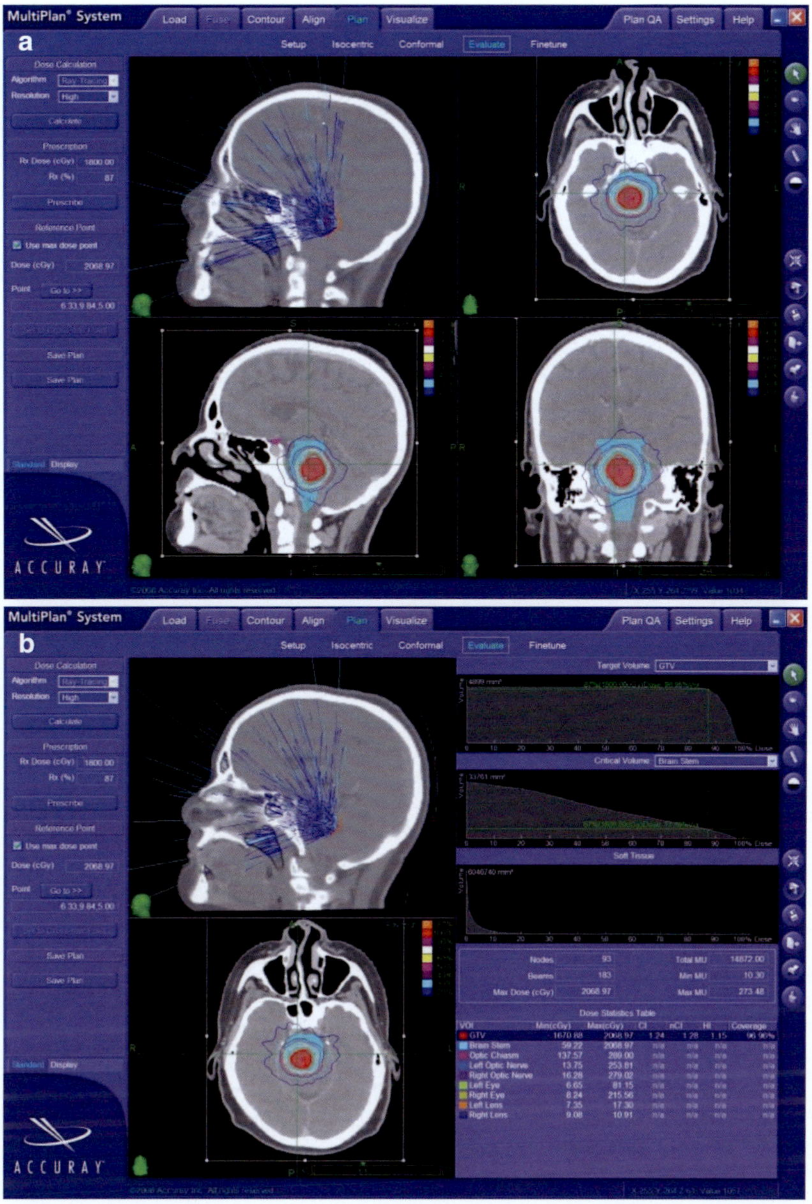

Fig. 34.2 (**a**) The CyberKnife® plan of the first FSRT of a patient with a pontine metastasis prescribed with 2 × 9 Gy. (**b**) The critical organ doses for the same plan

shown in transverse, coronal, and sagittal images in the right upper, left lower, and right lower corners, respectively, in Fig. 34.2a. In Fig. 34.2b, on the right, critical organ doses are shown. Maximum dose to the brainstem is 20 Gy in two fractions. After this first FSRT in May 2019, she underwent re-irradiation with 3 × 6 Gy FSRT to the progressed lesion in December 2019. The CyberKnife® plan and critical organ doses of the re-irradiation are shown in Fig. 34.3a, b. The maximum dose to the brainstem in re-irradiation was 20 Gy in three fractions.

34.5 Toxicity

Although SRS is an effective modality in the treatment of brainstem lesions, radiation-induced toxicity remains a critical concern. Radiation-related symptoms include headache, nausea and/

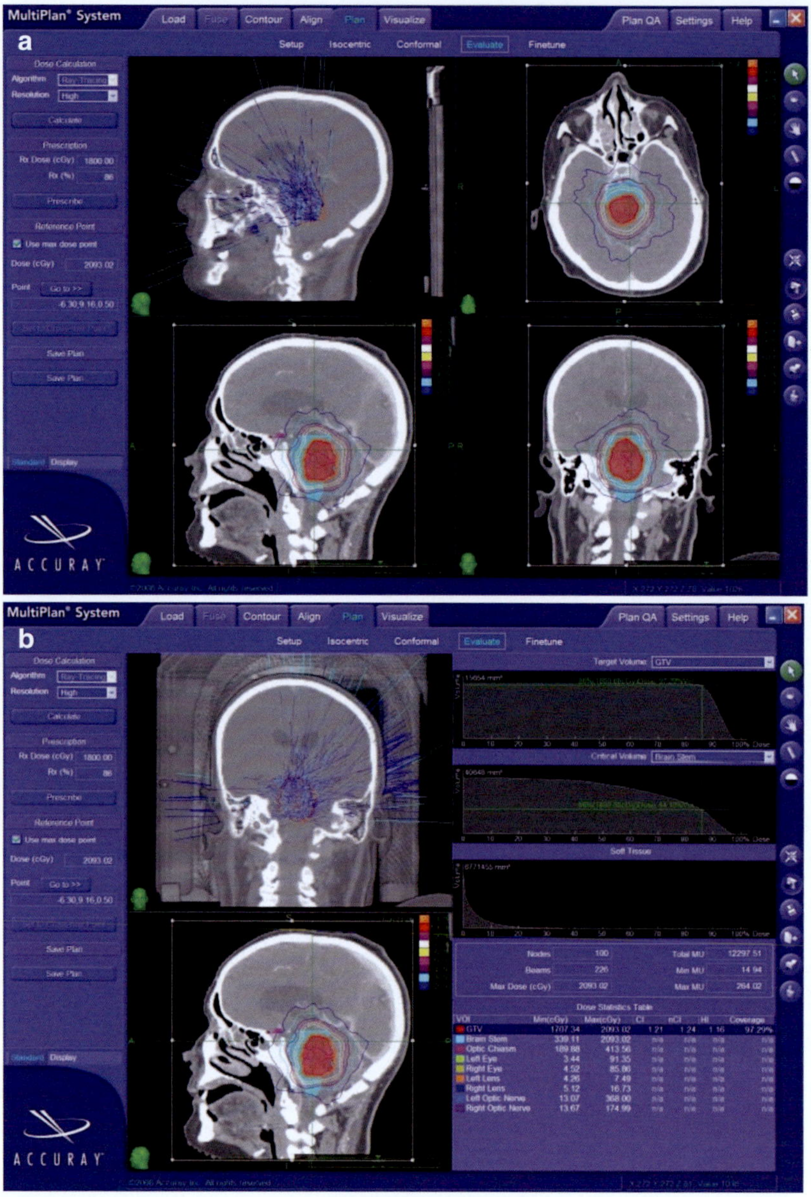

Fig. 34.3 (**a**) Re-irradiation of the same patient with 3 × 6 Gy 7 months after the first FSRT. (**b**) Critical organ doses for the re-irradiation plan

or vomiting, ataxia, and tinnitus developing due to an increased edema during SRS. The cranial neuropathies associated with SRS were also reported as a result of the dose received by the brainstem [76]. Although most toxicities are mild, severe to life-threatening toxicities have also been reported. In the Radiation Therapy Oncology Group's (RTOG) Protocol 0539 for intermediate- and high-risk meningiomas, a maximum dose of 60 Gy in 30 fractions was used for the brainstem during conventional RT to a volume of >0.03 cm^3 [77, 78].

On the other hand, the QUANTEC report [79] recommended a maximum dose of 64 Gy for conventional fractionated RT, and in the same report, a maximum point dose to the brainstem <12.5 Gy in a single fraction was, however, recommended.

Patel et al. [80] reviewed 29 retrospective studies on SRS of brainstem metastases and observed the rates of grade III or higher toxicity between 0% and 9.5%.

Median time leading to toxicity after SRS was 3 months, and 90% of toxicities occurred before 9 months and, as previously reported, 4.5 months for lesions in the cerebral parenchyma [81]. Although not clearly specified, the reason for the early onset of toxicity in the brainstem may be due to the lack of compressibility in the surrounding space for edema compared to the cerebral hemispheres. In the aforementioned review, median prescription dose in patients that encountered toxicity was 16 Gy, and only one third of them received >18 Gy [80]. Although an increasing risk of toxicity has been reported with higher doses, it can nonetheless develop at a wide range of doses [40]. A lower risk of toxicity development was reported for lesions located in the medulla oblongata (0.8%) compared to the lesions in the midbrain (2.8%) and pons (3%), but the authors associated the higher prevalence of toxicity in pontine lesions with the frequency of occurrence of brainstem metastasis in the pons [80].

In a study on SRS for vestibular schwannomas, a 3% complication rate [76] was reported in a single fraction of 14.2 Gy to the brainstem.

Xue et al. [82] claimed that only a very small volume of the brainstem can tolerate an extremely high dose without producing a severe clinical injury. Davidson et al. [83] retrospectively evaluated 107 patients with 114 lesions in the brainstem treated with Gamma Knife® and reported a 12% rate of delayed toxicity with a median latency of 6 months after a median dose of 16 Gy. They concluded that larger tumor volume and larger treatment volume significantly increased the incidence of delayed toxicity.

In animal studies it has been demonstrated that a shorter spinal cord length receiving the dose corresponds to a higher tolerance dose for a 50% complication rate [84].

In addition, a median dose of 20 Gy to the extreme small volumes of the human spinal cord in a single fraction via CyberKnife® was reported to be safe [85]. Considering the similar architecture of the spinal cord and brainstem and the higher radiation tolerance of the brainstem compared to the spinal cord, these data can also be valid for the brainstem.

The AAPM report 101 [86] recommends the threshold dose for <0.5 cm^3 of the brainstem (excluding the medulla) of 10 Gy, 18 Gy, and 23 Gy in one, three, and five fractions of stereotactic RT, respectively. The respective values for the maximum point doses are 15 Gy, 23.1 Gy, and 31 Gy. Timmerman [87], on the other hand, recommended the exact same doses for 1 cm^3 of the brainstem, except for 26 Gy as a threshold dose in five fractions.

34.6 Conclusion

SRS and FSRT are reasonable options for brainstem tumors that are generally not suitable for complete surgical resection. These modalities can apply a more homogeneous dose to the target volume inside the brainstem while preserving the normal brainstem tissues better compared to conventional techniques. However, as the brainstem itself is an extremely critical organ, extra caution should be given to the maximum and mean doses. SRS is the treatment of choice for its shorter total treatment time, but if the tolerance doses are not adequate, as in the case of large tumors, FSRT can be more feasible.

Practical Guide

Fractionation schemes (depending on the tolerance dose of the brainstem and optic apparatus)		10–25 Gy in a single fraction 18–30 Gy in 3–5 fractions
Brainstem tolerance dose (for 0.5–1 cm^3)	Threshold dose (Gy)	Maximum dose (Gy)
1 fraction	10	15
3 fractions	18	23
5 fractions	23–26	31

References

1. Pollack IF, Hoffman HJ, Humphreys RP, Becker L. The long-term outcome after surgical treatment of dorsally exophytic brain-stem gliomas. J Neurosurg. 1993;78:859–63.

2. Purohit B, Kamli AA, Kollias SS. Imaging of adult brainstem gliomas. Eur J Radiol. 2015;84:709–20.
3. Freeman CR, Krischer J, Sanford RA, et al. Hyperfractionated radiotherapy in brain stem tumors: results of a Pediatric Oncology Group study. Int J Radiat Oncol Biol Phys. 1988;15:311–8.
4. Freeman CR, Krischer J, Sanford RA, et al. Hyperfractionated radiation therapy in brain stem tumors. Results of treatment at the 7020 cGy dose level of Pediatric Oncology Group study #8495. Cancer. 1991;68:474–81.
5. Freeman CR, Krischer JP, Sanford RA, et al. Final results of a study of escalating doses of hyperfractionated radiotherapy in brain stem tumors in children: a Pediatric Oncology Group study. Int J Radiat Oncol Biol Phys. 1993;27:197–206.
6. Packer RJ, Boyett JM, Zimmerman RA, et al. Outcome of children with brain stem gliomas after treatment with 7800 cGy of hyperfractionated radiotherapy. A Childrens Cancer Group Phase I/II Trial. Cancer. 1994;74:1827–34.
7. Mandell LR, Kadota R, Freeman C, et al. There is no role for hyperfractionated radiotherapy in the management of children with newly diagnosed diffuse intrinsic brainstem tumors: results of a Pediatric Oncology Group phase III trial comparing conventional vs. hyperfractionated radiotherapy. Int J Radiat Oncol Biol Phys. 1999;43:959–64.
8. Marcus KJ, Dutton SC, Barnes P, et al. A phase I trial of etanidazole and hyperfractionated radiotherapy in children with diffuse brainstem glioma. Int J Radiat Oncol Biol Phys. 2003;55:1182–5.
9. Janssens GO, Jansen MH, Lauwers SJ, et al. Hypofractionation vs conventional radiation therapy for newly diagnosed diffuse intrinsic pontine glioma: a matched-cohort analysis. Int J Radiat Oncol Biol Phys. 2013;85:315–20.
10. Ramos A, Hilario A, Lagares A, et al. Brainstem gliomas. Semin Ultrasound CT MR. 2013;34:104–12.
11. Reyes-Botero G, Mokhtari K, Martin-Duverneuil N, et al. Adult brainstem gliomas. Oncologist. 2012;17:388–97.
12. Guillamo JS, Monjour A, Taillandier L, et al. Brainstem gliomas in adults: prognostic factors and classification. Brain. 2001;124:2528–39.
13. Salmaggi A, Fariselli L, Milanesi I, et al. Natural history and management of brainstem gliomas in adults. A retrospective Italian study. J Neurol. 2008;255:171–7.
14. Delattre JY, Krol G, Thaler HT, Posner JB. Distribution of brain metastases. Arch Neurol. 1988;45:741–4.
15. Matsumoto K, Tada E, Tamesa N, et al. Stereotactic brachytherapy for a cystic metastatic brain tumor in the midbrain. Case report. J Neurosurg. 1998;88:141–4.
16. Churilla TM, Chowdhury IH, Handorf E, et al. Comparison of local control of brain metastases with stereotactic radiosurgery vs surgical resection: a secondary analysis of a randomized clinical trial. JAMA Oncol. 2019;5:243–7.
17. Shin SM, Vatner RE, Tam M, et al. Resection followed by involved-field fractionated radiotherapy in the management of single brain metastasis. Front Oncol. 2015;5:206.
18. Brown PD, Ballman KV, Cerhan JH, et al. Postoperative stereotactic radiosurgery compared with whole brain radiotherapy for resected metastatic brain disease (NCCTG N107C/CEC.3): a multicentre, randomised, controlled, phase 3 trial. Lancet Oncol. 2017;18:1049–60.
19. Asher AL, Burri SH, Wiggins WF, et al. A new treatment paradigm: neoadjuvant radiosurgery before surgical resection of brain metastases with analysis of local tumor recurrence. Int J Radiat Oncol Biol Phys. 2014;88:899–906.
20. Atalar B, Modlin LA, Choi CY, et al. Risk of leptomeningeal disease in patients treated with stereotactic radiosurgery targeting the postoperative resection cavity for brain metastases. Int J Radiat Oncol Biol Phys. 2013;87:713–8.
21. Prabhu RS, Press RH, Patel KR, et al. Single-fraction stereotactic radiosurgery (SRS) alone versus surgical resection and SRS for large brain metastases: a multi-institutional analysis. Int J Radiat Oncol Biol Phys. 2017;99:459–67.
22. Szeifert GT, Timperley WR, Forster DMC, Kemeny AA. Histopathological changes in cerebral arteriovenous malformations following Gamma Knife radiosurgery. Prog Neurol Surg. 2007;20:212–9.
23. Fowler JF, Welsh JS, Howard SP. Loss of biological effect in prolonged fraction delivery. Int J Radiat Oncol Biol Phys. 2004;59:242–9.
24. Ling CC, Gerweck LE, Zaider M, Yorke E. Dose-rate effects in external beam radiotherapy redux. Radiother Oncol. 2010;95:261–8.
25. Kida Y, Mori Y. Stereotactic radiotherapy with fractionation for the lesions in and around the brainstem and optic nerve. Cureus. 2019;11:e6087.
26. Song CWPH, Griffin RJ, Levitt SH. Radiobiology of stereotactic radiosurgery and stereotactic body radiation therapy. In: Levitt SHPC, Poortmans P, editors. Technical basis of radiation therapy, practical clinical applications. Berlin, Heidelberg: Springer; 2012. p. 51–61.
27. Theeler BJ, Ellezam B, Melguizo-Gavilanes I, et al. Adult brainstem gliomas: correlation of clinical and molecular features. J Neurol Sci. 2015;353:92–7.
28. Boethius J, Ulfarsson E, Rahn T, Lippittz B. Gamma knife radiosurgery for pilocytic astrocytomas. J Neurosurg. 2002;97:677–80.
29. Hafez RF. Stereotaxic gamma knife surgery in treatment of critically located pilocytic astrocytoma: preliminary result. World J Surg Oncol. 2007;5:39.
30. Kano H, Niranjan A, Kondziolka D, et al. Stereotactic radiosurgery for pilocytic astrocytomas part 2: outcomes in pediatric patients. J Neuro-Oncol. 2009;95:219–29.
31. Kida Y, Kobayashi T, Mori Y. Gamma knife radiosurgery for low-grade astrocytomas: results of long-term follow up. J Neurosurg. 2000;93(Suppl 3):42–6.

32. Kihlstrom L, Lindquist C, Lindquist M, Karlsson B. Stereotactic radiosurgery for tectal low-grade gliomas. Acta Neurochir Suppl. 1994;62:55–7.
33. Liu JS, Foo D, Yeo TT, et al. Twenty-three years follow-up after low-dose gamma knife surgery of a brainstem juvenile pilocytic astrocytoma: a case report and review of the literature. Childs Nerv Syst. 2019;35:1227–30.
34. Phuong PC, Hung NQ, Ngoc TB, et al. Rotating gamma system irradiation: a promising treatment for low-grade brainstem gliomas. In Vivo. 2017;31:957–60.
35. Fuchs I, Kreil W, Sutter B, et al. Gamma Knife radiosurgery of brainstem gliomas. Acta Neurochir Suppl. 2002;84:85–90.
36. Winograd E, Rivers CI, Fenstermaker R, et al. The case for radiosurgery for brainstem metastases. J Neuro-Oncol. 2019;143:585–95.
37. Patel A, Mohammadi H, Dong T, et al. Brainstem metastases treated with Gamma Knife stereotactic radiosurgery: the Indiana University Health experience. CNS Oncol. 2018;7:15–23.
38. Murray L, Menard C, Zadeh G, et al. Radiosurgery for brainstem metastases with and without whole brain radiotherapy: clinical series and literature review. J Radiat Oncol. 2017;6:21–30.
39. Joshi R, Johnson MD, Maitz A, et al. Utility of graded prognostic assessment in evaluation of patients with brainstem metastases treated with radiosurgery. Clin Neurol Neurosurg. 2016;147:30–3.
40. Trifiletti DM, Lee CC, Kano H, et al. Stereotactic radiosurgery for brainstem metastases: an international cooperative study to define response and toxicity. Int J Radiat Oncol Biol Phys. 2016;96:280–8.
41. Voong KR, Farnia B, Wang Q, et al. Gamma knife stereotactic radiosurgery in the treatment of brainstem metastases: the MD Anderson experience. Neurooncol Pract. 2015;2:40–7.
42. Trifiletti DM, Lee CC, Winardi W, et al. Brainstem metastases treated with stereotactic radiosurgery: safety, efficacy, and dose response. J Neuro-Oncol. 2015;125:385–92.
43. Kilburn JM, Ellis TL, Lovato JF, et al. Local control and toxicity outcomes in brainstem metastases treated with single fraction radiosurgery: is there a volume threshold for toxicity? J Neuro-Oncol. 2014;117:167–74.
44. Peterson HE, Larson EW, Fairbanks RK, et al. Gamma knife treatment of brainstem metastases. Int J Mol Sci. 2014;15:9748–61.
45. Jung EW, Rakowski JT, Delly F, et al. Gamma knife radiosurgery in the management of brainstem metastases. Clin Neurol Neurosurg. 2013;115:2023–8.
46. Sengoz M, Kabalay IA, Tezcanli E, et al. Treatment of brainstem metastases with gamma-knife radiosurgery. J Neuro-Oncol. 2013;113:33–8.
47. Kawabe T, Yamamoto M, Sato Y, et al. Gamma knife surgery for patients with brainstem metastases. J Neurosurg. 2012;117(Suppl):23–30.
48. Li Y, Xu D, Zhang Z, et al. Gamma Knife surgery for brainstem metastases. J Neurosurg. 2012;117(Suppl):13–6.
49. Lin CS, Selch MT, Lee SP, et al. Accelerator-based stereotactic radiosurgery for brainstem metastases. Neurosurgery. 2012;70:953–8; discussion 958.
50. Yoo TW, Park ES, Kwon DH, Kim CJ. Gamma knife radiosurgery for brainstem metastasis. J Korean Neurosurg Soc. 2011;50:299–303.
51. Kelly PJ, Lin YB, Yu AY, et al. Linear accelerator-based stereotactic radiosurgery for brainstem metastases: the Dana-Farber/Brigham and Women's Cancer center experience. J Neuro-Oncol. 2011;104:553–7.
52. Valery CA, Boskos C, Boisserie G, et al. Minimized doses for linear accelerator radiosurgery of brainstem metastasis. Int J Radiat Oncol Biol Phys. 2011;80:362–8.
53. Hatiboglu MA, Chang EL, Suki D, et al. Outcomes and prognostic factors for patients with brainstem metastases undergoing stereotactic radiosurgery. Neurosurgery. 2011;69:796–806.. discussion 806
54. Koyfman SA, Tendulkar RD, Chao ST, et al. Stereotactic radiosurgery for single brainstem metastases: the Cleveland clinic experience. Int J Radiat Oncol Biol Phys. 2010;78:409–14.
55. Lorenzoni JG, Devriendt D, Massager N, et al. Brain stem metastases treated with radiosurgery: prognostic factors of survival and life expectancy estimation. Surg Neurol. 2009;71:188–95; discussion 195, 195–186.
56. Samblas JM, Sallabanda K, Bustos JC, et al. Radiosurgery and whole brain therapy in the treatment of brainstem metastases. Clin Transl Oncol. 2009;11:677–80.
57. Kased N, Huang K, Nakamura JL, et al. Gamma knife radiosurgery for brainstem metastases: the UCSF experience. J Neuro-Oncol. 2008;86:195–205.
58. Hussain A, Brown PD, Stafford SL, Pollock BE. Stereotactic radiosurgery for brainstem metastases: survival, tumor control, and patient outcomes. Int J Radiat Oncol Biol Phys. 2007;67:521–4.
59. Yen CP, Sheehan J, Patterson G, Steiner L. Gamma knife surgery for metastatic brainstem tumors. J Neurosurg. 2006;105:213–9.
60. Fuentes S, Delsanti C, Metellus P, et al. Brainstem metastases: management using gamma knife radiosurgery. Neurosurgery. 2006;58:37–42; discussion 37–42.
61. Shuto T, Fujino H, Asada H, et al. Gamma knife radiosurgery for metastatic tumours in the brain stem. Acta Neurochir. 2003;145:755–60.
62. Huang CF, Kondziolka D, Flickinger JC, Lunsford LD. Stereotactic radiosurgery for brainstem metastases. J Neurosurg. 1999;91:563–8.
63. Hall EJ, Brenner DJ. The radiobiology of radiosurgery: rationale for different treatment regimes for AVMs and malignancies. Int J Radiat Oncol Biol Phys. 1993;25:381–5.
64. Fowler JF. An arguing point? ESTRO. 2007;64:15.

65. Zhang J, Liu Q, Yuan Z, et al. Clinical efficacy of CyberKnife radiosurgery for adult brainstem glioma: 10 years experience at Tianjin CyberKnife center and review of the literature. Front Oncol. 2019;9:257.
66. Yin L, Zhang L. Correlation between MRI findings and histological diagnosis of brainstem glioma. Can J Neurol Sci. 2013;40:348–54.
67. Leeman JE, Clump DA, Wegner RE, et al. Prescription dose and fractionation predict improved survival after stereotactic radiotherapy for brainstem metastases. Radiat Oncol. 2012;7:107.
68. Nakamura M, Nishimura H, Mayahara H, et al. Investigation of the efficacy and safety of CyberKnife hypofractionated stereotactic radiotherapy for brainstem metastases using a new evaluation criterion: 'symptomatic control'. J Radiat Res. 2017;58:834–9.
69. Liu SH, Murovic J, Wallach J, et al. CyberKnife radiosurgery for brainstem metastases: management and outcomes and a review of the literature. J Clin Neurosci. 2016;25:105–10.
70. Fontanilla HP, Pinnix CC, Ketonen LM, et al. Palliative reirradiation for progressive diffuse intrinsic pontine glioma. Am J Clin Oncol. 2012;35:51–7.
71. Chidambaram S, Pannullo SC, Schwartz TH, Wernicke AG. Reirradiation of recurrent brain metastases: where do we stand? World Neurosurg. 2019;125:156–63.
72. Noel G, Proudhom MA, Valery CA, et al. Radiosurgery for re-irradiation of brain metastasis: results in 54 patients. Radiother Oncol. 2001;60:61–7.
73. Kim DH, Schultheiss TE, Radany EH, et al. Clinical outcomes of patients treated with a second course of stereotactic radiosurgery for locally or regionally recurrent brain metastases after prior stereotactic radiosurgery. J Neuro-Oncol. 2013;115:37–43.
74. Minniti G, Scaringi C, Paolini S, et al. Repeated stereotactic radiosurgery for patients with progressive brain metastases. J Neuro-Oncol. 2016;126:91–7.
75. Mayer R, Sminia P. Reirradiation tolerance of the human brain. Int J Radiat Oncol Biol Phys. 2008;70:1350–60.
76. Meeks SL, Buatti JM, Foote KD, et al. Calculation of cranial nerve complication probability for acoustic neuroma radiosurgery. Int J Radiat Oncol Biol Phys. 2000;47:597–602.
77. Rogers CL, Won M, Vogelbaum MA, et al. High-risk meningioma: initial outcomes from NRG oncology/RTOG 0539. Int J Radiat Oncol Biol Phys. 2020;106(4):790–9.
78. Rogers L, Zhang P, Vogelbaum MA, et al. Intermediate-risk meningioma: initial outcomes from NRG oncology RTOG 0539. J Neurosurg. 2018;129:35–47.
79. Mayo C, Yorke E, Merchant TE. Radiation associated brainstem injury. Int J Radiat Oncol Biol Phys. 2010;76:S36–41.
80. Patel A, Dong T, Ansari S, et al. Toxicity of radiosurgery for brainstem metastases. World Neurosurg. 2018;119:e757–64.
81. Shaw E, Scott C, Souhami L, et al. Single dose radiosurgical treatment of recurrent previously irradiated primary brain tumors and brain metastases: final report of RTOG protocol 90-05. Int J Radiat Oncol Biol Phys. 2000;47:291–8.
82. Xue J, Goldman HW, Grimm J, et al. Dose-volume effects on brainstem dose tolerance in radiosurgery. J Neurosurg. 2012;117(Suppl):189–96.
83. Davidson L, Zada G, Yu C, et al. Delayed toxicity from gamma knife radiosurgery to lesions in and adjacent to the brainstem. J Clin Neurosci. 2009;16:1139–47.
84. Bijl HP, van Luijk P, Coppes RP, et al. Dose-volume effects in the rat cervical spinal cord after proton irradiation. Int J Radiat Oncol Biol Phys. 2002;52:205–11.
85. Daly ME, Choi CY, Gibbs IC, et al. Tolerance of the spinal cord to stereotactic radiosurgery: insights from hemangioblastomas. Int J Radiat Oncol Biol Phys. 2011;80:213–20.
86. Benedict SH, Yenice KM, Followill D, et al. Stereotactic body radiation therapy: the report of AAPM Task Group 101. Med Phys. 2010;37:4078–101.
87. Timmerman RD. An overview of hypofractionation and introduction to this issue of seminars in radiation oncology. Semin Radiat Oncol. 2008;18:215–22.

Uveal Melanoma

35

David Roberge

35.1 Introduction

Ocular melanoma is a rare disease. The incidence in Canada is approximately 0.7 per 100,000 population per year. This represents approximately 385 cases per year in Canada, and 80 cases in the province of Quebec—many of which are referred to our institution [1]. The reported incidence in the United States and Europe is similar, but it is significantly lower in non-Caucasian populations, notably in Africa and Asia [2].

The treatment of choroidal melanoma is controversial because of the paucity of high-level evidence and the visual toxicity of available treatments [3]. Small tumors can be clinically indistinguishable from large nevi. Thus, for smaller tumors the question arises as to whether the patient's eyesight should be risked in the management of a lesion that might not have the potential for progression. Conversely, should a suspicious lesion be observed, potentially exposing the patient to a higher risk of metastatic disease? For medium-sized tumors, the main controversy is the choice of a treatment modality in the absence of convincing prospective comparisons. For large tumors in which visual outcomes are poor and distant metastases frequent, one can debate which patients should have eye preservation therapy and which should proceed with primary enucleation. For patients presenting with metastatic tumors and paucisymptomatic eyes, it remains undefined which patients benefit from immediate local therapy in the face of generally ineffective systemic therapy and limited life expectancy.

The Collaborative Ocular Melanoma Study (COMS) confirmed in a randomized trial that, for medium-sized tumors, organ preservation can be attempted without significantly impacting overall survival. Nonetheless, enucleation remains an option for patients with limited visual potential for whom follow-up would be burdensome. Although the COMS trial was limited to I-125 brachytherapy, its conclusions are assumed to apply to other local therapies known to produce similar levels of local tumor control. The main modalities are transpupillary thermotherapy, endoresection, plaque brachytherapy [4], proton beam therapy [5], and stereotactic radiation [6]. Treatment selection is based on tumor size, tumor location, patient preference, treatment availability, and, in large part, the treating physician's opinion. In our practice, we choose to offer plaque brachytherapy (iodine or ruthenium, depending on tumor thickness) to most patients with small- or medium-sized tumors but prefer

D. Roberge (✉)
Department of Radiation Oncology, Centre Hospitalier de L'Université de Montréal (CHUM), Montreal, Canada

Department of Radiology, Radiation Oncology and Nuclear Medicine, University of Montreal, Montreal, Canada
e-mail: david.roberge.chum@ssss.gouv.qc.ca

© Springer Nature Switzerland AG 2020
A. Conti et al. (eds.), *CyberKnife NeuroRadiosurgery*,
https://doi.org/10.1007/978-3-030-50668-1_35

stereotactic radiation for juxtapapillary tumors. In this context, the CyberKnife robotic radiosurgery platform allows delivery of a high dose of radiation to the small target volume within the eye with a steep dose gradient [7]. Similarly to other radiosurgery platforms, the CyberKnife can accurately target structures having a rigid relationship to the skull. However it does not intrinsically provide a means to track eye movement. Our practice is to achieve eye immobilization using an in-house system of which we have characterized the reproducibility [8]. Our experience with this system over the past years informs sections of this manuscript.

35.2 Staging

Local spread of uveal melanoma to other organs is rare, as is lymph node involvement. Most staging efforts center on the size of the primary tumor and its involvement of substructures of the eye. Until benefit is shown to adjuvant systemic therapy, the main purpose of local staging is to offer prognostic information and select a local management strategy. The tumor is principally characterized by ultrasound but MRI can contribute to the finding of extrascleral extension [9]. The eighth Edition of the American Joint Committee on Cancer proposes a detailed tumor staging system for ciliary body and choroid uveal melanoma. This system is based on the thickness of the tumor, the maximal basal diameter, involvement of the ciliary body, and extraocular extension. For example, a tumor of 12 mm or less in diameter and 3 mm or less in height as well as those of 9 mm or less in diameter and 6 mm or less in height are T1 tumors—T1a if there is no ciliary involvement or extraocular extension. At the other end of the spectrum, a T4 tumor is any tumor with more than 5 mm of extraocular extension, any tumor of more than 18 mm of basal diameter, more than 15 mm of height, or a tumor combining a diameter of more than 15 mm with a height of more than 12 mm. The complexity and relative clinical uselessness of this staging system might explain why the simpler COMS staging system remains in use. In this revised 3-tiered system, tumors are divided into small, medium, and large tumors [10].

- COMS Small: Diameter 5–16 mm and height 1–2.5 mm
- COMS Medium: Diameter >16 mm and height ≤2 mm or diameter <16 mm and height 2.6–10 mm
- COMS Large: Diameter >16 mm and height >2 mm or height >10 mm

When uveal melanoma spreads beyond the orbit, the disease will have a strong predilection for the liver. Median survival for patients with metastatic disease is less than 1 year and the evidence that any liver-directed therapy or systemic cancer therapy prolongs survival or improves quality of life is weak [11, 12]. Although approximately 10%, 25%, and 50% of patients with small, medium and large tumors will eventually develop metastatic disease, the yield of liver imaging at the time of diagnosis is low. Ultrasound, CT scan, and MRI imaging are reasonable means of screening for liver metastases (there is no good evidence of incremental benefit of total body FDG/PET imaging), but will all be more likely to yield incidental findings than true positive findings of metastatic disease [13].

35.3 Rationale for Radiation Therapy

Over a 12-year period from 1986 to 1998, 1317 patients with medium-sized uveal melanoma were randomized to enucleation or iodine-125 brachytherapy. These trial participants represented nearly half of eligible patients in the participating institutions. Mortality in both treatment arms was similar at 12 years with better visual outcomes in patients spared enucleation. In the brachytherapy arm, the 5-year actuarial risk of treatment failure was 10.3%, and the enucleation rate was 12.5% [14]. This high-level evidence has cemented radiotherapy as a standard treatment for medium-sized uveal melanoma. The body of evidence to support the treatment of smaller or larger tumors is mainly retrospective.

Based on reported outcomes of thousands of patients, the eye preservation rate is high and the local control favorable. Long-term preservation of visual acuity is disappointing—especially for larger tumors.

The evidence for non-radiation eye preservation therapies is limited. Transpupillary thermotherapy (TTT) may be a reasonable option for the smallest tumors [15]. However, when patients are poorly selected, the local control will be poor and the visual outcomes disappointing [16]. Endoresection is a more complex procedure which also aims to avoid late radiation toxicity (at the cost of more acute toxicity) for which the jury is still out [17].

35.4 Single-Fraction Radiosurgery

Radiosurgery devices had started to be used to treat small brain tumors with apparent satisfactory local control and limited toxicity when the first report of single-fraction radiosurgery for uveal melanoma was published in the late 1980s. At the time, radiosurgery programs used semi-invasive head immobilization, and fractionated treatments were rare and cumbersome [18]. The initial doses used were high by today's standard—60–90 Gy in a single fraction. As an illustration, in 1992–1993, the Sheffield group recruited 14 patients with uveal melanoma to be treated using the Leksell Gamma Knife. The typical treatment was 70 Gy in a single fraction using a spherical "shot." The entire procedure was accomplished in a single morning under retrobulbar anesthesia ("Akinesia of the globe was achieved by a standard retrobulbar injection of local anesthetic. Approximately 4 mL of a mixture lignocaine 2% and bupivacaine 0.5% was used followed by gentle massage of the globe for 5 min."). Although early tumor responses were seen, 13 of the 14 patients had serious adverse reactions leading the authors to the sober conclusion that "Several issues must be resolved before stereotactic radiosurgery can be accepted as a viable alternative method of treating intraocular tumors. An optimal dose which adequately treats the tumor with a minimum of side effects has yet to be established. Similarly, the value of fractionating the dose remains unknown." [19].

Over the ensuing decades, the single-fraction experience has grown, and the dose prescription refined to be closer to 20 Gy. Although outcomes are quite favorable in selected small series [20], the 5-year actuarial eye retention rate (73%) and local control (70.8%) in the largest series of 271 patients (treated with 18–22 Gy) may be lower than expected. As the experience remains limited and heterogeneous compared to the published series using 5 fractions, it is difficult to ascertain if the local control and enucleation rates are comparable. There is certainly a trend across radiation oncology away from ablative single-fraction treatments, whether it be in prostate cancer, lung cancer, or brain metastases. Irrespective of biological concerns, the margin required to ensure target coverage is inversely proportional to the number of fractions and make single-fraction treatments especially sensitive to various uncertainties. It remains that a single day procedure is convenient and while the jury remains out, single-fraction radiosurgery will continue to be appropriately used for selected patients in a small number of clinical programs.

35.5 Image-Guided Radiosurgery and Hypofractionated Radiotherapy

Although there is no prospective comparative data, the outcomes in a review of more than 10,000 patients treated with proton therapy at various institutions in North America and Europe compare favorably with I-125 plaque brachytherapy in terms of local control, eye preservation, and toxicity [21]. This is in keeping with a randomized study of 184 patients meeting broad eligibility criteria who were randomized to 70 CGE of helium ion therapy (5 fractions over 7–11 days) or 70 Gy to the tumor apex (changed mid-trial to 1 mm beyond the tumor apex) using I-125 plaque brachytherapy. In this trial, local control and enucleation-free survival were significantly improved in the particle therapy group. Although the difference was less marked when tumors <2 mm from the optic disc were excluded, it remained statistically significant [22]. These results must of course be interpolated with caution as helium ions are biologically different from photons/protons, and the brachytherapy treatments delivered in the trial were not in keeping with the standards of the COMS trial (85 Gy to a minimum of 6 mm from the plaque) or the recommendations of the American Brachytherapy Society (85 Gy to the tumor apex with a radial margin of 2–3 mm) [23].

As the dosimetry of stereotactic photon radiation does not appear meaningfully different from that of proton therapy, it is logical to transpose the proton experience to the more widely available and less costly stereotactic photon treatments [24, 25]. In this context, the best described treatment schedule is 50 Gy over 5 fractions. This dose is supported by a trial randomizing 188 patients with small- or medium-sized choroidal melanomas (<15 mm in diameter and <5 mm in height) to 50 CGE or 70 CGE of proton irradiation over the same 5 fractions. In this trial, the lower dose resulted in less visual field toxicity and similar oncological outcomes [26]. As photon plans tend to be less homogeneous than proton plans, the minimum dose to the tumor will anyways be closer to 60 Gy when the prescription to the PTV is 50 Gy.

Selected series of hypofractionated photon stereotactic radiation are presented in Table 35.1. No firm conclusions should be drawn as the series are relatively small and heterogeneous, but the outcomes are compatible with those reported for the same fractionation schemes delivered with proton beam irradiation.

35.6 Toxicity

Radiation treatment of uveal melanoma is highly focused and thus, the principal toxicity of radiosurgery will be from irradiation of the eye and globe. Other than enucleation as a consequence of painful glaucoma, the most common and most concerning toxicity will be the loss of visual acuity. This risk will depend on the location of the tumor within the eye and its size, but it is appropriate to expect less than half of patients with good pre-treatment visual acuity to maintain good visual acuity in the years following treatment. A greater risk will be seen in patients with large tumors and those with tumors encroaching on the macula or optic disc. Radiation retinopathy will be the most common related injury although patients may also suffer from cataract, secondary glaucoma, radiation maculopathy, optic neuropathy, retinal and vitreous hemorrhages, and retinal detachment. When enucleated eyes are examined, radiation injury is marked at the level of blood vessels which develop outpouchings, fusiform dilatation, and microaneurysms. Collateral circulation can be seen and vascular incompetence results in vascular leakage, edema, and retinal detachment. The capillary lumens also narrow leading to ischemia and infarction [30]. Elevated levels of vascular-

Table 35.1 Selected series of fractionated stereotactic radiotherapy

Authors	Year	*N*	Tumor size	Dose/fractions	Enucleation-free survival	Local control
Dunavoelgyi et al. [27]	2011	212	8% small 89% medium 3% large	50–70 Gy/5	79% 5-year 73% 10-year	96% 5-year 93% 10-year
Fernandes et al. [28]	2011	64	Median height 4.2 mm (range 1.5–11)	70 Gy/5	84% crude	95% crude (median f/u 37 months)
van den Bosch et al. [29]	2015	118	87% medium 13% large	50 Gy/5	84% crude (median f/u 4.7 years)	96% crude (subgroup with median f/u 32 months)

endothelial growth factor participate in mediating toxicity and anti-VEGF therapy may reduce vascular permeability and improve neovascular glaucoma, subretinal fluid, and retinal detachment. The benefit of *prophylactic* intravitreal anti-VEGF antibodies is controversial [31], but our institutional bias is in favor of early use of these drugs following radiotherapy of uveal melanoma.

Though retinal toxicity can be unavoidable, with optimization of the dose to the lachrymal gland and anterior chamber, the risk of xerophthalmia, keratitis, cataract, or loss of eyelashes can be minimized.

Beyond the orbit, there will be a low total body dose which likely incurs a small excess risk of malignancy. Although we have modeled the risk of extracranial second malignancies to exceed that of intracranial radiation-induced malignancies, it would not be possible to distinguish these tumors from the background risk of cancer. An aggressively behaving meningioma or an orbital sarcoma would be more readily attributed to radiotherapy [32]. The main factors to consider in the risk of radiation-induced malignancy would be patient age at the time of treatment, number of monitor units delivered and generation of the CyberKnife unit. As the absolute risk is modeled to be less than 1% at 20 years, it will typically not factor into the decision to treat.

Low total body doses will be more relevant in the case of pregnancy. The dose to the fetus can be in the order of magnitude of 0.5% of the prescription dose which is enough to be clinically relevant [33]. Pregnancy status should thus be ascertained in women of childbearing potential, and treatment of a pregnant woman should only proceed after considering the alternatives (ruthenium plaque brachytherapy, for example) and after implicating the radiation safety officer.

There are several reports of quality of life in patients irradiated for ocular melanoma. Few compare different radiation modalities, and the only high-level data is from the comparison of enucleation to iodine-125 brachytherapy [34, 35]. It is clear that, as with other patients diagnosed with cancer, the diagnosis of ocular melanoma is associated with a reduced quality of life. The main areas of impact relate to decreased vision, ocular discomfort, and mood disturbance. In the first years following treatment, patients benefit from a vision-preserving treatment but the benefit decreases after the second year as the incidence of radiation toxicity increases. Unsurprisingly, quality of life is better in those patients not suffering from secondary glaucoma. Depression is seen irrespective of treatment modality (although it may correlate with eyesight) and anxiety might be more likely to resolve with time in patients undergoing enucleation [36, 37].

35.7 Conclusion

Robotic stereotactic irradiation is one tool in the varied radiotherapy armamentarium for choroidal melanoma. It is a convenient and noninvasive treatment for which the results are in keeping with those of the more commonly described iodine plaque brachytherapy and proton beam therapy. Although prospective comparisons are not expected in the near future, continued accumulation and publication of retrospective evidence may further increase clinical adoption.

35.8 Practical Guide

Our treatment technique has previously been published [38]. The patient undergoes a 1.5 T planning MRI. Focused sequences are obtained with the patient fixing a dot within the coil (placed in the approximate position of the light used in our immobilization device). Three sequences are obtained, a thin slice T2 2D Turbo Spin Echo, a 3D 1 mm isotropic T2 series, and a gadolinium-enhanced 3D T1 isotropic series (Fig. 35.1). The patient is then immobilized supine in a thick (3.2 mm) thermoplastic mask with Kevlar reinforcement, a cutout for the eyes (in patients unable to see with the involved eye, immobilization is based on the seeing eye), and a wide base to support the camera system. The camera system is part of a custom immobilization device which provides a light for the patient

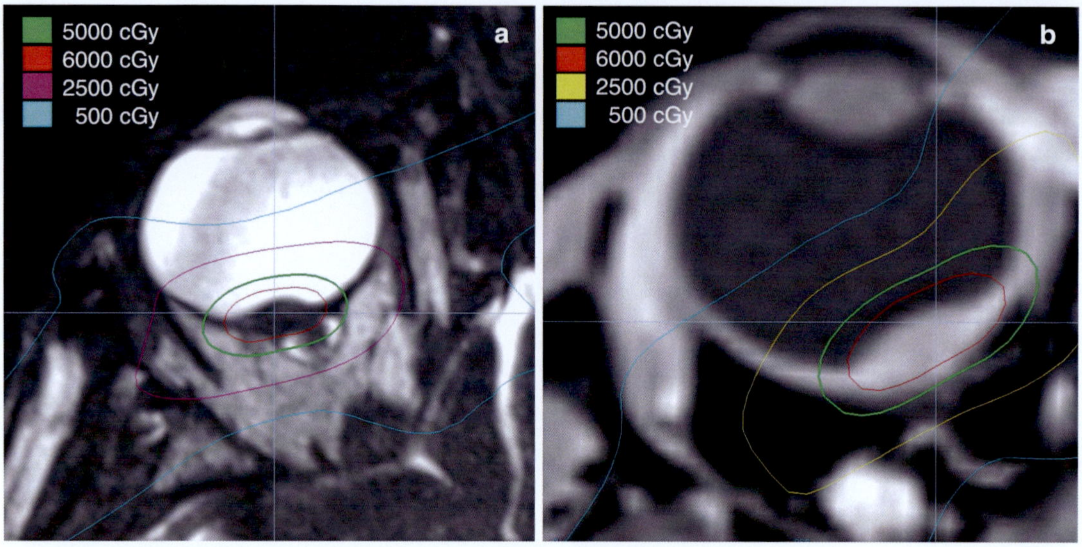

Fig. 35.1 Dosimetry for juxtapapillary choroidal melanoma treated in 5 fractions. (**a**) T2 MRI with 8.9 mm × 7.7 mm × 3.4 mm melanoma. (**b**) T1 contrast-enhanced MRI with 11.9 mm × 13.1 mm × 3.6 mm melanoma

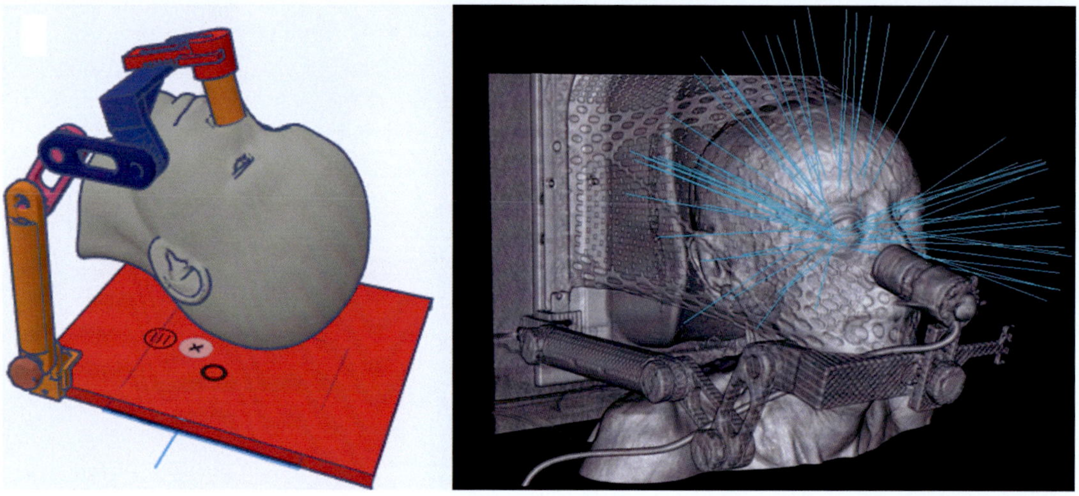

Fig. 35.2 3D-printed eye fixation device

to fix the position which can be recorded and reproduced (Fig. 35.2). The camera system allows for monitoring of patient compliance. The position of the iris is marked on a transparency overlaid on the screen linked to the monitoring camera. Simulation CT is acquired with 2 mm thick slices every 1 mm. The field of view is sufficient to visualize the entire immobilization device.

In the planning system, CT and MRI sequences are manually co-registered using the insertion and the optic nerve and lens as principal landmarks. The gross tumor volume (GTV) is segmented using both MRI sequences and fundus schema. The dimensions of the contoured volume are checked in relation to those measured on ocular ultrasound. A 2 mm planning target margin (PTV) is added, which is trimmed where it obviously extends beyond the sclera. Organs at risk contoured include ipsilateral lens, ipsilateral optic nerve, ipsilateral lachrymal gland, contralateral eye, immobilization device, and oral cavity. A shell structure is created 1.5 mm beyond the PTV. Collimator selection is a compromise between dose conformity and the treatment duration. The immobilization device is blocked with a 2–3 cm margin. In each case, the contralateral eye and oral cavity are either blocked or spared via strict optimization criteria. The plan is optimized so that the entire PTV is covered by at least 95% of the prescription dose and 99% is covered by 100% of the prescription isodose volume (typically this is 65–75% of the maximum dose). The conformity index (CI) is kept below 1.5, and the 25 Gy isodose volume is inspected for conformity. The lachrymal gland is optimized to a mean dose of less than 23.4 Gy. The entire contralateral eye (with a 1 cm margin) is kept below 2 Gy. When possible, the ipsilateral lens is kept under 2 Gy. Target coverage is prioritized over organs at risk, but a very steep gradient is created at the optic nerve in order to reduce the dose as much as possible without underdosing the PTV.

The dose is calculated considering tissue heterogeneity using a ray tracing algorithm. An independent monitor unit calculation is used to verify the plan. A dry run is performed with the mask, a head phantom and the immobilization device to identify potential collisions prior to the first fraction. The fractions are typically delivered every other day but can be delivered daily for patients in whom a 2-week treatment would be burdensome. During treatment, the position of the iris is monitored to be within the markings taken at simulation. Typically, the treatment is delivered in 1-min increments between which the patient can rest their eyes (Table 35.2).

Table 35.2 Planning constraints

Structure	Dose constraint
PTV	50 Gy in 5 fractions near minimum dose (99%)
GTV	100% >50 Gy (minimum dose typically >60 Gy)
Lachrymal gland	Mean dose <23.4 Gy
Ipsilateral lens	When possible, kept below 2 Gy
Optic nerve	A steep gradient is created near the nerve, but the nerve cannot be kept below 50 Gy for many juxtapapillary tumors
Contralateral eye	Maximum dose below 2 Gy
Oral cavity	Excluded from the beam path

References

1. New cases and age-standardized rate for primary cancer (based on the May 2015 CCR tabulation file), by cancer type and sex, Canada, provinces and territories. 2016 March 11. http://www5.statcan.gc.ca/cansim/a05.
2. Kaliki S, Shields CL. Uveal melanoma: relatively rare but deadly cancer. Eye (Lond). 2017;31(2):241–57.
3. Damato B. Does ocular treatment of uveal melanoma influence survival? Br J Cancer. 2010;103(3):285–90.
4. Lommatzsch PK, Lommatzsch R. Treatment of juxtapapillary melanomas. Br J Ophthalmol. 1991;75(12):715–7.
5. Seddon JM, et al. Uveal melanomas near the optic disc or fovea. Visual results after proton beam irradiation. Ophthalmology. 1987;94(4):354–61.
6. Emara K, et al. Stereotactic radiotherapy in the treatment of juxtapapillary choroidal melanoma: preliminary results. Int J Radiat Oncol Biol Phys. 2004;59(1):94–100.
7. Roberge D, et al. Treatment of medium-sized juxtapapillary melanoma with external Co-60 photon therapy. Radiother Oncol. 2005;74(1):71–3.
8. Iskanderani O, et al. Reproducibility of a noninvasive system for eye positioning and monitoring in stereotactic radiotherapy of ocular melanoma. Technol Cancer Res Treat. 2017;16(3):352–6.
9. Recsan Z, et al. MRI for the evaluation of scleral invasion and extrascleral extension of uveal melanomas. Clin Radiol. 2002;57(5):371–6.
10. Accuracy of diagnosis of choroidal melanomas in the Collaborative Ocular Melanoma Study. COMS report no. 1. Arch Ophthalmol. 1990;108(9):1268–73.

11. Khoja L, et al. Meta-analysis in metastatic uveal melanoma to determine progression-free and overall survival benchmarks: an international rare cancers initiative (IRCI) ocular melanoma study. Ann Oncol. 2019;30:1370.
12. Rantala ES, Hernberg M, Kivela TT. Overall survival after treatment for metastatic uveal melanoma: a systematic review and meta-analysis. Melanoma Res. 2019;29(6):561–8.
13. Strobel K, et al. Limited value of 18F-FDG PET/CT and S-100B tumour marker in the detection of liver metastases from uveal melanoma compared to liver metastases from cutaneous melanoma. Eur J Nucl Med Mol Imaging. 2009;36(11):1774–82.
14. Jampol LM, et al. The COMS randomized trial of iodine 125 brachytherapy for choroidal melanoma: IV. Local treatment failure and enucleation in the first 5 years after brachytherapy. COMS report no. 19. Ophthalmology. 2002;109(12):2197–206.
15. Shields CL, et al. Primary transpupillary thermotherapy for small choroidal melanoma in 256 consecutive cases: outcomes and limitations. Ophthalmology. 2002;109(2):225–34.
16. Harbour JW, et al. Transpupillary thermotherapy versus plaque radiotherapy for suspected choroidal melanomas. Ophthalmology. 2003;110(11):2207–14; discussion 2215.
17. Reichstein D, Karan K. Endoresection utilizing pars plana vitrectomy for benign and malignant intraocular tumors. Curr Opin Ophthalmol. 2019;30(3):151–8.
18. Roberge D, et al. Hypofractionated stereotactic radiotherapy for low grade glioma at McGill University: long-term follow-up. Technol Cancer Res Treat. 2006;5(1):1–8.
19. Rennie I, et al. The use of single fraction Leksell stereotactic radiosurgery in the treatment of uveal melanoma. Acta Ophthalmol Scand. 1996;74(6):558–62.
20. Sikuade MJ, et al. Outcomes of treatment with stereotactic radiosurgery or proton beam therapy for choroidal melanoma. Eye (Lond). 2015;29(9):1194–8.
21. Verma V, Mehta MP. Clinical outcomes of proton radiotherapy for uveal melanoma. Clin Oncol (R Coll Radiol). 2016;28(8):e17–27.
22. Mishra KK, et al. Long-term results of the UCSF-LBNL randomized trial: charged particle with helium ion versus Iodine-125 plaque therapy for choroidal and ciliary body melanoma. Int J Radiat Oncol Biol Phys. 2015;92(2):376–83.
23. Nag S, et al. The American Brachytherapy Society recommendations for brachytherapy of uveal melanomas. Int J Radiat Oncol Biol Phys. 2003;56(2):544–55.
24. Weber DC, et al. Proton beam radiotherapy versus fractionated stereotactic radiotherapy for uveal melanomas: a comparative study. Int J Radiat Oncol Biol Phys. 2005;63(2):373–84.
25. Daftari IK, et al. Newer radiation modalities for choroidal tumors. Int Ophthalmol Clin. 2006;46(1):69–79.
26. Gragoudas ES, et al. A randomized controlled trial of varying radiation doses in the treatment of choroidal melanoma. Arch Ophthalmol. 2000;118(6):773–8.
27. Dunavoelgyi R, et al. Local tumor control, visual acuity, and survival after hypofractionated stereotactic photon radiotherapy of choroidal melanoma in 212 patients treated between 1997 and 2007. Int J Radiat Oncol Biol Phys. 2011;81(1):199–205.
28. Fernandes BF, et al. Neovascular glaucoma after stereotactic radiotherapy for juxtapapillary choroidal melanoma: histopathologic and dosimetric findings. Int J Radiat Oncol Biol Phys. 2011;80(2):377–84.
29. van den Bosch T, et al. Risk factors associated with secondary enucleation after fractionated stereotactic radiotherapy in uveal melanoma. Acta Ophthalmol. 2015;93(6):555–60.
30. Groenewald C, Konstantinidis L, Damato B. Effects of radiotherapy on uveal melanomas and adjacent tissues. Eye (Lond). 2013;27(2):163–71.
31. Haji Mohd Yasin NA, et al. Choroidal melanoma treated with stereotactic fractionated radiotherapy and prophylactic intravitreal bevacizumab: the Dunedin Hospital experience. J Med Imaging Radiat Oncol. 2016;60(6):756–63.
32. Scaringi C, et al. Radiation-induced malignant meningioma following proton beam therapy for a choroidal melanoma. J Clin Neurosci. 2015;22(6):1036–7.
33. Pantelis E, et al. Radiation dose to the fetus during CyberKnife radiosurgery for a brain tumor in pregnancy. Phys Med. 2016;32(1):237–41.
34. Chabert S, Velikay-Parel M, Zehetmayer M. Influence of uveal melanoma therapy on patients' quality of life: a psychological study. Acta Ophthalmol Scand. 2004;82(1):25–31.
35. Miniati M, et al. Quality of life, depression, and anxiety in patients with uveal melanoma: a review. J Oncol. 2018;2018:5253109.
36. Moschos MM, et al. Depression in choroidal melanoma patients treated with proton beam radiotherapy. Anticancer Res. 2018;38(5):3055–61.
37. Melia M, et al. Quality of life after iodine 125 brachytherapy vs enucleation for choroidal melanoma: 5-year results from the collaborative ocular melanoma study: COMS QOLS report no. 3. Arch Ophthalmol. 2006;124(2):226–38.
38. Beliveau-Nadeau D, Callejo S, Roberge D. Technique for robotic stereotactic irradiation of choroidal melanoma. Cureus. 2016;8(4):e582.

Pediatric Radiosurgery

36

Valérie Bernier-Chastagner

36.1 Introduction

All the pathologies described in the previous chapters could theoretically concern the pediatric area. However, the frequency of indications is widely different from those of adults. Most of the publications available concern Gamma Knife, and the recent use of CyberKnife does not allow the retrieval of the same long-term follow-up data. Even if it is reasonable to assume that results are not different, publications on large series are still awaited.

Pediatric specificity is strongly correlated with the fact that the organs of young patients are growing at different rates according to age and organs. Thus, the patient's integrity and quality of life must be fully preserved for several decades in the case of treatment. More than ever, the radiotherapist's compromise between effectiveness and tolerance must be respected.

The advantages of radiosurgery are undeniably the strong dose gradients with rapid protection of organs at risk (OAR) near the target volume, due to the rapid fall-off of the dose. The intersecting multiple beams create an extremely compliant radiation therapy plan.

The first systems such as Gamma Knife used a stereotactic frame anchored to the skull. A general anesthesia is usually mandatory despite the high accuracy, because of the discomfort for young patients. Moreover, the thin skull does not permit the fixation of frame-based systems in infants and very young children [1].

The CyberKnife robotic radiosurgical system achieves precision and accuracy in competition with frame-based systems [2–4]. The advantage of a frameless system is a real asset for children, offering the possibility of a hypofractionated treatment and reducing the need for general anesthesia.

Hypofractionation was first tested in adults and then in children. Many publications have shown the feasibility of the CyberKnife technique [5–8]. But series are inhomogeneous, with few cases and various pathologies.

The disadvantages are the limitation to small target volumes, the high dose of radiation per fraction, more deleterious for growing tissue, the absence of sufficient follow-up to appreciate the very long-term (beyond 20-40 years) efficacy, and the toxicity profile. Concerning the risk of second cancers, Wolf et al. [9] reported the incidence of malignancies after stereotactic radiosurgery with Gamma Knife, with a 10-year risk of malignant intracranial tumor rate <1%, similar to the spontaneous risk in the general population. The low risk during the first decade increases thereafter, implying the need for a follow-up of at least 20 years.

V. Bernier-Chastagner (✉)
Department of Radiation Oncology, Institut de Cancérologie de Lorraine, 54500 Vandoeuvre les Nancy, France
e-mail: v.bernier@nancy.unicancer.fr; http://www.icl-lorraine.fr

© Springer Nature Switzerland AG 2020
A. Conti et al. (eds.), *CyberKnife NeuroRadiosurgery*, https://doi.org/10.1007/978-3-030-50668-1_36

36.2 Indications and Results

36.2.1 Pituitary Adenomas

Pituitary adenomas comprise about 3% of all intracranial tumors in childhood. The literature reported only Gamma Knife experience [10, 11] with a local control rate between 87% and 95% at 4 years and about 64% at 10 years. The isodose lines were usually 50%, with median margin single dose of 25 Gy (12.9–27.1 Gy). The maximum dose to the optic apparatus was kept below 8 Gy. The predictive factors for endocrine remission were age <15 years and higher margin dose.

36.2.2 Arteriovenous Malformations (AVMs)

AVMs represent the etiology of about 50% of spontaneous intracranial hemorrhages in the pediatric population [12]. Compared to adults, and due to a long life expectancy, the pediatric population has a higher lifetime risk of AVM rupture. Therefore, pediatric AVMs are more likely to be treated with interventional approaches than with conservative management. In a retrospective study concerning two matched cohorts (315 pediatric vs 315 adult) treated by Gamma Knife between 1987 and 2014, Chen et al. didn't find differences in terms of efficacy, outcomes, and toxicities [13]. Female sex, smaller AVM volume, and deep venous drainage seem to be independent predictors of hemorrhagic presentation in the pediatric population [14].

Hasegawa et al. [15] reported the analysis of the largest series of 189 pediatric AVM patients, treated with a median marginal single dose of 20 Gy with Gamma Knife. MRI was performed at 3-month intervals for the first year, at 6-month intervals for the second and third years, and then annually. Angiography was performed 3 years after treatment and until obtainment of nidus obliteration. The actuarial 5, 8, and 10 years nidus obliteration rates after single fraction, with a mean follow-up period of 136 months, were 66%, 74%, and 77%, respectively. This nidus obliteration rate depends on the Spetzler-Martin grade of the AVM (65% for grade I to 34% for grade IV, at 5 years). In the multivariate analysis, pre-treatment embolization and marginal margin (cutoff 21.8 Gy) were significantly associated with nidus obliteration.

The annual pre-treatment hemorrhage rate was 8.1%. The annual post-treatment hemorrhage rate during the latency period was 2.8%, increased for patients with pre-treatment hemorrhage (2.9% versus 2.4%), and there was no hemorrhage after nidus obliteration. The cumulative hemorrhage rates after treatment were 3.3% at 3 years to 11.9% at 10 years.

In multivariate analysis, the Spetzler-Martin grade was the only significant factor for hemorrhage after treatment. A combined therapy with endovascular embolization should be considered to reduce the rate of bleeding during the latency period.

Hasegawa T et al. [16] also reported the long-term toxicity for 201 pediatric patients, treated by GK between 1991 and 2014. With a median follow-up of 136 months, the incidences of cyst formation, chronic encapsulated hematoma, and radiation-induced tumor were, respectively, 4%, 4%, and 1%. In the multivariate analysis, large nidus volume alone was a significant factor for late adverse radiation effects.

With a mean follow-up period of 130 months, pediatric patients, compared to adult patients, were significantly more likely to have bleeding before treatment and less after; nidus obliteration occurred earlier, but the final rate was the same as in adult patients. This suggests that the sensitivity of the vessels to radiation is greater for pediatric patients [17].

36.2.3 Re-Irradiation

The strong dose gradient, with a protection of organs at risk in proximity of the target volume, is a real advantage to consider re-irradiation. We know that some cerebral re-irradiation is feasible, safe, and effective [18–21], especially for high-grade glioma, medulloblastoma, germinoma, and ependymoma. Rao et al. [18] evaluated the toxicity and outcomes of 67 pediatric patients, including four stereotactic radiosurgeries. The crucial risk of cerebral radionecrosis occurred at cumulative dose >100 Gy when total dose was normalized in 2 Gy fractions.

The French Society of children's cancer reported the experience of full dose re-irradiation for local intracranial ependymoma recurrences [22]. With a median follow-up of 37 months, the median local recurrence-free survival was 31 months, and the overall survival rate was 70% at 1 year and 48% at 3 years, confirming the interest of re-irradiation in this pathology. No toxicity >2 was reported. For the brainstem, the authors recommended a V59 beyond 10 cm^3 and a maximal dose of 64 Gy. The cumulative dose to any point of the brain should be less than 115 Gy.

36.2.4 Metastasis

Used in the adult population with randomized evidence, the use of stereotactic irradiation for metastasis in the pediatric population is increasing.

Chandy et al. reported the retrospective series of 18 treated lesions, with an excellent local control (78.6% at 1 year and 57.1% at 2 years) and no toxicity, achieving to ameliorate the quality of life in those palliative situations. The European FaR-RMS protocol that opened at the beginning of 2020 includes the possibility of stereotactic treatment of metastasis for rhabdomyosarcomas.

36.3 Conclusion

The use of stereotactic irradiation is increasing in the pediatric population and should be considered as an option in selected cases. The CyberKnife system offers the possibility to fractionate the dose, which is less deleterious for growing tissues, and opens the field of treatment possibilities, on primary tumors, for recurrences, metastasis, or re-irradiation. The dose escalation is also a new field of research.

There is an evident lack of publication on homogeneous and large pediatric series, to confirm the promising results. Database, like the French PediaRT base, is a precious tool to increment the outcomes.

References

1. Giller CA, et al. Robotically guided radiosurgery for children. Pediatr Blood Cancer. 2005;45:304–10.
2. Adler JR, Murphy MJ, Chang SD, et al. Image-guided robotic radiosurgery. Neurosurgery. 1999;44:1299–307.
3. Chang SD, Main W, Martin DP, et al. An analysis of the accuracy of the Cyberknife: a robotic frameless stereotactic radiosurgical system. Neurosurgery. 2003;52:140–7.
4. Murphy MJ, Cox RS. The accuracy of dose localization for an image-guided frameless radiosurgery system. Med Phys. 1996;23:2043–9.
5. Aggarwal R, Yeung D, Kumar P, et al. Efficacy and feasibility of stereotactic radiosurgery in the primary management of unfavourable pediatric ependymoma. Radiother Oncol. 1997;43:269–73.
6. Kalapurakal JA, Kepka A, Bista T, et al. Fractionated stereotactic radiotherapy for pediatric brain tumors: the Chicago Children's experience. Child's Nerv System. 2000;16:296–300.
7. Saran FH, Baumert BG, Khoo VS, et al. Stereotactically guided conformal radiotherapy for progressive low-grade gliomas of childhood. Int J Radiat Oncol Biol Phys. 2002;53:43–51.
8. Weprin BE, Hall WA, Cho KH, Sperduto PW, Gerbi BJ, Moertel C. Stereotactic radiosurgery in pediatric patients. Pediatr Neurol. 1996;15(3):193–9.
9. Wolf A, Naylor K, Tam M, et al. Risk of radiation-associated intracranial malignancy after stereotactic radiosurgery: a retrospective, multicentre, cohort study. Lancet Oncol. 2019 Jan;20(1):159–64.
10. Mehta GU, Ding D, Patibandla MR, et al. Stereotactic radiosurgery for Cushing disease: results of an international, multicenter study. J Clin Endocrinol Metab. 2017;102:4284–91.
11. Shrivastava A, Nasser M, Xu Z, Liscak R, et al. Outcomes after Gamma Knife stereotactic radiosurgery in pediatric patients with Cushing disease or acromegaly: a multi-institutional study. World Neurosurg. 2019;125:1104–13.
12. Meyer-Heim AD, Boltshauser E. Spontaneous intracranial haemorrhage in children: aetiology, presentation and outcome. Brain and Development. 2003;25:416–21.
13. Chen CJ, Ding D, et al. Stereotactic radiosurgery for pediatric versus adult brain arteriovenous malformations. Stroke. 2018;49:1939–45.
14. Ding D, Starke RM, Kano H, Mathieu D, et al. International multicentre cohort study of pediatric brain arteriovenous malformations. Part 1: predictors of hemorrhagic presentation. J Neurosurg Pediatr. 2017;19:127–35.

15. Hasegawa T, Kato T, Naito T, et al. Long -term outcomes for pediatric patients with brain AVMs treated with Gammaknife radiosurgery, part 1: analysis of nidus obliteration rates and related factors. World Neurosurg. 2019;126:1518–25.
16. Hasegawa T, Kato T, Naito T, et al. Long-term outcomes for pediatric patients with brain AVMs treated with Gammaknife radiosurgery, part 2: the incidence of cyst formation, encapsulated hematoma and radiation-induced tumor. World Neurosurg. 2019;126:1526–36.
17. Hasegawa H, Hanakita S, Shin M, et al. Comparison of the long-term efficacy and safety of Gammaknife radiosurgery for AVMs in pediatric and adult patients. Neurol Med Chir (Tokyo). 2018;58:231–9.
18. Rao AD, Rashid AS, Chen Q, et al. Reirradiation for recurrent pediatric central nervous system malignancies: a multi-institutional review. Int J Radiat Oncol Biol Phys. 2017;99(3):634–41.
19. Gultekin M, Cengiz M, Sezen D, et al. Re-irradiation of pediatric tumors using hypofractionated stereotactic radiotherapy. Technol Cancer Res Treatment. 2017;16(2):195–202.
20. Tsang DS, Laperriere NJ. Re-irradiation for paediatric tumours. Clin Oncol. 2019;31:191–8.
21. Chandy E, Taylor H, Gaito S, et al. Hypofractionated stereotactic ablative radiotherapy for recurrent or oligometastatic tumours in children and young adults. Clin Oncol. 2019;32:316. https://doi.org/10.1016/j.clon.2019.11.005.
22. Regnier E, Laprie A, Bernier V, et al. Re-irradiation of locally recurrent pediatric intracranial ependymoma: experience of the French Society of Children's Cancer. Radiother Oncol. 2019;132:1–7.

Immunotherapy and Radiosurgery

37

Ravi Medikonda and Michael Lim

37.1 CNS Immune Privilege

Historically, the central nervous system (CNS) has been considered an immunoprivileged organ. This notion arose from early experiments showing homologous tissue grafts to the brain did not generate a strong immune reaction compared to grafts in non-CNS sites [1]. This view was strengthened by characterization of the blood-brain barrier (BBB) [2], an apparent lack of conventional brain lymphatic drainage pathways, relatively low major histocompatibility compex (MHC) class I and II expression in the CNS [3], and the absence of professional antigen-presenting cells (APCs) in the brain [4].

The concept of CNS immune privilege has been called into question as further experimentation has revealed several inconsistencies [5]. Medawar et al. showed that tissue grafted to the brain is rejected if a subcutaneous graft is first placed, suggesting foreign antigens can be recognized in the CNS after priming in peripheral tissue [6]. Furthermore, it has been demonstrated that the BBB permeability to immune cells can be modulated by inflammatory chemokines such as CXCL9, CXCL10, and CXCL11 interacting with the CXCR3 receptor on lymphocytes [7–9]. More recently, a novel route of lymphatic draining was discovered from the brain to the deep cervical lymph nodes, allowing CNS antigens to be presented to the B and T cells at these lymph nodes [10, 11]. Several studies have indicated that the CNS is subject to active immunosurveillance and can mount a strong immune response against foreign antigens [12, 13]. Together, these data suggest significant interactions between the CNS and immune system. Given the success of novel immunotherapies for non-CNS malignancies, there has been an interest in applying immunotherapy to malignancies within the CNS.

37.2 Immunotherapy for Non-CNS Malignancies

The concept of cancer immunosurveillance was first established by Burnet and Thomas who found that immune cells are capable of detecting and killing malignant cells [14–16]. However, the immune system is often unable to eliminate all of the malignant cells, and the cells that survive often develop resistance to immune attack via the process of immunoediting [17]. Various mechanisms of immune escape have been extensively characterized, and the concept of cancer immunosurveillance has given rise to the field of cancer immunotherapy [18–20].

Immunotherapies have been best established for tumors outside the CNS [20–22]. Trastuzumab, an anti-HER2 monoclonal antibody, has become the gold standard in treatment of HER2-positive

R. Medikonda · M. Lim (✉)
Department of Neurosurgery, The Johns Hopkins University School of Medicine, Baltimore, MD, USA
e-mail: rmediko1@jhmi.edu; mlim3@jhmi.edu

© Springer Nature Switzerland AG 2020
A. Conti et al. (eds.), *CyberKnife NeuroRadiosurgery*,
https://doi.org/10.1007/978-3-030-50668-1_37

breast cancer [23]. Sipuleucel-T is an autologous therapeutic cancer vaccine developed for metastatic castrate-resistant prostate cancer and has shown to improve median survival by 4 months [24]. Talimogene laherparepvec, an oncolytic virus therapy, has shown to significantly improve median overall survival in melanoma patients [25]. Chimeric antigen receptor (CAR) T cell therapy targeting CD22 has shown significant efficacy for refractory acute lymphoblastic lymphoma and diffuse large B cell lymphoma [26].

Immune checkpoint inhibitors are among the most promising antitumor immunotherapy approaches. Immune checkpoints are inhibitory pathways intrinsic to the immune system designed to maintain self-tolerance and modulate the duration and amplitude of the immune response to foreign antigens [27]. Ipilimumab, an anti-cytotoxic T-lymphocyte-associated protein-4 (CTLA-4) checkpoint inhibitor, has been established for the treatment of metastatic melanoma following several clinical trials showing survival benefit and is currently under investigation for a variety of other cancers such as renal cell carcinoma, non-small cell lung cancer (NSCLC), and prostate cancer [28–31]. CTLA-4 is an inhibitory receptor that is upregulated on T cells during activation, and it is responsible for downregulating the effector T cell phenotype associated with the antitumor immune response [32, 33]. Another set of checkpoint inhibitors have been developed to target the programmed cell death protein-1 (PD-1)/programmed death-ligand 1 (PD-L1) axis. PD-1 is a cell surface receptor shown to inhibit effector T cell activity upon binding to its ligand PD-L1 via upregulation of T regulatory cells which promote exhaustion of already differentiated effector T cells [34, 35]. The inflammatory stimuli present in tumor microenvironments upregulate expression of PD-L1, which acts to suppress effector T cells expressing the PD-1 receptor [36, 37]. Nivolumab, an anti-PD-1 checkpoint inhibitor, has shown significant benefit for many solid tumors such as NSCLC, renal cell carcinoma (RCC), advanced melanoma, urothelial carcinoma, and hepatocellular carcinoma [38–42]. Pembrolizumab is another anti-PD-1 checkpoint inhibitor that is currently approved for the treatment of several solid tumors such as metastatic melanoma, metastatic NSCLC, gastric cancer, and urothelial carcinoma [43–47]. Targeting the PD-L1 ligand has also shown promise for several solid tumors, and two anti-PD-L1 checkpoint inhibitors atezolizumab and durvalumab have been approved for specific indications in metastatic urothelial carcinoma and NSCLC [48–50].

Historically, there has been limited data on the efficacy of checkpoint inhibitors for patients with brain metastases as these patients were excluded from the initial trials testing these checkpoint inhibitors. However, advances in treating metastatic brain disease have led to an improvement in survival and quality of life, and there is newfound interest in studying the role of checkpoint inhibitors in patients with brain metastasis.

37.3 Current Treatment Options for Brain Metastasis and the Role of Immunotherapy

The management of patients with metastatic brain disease depends on prognostic factors such as age, KPS, and number of intracranial metastases, and the current standard of care includes radiation therapy with or without surgical resection [51–54]. Despite these treatment options, the median overall survival for patients with brain metastasis remains at a few months [55–57]. Given the poor prognosis, there have been significant efforts to improve survival through novel therapeutic strategies such as immunotherapy.

One of the first immunotherapy strategies studied for brain metastasis was IL-2 therapy [58]. Guirguis et al. conducted a retrospective study in 1069 patients to determine the efficacy of recombinant high-dose IL-2 therapy for brain metastasis from melanoma or RCC [59]. They found that IL-2 had an 18.5% clinical response rate in patients previously treated for brain metastasis, but only a 5.6% clinical response rate in patients with previously untreated brain metastasis. Unfortunately, some patients experienced significant toxicity including fluid retention and capillary leak syndrome leading to increased peri-tumoral edema. The authors concluded that IL-2 therapy is efficacious for carefully selected patients with brain metastasis. Chandar et al. prospectively evaluated the safety and efficacy of high-dose IL-2 therapy for 18 patients with brain metastasis from RCC and found that no patients had complete response, 1/18 patients had partial response, and 4/18 patients had stable disease [60]. Adverse neurological events were reported in 6/18 patients and included confusion, mental fatigue, rigors, and anxiety. The authors concluded that IL-2 therapy may be therapeutic for carefully selected patients with brain metastasis.

Given the lackluster results with IL-2 therapy, clinicians shifted focus to immune checkpoint inhibitors such as ipilimumab, an anti-CTLA-4 monoclonal antibody, and nivolumab or pembrolizumab, both anti-PD-1 monoclonal antibodies, for the treatment of brain metastasis [58]. A phase 2 open-label trial studied the safety and efficacy of ipilimumab in 72 patients with melanoma brain metastasis [61]. The trial found ipilimumab to have some activity especially in patients with small and asymptomatic brain metastasis. Di Giacomo et al. conducted a phase 2 single-arm clinical trial in 86 patients with melanoma brain metastasis studying the combination of ipilimumab with fotemustine chemotherapy [62]. They found this combination to have clinical activity in patients with melanoma brain metastasis both at 1-year follow-up and 3-year follow-up [63]. Efficacy of anti-PD-1 therapy for patients with NSCLC or melanoma with brain metastasis was studied in a phase 2 open-label clinical trial [64]. Thirty-six patients were enrolled in this study, half with melanoma and half with NSCLC. Brain metastasis response was achieved in 4 out of 18 patients with melanoma and 6 of 18 patients with NSCLC. Given the purported benefit of anti-CTLA-4 therapy and anti-PD-1 therapy, combination of both checkpoint inhibitors was also evaluated. A multi-center, phase 2 clinical trial was conducted by Tawbi et al. to study the efficacy of combined ipilimumab and nivolumab therapy for melanoma patients with asymptomatic, untreated brain metastasis [65]. Ninety-four patients were enrolled in this trial, and the intracranial complete response rate was 26% and partial response rate was 30%. However, grade 3 or 4 adverse events were reported in 55% of patients, and one patient died of immune-related myocarditis. The authors concluded that this combination therapy has meaningful intracranial efficacy. Long et al. compared combination of nivolumab and ipilimumab or nivolumab alone for 79 melanoma patients with brain metastasis in a phase 2 clinical trial. Intracranial response was detected in 46% of the patients receiving combination therapy and in 20% of the patients receiving nivolumab alone [66]. However, there was significantly greater toxicity associated with the combination therapy than with nivolumab monotherapy as 63% of patients had a grade 3 or higher adverse event in the combination arm while 16% of patients had a grade 3 or higher adverse event in the nivolumab monotherapy arm. The authors concluded that combination therapy of nivolumab and ipilimumab should be considered as first-line therapy in this patient population. Further clinical trials are underway studying the role of checkpoint inhibitors for metastatic brain cancer (Table 37.1).

37.4 Current Treatment Options for Glioblastoma and the Role of Immunotherapy

The current standard of care for glioblastoma is maximal surgical resection with radiotherapy and concomitant and adjuvant temozolomide (TMZ) [67]. Despite standard of care, recurrence is common, and median overall survival is 14.6 months [67, 68]. Treatment options following disease recurrence are limited in efficacy and include further surgical resection if possible and bevacizumab, an anti-vascular endothelial growth factor monoclonal antibody [69]. Poor prognosis is attributed in part to resistance to chemotherapy and radiotherapy mediated through glioma stem cells [70, 71]. It is thought that glioma stem cells resist treatment via intrinsic genetic heterogeneity as well as cellular signaling-mediated resistance pathways [72–74]. Given the poor prognosis, there has been interest in developing novel therapies to supplement the current standard of care.

Table 37.1 Clinical trials evaluating immunotherapy for intracranial metastasis

Trial	*n*	Primary tumor	Treatment	Phase	Key findings
Margolin et al.	72	Melanoma	Ipilimumab	2	Ipilimumab has some activity especially in patients with small or asymptomatic brain metastasis
Di Giacomo et al.	86	Melanoma	Ipilimumab + fotemustine	2	Ipilimumab + fotemustine has clinical activity at 1-year and 3-year follow-up
Goldberg et al.	36	NSCLC or melanoma	Pembrolizumab	2	Brain metastasis response in 4 of 18 melanoma patients and 6 of 18 NSCLC patients
Tawbi et al.	94	Melanoma	Nivolumab + ipilimumab	2	26% complete intracranial response rate and 30% partial response rate. However, grade 3 or 4 adverse events reported in 55% of patients
Long et al.	79	Melanoma	Nivolumab + ipilimumab	2	46% intracranial response rate with combination therapy 20% intracranial response rate with nivolumab monotherapy 63% of patients with combination therapy had grade 3+ toxicity compared to 16% in patients with nivolumab monotherapy

NSCLC Non-small cell lung cancer, *n* Number of patients enrolled

It has been discovered that GBM induces profound intratumoral and systemic immunosuppression [75]. Several mechanisms have been proposed to explain GBM immunosuppression including T cell dysfunction [76], secretion of immunosuppressive cytokines [77], and upregulation of immune checkpoints such as PD-L1 [78]. The discovery that glioblastoma induces immunosuppression coupled with a better understanding of CNS immune privilege has led to an interest in applying immunotherapy to potentially reverse GBM immunosuppression and improve overall survival for GBM patients.

Immune checkpoint blockade has garnered attention for GBM after showing promise for other solid tumors and in preclinical GBM models. Fecci et al. showed that CTLA-4 blockade restores normal CD4 T cell counts and abrogates increases in T regulatory cells in a murine model [79]. Reardon et al. showed that immune checkpoint blockade via anti-CTLA-4 alone, anti-PD-L1 alone, or combination therapy in mice improves tumor-free survival, promotes tumor-specific memory response, and increases the number of natural killer cells and activated CD8 T cells [80]. Furthermore, several case reports have suggested anti-PD-1 therapy can be effective in the management of GBM. Two siblings treated with nivolumab for recurrent multifocal childhood GBM had clinical and radiological response [81]. Another patient presenting with hypermutated GBM showed radiographic response to pembrolizumab [82]. Given the preclinical success and early case reports, there have been attempts to incorporate immune checkpoint therapy into routine care for GBM patients. CheckMate 143 was the first major randomized clinical trial evaluating the safety and efficacy of nivolumab for recurrent GBM [83]. Preliminary results from a phase I safety cohort study showed that all nivolumab-related adverse events were grade 1 or grade 2. However, eight out of ten patients receiving a combination of nivolumab and ipilimumab had grade 3 or 4 toxicity, leading to discontinuation of the combination regimen in five of ten patients [83]. Initial results presented in a WFNOS 2017 abstract described no significant survival advantage for nivolumab compared to bevacizumab in recurrent GBM, and this arm of the trial was subsequently closed. Combination of nivolumab with current standard of care is still being evaluated in the CheckMate 143 trial, and exploratory cohorts suggest an acceptable safety profile. The CheckMate 498 trial (NCT02617589) is evaluating nivolumab as an alternative to TMZ in the standard of care paradigm for patients with MGMT promoter-unmethylated tumors. CheckMate 548 (NCT02667587) is evaluating nivolumab in addition to current standard of care for MGMT promoter-methylated tumors.

Other immunotherapy strategies also being investigated for GBM include peptide vaccines, oncolytic virus therapy, and CAR T cell therapy. GBM vaccines are being developed with an aim to strengthen the response of the adaptive immune system against glioma cells. Rindopepimut is a peptide vaccine that targets a constitutively active EGFR variant exclusive to GBM cells found in 25–30% of patients [84]. Although phase 2 studies of rindopepimut suggested improved median overall survival compared to historical controls, a multi-center phase 3 trial was prematurely concluded after interim analysis found no significant difference in overall survival. An IDH1 peptide vaccine is under clinical investigation in two clinical trials (NCT02193347 and NCT02454643) with IDH1-mutated glioma patients. Oncolytic virus therapies are also being studied for GBM. PVSRIPO is a recombinant oncolytic poliovirus that showed a higher survival rate at 24 and 36 months than historical controls [85]. Vocimagene amiretrorepvec, a murine leukemia virus, is currently being investigated in a phase 2/3 clinical trial (NCT02414165). Several adenovirus-based oncolytic virus therapies are currently being investigated in early phase clinical trials (NCT02798406, NCT02197169). A measles virus-based oncolytic virus therapy is also under investigation in a phase I clinical trial (NCT00390299) after encouraging preclinical data showed improved survival in murine GBM model [86]. Several clinical trials are studying herpes simplex virus-based oncolytic virus therapies (NCT00028158, NCT00157703, NCT02457845, NCT02031965, NCT02062827). Chimeric antigen receptor (CAR) T cells are genetically modified T cells with antigen receptors specific to tumor-associated antigens. CAR T cells do not require co-stimulatory MHC activation. A case report by Brown et al. reported a patient with recurrent GBM who received CAR T cells targeting the tumor-associated antigen IL13Rα2. The CAR T cells were administered over the course of 220 days either directly into the resected tumor cavity or into the ventricular system. The patient did not have any grade 3 or higher toxicities and showed clinical and radiographic response with regression of all intracranial and spinal tumors. This clinical response lasted for 7.5 months after starting CAR T cell treatment. Further research is warranted for CAR T cell therapy as a phase I clinical study demonstrated treatment safety and evidence of tumor infiltration by CAR T cells. However, a survival benefit was not detected [87].

37.5 Immunotherapy May Synergize with Radiation Therapy and Chemotherapy

There has been significant interest in integrating immunotherapy into the current standard of care for brain metastasis and GBM. The use of hyperfractionated radiation therapy has been shown to induce systemic immunosuppression which may abrogate the effect of subsequent immunotherapy. Grossman et al. followed the CD4 T cell count every month for 1 year in 96 patients with newly diagnosed high-grade glioma receiving hyperfractionated radiation, TMZ, and glucocorticoids [88]. The median CD4 count before radiation and TMZ treatment was 664 cells/mm^3. The CD4 count reached a nadir at 2 months after initiating treatment with fewer than 200 cells/mm^3 in 40% of patients and fewer than 300 cells/mm^3 in 73% of patients. The median overall survival in patients with a CD4 count less than 200 cells/mm^3 at 2 months after starting treatment was 13.1 months compared to 19.7 months in patients with a CD4 count greater than 200 cells/mm^3 at 2 months. Subsequent studies by the same researchers in patients with pancreatic cancer found that the patients who received stereotactic body radiation had less severe lymphopenia compared to the patients that received conventional hyperfractionated radiotherapy [89]. Together, these studies suggest that conventional hyperfractionated radiation therapy may not synergize with immunotherapy. Bamashmos et al. retrospectively studied the absolute lymphocyte count (ALC) in 231 newly diagnosed GBM patients receiving concurrent temozolomide chemoradiation [90]. They found that patients with a low ALC at 4 weeks after starting chemoradiation had higher mortality, although there was no difference in progression-free survival.

TMZ has also been shown to have a myelosuppressive effect initially reported by Brock et al. in in a phase 1 trial of TMZ [91]. Subsequent studies have shown that bone marrow is particularly vulnerable to TMZ, and it has been observed that TMZ selectively targets monocytes, but not dendritic cells [92]. In a preclinical murine GBM model, Mathios et al. showed that systemic chemotherapy induces systemic and intratumoral lymphodeletion and decreases immune memory in long-term survivors. However, local chemotherapy with anti-PD1 immunotherapy elicited a robust antitumor immune response and showed a survival benefit [93]. These results suggest that systemic chemotherapy and conventional hyperfractionated radiation therapy may not synergize with immunotherapy, but stereotactic radiosurgery (SRS) and local chemotherapy may synergize well with immunotherapy.

Indeed, there is a significant body of literature on the synergy between radiation therapy and immunotherapy. It has been shown that radiation therapy can enhance the immune system's antitumor response through a varicty of mechanisms. One of these mechanisms has been well defined as the abscopal effect. First described by Mole in 1953, the abscopal effect refers to shrinkage of an untreated metastasis after local radiation treatment of a tumor [94]. Demaria et al. demonstrated that the abscopal effect is at least in part mediated by the immune system [95]. In their study, mice with mammary carcinoma in both flanks were treated daily with a dendritic cell growth factor Flt3-L, and one of the two flank tumor sites received radiation. They found that the mice treated with radiation alone demonstrated a decrease in the size of the irradiated tumor only, and mice receiving Flt3-L only showed no decrease in tumor size. However, mice receiving both the dendritic cell growth factor and radiation therapy demonstrated a decrease in tumor size of both radiated and non-radiated flank tumors. The abscopal effect has been extensively studied for a variety of tumors [96–99]. In addition to the abscopal effect, synergy between radiation therapy and immunotherapy may be mediated through increased antigen presentation and development of neoantigens. Reits et al. demonstrated that radiation therapy can upregulate the expression of MHC I on tumor cells and increase antigen presentation to T cells that mediate the antitumor response [100]. Furthermore, it is thought that DNA damage induced by radiation therapy may increase the mutational load of the tumor and lead to the development of neoantigens for immune recognition [101].

Numerous preclinical studies have evaluated the combination of radiation therapy with immunotherapy for CNS tumors. Zeng et al. treated mice with either radiation therapy only, anti-PD-1 only, or a combination of radiation therapy and anti-PD-1 [102]. The median survival in the combination arm was nearly twice the median survival in the other arms. Furthermore, long-term survival was only seen in the combination treatment arm. Kim et al. studied combination therapy with anti-PD-1, another immune checkpoint inhibitor anti-TIM-3, and focal radiation in a murine glioma model. They found that anti-TIM-3 plus SRS improved survival more than anti-TIM-3 alone. Furthermore, combination of anti-TIM-3 with anti-PD-1 and SRS resulted in 100% overall survival [103]. Belcaid et al. evaluated combination radiation therapy with 4-1BB activation and CTLA-4 blockade in a murine glioma model [104]. 4-1BB is a co-stimulatory signal expressed by activated T lymphocytes that promotes cytotoxic function. Mice were treated with a triple therapy of 4-1BB agonist antibodies, CTLA-4 inhibiting antibodies, and focal radiation therapy to an intracranial GL261 tumor. Mice that received triple therapy had a median survival of 67 days compared to 24 days in mice that only received focal radiation. Furthermore, depletion of CD4 and CD8 T tumor infiltrating lymphocytes abrogated the efficacy of triple therapy. Patel et al. studied combination therapy of anti-GITR antibody with SRS in a murine GBM model. GITR is a co-stimulatory molecule constitutively expressed on regulatory T cells and effector T cells upon activation. Mice were treated with either SRS only, anti-GITR only or combination therapy. It was found that combination therapy improved survival more than either therapy alone.

The findings of these preclinical studies have encouraged several clinical trials currently underway to evaluate the combination of SRS and immunotherapy in GBM patients. CheckMate 548 is a multi-center phase 3 randomized trial studying temozolomide plus radiation therapy combined with nivolumab or placebo for newly diagnosed GBM (NCT02667587). CheckMate 498 is another multi-center phase 3 randomized trial studying nivolumab versus temozolomide, each in combination with radiation therapy for newly diagnosed GBM patients (NCT02617589). A phase 1 trial is currently studying the safety profile of radiation therapy combined with temozolomide and pembrolizumab for newly diagnosed GBM patients (NCT02530502). Another phase 1 trial is studying hypofractionated stereotactic radiation with pembrolizumab and bevacizumab for patients with recurrent high-grade glioma (NCT02313272) (Table 37.2).

Combination of radiation therapy with immunotherapy has also garnered attention for the management of brain metastasis. Silk et al. conducted a retrospective review of ipilimumab and radiation therapy in 33 patients with melanoma brain metastasis [105]. The median survival in the cohort that received ipilimumab and radiation therapy was 18.3 months, and the median survival in the cohort of patients that only received radiation therapy was 5.3 months. The authors concluded that combination of ipilimumab with radiation therapy significantly reduced the risk of death in patients with melanoma brain metastasis. Minniti et al. retrospectively evaluated 80 patients with melanoma brain metastasis receiving SRS in combination with either ipilimumab or nivolumab [106]. Forty-five patients in this study received SRS plus ipilimumab, and 35 received SRS plus nivolumab. Twelve-month intracranial progression-free survival (PFS) rates were 42% with SRS plus nivolumab and 17% with SRS plus ipilimumab. Furthermore, it was found that multi-fraction SRS (3 × 9 Gy) had better intracranial PFS than single-fraction SRS. Gabani et al. used the National Cancer Database to retrospectively evaluate patients with melanoma brain metastasis receiving either radiation therapy alone or radiation therapy plus immunotherapy [107]. They found that patients receiving SRS were more likely to receive immunotherapy than patients being treated with WBRT. The median overall survival was significantly higher for the radiation therapy plus immunotherapy cohort than the radiation therapy only cohort (11.1 versus 6.2 months). Anderson et al. retrospectively analyzed 21 patients receiving concurrent radiation therapy with pembrolizumab for brain metastasis [108]. It was found that 70% of patients had a complete or partial response to combination therapy at the first scheduled follow-up MRI. The authors concluded that concurrent pembrolizumab with brain RT is safe and particularly effective in reducing the size of brain metastasis at first follow-up MRI. Lehrer et al. conducted an international meta-analysis of patients treated for brain metastasis with SRS and immune checkpoint inhibitors either concurrently or non-concurrently [109]. Among the 534 patients evaluated in this meta-analysis, the 1-year overall survival was 64.6% for concurrent radiation and immune checkpoint inhibitor therapy and 51.6% in non-concurrent therapy. The local control and regional brain control at 1 year were also significantly higher with concurrent therapy. The results from these retro-

Table 37.2 Current clinical trials evaluating immunotherapy plus radiation therapy for GBM

Trial	Treatment	Primary vs recurrent	MGMT status	Phase
CheckMate 498	Nivolumab + RT vs TMZ + RT	Primary	Negative	2
CheckMate 548	Nivolumab + TMZ + RT vs TMZ + RT	Primary	Positive	2
NCT02530502	Pembrolizumab + RT + TMZ	Primary	Both	1
NCT02313272	Pembrolizumab + HFSRT + bevacizumab	Recurrent	Both	1

MGMT Methylguanine-DNA methyltransferase, *RT* Radiation therapy, *TMZ* Temozolomide, *HFSRT* Hypofractionated stereotactic radiation therapy spective studies have warranted several clinical trials to evaluate the role of combining immunotherapy with SRS for NSCLC and melanoma brain metastasis (NCT02696993, NCT02097732, NCT01703507).

37.6 Conclusions

For many years, the prognosis of patients with GBM or brain metastasis has been poor despite significant advances in surgery, radiation therapy, and chemotherapy. The advent of immunotherapy revolutionized the management of a variety of non-CNS tumors, and with the recent advances in our understanding of CNS immune privilege, there is hope that immunotherapy may improve outcomes for patients with CNS tumors. Early preclinical data and case reports of immunotherapy for GBM are promising. However, it has been challenging to translate these results into the clinical setting, and several clinical trials are still currently underway. The application of immunotherapy to the management of brain metastasis is still in its infancy. Whereas immune checkpoint inhibitors are beginning to be characterized in the management of brain metastasis, other immunotherapy approaches including CAR T cells, oncolytic viruses, and tumor vaccines are still poorly characterized. Immune checkpoint inhibitors have shown efficacy especially in clinical trials for melanoma and NSCLC brain metastasis and more so when used in combination. However, dual immune checkpoint therapy was associated with significantly higher toxicity than monotherapy. Checkpoint inhibitors continue to be evaluated in several clinical trials in combination with other therapies.

There is significant interest in integrating immunotherapy with the current standard of care for GBM and brain metastasis, specifically stereotactic radiosurgery. The abscopal effect has been described extensively in the literature, and preclinical studies suggest immunotherapy may synergize with radiation therapy for CNS tumors. Furthermore, several retrospective studies suggest that the combination of immunotherapy with radiation therapy may improve outcomes in patients with brain metastasis. Given these findings, several clinical trials are underway in both GBM patients and brain metastasis patients to evaluate this combinatorial approach. The results from these clinical trials will guide the standard of care for patients with these deadly CNS tumors.

References

1. Barker CF, Billingham RE. Immunologically privileged sites. Adv Immunol. 1977;25:1–54.
2. Banks WA. The blood-brain barrier in neuroimmunology: tales of separation and assimilation. Brain Behav Immun. 2015;44:1–8. https://doi.org/10.1016/j.bbi.2014.08.007.
3. Perry VH. A revised view of the central nervous system microenvironment and major histocompatibility complex class II antigen presentation. J Neuroimmunol. 1998;90(2):113–21.
4. Forrester JV, McMenamin PG, Dando SJ. CNS infection and immune privilege. Nat Rev Neurosci. 2018;19(11):655–71. https://doi.org/10.1038/s41583-018-0070-8.
5. Galea I, Bechmann I, Perry VH. What is immune privilege (not)? Trends Immunol. 2007;28(1):12–8. https://doi.org/10.1016/j.it.2006.11.004.
6. Medawar PB. Immunity to homologous grafted skin; the fate of skin homografts transplanted to the brain, to subcutaneous tissue, and to the anterior chamber of the eye. Br J Exp Pathol. 1948;29(1):58–69.
7. Gorbachev AV, Kobayashi H, Kudo D, Tannenbaum CS, Finke JH, Shu S, Farber JM, Fairchild RL. CXC chemokine ligand 9/monokine induced by IFN-γ production by tumor cells is critical for T cell-mediated suppression of cutaneous tumors. J Immunol. 2007;178(4):2278–86. https://doi.org/10.4049/jimmunol.178.4.2278.
8. Klein RS, Izikson L, Means T, Gibson HD, Lin E, Sobel RA, Weiner HL, Luster AD. IFN-inducible protein 10/CXC chemokine ligand 10-independent induction of experimental autoimmune encephalomyelitis. J Immunol. 2004;172(1):550–9. https://doi.org/10.4049/jimmunol.172.1.550.
9. Groom JR, Luster AD. CXCR3 in T cell function. Exp Cell Res. 2011;317(5):620–31. https://doi.org/10.1016/j.yexcr.2010.12.017.
10. Louveau A, Smirnov I, Keyes TJ, Eccles JD, Rouhani SJ, Peske JD, Derecki NC, Castle D, Mandell JW, Lee KS, Harris TH, Kipnis J. Structural and functional features of central nervous system lymphatic vessels. Nature. 2015;523(7560):337–41. https://doi.org/10.1038/nature14432.
11. Aspelund A, Antila S, Proulx ST, Karlsen TV, Karaman S, Detmar M, Wiig H, Alitalo K. A dural lymphatic vascular system that drains brain

interstitial fluid and macromolecules. J Exp Med. 2015;212(7):991–9. https://doi.org/10.1084/jem.20142290.
12. Ransohoff RM, Engelhardt B. The anatomical and cellular basis of immune surveillance in the central nervous system. Nat Rev Immunol. 2012;12(9):623–35. https://doi.org/10.1038/nri3265.
13. Mayo L, Quintana FJ, Weiner HL. The innate immune system in demyelinating disease. Immunol Rev. 2012;248(1):170–87. https://doi.org/10.1111/j.1600-065X.2012.01135.x.
14. Burnet FM. The concept of immunological surveillance. Prog Exp Tumor Res. 1970;13:1–27.
15. Burnet M. Cancer: a biological approach. III. Viruses associated with neoplastic conditions. IV. Practical applications. Br Med J. 1957;1(5023):841–7. https://doi.org/10.1136/bmj.1.5023.841.
16. Thomas L. Cellular and humoral aspects of the hypersensitive states. Acta Med Scand. 1961;170(1):128. https://doi.org/10.1111/j.0954-6820.1961.tb00220.x.
17. Dunn GP, Bruce AT, Ikeda H, Old LJ, Schreiber RD. Cancer immunoediting: from immunosurveillance to tumor escape. Nat Immunol. 2002;3(11):991–8. https://doi.org/10.1038/ni1102-991.
18. Prestwich RJ, Errington F, Hatfield P, Merrick AE, Ilett EJ, Selby PJ, Melcher AA. The immune system—is it relevant to cancer development, progression and treatment? Clin Oncol (R Coll Radiol). 2008;20(2):101–12. https://doi.org/10.1016/j.clon.2007.10.011.
19. Sharma P, Hu-Lieskovan S, Wargo JA, Ribas A. Primary, adaptive, and acquired resistance to cancer immunotherapy. Cell. 2017;168(4):707–23. https://doi.org/10.1016/j.cell.2017.01.017.
20. Zhang H, Chen J. Current status and future directions of cancer immunotherapy. J Cancer. 2018;9(10):1773–81. https://doi.org/10.7150/jca.24577.
21. Subramaniam DS, Liu SV, Giaccone G. Novel approaches in cancer immunotherapy. Discov Med. 2016;21(116):267–74.
22. Voena C, Chiarle R. Advances in cancer immunology and cancer immunotherapy. Discov Med. 2016;21(114):125–33.
23. Maximiano S, Magalhaes P, Guerreiro MP, Morgado M. Trastuzumab in the treatment of breast cancer. BioDrugs. 2016;30(2):75–86. https://doi.org/10.1007/s40259-016-0162-9.
24. Gardner TA, Elzey BD, Hahn NM. Sipuleucel-T (Provenge) autologous vaccine approved for treatment of men with asymptomatic or minimally symptomatic castrate-resistant metastatic prostate cancer. Hum Vaccin Immunother. 2012;8(4):534–9. https://doi.org/10.4161/hv.19795.
25. Andtbacka RH, Kaufman HL, Collichio F, Amatruda T, Senzer N, Chesney J, Delman KA, Spitler LE, Puzanov I, Agarwala SS, Milhem M, Cranmer L, Curti B, Lewis K, Ross M, Guthrie T, Linette GP, Daniels GA, Harrington K, Middleton MR, Miller WH Jr, Zager JS, Ye Y, Yao B, Li A, Doleman S, VanderWalde A, Gansert J, Coffin RS. Talimogene laherparepvec improves durable response rate in patients with advanced melanoma. J Clin Oncol. 2015;33(25):2780–8. https://doi.org/10.1200/jco.2014.58.3377.
26. Pehlivan KC, Duncan BB, Lee DW. CAR-T cell therapy for acute lymphoblastic leukemia: transforming the treatment of relapsed and refractory disease. Curr Hematol Malig Rep. 2018;13(5):396–406. https://doi.org/10.1007/s11899-018-0470-x.
27. Pardoll DM. The blockade of immune checkpoints in cancer immunotherapy. Nat Rev Cancer. 2012;12(4):252–64. https://doi.org/10.1038/nrc3239.
28. McDermott D, Haanen J, Chen TT, Lorigan P, O'Day S. Efficacy and safety of ipilimumab in metastatic melanoma patients surviving more than 2 years following treatment in a phase III trial (MDX010-20). Ann Oncol. 2013;24(10):2694–8. https://doi.org/10.1093/annonc/mdt291.
29. Weber J, Thompson JA, Hamid O, Minor D, Amin A, Ron I, Ridolfi R, Assi H, Maraveyas A, Berman D, Siegel J, O'Day SJ. A randomized, double-blind, placebo-controlled, phase II study comparing the tolerability and efficacy of ipilimumab administered with or without prophylactic budesonide in patients with unresectable stage III or IV melanoma. Clin Cancer Res. 2009;15(17):5591–8. https://doi.org/10.1158/1078-0432.ccr-09-1024.
30. Weber JS, Amin A, Minor D, Siegel J, Berman D, O'Day SJ. Safety and clinical activity of ipilimumab in melanoma patients with brain metastases: retrospective analysis of data from a phase 2 trial. Melanoma Res. 2011;21(6):530–4. https://doi.org/10.1097/CMR.0b013e32834d3d88.
31. Hodi FS, O'Day SJ, McDermott DF, Weber RW, Sosman JA, Haanen JB, Gonzalez R, Robert C, Schadendorf D, Hassel JC, Akerley W, van den Eertwegh AJ, Lutzky J, Lorigan P, Vaubel JM, Linette GP, Hogg D, Ottensmeier CH, Lebbe C, Peschel C, Quirt I, Clark JI, Wolchok JD, Weber JS, Tian J, Yellin MJ, Nichol GM, Hoos A, Urba WJ. Improved survival with ipilimumab in patients with metastatic melanoma. N Engl J Med. 2010;363(8):711–23. https://doi.org/10.1056/NEJMoa1003466.
32. Qureshi OS, Zheng Y, Nakamura K, Attridge K, Manzotti C, Schmidt EM, Baker J, Jeffery LE, Kaur S, Briggs Z, Hou TZ, Futter CE, Anderson G, Walker LSK, Sansom DM. Trans-endocytosis of CD80 and CD86: a molecular basis for the cell-extrinsic function of CTLA-4. Science. 2011;332(6029):600–3. https://doi.org/10.1126/science.1202947.
33. Krummel MF, Allison JP. CD28 and CTLA-4 have opposing effects on the response of T cells to stimulation. J Exp Med. 1995;182(2):459–65. https://doi.org/10.1084/jem.182.2.459.

34. Latchman Y, Wood CR, Chernova T, Chaudhary D, Borde M, Chernova I, Iwai Y, Long AJ, Brown JA, Nunes R, Greenfield EA, Bourque K, Boussiotis VA, Carter LL, Carreno BM, Malenkovich N, Nishimura H, Okazaki T, Honjo T, Sharpe AH, Freeman GJ. PD-L2 is a second ligand for PD-1 and inhibits T cell activation. Nat Immunol. 2001;2(3):261–8. https://doi.org/10.1038/85330.
35. Freeman GJ, Long AJ, Iwai Y, Bourque K, Chernova T, Nishimura H, Fitz LJ, Malenkovich N, Okazaki T, Byrne MC, Horton HF, Fouser L, Carter L, Ling V, Bowman MR, Carreno BM, Collins M, Wood CR, Honjo T. Engagement of the Pd-1 Immunoinhibitory receptor by a novel B7 family member leads to negative regulation of lymphocyte activation. J Exp Med. 2000;192(7):1027–34. https://doi.org/10.1084/jem.192.7.1027.
36. Topalian Suzanne L, Drake Charles G, Pardoll Drew M. Immune checkpoint blockade: a common denominator approach to cancer therapy. Cancer Cell. 2015;27(4):450–61. https://doi.org/10.1016/j.ccell.2015.03.001.
37. Hargadon KM, Johnson CE, Williams CJ. Immune checkpoint blockade therapy for cancer: an overview of FDA-approved immune checkpoint inhibitors. Int Immunopharmacol. 2018;62:29–39. https://doi.org/10.1016/j.intimp.2018.06.001.
38. Paz-Ares L, Horn L, Borghaei H, Spigel DR, Steins M, Ready N, Chow LQM, Vokes EE, Felip E, Holgado E, Barlesi F, Kohlhaeufl M, Rodriguez O, Burgio MA, Fayette J, Gettinger SN, Harbison C, Dorange C, Finckenstein FG, Brahmer JR. Phase III, randomized trial (CheckMate 057) of nivolumab (NIVO) versus docetaxel (DOC) in advanced non-squamous cell (non-SQ) non-small cell lung cancer (NSCLC). J Clin Oncol. 2015;33(18_suppl):LBA109-LBA109. https://doi.org/10.1200/jco.2015.33.18_suppl.lba109.
39. Motzer RJ, Escudier B, McDermott DF, George S, Hammers HJ, Srinivas S, Tykodi SS, Sosman JA, Procopio G, Plimack ER, Castellano D, Choueiri TK, Gurney H, Donskov F, Bono P, Wagstaff J, Gauler TC, Ueda T, Tomita Y, Schutz FA, Kollmannsberger C, Larkin J, Ravaud A, Simon JS, Xu LA, Waxman IM, Sharma P. Nivolumab versus everolimus in advanced renal-cell carcinoma. N Engl J Med. 2015;373(19):1803–13. https://doi.org/10.1056/NEJMoa1510665.
40. Weber JS, D'Angelo SP, Minor D, Hodi FS, Gutzmer R, Neyns B, Hoeller C, Khushalani NI, Miller WH, Lao CD, Linette GP, Thomas L, Lorigan P, Grossmann KF, Hassel JC, Maio M, Sznol M, Ascierto PA, Mohr P, Chmielowski B, Bryce A, Svane IM, Grob J-J, Krackhardt AM, Horak C, Lambert A, Yang AS, Larkin J. Nivolumab versus chemotherapy in patients with advanced melanoma who progressed after anti-CTLA-4 treatment (CheckMate 037): a randomised, controlled, open-label, phase 3 trial. Lancet Oncol. 2015;16(4):375–84. https://doi.org/10.1016/S1470-2045(15)70076-8.
41. Sharma P, Retz M, Siefker-Radtke A, Baron A, Necchi A, Bedke J, Plimack ER, Vaena D, Grimm M-O, Bracarda S, Arranz JÁ, Pal S, Ohyama C, Saci A, Qu X, Lambert A, Krishnan S, Azrilevich A, Galsky MD. Nivolumab in metastatic urothelial carcinoma after platinum therapy (CheckMate 275): a multicentre, single-arm, phase 2 trial. Lancet Oncol. 2017;18(3):312–22. https://doi.org/10.1016/S1470-2045(17)30065-7.
42. El-Khoueiry AB, Sangro B, Yau T, Crocenzi TS, Kudo M, Hsu C, Kim T-Y, Choo S-P, Trojan J, Welling TH, Meyer T, Kang Y-K, Yeo W, Chopra A, Anderson J, dela Cruz C, Lang L, Neely J, Tang H, Dastani HB, Melero I. Nivolumab in patients with advanced hepatocellular carcinoma (CheckMate 040): an open-label, non-comparative, phase 1/2 dose escalation and expansion trial. Lancet. 2017;389(10,088):2492–502. https://doi.org/10.1016/S0140-6736(17)31046-2.
43. Robert C, Schachter J, Long GV, Arance A, Grob JJ, Mortier L, Daud A, Carlino MS, McNeil C, Lotem M, Larkin J, Lorigan P, Neyns B, Blank CU, Hamid O, Mateus C, Shapira-Frommer R, Kosh M, Zhou H, Ibrahim N, Ebbinghaus S, Ribas A. Pembrolizumab versus Ipilimumab in advanced melanoma. N Engl J Med. 2015;372(26):2521–32. https://doi.org/10.1056/NEJMoa1503093.
44. Ribas A, Puzanov I, Dummer R, Schadendorf D, Hamid O, Robert C, Hodi FS, Schachter J, Pavlick AC, Lewis KD, Cranmer LD, Blank CU, O'Day SJ, Ascierto PA, Salama AKS, Margolin KA, Loquai C, Eigentler TK, Gangadhar TC, Carlino MS, Agarwala SS, Moschos SJ, Sosman JA, Goldinger SM, Shapira-Frommer R, Gonzalez R, Kirkwood JM, Wolchok JD, Eggermont A, Li XN, Zhou W, Zernhelt AM, Lis J, Ebbinghaus S, Kang SP, Daud A. Pembrolizumab versus investigator-choice chemotherapy for ipilimumab-refractory melanoma (KEYNOTE-002): a randomised, controlled, phase 2 trial. Lancet Oncol. 2015;16(8):908–18. https://doi.org/10.1016/S1470-2045(15)00083-2.
45. Herbst RS, Baas P, Kim D-W, Felip E, Pérez-Gracia JL, Han J-Y, Molina J, Kim J-H, Arvis CD, Ahn M-J, Majem M, Fidler MJ, de Castro G, Garrido M, Lubiniecki GM, Shentu Y, Im E, Dolled-Filhart M, Garon EB. Pembrolizumab versus docetaxel for previously treated, PD-L1-positive, advanced non-small-cell lung cancer (KEYNOTE-010): a randomised controlled trial. Lancet. 2016;387(10,027):1540–50. https://doi.org/10.1016/S0140-6736(15)01281-7.
46. Bellmunt J, de Wit R, Vaughn DJ, Fradet Y, Lee J-L, Fong L, Vogelzang NJ, Climent MA, Petrylak DP, Choueiri TK, Necchi A, Gerritsen W, Gurney H, Quinn DI, Culine S, Sternberg CN, Mai Y, Poehlein CH, Perini RF, Bajorin DF. Pembrolizumab as second-line therapy for advanced urothelial car-

cinoma. N Engl J Med. 2017;376(11):1015–26. https://doi.org/10.1056/NEJMoa1613683.
47. Fuchs CS, Doi T, Jang RW, Muro K, Satoh T, Machado M, Sun W, Jalal SI, Shah MA, Metges J-P, Garrido M, Golan T, Mandala M, Wainberg ZA, Catenacci DV, Ohtsu A, Shitara K, Geva R, Bleeker J, Ko AH, Ku G, Philip P, Enzinger PC, Bang Y-J, Levitan D, Wang J, Rosales M, Dalal RP, Yoon HH. Safety and efficacy of pembrolizumab monotherapy in patients with previously treated advanced gastric and gastroesophageal junction cancer: phase 2 clinical KEYNOTE-059 trial pembrolizumab in advanced gastric and gastroesophageal junction cancer pembrolizumab in advanced gastric and gastroesophageal junction cancer. JAMA Oncol. 2018;4(5):e180013. https://doi.org/10.1001/jamaoncol.2018.0013.
48. Powles T, Durán I, van der Heijden MS, Loriot Y, Vogelzang NJ, De Giorgi U, Oudard S, Retz MM, Castellano D, Bamias A, Fléchon A, Gravis G, Hussain S, Takano T, Leng N, Kadel EE, Banchereau R, Hegde PS, Mariathasan S, Cui N, Shen X, Derleth CL, Green MC, Ravaud A. Atezolizumab versus chemotherapy in patients with platinum-treated locally advanced or metastatic urothelial carcinoma (IMvigor211): a multicentre, open-label, phase 3 randomised controlled trial. Lancet. 2018;391(10,122):748–57. https://doi.org/10.1016/S0140-6736(17)33297-X.
49. Rittmeyer A, Barlesi F, Waterkamp D, Park K, Ciardiello F, von Pawel J, Gadgeel SM, Hida T, Kowalski DM, Dols MC, Cortinovis DL, Leach J, Polikoff J, Barrios C, Kabbinavar F, Frontera OA, De Marinis F, Turna H, Lee J-S, Ballinger M, Kowanetz M, He P, Chen DS, Sandler A, Gandara DR. Atezolizumab versus docetaxel in patients with previously treated non-small-cell lung cancer (OAK): a phase 3, open-label, multicentre randomised controlled trial. Lancet. 2017;389(10,066):255–65. https://doi.org/10.1016/S0140-6736(16)32517-X.
50. Antonia SJ, Villegas A, Daniel D, Vicente D, Murakami S, Hui R, Yokoi T, Chiappori A, Lee KH, de Wit M, Cho BC, Bourhaba M, Quantin X, Tokito T, Mekhail T, Planchard D, Kim Y-C, Karapetis CS, Hiret S, Ostoros G, Kubota K, Gray JE, Paz-Ares L, de Castro CJ, Wadsworth C, Melillo G, Jiang H, Huang Y, Dennis PA, Özgüroğlu M. Durvalumab after chemoradiotherapy in stage III non–small-cell lung cancer. N Engl J Med. 2017;377(20):1919–29. https://doi.org/10.1056/NEJMoa1709937.
51. Patchell RA, Tibbs PA, Walsh JW, Dempsey RJ, Maruyama Y, Kryscio RJ, Markesbery WR, Macdonald JS, Young B. A randomized trial of surgery in the treatment of single metastases to the brain. N Engl J Med. 1990;322(8):494–500. https://doi.org/10.1056/nejm199002223220802.
52. Order SE, Hellman S, Von Essen CF, Kligerman MM. Improvement in quality of survival following whole-brain irradiation for brain metastasis. Radiology. 1968;91(1):149–53. https://doi.org/10.1148/91.1.149.
53. Markesbery WR, Brooks WH, Gupta GD, Young AB. Treatment for patients with cerebral metastases. Arch Neurol. 1978;35(11):754–6. https://doi.org/10.1001/archneur.1978.00500350058012.
54. Cairncross JG, Kim J-H, Posner JB. Radiation therapy for brain metastases. Ann Neurol. 1980;7(6):529–41. https://doi.org/10.1002/ana.410070606.
55. Yuan Y, Shi Q, Li M, Nagamuthu C, Andres E, Davis FG. Canadian brain cancer survival rates by tumour type and region: 1992-2008. Can J Public Health. 2016;107(1):e37–42. https://doi.org/10.17269/cjph.107.5209.
56. Weltman E, Salvajoli JV, Brandt RA, de Morais HR, Prisco FE, Cruz JC, de Oliveira Borges SR, Wajsbrot DB. Radiosurgery for brain metastases: a score index for predicting prognosis. Int J Radiat Oncol Biol Phys. 2000;46(5):1155–61. https://doi.org/10.1016/S0360-3016(99)00549-0.
57. Lagerwaard F, Levendag P, Nowak PCM, Eijkenboom WH, Hanssens PJ, Schmitz PM. Identification of prognostic factors in patients with brain metastases: a review of 1292 patients. Int J Radiat Oncol Biol Phys. 1999;43(4):795–803. https://doi.org/10.1016/S0360-3016(98)00442-8.
58. Chukwueke U, Batchelor T, Brastianos P. Management of brain metastases in patients with melanoma. J Oncol Pract. 2016;12(6):536–42. https://doi.org/10.1200/JOP.2016.011882.
59. Guirguis LM, Yang JC, White DE, Steinberg SM, Liewehr DJ, Rosenberg SA, Schwartzentruber DJ. Safety and efficacy of high-dose interleukin-2 therapy in patients with brain metastases. J Immunother (Hagerstown, MD: 1997). 2002;25(1):82–7.
60. Chandar A, Silk AW, Clark JI, Daniels GA, McDermott DF, Morse M, Wong MKK, Stein M, Mehnert J, Danish S, Aung S, Kaufman HL. Efficacy and safety of high-dose interleukin-2 treatment in patients with a history of brain metastases from renal cell carcinoma. J Immunother Cancer. 2015;3(Suppl 2):P129. https://doi.org/10.1186/2051-1426-3-S2-P129.
61. Margolin K, Ernstoff MS, Hamid O, Lawrence D, McDermott D, Puzanov I, Wolchok JD, Clark JI, Sznol M, Logan TF, Richards J, Michener T, Balogh A, Heller KN, Hodi FS. Ipilimumab in patients with melanoma and brain metastases: an open-label, phase 2 trial. Lancet Oncol. 2012;13(5):459–65. https://doi.org/10.1016/s1470-2045(12)70090-6.
62. Di Giacomo AM, Ascierto PA, Pilla L, Santinami M, Ferrucci PF, Giannarelli D, Marasco A, Rivoltini L, Simeone E, Nicoletti SV, Fonsatti E, Annesi D, Queirolo P, Testori A, Ridolfi R, Parmiani G, Maio M. Ipilimumab and fotemustine in patients with advanced melanoma (NIBIT-M1): an open-label, single-arm phase 2 trial. Lancet Oncol.

2012;13(9):879–86. https://doi.org/10.1016/s1470-2045(12)70324-8.
63. Di Giacomo AM, Ascierto PA, Queirolo P, Pilla L, Ridolfi R, Santinami M, Testori A, Simeone E, Guidoboni M, Maurichi A, Orgiano L, Spadola G, Del Vecchio M, Danielli R, Calabro L, Annesi D, Giannarelli D, Maccalli C, Fonsatti E, Parmiani G, Maio M. Three-year follow-up of advanced melanoma patients who received ipilimumab plus fotemustine in the Italian network for tumor biotherapy (NIBIT)-M1 phase II study. Ann Oncol. 2015;26(4):798–803. https://doi.org/10.1093/annonc/mdu577.
64. Goldberg SB, Gettinger SN, Mahajan A, Chiang AC, Herbst RS, Sznol M, Tsiouris AJ, Cohen J, Vortmeyer A, Jilaveanu L, Yu J, Hegde U, Speaker S, Madura M, Ralabate A, Rivera A, Rowen E, Gerrish H, Yao X, Chiang V, Kluger HM. Pembrolizumab for patients with melanoma or non-small-cell lung cancer and untreated brain metastases: early analysis of a non-randomised, open-label, phase 2 trial. Lancet Oncol. 2016;17(7):976–83. https://doi.org/10.1016/s1470-2045(16)30053-5.
65. Tawbi HA, Forsyth PA, Algazi A, Hamid O, Hodi FS, Moschos SJ, Khushalani NI, Lewis K, Lao CD, Postow MA, Atkins MB, Ernstoff MS, Reardon DA, Puzanov I, Kudchadkar RR, Thomas RP, Tarhini A, Pavlick AC, Jiang J, Avila A, Demelo S, Margolin K. Combined nivolumab and ipilimumab in melanoma metastatic to the brain. N Engl J Med. 2018;379(8):722–30. https://doi.org/10.1056/NEJMoa1805453.
66. Long GV, Atkinson V, Lo S, Sandhu S, Guminski AD, Brown MP, Wilmott JS, Edwards J, Gonzalez M, Scolyer RA, Menzies AM, McArthur GA. Combination nivolumab and ipilimumab or nivolumab alone in melanoma brain metastases: a multicentre randomised phase 2 study. Lancet Oncol. 2018;19(5):672–81. https://doi.org/10.1016/s1470-2045(18)30139-6.
67. Stupp R, Mason WP, van den Bent MJ, Weller M, Fisher B, Taphoorn MJ, Belanger K, Brandes AA, Marosi C, Bogdahn U, Curschmann J, Janzer RC, Ludwin SK, Gorlia T, Allgeier A, Lacombe D, Cairncross JG, Eisenhauer E, Mirimanoff RO. Radiotherapy plus concomitant and adjuvant temozolomide for glioblastoma. N Engl J Med. 2005;352(10):987–96. https://doi.org/10.1056/NEJMoa043330.
68. Alifieris C, Trafalis DT. Glioblastoma multiforme: pathogenesis and treatment. Pharmacol Ther. 2015;152:63–82. https://doi.org/10.1016/j.pharmthera.2015.05.005.
69. Cohen MH, Shen YL, Keegan P, Pazdur R. FDA drug approval summary: bevacizumab (Avastin) as treatment of recurrent glioblastoma multiforme. Oncologist. 2009;14(11):1131–8. https://doi.org/10.1634/theoncologist.2009-0121.
70. Lathia JD, Mack SC, Mulkearns-Hubert EE, Valentim CL, Rich JN. Cancer stem cells in glioblastoma. Genes Dev. 2015;29(12):1203–17. https://doi.org/10.1101/gad.261982.115.
71. Bao S, Wu Q, McLendon RE, Hao Y, Shi Q, Hjelmeland AB, Dewhirst MW, Bigner DD, Rich JN. Glioma stem cells promote radioresistance by preferential activation of the DNA damage response. Nature. 2006;444(7120):756–60. https://doi.org/10.1038/nature05236.
72. Chen J, Li Y, Yu TS, McKay RM, Burns DK, Kernie SG, Parada LF. A restricted cell population propagates glioblastoma growth after chemotherapy. Nature. 2012;488(7412):522–6. https://doi.org/10.1038/nature11287.
73. Osuka S, Sampetrean O, Shimizu T, Saga I, Onishi N, Sugihara E, Okubo J, Fujita S, Takano S, Matsumura A, Saya H. IGF1 receptor signaling regulates adaptive radioprotection in glioma stem cells. Stem Cells (Dayton, OH). 2013;31(4):627–40. https://doi.org/10.1002/stem.1328.
74. Osuka S, Van Meir EG. Overcoming therapeutic resistance in glioblastoma: the way forward. J Clin Invest. 2017;127(2):415–26. https://doi.org/10.1172/JCI89587.
75. Razavi S-M, Lee KE, Jin BE, Aujla PS, Gholamin S, Li G. Immune evasion strategies of glioblastoma. Front Surg. 2016;3:11. https://doi.org/10.3389/fsurg.2016.00011.
76. Roszman T, Elliott L, Brooks W. Modulation of T-cell function by gliomas. Immunol Today. 1991;12(10):370–4. https://doi.org/10.1016/0167-5699(91)90068-5.
77. Hao C, Parney IF, Roa WH, Turner J, Petruk KC, Ramsay DA. Cytokine and cytokine receptor mRNA expression in human glioblastomas: evidence of Th1, Th2 and Th3 cytokine dysregulation. Acta Neuropathol. 2002;103(2):171–8. https://doi.org/10.1007/s004010100448.
78. Wei B, Wang L, Zhao X, Du C, Guo Y, Sun Z. The upregulation of programmed death 1 on peripheral blood T cells of glioma is correlated with disease progression. Tumour Biol. 2014;35(4):2923–9. https://doi.org/10.1007/s13277-013-1376-9.
79. Fecci PE, Ochiai H, Mitchell DA, Grossi PM, Sweeney AE, Archer GE, Cummings T, Allison JP, Bigner DD, Sampson JH. Systemic CTLA-4 blockade ameliorates glioma-induced changes to the CD4+ T cell compartment without affecting regulatory T-cell function. Clin Cancer Res. 2007;13(7):2158–67. https://doi.org/10.1158/1078-0432.ccr-06-2070.
80. Reardon DA, Gokhale PC, Klein SR, Ligon KL, Rodig SJ, Ramkissoon SH, Jones KL, Conway AS, Liao X, Zhou J, Wen PY, Van Den Abbeele AD, Hodi FS, Qin L, Kohl NE, Sharpe AH, Dranoff G, Freeman GJ. Glioblastoma eradication following immune checkpoint blockade in an orthotopic, immunocompetent model. Cancer Immunol Res. 2016;4(2):124–35. https://doi.org/10.1158/2326-6066.cir-15-0151.

81. Bouffet E, Larouche V, Campbell BB, Merico D, de Borja R, Aronson M, Durno C, Krueger J, Cabric V, Ramaswamy V, Zhukova N, Mason G, Farah R, Afzal S, Yalon M, Rechavi G, Magimairajan V, Walsh MF, Constantini S, Dvir R, Elhasid R, Reddy A, Osborn M, Sullivan M, Hansford J, Dodgshun A, Klauber-Demore N, Peterson L, Patel S, Lindhorst S, Atkinson J, Cohen Z, Laframboise R, Dirks P, Taylor M, Malkin D, Albrecht S, Dudley RWR, Jabado N, Hawkins CE, Shlien A, Tabori U. Immune checkpoint inhibition for hypermutant glioblastoma multiforme resulting from germline biallelic mismatch repair deficiency. J Clin Oncol. 2016;34(19):2206–11. https://doi.org/10.1200/JCO.2016.66.6552.
82. Johanns TM, Miller CA, Dorward IG, Tsien C, Chang E, Perry A, Uppaluri R, Ferguson C, Schmidt RE, Dahiya S, Ansstas G, Mardis ER, Dunn GP. Immunogenomics of hypermutated glioblastoma: a patient with germline POLE deficiency treated with checkpoint blockade immunotherapy. Cancer Discov. 2016;6(11):1230–6. https://doi.org/10.1158/2159-8290.cd-16-0575.
83. Sampson JH, Vlahovic G, Sahebjam S, Omuro AMP, Baehring JM, Hafler DA, Voloschin AD, Paliwal P, Grosso J, Coric V, Cloughesy TF, Lim M, Reardon DA. Preliminary safety and activity of nivolumab and its combination with ipilimumab in recurrent glioblastoma (GBM): CHECKMATE-143. J Clin Oncol. 2015;33(15_suppl):3010. https://doi.org/10.1200/jco.2015.33.15_suppl.3010.
84. Weller M, Kaulich K, Hentschel B, Felsberg J, Gramatzki D, Pietsch T, Simon M, Westphal M, Schackert G, Tonn JC, von Deimling A, Davis T, Weiss WA, Loeffler M, Reifenberger G. Assessment and prognostic significance of the epidermal growth factor receptor vIII mutation in glioblastoma patients treated with concurrent and adjuvant temozolomide radiochemotherapy. Int J Cancer. 2014;134(10):2437–47. https://doi.org/10.1002/ijc.28576.
85. Desjardins A, Gromeier M, Herndon JE, Beaubier N, Bolognesi DP, Friedman AH, Friedman HS, McSherry F, Muscat AM, Nair S, Peters KB, Randazzo D, Sampson JH, Vlahovic G, Harrison WT, McLendon RE, Ashley D, Bigner DD. Recurrent glioblastoma treated with recombinant poliovirus. N Engl J Med. 2018;379(2):150–61. https://doi.org/10.1056/NEJMoa1716435.
86. Phuong LK, Allen C, Peng KW, Giannini C, Greiner S, TenEyck CJ, Mishra PK, Macura SI, Russell SJ, Galanis EC. Use of a vaccine strain of measles virus genetically engineered to produce carcinoembryonic antigen as a novel therapeutic agent against glioblastoma multiforme. Cancer Res. 2003;63(10):2462–9.
87. O'Rourke DM, Nasrallah MP, Desai A, Melenhorst JJ, Mansfield K, Morrissette JJD, Martinez-Lage M, Brem S, Maloney E, Shen A, Isaacs R, Mohan S, Plesa G, Lacey SF, Navenot JM, Zheng Z, Levine BL, Okada H, June CH, Brogdon JL, Maus MV. A single dose of peripherally infused EGFRvIII-directed CAR T cells mediates antigen loss and induces adaptive resistance in patients with recurrent glioblastoma. Sci Transl Med. 2017;9(399):eaaa0984. https://doi.org/10.1126/scitranslmed.aaa0984.
88. Grossman SA, Ye X, Lesser G, Sloan A, Carraway H, Desideri S, Piantadosi S. Immunosuppression in patients with high-grade gliomas treated with radiation and temozolomide. Clin Cancer Res. 2011;17(16):5473–80. https://doi.org/10.1158/1078-0432.ccr-11-0774.
89. Wild AT, Herman JM, Dholakia AS, Moningi S, Lu Y, Rosati LM, Hacker-Prietz A, Assadi RK, Saeed AM, Pawlik TM, Jaffee EM, Laheru DA, Tran PT, Weiss MJ, Wolfgang CL, Ford E, Grossman SA, Ye X, Ellsworth SG. Lymphocyte-sparing effect of stereotactic body radiation therapy in patients with unresectable pancreatic cancer. Int J Radiat Oncol Biol Phys. 2016;94(3):571–9. https://doi.org/10.1016/j.ijrobp.2015.11.026.
90. Saeed Bamashmos A, Ali A, Barnett A, Sagar S, Rybicki LA, Barnett GH, Mohammadi AM, Angelov L, Chao ST, Murphy ES, Suh JH, Yu JS, Peereboom DM, Stevens G, Ahluwalia MS, Wei W. Absolute lymphocyte count in patients with glioblastoma treated with temozolomide chemoradiation. J Clin Oncol. 2019;37(15_suppl):e13564. https://doi.org/10.1200/JCO.2019.37.15_suppl.e13564.
91. Brock CS, Newlands ES, Wedge SR, Bower M, Evans H, Colquhoun I, Roddie M, Glaser M, Brampton MH, Rustin GJS. Phase I trial of temozolomide using an extended continuous oral schedule. Cancer Res. 1998;58(19):4363–7.
92. Briegert M, Kaina B. Human monocytes, but not dendritic cells derived from them, are defective in base excision repair and hypersensitive to methylating agents. Cancer Res. 2007;67(1):26–31. https://doi.org/10.1158/0008-5472.CAN-06-3712.
93. Mathios D, Kim JE, Mangraviti A, Phallen J, Park CK, Jackson CM, Garzon-Muvdi T, Kim E, Theodros D, Polanczyk M, Martin AM, Suk I, Ye X, Tyler B, Bettegowda C, Brem H, Pardoll DM, Lim M. Anti-PD-1 antitumor immunity is enhanced by local and abrogated by systemic chemotherapy in GBM. Sci Transl Med. 2016;8(370):370ra180. https://doi.org/10.1126/scitranslmed.aag2942.
94. Mole RH. Whole body irradiation; radiobiology or medicine? Br J Radiol. 1953;26(305):234–41. https://doi.org/10.1259/0007-1285-26-305-234.
95. Demaria S, Ng B, Devitt ML, Babb JS, Kawashima N, Liebes L, Formenti SC. Ionizing radiation inhibition of distant untreated tumors (abscopal effect) is immune mediated. Int J Radiat Oncol Biol Phys. 2004;58(3):862–70. https://doi.org/10.1016/j.ijrobp.2003.09.012.
96. Ohba K, Omagari K, Nakamura T, Ikuno N, Saeki S, Matsuo I, Kinoshita H, Masuda J, Hazama H, Sakamoto I, Kohno S. Abscopal regression of hepatocellular carcinoma after radiotherapy for bone metastasis. Gut. 1998;43(4):575–7. https://doi.org/10.1136/gut.43.4.575.

97. Ehlers G, Fridman M. Abscopal effect of radiation in papillary adenocarcinoma. Br J Radiol. 1973;46(543):220–2. https://doi.org/10.1259/0007-1285-46-543-220.
98. Rees GJ, Ross CM. Abscopal regression following radiotherapy for adenocarcinoma. Br J Radiol. 1983;56(661):63–6. https://doi.org/10.1259/0007-1285-56-661-63.
99. Antoniades J, Brady LW, Lightfoot DA. Lymphangiographic demonstration of the abscopal effect in patients with malignant lymphomas. Int J Radiat Oncol Biol Phys. 1977;2(1–2):141–7. https://doi.org/10.1016/0360-3016(77)90020-7.
100. Reits EA, Hodge JW, Herberts CA, Groothuis TA, Chakraborty M, Wansley EK, Camphausen K, Luiten RM, de Ru AH, Neijssen J, Griekspoor A, Mesman E, Verreck FA, Spits H, Schlom J, van Veelen P, Neefjes JJ. Radiation modulates the peptide repertoire, enhances MHC class I expression, and induces successful antitumor immunotherapy. J Exp Med. 2006;203(5):1259–71. https://doi.org/10.1084/jem.20052494.
101. Germano G, Lamba S, Rospo G, Barault L, Magri A, Maione F, Russo M, Crisafulli G, Bartolini A, Lerda G, Siravegna G, Mussolin B, Frapolli R, Montone M, Morano F, de Braud F, Amirouchene-Angelozzi N, Marsoni S, D'Incalci M, Orlandi A, Giraudo E, Sartore-Bianchi A, Siena S, Pietrantonio F, Di Nicolantonio F, Bardelli A. Inactivation of DNA repair triggers neoantigen generation and impairs tumour growth. Nature. 2017;552(7683):116–20. https://doi.org/10.1038/nature24673.
102. Zeng J, See AP, Phallen J, Jackson CM, Belcaid Z, Ruzevick J, Durham N, Meyer C, Harris TJ, Albesiano E, Pradilla G, Ford E, Wong J, Hammers HJ, Mathios D, Tyler B, Brem H, Tran PT, Pardoll D, Drake CG, Lim M. Anti-PD-1 blockade and stereotactic radiation produce long-term survival in mice with intracranial gliomas. Int J Radiat Oncol Biol Phys. 2013;86(2):343–9. https://doi.org/10.1016/j.ijrobp.2012.12.025.
103. Kim JE, Patel MA, Mangraviti A, Kim ES, Theodros D, Velarde E, Liu A, Sankey EW, Tam A, Xu H, Mathios D, Jackson CM, Harris-Bookman S, Garzon-Muvdi T, Sheu M, Martin AM, Tyler BM, Tran PT, Ye X, Olivi A, Taube JM, Burger PC, Drake CG, Brem H, Pardoll DM, Lim M. Combination therapy with anti-PD-1, anti-TIM-3, and focal radiation results in regression of murine gliomas. Clin Cancer Res. 2017;23(1):124–36. https://doi.org/10.1158/1078-0432.ccr-15-1535.
104. Belcaid Z, Phallen JA, Zeng J, See AP, Mathios D, Gottschalk C, Nicholas S, Kellett M, Ruzevick J, Jackson C, Albesiano E, Durham NM, Ye X, Tran PT, Tyler B, Wong JW, Brem H, Pardoll DM, Drake CG, Lim M. Focal radiation therapy combined with 4-1BB activation and CTLA-4 blockade yields long-term survival and a protective antigen-specific memory response in a murine glioma model. PLoS One. 2014;9(7):e101764. https://doi.org/10.1371/journal.pone.0101764.
105. Silk AW, Bassetti MF, West BT, Tsien CI, Lao CD. Ipilimumab and radiation therapy for melanoma brain metastases. Cancer Med. 2013;2(6):899–906. https://doi.org/10.1002/cam4.140.
106. Minniti G, Anzellini D, Reverberi C, Cappellini GCA, Marchetti L, Bianciardi F, Bozzao A, Osti M, Gentile PC, Esposito V. Stereotactic radiosurgery combined with nivolumab or Ipilimumab for patients with melanoma brain metastases: evaluation of brain control and toxicity. J Immunother Cancer. 2019;7(1):102. https://doi.org/10.1186/s40425-019-0588-y.
107. Gabani P, Fischer-Valuck BW, Johanns TM, Hernandez-Aya LF, Keller JW, Rich KM, Kim AH, Dunn GP, Robinson CG, Chicoine MR, Huang J, Abraham CD. Stereotactic radiosurgery and immunotherapy in melanoma brain metastases: patterns of care and treatment outcomes. Radiother Oncol. 2018;128(2):266–73. https://doi.org/10.1016/j.radonc.2018.06.017.
108. Anderson ES, Postow MA, Wolchok JD, Young RJ, Ballangrud A, Chan TA, Yamada Y, Beal K. Melanoma brain metastases treated with stereotactic radiosurgery and concurrent pembrolizumab display marked regression; efficacy and safety of combined treatment. J Immunother Cancer. 2017;5(1):76. https://doi.org/10.1186/s40425-017-0282-x.
109. Lehrer EJ, Peterson J, Brown PD, Sheehan JP, Quinones-Hinojosa A, Zaorsky NG, Trifiletti DM. Treatment of brain metastases with stereotactic radiosurgery and immune checkpoint inhibitors: an international meta-analysis of individual patient data. Radiother Oncol. 2019;130:104–12. https://doi.org/10.1016/j.radonc.2018.08.025.

Part VI

Spine

Spinal Metastases

38

Souvik Roy, Nitin Agarwal, Steven A. Burton, John C. Flickinger, and Peter Gerszten

38.1 Introduction

Spinal metastases are the most common type of spinal tumors and are most often found in the extradural space [1]. Nearly 5–10% of all cancer patients will develop spinal metastases, the majority of which undergo palliative radiation therapy [2]. There are over 180,000 new cases of spinal metastases diagnosed in North America each year; these include 20,000 clinical cases of spinal cord compression [3, 4]. Spinal metastases reflect the end stages of cancer pathology and are understandably debilitating with a high risk for surgical morbidity. Up to 90% of patients present with spine bone pain and tumor extension threatening neurological integrity [5]. As curing the underlying pathology is unlikely, the goals of therapy include pain relief, slowing local disease progression, avoiding pathological fractures, and alleviating neurological compromise. The role of radiation therapy in the treatment of malignant spine tumors is well established [6–10].

Radiotherapy offers a noninvasive option which is often utilized as first-line therapy. Surgery is usually reserved for cases that involve spinal instability or subluxation or for isolated lesions which have intractable pain or persistent neurological deficits. Given the high surgical morbidity, minimally invasive surgical procedures are currently being explored as potential elements of treatment. Radiotherapy may also be used for multimodal treatment in conjugation with chemotherapy, hormonal therapy, radionuclide therapy, or surgical decompression and stabilization followed by radiotherapy [11, 12].

38.2 The Development of Spine Radiosurgery Technique

The radiation doses that are capable of being delivered via conventional radiotherapy are fundamentally limited by the relatively low tolerance of the spinal cord to radiation. External beam radiotherapy does not have the precision necessary to deliver large single-fraction doses of radiation in the context of spinal metastases due to the presence of local radiosensitive structures. This means that the radiation doses available with conventional radiotherapy would be below the optimal therapeutic dose lest

S. Roy · N. Agarwal
Department of Neurological Surgery, University of Pittsburgh Medical Center, Pittsburgh, PA, USA

S. A. Burton
Department of Radiation Oncology, University of Pittsburgh Medical Center, Pittsburgh, PA, USA

J. C. Flickinger · P. Gerszten (✉)
Department of Neurological Surgery, University of Pittsburgh Medical Center, Pittsburgh, PA, USA

Department of Radiation Oncology, University of Pittsburgh Medical Center, Pittsburgh, PA, USA
e-mail: gersztenpc@upmc.edu

© Springer Nature Switzerland AG 2020
A. Conti et al. (eds.), *CyberKnife NeuroRadiosurgery*,
https://doi.org/10.1007/978-3-030-50668-1_38

they exacerbate neurological deficits [8, 13]. For this reason, spine tumors often progress or recur after cEBRT is attempted. Precise confinement of radiation dose to treatment volume is crucial as it increases the likelihood of tumor control while minimizing the risk of spinal cord injury [14].

Spine SRS was first published in 1995 using a stereotactic frame-based system resembling the earlier techniques of intracranial stereotactic radiosurgery [15]. Although this technique was feasible, it was both invasive and impractical. CyberKnife (Accuray Incorporated, Sunnyvale, CA), introduced in 1994, was the first prototype for an image-guided robotic radiosurgical delivery system. This new system was able to overcome the limitations of stereotactic frame placement. The system utilized multibeam delivery and a dynamic X-ray targeting system accounting for real-time inter- and intrafraction motion to achieve a 1-mm accuracy of dose delivery with a steep dose falloff. This allowed safe dose escalation for spine tumors which were in close proximity to the spinal cord without invasive immobilization schemes.

Recent advances have allowed the radiosurgical treatment of malignant lesions in the paraspinal region, vertebral bodies, as well as intradural and intramedullary locations. These advancements include improvements to imaging technology for localization and treatment planning along with the advent of intensity-modulated radiotherapy (IMRT). This increase in accuracy and ability to deliver a tumoricidal dose has improved the responses of malignant spine tumors previously considered resistant to cEBRT [9].

Regardless of the technology utilized, target delineation and tumor contouring are crucial for SRS treatment. Failure in this aspect may result in tumor progression or spinal cord injury. Lesions are optimally visualized with magnetic resonance imaging (MRI). However, treatment simulation is based on computed tomography (CT) [16]. Radiosurgery systems have MRI and CT fusion algorithms to identify targets and assist in treatment planning. Positron emission tomography may also be used to identify tumor targets more precisely. CT with myelography may be useful in delineating the spinal cord for cases with previous spinal instrumentation.

Gross tumor volume (GTV) must be precisely contoured to identify the tumor visualized on MRI and CT. T1-weighted and T2-weighted short tau inversion recovery is the best imaging modality for tumor identification. The fact that tumors have variable intensities on T2-weighted images makes that modality useless for tumor delineation. However, axial T2-weighted images are the best way to assess spinal cord impingement. Based on the GTV contour, a clinical target volume (CTV) is drawn to account for microscopic disease outside the GTV contour. In contrast to intracranial metastases, vertebral body tumors are thought to have an infiltrative penumbra which puts the entire vertebra at risk. Therefore, if the GTV involves a small part of the vertebral body, the CTV will involve the entire vertebral body [6, 17]. The International Spine Radiosurgery Consortium recommends a CTV expansion of the entire vertebral body for an anterior lesion or the entire spinous process and bilateral lamina for a posterior lesion [17]. A retrospective single-institution study found that contouring the entire vertebral body may reduce the risk of recurrence and improve symptom relief over partial vertebral body coverage [18].

38.3 Radiation Dose Selection

Tumor dosing is guided by tumor histology along with radiation tolerance of radiosensitive structures such as the spinal cord and cauda equina [19]. It is crucial to keep treatment history in mind as previous radiation quantity to normal tissue must be taken into account. There has been no large experiment to date with spine radiosurgery or hypofractionated radiotherapy that has fully developed the optimal doses for radiosurgical treatment of spinal metastases. Therefore, dose and fractionation differ by institution [20]. Currently, centers that are capable of using intensity-modulated, near-simultaneous, CT image-guided stereotactic radiotherapy techniques use doses from 6 to 30 Gy in one to five fractions [21–27]. That being said, there is no experience precisely outlining the radiation tolerance of radiosensitive structures within the spinal canal. The atlas curated by Kong et al. defines the maximal radiation before spinal cord damage as 18 Gy in three fractions for stereotactic radiation therapy [28]. Sahgal et al. compared the dosing of five patients with radiation-induced myelopathy to a control group and recommended a maximum of 10 Gy to a single point [29]. A later study by Sahgal et al. found that stereotactic radiosurgery given at least 5 months after conventional radiotherapy appears to be safe, provided that the total biologically effective doses do not exceed 70 Gy [30].

At our institution, maximum tumor dose is maintained at 12.5–22.5 Gy (mean 20 Gy) delivered in a single fraction. Tumor dose is prescribed to the 80% isodose line. A maximum dose of 20 Gy with 16 Gy to the tumor margin appears to offer good tumor control without radiation-induced injury. During each spine radiosurgery case at our institution, either the spinal cord or cauda equina is outlined as a critical structure. The maximal spinal cord dose for treatment planning calculations is 10 Gy. This limit is raised to 11 Gy for the cauda equina as we outline the entire spinal canal which is not the actual neural tissue [31]. A limit for maximal dose is set for 2 Gy for each kidney and 8 Gy for the bowel. This is particularly important if patients have received previous radiotherapy or chemotherapy in these regions.

38.4 Radiation Fractionation

Since the introduction of spine SRS, several clinical trials have compared the efficacy of various dose-fractionation protocols for spinal bone metastases [32–39]. These studies have demonstrated the clinical efficacy of single-fraction therapy [32, 40]. Bruner et al. conducted a Radiation Therapy Oncology Group phase III trial comparing 8 Gy in a single fraction to 30 Gy over 10 fractions in 363 patients. They found no differences in health-related quality of life and health utility indexes between the two arms over a follow-up period of 3 months [32]. Wu et al. conducted a meta-analysis of 16 trials comparing single to a variety of multiple fraction schedules and found no differences in pain relief along with no dose–response relationship [39]. These trials often used 8 Gy for their single-fraction arms. Given our institution's extensive experience with the intracranial radiosurgery principles involved with the Leksell Gamma Knife, we initially elected to utilize a single-fraction radiosurgery technique for spinal metastases as well. The single-fraction paradigm offers convenience to the patient while minimizing interference with other cancer treatment schedules. We have continued to prefer the single-fraction paradigm at our institution due to good clinical response with a lack of adverse effects on normal tissue.

Additionally, single-fraction radiosurgery carries the benefit that treatment can be completed in a single day rather than over several weeks. This can have a positive impact on patients who have an otherwise limited life expectancy. Furthermore, single-fraction radiosurgery is convenient for patients as they may need to travel large distances to reach a spine radiosurgery center.

38.5 Clinical Outcomes

Spine radiosurgery tends to have good outcomes in terms of tumor control, as stated previously. In terms of symptom control, outcomes are equally as promising. In the aforementioned cohort of our first 500 patients, 86% of patients who were experiencing pain symptoms saw a relief in the 12 months after receiving radiosurgery [41]. Similar success rates were seen by Degen et al. when using visual analogue scales (VAS) and the 12-item Short Form Health Survey (SF-12) [42]. These results have been corroborated with other centers as well [22, 26, 43–45]. Back and leg pain usually decreases within weeks after radiosurgery and can rarely decrease within days. Spine radiosurgery can also alleviate radicular pain caused by tumor compression on adjacent nerve roots.

In terms of treating progressive neurological deficits, radiosurgery is employed when open surgical intervention is determined to be contraindicated by the surgical team. Therefore, the outcomes are more mixed as providers may choose differing approaches. Interestingly, pain is a more common aspect of metastatic spinal tumors than progressive neurological deficits. Our center saw at least some improvement in 36 of 42 patients (86%) [11]. Yamada et al. reported a 90% and 92% palliation rate when examining weakness and paresthesias, respectively. However, Degen et al. prospectively examined a group of 52 patients receiving spine radiosurgery for metastases and found deficit improvement in 16, no change in 24, and worsening in 11. Of those who worsened, five were the result of an irradiated tumor. However, overall quality of life scores were maintained throughout their 1-year follow-up period [42]. Of note, intermedullary spinal metastases tend to have significantly poorer outcomes [46].

There are several theoretical benefits to utilizing radiosurgery as a primary treatment for spine metastases. Early treatment of metastatic lesions may avoid the emergence of neurologic symptoms which may be irreversible or debilitating. Additionally, it may also remove the need for extensive spinal surgeries for decompression or fixation in a debilitated patient population. Conformation radiosurgery avoids irradiating large segments of the spinal column which can have a negative effect on bone marrow reserve. Preserving bone marrow is important for patient groups receiving chemotherapy. Within the goals of palliation, a minimally invasive procedure with few complications can be more preferable to patients than a treatment modality with significant side effects [47–49].

Table 38.1 Candidate lesions for spine radiosurgery

Well-circumscribed lesions
Minimal spinal cord compromise
Radioresistant lesions that would benefit from a radiosurgical boost
Residual tumor after surgery
Previously irradiated lesions precluding further external beam irradiation
Recurrent surgical lesions
Lesions requiring difficult surgical approaches
Relatively short life expectancy as an exclusion criterion for open surgical intervention
Lesions not requiring open spinal stabilization techniques

Table 38.2 Indications for radiosurgery for spinal metastases

Pain
Primary treatment modality
Prevention of tumor progression
Radiation boost
Progressive neurologic deficit
Residual tumor after surgery
Postsurgical treatment of residual disease
Postsurgical tumor progression

38.6 Indications

Spine radiosurgery was initially introduced into the treatment paradigm for spinal tumors for a subset of oncology patients who did not meet the criteria for other forms of therapy including conventional radiotherapy and open surgical techniques. The indications for spinal radiosurgery at our center have evolved over time and will continue to do so. This is a similar path that intracranial radiosurgery took to evolve its indications. The ideal lesion for spinal radiosurgery is well circumscribed with minimal spinal cord compromise. Other important characteristics are listed in Table 38.1. The current indications for radiosurgery for spinal metastases are listed in Table 38.2. These indications can be grouped into one of three treatment paradigms: (1) primary definitive therapy for previously unirradiated tumors, (2) salvage radiosurgery for recurrent or progressive tumors having failed prior to cEBRT, and (3) postoperative radiosurgery after surgical intervention with or without spinal stabilization [50].

38.6.1 Radiosurgery as Neoadjuvant or Definitive Therapy

Spinal SRS as definitive local treatment of malignant spine tumors is one of the technology's most significant applications. However, this approach is restricted to tumors that involve the vertebrae alone or with minimal epidural abutment. High-grade spinal cord compression is a relative contraindication to exclusively administering SRS treatment. The precise application of

SRS varies with tumor histology. Radiosensitive tumors such as hematological malignancies can be treated with hypofractionated regimens. However, the greatest application of SRS is for tumors such as renal cell, melanoma, and sarcoma which maybe radioresistant prior to cEBRT. Large series reporting outcomes after cEBRT for spinal metastases only show good tumor control when they do not stratify for radiosensitivity of tumors. Marked difference between tumor responses can be seen when stratifying for tumor radiosensitivity [2]. Maranzano et al. found that radioresistant tumors such as hepatocellular carcinoma had only a 20% response rate at 1–3 months. On the other hand, radiosensitive tumors had an 80% response rate at 16 months [51]. In contrast to the poor responses seen to cEBRT regimens, several series show response rates greater than 90% with long-term follow-up when using SRS [34–37, 51–55]. The largest published series to date reported outcomes of a prospective cohort of 500 cases in 393 patients who had a variety of primary tumor histologies and were treated with single-fraction radiosurgery at all spine levels [41]. Maximum tumor dose was 12.5–22.5 Gy (mean 20 Gy). Pain and radiographic tumor controls were 86% and 90%, respectively, with a median follow-up of 21 months. Furthermore, Chao et al. used a cohort of 174 patients receiving a mean dose of 14 Gy to establish a prognostic index using recursive partitioning analysis. There was wide variability in survival. However, the index was found to be predictive for overall survival and may be useful in predicting which patients may benefit most from radiosurgery [56, 57].

A systematic review conducted in 2009 looked at 29 case series for both conventional radiotherapy and radiosurgery for metastatic spine disease. A Guyatt analysis showed that radiosurgery for metastatic spine disease was both safe and effective with durable symptomatic response and local control for radioresistant histologies, regardless of prior fractionated radiotherapy [2]. It is also important to note that the neoadjuvant approach advantageously allows the ability to carefully delineate tumor volumes in unviolated tissue planes.

38.6.2 Radiosurgery in the Setting of Reirradiation

Spine radiosurgery is frequently used to halt radiographic tumor progression which can be seen after conventional irradiation treatment fails. This often means that conventional irradiation is no longer appropriate as the maximal spinal cord dose has been reached. Many tumors treated with cEBRT may recur, or symptoms may persist requiring further palliative intervention [39]. Surgical intervention can be constrained by radiation-induced hypoxia and fibrosis which impair wound healing. In this role, spine radiosurgery can be employed as a “salvage” technique to avoid significant reirradiation of the spinal cord by conventional methods [58, 59].

The most frequent application of SRS is in treating tumor recurrence after prior cEBRT. In this situation, it is important to keep in mind the radiation tolerance of the spinal cord. Sahgal et al. reviewed the timing and safe dosage for reirradiation SRS of the spine and described the impact of dose “hot spots” on spinal cord tolerance [30, 60]. Within our first 500 patients treated with single-fraction SRS, 344 (69%) had received previous external beam radiation and had no cases of spinal cord toxicity [41].

Choi et al. reported a series of 51 cases involving recurrent tumors in close proximity to a previously irradiated spinal cord. Forty-one (80%) of the lesions were treated with a multisession schedule of two to five fractions with a median marginal target dose of 20 Gy. Five of the 13 failures were due to underdosing an epidural tumor out of concern for spinal toxicity. One patient had grade 4 spinal cord toxicity 6 months after receiving SRS in two fractions. The patient should be treated in multiple sessions if the single-session option would expose the spinal cord to more than 70% of the prescription dose. This is partially because interfraction tumor reoxygenation and cell reassortment might increase tumor kill by minimizing hypoxia-induced radiation resistance and cell cycle-specific radiation sensitivity, respectively [61].

Finally, Damast et al. specifically examined causes of local failure after SRS for recurrent spine metastases in a series of 92 patients who received either five 4 Gy fractions or 6 Gy fractions [62]. Forty-eight percent of the patients had received decompressive surgery prior to SRS treatment. No patients had radiation-induced myelopathy. The only treatment characteristic found to significantly impact incidence of local failure was total dose. Therefore, a significant decrease in local failure rate was observed by utilizing a higher treatment dose while avoiding myelopathy. The ultimate goal of reirradiation SRS should be high-dose single-fraction treatment which may offer better local tumor control rates than hypofractionated schedules.

38.6.3 Radiosurgery as a Postsurgical Adjuvant Treatment

Within this multimodality treatment paradigm, surgery is utilized as an initial step to treat tumors with high-grade spinal cord or cauda equina compression or for cases of significant spinal instability. The majority of solid tumors do not respond to conventional radiotherapy techniques and can benefit from decompressive surgery or stabilization surgery. Radiation alone cannot stabilize an unstable spine [63].

Prospective randomized trial data have established the utility of direct surgical decompression for patients with metastatic spinal disease with symptomatic cord compression. Surgery plus postoperative cEBRT was shown to improve ambulation rates, bowel bladder continence, narcotic utilization, and overall survival over conventional radiotherapy alone. Although surgery is essential for resolving spinal instability, effective radiation therapy is critical for local tumor control [64].

For this purpose, adjuvant SRS tends to have improved tumor control compared to adjuvant cEBRT. Klekamp et al. reported the local tumor control rates for 106 patients undergoing decompressive surgery with adjuvant cEBRT. The recurrence rates derived from Kaplan-Meier curves were 58% after 6 months and 96% after 4 years [65]. Essentially, this meant that tumors would recur in almost all patients given that the patient lived long enough. This recurrence rate is very different from the recurrence rates allowed by adjuvant SRS. Rock et al. evaluated this treatment combination in a prospective cohort of 18 patients and found a tumor control rate of 94% [66].

The theoretical rationale behind multimodal therapy is that one can potentially perform a less aggressive tumor resection with the expectation that tumor control can be taken care of by high-dose radiation therapy. This is particularly useful for aggressive, radioresistant tumors such as renal cell carcinoma for which en bloc removal was thought to be essential for local tumor control. With multimodal treatment, the focus of surgery shifts to epidural decompression and instrumented stabilization which is termed "separation surgery." This approach was utilized by Laufer et al. to treat 186 patients. For this cohort, 1-year actuarial control was 84% for all patients and 96% for those who received high-dose radiosurgery of 24–30 Gy in three fractions [67]. Given the steep falloff gradient achieved with SRS, high-dose radiosurgery can be offered early in the postoperative period, as early as 1 week.

Moulding et al. reported on 21 patients who underwent separation surgery and posterior segmental instrumentation for radioresistant tumors. GTV for SRS was based upon the preoperative tumor volume rather than the postoperative residual tumor. Additionally, the spinal cord and thecal sac contours were visualized using CT myelography which provides excellent anatomic detail even in the presence of implant artifacts. The GTV received 24 Gy in 16 patients and 18–21 Gy in 5 patients. Overall local tumor control was 81% with a 1-year failure rate of 9.5%. The local tumor control rate was significantly higher in the group receiving 24 Gy versus a lower dose of radiation (94% versus 60%, respectively) [12]. When utilizing postoperative radiosurgery, difficulties exist in delineating the interface between the spinal cord and residual tumor which can be exacerbated by implant artifacts. As a result, this location is a potential site of underdosing or even overdosing. To overcome this concern, CT myelography or PET/CT imaging may be useful

[68, 69]. However, PET/CT may be falsely positive immediately after surgery.

38.7 Contraindications

Contraindications to spinal radiosurgery include overt spinal instability, neurological deficit due to bony compression, and previous radiation summating to tolerance doses of neural elements. The spinal instability neoplastic score (SINS) can be utilized to assess the degree of instability and determine if fixation is necessary in addition to radiosurgery [70]. More clinical experience and research are necessary to fully elucidate the absolute contraindications to spinal radiosurgery (Figs. 38.1 and 38.2).

38.8 Complications

As mentioned earlier the feared complication of spine radiosurgery is radiation-induced spinal cord injury [71]. We found no clinically detectable signs of acute or subacute radiation-induced spinal cord injury in our series of metastatic spinal lesions during a follow-up of >60 months [41]. The lack of clinical signs was corroborated by the lack of radiographic signs on MRI imaging. A multicenter summation of 1075 patient cases found only six cases of radiation-induced myelopathy at a mean follow-up of 6.3 months (range 2–9 months) [14]. These cases occurred over a spectrum of dose parameters which prevents the identification of a specific dose toxicity. Yamada et al. utilized a maximum dose con-

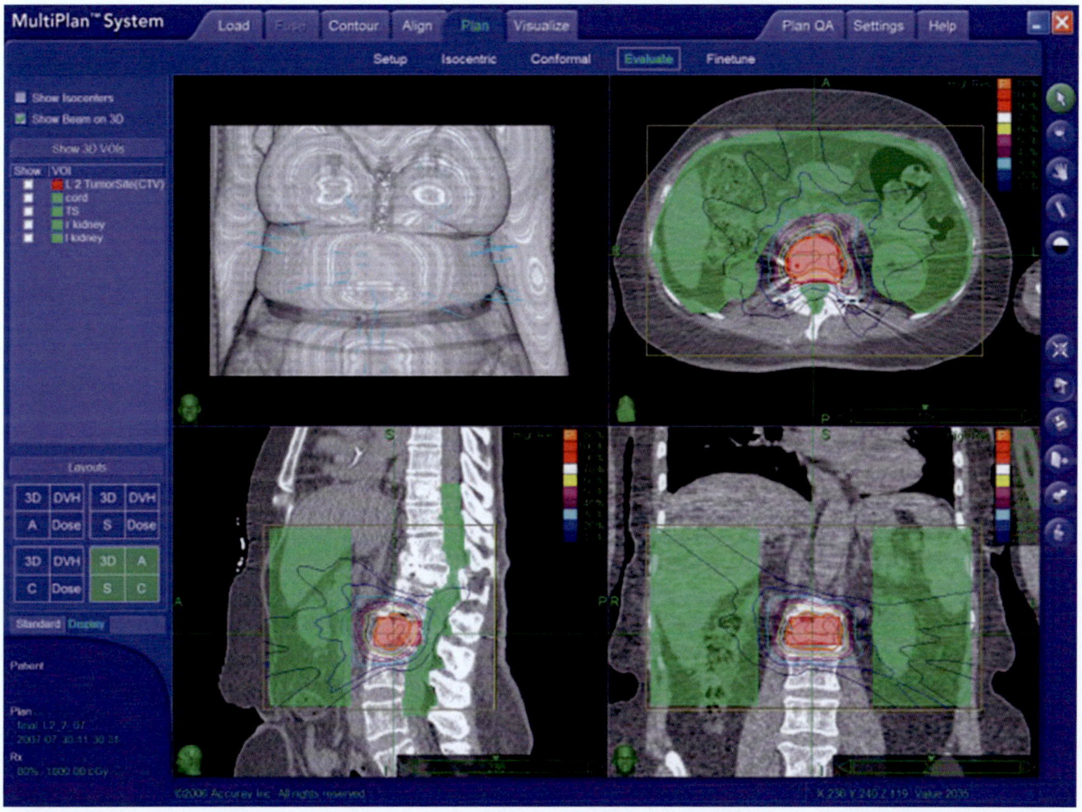

Fig. 38.1 Case example of a 53-year-old woman with breast cancer with a symptomatic L2 metastasis. She had undergone conventional fractionated radiotherapy to the lesion with only temporary pain relief. It was decided to treat the lesion with SRS. The GTV was treated with a prescribed dose of 18 Gy to the 80% isodose line delivered in a single fraction. The GTV was 43 cm^3 and the D_{max} was 22.5 Gy. Critical structures for dose avoidance included the cauda equina, kidneys, and bowel. The maximum dose to a single voxel of the cauda equina was less than 12 Gy

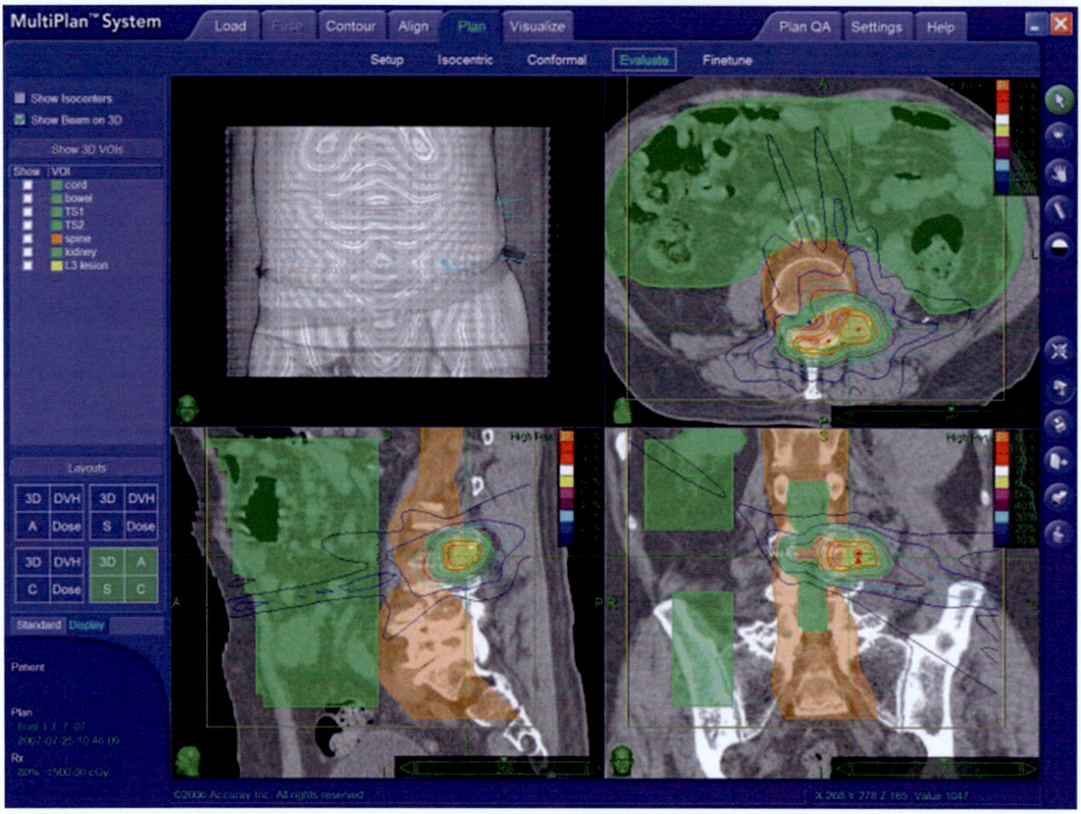

Fig. 38.2 A 70-year-old man with prostate cancer with a symptomatic L3 metastasis. He had undergone conventional fractionated radiotherapy to the lesion with only temporary relief of his pain. The GTV was treated with a prescribed dose of 15 Gy to the 80% isodose line delivered in a single fraction. The GTV was 37 cm^3 and the D_{max} was 22.5 Gy. Critical structures for dose avoidance included the cauda equina, kidneys, and bowel. The maximum dose to a single voxel of the cauda equina was 10 Gy

straint of 14 Gy to any portion of the spinal cord and noted no cases of toxicity [72].

Immediate complications of the radiosurgical procedure itself are usually mild and self-limited. They included transient esophagitis, dysphagia, paresthesias, diarrhea, and flare-ups of tumor-related pain. These complications may be treated prophylactically with single dose steroids. Given the nature of radiosurgery, adjacent levels are not included in the radiation field. One possible concern is that tumors may metastasize to adjacent vertebra. However in our case series, we found the rate of this occurrence to be 5%, justifying the decision not to irradiate other levels [73].

Finally, there is discussion in the literature of the risk of vertebral compression fractures (VCF) after SRS. Radiation therapy is known to predispose patients to spontaneous bony fractures in a dose-dependent manner, and vertebrae containing lesions may already be at risk of fracture. However, there is uncertainty regarding the natural history of metastatic spine disease, and the risk associated with SRS may be overstated in the literature. The actual association between SRS and post-treatment VCFs is still unclear [74].

38.9 Conclusion

Spine radiosurgery is an effective, minimally invasive therapy for palliation of spinal metastases and can be used as a first-line option. Pain is the most common indication for radiosurgery, and outcomes show that spine radiosurgery can achieve effective pain control. Spine SRS has been shown to achieve better tumor control than cEBRT and has begun changing treatment paradigms for spinal metastases. Dosing still varies by institution; however, single-fraction radiation is a convenient schedule which has been shown to be as effective as multi-fraction. Radiation-induced damage of healthy spinal tissue is the largest concern when considering radiosurgery; however, the CyberKnife system minimizes this risk. As confidence and experience with spinal radiosurgery as a postoperative adjuvant have increased, surgical resection has become less aggressive, and a minimally invasive paradigm has prevailed.

References

1. Agarwal N. Neurosurgery fundamentals. New York: Thieme; 2019.
2. Gerszten PC, Mendel E, Yamada Y. Radiotherapy and radiosurgery for metastatic spine disease: what are the options, indications, and outcomes? Spine. 2009;34:S78–92.
3. McClelland S III, Kim E, Passias PG, Murphy JD, Attia A, Jaboin JJ. Spinal stereotactic body radiotherapy in the United States: a decade-long nationwide analysis of patient demographics, practice patterns, and trends over time. J Clin Neurosci. 2017;46:109–12.
4. Sahgal A, Atenafu EG, Chao S, Al-Omair A, Boehling N, Balagamwala EH, et al. Vertebral compression fracture after spine stereotactic body radiotherapy: a multi-institutional analysis with a focus on radiation dose and the spinal instability neoplastic score. J Clin Oncol Off J Am Soc Clin Oncol. 2013;31:3426–31.
5. Klimo P, Schmidt MH. Surgical management of spinal metastases. Oncologist. 2004;9:188–96.
6. Bilsky MH, Angelov L, Rock J, Weaver J, Sheehan J, Rhines L, et al. Spinal radiosurgery: a neurosurgical perspective. J Radiosurg SBRT. 2011;1:47.
7. Black P. Spinal metastasis: current status and recommended guidelines for management. Neurosurgery. 1979;5:726–46.
8. Faul CM, Flickinger JC. The use of radiation in the management of spinal metastases. J Neuro-Oncol. 1995;23:149–61.
9. Guckenberger M, Mantel F, Gerszten PC, Flickinger JC, Sahgal A, Létourneau D, et al. Safety and efficacy of stereotactic body radiotherapy as primary treatment for vertebral metastases: a multi-institutional analysis. Radiat Oncol. 2014;9:226.
10. Husain ZA, Sahgal A, De Salles A, Funaro M, Glover J, Hayashi M, et al. Stereotactic body radiotherapy for de novo spinal metastases: systematic review: International Stereotactic Radiosurgery Society practice guidelines. J Neurosurg Spine. 2017;27:295–302.
11. Gerszten PC, Burton SA. Radiosurgery of spinal metastases. In: Gerszten PC, Ryu S, editors. Spine radiosurgery. New York: Thieme; 2009. p. 82–90.
12. Moulding HD, Elder JB, Lis E, Lovelock DM, Zhang Z, Yamada Y, et al. Local disease control after decompressive surgery and adjuvant high-dose single-fraction radiosurgery for spine metastases. J Neurosurg Spine. 2010;13:87–93.
13. Ryu SI, Chang SD, Kim DH, Murphy MJ, Le Q-T, Martin DP, et al. Image-guided hypo-fractionated stereotactic radiosurgery to spinal lesions. Neurosurgery. 2001;49:838–46.
14. Gibbs IC, Patil C, Gerszten PC, Adler JR Jr, Burton SA. Delayed radiation-induced myelopathy after spinal radiosurgery. Neurosurgery. 2009;64:A67–72.
15. Hamilton AJ, Lulu BA, Fosmire H, Stea B, Cassady JR. Preliminary clinical experience with linear accelerator-based spinal stereotactic radiosurgery. Neurosurgery. 1995;36:311–9.
16. Thibault I, Chang EL, Sheehan J, Ahluwalia MS, Guckenberger M, Sohn M-J, et al. Response assessment after stereotactic body radiotherapy for spinal metastasis: a report from the SPIne response assessment in neuro-oncology (SPINO) group. Lancet Oncol. 2015;16:e595–603.
17. Cox BW, Spratt DE, Lovelock M, Bilsky MH, Lis E, Ryu S, et al. International Spine Radiosurgery Consortium consensus guidelines for target volume definition in spinal stereotactic radiosurgery. Int J Radiat Oncol Biol Phys. 2012;83:e597–605.
18. Patel VB, Wegner RE, Heron DE, Flickinger JC, Gerszten P, Burton SA. Comparison of whole versus partial vertebral body stereotactic body radiation therapy for spinal metastases. Technol Cancer Res Treat. 2012;11:105–15.
19. Sahgal A, Weinberg V, Ma L, Chang E, Chao S, Muacevic A, et al. Probabilities of radiation myelopathy specific to stereotactic body radiation therapy to guide safe practice. Int J Radiat Oncol Biol Phys. 2013;85:341–7.
20. Toussaint A, Richter A, Mantel F, Flickinger JC, Grills IS, Tyagi N, et al. Variability in spine radiosurgery treatment planning–results of an international multi-institutional study. Radiat Oncol. 2016; 11:57.
21. Bilsky MH, Yamada Y, Yenice KM, Lovelock M, Hunt M, Gutin PH, et al. Intensity-modulated stereotactic radiotherapy of paraspinal tumors: a preliminary report. Neurosurgery. 2004;54:823–31.
22. Chang EL, Shiu AS, Mendel E, Mathews LA, Mahajan A, Allen PK, et al. Phase I/II study of stereotactic body radiotherapy for spinal metastasis and its pattern of failure. J Neurosurg Spine. 2007;7:151–60.
23. Joaquim AF, Ghizoni E, Tedeschi H, Pereira EB, Giacomini LA. Stereotactic radiosurgery for spinal metastases: a literature review. Einstein (São Paulo). 2013;11:247–55.

24. Klish MD, Watson GA, Shrieve DC. Radiation and intensity-modulated radiotherapy for metastatic spine tumors. Neurosurg Clin. 2004;15:481–90.
25. Rock JP, Ryu S, Yin F-F. Novalis radiosurgery for metastatic spine tumors. Neurosurg Clin N Am. 2004;15:503–9.
26. Ryu S, Fang Yin F, Rock J, Zhu J, Chu A, Kagan E, et al. Image-guided and intensity-modulated radiosurgery for patients with spinal metastasis. Cancer. 2003;97:2013–8.
27. Yamada Y, Lovelock DM, Yenice KM, Bilsky MH, Hunt MA, Zatcky J, et al. Multifractionated image-guided and stereotactic intensity-modulated radiotherapy of paraspinal tumors: a preliminary report. Int J Radiat Oncol Biol Phys. 2005;62:53–61.
28. Kong F-M, Ritter T, Quint DJ, Senan S, Gaspar LE, Komaki RU, et al. Consideration of dose limits for organs at risk of thoracic radiotherapy: atlas for lung, proximal bronchial tree, esophagus, spinal cord, ribs, and brachial plexus. Int J Radiat Oncol Biol Phys. 2011;81:1442–57.
29. Sahgal A, Ma L, Gibbs I, Gerszten PC, Ryu S, Soltys S, et al. Spinal cord tolerance for stereotactic body radiotherapy. Int J Radiat Oncol Biol Phys. 2010;77:548–53.
30. Sahgal A, Ma L, Weinberg V, Gibbs IC, Chao S, Chang U-K, et al. Reirradiation human spinal cord tolerance for stereotactic body radiotherapy. Int J Radiat Oncol Biol Phys. 2012;82:107–16.
31. Gerszten PC, Quader M, Novotny J Jr, Flickinger JC. Prospective evaluation of spinal cord and cauda equina dose constraints using cone beam computed tomography (cbct) image guidance for spine radiosurgery. J Radiosurg SBRT. 2011;1:197.
32. Bruner D, Winter K, Hartsell W, Konski A, Curran W, Roach M, et al. Prospective health-related quality of life valuations (utilities) of 8 Gy in 1 fraction vs 30 Gy in 10 fractions for palliation of painful bone metastases: Preliminary results of RTOG 97–14. Int J Radiat Oncol Biol Phys. 2004;60:S142.
33. Hartsell WF, Scott CB, Bruner DW, Scarantino CW, Ivker RA, Roach M III, et al. Randomized trial of short-versus long-course radiotherapy for palliation of painful bone metastases. J Natl Cancer Inst. 2005;97:798–804.
34. Maranzano E, Latini P, Perrucci E, Beneventi S, Lupattelli M, Corgna E. Short-course radiotherapy (8 Gy× 2) in metastatic spinal cord compression: an effective and feasible treatment. Int J Radiat Oncol Biol Phys. 1997;38:1037–44.
35. Rades D, Fehlauer F, Stalpers LJA, Wildfang I, Zschenker O, Schild SE, et al. A prospective evaluation of two radiotherapy schedules with 10 versus 20 fractions for the treatment of metastatic spinal cord compression. Cancer. 2004;101:2687–92.
36. Rades D, Karstens JH, Alberti W. Role of radiotherapy in the treatment of motor dysfunction due to metastatic spinal cord compression: comparison of three different fractionation schedules. Int J Radiat Oncol Biol Phys. 2002;54:1160–4.
37. Rades D, Stalpers L, Veninga T, Schulte R, Hoskin PJ, Obralic N, et al. Evaluation of five radiation schedules and prognostic factors for metastatic spinal cord compression. J Clin Oncol. 2005;23:3366–75.
38. Redmond KJ, Sahgal A, Foote M, Knisely J, Gerszten PC, Chao ST, et al. Single versus multiple session stereotactic body radiotherapy for spinal metastasis: the risk–benefit ratio. Future Oncol. 2015;11:2405–15.
39. Wu JS-Y, Wong R, Johnston M, Bezjak A, Whelan T. Meta-analysis of dose-fractionation radiotherapy trials for the palliation of painful bone metastases. Int J Radiat Oncol Biol Phys. 2003;55:594–605.
40. Chow E, Van Der Linden YM, Roos D, Hartsell WF, Hoskin P, Wu JSY, et al. Single versus multiple fractions of repeat radiation for painful bone metastases: a randomised, controlled, non-inferiority trial. Lancet Oncol. 2014;15:164–71.
41. Gerszten PC, Burton SA, Ozhasoglu C, Welch WC. Radiosurgery for spinal metastases: clinical experience in 500 cases from a single institution. Spine. 2007;32:193–9.
42. Degen JW, Gagnon GJ, Voyadzis J-M, McRae DA, Lunsden M, Dieterich S, et al. CyberKnife stereotactic radiosurgical treatment of spinal tumors for pain control and quality of life. J Neurosurg Spine. 2005;2:540–9.
43. Benzil DL, Saboori M, Mogilner AY, Rocchio R, Moorthy CR. Safety and efficacy of stereotactic radiosurgery for tumors of the spine. J Neurosurg. 2004;101:413–8.
44. Milker-Zabel S, Zabel A, Thilmann C, Schlegel W, Wannenmacher M, Debus J. Clinical results of retreatment of vertebral bone metastases by stereotactic conformal radiotherapy and intensity-modulated radiotherapy. Int J Radiat Oncol Biol Phys. 2003;55:162–7.
45. Ryken TC, Meeks SL, Pennington EC, Hitchon P, Traynelis V, Mayr NA, et al. Initial clinical experience with frameless stereotactic radiosurgery: analysis of accuracy and feasibility. Int J Radiat Oncol Biol Phys. 2001;51:1152–8.
46. Sung W-S, Sung M-J, Chan JH, Manion B, Song J, Dubey A, et al. Intramedullary spinal cord metastases: a 20-year institutional experience with a comprehensive literature review. World Neurosurg. 2013;79:576–84.
47. Adler JR Jr, Colombo F, Heilbrun MP, Winston K. Toward an expanded view of radiosurgery. Neurosurgery. 2004;55:1374–6.
48. Coste-Manière È, Olender D, Kilby W, Schulz R. Robotic whole body stereotactic radiosurgery: clinical advantages of the CyberKnife® integrated system. Int J Med Robot. 2005;1:28–39.
49. Gerszten PC, Burton SA. Clinical assessment of stereotactic IGRT: spinal radiosurgery. Med Dosim. 2008;33:107–16.
50. Jabbari S, Gerszten PC, Ruschin M, Larson DA, Lo SS, Sahgal A. Stereotactic body radiotherapy for spinal metastases: practice guidelines, outcomes, and risks. Cancer J. 2016;22:280–9.

51. Maranzano E, Latini P. Effectiveness of radiation therapy without surgery in metastatic spinal cord compression: final results from a prospective trial. Int J Radiat Oncol Biol Phys. 1995;32:959–67.
52. Bishop AJ, Tao R, Rebueno NC, Christensen EN, Allen PK, Wang XA, et al. Outcomes for spine stereotactic body radiation therapy and an analysis of predictors of local recurrence. Int J Radiat Oncol Biol Phys. 2015;92:1016–26.
53. Gerszten PC, Burton SA, Ozhasoglu C, Vogel WJ, Welch WC, Baar J, et al. Stereotactic radiosurgery for spinal metastases from renal cell carcinoma. J Neurosurg Spine. 2005;3:288–95.
54. Gerszten PC, Burton SA, Quinn AE, Agarwala SS, Kirkwood JM. Radiosurgery for the treatment of spinal melanoma metastases. Stereotact Funct Neurosurg. 2005;83:213–21.
55. Gerszten PC, Burton SA, Welch WC, Brufsky AM, Lembersky BC, Ozhasoglu C, et al. Single-fraction radiosurgery for the treatment of spinal breast metastases. Cancer. 2005;104:2244–54.
56. Chao ST, Koyfman SA, Woody N, Angelov L, Soeder SL, Reddy CA, et al. Recursive partitioning analysis index is predictive for overall survival in patients undergoing spine stereotactic body radiation therapy for spinal metastases. Int J Radiat Oncol Biol Phys. 2012;82:1738–43.
57. Tang C, Hess K, Bishop AJ, Pan HY, Christensen EN, Yang JN, et al. Creation of a prognostic index for spine metastasis to stratify survival in patients treated with spinal stereotactic radiosurgery: secondary analysis of mature prospective trials. Int J Radiat Oncol Biol Phys. 2015;93:118–25.
58. Hashmi A, Guckenberger M, Kersh R, Gerszten PC, Mantel F, Grills IS, et al. Re-irradiation stereotactic body radiotherapy for spinal metastases: a multi-institutional outcome analysis. J Neurosurg Spine. 2016;25:646–53.
59. Sahgal A, Ames C, Chou D, Ma L, Huang K, Xu W, et al. Stereotactic body radiotherapy is effective salvage therapy for patients with prior radiation of spinal metastases. Int J Radiat Oncol Biol Phys. 2009;74:723–31.
60. Sahgal A, Ma L, Fowler J, Weinberg V, Gibbs I, Gerszten P, et al. Impact of dose hot spots on spinal cord tolerance following stereotactic body radiotherapy: a generalized biological effective dose analysis. Technol Cancer Res Treat. 2012;11:35–40.
61. Choi CY, Adler JR, Gibbs IC, Chang SD, Jackson PS, Minn AY, et al. Stereotactic radiosurgery for treatment of spinal metastases recurring in close proximity to previously irradiated spinal cord. Int J Radiat Oncol Biol Phys. 2010;78:499–506.
62. Damast S, Wright J, Bilsky M, Hsu M, Zhang Z, Lovelock M, et al. Impact of dose on local failure rates after image-guided reirradiation of recurrent paraspinal metastases. Int J Radiat Oncol Biol Phys. 2011;81:819–26.
63. Bilsky MH, Laufer I, Burch S. Shifting paradigms in the treatment of metastatic spine disease. Spine. 2009;34:S101–7.
64. Sahgal A, Bilsky M, Chang EL, Ma L, Yamada Y, Rhines LD, et al. Stereotactic body radiotherapy for spinal metastases: current status, with a focus on its application in the postoperative patient: a review. J Neurosurg Spine. 2011;14:151–66.
65. Klekamp J, Samii H. Surgical results for spinal metastases. Acta Neurochir. 1998;140:957–67.
66. Rock JP, Ryu S, Shukairy MS, Yin F-F, Sharif A, Schreiber F, et al. Postoperative radiosurgery for malignant spinal tumors. Neurosurgery. 2006;58:891–8.
67. Laufer I, Iorgulescu JB, Chapman T, Lis E, Shi W, Zhang Z, et al. Local disease control for spinal metastases following "separation surgery" and adjuvant hypofractionated or high-dose single-fraction stereotactic radiosurgery: outcome analysis in 186 patients. J Neurosurg Spine. 2013;18:207–14.
68. Gwak H-S, Youn S-M, Chang U, Lee D, Cheon GJ, Rhee C, et al. Usefulness of 18F-fluorodeoxyglucose PET for radiosurgery planning and response monitoring in patients with recurrent spinal metastasis. Minim Invasive Neurosurgery. 2006;49:127–34.
69. Thariat J, Castelli J, Chanalet S, Marcie S, Mammar H, Bondiau P-Y. CyberKnife stereotactic radiotherapy for spinal tumors: value of computed tomographic myelography in spinal cord delineation. Neurosurgery. 2009;64:A60–6.
70. Fourney DR, Frangou EM, Ryken TC, DiPaola CP, Shaffrey CI, Berven SH, et al. Spinal instability neoplastic score: an analysis of reliability and validity from the spine oncology study group. J Clin Oncol. 2011;29:3072–7.
71. Ling DC, Flickinger JC, Burton SA, Heron DE, Quinn AE, Bejjani GK, et al. Long-term outcomes after stereotactic radiosurgery for spine metastases: radiation dose–response for late toxicity. Int J Radiat Oncol Biol Phys. 2018;101:602–9.
72. Yamada Y, Bilsky MH, Lovelock DM, Venkatraman ES, Toner S, Johnson J, et al. High-dose, single-fraction image-guided intensity-modulated radiotherapy for metastatic spinal lesions. Int J Radiat Oncol Biol Phys. 2008;71:484–90.
73. Ryu S, Rock J, Rosenblum M, Kim JH. Patterns of failure after single-dose radiosurgery for spinal metastasis. J Neurosurg. 2004;101:402–5.
74. Jawad MS, Fahim DK, Gerszten PC, Flickinger JC, Sahgal A, Grills IS, et al. Vertebral compression fractures after stereotactic body radiation therapy: a large, multi-institutional, multinational evaluation. J Neurosurg Spine. 2016;24:928–36.

39 Re-irradiation of Spinal Metastases

Emma M. Dunne, Timothy K. Nguyen, Jay Detsky, Mitchell Liu, Simon S. Lo, and Arjun Sahgal

39.1 Introduction

It is estimated that 40% of all patients with cancer will develop metastatic disease to the spine [1]. Local control (LC) is important given the increased risk of debilitating complications associated with progression, including neurological compromise and spinal cord compression upon disease progression. Though conventional external beam radiation therapy (cEBRT) can effectively palliate painful bone metastases at doses of 8 Gy in a single fraction (SF) or 20–30 Gy in multiple fractions (MF), controversy still reigns over the optimal dose fractionation to achieve durable pain control in this setting [2]. Randomised trials of dose fractionation schedules with cEBRT have reported rates of pain relief ranging between 50 and 85% and complete response rates to pain of up to 20%. Caution is warranted interpreting these results due to the different definitions of pain relief and pain measurement scales used throughout the trials. Notably, reirradiation rates of up to 42% and 24% with SF and MF regimens respectively were reported due to persistent or recurrent pain [3, 4]. In an era where novel targeted therapies and improved systemic treatment have been shown to extend patient survival, the role of durable salvage reirradiation is gaining traction with a particular focus on determining the optimal dose and radiotherapy delivery platform.

E. M. Dunne (✉)
BC Cancer—Vancouver Centre, Vancouver, BC, Canada

T. K. Nguyen · J. Detsky · A. Sahgal
Department of Radiation Oncology, Sunnybrook Health Sciences Centre and University of Toronto, Toronto, ON, Canada
e-mail: timothy.nguyen@lhsc.on.ca; jay.detsky@sunnybrook.ca; Arjun.Sahgal@sunnybrook.ca

M. Liu
Department of Radiation Oncology, BC Cancer—Vancouver Centre, Vancouver, BC, Canada
e-mail: mliu@bccancer.bc.ca

S. S. Lo
Department of Radiation Oncology, University of Washington, Seattle, WA, USA
e-mail: simonslo@uw.edu

39.2 The Evidence for Reirradiation

In 2014, the Symptom Control (SC) phase III randomised controlled non-inferiority trial led by the Canadian Clinical Trials Group (CCTG) sought to compare the efficacy of 8 Gy in a SF or 20 Gy in MF (1:1 randomisation) in a group of patients who previously received palliative cEBRT for painful bone metastases. Of the 850 patients randomly assigned, 28% had metastatic disease to the spine. Previous spine cEBRT doses included 6, 7, or 8 Gy in a SF or MF regimens including 4.5 Gy × 4 fractions and 5 Gy × 5 fractions

© Springer Nature Switzerland AG 2020
A. Conti et al. (eds.), *CyberKnife NeuroRadiosurgery*,
https://doi.org/10.1007/978-3-030-50668-1_39

(biologically equivalent doses (BED) ≤ 60 Gy_2). Sixty-six percent had previously been treated with 8 Gy in a SF. Patients with spinal cord compression or who had received initial doses higher that those stated above were excluded. On intention to treat analysis, 8 Gy in SF was shown to be non-inferior and less toxic than 20 Gy in 5 fractions. However, only 28% of patients in the 8 Gy SF cohort and 32% of patients in the 20 Gy MF cohort had any pain response, and a complete response was only achieved in 8% in the SF and 7% in the MF cohorts, respectively. Of those patients receiving 8 Gy in SF, there were seven (2%) cases of spinal cord or cauda equina compression reported, compared to two (<1%) cases in the MF group. There were no cases of reirradiation radiation myelopathy [5].

One of the major challenges associated with salvage spine irradiation is delivering an effective dose of radiation to the disease while respecting the cumulative tolerance of the spinal cord. Stereotactic body radiation therapy (SBRT) is a technique designed to accomplish such a goal. SBRT allows for the delivery of high biologically equivalent doses of radiation to the vertebral target with sub-millimetre precision while controlling the differential dose exposure to the critical neural tissues (CNT). Thesc include the spinal cord and thecal sac and other dose limiting organs at risk (OAR). Though results of randomised trials are forthcoming in patients with painful and previously un-irradiated spinal metastases (de novo), mature single and multi-institution studies have reported that SBRT is feasible, safe and effective with prospective series reporting an actuarial LC rate of approximately 88% at 18 months [6, 7].

39.2.1 Salvage SBRT Following cEBRT

As early as 2009, Sahgal et al. demonstrated a 1-year LC rate of 82% and a median overall survival (OS) of 21 months in a cohort of 37 vertebral metastases reirradiated with a median total SBRT dose of 24 Gy in 3 fractions prescribed to the 60% isodose line [8]. The median initial cEBRT dose in this group of patients was 36 Gy in 14 fractions. More recently, following a median initial 30 Gy in 10 fractions of cEBRT and a median time interval to reirradiation of 13.5 months, Hashmi et al. reported on a multi-institutional pooled analysis of 215 patients with 247 vertebral metastases salvaged with spine SBRT. Sixty percent of the vertebral metastases were treated with SF SBRT, receiving a median dose of 16.6 Gy [equivalent dose in 2 Gy fractions (EQD2) using an α/β of 10 for tumour (EQD2/10) of 36.8 Gy] and 40% with MF SBRT receiving a median dose/fractionation of 24 Gy in 3 fractions (EQD2/10 of 36 Gy). Results showed a 1-year LC rate of 83% and a 1-year OS of 48%. On a multivariate analysis, SF SBRT was a significant predictive factor for better local control, while a Karnofsky Performance Status (KPS) <70 was a significant prognostic factor for worse survival. Even with the high proportion of salvage treatments delivered in a SF, the vertebral compression fracture (VCF) rate was 4.5%, and no cases of radiation myelopathy were recorded.

Additional evidence of the safety and efficacy of salvage SBRT has been shown by the group at the Sunnybrook Health Sciences Centre, who retrospectively reviewed 43 patients with 83 spinal segments treated with reirradiation SBRT. In this cohort, 6/83 segments were initially treated with SBRT, 60/83 received one prior course of cEBRT, 17/83 with two prior courses of cEBRT, and 1/83 segments had received three prior courses of cEBRT prior to salvage treatment. The majority of patients (65%) had epidural disease at the time of salvage SBRT. With a median follow-up time of 12.4 months (range 0.5–52.4 months), the median overall survival was 13.2 months. Failure occurred in 15/83 segments (18%) with actuarial local failure (LF) rates of 7%, 14% and 19% at 6 months, 12 months and 24 months respectively. The crude risk of VCF was 4%, and no cases of radiation myelopathy were observed. Currently, the most common salvage SBRT dose used by this group is 30 Gy in 4 fractions. In cases with no epidural disease, a 16.2 Gy maximum point dose constraint to the CNT is applied. This dose limit is based on recommendations by Sahgal et al., also endorsed by the Hypofractionated Treatment Effects in the Clinic (HyTEC) spinal cord group. As no cases of radiation myelopathy have been observed using these constraints, the group has applied an increase of 20% to the D_{max} constraint in cases where epidural disease is present (allowing 19.4 Gy to the thecal sac or cord planning organ at risk volume (PRV) as a point dose). They do, however, recognise that this is based on limited clinical experience and in time will report their outcomes to help validate this practice [9]. In non-complicated cases, those reirradiated with 24 Gy in 2 fractions have a 2 fraction retreatment spinal cord PRV/thecal sac dose limit applied as recommended by Sahgal et al. and HyTEC (12.2 Gy as point maximum). This dose limit is increased by 20% when epidural disease is present (14.6 Gy as a point maximum).

Between 2009 and 2017, the literature is limited to eight retrospective reviews, one prospective study, one phase I/II trial and one multi-institutional pooled analysis specific to outcomes following salvage SBRT [8, 10–19] (Table 39.1). Vast heterogeneity exists between these studies with respect to patient inclusion, treatment planning (including target and OAR delineation), dose constraints and the dose/fractionation delivered. Despite the variation in the definitions of LC, lack of detail with respect to adverse effects and the cumulative dose exposure, and retrospective nature of the data, one can conclude that SBRT given as salvage following cEBRT failure is effective with durable LC rates ranging from a median of 66 to 93% at 1 year. With respect to dose, there is a wide range of practice from 16 Gy in 1 fraction to

Table 39.1 Summary of reirradiation spine stereotactic body radiation therapy (SBRT) studies to date

Reference	Study type	Targets treated	Median retreatment total dose/ fraction	Median FU in months (range)	Time to reirradiation (months)	Local control (1 year)	Overall survival
Sahgal et al. (2009) [8]	Retrospective	37	24 Gy/3	7 (1–48)	11	82%	Median, 21 months
Choi et al. (2010) [10]	Retrospective	51	20 Gy/2	7 (4–27)	19	73%	1 year, 68%
Damast et al. (2011) [11]	Retrospective	92	30 Gy/5	12.1 (0.2–63.6)	43–45% of RI <12 months	66%	Median, 13.6 months
Mahadevan (2011) [12]	Retrospective	81	30 Gy/5	12 (4–36)	20	93%	Median, 11 months
Chang et al. (2012) [13]	Retrospective	54	20.6 Gy/1	17.3	24.5	81%	Median, 11 months
Thibault et al. (2014) [14]	Retrospective	11	24 Gy/2	12.3 (1.2–55.4)	–	83%	–
Thibault et al. (2015) [15]	Retrospective	56	30 Gy/4	6.8 (0.9–39) N	12.9	81%	Median, 10 months
Boyce-Fappiano et al. (2017)	Retrospective	237	16 Gy/1	4	10.2	71%	–
Ahmed et al. (2012) [16]	Prospective	22	24 Gy/3	8.2	–	83%	1 year, 28%
Garg et al. (2011) [17]	Prospective phase I/II	63	27 Gy/3	17.6 (0.9–67.5)	–	68%	1 year, 76%
Hashmi et al. (2016) [18]	Multi institutional pooled analysis	247	16.6 Gy/1 24 Gy/3[a]	8.1	13.5	83%	1 year, 48%

FU follow-up, *RI* reirradiation
Median dose/fractionation is provided for both single and multiple fraction SBRT
[a]Pooled analysis

30 Gy in 5 fractions (median 16 Gy/1 fraction), and no conclusions can be made as to optimal practice.

39.2.2 Salvage SBRT Following Initial SBRT

Unsurprisingly, there is a dearth of evidence in the literature regarding the management of infield failures following an initial course of SBRT and subsequent treatment with salvage SBRT. This is a challenging situation requiring careful consideration of the previously delivered dose to the CNT. The only significant published series of salvage SBRT following initial SBRT failures was reported by Thibault et al. [15]. They reported on 56 spinal metastases in 40 patients, of which 24 (42.9%) had been initially irradiated with cEBRT, followed by SBRT and followed by a second course of salvage SBRT. For those receiving SBRT as their first course of treatment, the median total dose and number of fractions was 24 Gy (range 20–35 Gy) and 2 (range 1–5 fractions), respectively. The median time from the first SBRT course to local tumour progression was 11.7 months (range 2.1–41.9 months). The median total dose and number of fractions to the CNT before the second salvage SBRT treatment was 22.5 Gy (range 20–30 Gy) and 5 (range 5–40 fractions), respectively. The median salvage SBRT total dose and number of fractions was 30 Gy in 4 fractions. The median OS after salvage second SBRT was 10 months (95% CI, 6.4–13.5 months) with univariate analysis predicting for longer OS in those patients with a longer time interval between the first and second salvage SBRT and those with oligometastatic disease, though what specifically defined the oligometastatic state wasn't detailed in the paper. Although the radiographic actuarial LC at 1 year was encouraging at 81%, the 6-month and 1-year OS rates were 71.8% and 48%, respectively. The poor survival rates are likely a reflection of the limited life expectancy at this stage of treatment. Importantly, no VCF, radiation myelopathy or grade 3 toxicities were observed.

39.3 International Spine Radiosurgery (ISRS) Recommendations for Salvage SBRT

In 2017, the International Spine Radiosurgery Society (ISRS) recommended SBRT for reirradiation following either cEBRT or to salvage SBRT failures with the intent of improving LC or optimising pain control [20]. However, the recommendations advise that in those patients with high-grade epidural disease, malignant epidural spinal cord compression or mechanical instability of the spine, it is imperative that there is a surgical consultation prior to commencing salvage SBRT. The ISRS based their recommendations on nine eligible studies (three non-randomised prospective case series, one phase I/II prospective study and four retrospective series) identified in the literature from 2005 to 2015. The systematic review did, however, find that the median survival following salvage SBRT ranged from 10 to 22.5 months negating the concept that the perceived poor prognosis of this population should prevent treatment with this technique.

Though this field shows great promise, in the absence of high-level evidence, randomised trials are needed to determine exactly which patients receive the most benefit from this technique and how best salvage SBRT can be optimised to minimise infield failures, improve LC rates and minimise toxicity. Until further data is available, we would suggest adhering to the following recommendations endorsed by the ISRS when considering salvage spine SBRT in patients:

1. Following cEBRT, retreatment with SBRT is a recommended therapeutic option in suitable patients based on multidisciplinary assessment (Level III evidence).
2. Following SBRT, retreatment with SBRT is a treatment option in suitable patients based on multidisciplinary assessment (Level III evidence).
3. For patients with clinical features concerning for malignant epidural spinal cord compression, mechanical instability or baseline verte-

bral body compression fracture, the radiation oncologist should consult a spine surgeon before the patient undergoes SBRT.

39.4 Toxicity Assessment

39.4.1 Radiation-Induced Myelopathy

Although the risk of radiation-induced myelopathy has been reported at <5% in the literature, it remains the most morbid and feared complication of SBRT due to its debilitating consequences [21–24]. A study by Sahgal et al. has suggested safe maximum point doses to the thecal sac in the reirradiation setting, advising a minimum time to reirradiation of at least 5 months [25]. This data was based on the comparison of five cases of grade 4 reirradiation radiation-induced myelopathy, as per the RTOG/EORTC Late Radiation Morbidity Scoring System, and a control group of 14 patients (16 spinal segments) receiving salvage SBRT. In the first course of radiation, the thecal sac received an $EQD2_2$ D_{max} ranging from 18.3 to 52.5 Gy. The salvage SBRT thecal sac $EQD2_2$ D_{max} ranged from 44.1 to 104.9 Gy. In the group that developed radiation myelopathy, the cumulative $EQD2_2$ of the thecal sac was 99.6 Gy (range 77.2–154.9 Gy) with a median dose of 61.7 Gy (range 44.1–104.9 Gy), compared with a cumulative thecal sac $EQD2_2$ of 52.4 Gy (range 39.1–111.2 Gy) and median dose of 12.5 Gy (range 1.9–58.7 Gy) in the non-radiation myelopathy group. Based on this study, recommendations by Sahgal et al. have been made for SBRT delivered in 1–5 fractions in the setting of reirradiation as follows:

1. The cumulative thecal sac $EQD2_2$ D_{max} should not exceed 70 Gy.
2. The reirradiation SBRT thecal sac $EQD2_2$ D_{max} should not exceed 25 Gy.
3. The reirradiation SBRT thecal sac $EQD2_2$ D_{max} to cumulative $EQD2_2$ D_{max} ratio should not exceed 0.5.
4. The minimum time interval to reirradiation should be at least 5 months.

The recently published HyTEC report [25] provides recommendations for dose limits in the reirradiation setting based on dose, volume and outcome data since the publication of the American Association of Physicists in Medicine Task Group 101 (TG101) report [26]. Table 39.1 details maximal spinal cord doses for reirradiation associated with a low risk of radiation myelopathy [25, 26].

39.4.2 Vertebral Compression Fracture (VCF)

The most common late toxicity from SBRT is radiation-induced VCF, a condition that can result in significant pain, destabilisation or neurological compromise requiring surgical intervention. Rates of VCF after SBRT have been reported at up to 40% [27–29]. Previously reported data have suggested that higher doses per fraction may result in an increased risk of VCF, with rates of 21% following 18 Gy/1 fraction and 36–39% following 24 Gy/1 fraction, suggesting fractionated SBRT may mitigate this risk [30]. A multi-institutional analysis of 252 patient with 410 spinal segments conducted by Sahgal et al. identified three Spinal Instability Neoplastic Score (SINS) criteria predictive for VCF, prior to VCF, lytic lesion type and spinal misalignment. The remaining three SINS criteria, mechanical pain, posterolateral involvement of spinal elements and location, were not associated with an increased risk of VCF in this analysis. Additionally, this study found that the overall SINS score was not predictive of VCF risk. The authors suggested that SINS only comprises one component of the overall risk stratification given the increased risk of this complication in those patients treated with higher dose per fraction SBRT [30].

The Sahgal et al. study excluded patients who received surgical or salvage radiotherapy to the treated vertebral segment following initial spine SBRT. Unsurprisingly, limited data exists on the rates of VCF in the reirradiation setting. Rates between 0 and 22% have been described by Myrehaug et al. who summarised ten studies specific to salvage spine SBRT, noting that only six studies reported VCF as an adverse event [31]. Furthermore, of the nine eligible studies the ISRS based their reirradiation treatment recommendations on, only three articles specifically reported on VCF rates. It was observed that of the 186 spinal segments included in the ISRS analysis, 12% of spinal metastases treated with salvage SBRT developed VCF, with the highest risk of fracture in those patients treated with single fraction SBRT [19]. In the only study detailing outcomes of salvage SBRT following initial SBRT failure, Thibault et al. reported no incidences of VCF [15] which likely reflects a bias associated with patient selection.

39.4.3 Tolerance of Other Organ-at-Risk (OAR) Structures in the Reirradiation Setting

As discussed, more evidence is emerging to sufficiently guide safe maximum point doses to the thecal sac and spinal cord in the reirradiation setting [25, 26]. However, further consideration needs to be given to the utility of current constraints for OARs such as the oesophagus, trachea, major vessels and bowel to ensure that dose to these structures does not significantly exceed conventional dose limits with unknown consequences. Further understanding about the kinetics of normal tissue recovery and dose volume effect is a key area for further research.

39.5 Treatment Planning Overview

39.5.1 Patient Selection

Careful patient selection remains paramount when deciding on who should receive salvage SBRT. Various survival models have been developed to aid patient selection when treating with spine SBRT in the de novo setting. In 2013, the Neurologic Oncologic Mechanical and Systemic (NOMS) decision framework was developed to aid optimal patient selection and management of patients presenting with metastatic disease to the spine [32]. In 2017, the International Spine Oncology Consortium developed two algorithms to aid patient selection. Central to this framework was the importance of the patient's performance status, the systemic burden of disease and the potential systemic and therapeutic treatment options available to the patient when deciding who and how a patient should be treated. Other factors such as the amount of epidural disease present, tumour histology, specifically radiosensitivity, radioresponsiveness (rapid vs slow), the vascularity of the tumour and mechanical stability were considered before deciding on optimal management of these patients, including treating with SBRT [33].

No such frameworks exist when choosing patients for salvage SBRT, highlighting the importance of multidisciplinary input and collaboration between radiation oncologists, spine surgeons and medical oncologists. Furthermore, in the absence of biomarkers or next generation molecular diagnostics, we cannot determine with certainty which patients with limited disease burden or oligometastatic disease would benefit from being treated with salvage SBRT following initial treatment failure. However, extrapolating from the studies detailed above, an ideal patient would adhere to the ISRS recommendations and have a life expectancy ≥3 months, a favourable KPS ≥ 70, limited disease burden with an interval of greater than 5 months between previous irradiation and salvage SBRT treatment. The presence of brain metastases and neurological compromise (Asia Impairment A to D) significantly limits life expectancy and can also guide patient selection for salvage SBRT.

39.5.2 Technical Requirements for Spine SBRT

The recommended practice for the delivery of spine SBRT in any setting, including in the setting of reirradiaton, is using near-rigid body immobilisation to improve the patients' stability and reduce intrafraction motion [34]. Typical practice also requires the use of a five-point thermoplastic head and shoulder mask when treating any lesion located in the cervical or upper thoracic spine. For lesions below T4, the dual vacuum system (BodyFIX, Elekta AB) can provide excellent immobilisation [35]. Computer tomography (CT) simulation scan requires 1 mm thin slices, and images are fused with volumetric thin-sliced axial T1 and T2 magnetic resonance images which include at least one vertebral level above and below the target volume. For CyberKnife, the patient is placed in a vacuum cushion or cradle, and X-Sight Spine is used for near real-time tracking to facilitate adjustment of intrafractional positional deviations.

39.5.3 Target and Critical Neural Element Delineation

Accurate target and CNT delineation is paramount when treating with SBRT given the steep dose gradient deliberately used by this technique to ensure maximum dose to the tumour while maintaining the spinal cord or thecal sac within acceptable and safe constraints. Target volume and OAR delineation variability remains a large source of dosimetric uncertainty in planning. For example, over contouring the CNT could limit the prescription dose to the disease and underdose the epidural space which is the most common site of recurrence following treatment with SBRT. Conversely, underestimating the CNT volumes could result in the spinal cord or thecal sac receiving a dose beyond tolerance which would risk catastrophic neurological consequences.

Variability in clinical target volume (CTV) contouring is the single largest uncertainty throughout the planning process. Eliminating variation completely is not feasible, but the utility of consensus contouring guidelines has been recognised as a way to decrease uncertainty and standardise practice, allowing meaningful comparisons of outcomes and to understand patterns of recurrence across clinical networks and institutions. There have been published consensus guidelines on CTV delineation when treating metastatic disease to the spine in the de novo and post-operative setting [36, 37] and now the sacrum [38].

Accurate delineation of the spinal cord or thecal sac remains challenging with no reference or gold standard contour available specifically for thecal sac delineation. The first step to accurate identification of the CNT is fusing thin sliced, volumetric T1-weighted and T2-weighted axial non-contrast enhanced magnetic resonance imaging (MRI) sequences to the treatment planning computed tomography (CT). The spinal cord needs to be defined on the MRI images [36–39], and typically we use the T1-weighted images although many centres use the T2. What is important is consistency in the approach and it is critical to ensure accuracy of fusions to minimise the level of uncertainty. Some centres use a CT myelogram to visualise the spinal cord and thecal sac, particularly in the setting of post-operative spine SBRT where significant artefacts as a result of the implanted hardware can impact accurate visualisation of the neural elements [39, 40]. This invasive procedure is best performed right before the patient is appropriately immobilised and simulated in the treatment position. These two methods are the only means by which the spinal cord can be visualised for spine SBRT and CT alone should never be used to define the spinal cord. Caution has to be taken to select the right window level for the CT myelogram for cord contouring to avoid underestimation of the cord extent (Fig. 39.1). Planning organ-at-risk margins (PRV) of 1–2 mm are typically applied to the spinal cord to mitigate setup errors which may alter the spinal cord position during treatment. In some cases, the thecal sac or the spinal canal, without an applied PRV, is used as a surrogate for the spinal cord. Caution is warranted when using the spinal canal as an avoidance structure in order to restrict the dose received by the spinal cord, as given the purposefully steep dose gradient between the cord and the tumour, there is a risk of significantly compromising the dose to the disease, and again not recommended.

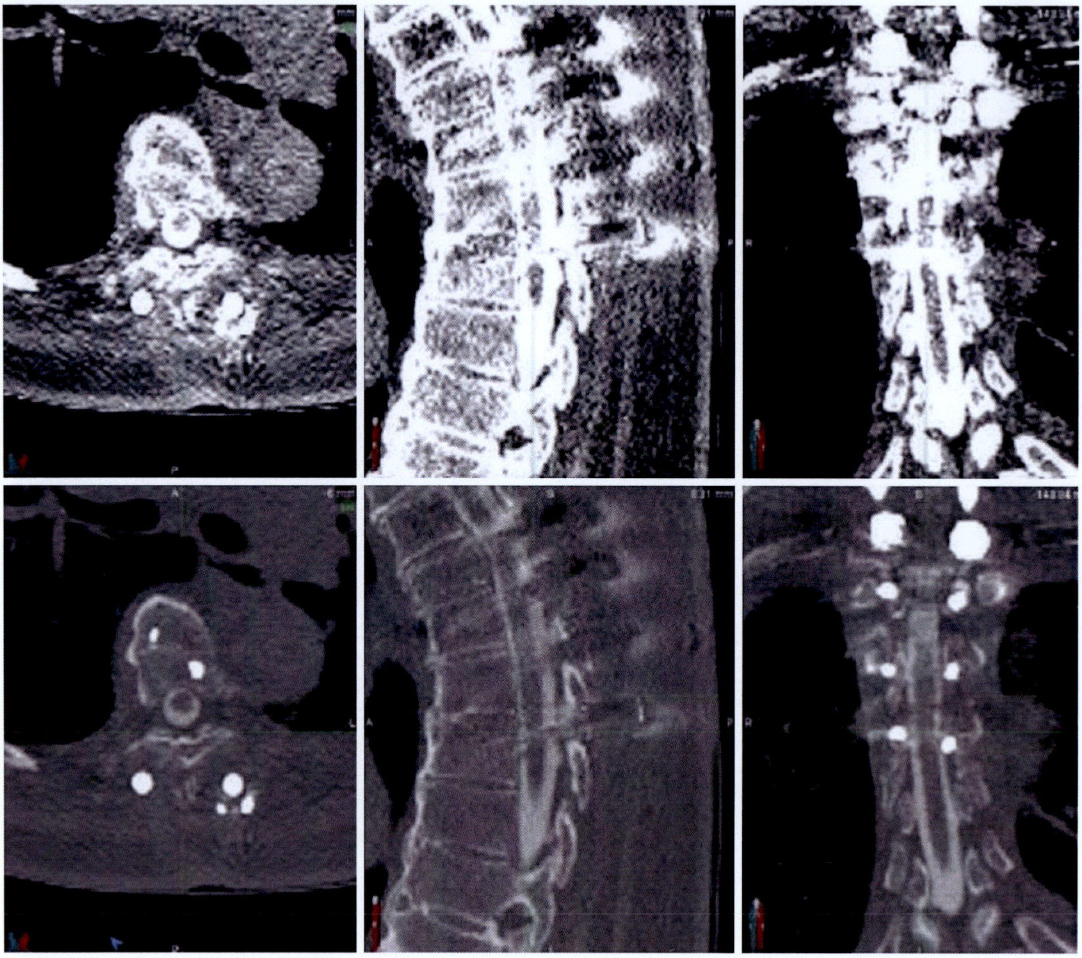

Fig. 39.1 CT myelogram to aid delineation of the spinal cord. Upper panel, incorrect windowing with small cord contours; lower panel, correct windowing showing true cord contours

39.5.4 Dose/Fractionation

No consensus exists on the optimal dose/fractional schedule when treating metastatic disease to the spine in the de novo setting. Commonly used schedules include 16–24 Gy/1 fraction, 24–27 Gy/2–3 fractions and 30–35 Gy in 5 fractions [7, 13, 21, 22, 41–44]. A phase III multicentre randomised study comparing 27 Gy in 3 fractions in three consecutive days or 24 Gy in a SF is underway and due to complete in 2020 (NCT01223248). It must be noted however that this trial is not specific for metastatic disease to the spine, with disease to bone, soft tissue and lymph nodes additionally included in the study [45]. Current practice at the Sunnybrook Health Sciences Center of the University of Toronto is to deliver 24 Gy in 2 fractions, or 30 Gy in 4 fractions for complex large volume cases, while adhering to published limits to the CNT. In those patients eligible for salvage SBRT following initial treatment with SBRT, a dose of 30 Gy in 4 fractions with a spinal cord D_{max} of 16.2 Gy is practiced (Table 39.2) [25, 26, 31]. Recognising the paucity of data and the inherent limitations with published studies, the dose/fraction and spinal cord tolerances are recommendations based mainly on clinical experience. The authors recognise that there may be situations where these constraints need to be exceeded. However, that is

Table 39.2 Spinal cord dose constraints in the reirradiation setting for a probability of radiation myelopathy (RM) <5% [25, 26]

Prior cEBRT Dose and fractionation	Dmax limit (Gy) 1 fraction	Dmax limit (Gy) 2 fractions	Dmax limit (Gy) 3 fractions	Dmax limit (Gy) 4 fractions	Dmax limit (Gy) 5 fractions
20 Gy in 5 fractions	9	12.2	14.5	16.2	18
30 Gy in 10 fractions	9	12.2	14.5	16.2	18
40 Gy in 20 fractions	N/A	12.2	14.5	16.2	18
45 Gy in 25 fractions	N/A	12.2	14.5	16.2	18
50 Gy in 25 fractions	N/A	11	12.5	14	15.5

up to the individual physician to determine based on the particular nuances of their own practice.

39.6 Response Assessment

39.6.1 Pattern of Failure Analysis

Early pattern of failure analyses suggested that local tumour progression following spine SBRT occurred most commonly in the bone adjacent to gross tumour volume (GTV), when the entire vertebral body was not included in the CTV. As such, even prior to the publication of the consensus guidelines on spine SBRT in 2012, there was a trend toward including the entire vertebral body as the target volume and associated bony anatomy at risk [42, 44, 46]. With the increasing practice of reporting patterns of failure, it has become clear that the most common site of failure is now epidural. Sahgal et al. was one of the first to describe this finding. In a series of 39 patients with 60 metastases (37 of which had previous radiation), 6/8 (75%) failures occurred at a distance ≤1 mm from the tumour to the thecal sac [8]. Within the salvage reirradiation setting, Thibault et al. reported that 11/13 (85%) spinal segments that progressed locally did so within the epidural space, 9 (69%) within the bone segment and 6 (46%) within the paraspinal tissue. They suggested that as treatment in this group was delivered following initial salvage SBRT, with some cases additionally having cEBRT prior to the first course of SBRT, that inherent radioresistance may have contributed to the pattern observed. Furthermore, given the salvage setting and the risk of toxicity to the spinal cord, the more conservative fractionated SBRT dosing employed (median salvage SBRT total dose was 30 Gy in 4 fractions) could have also contributed to this outcome [15]. Recently, Detsky et al. observed that bulky paraspinal disease was a factor predictive of local failure specific to their salvage SBRT series and suggested radioresistance may be causal given that dose is typically not compromised within the paraspinal tissues as compared to the epidural space [9].

Regarding epidural disease and its relation to SBRT outcomes, we have learned much from the post-operative spine SBRT experience, as in that situation most patients are operated for malignant epidural spinal cord compression. It is now clear that a post-operative epidural Bilsky grade of 0 or 1 versus Bilsky 2 or 3 is a significant predictor of local control. This finding was first reported by Al-Omair et al. in a series of 80 patients treated with post-operative SBRT to a dose of 18–26 Gy in 1–2 fractions [47]. Fifteen of the 21 failures (71%) were within the epidural space. In 48 of the 80 patients who presented with high-grade epidural disease, there was a significant improvement in local control when high-grade preoperative epidural disease (Bilsky grade 2) was surgically downgraded to Bilsky 0/1. This finding was validated by a post-operative spine SBRT recent series by Alghamdi et al. [48].

Failure at the epidural space is likely due to the relative underdosing required to respect the tolerance of the spinal cord, not fully encompassing the epidural space in the target volume, or simply that these patients have aggressive disease [49]. Dosimetric aspects of optimising the separation between the spinal cord and disease have been studied by Jakobovic et al. [50]. They performed a treatment planning study that indicated that with increasing distance between the disease and the spinal cord, dosimetric gains are observed in particular for the D_{min}. This is relevant as it has been shown that D_{min} may be predictive of local failure [51]. This is sensible and expected; however, they did observe patient-specific gains which are a reflection of the complex anatomy. Ultimately, the group is developing software to feedback to the surgeon in real-time intraoperatively, when sufficient decompression has been performed to optimise the post-operative SBRT treatment plan. With respect to contouring and ensuring sufficient coverage in the target volume, post-operative spine SBRT contouring consensus guidelines have been published, and essentially, at Sunnybrook, most patients are treated with a donut CTV to avoid missing epidural disease. This practice has observed to be safe and efficacious [47, 48]. Furthermore, they practice a 5 mm craniocaudal margin along the epidural space to ensure they do not miss disease extension in that plane to reduce the risk of marginal epidural progression.

39.6.2 Radiographic Response

Under the direction of the Response Assessment in Neuro-Oncology (RANO) working group, the SPine response assessment in Neuro-Oncology (SPINO) committee was formed, which was comprised of 13 international experts in spine SBRT. Members of this group completed a survey which led to the development of recommendations for the use of imaging in treatment planning and response assessment for spine SBRT [52]. They recommended that radiographic response after spine SBRT should be assessed with a spine MRI every 2–3 months for the first 12–18 months and then every 3–6 months thereafter. Imaging should be interpreted by both a radiologist and radiation oncologist to determine if there is local control, local progression or pseudoprogression (Fig. 39.2). Local control was defined as "the absence of progression within the treated area on serial imaging" based on MRI at least 6 weeks apart. Local progression was defined as a "gross unequivocal increase in tumour volume or linear dimension, [or] any new or progressive tumour within the epidural space, [or] neurological deterioration attributable to pre-existing epidural disease with equivocal increase in epidural disease on MRI." Importantly, these guidelines note that traditional RECIST criteria are not ideal in this setting, and dedicated radiographic response assessment criteria are needed.

A second report from the SPINO group provided international consensus recommendations on measuring physician-reported and patient-reported outcomes after spine SBRT that include mechanical spinal stability, quality of life, neurological status and physical function [53]. For physician-reported outcomes, these guidelines recommended using the Spinal Instability in Neoplastic Score (SINS) for spinal stability and the Bilsky grade to characterize the extent of epidural disease. For neurological status, the American Impairment Scale (AIS), the Medical Research Council (MRC) scale and the 10-m walk test were the preferred measurement tools, and the Karnofsky performance status (KPS) and Eastern Cooperative Oncology Group (ECOG) scale recommended for performance status assessment and survival prediction. For patient-reported outcomes, pain should be measured using the Brief Pain Inventory (BPI), and recommended tumour-specific scales included the MD Anderson Symptom Inventory (MDASI) and the Spine Oncology Study Group Outcomes Questionnaire (SOSGOQ) with the latter created specifically for patients with spine metastases.

Radiographic assessment after spine SBRT can be confounded by pseudoprogression: a phenomenon where a transient increase in intraosseous tumour volume is observed following SBRT that can mimic true local progression. Amini et al.

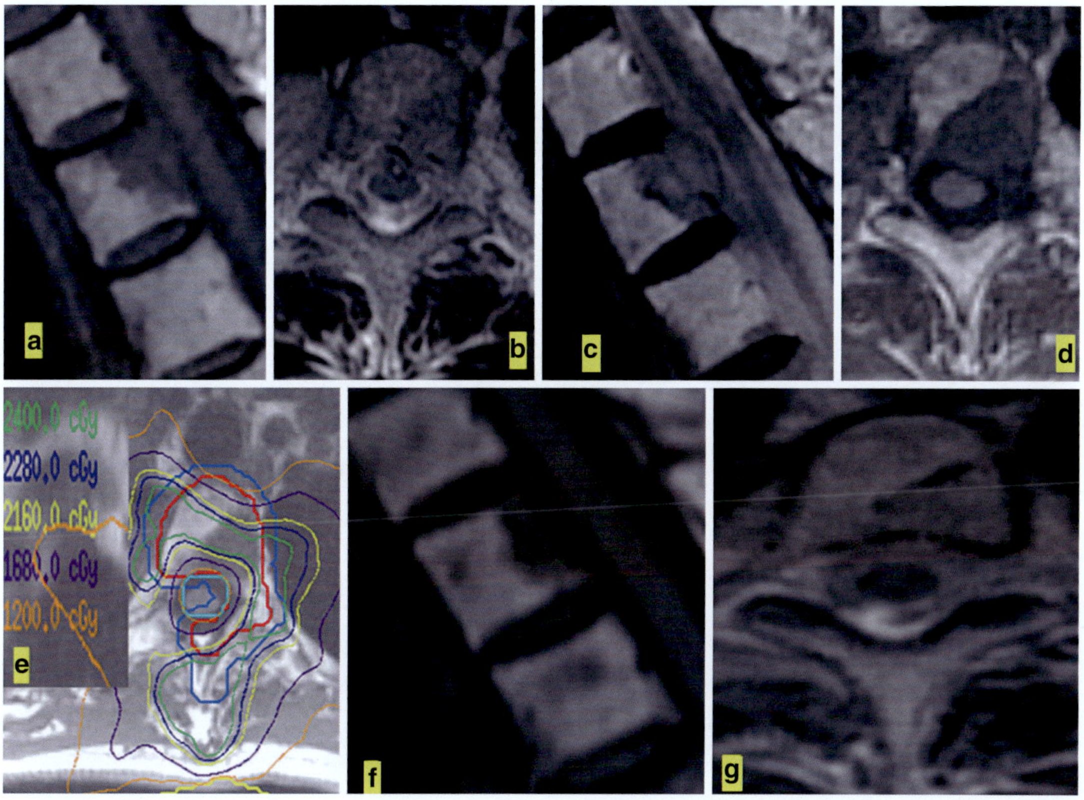

Fig. 39.2 A patient with RCC found to have a metastasis in the left T2 vertebral body. (**a**) Sagittal T1 MRI, (**b**) axial T2 MRI prior to upfront SBRT. One year later there was progression with new left anterolateral epidural disease as seen in the (**c**) sagittal T2 MRI and (**d**) axial T1 MRI. Reirradiation SBRT to 24 Gy in three fractions was delivered (**e**), and 3 months significant regression of epidural disease was demonstrated on (**f**) sagittal T1 MRI and (**g**) axial T2 MRI

first reported the incidence of pseudoprogression at 14% based on a subset of 37 lesions from a larger phase I/II prospective clinical trial [54]. In all these cases, the lesions returned to their baseline size within 23–52 weeks from the MRI scan in which it was first observed. Similarly, in a retrospective series of 49 spinal segments treated with CyberKnife SBRT, the incidence of pseudoprogression was 18% and was diagnosed at a median of 5 months (range 3–9 months) from treatment completion [55]. Maralani et al. reported a higher incidence of pseudoprogression of 37%, which was observed in a retrospective series of 37 spinal segments and 31 patients with prostate cancer or renal cell carcinoma [56]. This study was unique in that it also reported the mean time to maximum pseudoprogression changes (137 ± 21 days), and the mean time until pseudoprogression could be differentiated from true progression on subsequent MRI (245 ± 24 days). Based on multivariable analyses, earlier time to lesion enlargement, enlargement within the 80% isodose line and lytic lesions (as opposed to sclerotic) was associated with an increased risk of pseudoprogression [52, 53].

39.7 Conclusion

As SBRT is emerging as the standard of care for treating patients with metastatic disease to the spine, particularly in the reirradiation setting, we hope this chapter provides a comprehensive critical review of the current evidence available and guidelines to aid practice when treating this challenging group of patients.

References

1. Holman PJ, Suki D, McCutcheon I, Wolinsky J, Rhines LD, Gokaslan ZL. Surgical management of metastatic disease of the lumbar spine: experience with 139 patients. J Neurosurg Spine. 2005;2(5):550–63.
2. Steenland E, Leer J, van Houwelingen H, Post WJ, van den Hout WB, Kievit J, et al. The effect of a single fraction compared to multiple fractions on painful bone metastases: a global analysis of the Dutch Bone Metastasis Study. Radiother Oncol. 1999;52(2):101–9.
3. Chow E, Harris K, Fan G, Tsao M, Sze WM. Palliative radiotherapy trials for bone metastases: a systematic review. JCO. 2007;25(11):1423–36.
4. Chow E, Hoskin PJ, Wu J, Roos D, van der Linden Y, Hartsell W, et al. A phase III international randomized trial comparing single with multiple fractions for re-irradiation of painful bone metastases: National Cancer Institute of Canada Clinical Trials Group (NCIC CTG) SC 20. Clin Oncol. 2006;18(2): 125–8.
5. Chow E, van der Linden Y, Roos D, Hartsell WF, Hoskin P, Wu J, et al. Single versus multiple fractions of repeat radiation for painful bone metastases: a randomised, controlled, non-inferiority trial. Lancet Oncol. 2014;15(2):164–71.
6. Husain ZA, Sahgal A, De Salles A, Funaro M, Glover J, Hayashi M, et al. Stereotactic body radiotherapy for de novo spinal metastases: systematic review. J Neurosurg Spine. 2017;27(3):295–302.
7. Garg AK, Shiu AS, Yang J, Wang X, Allen P, Brown BW, et al. Phase 1/2 trial of single-session stereotactic body radiotherapy for previously unirradiated spinal metastases. Cancer. 2012;118(20):5069–77.
8. Sahgal A, Ames C, Chou D, Ma L, Huang K, Xu W, et al. Stereotactic body radiotherapy is effective salvage therapy for patients with prior radiation of spinal metastases. Int J Radiat Oncol Biol Phys. 2009;74:723–31.
9. Detsky J, Nguyen T, Soliman H, Tseng CL, Myrehaug SD, Sahgal A. Re-irradiation with spine stereotactic body radiotherapy in a heavily pre-treated cohort. Int J Radiat Oncol Biol Phys. 2019;105(1 Supp 1):E122. https://doi.org/10.1016/j.ijrobp.2019.06.2240.
10. Choi CY, Adler JR, Gibbs IC, Chang SD, Jackson PS, Minn AY, et al. Stereotactic radiosurgery for treatment of spinal metastases recurring in close proximity to previously irradiated spinal cord. Int J Radiat Oncol Biol Phys. 2010;78(2):499–506.
11. Damast S, Wright J, Bilsky M, Hsu M, Zhang Z, Lovelock M, et al. Impact of dose on local failure rates after image-guided reirradiation of recurrent paraspinal metastases. Int J Radiat Oncol Biol Phys. 2011;81(3):819–26.
12. Mahadevan A, Floyd S, Wong E, Jeyapalan S, Groff M, Kasper E. Stereotactic body radiotherapy reirradiation for recurrent epidural spinal metastases. Int J Radiat Oncol Biol Phys. 2011;81(5):1500–5.
13. Chang UK, Cho WI, Kim MS, Cho CK, Lee DH, Rhee CH. Local tumor control after retreatment of spinal metastasis using stereotactic body radiotherapy; comparison with initial treatment group. Acta Oncol. 2012;51(5):589–95.
14. Thibault I, Al-Omair A, Masucci GL, Masson-Cote L, Lochray F, Korol R, et al. Spine stereotactic body radiotherapy for renal cell cancer spinal metastases: analysis of outcomes and risk of vertebral compression fracture. J Neurosurg Spine. 2014;21(5): 711–8.
15. Thibault I, Campbell M, Tseng CL, Atenafu EG, Letourneau D, Yu E, et al. Salvage stereotactic body radiotherapy (SBRT) following in-field failure of initial SBRT for spinal metastases. Int J Radiat Oncol Biol Phys. 2015;93:353–60.
16. Ahmed KA, Stauder MC, Miller RC, Bauer HJ, Rose PS, Olivier KR, et al. Stereotactic body radiation therapy in spinal metastases. Int J Radiat Oncol Biol Phys. 2012;82(5):e803–9.
17. Garg AK, Wang XS, Shiu AS, Allen P, Yang J, McAleer MF, et al. Prospective evaluation of spinal reirradiation by using stereotactic body radiation therapy: the University of Texas MD Anderson Cancer Center experience. Cancer. 2011;117(15):3509–16.
18. Hashmi A, Guckenberger M, Kersh R, Gerszten PC, Mantel F, Grills IS, et al. Re-irradiation stereotactic body radiotherapy for spinal metastases: a multi-institutional outcome analysis. J Neurosurg Spine. 2016;25(5):646–53.
19. Boyce-Fappiano D, Elibe E, Zhao B, Salim Siddiqui M, Lee I, Rock J, et al. Reirradiation of the spine with stereotactic radiosurgery: efficacy and toxicity. Pract Radiat Oncol. 2017;7:e409–17.
20. Myrehaug S, Sahgal A, Hayashi M, Levivier M, Ma L, Martinez R, et al. Reirradiation spine stereotactic body radiation therapy for spinal metastases: systematic review. J Neurosurg Spine. 2017;27:1–8.
21. Guckenberger M, Mantel F, Gerszten PC, Flickinger JC, Sahgal A, Létourneau D, et al. Safety and efficacy of stereotactic body radiotherapy as primary treatment for vertebral metastases: a multi-institutional analysis. Radiat Oncol. 2014;9(1):226.
22. Folkert MR, Bilsky MH, Tom AK, Oh JH, Alektiar KM, Laufer I, et al. Outcomes and toxicity for hypofractionated and single-fraction image-guided stereotactic radiosurgery for sarcomas metastasizing to the spine. Int J Radiat Oncol Biol Phys. 2014;88(5):1085–91.
23. Hall WA, Stapleford LJ, Hadjipanayis CG, Curran WJ, Crocker I, Shu H-KG. Stereotactic body radiosurgery for spinal metastatic disease: an evidence-based review. Int J Surg Oncol. 2011;2011:979214.
24. Sahgal A, Weinberg V, Ma L, Chang E, Chao S, Muacevic A, et al. Probabilities of radiation myelopathy specific to stereotactic body radiation therapy to guide safe practice. Int J Radiat Oncol Biol Phys. 2013;85(2):341.
25. Sahgal A, Ma L, Weinberg V, Gibbs IC, Chao S, Chang EL, et al. Reirradiation human spinal cord tol-

erance for stereotactic body radiotherapy. Int J Radiat Oncol Biol Phys. 2012;82(1):107–16.
26. Sahgal A, Chang JH, Ma L, Marks LB, Milano MT, Medin P, et al. Spinal cord dose tolerance to stereotactic body radiation therapy. Int J Radiat Oncol Biol Phys. 2019. https://doi.org/10.1016/j.ijrobp.2019.09.038. [Epub ahead of print].
27. Rose PS, Laufer I, Boland PJ, Hanover A, Bilsky MH, et al. Risk of fracture after single fraction image-guided intensity-modulated radiation therapy to spinal metastases. J Clin Oncol. 2009;27:5075–9.
28. Cunha MV, Al-Omair A, Atenafu EG, Masucci GL, Letourneau D, Korol R, et al. Vertebral compression fracture (VCF) after spine stereotactic body radiation therapy (SBRT): analysis of predictive factors. Int J Radiat Oncol Biol Phys. 2012;84:e343–9.
29. Boehling NS, Grosshans DR, Allen PK, McAleer MF, Burton AW, Azeem S, et al. Vertebral compression fracture risk after stereotactic body radiotherapy for spinal metastases. J Neurosurg Spine. 2012;16:379–86.
30. Sahgal A, Whyne CM, Ma L, Larson DA, Fehlings MG. Vertebral compression fracture after stereotactic body radiotherapy for spinal metastases. Lancet Oncol. 2013;14(8):e320.
31. Myrehaug S, Soliman H, Tseng C, Heyn C, Sahgal A. Re-irradiation of vertebral body metastases: treatment in the radiosurgery era. Clin Oncol (R Coll Radiol). 2018;30(2):85–92.
32. Laufer I, Rubin DG, Lis E, Cox BW, Stubblefield MD, Yamada Y, et al. The NOMS framework: approach to the treatment of spinal metastatic tumors. Oncologist. 2013;18(6):744–51.
33. Spratt DE, Beeler WH, de Moraes FY, Rhines LD, Gemmete JJ, Chaudhary N, et al. An integrated multidisciplinary algorithm for the management of spinal metastases: an International Spine Oncology Consortium report. Lancet Oncol. 2017;18(12):e720.
34. Ma L, Sahgal A, Hossain S, et al. Nonrandom Intrafraction target motions and general strategy for correction of spine stereotactic body radiotherapy. Int J Radiat Oncol. 2009;75(4):1261–5. https://doi.org/10.1016/j.ijrobp.2009.04.027.
35. Li W, Sahgal A, Foote M, Millar B, Jaffray DA, Letourneau D. Impact of immobilization on Intrafraction motion for spine stereotactic body radiotherapy using cone beam computed tomography. Int J Radiat Oncol Biol Phys. 2012;84(2):520–6.
36. Cox BW, Spratt DE, Lovelock M, Bilsky MH, Lis E, Ryu S, et al. International spine radiosurgery consortium consensus guidelines for target volume definition in spinal stereotactic radiosurgery. Int J Radiat Oncol Biol Phys. 2012;83(5):e597–605.
37. Redmond KJ, Robertson S, Lo SS, Soltys SG, Ryu S, McNutt T, et al. Consensus contouring guidelines for postoperative stereotactic body radiation therapy for metastatic solid tumor malignancies to the spine. Int J Radiat Oncol Biol Phys. 2017;97(1):64–74.
38. Dunne EM, Sahgal A, Lo SS, Bergman A, Kosztyla R, Dea N, et al. International consensus recommendations for target volume delineation specific to sacral metastases and spinal stereotactic body radiation therapy (SBRT). Radiother Oncol. 2019;145:21–9. https://doi.org/10.1016/j.radonc.2019.11.026. [Epub ahead of print].
39. Redmond KJ, Lo SS, Soltys SG, Yamada Y, Barani IJ, Brown PD, et al. Consensus guidelines for postoperative stereotactic body radiation therapy for spinal metastases: results of an international survey. J Neurosurg Spine. 2017;26:299–306.
40. Benedict SH, Yenice KM, Followill D, Galvin JM, Hinson W, Kavanagh B, et al. Stereotactic body radiation therapy: the report of AAPM Task Group 101. Med Phys. 2010;37:4078–101.
41. Yamada Y, Bilsky MH, Lovelock DM, Venkatraman ES, Toner S, Johnson J, et al. High-dose, single-fraction image-guided intensity-modulated radiotherapy for metastatic spinal lesions. Int J Radiat Oncol Biol Phys. 2008;71(2):484–90.
42. Ryu S, Rock J, Rosenblum M, Kim JH. Patterns of failure after single-dose radiosurgery for spinal metastasis. J Neurosurg. 2004;101(Suppl 3):402.
43. Staehler M, Haseke N, Nuhn P, Tüllmann C, Karl A, Siebels M, et al. Simultaneous anti-angiogenic therapy and single-fraction radiosurgery in clinically relevant metastases from renal cell carcinoma. BJU Int. 2011;108(5):673–8.
44. Chang EL, Shiu AS, Mendel E, Mathews LA, Mahajan A, Allen PK, et al. Phase I/II study of stereotactic body radiotherapy for spinal metastasis and its pattern of failure. J Neurosurg Spine. 2007;7(2):151–60.
45. A Phase III randomized study comparing two dosing schedules for hypofractionated image guided radiation therapy in patients with metastatic cancer. National Library of Medicine (US). 2017. [Internet]. 2019 Dec 09 [cited 2019 Dec 09]. https://clinicaltrials.gov/show/NCT01223248.
46. Patel VB, Wegner RE, Heron DE, Flickinger JC, Gerszten P, Burton SA. Comparison of whole versus partial vertebral body stereotactic body radiation therapy for spinal metastases. Technol Cancer Res Treat. 2012;11(2):105–15.
47. Al-Omair A, Masucci L, Masson-Cote L, Campbell M, Atenafu EG, Parent A, et al. Surgical resection of epidural disease improves local control following postoperative spine stereotactic body radiotherapy. Neurooncology. 2013;15(10):1413–9.
48. Alghamdi M, Sahgal A, Soliman H, Myrehaug S, Yang VXD, Das S, et al. Postoperative stereotactic body radiotherapy for spinal metastases and the impact of epidural disease grade. Neurosurgery. 2019;85(6):E1111–8. https://doi.org/10.1093/neuros/nyz349.
49. Redmond KJ, Sciubba D, Khan M, Gui C, Lo SL, Gokaslan ZL, Leaf B, Kleinberg L, Grimm J, Ye X, Lim M. A phase 2 study of post-operative stereotactic body radiation therapy (SBRT) for solid tumor spine metastases. Int J Radiat Oncol Biol

Phys. 2019;106:261. https://doi.org/10.1016/j.ijrobp.2019.10.011. . [Epub ahead of print].

50. Jakubovic R, Ruschin M, Tseng CL, Pejović-Milić A, Sahgal A, Yang VXD. Surgical resection with radiation treatment planning of spinal tumors. Neurosurgery. 2019;84(6):1242–50. https://doi.org/10.1093/neuros/nyy176.
51. Lovelock DM, Zhang Z, Jackson A, Keam J, Bekelman J, Bilsky M, Lis E, Yamada Y. Correlation of local failure with measures of dose insufficiency in the high-dose single-fraction treatment of bony metastases. Int J Radiat Oncol Biol Phys. 2010;77(4):1282–7. https://doi.org/10.1016/j.ijrobp.2009.10.003. Epub 2010 Mar 28.
52. Thibault I, Chang EL, Sheehan J, et al. Response assessment after stereotactic body radiotherapy for spinal metastasis: a report from the SPIne response assessment in Neuro-Oncology (SPINO) group. Lancet Oncol. 2015;16(16):e595–603. https://doi.org/10.1016/S1470-2045(15)00166-7.
53. Laufer I, Lo SS, Chang EL, et al. Population description and clinical response assessment for spinal metastases: part 2 of the SPIne response assessment in Neuro-Oncology (SPINO) group report. Neuro-Oncology. 2018;20(9):1215–24. https://doi.org/10.1093/neuonc/noy047.
54. Amini B, Beaman CB, Madewell JE, et al. Osseous pseudoprogression in vertebral bodies treated with stereotactic radiosurgery: a secondary analysis of prospective phase I/II clinical trials. AJNR Am J Neuroradiol. 2016;37(2):387–92. https://doi.org/10.3174/ajnr.A4528.
55. Bahig H, Simard D, Letourneau L, et al. A study of pseudoprogression after spine stereotactic body radiation therapy. Int J Radiat Oncol Biol Phys. 2016;96(4):848–56. https://doi.org/10.1016/j.ijrobp.2016.07.034.
56. Jabehdar Maralani P, Winger K, Symons S, et al. Incidence and time of onset of osseous pseudoprogression in patients with metastatic spine disease from renal cell or prostate carcinoma after treatment with stereotactic body radiation therapy. Neurosurgery. 2018;84:1–8. https://doi.org/10.1093/neuros/nyy075.

Radiosurgery for Benign Spinal Tumors

40

Scott G. Soltys, Antonio Meola, Alex Chin, and Erqi Pollom

40.1 Introduction

Tumors in the spine can be categorized by their anatomical location in relation to the dura and the spinal cord. Prior to histological confirmation, this location correlates with the tumor histology [1]. Tumors external to the thecal sac are in the extradural compartment and include the primary bone tumors such as chordoma and sarcomas, as well as secondary tumors such as spinal metastases. Tumors within the thecal sac are divided into intramedullary tumors within the spinal cord and extramedullary tumors external to the cord. Common intramedullary tumors include ependymomas, astrocytomas, hemangioblastomas, metastases, and non-malignant entities such as spinal cord arteriovenous malformations. Benign tumors of the spine are primarily within the intradural extramedullary spinal compartment. Histologies include meningiomas, schwannomas, and neurofibromas.

S. G. Soltys (✉) · A. Chin · E. Pollom
Department of Radiation Oncology, Stanford University, Stanford Cancer Institute, Stanford, CA, USA
e-mail: sgsoltys@stanford.edu; alexchin@stanford.edu; erqiliu@stanford.edu

A. Meola
Department of Neurosurgery, Stanford University, Stanford Cancer Institute, Stanford, CA, USA
e-mail: ameola@stanford.edu

© Springer Nature Switzerland AG 2020
A. Conti et al. (eds.), *CyberKnife NeuroRadiosurgery*, https://doi.org/10.1007/978-3-030-50668-1_40

The treatment of benign spinal tumors has paralleled the treatment of benign intracranial tumors. Historically, surgical resection was the only treatment option for benign brain tumors and thus the standard of care. With time, large series have reported outcomes of long-term durable tumor control with radiotherapy and stereotactic radiosurgery (SRS) [2–7]. For intracranial meningiomas and schwannomas, radiotherapy and SRS are standard-of-care alternatives to surgical resection and are often the preferred treatment option for some patients [8, 9].

Similarly, in the past, open surgical resection has been considered the only treatment option for benign spinal tumors [1]. However, the radiosurgical techniques developed for the treatment of benign intracranial tumors have been applied to benign spinal tumors. The most recent large series reporting outcomes of SRS suggest long-term durable tumor control for benign histologies in the spine [10], similar to intracranial outcomes.

In this chapter, we review the latest reported outcomes of SRS for benign spinal tumors, provide treatment recommendations, and highlight the potential toxicities of high-dose SRS near the spinal cord.

40.2 Rationale for Irradiation of Benign Tumors in the Brain and Spine

Therapeutic irradiation for benign tumors can be classified as conventionally fractionated radiotherapy, typically 1.8–2.0 Gy per day over approximately 5–6 weeks or as SRS. SRS was defined by the AANS/CNS/ASTRO consensus report [11] as stereotactic irradiation in 1 to 5 fractions. We term treatments performed in 1 fraction as "SRS," with 2–5 fractions considered hypofractionated SRS (fSRS).

A standard of care for benign histologies, including meningiomas, is surgical resection. Tumor control outcomes following surgical resection of a meningioma correlate with extent of resection (commonly characterized by the Simpson Grade [12]), the histological grade, and the length of follow-up. Similar to most radiotherapy series, reports of long-term outcomes of greater than 10 years following resection are lacking. Given the slow growth of benign tumors such as meningiomas, the clinician must be cautious in applying results with follow-up of 5 years or less. For example, a series of 51 patients with intracranial meningiomas treated with surgical resection highlights that tumor recurrence can occur beyond 10–20 years [13]. For Simpson grade 1–2 tumors (representing an imaging-defined gross-tumor resection), the crude incidence of tumor recurrence was 13% at 10 years and 38% at 25 years. For a Simpson grade 3 resection (also denoting a gross-tumor resection), tumor recurrence was 33% at 10 years and 42% at 25 years.

Large series highlight that both radiotherapy and SRS provide long-term, durable local tumor control for benign histologies and are standard-of-care alternatives to surgical resection in appropriately selected patients. For example, a series of 507 patients with intracranial meningiomas reported outcomes of conventionally fractionated irradiation to a median dose of 57.6 Gy in 1.8 Gy fractions [3]. With a median follow-up of 107 months, local tumor control for grade I meningiomas was 91% at 10 years and approximately 85% at 15 years.

A large series reported the outcomes of SRS in 4565 patients with 5300 presumed grade I meningiomas [5]. For a median tumor volume of 4.8 cm^3 treated with a median dose of 14 Gy, the median imaging follow-up was 63 months. The local tumor control was 95% at 5 years and 89% at 10 years. A single-institution series of SRS in 251 patients treated with a median SRS dose of 16 Gy reported a 10-year local control rate of 99% [6]. However, the mean follow-up in this series was 63 months; thus the confidence intervals are large when reporting 10-year data. Additionally, the tumor recurrences in this report occurred at 28, 145, and 150 months and highlight the need for longer-term data.

For meningiomas, if one can achieve similar tumor doses for spinal meningiomas as the above reports of 14–16 Gy in a single fraction for intracranial meningiomas [7], then one may reasonably expect similar durable tumor control. A unique challenge of spinal SRS, however, is the tolerance of the spinal cord, as discussed below. The clinician may be tempted to give lower tumor doses for spinal meningiomas given the proximity of the spinal cord. No clear SRS dose response data exist for meningiomas, but some reports may provide guidance. Outcomes of 189 grade I parasellar meningiomas with a median follow-up of 71 months found a 10-year local control of 96% if treated with SRS of 16 Gy or more compared to 82% if less than 16 Gy. An early report in 1993 of 20 patients treated with SRS suggested tumor growth in 40% of tumors treated with 10 Gy or less compared to 0% with 12 Gy or more [14]. The role of fractionation to separate the tumor control from the normal tissue complication probability curves remains unexplored.

A similar rationale exists for the treatment of spinal schwannomas. SRS provides long-term, durable tumor control for intracranial vestibular and non-vestibular schwannomas [9]. Durable tumor control was reported in 440 patients with vestibular schwannomas treated with a median of 12.8 Gy in 1 fraction [2]. With a median follow-up of 12.5 years, the 10-year tumor control was 92%, with no tumor progression after 10 years.

The invention of the CyberKnife frameless stereotactic radiosurgical system allowed the stereotactic principles pioneered for intracranial tumors to be applied in the spine.

40.3 Single Fraction or Hypofractionated Stereotactic Radiosurgery for Benign Spinal Tumors

Early reports applying intracranial SRS techniques to benign spinal tumors found promising, but preliminary, outcomes. An early report of 51 patients treated from 1999 to 2005 reported a 100% control with a mean follow-up of 36 months [15]. This report highlighted the risks of SRS near the spinal cord, as 1 patient developed radiation myelopathy, as discussed further below.

A similar early report of single fraction SRS to 73 tumors from 2001 to 2006 noted 100% tumor control with a median follow-up time of 37 months [16]. Three patients in this early report had SRS-associated spinal cord injury. Later, the same authors published a series of 45 benign tumors, again with 100% tumor control with a median single-fraction dose of 16 Gy and a short median follow-up of 32 months [17]. Subsequent series found similar outcomes, with short-term local control of over 90% at 3–5 years (see Table 40.1) with SRS/fSRS for benign spinal tumors.

The largest series on spinal SRS for benign tumors was an update of the previously reported Stanford experience [10]. From 1999 to 2016, 120 patients with 149 benign spinal tumors received SRS. Tumors included 39 meningiomas, 26 neurofibromas, and 84 schwannomas. The most common dose/fractionation regimens were 16 Gy in 1 fraction, 20 Gy in 2 fractions, and 21 in 3 fractions. With 49 months median follow-up (range 25–103 months), including 24 patients with greater than 10 years of follow-up, the cumulative incidence, with death as a competing risk, of tumor progression was 5% at 5 years and 12% at 10 years for all patients. Ten tumors had been previously irradiated prior to SRS, which presumably led to lower prescription doses due to consideration of spinal cord tolerance and/or selecting tumors that are potentially more resistant to irradiation. Omitting these 10 tumors that had been previously irradiated, local tumor progression was 2% at 5 years and 8% at 10 years. Considering symptom response in these 149 tumors, 71% had symptoms prior to SRS, most commonly pain in 55% [10]. Overall, pain improved in 36%, was stable in 53%, and worsened in 11%.

The median doses in this series were 18 Gy in 1 fraction for schwannomas and 20 Gy in 2 fractions for meningiomas and neurofibromas. The median corresponding single fraction equivalent doses using the linear-quadratic model were 13.0 Gy_3 for meningiomas, 14.3 Gy_3 for neurofibromas, and 13.9 Gy_3 for schwannomas, doses comparable to single-fraction doses for intracranial SRS treatments.

As noted for intracranial SRS, lower doses may not provide durable local control. A series studied the impact of dose and tumor control for benign spinal tumors [22]. In 38 patients with 47 tumors, local tumor control at 5 years was 76% overall and 73% for $BED_{10Gy} < 30Gy$ and 83% for $BED_{10Gy} > 30$ ($p = 0.5$). Of note, a BED_{10Gy} of 30 is approximately equal to a single-fraction SRS dose of 13 Gy. Possible contributing factors for poor local control in this series include that 19% had previously been irradiated, 17% had hemangioblastomas (which have worse local control outcomes [23]), and 52% had prior surgery. However, the authors reported no significant differences in tumor control in these small sub-groups and conclude that dose de-escalation might be a reasonable approach. On the contrary, we feel that local tumor control of only 76% at 5 years is much lower than expected for benign histologies in the spine (see Table 40.1) or the brain, as noted above, and we would not recommend these low doses.

Overall, these results of SRS for benign spinal tumors are similar to the results of SRS for benign intracranial tumors. Although we await further maturation of these data, with even longer-term outcomes, we consider SRS a standard of care for spinal benign tumors if appropriate doses can be delivered.

Table 40.1 Selected publications of spinal SRS for benign tumors. Series were included if they contained more than 20 tumors

Publication	Total number of tumors	Tumor types	Treatment technique	Median dose in Gy (range)	Median number of fractions (range)	Median Follow-up in months (range)	Tumor control rate
Gerszten (2008) [16]	73	13 meningiomas 35 schwannomas 25 neurofibromas	CyberKnife	22 (15–25)	Not recorded (1–3)	37 (8–71)	100% at 3 and 5 years
Selch (2009) [18]	25	8 schwannomas 8 neurofibromas	LINAC	12	1	18 (12–58)	100% at 3 years
Chang (2011) [19]	30	5 schwannomas 2 meningiomas 8 hemangioblastomas 15 neurogenic tumors	CyberKnife	Not recorded (13–23)	Not recorded (1–5)	36 (12–84)	90% at 3 years
Gerszten (2012) [17]	45	10 meningiomas 16 schwannomas 14 neurofibromas 5 others (ganglioglioma, hemangioma, giant cell tumor, aneurysmal bone cyst)	LINAC (Synergy S)	16 (12–24)	Not recorded (1–3)	32 (3–55)	100% at 3 years
Marchetti (2013) [20]	21	11 meningiomas 9 schwannomas 1 neurofibroma	CyberKnife	24 (10–25)	Not recorded (1–6)	43 (32–73)	100% at 3 and 5 years
Shin (2015) [21]	92	69 schwannomas 23 NF1	LINAC (Novalis)	13 (range N/A)	1 (range N/A)	44	95% local control
Kalash (2018) [22]	47	18 meningiomas 18 schwannomas 10 hemangioblastomas 1 Paraganglioma	CyberKnife [11] Synergy S [19] Truebeam [15]	13 (12–17)	1 (1–3)	54 (1–133)	76% at 5 years 73% for $BED_{10Gy} < 30$ 83% for $BED_{10Gy} > 30$
Chin (2019) [10]	149	39 meningiomas 84 schwannomas 26 neurofibromas	CyberKnife	Single-fraction equivalent dose: 13.0–14.3Gy	1–2	49 (3–216)	Radiation naive tumors: 95% at 3 and 92% at 10 years

40.4 Toxicity

The primary concern with spinal SRS, particularly for a benign tumor, is radiation-associated myelopathy, which may cause sensory deficits, weakness, or paralysis. During the initial period of adoption of spinal SRS in the 1990s, data were limited regarding the tolerance of the spinal cord to high-dose single-fraction or hypofractionated treatments.

Early analyses of reported cases of myelopathy attempted to provide guidance on spinal cord dose constraints [24]. Updated series [25] provide dose guidelines for the thecal sac which is often several millimeters distant to the spinal cord. Given the steep dose gradient seen in modern spinal SRS plans, how these thecal sac dose constraints correlate with true spinal cord dose constraints is unclear.

In the largest series analyzing spinal cord dosimetry in SRS plans for metastases, spinal myelopathy was reported in 2 of 259 SRS treatment plans (a 0.78% risk) [26]. The maximum spinal cord doses in these two patients were greater than 13.3 Gy in 1 fraction. However, there were 194 other SRS plans that had higher spinal cord doses. Of note, these patients were treated with an established immobilization system and cone-beam CT image guidance. It is unknown if these data using cone-beam CT-based management are applicable to the more frequent intrafraction imaging and motion management of the CyberKnife.

On RTOG 0631, the spinal cord dose constraints for spinal metastases were a V14Gy of 0.03 cm^3 and a V10Gy of 0.35 cm^3, with a volume of cord receiving greater than 10 Gy of less than 10% of the spinal cord, defined as 5–6 mm above/below the tumor site. A recent systematic review [27] modeled a myelopathy risk of 1–5% for maximum spinal cord doses of 12.4–14.0 Gy in 1 fraction, 17.0 Gy in 2 fractions, 20.3 Gy in 3 fractions, 23.0 Gy in 4 fractions, and 25.3 Gy in 5 fractions.

As the doses needed to control benign tumors are lower than those used for spinal metastatic disease, the spinal cord constraints for benign tumors may be lower than that for malignant disease. When treating benign tumors, we typically limit the cord maximum dose to 10–12 Gy in 1 single fraction, which is 2–4 Gy lower than standard RTOG 0631 spinal cord constraints for metastatic disease (see Table 40.2).

Spinal myelopathy following SRS for a benign spinal tumor was reported in 2006 [15]. This patient had a 7.6 cm^3 C7-T1 meningioma targeted with 24 Gy in three fractions. She developed posterior column dysfunction associated with cord enhancement and edema. Dosimetry found that 1.7 cm^3 of the spinal cord received more than the prescription dose of 24 Gy. Improvements in dosimetry and awareness of the doses associated with a risk of myelopathy have led to no further instances of myelopathy, as seen in the largest

Table 40.2 Typical treatment planning parameters utilized at the authors' institution

Imaging for benign spinal SRS treatment planning		
Imaging type	**Sequence parameters**	**Purpose**
CT with contrast	1.0–2.0 mm slice thickness	Tumor delineation Treatment plan dosimetry
MRI	T1 pre-contrast, no fat saturation 2 mm slice thickness	Tumor interface with paraspinous fat Assess if tumor invades into bone
	T1 post-contrast, fat saturation 2 mm slice thickness	Tumor delineation
	FIESTA/CISS 1–2 mm slice thickness Typically, a sagittal acquisition	Delineation of spinal cord/cauda equina Delineation of medial border of tumor if abutting the thecal sac
Treatment planning parameters		
Target or structure	**Recommended dose or dose constraint**	**Notes**
Meningioma	14–16 Gy in 1 fraction 21–24 Gy in 3 fractions	This dosing is similar to doses for intracranial meningiomas
Schwannoma/ neurofibroma	13–16 Gy in 1 fraction 21–24 Gy in 3 fractions	This dosing is slightly higher than doses for vestibular schwannomas, as there is no concern for hearing preservation or facial nerve toxicity
Spinal cord	10 Gy Dmax, V8 0.3 cc in 1 fraction 12 Gy Dmax, V10 0.3 cc in 1 fraction 20 Gy Dmax, V17 0.3 cc in 3 fractions	This constraint is lower than the RTOG 0631 constraint that is typically used for metastatic/ malignant disease
Cauda equina	16 Gy Dmax in 1 fraction 21 Gy Dmax in 3 fractions	

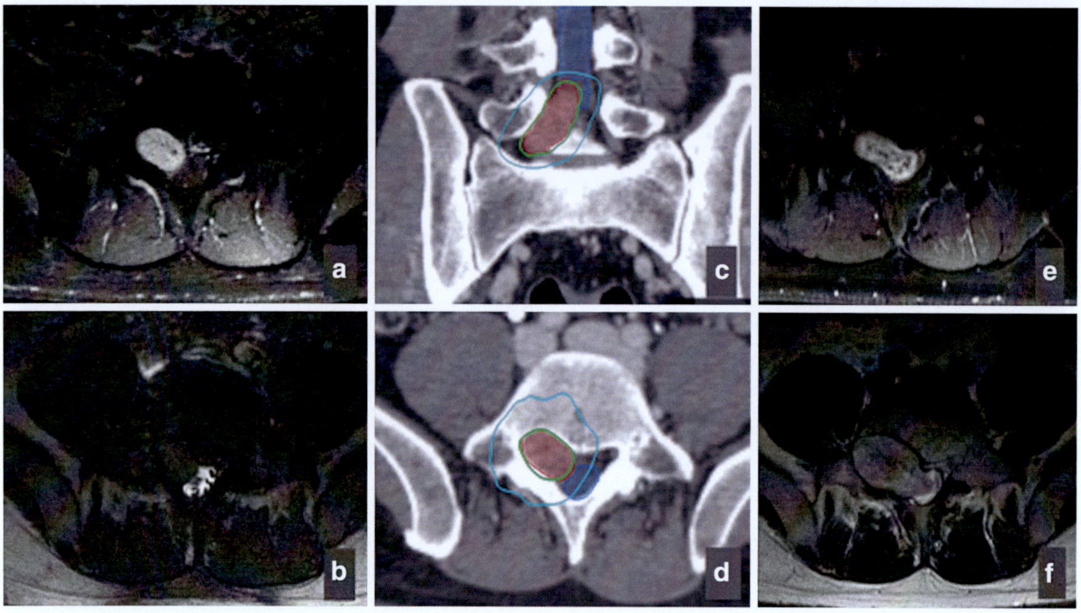

Fig. 40.1 A representative spinal SRS plan for treatment of a right L5 schwannoma. This 40-year-old man presented with right L5 paresthesias. Imaging revealed a right L5 schwannoma, which progressed on observation over 1 year (Panel **a**, pre-treatment T1 post-contrast fat-saturated MRI. Panel **b**, pre-treatment CISS MRI). He chose treatment with SRS rather than open surgical resection (Panels **c** and **d**, SRS Plan. The 14 Gy isodose line is green. Cyan is the 7 Gy line. The PTV is in red, cauda in blue). His 5.5 cm^3 schwannoma received 14 Gy in 1 fraction to the 85% isodose line. Tumor coverage was 95% at 14 Gy. The tumor was undercovered near the cauda equina nerve roots, but the tumor minimum dose was 12.2 Gy, an adequate dose to control schwannomas. Cauda equina maximum dose was 15.1 Gy. In his first follow-up 6 months after SRS, the tumor had post-SRS enlargement typical of schwannomas, with greater compression of the thecal sac and the contralateral nerve roots of the cauda equina (Panel **e**, post-SRS T1 post-contrast fat-saturated MRI; Panel **f**, post-SRS CISS MRI). He had no change in his baseline paresthesias, so no surgery was performed for this post-SRS tumor enlargement. Four years later, his tumor is slightly smaller, and he remains stable with no progression of symptoms. This example highlights the need for appropriate case selection in benign tumor spinal SRS. Post-SRS swelling of schwannomas may lead to neurologic compression and the potential need for surgical decompression for symptomatic tumor swelling. Had this tumor been over the spinal cord rather than the cauda, this tumor enlargement post-SRS would likely have been symptomatic

series of benign spinal SRS [10]. With MR imaging that can better define the spinal cord [28] than in the past, combined with the better dosimetry capable with modern radiosurgical planning [29, 30], the risk of spinal myelopathy is very low, particularly with the lower doses needed for benign compared to metastatic disease.

A common argument for the advantage of surgical resection of a benign intracranial tumor is that intracranial SRS poses a risk of secondary malignancy. Recent data provide guidance on how to weigh the risk of surgical morbidity and mortality [31, 32] (from surgical complications, anesthesia risks, pulmonary emboli, infection) with that of SRS-associated secondary malignancy. A 5-institution review of 4905 patients found an incidence of SRS-associated malignancy of 6.8 per 100,000 patient-years or a cumulative incidence of 0.00045% over 10 years [33]. This risk is similar to the risk of developing a malignancy in those that have not received SRS, per the Central Brain Tumor Registry of the United States. Similar results were seen in a series of 11,264 patient-years of follow-up, where no SRS-induced malignancies were reported [34]. The authors concluded that “the risk of radiation-induced tumors or malignant transformation after SRS is very low and should not be used as a justification for choosing alternative treatment approaches such as surgery”. It is unknown if these risks for intracranial SRS are applicable to spine SRS. In the largest series reported, no secondary malignancy was seen with 776 patient-years of follow-up [10].

Finally, with SRS for spinal schwannomas, appropriate patient selection is important, as following SRS, schwannomas will enlarge. In vestibular schwannomas, post-SRS swelling occurs in over 50% of treatments [35]. Similar post-treatment enlargement is seen in spinal schwannomas. In the series by Chin et al., a patient with a spinal schwannoma causing cord compression initially declined open surgical resection [10]. She was treated with SRS and developed acute worsening of compressive symptoms 2 weeks later which required urgent surgical resection. A similar example of post-SRS schwannoma enlargement is seen in Fig. 40.1. Patients with a schwannoma in the neuroforamen or adjacent to the spinal cord should be consented for the potential need for post-SRS resection should they become symptomatic from tumor swelling.

40.5 Conclusion

SRS for benign intracranial tumors is a standard of care, demonstrated by large series reporting high tumor control rates and low toxicity. The adoption of stereotactic principles for benign spinal tumors was initially limited by lack of long-term data. Recent series with larger patient numbers and longer follow-up report benign spinal tumor control rates comparable to intracranial tumor control rates. In previously unirradiated tumors, with doses of 14–16 Gy in a single fraction or equivalent, local progression following SRS for benign spinal tumors was reported in 2% at 5 years and 8% at 10 years [10]. Following SRS, patients reported that pain improved in 36%, was stable in 53%, and worsened in 11%. Appropriate patient selection for SRS is critical, particularly in patients with schwannoma, given their risk of post-SRS enlargement. Recommended single-fraction and hypofractionated doses and constraints to spinal cord are presented below. As for benign intracranial tumors, SRS for benign spinal tumors is a standard of care and an alternative to surgical resection.

40.6 Practical Guide

Representative treatment planning guidelines, tumor doses, organ-at-risk doses, and case examples for benign spinal tumor stereotactic radiosurgery follow.

References

1. Chamberlain MC, Tredway TL. Adult primary intradural spinal cord tumors: a review. Curr Neurol Neurosci Rep. 2011;11(3):320–8.
2. Hasegawa T, Kida Y, Kato T, Iizuka H, Kuramitsu S, Yamamoto T. Long-term safety and efficacy of stereotactic radiosurgery for vestibular schwannomas: evaluation of 440 patients more than 10 years after treatment with gamma knife surgery. J Neurosurg. 2013;118(3):557–65.
3. Combs SE, Adeberg S, Dittmar J-O, Welzel T, Rieken S, Habermehl D, et al. Skull base meningiomas: long-term results and patient self-reported outcome in 507 patients treated with fractionated stereotactic radiotherapy (FSRT) or intensity modulated radiotherapy (IMRT). Radiother Oncol. 2013;106(2):186–91.
4. Combs SE, Engelhard C, Kopp C, Wiedenmann N, Schramm O, Prokic V, et al. Long-term outcome after highly advanced single-dose or fractionated radiotherapy in patients with vestibular schwannomas—pooled results from 3 large German centers. Radiother Oncol. 2015;114(3):378–83.
5. Santacroce A, Walier M, Régis J, Liščák R, Motti E, Lindquist C, et al. Long-term tumor control of benign intracranial meningiomas after radiosurgery in a series of 4565 patients. Neurosurgery. 2012;70(1):32–9.
6. Pollock BE, Stafford SL, Link MJ, Garces YI, Foote RL. Single-fraction radiosurgery for presumed intracranial meningiomas: efficacy and complications from a 22-year experience. Int J Radiat Oncol Biol Phys. 2012;83(5):1414–8.
7. Rogers L, Barani I, Chamberlain M, Kaley TJ, McDermott M, Raizer J, et al. Meningiomas: knowledge base, treatment outcomes, and uncertainties. A RANO review. J Neurosurg. 2015;122(1):4–23.
8. Goldbrunner R, Minniti G, Preusser M, Jenkinson MD, Sallabanda K, Houdart E, et al. EANO guidelines for the diagnosis and treatment of meningiomas. Lancet Oncol. 2016;17(9):e383–91.
9. Tsao MN, Sahgal A, Xu W, De Salles A, Hayashi M, Levivier M, et al. Stereotactic radiosurgery for vestibular schwannoma: international stereotactic radiosurgery society (ISRS) practice guideline. J Radiosurg SBRT. 2017;5(1):5–24.
10. Chin AL, Fujimoto D, Kumar KA, Tupper L, Mansour S, Chang SD, et al. Long-term update of stereotactic radiosurgery for benign spinal tumors. Neurosurgery. 2019;85(5):708–16.

11. Barnett GH, Linskey ME, Adler JR, Cozzens JW, Friedman WA, Heilbrun MP, et al. Stereotactic radiosurgery—an organized neurosurgery-sanctioned definition. J Neurosurg. 2007;106(1):1–5.
12. Simpson D. The recurrence of intracranial meningiomas after surgical treatment. J Neurol Neurosurg Psychiatry. 1957;20(1):22–39.
13. Pettersson-Segerlind J, Orrego A, Lönn S, Mathiesen T. Long-term 25-year follow-up of surgically treated parasagittal meningiomas. World Neurosurg. 2011;76(6):564–71.
14. Ganz JC, Backlund E-O, Thorsen FA. The results of gamma knife surgery of meningiomas, related to size of tumor and dose. Stereotact Funct Neurosurg. 1993;61(1):23–9.
15. Dodd RL, Ryu M-R, Kamnerdsupaphon P, Gibbs IC, Chang SD, Adler JR. CyberKnife radiosurgery for benign intradural extramedullary spinal tumors. Neurosurgery. 2006;58(4):674–85.
16. Gerszten PC, Burton SA, Ozhasoglu C, McCue KJ, Quinn AE. Radiosurgery for benign intradural spinal tumors. Neurosurgery. 2008;62(4):887–96.
17. Gerszten PC, Chen S, Quader M, Xu Y, Novotny J, Flickinger JC. Radiosurgery for benign tumors of the spine using the Synergy S with cone-beam computed tomography image guidance. J Neurosurg. 2012;117 Suppl:197–202.
18. Selch MT, Lin K, Agazaryan N, Tenn S, Gorgulho A, DeMarco JJ, et al. Initial clinical experience with image-guided linear accelerator-based spinal radiosurgery for treatment of benign nerve sheath tumors. Surg Neurol. 2009;72(6):668–74.
19. Chang U-K, Rhee CH, Youn SM, Lee DH, Park SQ. Radiosurgery using the Cyberknife for benign spinal tumors: Korea Cancer Center Hospital experience. Journal of Neurooncol. 2011;101(1):91–9.
20. Marchetti M, De Martin E, Milanesi I, Fariselli L. Intradural extramedullary benign spinal lesions radiosurgery. Medium- to long-term results from a single institution experience. Acta Neurochir. 2013;155(7):1215–22.
21. Shin D-W, Sohn M-J, Kim H-S, Lee D-J, Jeon SR, Hwang YJ, et al. Clinical analysis of spinal stereotactic radiosurgery in the treatment of neurogenic tumors. J Neurosurg Spine. 2015;23(4):429–37.
22. Kalash R, Glaser SM, Flickinger JC, Burton S, Heron DE, Gerszten PC, et al. Stereotactic body radiation therapy for benign spine tumors: is dose de-escalation appropriate? J Neurosurg Spine. 2018;29(2):220–5.
23. Daly ME, Choi CYH, Gibbs IC, Adler JR Jr, Chang SD, Lieberson RE, et al. Tolerance of the spinal cord to stereotactic radiosurgery: insights from hemangioblastomas. Int J Radiat Oncol Biol Phys. 2011;80(1):213–20.
24. Gibbs IC, Patil C, Gerszten PC, Adler JR, Burton SA. Delayed radiation-induced myelopathy after spinal radiosurgery. Neurosurgery. 2009;64(suppl_2):A67–72.
25. Sahgal A, Weinberg V, Ma L, Chang E, Chao S, Muacevic A, et al. Probabilities of radiation myelopathy specific to stereotactic body radiation therapy to guide safe practice. Int J Radiat Oncol Biol Phys. 2013;85(2):341–7.
26. Katsoulakis E, Jackson A, Cox B, Lovelock M, Yamada Y. A detailed dosimetric analysis of spinal cord tolerance in high-dose spine radiosurgery. Int J Radiat Oncol Biol Phys. 2017;99(3):598–607.
27. Sahgal A, Chang JH, Ma L, Marks LB, Milano MT, Medin P, et al. Spinal cord dose tolerance to stereotactic body radiation therapy. Int J Radiat Oncol Biol Phys. 2019;S0360-3016(19)33862-3.
28. Li Z, Chen YA, Chow D, Talbott J, Glastonbury C, Shah V. Practical applications of CISS MRI in spine imaging. Eur J Radiol Open. 2019;6:231–42.
29. Blanck O, Wang L, Baus W, Grimm J, Lacornerie T, Nilsson J, et al. Inverse treatment planning for spinal robotic radiosurgery: an international multi-institutional benchmark trial. J Appl Clin Med Phys. 2016;17(3):313–30.
30. Furuya T, Phua JH, Ruschin M, Tanaka H, Nihei K, Pinnaduwage D, et al. Assessing functionality and benefits of comprehensive dose volume prescriptions: an international, multi-institutional, treatment planning study in spine stereotactic body radiation therapy. Pract Radiat Oncol. 2019;9(1):9–15.
31. Raco A, Pesce A, Toccaceli G, Domenicucci M, Miscusi M, Delfini R. Factors leading to a poor functional outcome in spinal meningioma surgery: remarks on 173 cases. Neurosurgery. 2017;80(4):602–9.
32. Ansari SF, Terry C, Cohen-Gadol AA. Surgery for vestibular schwannomas: a systematic review of complications by approach. Neurosurg Focus. 2012;33(3):E14.
33. Wolf A, Naylor K, Tam M, Habibi A, Novotny J, Liščák R, et al. Risk of radiation-associated intracranial malignancy after stereotactic radiosurgery: a retrospective, multicentre, cohort study. Lancet Oncol. 2019;20(1):159–64.
34. Pollock BE, Link MJ, Stafford SL, Parney IF, Garces YI, Foote RL. The risk of radiation-induced tumors or malignant transformation after single-fraction intracranial radiosurgery: results based on a 25-year experience. Int J Radiat Oncol Biol Phys. 2017;97(5):919–23.
35. van de Langenberg R, Dohmen AJC, de Bondt BJ, Nelemans PJ, Baumert BG, Stokroos RJ. Volume changes after stereotactic LINAC radiotherapy in vestibular schwannoma: control rate and growth patterns. Int J Radiat Oncol Biol Phys. 2012;84(2):343–9.

Intradural Spinal Lesions

41

Kita Sallabanda, Morena Sallabanda, and Peter Gerszten

41.1 Introduction

The treatment of spinal lesions has significantly evolved over the past decade with the introduction of minimally invasive surgical techniques and advances made in spinal radiosurgery [1–8]. The goal for the successful treatment of spinal pathology consists of effective pain control, maintenance and recovery of neurological function and ambulation, spinal stability, reduction of the risk of bleeding, and preserving quality of life.

The methods to treat spinal tumors include surgery, radiotherapy, radiotherapy plus systemic therapy or surgical decompression, and/or stabilization followed by radiotherapy (i.e., separation surgery) [1–3]. For benign spinal lesions, surgery represents the first option to alleviate neurological deficits and intractable pain. Nevertheless, recurrence is common after surgery. Furthermore, there are circumstances in which surgery is contraindicated due to clinical conditions or technical aspects (i.e., tumors massively extending beyond the limits of the spinal column). In such cases, radiosurgery may represent a safe and effective treatment option [1, 2, 4, 5, 9–12].

Three-dimensional treatment planning with multiple carefully shaped fixed fields allows to conform dose distribution to the target volume, as well as to minimize the dose to the surrounding critical tissues.

In recent years, intensity-modulated radiation therapy (IMRT) and volumetric-modulated arc therapy (VMAT) have increasingly gained importance for the delivery of conformal therapy. Nevertheless, conventionally fractionated or hypofractionated radiotherapy should be considered a "palliative" treatment of spinal metastases due to the dose limitation by the close proximity of the spinal cord.

Radiotherapy allows, to some extent, improvement in pain, neurological deficit, and functional outcome [13], especially in radiosensitive tumors [14]. Recent advances in imaging technology have allowed the safe delivery of high-dose radiation to spinal tumors lying in close proximity to the spinal cord by introducing the concept of image-guided spinal radiosurgery. Image-guided stereotactic radiosurgery allows the delivery of "ablative" doses in a single or in a limited [1–5, 7, 9, 11, 12, 15, 16] number of fractions because of a higher level of conformality as compared to FSRT and IMRT.

K. Sallabanda (✉)
Universidad Complutense de Madrid, Madrid, Spain

Hospital Clinico San Carlos, Madrid, Spain

Genesiscare, Madrid, Spain
e-mail: kitasall@ucm.es

M. Sallabanda
Genesiscare, Madrid, Spain
e-mail: morena.sallabanda@genesicare.com

P. Gerszten
Department of Neurological Surgery, Presbyterian University Hospital, Pittsburgh, PA, USA
e-mail: gersztenpc@upmc.edu

© Springer Nature Switzerland AG 2020
A. Conti et al. (eds.), *CyberKnife NeuroRadiosurgery*,
https://doi.org/10.1007/978-3-030-50668-1_41

A substantial body of data supports a number of benefits of spine SRS over conventionally fractionated external beam radiotherapy (cEBRT), including several evidence-based reviews [1]. According to a multidisciplinary spine oncology study group's recommendations, the current indications for spine SRS can be grouped into three general categories:

1. Primary definitive therapy for previously unirradiated tumors
2. Salvage radiosurgery for recurrent or progressive tumors having failed prior cEBRT
3. Postoperative radiosurgery after surgical intervention with or without spinal stabilization

In regard to treatment modalities, we have progressively transferred our radiosurgery experience from the brain to the spinal cord, including the implementation of the concepts of single-session treatment or hypofractionation, maximum dose conformity, maximum dose concentration, and maximum protection of the organs at risk (OAR).

It was Alan Hamilton et al. who, in 1996, reported that column SRS was possible with a linear accelerator using a frame-based technique [17]. Since then, several frame-based techniques have been used with limited success, but with the introduction of the CyberKnife system (Accuray Inc., Sunnyvale, CA) and its sophisticated image guide, we had the opportunity to treat a large patient population [1, 4, 15]. The main limitation to the application of spinal radiosurgery remains the limited tolerance of the spinal cord to high-dose irradiation.

Sahgal et al. evaluated the dosimetric data in five cases of radiation-induced myelopathy after SRS to spine tumors and reported on the recommended spinal cord tolerance for stereotactic body radiotherapy. The study concluded that for single-fraction SRS, 10 Gy to a maximum point dose to any part of the spinal cord is extremely safe. A risk of myelopathy of 5% or less was observed when limiting the thecal sac D_{max} volume dose to 12.4 Gy in a single fraction, 17.0 Gy in two fractions, 20.3 Gy in three fractions, 23.0 Gy in four fractions, and 25.3 Gy in five fractions [18]. Here we describe our experience with the treatment of benign spinal lesions including intramedullary vascular malformations.

41.2 Radiosurgery for Benign Spinal Tumors

The most common benign tumors of the spine include meningiomas, schwannomas, and neurofibromas. These tumors are intradural and extramedullary in location. The primary treatment for these tumors is microsurgical resection without adjuvant radiotherapy. However, patients with multifocal disease, advanced age, and poor performance status may be poor surgical candidates, and for these, radiosurgery has been explored as a therapeutic option.

Several recent publications have now documented both the safety and long-term efficacy of radiosurgery for benign spinal tumors. As patients with benign spinal lesions have prolonged life expectancies compared to their malignant counterparts, the potential for delayed radiation myelopathy is a special concern [19–21].

41.2.1 Meningioma

Spinal meningiomas comprise 10% of all meningiomas and 25% of all spinal tumors. Most spinal meningiomas occur in the thoracic region, posterolateral to the spinal cord. Gross total surgical resection optimizes the outcomes [22].

Sachdev et al. [11] reported on a series of 32 spinal meningiomas with a mean volume of 3.03 cm^3 (0.14–11.05 cm^3) that were treated with a median dose of 20.57 Gy (16–30 Gy) over a median of 2 fractions. Approximately 50% of lesions had previously been resected, and the remaining were diagnosed by imaging. At median follow-up of 29 months, all treated meningiomas were either stable (47%) or decreased (53%) in size. Overall, 91% of meningiomas had stable or improved neurological symptoms, with 57% of patients reporting pain improvement and 43% reporting minimal change. One case of late-onset transient myelitis was observed 9 months after treatment for a 7.6 cm^3 recurrent meningioma treated to 24 Gy over three fractions [11].

From the series published by our institution [2], 13 spinal meningiomas were treated using a single-fraction technique (mean dose 21 Gy, mean tumor volume 4.9 cm^3). Eleven of 13 patients had radiosurgery as an adjunctive treatment for residual or recurrent tumor following open surgical resection. Radiographic tumor control was demonstrated in all cases with a median follow-up of 17 months [2].

In a study by Lee et al. [23], patients treated with SRS (seven patients; median dose 15 Gy) or SBRT (four patients; median dose 26 Gy in three fractions) were followed for a median of 46.9 months. All lesions were controlled locally, with an average volume reduction of 29.7%. No statistically significant changes in enhancement patterns or T2 signal intensity were found [23].

Overall, spinal meningiomas treated with SBRT to doses ranging from 14 to 16 Gy in one fraction to 25–30 Gy in five fractions achieved excellent local control rates comparable to surgical outcomes after gross total resection.

The results also showed parallel results observed with fractional radiotherapy and single-fraction radiosurgery for intracranial meningiomas. SBRT also appears to provide pain improvement in up to 30% of patients over the initial weeks to months. However, motor deficits rarely improve.

One important point in the treatment of spinal meningiomas is that the vascularization comes from the anterior part of the spine and we can use a technique of partial coverage of the tumor volume by applying the concept of irradiation of the dural attachment of the tumor and the following bystander effect. We have one series of five patients treated with this technique with very good results and a local control at 5 years of 100% (Fig. 41.1).

41.2.2 Schwannoma

Spinal schwannomas comprise a third of spinal neoplasms and arise from the posterolaterally placed dorsal nerve root. Given their posterior position relative to the spinal cord or cauda equina, their removal by laminectomy approach represents a safe and effective option [24].

Klekamp and Samii [25] found that the recurrence rate was 10.7% after 5 years and 28.2% after 10 and 15 years after surgery. Similarly, the authors of other studies reported a recurrence rate of ~10% at a mean of 4.1–4.3 years after surgery.

In the largest published series from Sachdev et al. [11], 47 spinal schwannomas were treated with CyberKnife-based SBRT, for which 11%, 21%, and 7% of lesions from the entire cohort (103 spinal lesions) had associated diagnoses of neurofibromatosis type 1 (NF1), NF2, and schwannomatosis, respectively. The mean tumor

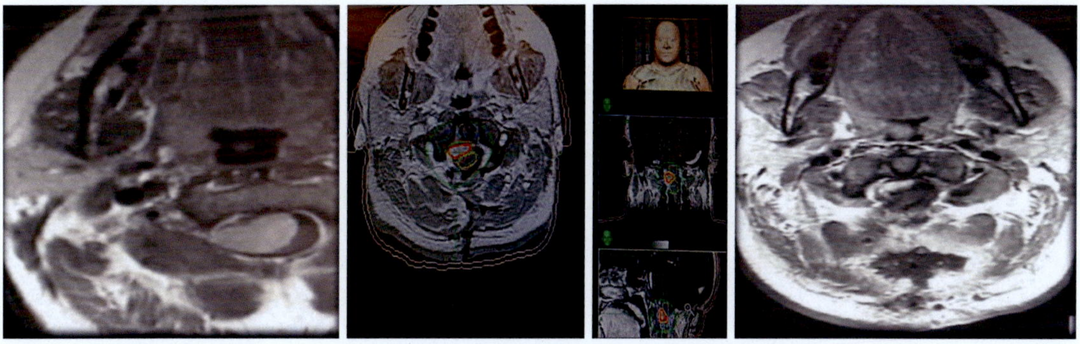

Fig. 41.1 A case of a 48-year-old woman with severe tetraparesis. A large meningioma was identified at C1–C2 level. After Simpson grade 2 resection, the tumor regrew, and we performed SRS with a partial covering technique. The treatment was delivered in a single dose. The dose selected was 14 Gy at the 82% isodose line with a 15 Gy dose at the tumor attachment and a maximum point dose of 14 Gy at the spinal cord

volume was 6.18 cm^3 (0.05–54.52 cm^3), and the mean dose delivered was 18.74 Gy (14–24 Gy) over a median of 1 fraction (range: 1–4 fractions). At a median follow-up of 29 months (range 6–87 months), a single lesion progressed 73 months after treatment after receiving 18 Gy in three fractions, resulting in a crude control rate of 98%. Radiographic regression was noted in 47% of patients, half of which decreased to less than half the original tumor size. Pain was improved in 54% of patients and progressed in only 14%. Salvage surgery was attempted in four patients (only one of which experienced radiographic progression), with subsequent symptomatic improvement in three of four patients and no change in symptoms in one patient. No late spinal cord toxicities were noted [11].

A Korean series by Shin et al. [16] reported outcomes of 54 patients with benign tumors (47 spinal schwannomas), who presented with pain (63%) or neurological symptoms (24%) [26]. Most were treated with single-fraction SBRT (72%) to a median of 13 Gy as primary therapy, and the remaining patients received a combination of surgery and SBRT. Radiographic control rate was 95%, with 55% of lesions showing regression. Transient swelling was noted in 20% of lesions at a median time from SRS of 8 months (range 5.1–44.3), and tumoral enhancement suggestive of necrosis was noted in 69%. Neither finding was significantly associated with local control ($p = 0.253$ and $p = 0.067$, respectively). Overall, significant improvements in pain scores were noted at a median of 8.1 months, and all patients with neurological symptoms improved after combined surgery and SBRT [16].

From our institution, 35 spinal schwannomas treated with CyberKnife-based SBRT, with overall rates of NF1 and NF2 being 29% and 12%, respectively [7]. The mean tumor volume was 11.0 cm^3 (1.0–47.7 cm^3), and 59% of lesions were located in the cervical spine. Prescription doses ranged from 17.5 to 25 Gy in 1–5 fractions (Fig. 41.2). At a median follow-up of 37 months (8–71 months; all patients), the radiographic control rate was 100%. Among initially symptomatic patients, 82.4% noted improvement in pain, and 80% had improvement (60%) or stabilization (20%) of neurological symptoms. Three patients ultimately underwent salvage surgery for progressive symptoms. Two patients experienced transient myelopathy, with subsequent return of strength [7]. Published series demonstrate local control rates after single-fraction SBRT for schwannomas, ranging above 90–95%. An excellent improvement in symptomatic pain has also been demonstrated over the months, with only a minority of patients experiencing persistent symptoms requiring salvage surgery.

41.2.3 Neurofibroma

Neurofibromas are benign nerve-sheath tumors that may arise from either peripheral or spinal nerve roots. Neurofibromas of the spine are often multiple, predominate in the cervical region, and are commonly associated with NF1. Neurofibromas are less common than schwannomas, constituting only of 3.5% of primary spinal tumors. These tumors grow both intra- and extradurally.

Surgical eradication of these tumors usually requires sectioning of the originating nerve root in order to completely resect the lesion.

Sahgal et al. [27] reported a series of 11 treated neurofibromas (mean dose 21 Gy delivered in 3 fractions, mean tumor volume 6.0 cm^3). Radiographic control was documented in nine patients. Three patients had NF1, and two of these suffered progression. In the published series from Stanford, nine neurofibromas in seven patients with NF1 (mean dose 10.6 Gy, mean tumor volume 4.31 cm^3) were treated with radiosurgery, and tumor stabilization on imaging was documented in six of seven (86%) patients. After a mean follow-up of 20 months, half of the patients described an improvement in symptoms after radiosurgery, and half of the patients documented a worsening in pain, weakness, or numbness at their last follow-up. However, the tumors were radiographically stable in all patients [28].

The authors caution that the role of radiosurgery for neurofibromas remains unclear, particularly considering that a significant number of the NF1 patients were myelopathic at presentation. They also state that the most realistic and attain-

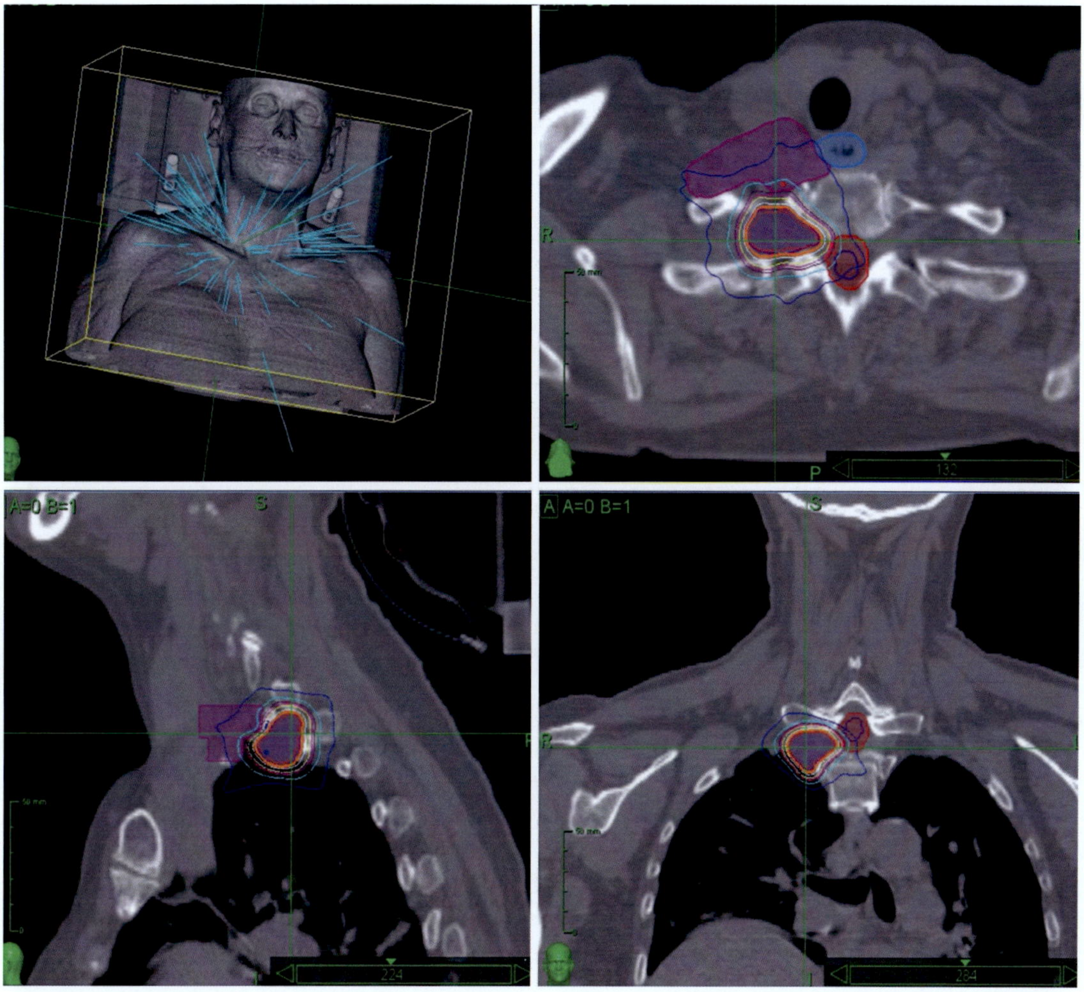

Fig. 41.2 CyberKnife treatment plan of a right T1 schwannoma treated with a dose of 18 Gy in three fractions of 6 Gy. The tumor growth was controlled 8 years later

able goal of neurofibroma treatment in myelopathic patients is tumor control without significant expectations of symptomatic improvement.

Our institution published an experience with 25 neurofibroma cases (mean dose 21.3 Gy, mean tumor volume 12.6 cm^3) [2]. Similar to the Stanford experience, no patient had evidence of radiographic tumor progression on follow-up. Twenty-one of these patients had NF1 and nine had NF2. Radiosurgery improved discomfort in 8 of 13 patients (61.5%) treated for pain. All patients without any improvement in pain had NF1 [2].

These findings echo the outcomes from the Stanford series which found that pain control in spinal neurofibromas associated with NF1 responds less well to radiosurgery [7, 28]. Poorer microsurgical results for neurofibromas have also been observed in patients with NF1. The multiplicity of neurofibromas in NF1 may be partially to blame as this factor makes identifying the symptomatic neurofibroma that needs treatment more difficult.

Furthermore, given that many of the patients with neurofibromas have multiple lesions along their spine, it can often be difficult to determine whether symptom progression is due to the treated lesion or from any of the other neurofibroma lesions within the spine.

Moreover, the infiltrating nature of neurofibromas, in contrast to the other benign extramedullary intradural spinal tumors, can cause more irreversible neural damage and increase the susceptibility of the native nerve root to injury from both microsurgical and radiosurgical treatments. Finally, future genomic investigations may reveal that intrinsic genetic differences in neurofibromas associated with NF1 predisposed to a weaker radiobiological response.

41.3 Spinal AVMs

Intramedullary arteriovenous malformations (AVMs) are rare, representing around 20% of all spinal AVMs (being spinal AVMs 16% of all central nervous system AVMs). Spinal AVMs are associated with a high morbidity and can very seriously affect the quality of life of patients. Intramedullary AVMs can be classified according to the angioarchitecture of the angiomatous nidus, the relationship with the arteries, and the distribution of the nidus around the medulla [26].

There are four types of Intramedullary AVMs:

- Type I: dural arteriovenous fistulas
- Type II: so-called glomus lesions, characterized by a compact intramedullary nidus
- Type III: so-called juveniles, which are characterized by a large and diffuse nidus with a wide extramedullary and paraspinal component
- Type IV (a-b-c): perimedullary arteriovenous fistulas

Type I and IV are treated by surgery, endovascular embolization, or a combination of both. Type III are very difficult to treat, and multiple endovascular treatments are necessary to reduce symptoms. Type II AVMs, featured by a compact and not very large nidus or glomus lesions, represent a potential target for radiosurgery. The principles of radiosurgery to spinal AVMs derive from those developed for the treatment of cerebral AVMs. The limiting factor to the application of radiosurgery to intramedullary AVMs is represented by the high sensitivity to radiation of the spinal cord. Furthermore, the spinal cord is a moving organ.

The use of image guidance gave us the opportunity to attempt the application of radiosurgery to such complicated lesions. By image guidance, it is actually possible to track the spine during the treatment delivery and to target the moving organ by using a robotic arm correction.

There is very limited experience in the treatment of these lesions in the literature [6, 29–35]; we treated five patients with spinal vascular malformations, three intramedullary, one lumbar, and one radicular AVMs. Two of the intramedullary malformations were type II AVMs, and one was a cavernous angioma. After 3 years of follow-up, the AVMs were obliterated (Fig. 41.3), whereas the cavernous angioma reduced in size without post-treatment bleeding, Zabramski grade 3. For the treatment of these lesions, we used three fractions of 7 Gy each for the type II AVMs and 13 Gy in single fraction for the cavernoma.

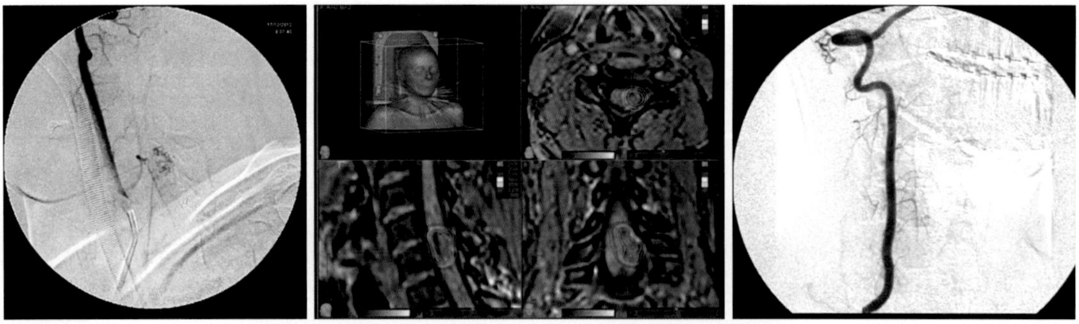

Fig. 41.3 (Left) Pre-treatment angiogram showing a type II intramedullary arteriovenous malformation (AVM) at the cervical level. (Middle) The CyberKnife treatment plan. (Right) The angiogram obtained 30 months after treatment shows the obliteration of the AVM

Based on this experience, we suggest that CyberKnife robotic radiosurgery represents a good option in the treatment of type II spinal AVMs, and our results mirror few other reports available in the literature (39, 40). However, the potential for radiation injury to the spinal cord after radiosurgery may depend on a range of critical variables including, but not limited to, spinal cord region (cervical versus thoracic), AVM volume, length of involved spinal cord, and the details of AVM blood supply.

There some other suggestions that we can derive from this experience concerning the dose constraints we adopted.

We suggest to keep the V10 <0.455 cm^3 in single fraction, <0.71 cm^3 in two fractions, and <0.788 cm^3 in three fractions and the V12 <0.286 cm^3 in single fraction, 0.557 cm^3 in two fractions, and <0.579 cm^3 in three fractions [21].

References

1. Conti A, Acker G, Kluge A, Loebel F, Kreimeier A, Budach V, et al. Decision making in patients with metastatic spine. The role of minimally invasive treatment modalities. Front Oncol. 2019;9:915.
2. Gerszten PC, Burton SA, Ozhasoglu C, McCue KJ, Quinn AE. Radiosurgery for benign intradural spinal tumors. Neurosurgery. 2008;62(4):887–96.
3. Moussazadeh N, Laufer I, Yamada Y, Bilsky MH. Separation surgery for spinal metastases: effect of spinal radiosurgery on surgical treatment goals. Cancer Control. 2014;21(2):168–74.
4. Gerszten PC, Mendel E, Yamada Y. Radiotherapy and radiosurgery for metastatic spine disease. Spine. 2009;34(Supplement):S78–92.
5. Gerszten PC, Monaco EA. Complete percutaneous treatment of vertebral body tumors causing spinal canal compromise using a transpedicular cavitation, cement augmentation, and radiosurgical technique. Neurosurg Focus. 2009;27(6):E9.
6. Ghobrial GM, Maulucci CM, Dalyai RT, Chalouhi N, Rosenwasser RH, Harrop JS. Radiosurgery for spinal intramedullary arteriovenous malformations: a literature review. J Neurol Surg A Cent Eur Neurosurg. 2015;76(5):392–8.
7. Sahgal A, Chou D, Ames C, Ma L, Lamborn K, Huang K, et al. Image-guided robotic stereotactic body radiotherapy for benign spinal tumors: the University of California San Francisco preliminary experience. Technol Cancer Res Treat. 2007;6(6):595–603.
8. Yamada Y. Faculty of 1000 evaluation for Tolerance of the spinal cord to stereotactic radiosurgery: insights from hemangioblastomas. F1000—Post-publication peer review of the biomedical literature: F1000 Research Ltd; 2011.
9. Faul CM, Flickinger JC. The use of radiation in the management of spinal metastases. J Neurooncol. 1995;23(2):149–61.
10. Kim YH, Fayos JV. Radiation tolerance of the cervical spinal cord. Radiology. 1981;139(2):473–8.
11. Sachdev S, Dodd RL, Chang SD, Soltys SG, Adler JR, Luxton G, et al. Stereotactic radiosurgery yields long-term control for benign intradural, extramedullary spinal tumors. Neurosurgery. 2011;69(3):533–9.
12. Sundaresan N, Digiacinto GV, Hughes JEO, Cafferty M, Vallejo A. Treatment of neoplastic spinal cord compression: results of a prospective study. Neurosurgery. 1991;29(5):645–50.
13. Cole JS, Patchell RA. Metastatic epidural spinal cord compression. Lancet Neurol. 2008;7(5):459–66.
14. Maranzano E, Latini P. Effectiveness of radiation therapy without surgery in metastatic spinal cord compression: final results from a prospective trial. Int J Radiat Oncol Biol Phys. 1995;32(4):959–67.
15. Muacevic A, Staehler M, Drexler C, Wowra B, Reiser M, Tonn J-C. Technical description, phantom accuracy, and clinical feasibility for fiducial-free frameless real-time image-guided spinal radiosurgery. J Neurosurg Spine. 2006;5(4):303–12.
16. Shin D-W, Sohn M-J, Kim H-S, Lee D-J, Jeon SR, Hwang YJ, et al. Clinical analysis of spinal stereotactic radiosurgery in the treatment of neurogenic tumors. J Neurosurg Spine. 2015;23(4):429–37.
17. Hamilton AJ, Lulu BA, Fosmire H, Stea B, Cassady JR. Preliminary clinical experience with linear accelerator-based spinal stereotactic radiosurgery. Neurosurgery. 1995;36(2):311–9.
18. Sahgal A, Ma L, Weinberg V, Gibbs IC, Chao S, Chang UK, et al. Reirradiation human spinal cord tolerance for stereotactic body radiotherapy. Int J Radiat Oncol Biol Phys. 2012;82(1):107–16.
19. Hashmi A, Guckenberger M, Kersh R, Gerszten PC, Mantel F, Grills IS, et al. Re-irradiation stereotactic body radiotherapy for spinal metastases: a multi-institutional outcome analysis. J Neurosurg Spine. 2016;25(5):646–53.
20. Sahgal A, Ames C, Chou D, Ma L, Huang K, Xu W, et al. Stereotactic body radiotherapy is effective salvage therapy for patients with prior radiation of spinal metastases. Int J Radiat Oncol Biol Phys. 2009;74(3):723–31.
21. Daly ME, Choi CYH, Gibbs IC, Adler JR, Chang SD, Lieberson RE, et al. Tolerance of the spinal cord to stereotactic radiosurgery: insights from hemangioblastomas. Int J Radiat Oncol Biol Phys. 2011;80(1):213–20.
22. Peker S, Pamir MN. Management of superior sagittal sinus invasion in parasagittal meningiomas. In: Meningiomas: Elsevier; 2010. p. 365–71.

23. Lee ME, Hwang YJ, Sohn MJ, Lee BH, Kim SY. Assessment of the treatment response of spinal meningiomas after radiosurgery focusing on serial MRI findings. Jpn J Radiol. 2015;33(9):547–58.
24. Seppälä MT, Haltia MJJ, Sankila RJ, Jääskeläinen JE, Heiskanen O. Long-term outcome after removal of spinal schwannoma: a clinicopathological study of 187 cases. J Neurosurg. 1995;83(4):621–6.
25. Klekamp J, Samii M. Surgery of spinal nerve sheath tumors with special reference to neurofibromatosis. Neurosurgery. 1998;42(2):279–89.
26. Kim LJ, Spetzler RF. Classification and surgical management of spinal arteriovenous lesions: arteriovenous fistulae and arteriovenous malformations. Neurosurgery. 2006;59(suppl_5):S3-195–201.
27. Sahgal A, Ma L, Gibbs I, Gerszten P, Ryu S, Weinberg V, et al. 104 spinal cord tolerance for stereotactic body radiotherapy. Radiother Oncol. 2009;92:S33.
28. Dodd RL, Ryu M-R, Kamnerdsupaphon P, Gibbs IC, Chang SD, Adler JR. CyberKnife radiosurgery for benign intradural extramedullary spinal tumors. Neurosurgery. 2006;58(4):674–85.
29. Cho WS, Wang KC, Phi JH, Lee JY, Chong S, Kang HS, et al. Pediatric spinal arteriovenous malformations and fistulas: a single institute's experience. Childs Nerv Syst. 2016;32(5):811–8.
30. Kalani MA, Choudhri O, Gibbs IC, Soltys SG, Adler JR, Thompson PA, et al. Stereotactic radiosurgery for intramedullary spinal arteriovenous malformations. J Clin Neurosci. 2016;29:162–7.
31. Mori Y, Hashizume C, Tsugawa T, Kato S, Shibamoto Y. Stereotactic radiotherapy for intramedullary spinal arteriovenous malformations. Cureus. 2018;10(7):e2908.
32. Potharaju M, John R, Venkataraman M, Gopalakrishna K, Subramanian B. Stereotactic radiosurgery results in three cases of intramedullary spinal cord arteriovenous malformations. Spine J. 2014;14(11):2582–8.
33. Rashad S, Endo T, Ogawa Y, Sato K, Endo H, Matsumoto Y, et al. Stereotactic radiosurgery as a feasible treatment for intramedullary spinal arteriovenous malformations: a single-center observation. Neurosurg Rev. 2017;40(2):259–66.
34. Sussman ES, Adler JR, Dodd RL. Radiosurgical ablation of spinal cord arteriovenous malformations. Handb Clin Neurol. 2017;143:175–87.
35. Zhan PL, Jahromi BS, Kruser TJ, Potts MB. Stereotactic radiosurgery and fractionated radiotherapy for spinal arteriovenous malformations—a systematic review of the literature. J Clin Neurosci. 2019;62:83–7.

Part VII

Vascular Lesions

Cerebral Arteriovenous Malformations

42

Umberto Fornezza, Zeno Perini, Gianpaolo Zambon, Mariano Zanusso, and Leopoldo S. Casentini

42.1 Introduction

The oldest description of extracranial MAV is dated 1500 BC. in Egypt and more precisely to the Ebers Papyrus. In the second half of the nineteenth century, Virchow defined these lesions as *of congenital origin*, and, as we know, this definition is still the subject of debate in the modern scientific world [1]. In the following centuries, scientists such as Dandy, Cushing, Olivecrona, and above all Yasargil, Drake, Spetzler, and De Oliveira determined important paths in the surgical treatment of these complicated but fascinating lesions. In 1960, Luessenhop carried out the first embolizations with the use of methacrylate, a method subsequently extended by Newton (1968) and Doppman (1971) for the treatment of spinal malformations [2–5].

Arteriovenous malformations (AVMs) are congenital vascular anomalies composed of an abnormal tangle of abnormal blood vessels. In these vessels, blood is shunt directly from the arterial to the venous compartment without capillaries to control the pressure boost. This vicious circle combined with the abnormal composition of the vessels can lead both to the formation of intra- and extranidal aneurysms and to a deterioration of the vessel walls increasing risk of rupture and intracranial/intracerebral hemorrhage [2, 6–8].

Single session stereotactic focused irradiation has been considered for a long time one of most effective treatment for selective cerebral AVMs. Early clinical experience was based on the use of Gamma Units and was described in 1970 by Steiner and colleagues, followed by other radiosurgical techniques, cyclotron, LINACs, and CyberKnife [6, 9–12]. Perhaps, the first observation of the beneficial effect on these lesions was accomplished by Harvey Cushing as early as 1928 when he performed brain surgery for an AVM of a patient who received, a few years earlier, non-stereotaxic X-ray irradiation [4]. Since its introduction in clinical practice, radiosurgical procedures have proven to be safe and reliable. Early results, in terms of complete AVM obliteration, appeared to be unrelated to the device used but strongly influenced by the nidus volume and adsorbed peripheral dose.

The morphology and size of the nidus are not the only two indices needed for a therapeutic strategy, but they are the two fundamental parameters in the development phase of radiosurgical treatment planning [10, 13]. In general, the following factors are evaluated when a patient is seen with an AVM: age, related medical condition,

U. Fornezza · Z. Perini (✉) · G. Zambon
M. Zanusso · L. S. Casentini
Department of Neurosurgery, City Hospital, Vicenza, Italy
e-mail: umberto.fornezza@aulss8.veneto.it; giampaolo.zambon@aulss8.veneto.it; mariano.zanusso@aulss8.veneto.it; leopoldosilviocasentini@aulss8.veneto.it

© Springer Nature Switzerland AG 2020
A. Conti et al. (eds.), *CyberKnife NeuroRadiosurgery*,
https://doi.org/10.1007/978-3-030-50668-1_42

any prior hemorrhagic events, location of the malformation, onset symptoms (headache, seizures, local neurologic deficits), angioarchitecture of the AVM (e.g., compact or diffuse nidus), surgical risks, presence of a proximal or intranidal aneurysm, and the surgical experience of the center [14].

In fact, surgical removal is an important option for patients with lobar vascular malformations of suitable size, especially at centers of excellence with extensive AVM experience. Incomplete removal requires adjuvant management, including radiosurgery [8, 15, 16].

In our experience, in all patients the treatment planning was performed with the aid of 3D reconstruction of the AVM nidus based on angiographic data. It is commonly accepted that 3D angiography remains the gold standard for characterization of AVMs, and in our opinion this exam is mandatory to perform a correct and useful treatment planning [13, 17–20].

Over the years, many aspects of AVM radiosurgical treatment have been studied. In fact, literature reports numerous articles describing encouraging results for lower Spetzler-Martin grades and lower Pollock-Flickinger scores [21–23]. Although radiosurgery for AVMs has proven to be effective, some limitations, drawbacks, questions, and challenges are still present.

The first problem is the risk of hemorrhage during the "latency period" between treatment and complete AVM obliteration. The use of radiosurgery reduces the risks associated with surgery, and this is the most important benefit [7, 8]. Nonetheless, it is equally true that the latency period required to achieve complete obliteration of the AVMs is the most important limitation of the method and the risk of hemorrhage for months to years after irradiation restricts the general application of radiosurgery as a first treatment option.

Furthermore, the difficulty of delineating the real target to irradiate represents another problem for radiosurgery, especially after embolization [24].

Another "open" question remains when we can determine that the biological effects of irradiation on the AVM are effectively over.

In regard to large AVMs, (this issue will be addressed in an another chapter) the role of radiosurgery versus microsurgery and embolization is under discussion because of the increased probability of volume-related adverse effects with a standard dose delivered in a single session.

Staged procedures are used for larger vascular malformations or in patients with a diffused nidus close to the optic pathway, brainstem, and functional areas [8, 25–28].

These are some of the main problems we encounter daily in the management of the patients affected by AVMs.

42.2 Vicenza Neurosurgery Department Experience

In the Vicenza Neurosurgery Department, radiosurgery was introduced in 1982 thanks to Dr. Federico Colombo (Fig. 42.1a) and the treatment of cerebral arteriovenous malformations (AVMs) 2 years later [29].

Until December 2002, 642 patients were treated with our LINAC based, converging arc technique (Fig. 42.1b) [9, 10, 13, 20]. Median follow-up was 120 months. Of these patients, 418 control angiographies were examined 24 months after treatment. Complete obliteration was influenced by AVM size ranging from 96% in small (S)-less than 15 mm in diameter to 33% in large (L)—over 25 mm in diameter. Twenty-eight patients suffered for complications due to radiation, ranging from mild transitory neurological complaints (14 patients) to permanent disabling ones (13 patients). One patient developed a radio-induced tumor 13 years after treatment of an AVM and eventually died.

The most frequent complication was related to bleeding during the latency period before complete obliteration. Forty-eight cerebral hemorrhages were observed. Nine patients died as a consequence of AVM rupture (1.4%). Emergency surgery was performed on nine patients. Fifteen patients suffered from stabilized neurological symptoms (2.3%). Poisson regression was used to estimate hemorrhage rates per 100 person/year of follow-up.

During this period, significant hardware changes were introduced in the procedure. The

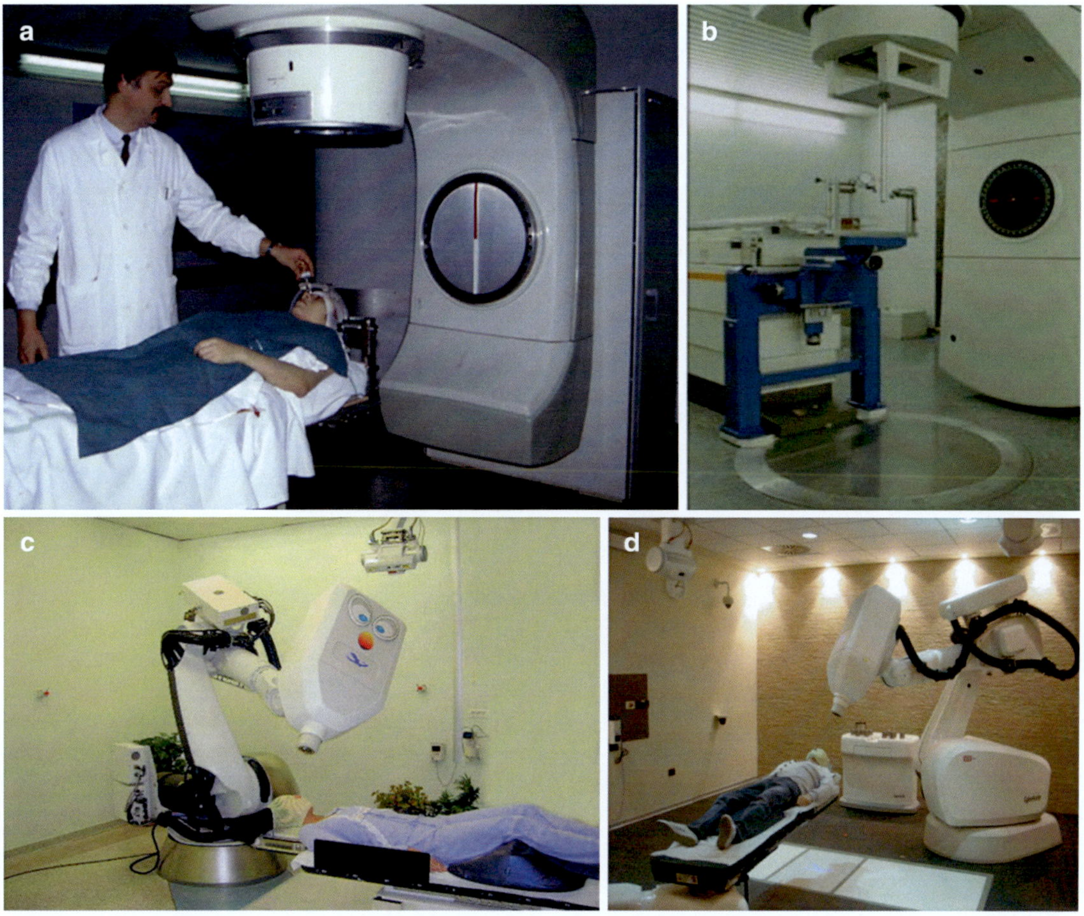

Fig. 42.1 (**a**) Dr. Colombo in 1982 during one of the first radiosurgical treatments with LINAC; (**b**) The LINAC (Siemens 6 MV®) in use in Vicenza from 1984 to 2002; (**c**) The first CyberKnife (Accuray Inc., Sunnyvale, CA®) in Vicenza arrived in 2003; (**d**) The current CyberKnife Unit in use at our department and established in 2012

first MV LINAC was substituted, after 213 patients, by a 6MV employed up to 2002 (Fig. 42.1b). Stereotactic biplanar angiography was the only morphological data base for target determination up to the early 1990s. Subsequently, computerized treatment planning systems were employed for combining stereotactic angiography with CT and/or MRI scans [10, 13, 30].

From January 2003, a frameless, image-guided robotic radiosurgery apparatus, the CyberKnife (Accuray Inc., Sunnyvale CA) [10] has been utilized for AVMs (Fig. 42.1c, d).

In a first step, contrast-enhanced CT images of the intended target are acquired. The imaging modality that has to be fused to CT is three-dimensional rotational angiography (3DRA). The maximization of the entropy correlation coefficient, a mutual information maximization algorithm [10, 20], is employed to calculate the parameters of a transformation that takes into account rotation, translation, linear scaling, and angular deformation [31, 32].

42.2.1 Image Registration Procedure

The CyberKnife stereotactic radiosurgery treatment planning system is based on CT for image guidance. Nevertheless, optimal target outlining requires information provided by multimodality images, such as angiography for AVMs or MR imaging for lesions close to critical structures.

Before the CT scan, a thermoplastic mask was prepared and used to minimize patient movement during scanning. Computed tomography images were then acquired by a standard scanner. The number of slices ranged from 90 to 120 depending on the target location, and slice thickness range was 1.25–2 mm. The modalities that could be co-registered to CT were 3D rotational angiography, MR imaging, PET, fMR imaging, and tractography (DTI) [10, 20, 33–36].

An angiographic study (3D rotational angiography) was administered the same day as CT examination. For image registration, the maximization of the entropy correlation coefficient was selected among mutual information maximization algorithms.

The image registration procedure was adapted for co-registration of fMR imaging. The aim of this procedure was to fuse a brain functional map to the CT volume used for treatment planning. This image set was used to contour critical structures located close to the AVM nidus so that dose constraints could be imposed.

Functional MR imaging studies were performed prior to the radiation procedure (usually a few days before) and prepared for registration to the CT scan. Functional maps were obtained by means of the blood oxygen level-dependent effect using EPI MR volumes and the statistical parametric mapping software package. We evaluated motor functions by means of hand, foot, and tongue movements, and evaluated language-cognitive functions using category generation, letter generation, simple question, and verbal generation tasks. For motor studies, functional maps were obtained by means of a t-test analysis with a probability value of 0.05 (false-positive corrected) and a minimum number of 20 adjacent voxels to define an activation cluster. For language-cognitive functions, the values used were $p = 0.001$ (uncorrected for false positives) and 40 voxels. These differences in activation analysis were required due to the spread of activated areas in language-related tasks. To improve the signal-to-noise ratio, images were filtered by means of convolution with a Gaussian kernel with full width at half maximum of 8 mm. Due to its negative effect on spatial resolution, this preprocessing step was factored in when evaluating proximity of an activated area to the target volume in radiosurgery [10].

Registration of the functional maps to the CT volume was performed using the following steps: (1) rigid registration of all MR imaging data sets to the first volume acquired to compensate for patient movement during the acquisition procedure; (2) Gaussian filtering; (3) statistical analysis and generation of the functional map; (4) affine registration of the EPI volumes to a T2-weighted MR imaging volume; (5) rigid registration of the T2-weighted MR imaging volume to the CT volume; and (6) application of the combined transformation defined in Steps 4 and 5 to the functional map obtained in Step 3. This final step allowed the functional map to be spatially registered to the CT volume used for treatment planning and therefore made it possible to define regions of interest corresponding to functional areas, within which dose constraints could be imposed during the optimization process.

When 3DRA datasets are fused to CT, visualization of anatomy, target outline, and critical structures on both modalities is enabled. For a better definition of AVM nidus, embolic material can be subtracted. Automatic delineation of AVM nidus contour is performed slice by slice on axial sections of 3DRA. The same image fusion procedure can be employed for implementing fMR and tractography (DTI) in the treatment planning procedures. Based on these functional images, automatic contouring of critical regions (motor strip, language cortical areas, occipital visual cortex, etc.) can be initiated (Fig. 42.2). For image registration, the maximization of the entropy correlation coefficient was selected among mutual information maximization algorithms [1, 10, 37–40].

42.2.1.1 Materials and Methods

From January 30, 2003, to December 31, 2018, 550 patients (303 males and 247 females) affected by cerebral AVMs and 1 by a spinal AVM were treated.

At treatment, age ranged from 9 to 74 years (mean 36 years, median 34). Symptoms of appearance were bleeding in 297, epilepsy in 87,

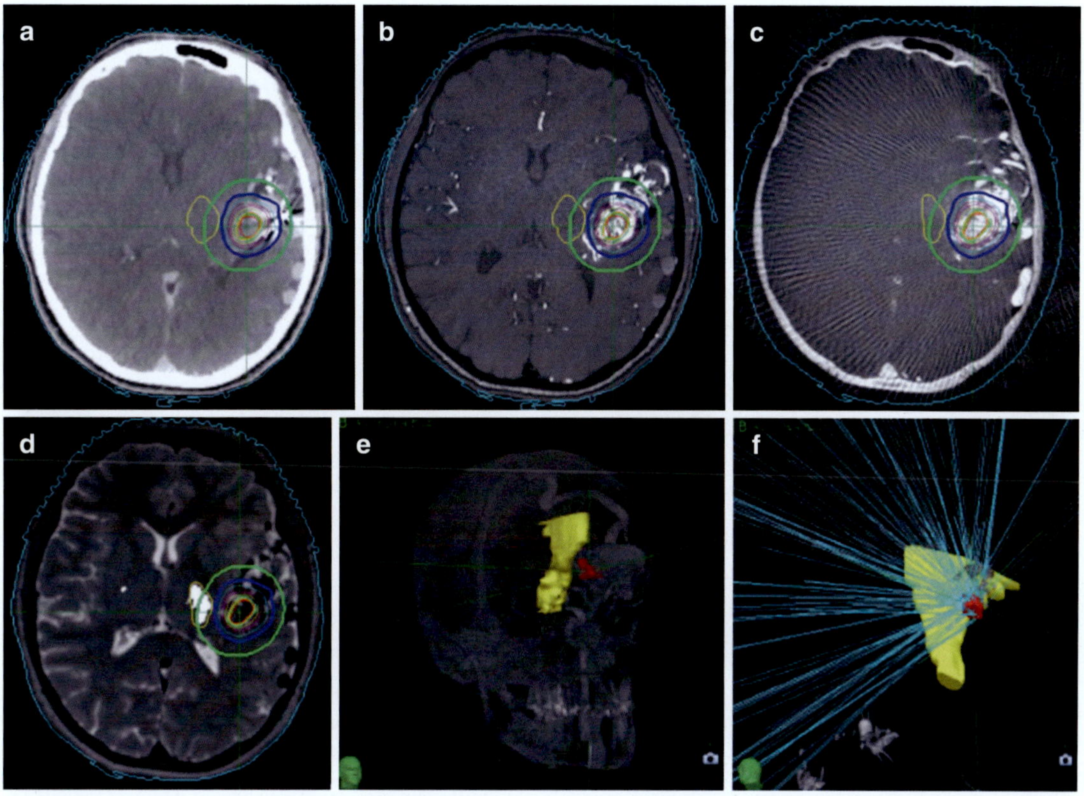

Fig. 42.2 (**a**) Contrast-enhanced CT scan. The embolizing material is also shown; (**b**) contrast-enhanced MRI; (**c**) angiography; (**d**) fMRI (in yellow the corticospinal tract); (**e**, **f**) the 3D angiographic images; in the picture f with treatment beams

neurological deterioration in 92, and headache in 45. In ten patients, the AVM was revealed by examinations made for non-related diseases or trauma.

Of this court, 378 of them had a follow-up superior to 36 months: 193 (51%) had bleeding as symptom of onset, 71 seizures, 54 neurological deterioration, 46 headache, and 14 were an occasional finding. 45% of the lesions were located in the basal ganglia and 25% in the corpus callosum. Before radiosurgery, 23 patients had undergone unsuccessful attempts of surgical removal; 187 patients incomplete AVM embolization and 54 radiosurgical treatment. Using the *Spetzler-Martin* grading system, we treated 69 patients with grade II, 242 with grade III, 54 grade IV, and only 13 with grade V. The *Pollock-Flickinger score* ranged from 0.41 to 6.2. AVM irradiated volume varied from 0.1 to 56 mL (mean 4.7 mL). Maximum radiation doses ranged between 22.5 and 30 Gy (mean 25.1 Gy). The borders of target volume were encompassed by isodose lines ranging from 70 to 85%.

In cerebral AVMs with target volumes smaller than 8 mL, radiation was delivered in single session (358). In 18 patients with target volumes larger than 8 mL, in 3 patients with brainstem or optic pathways AVMs, and in one patient with a spinal AVM, the radiation dose was delivered in 2 fractions, 8–30 days apart (dose staged strategy). In 11 patients with large AVMs, we adopted volume staged strategy considering two distinct portions of the nidus and irradiated in a separate session with a 2–14 months interval between the two procedures.

Follow-up protocol consisted in MRI images with contrast enhancement and MR angiographic sequences at 6, 12, 18, 24, and 36 months after radiosurgery. An angiogram was performed when MRI suggested a sensible reduction or complete

Table 42.1 Summary of patients of our cohort

Sex	
Female	185
Male	193
Age	
Range	9–74
Mean	36
Median	34
Symptoms	
Bleeding	193
Headache	46
Seizure	71
Focal neurological deficit	54
AVM location	
Supratentorial	223
Thalamus and basal ganglia	70
Brainstem	20
Cerebellum	26
Corpus callosum	39
Deep draining vein	229
SM grade	
II	69
III	242
IV/V	67
PF-score	
Range	0.41–6.2
Margin dose	
Range	75–80%
Maximum dose	
Range	22.5–30 Gy
Mean	25.1

obliteration of the nidus. In case of stable remnant, we repeated irradiation or suggested alternative treatments. Total follow-up was 192 months (Table 42.1).

42.2.2 Results

In 70% of patients (265/378), a complete follow-up was obtained and the obliteration rate was about 90% (238/265). MRI showed a significant nidus reduction in 95% of patients (252/265). In these patients, control angiography was performed in 100: complete obliteration of the malformation was confirmed in 88 out of 100 (in 12 very tiny AVM remnants persisted, to be verified at later follow-up). Three-year complete obliteration rate is 90% (Figs. 42.2 and 42.3).

Hemorrhage after treatment was observed in 27 patients (7%) from 2 days to 10 months after radiosurgery. All of them had suffered previous hemorrhages. Two died and two reported permanent neurological deficits.

Radionecrosis was observed in 48 patients (13%), but only three were symptomatic: one demonstrated transitory dysphasia, in one motor deficit worsened and one manifested VI cranial nerve paresis. Forty-five patients with MRI positive for likely radionecrosis were clinically silent.

Ultimately, in eight patients, radiosurgery failed: two patients underwent microsurgery, three endovascular treatment, and three repeated radiosurgery.

42.3 CyberKnife Radiosurgery of Unruptured Cerebral AVMs. Retrospective Analysis of 220 Cases Treated Between 2003 and 2014

The natural course of cerebral arteriovenous malformations (AVMs) has long been a matter of study and debate. Bleeding is the most common and dangerous manifestation of AVMs, but other signs can lead to the detection of AVMs: seizures, headaches, transient neurological deficits, or symptoms. In some cases, the malformation is found accidentally when neuroradiological examinations are performed for unrelated pathology or trauma [8, 15, 25].

The recent Randomized Trial of Unruptured Brain Arteriovenous Malformations (ARUBA) [25] analyzing the outcome of a series of 223 patients with unruptured AVMs showed that the risk of stroke and death is higher in the group of 114 patients interventionally treated (5 with surgical procedure, 30 with embolization, 31 with radiotherapy, 48 with multimodal therapy) compared to 109 patients only medically managed.

The aim of our study is to review the results of stereotactic radiosurgery (SRS) with CyberKnife on 220 patients with unruptured AVMs between

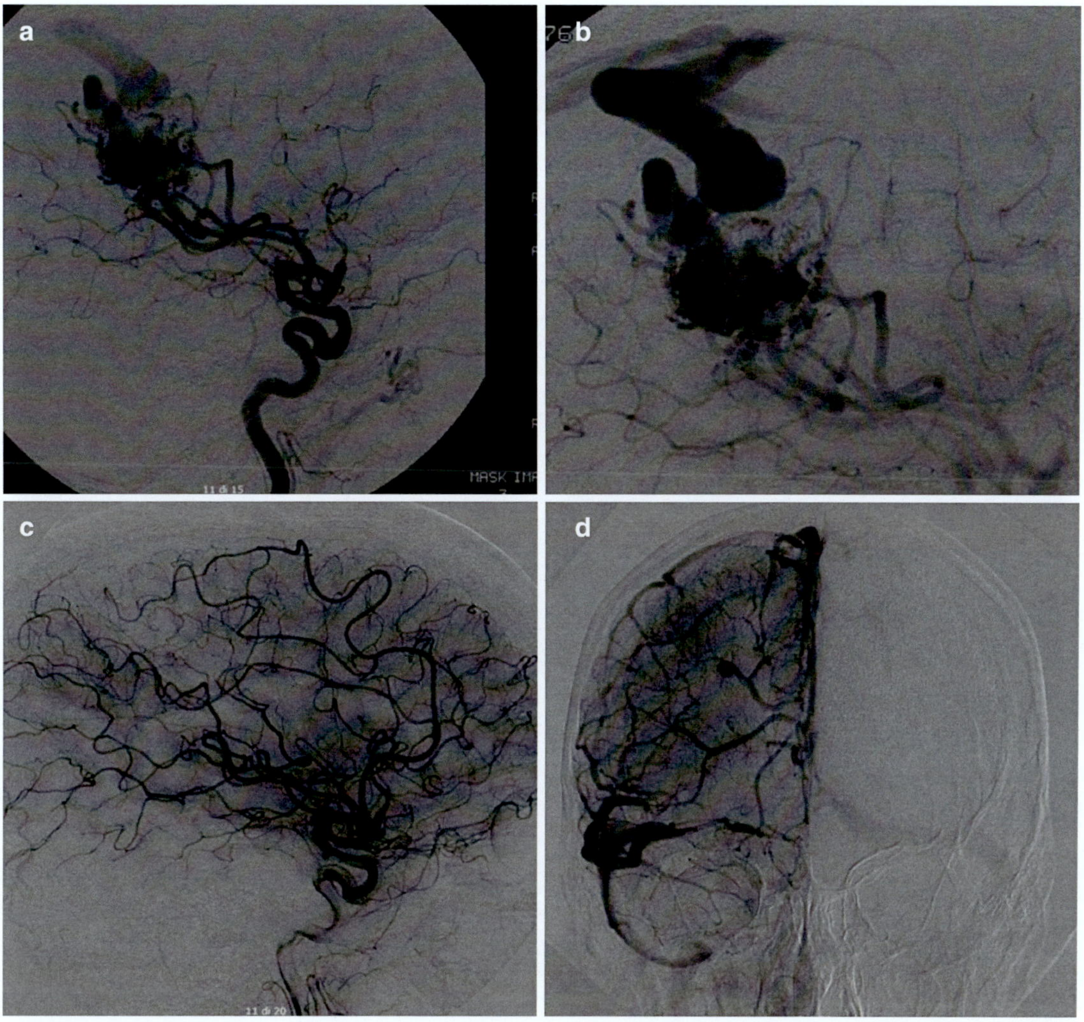

Fig. 42.3 (**a**, **b**) Female, 30 years old. Right parietal AVM. Treatment: conformal, 7.5 mm collimator, 25.33Gy (80%. 20.26Gy) single session (2006). Pollock-Flickinger score: 1.25. (**c**, **d**) Control angiogram obtained 36 months later showing the complete obliteration of the AVM

2003 and 2014 at the Vicenza Hospital, to evaluate the obliteration rate and the incidence of complications after CyberKnife SRS (Fig. 42.4).

42.3.1 Materials and Methods

Between February 2003 and December 2014, 220 patients with unruptured brain AVMs were treated with CyberKnife at the Vicenza Hospital (Fig. 42.5). Seventy-seven patients presented seizures, 55 headache, 39 visual disturbances, 18 trauma or incident, 17 paresthesias, and 15 focal neurological deficit (in one patient both focal deficit and paresthesias were considered). There were 115 males and 105 females with a median age of 38 years (mean 39.1 years, range 12–81 years).

One hundred seventeen patients came to our observation after previous treatment: 100 embolization, 12 LINAC SRS, 3 surgical approach, and 2 embolization plus LINAC SRS.

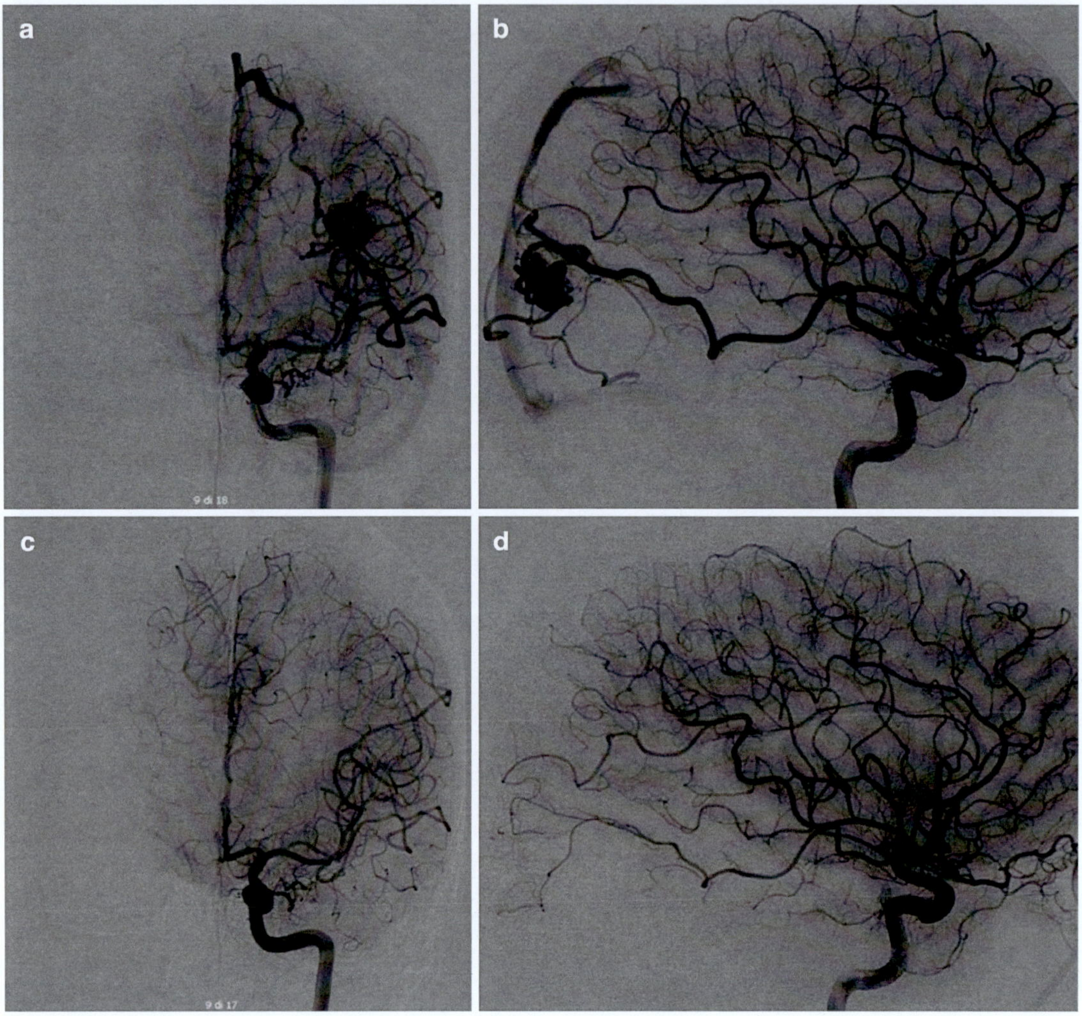

Fig. 42.4 (**a**, **b**) Female, 39 years old, left occipital AVM. Treatment: conformal 7.5 mm collimator, 25.33 Gy (75%, 19 Gy) single session (2009). P-F score: 1.36. (**c**, **d**) Control angiogram obtained 36 months later showing the complete obliteration of the AVM

42.3.2 AVM Characteristics

The localization was mainly lobar: 166 of which 23 frontal, 57 temporal, 48 parietal, 38 occipital, 27 brainstem and basal ganglia, 15 cerebellar, and 12 corpus callosum. The nidus volumes ranged from 0.1 to 23 mL (median 1.58 mlL, mean 2.28 mL). The Martin-Spetzler grading at the first CyberKnife SRS was grade I in 3 patients, grade II in 66 patients, grade III in 116 patients, grade IV in 27 patients, and grade V in 8 patients. The median Pollock-Flickinger score was 1.26 (mean 1.35, range 0.23–4.63).

Forty-nine cases required a second CyberKnife SRS and 10 a third CyberKnife SRS. The median nidus volume at the second treatment was 1.6 mL (mean 2.1 mL, range 0.15–9.6 mL); at the third treatment, median and mean volume were 1.6 mL (range 0.6–3.45 mL).

42.3.3 Treatment Planning

The median margin dose at the first treatment was 18.99 Gy (mean 18.81 Gy, range 10.5–22.5 Gy); the median maximum dose was

25.33 Gy (mean 25.01 Gy, range 15–30 Gy). At the second SRS, the parameters were median margin dose was 18.5 Gy (mean 18.42 Gy, range 14–20 Gy); median maximum dose was 24.66 Gy (mean 24.28 Gy, range 20–26 Gy). In the third treatment, median margin dose was 17.75 Gy (mean 16.97 Gy, range 12–19.5 Gy); median maximum dose was 23.66 Gy (mean 22.56 Gy, range 16–26 Gy).

42.3.4 Follow-Up

MRIs were performed at 6-month intervals for the first 2 years after SRS and yearly afterward. Further MRI control was performed in patients with side effects. When MRI revealed obliteration of the AVM, an angiography was planned, generally after at least 24 months after SRS. If the nidus was still evident at the angiographic

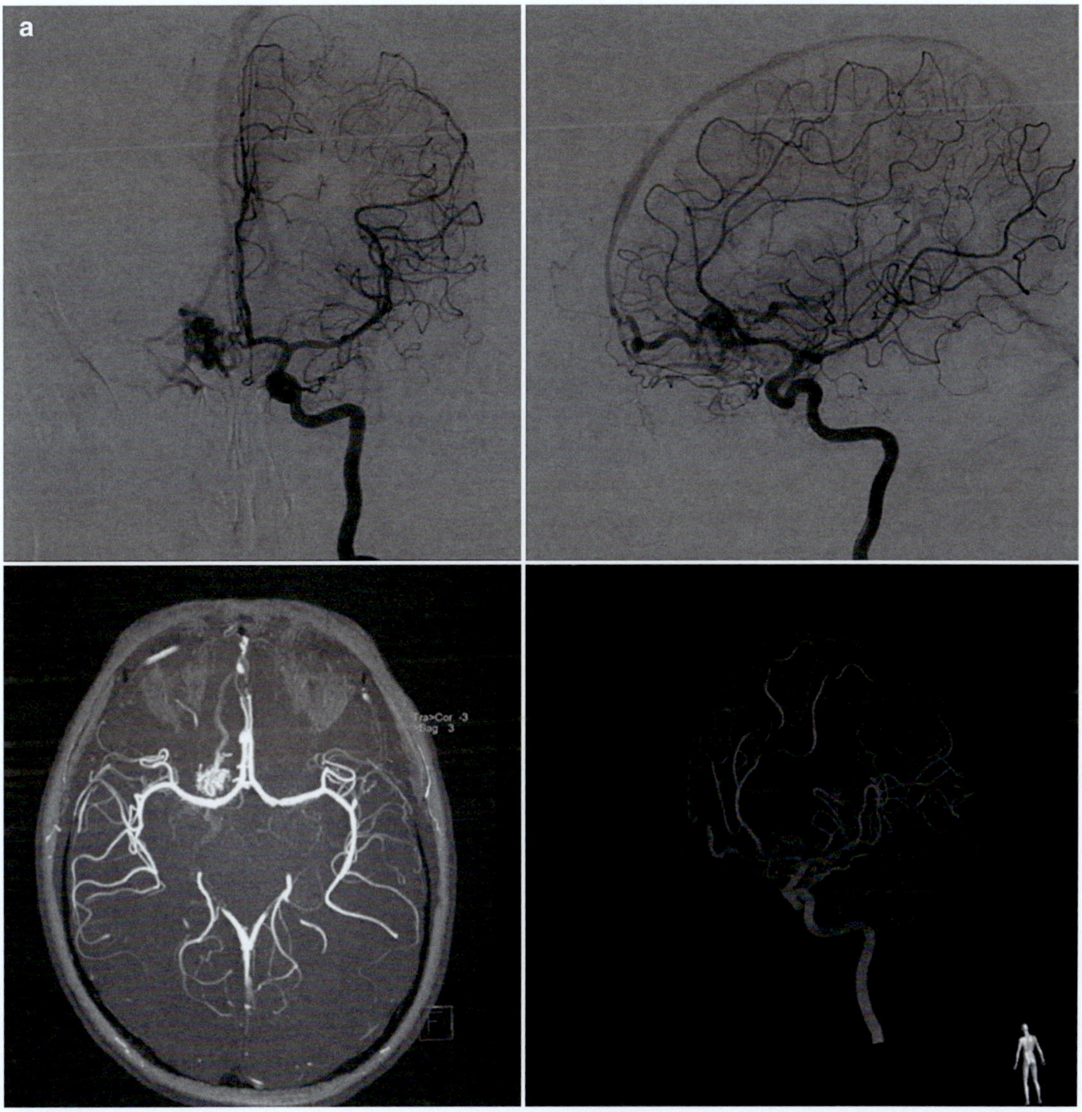

Fig. 42.5 An example of treated unruptured AVM. Female, 53 years old. Dose 25.33 Gy (isodose 75%, 19 Gy) in single fraction. (**a**) Imaging we used for planning. Angiography (anterior and oblique); MR angiogram; 3D angiography reconstruction. (**b**) Details of treatment planning (DVH and doses)

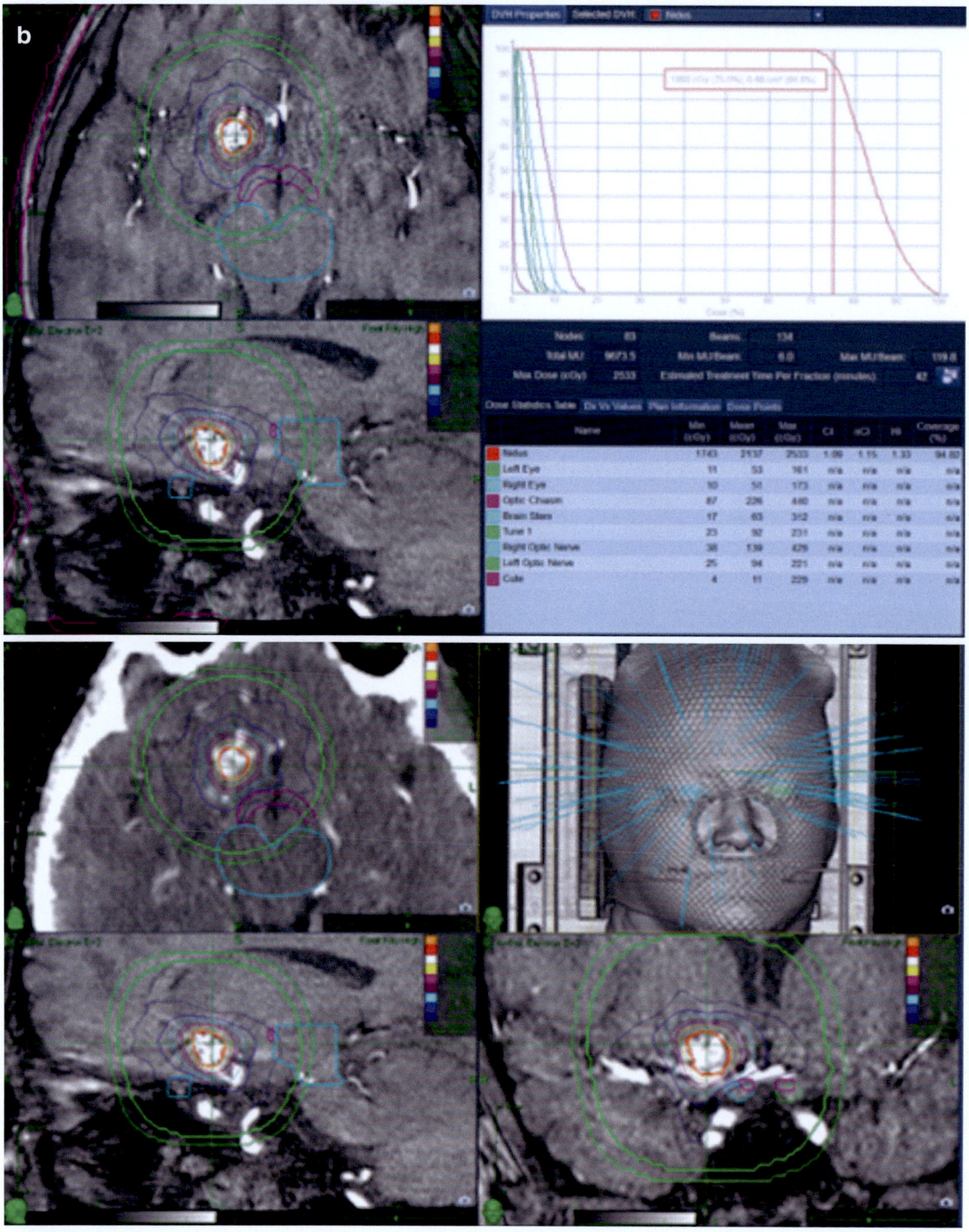

Fig. 42.5 (continued)

control at 36 months, a second treatment was planned. If the residual appeared unchanged, with the exception of significant progression of obliteration, the SRS was delayed 1 or 2 years. AVMs reduction at angiography showed various patterns: nidus shrinkage was more or less evident, flow reduction and later appearance of draining vein, steal decrease with visualization of normal arterial vessels, reduction of number and/or dimensions of draining vein(s), visualization of remaining nidi in staged SRS, and lower "density" of the treated nidus.

Subsequent to angiographically demonstrated obliteration, MRI every 5 years was advised.

Twenty-one patients did not return for follow-up within the first 12 months; 199 patients received follow-up after over 12 months (129 angiographies, 68 MRI, 2 CT).

One hundred thirty-eight patients underwent follow-up more than 36 months later. In 84 patients (61%) obliteration was seen: in 80 patients by angiography and in 4 by MRI. In 10 patients, obliteration was obtained with a second CyberKnife SRS.

42.3.5 Complications

We observed radionecrosis/MRI changes in 15 cases, hemorrhage in 12/205 cases, seizures in 6 cases, paresthesia in 3, visual disturbances in 4, headache in 1, aphasia in 1, and confusion and transient amnesia in 2.

42.3.6 Results

At the end of the study, 21 patients underwent a follow-up of less than 12 months. Of the 199 patients with follow-up longer than 12 months, 157 received angiographic control, 47 only MRI, and 2 only CT.

Follow-up longer than 36 months was performed on 138 patients. As previously reported, only this group has been considered for statistical analysis of factors determining successful treatment, while other data such as clinical complication or imaging changes are referred to all of the 220 patients.

Obliteration was achieved in 84/138 patients (61%): the median obliteration time was 79 months (IC 95%: 51–96), median obliteration time after single SRS was 42 months (IC 95%: 37–48 months), median after second SRS was 101 months (IC 95%: 88—not valuable months), and median after third SRS was not valuable (IC 95% not valuable).

Bleeding risk was calculated at about 1.8–2% per year for the first 5 years (equal to the natural history of the disease) and significantly decreasing after this time to 0.2% per year.

Although in this cohort of study the percentage of healing was not high, the limited appearance of side effects suggests that his method offers an important chance of recovery to the patient.

42.4 Discussion

In this work, we performed an update of the experience reported by Colombo et al. [10]. The parameters of treatment used are substantially the same. In fact, we performed angioMR for the treatment in only three patients (one patient refused angiography and two had liver failure). All other patients underwent 3D angiography.

As seen previously in 2009, no differences on the percentage of obliteration were recorded between previously embolized patients and those who did not receive endovascular treatments, refuting the Pittsburgh study group findings [8]. In fact, Lunsford et al. (2016) underline that in a group of 120 previously embolized patients, the ratio between those who undergo complete obliteration of the malformation is significantly lower than those who have not been embolized. The authors affirm that this data is due to the latency time required by radiosurgery where in previously embolized patients, it can allow a recanalization of the malformation "thus canceling" the effect of irradiation. This thesis would be supported by the fact that in a study that compares 47 previously embolized patients with 47 others treated with radiosurgery alone. The percentage of obliteration was clearly in favor of the latter [41]. The results obtained in our series could be explained by the improvement in target definition with our procedure with 3D angiography and subtraction of embolic material.

With regard to hemorrhage risk, we observed 27 patients (7%) with bleeding from 2 days to 10 months after radiosurgery. All of them had previous hemorrhages, two died, and two reported permanent neurological deficits. Instead, in our unruptured AVMs experience, we reported a bleeding risk related to radiosurgery which is substantially equal to the risk linked to the natural history of the disease (2% per year) for the first 5 years after treatment and which then drops significantly in the following years. In our opinion, the interesting point in this retrospective observational study does not lie in the number of obliterations obtained (60% c.a.), but the scarcity of side effects observed in the treated patients, even those treated several times. Moreover, this "low" cure rate must be compared to the treated lesions volumes which were in a range equal to 1.13–23 mL (mean 2.28 mL).

An interesting study, edited by Yen et al. in 2011 with Gamma Knife radiosurgery in 1204 patients, showed a hemorrhage rate of 2% in the pre-radiosurgery period and a hemorrhage rate of 2.5% after treatment [7]. The same results were obtained by Colombo et al. in 2009 [10]. Slightly higher rates were reported in a study of 2004 by Choi et al. [42] on 214 patients, in which the hemorrhage rate calculated was 3.2%. Two recent studies by Arslan et al. 2017 and Matthieu et al., published in April 2018 [6, 11], reported hemorrhage risk after radiosurgery as 3.5% and 4.7%, respectively. The cohorts were composed of 199 patients in the first study and of 57 patients (for a total of 64 treatments) in the second study. A very low hemorrhage risk was reported in the study published by the group of the Burdenko neurosurgical institute of Moscow (2015) [43] where the risk after radiosurgery is equal to 0.07% in a cohort of 93 patients and with a percentage of obliteration equal to 80.6%.

In our study, we also investigated a small cohort of pediatric patients where only 24/38 had a follow-up longer than 3 years. Most of them had SM grade II–III (19/24). The main onset symptom was hemorrhage (79%), three patients underwent to surgery, and five of them were embolized before treatment. The angiographic obliteration rate was 62% (15/24) with a significant nidus reduction in nine of the remaining patients. During follow-up (up to 189 months), we observed one case of radionecrosis, three hemorrhages, and seven patients underwent repeated radiosurgery. An interesting study on this topic was edited by Umansky et al. (Department of Neurosurgery of Tel Aviv) in 2018 [44]. In a cohort of 14 pediatric patients (mean age 17.3 years) who had combined treatment, radiosurgery after embolization was analyzed. A significant reduction of the nidus, thanks to the onyx, 10/14 completely confirmed closures (71%) and 2/14 significant flow reductions were observed. Although not clinically relevant, two cases of edema were observed.

In conclusion it is our opinion that the long-term effects in the pediatric population are the primary issue in radiosurgical treatment and represent one of the open questions and challenges in the management of AVMs [48].

References

1. Kakizawa Y, Nagashima H, Oya F, Ito K, Tanaka Y, Hongo K, Kobayashi S. Compartments in arteriovenous malformation nidi demonstrated with rotational three-dimensional digital subtraction angiography by using selective microcatheterization. J Neurosurg. 2002;96(2):770–4.
2. Hunter W. Observation in arteriovenous malformations. London: Royal Society of Medicine; 1762.
3. Krayenbuhl H, Yasargil MG. Das hirnaneuriysma. JR Geigy SA: Basel; 1958.
4. Nicolato A, Foroni R, Lupidi F, Gerosa M. La GammaKnife nel trattamento delle malformazioni artero-venose cerebrali. New Magazine Edizioni; 2006
5. Virchow R. Die Krankhaften Geschwulste. Berlin: Hirschwald Publications; 1867. p. 456.
6. Arslan I, Tezcanli E, Yilmaz M, Cizmeli O, Sengoz M, Peker S. Gamma knife radiosurgery for arteriovenous malformations: clinical series of 199 patients. Turk Neurosurg. 2017;27(2):301–8.
7. Chung PY, Sheehan JP, Lucia S, David S. Hemorrhage risk of cerebral arteriovenous malformations before and during the latency period after gamma knife raidosurgery. Stroke. 2011;42:1691–6.
8. Dade LL, Sheehan Jason P. Intracranial stereotactic radiosurgery. 2nd ed. New York: Thieme Medical Publishers; 2016.
9. Colombo F. Linear accelerator radiosurgery of arteriovenous malformations: Vicenza experience. In: Smee R, editor. Proceedings of the 4th Congress of International Stereotactic Radiosurgery Society, the Price of Wales Hospital Press, Sydney; 1999. p. 47 (abs).
10. Colombo F, Cavedon C, Casentini L, Francescon P, Causin F, Pinna V. Early results of CyberKnife radio-

surgery for arteriovenous malformations. J Neurosurg. 2009;111:807–919.
11. Raboud M, Tuleasca C, Maeder P, Schiappacasse L, Marguet M, Daniel RT, Levivier M. GammaKnife radiosurgery for arteriovenous malformations: general principles and preliminary results in a Swiss cohort. Swiss Med Wkly. 2018;148:w14602.
12. Steiner L, Prasad D, Lindquist C, Steiner M. Clinical aspects of gamma knife stereotactic radiosurgery. In: Gildenberg PL, Tasker RR, editors. Textbook of stereotactic and functional neurosurgery. New York: McGrow-Hill; 1998. p. 763–803.
13. Colombo F, Casentini L, Cavedon C, Francescon P, Causin F, Pinna V. Arteriovenous malformations radiosurgery: evolution of the technique. In: Kondziolka D, editor. Radiosurgery, vol. 6. Basel: Karger; 2006. p. 1–11.
14. Karlsson B, Lindquist C, Steiner L. Prediction of obliteration after gamma knife surgery for cerebral arteriovenous malformations. Neurosurgery. 1997;40:425–31.
15. Narayanan M, Atwal GS, Nakaji P. Multimodality management of cerebral arteriovenous malformations. Handb Clin Neurol. 2017;143:85–96.
16. Tsuji A, Nozaki K. A prospective and retrospective study of cerebral AVM treatment strategies 1990-2014. Acta Neurochir Suppl. 2016;123:135–9.
17. Anoop H, Jillian M, Santanu C, John S, Janos S, Daniela I, Shawn M. Dinamic CT angiography for cyberknife radiosurgery planning of intracranial arteriovenous malformations: a technical/feasibility report. Radiol Oncol. 2015;49(2):192–9.
18. Bridcut RR, Winder RJ, Workman A, Flynn P. Assessment of distortion in three dimensional rotational angiography system. Br J Radiol. 2002;75:226–70.
19. Chang SD, Murphy MJ, Martin DP, Hancock SL, Doty JR, Adler JR. Image guided robotic radiosurgery: clinical and radiographic results with the Cyberknife. In: Kondziolka D, editor. Radiosurgery 1999, vol. 3. Basel: Karger; 2000. p. 23–33.
20. Colombo F, Cavedon C, Francescon P, Casentini L, Fornezza U, Castellan L, Causin F, Perini S. Three-dimensional angiography for radiosurgical treatment planning for arteriovenous malformations. J Neurosurg. 2003;98:536–43.
21. Pollock BE, Flickinger JC. A proposed radiosurgery—based grading system for arteriovenous malformations. J Neurosurg. 2002;96:79–85.
22. Pollock BE. Gamma knife radiosurgery of arteriovenous malformations: long-term outcomes and late effects. Prog Neurol Surg. 2019;34:238–47.
23. Spetzler RF, Martin NA. A proposed grading system for arteriovenous malformations. J Neurosurg. 1986;65:476–83.
24. Todnem N, Ward A, Nahhas M, Vender JR, Alleyne CH, Rahimi SY. A retrospective cohort analysis of hemorrhagic arteriovenous malformations treated with combined endovascular embolization and gamma knife stereotactic radiosurgery. World Neurosurg. 2019;122:e713–22.
25. Mohr JP, Parides MK, Stapf C, Moquete E, Moy CS, Overbey JR, Al-Shahi Salman R, Vicaut E, Young WL, Houdart E, Cordonnier C, Stefani MA, Hartmann A, von Kummer R, Biondi A, Berkefeld J, Klijn CJ, Harkness K, Libman R, Barreau X, Moskowitz AJ, International ARUBA Investigators. Medical management with or without interventional therapy for unruptured brain arteriovenous malformations (ARUBA): a multicentre, non-blinded, randomised trial. Lancet. 2014;383(9917):614–21.
26. Russel D, Peck T, Ding D, Chen C-J, Taylor DG, Stark RM, Lee C-C, Sheehan JP. Sterotactic radiosurgery alone or combined with embolization for brain arteriovenous malformations: a systematic review and meta-analysis. J Neurosurg. 2018;128: 1338–48.
27. Kano H, Kondziolka D, Flickinger JC, Park KJ, Iyer A, Yang HC, et al. Stereotactic radiosurgery for arteriovenous malformations after embolization: a case-control study. J Neurosurg. 2012;117:265–75.
28. Russell D, Peck T, Ding D, Chen CJ, Taylor DG, Starke RM, Lee CC, Sheehan JP. Stereotactic radiosurgery alone or combined with embolization for brain arteriovenous malformations: a systematic review and meta-analysis. J Neurosurg. 2018;128(5):1338–48.
29. Colombo F, Benedetti A, Pozza F, Marchetti C, Chierego G. Linear accelerator radiosurgery of cerebral arteriovenous malformations. Neurosurgery. 1989;24:833–44.
30. Colombo F, Pozza F, Chierego G, Casentini L, De Luca GP, Francescon P. Linear accelerator radiosurgery of cerebral arteriovenous malformations: an update. Neurosurgery. 1994;34(1):14–21.
31. Kooy HM, Bellerive MR, Loeffler JS. Technical concepts of linac radiosurgery dosimetry. In: Gildenberg PL, Tasker RR, editors. Textbook of stereotactic and functional neurosurgery. New York: McGrow-Hill; 1998. p. 687–704.
32. Meeks SL, Bova FJ, Friedman WA. Technical aspects of radiation physics. In: Gildenberg PL, Tasker RR, editors. Textbook of stereotactic and functional neurosurgery. New York: McGrow-Hill; 1998. p. 649–68.
33. Friedman L. Comment on Kondziolka D, Lunsford LD, Kanal E, Talagala L: Stereotactic magnetic resonance angiography for targeting in arteriovenous malformations radiosurgery. Neurosurgery. 1994;33(4):591.
34. Adler JR, Murphy MJ, Chang SD, Hancock SL. Image-guided robotic radiosurgery. Neurosurgery. 1999;44:1299–307.
35. Blatt DR, Friedman WA, Bova FJ. Modifications based on computer tomography imaging in planning the radiosurgical treatment of arteriovenous malformations. Neurosurgery. 1993;33(4):588–96.
36. Van Herk M, Kooy HM. Automatic three-dimensional correlation of CT-CT, CT-MRI, and CT-Spect using chamfer matching. Med Phys. 1994;21:1163–78.

37. Holupta EJ, Kooy HM. A geometrical algorithm for medical image correlations. Med Phys. 1992;19:433–8.
38. Kondziolka D, Lunsford LD, Kanal E, Talagala L. Stereotactic magnetic resonance angiography for targeting in arteriovenous malformations radiosurgery. Neurosurgery. 1994;33(4):585–91.
39. Stancanello J, Cavedon C, Francescon P, Cerveri P, Ferrigno G, Colombo F, Perini S. Development and validation of a CT-3D rotational angiography registration method for AVM radiosurgery. Med Phys. 2004;31:1363–71.
40. Spiegelman R, Friedman WA, Bova FJ. Limitation of angiographic target localization in planning radiosurgical treatment. Neurosurgery. 1992;30(4):619–23.
41. Börcek AÖ, Çeltikçi E, Aksoğan Y, Rousseau MJ. Clinical outcomes of stereotactic radiosurgery for cerebral arteriovenous malformations in pediatric patients: systematic review and meta-analysis. Neurosurgery. 2019;85(4):E629–40.
42. Seok KC, Young JL, Jun SK, Bong AR, Gook KK, Tae SK. Post-treatment bleeding of cerebral arteriovenous malformations after gamma knife radiosurgery. J Korean Neurosurg Soc. 2004;36:363–8.
43. Shekhtman OD, Maryashev SA, Eliava SS, Yakovlev SB, Golanov AV, Shishkina LV, Pilipenko YV, Okishev DN, Bocharov AV, Bukharin EY, Mikeladze KG, Kisar'ev SA, Vinogradov EV, Kaftanov AN, Konovalov AN. Combined Treatment of Cerebral Arteriovenous Malformations. Experience of the Burdenko Neurosurgical Institute. Zh Vopr Neirokhir Im N N Burdenko. 2015;79(4):4–18.
44. Umansky D, Corn BW, Strauss I, Shtraus N, Costantini S, Frolov V, Maimon S, Kanner AA. Combined treatment approach to cerebral arteriovenous malformation in pediatric patients: stereotactic radiosurgery to partially Onyx-embolized AVM. Childs Nerv Syst. 2018;34(11):2269–74.
45. Ching-Jen C, Kathryn NK, Dale D, Hideyuki K, David M, Kondziolka D, Caleb F, Rafael RM, Grills IS, Gene HB, Lunsford LD, Sheehan JP. Stereotactic radiosurgery for arteriovenous malformations of the basal ganglia and thalamus: an international multicenter study. J Neurosurg. 2019;132(1):122–31.
46. Ryu SI, Chang SD, Kim DH, Murphy MJ, Le QT, Martin DP, Adler JR. Image guided hypo-fractionated stereotactic radiosurgery of spinal lesions. Neurosurgery. 2001;49(4):838–46.
47. Sinclair J, Chang SD, Gibbs IC, Adler JR. Multisession Cyberknife radiosurgery for intramedullary spinal cord arteriovenous malformations. Neurosurgery. 2006;58(6):1081–9.
48. Meling TR, Patet G. What is the best therapeutic approach to a pediatric patient with a deep-seated brain AVM? Neurosurg Rev. 2019;42(2):409–16.

Large Arteriovenous Malformations

43

Umberto Fornezza, Zeno Perini, Leopoldo S. Casentini, Gianpaolo Zambon, and Mariano Zanusso

43.1 Introduction

Radiosurgery for large cerebral arteriovenous malformations (AVMs) has been the subject of discussion for many years for a lower obliteration rate achieved and a greater probability of volume-related to adverse effects with a standard dose delivered in a single session. Therefore, a multimodal therapeutic approach for large AVMs including partial surgical resection followed by stereotactic radiosurgery, embolization and radiosurgery, embolization, or surgical removal after radiosurgery failure, especially in case of re-bleeding during the latency period after irradiation, has been widely established [1, 2]. The management of large AVMs remains challenging due to the worse natural history compared to other AVMs. In fact, the risk of bleeding in these malformations persists in the latency period at a rate higher than the expected rate for smaller AVMs [3]. Colombo and colleagues [4] presented in 2009 a series of 279 patients with cerebral AVMs treated using CyberKnife radiosurgery. A statistical evaluation of factors associated with successful treatment confirmed a strong negative correlation with AVM volume—a well-established relationship [5–7]—and with the Pollock-Flickinger score [8]. Contradicting other authors' experience [6, 8], a surprising significantly strong negative correlation with Spetzler-Martin grade was also found [9]. This fact was explained by the dependence of Spetzler-Martin (SM) grading system for determining AVM volume, combined with the wider range of grade variability (from II to V).

Recently Patibanda and colleagues [10] have presented an international multicenter study regarding Stereotactic radiosurgery (SRS) for Spetzler-Martin Gade IV and V: a retrospectively study of 233 patients with SM Grade IV and V treated with single-shot SRS with a mean nidus volume of 9.7 cm^3.

The statistical analyses performed to identify factors associated with post-SRS outcome established that only larger AVM diameter was an independent predictor of unfavorable outcome. The rate of favorable outcome was significantly lower for unruptured SM Grade IV and V AVMs compared with ruptured ones. Another finding was that prior embolization was a negative independent predictor of obliteration as it may obscure the final target volume, reducing the successful obliteration rate [11].

Nidus architecture represents a further important factor for outcome [12]. Although it is very difficult to quantify the rate of diffuse and

U. Fornezza · Z. Perini (✉) · L. S. Casentini
G. Zambon · M. Zanusso
Department of Neurosurgery, City Hospital, Vicenza, Italy
e-mail: umberto.fornezza@aulss8.veneto.it; leopoldosilviocasentini@aulss8.veneto.it; giampaolo.zambon@aulss8.veneto.it; mariano.zanusso@aulss8.veneto.it

© Springer Nature Switzerland AG 2020
A. Conti et al. (eds.), *CyberKnife NeuroRadiosurgery*, https://doi.org/10.1007/978-3-030-50668-1_43

compact nidus architecture, the compact lesions have an overall higher response rate at lower doses.

The margin dose absorbed by large AVMs nidus seems to be one of the factors significantly associated with the obliteration rate [13]. In a recent clinical article, Khano et al. [7] reported that the outcomes of volume-staged radiosurgery for large AVMs can be improved with a margin dose higher than 17 Gy and the 20 Gy volume with more than 63% of the target volume. By a radiosurgical point of view, two distinct therapeutic strategies have therefore been considered: dose-staged and volume- staged radiosurgery. Both dose- and volume-staged SRS were used in order to reduce adverse radiation effects on surrounding brain tissue including radiation necrosis, cerebral edema, cyst formation, and the onset of neurological deficits or seizure.

Dose-staging was described as either hypofractionated stereotactic radiotherapy (hFSRT) by administering several small radiation doses to the entire nidus over a period of some weeks or repeating radiosurgery with initial high dose and another lower dose after several months or years.

Volume staging radiosurgery is performed by irradiation of separate portions of the AVM over time.

Usually, the nidus is divided into two or more volumes and treated in separate sessions with a 2- to 12-month interval between the two procedures [3, 14]. This strategy reduces the dose absorbed by normal nervous tissue and then toxicity while maintaining adequate prescription doses.

In a systematic review of literature, Moosa et al. [15] observed that volume-staging radiosurgery provides higher obliteration rates compared with dose-staged radiosurgery. The mean complete obliteration rates for the dose- and volume-staged groups were 22.8% and 47.5%, respectively.

The mean complications rates were 13.5% and 13.6%, respectively, in the dose- and volume-staged groups.

43.2 Vicenza Neurosurgery Department Experience

At the Vicenza Neurosurgery Department, cerebral AVM radiosurgery was introduced in 1984 with LINAC-based, converging arc technique [5, 16–19]. Until December 2002, 642 patients were treated. In this series, complete obliteration was influenced by AVM size (6, 7) ranging from 96% in small (S), less than 15 mm in diameter, to 33% in large (L), over 25 mm in diameter.

During these years, we considered the bleeding risk in the latency period as the main problem. An increase of bleeding risk in large and inhomogeneously irradiated AVMs seemed possible [5, 17].

In January 2003, we started using a frameless, image-guided robotic radiosurgery system, the CyberKnife (Accuray Inc., Sunnyvale CA) [4, 20–23], for the treatment of AVMs.

The CyberKnife stereotactic radiosurgery treatment planning system is based on CT for target definition. Optimal AVM nidus target outlining often requires information provided by multimodal images such as angiography, MRI, MR angiogram (3D TOF MRI), functional MRI (fMRI), and diffusion-weighted imaging (DTI) tractography.

Treatment planning was performed using 3D rotational angiography registered to CT scan, which represents the reference imaging modality of CyberKnife for skull position tracking and intrafraction motion correction by the robotic system [24, 25].

Before the radiation procedures, all patients underwent MRI which proved to be very useful for modelling the surrounding structures [26, 27]. Automatic target delineation was routinely utilized, whereas automatic delineation of critical structures (motor cortex, language areas, and cortical spinal tract) was used in critically located AVMs.

Moreover, after 2010, we also performed ANGIO MRI (3D TOF MRI) which proved to be very useful in cases of previous embolization to identify the nidus patent vessel in axial reconstructions.

The image registration procedure was adapted for co-registration of functional MR imaging and also DTI tractography. This image set was used to contour functional structures located close to the AVM nidus so that dose constraints could be imposed.

Once the recorded datasets have been imported into the CyberKnife treatment planning system, the contour of the AVM nidus can be performed, portion by portion, on axial sections of CT scans and other co-registered neuroradiological studies, using an automatic contouring tool with an appropriate threshold on voxel values that delineate the boundaries of the target and reconstruct the volume of the nidus in 3D space. In our experience [4], the fMRI imaging-based treatment planning was also able to reduce the maximal dose absorbed by critical structures of 19–58% (mean 39%).

43.3 Clinical Experience

From January 30th 2003 to December 2018, 29 patients (18 men and 11 women) affected by large cerebral AVMs (nidus >8 mL) were treated (Fig. 43.1). Age ranged from 14 to 68 years (mean 38 years). Symptoms onset were bleeding in 23, epilepsy in 4, and headache in 2.

AVM-irradiated volume varied from 8.2 to 56 mL (mean 13.3 mL). The maximum radiation

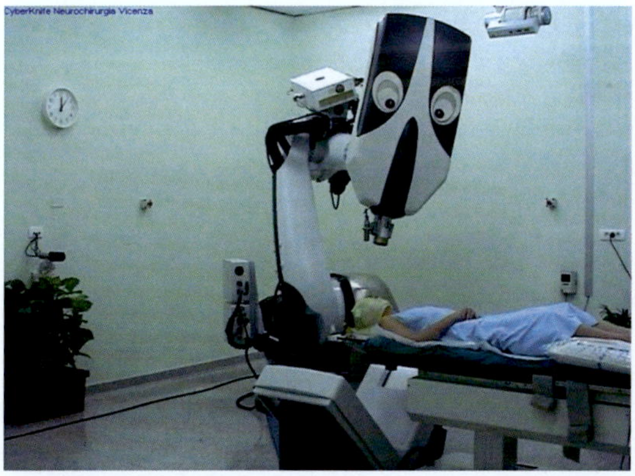

Fig. 43.1 First radiosurgical procedure using CyberKnife for AVM at Vicenza Neurosurgery Department in January 30, 2003

dose delivered ranged from 24 to 30 Gy (mean 25.1 Gy). The borders of target volume were encompassed by isodose surfaces from 70 to 85%.

Before radiosurgery, two patients underwent unsuccessful attempts of surgical removal. Sixteen patients underwent incomplete AVM embolization. Six patients were re-treated after a first radiosurgery (4 with CyberKnife and 2 with LINAC, before 2003), and two patients underwent three radiosurgical treatments (1 CyberKnife and 1 LINAC and CyberKnife).

Our follow-up included MRI and MRA imaging at 6, 12, 24, and 36 months after treatment. 3D angiography was obtained at 24 months until 2008. Subsequently, we preferred to perform angiography at 36 months in patients with bleeding onset or when MR studies suggested the complete obliteration.

If a residual AVM nidus was disclosed, but significantly reduced, we only repeated the angiography at annual intervals until the complete elimination was achieved. If AVM remained unchanged, suggesting that the process was over, radiosurgery was then repeated.

Until June 2008, in 18 patients with target volumes larger than 8 mL, we used the dose-staged radiosurgery strategy. The radiation dose was given to the entire nidus in two equal fractions, 8–30 days apart (Figs. 43.2 and 43.3).

Later, we adopted the volume-staged radiosurgery approach in 11 patients in order to increase the rate of complete AVM obliteration (Fig. 43.4). Four patients we had treated before 2008 with the dose-staged technique were subsequently re-irradiated with the volume-staged strategy. In three of these patients with hemorrhagic onset, we obtained, after the first irradiation, only a partial obliteration at a 36 months angiography. In one patient, with clinical onset consisting of epileptic seizure, a significant result was not evident at a 5 years angiography. Overall, 15 patients were treated with volume dose strategy. For all patients who underwent volume-staging after 2008, the nidus was divided into two volumes and treated in separate sessions with a 6- to 13-month interval between the two procedures.

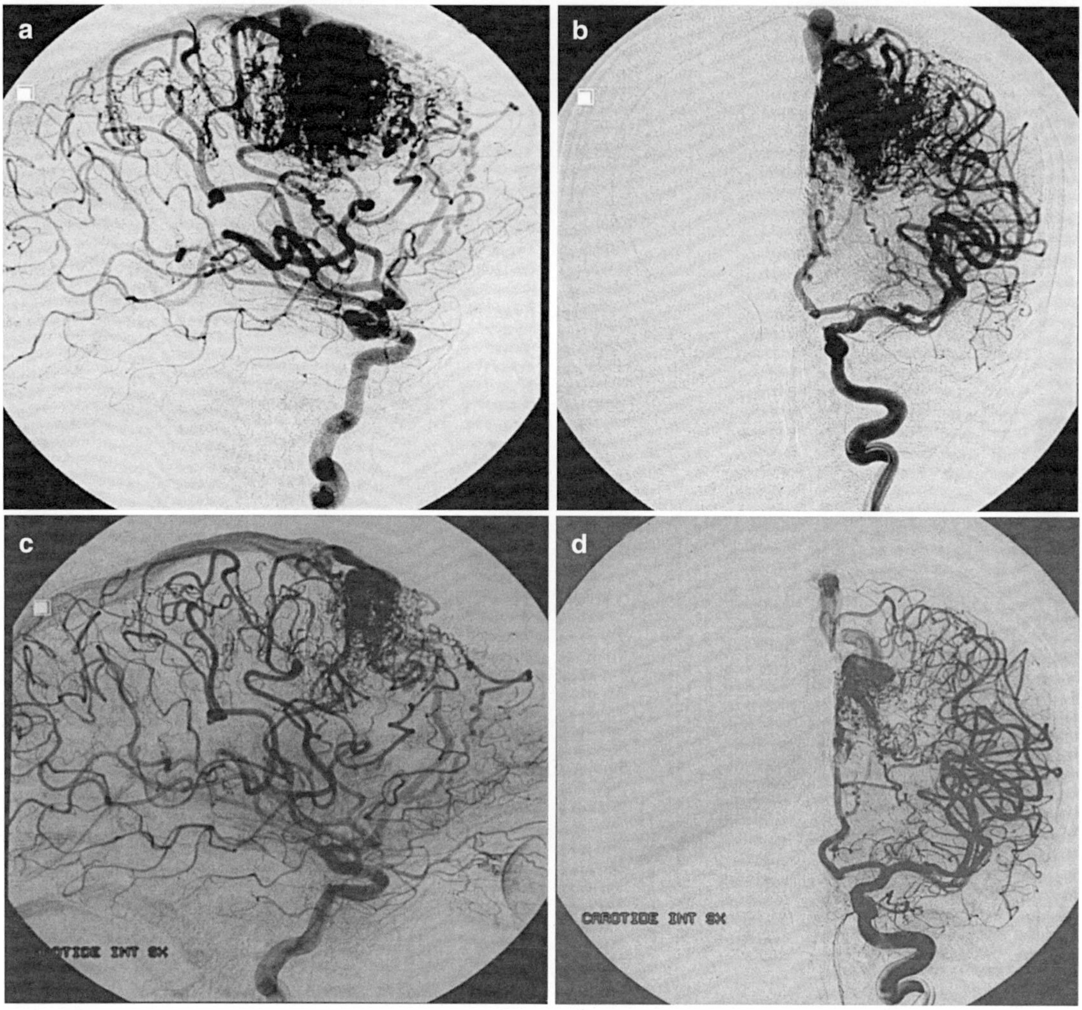

Fig. 43.2 A 30-year old patient with a left frontal AVM. Volume 15.5 cm^3 (**a**, **b**). Clinical onset with hemorrhage, three previous embolizations. Treated by dose-staging. Total dose 25 Gy, in two fractions 30 days apart. Six-year control angiography (**c**, **d**): only a reduction of nidus volume was evident. Patient refused other treatments

43.4 Results

Follow-up ranged from 12 to 192 months. In 25 of 29 patients, we had a follow-up longer than 36 months with angiographic findings. One patient suffered from a transient complication after irradiation, requiring temporary corticoid medication. Three bleeding recurrences were observed, fatal in one case. The other two patients presented a rapid recovery of clinical symptoms, and the control angiographies at 4 years and 6 years, respectively, showed a complete obliteration of the malformation.

In the group of patients treated with the dose-staged strategy, 17 underwent angiography.

One patient died of a hemorrhage. In 6 patients (35.3%) full AVM obliteration was highlighted.

Considering the other group of 11 patients irradiated with the volume-staged strategy, only 8 of them underwent control angiographies. In 3 patients, the angiography has not been performed yet, but a significant reduction in the volume of the nidus is evident on MRI and MRA. To date, complete obliteration has only been demonstrated in three (37.5%) patients.

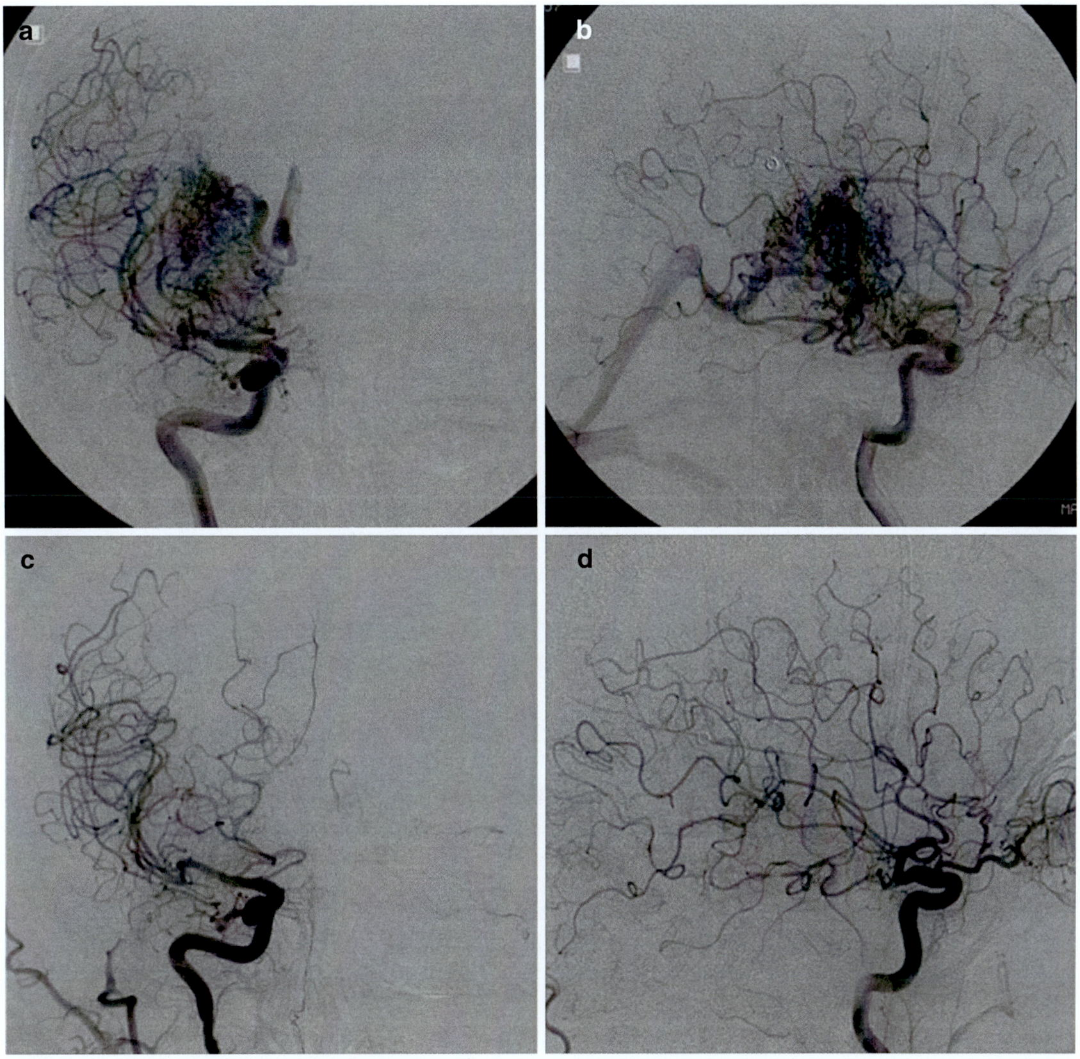

Fig. 43.3 A 54-year old patient with a right basal ganglia AVM (**a**, **b**). Volume 9.3 cm^3. Clinical onset with hemorrhage. Left hemiparesis. Treated by dose-staging. Total dose 26 Gy, two fractions 30 days apart. The 36 months angiography (**c**, **d**) demonstrated the complete obliteration of the AVM

In four patients undergoing re-irradiation with the volume-staged approach after the previous dose-staged strategy, a complete obliteration was achieved in three patients and confirmed by angiography obtained 4 years after irradiation.

Therefore, in a total number of 12 patients undergoing staged-volume therapy and follow-up angiography, six angiographical cures (50%) were achieved. Out of six patients with unruptured AVMs, only two complete obliterations were obtained: the first patient in the group of the volume-staged strategy after 6 years, and the other one in the group of the combined dose-staged/volume-staged, 11 years after the first irradiation. Noteworthy, no clinical toxicity was observed in these six patients.

We did not find a significant difference in the percentage of obliteration between previously embolized (34.6%) and not embolized patients (35.2%).

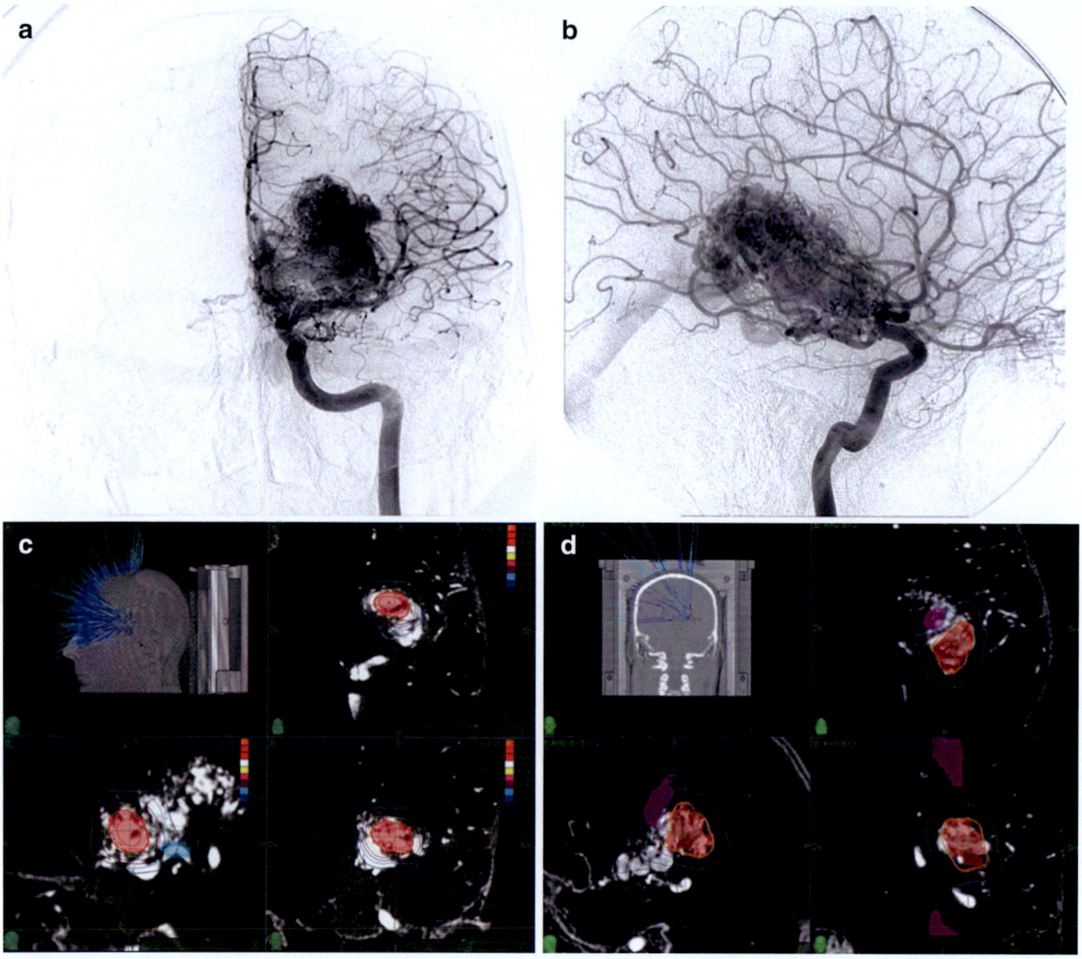

Fig. 43.4 A 17-year old patient with a left basal ganglia AVM (SM grade IV). Two previous intracranial hemorrhages. Right hemiparesis. Volume 8.5 mL (**a**, **b**). Treated by volume-staging. The anterior compartment (**c**) was treated by conformal planning and a total dose of 25 Gy in single session. After 12 months (**d**), the posterior compartment was treated. Conformal planning. Total dose 25 Gy in a single session

43.5 Discussion

Angiography represents the optimal investigation for the diagnosis and treatment of cerebral AVMs [7, 17] and the only available dataset for the radiosurgery treatment planning of AVM until the introduction of CT scan and MRI and development of 3D planning. Nevertheless, we consider CT angiogram and MR angiogram insufficient and the use of three-dimensional rotational angiography study [2, 25, 28, 29] essential for treating AVMs. Kakizawa et al. [30] have suggested that the information obtained by 3D angiography would improve the treatment strategy of large and complex AVMs.

In embolized patients, subtracted 3D angiography data sets can be utilized in order to exclude embolic material from intended target, reducing the final target volume and total radiation dose to deliver [22]. Another advantage of using 3D angiography for treatment planning is the ability to three-dimensionally reconstruct any selected isodose surface and relative volume coverage by shifting the viewpoint at the operator's choice. In our opinion, this option represents an important step in the iterative procedure aimed at finding the optimal solution for the treatment.

In general, changes induced by the volumetric critical evaluation of angiographic data result in the decrease of the final target volume. We think that a possible explanation is that angiographic data permit a better discrimination between nidus shunt vessels and large draining veins, often obscuring the target outline.

Contrary to previous experience [11, 13], in the series of our patients, a previous embolization seemed to be irrelevant to the outcome even in large AVMs. The fact that we were able to achieve the same obliteration rate could be explained by the improvement in the definition of the target that involves the use of the procedure with 3D rotational angiography and the subtraction of the embolic material [25].

We have not found a significant difference, in terms of obliteration, between patients treated with by the volume-staging and those by the dose-staging strategy (35.6% vs. 33.3%), but in the case of staged-dose failure, re-irradiation by the staged-volume strategy can be proposed after an appropriate time interval.

As for the large unruptured AVMs, we treated only six patients and obtained two complete obliterations. Based on this experience, together with surgical and endovascular literature evidence, we believe that radiosurgical treatment in large unruptured AVMs is only to be proposed in rare cases. In patients with non-critical AVMs located and with compact angioarchitecture, healing is indeed possible even after many years and with repeated irradiation. On the other hand, a radiosurgical treatment in large AVMs with extensive and inhomogeneous nidus and in a critical position does not seem justified because of potential side effects.

References

1. Deruty R, Pelissou-Guyotat I, Amat D, Mottolese C, Bascoulergue Y, Turjman F, et al. Complications after multidisciplinary treatment of cerebral arteriovenous malformations. Acta Neurochir. 1996;138:119–31.
2. Gobin YP, Laurent A, Merienne L, Schlienger M, Aymard A, Houdart E, et al. Treatment of brain arteriovenous malformations by embolization and radiosurgery. J Neurosurg. 1996;85:19–28.
3. Seymour ZA, Snced PK, Gupta N, Lawton MT, Molinaro AM, Young W, et al. Volume-staged radiosurgery for large arteriovenous malformations: an evolving paradigm. J Neurosurg. 2016;124(1):163–74.
4. Colombo F, Cavedon C, Casentini L, Francescon P, Causin F, Pinna V. Early results of CyberKnife radiosurgery for arteriovenous malformations. J Neurosurg. 2009;111:807–919.
5. Colombo F, Pozza F, Chierego G, Casentini L, De Luca GP, Francescon P. Linear accelerator radiosurgery of cerebral arteriovenous malformations: an update. Neurosurgery. 1994;34(1):14–21.
6. Friedman WA, Bova FJ, Mendenhall WM. Linear accelerator radiosurgery for arteriovenous malformations: the relationship of size and volume. J Neurosurg. 1995;82:180–9.
7. Kano H, Flickinger JC, Nakamura A, Jacobs RC, Tonetti DA, Lehocky C, Park KJ, Yang H, Niranjan A, Lunsford D. How to improve obliteration rates during volume-staged stereotactic radiosurgery for arteriovenous malformations. J Neurosurg. 2019;130: 1809–16.
8. Pollock BE, Flickinger JC. A proposed radiosurgery—based grading system for arteriovenous malformations. J Neurosurg. 2002;96:79–85.
9. Spetzler RF, Martin NA. A proposed grading system for arteriovenous malformations. J Neurosurg. 1986;65:476–83.
10. Patibandla MR, Ding D, Kano H, Xu Z, Lee JY, et al. Stereotactic radiosurgery for Spetler-Martin grade IV and V arteriovenous malformations: an international multicenter study. J Neurosurg. 2018;129:498–507.
11. Kano H, Kondziolka D, Flickinger JC, Park KJ, Iyer A, Yang HC, et al. Stereotactic radiosurgery for arteriovenous malformations after embolization: a case-control study. J Neurosurg. 2012;117:265–75.
12. Seymour ZA, Chan JW, Sneed PK, Hano H, Lehocky CA, Jacobs RC, Ye H, et al. Dose response and architecture in volume staged radiosurgery for large arteriovenous malformations: a multi-institutional study. Radiother Oncol. 2020;144:180–8.
13. Karlsson B, Lindquist C, Steiner L. Prediction of obliteration after gamma knife surgery for cerebral arteriovenous malformations. Neurosurgery. 1997;40:425–31.
14. Kakizawa Y, Nagashima H, Oya F, Ito K, Tanaka Y, Hongo K, Kobayashi S. Compartments in arteriovenous malformation nidi demonstrated with rotational three-dimensional digital subtraction angiography by using selective microcatheterization. J Neurosurg. 2002;96(2):770–4.
15. Moosa S, Chen CJ, Ding D, Lee CC, Chivukula S, Starke RM, Yen CP, Xu Z, Sheehan JP. Volume-staged versus dose-staged radiosurgery outcomes for large intracranial arteriovenous malformations. Neurosurg Focus. 2014;37(3):E18.
16. Blatt DR, Friedman WA, Bova FJ. Modifications based on computer tomography imaging in planning

the radiosurgical treatment of arteriovenous malformations. Neurosurgery. 1993;33(4):588–96.
17. Colombo F, Benedetti A, Pozza F, Marchetti C, Chierego G. Linear accelerator radiosurgery of cerebral arteriovenous malformations. Neurosurgery. 1989;24:833–44.
18. Colombo F. Linear accelerator radiosurgery of arteriovenous malformations: Vicenza experience. In: Smee R, editor. Proceedings of the 4th Congress of International Stereotactic Radiosurgery Society, the Price of Wales Hospital Press, Sydney; 1999. p. 47 (abs).
19. Kooy HM, Bellerive MR, Loeffler JS. Technical concepts of linac radiosurgery dosimetry. In: Gildenberg PL, Tasker RR, editors. Textbook of stereotactic and functional neurosurgery. New York: Mc Grow-Hill; 1998. p. 687–704.
20. Adler JR, Murphy MJ, Chang SD, Hancock SL. Image-guided robotic radiosurgery. Neurosurgery. 1999;44:1299–307.
21. Chang SD, Murphy MJ, Martin DP, Hancock SL, Doty JR, Adler JR. Image guided robotic radiosurgery: clinical and radiographic results with the Cyberknife. In: Kondziolka D, editor. Radiosurgery 1999, vol. 3. Basel: Karger; 2000. p. 23–33.
22. Colombo F, Casentini L, Cavedon C, Francescon P, Causin F, Pinna V. Arteriovenous malformations radiosurgery: evolution of the technique. In: Kondziolka D, editor. Radiosurgery, vol. 6. Basel: Karger; 2006. p. 1–11.
23. Holupta EJ, Kooy HM. A geometrical algorithm for medical image correlations. Med Phys. 1992;19:433–8.
24. Bridcut RR, Winder RJ, Workman A, Flynn P. Assessment of distortion in three dimensional rotational angiography system. Br J Radiol. 2002;75:226–70.
25. Colombo F, Cavedon C, Francescon P, Casentini L, Fornezza U, Castellan L, Causin F, Perini S. Three-dimensional angiography for radiosurgical treatment planning for arteriovenous malformations. J Neurosurg. 2003;98:536–43.
26. Friedman L. Comment on Kondziolka D, Lunsford LD, Kanal E, Talagala L: Stereotactic magnetic resonance angiography for targeting in arteriovenous malformations radiosurgery. Neurosurgery. 1994;33(4):591.
27. Kondziolka D, Lunsford LD, Kanal E, Talagala L. Stereotactic magnetic resonance angiography for targeting in arteriovenous malformations radiosurgery. Neurosurgery. 1994;33(4):585–91.
28. Meeks SL, Bova FJ, Friedman WA. Technical aspects of radiation physics. In: Gildenberg PL, Tasker RR, editors. Textbook of stereotactic and functional neurosurgery. New York: Mc Grow-Hill; 1998. p. 649–68.
29. Stancanello J, Cavedon C, Francescon P, Cerveri P, Ferrigno G, Colombo F, Perini S. Development and validation of a CT-3D rotational angiography registration method for AVM radiosurgery. Med Phys. 2004;31:1363–71.
30. Friedman WA, Bova FJ, Bollampally S, Bradshaw P. Analysis of factors preditictive of success or complications in arterio-venous malformation radiosurgery. Neurosurgery. 2003;52:296–308.

44 Cyberknife Radiosurgery for Cerebral Cavernous Malformations

François Nataf

44.1 Introduction

Cavernous malformations (CM), also known as cavernous angiomas or cavernomas, are low-flow vascular malformations of the central nervous system (brain and spinal cord) that consist of clusters of dilated vascular sinusoidal channels (known as "caverns") lined with a thin endothelium without intervening tight junctions, lacking smooth muscle and elastic layers. Sinusoid vessels have varying diameters and wall thickness and are associated with thrombotic phenomena at various stages of organization. Clots often fill the thin-walled vascular cavernous vessels. Chronic hemorrhagic stigmata such as hemosiderin and gliosis often surround CM in adjacent neuroglial parenchyma [1–3].

Cerebral CM (CCM) are grossly distinct from the adjacent brain, with no neural tissue inside and have a lobulated appearance sometimes resembling a mulberry [4, 5]. They can be associated with the development of venous anomalies in 33% of cases [6].

CCMs classically cannot be seen on subtraction angiography (thus termed as angiographically occult vascular malformations) but may sometimes be visualized on late venous phases on angiograms, signing a low-flow circulation.

F. Nataf (✉)
Department of Neurosurgery, Centre Hospitalier Sainte-Anne, GHU Paris, Paris, France
e-mail: f.nataf@ghu-paris.fr

CCMs are prone to hemorrhage [7], so multiple small hemorrhage events lead to a pathognomic "popcorn-like" appearance on magnetic resonance imaging (MRI) because of hemosiderin staining from blood products at various stages [3, 7, 8].

CCMs are rather frequent, occurring in about 0.2–0.4% of the population [9–12]. They are the second most common type of vascular malformation of the CNS, accounting for about 10–15% of cerebral vascular malformations [13]. About half of them are located in cerebral hemispheres, 35% in the brainstem and about 15% in deep locations or the cerebellum [14]. These lesions can be sporadic or inherited as an autosomal dominant trait. When inherited, cavernomas are multiple (cavernomatosis) [15]. They also may appear de novo, especially after radiation therapy [16].

Clinically, hemorrhage is the most common sign (about 35%) followed by seizures (about 20%) and focal neurological deficit (about 15%). Incidental discovery occurs in about 30% of cases [4, 14, 17–19].

Annual rate of bleeding is about 0.3% in non-brainstem CMs and seems to be higher in brainstem CM [20]. Annual rate of rebleeding seems to be higher in brainstem lesions and in women [20]. However, brainstem location is not found as a risk factor in all series [18]. Rate of re-hemorrhage within 2 years of the first hemorrhage is higher than after 2 years [18, 20]. Post-hemorrhage full recovery or minimal disability is

© Springer Nature Switzerland AG 2020
A. Conti et al. (eds.), *CyberKnife NeuroRadiosurgery*,
https://doi.org/10.1007/978-3-030-50668-1_44

recorded as about 80%/person/year, and mortality after bleeding is about 2.2%.

Surgical resection is the gold standard of interventional treatment when required. Indications for surgery include multiple hemorrhages, neurological deficit, and progressive seizures, if its location involves acceptable surgical risk. But in the case of functional areas or deep-seated lesions, especially the brainstem, surgical risk may be higher than the proper risk of the CM with a rate of permanent neurological deficits ranging from 10.8 to 36% [21–24].

In these cases, radiosurgery (RS) has been proposed as an alternative treatment for cavernoma, with the same rationale than for arteriovenous malformations. Many series on the topic have been published (Table 44.1), substantially showing a decreased risk of hemorrhage compared with that of untreated patients. Nonetheless, many controversies still remain.

CONTROVERSIES:

The first controversy concerns the mechanism of action of RS since CCM lacks elastic and smooth muscular layers which are affected by high-dose radiations in arteriovenous malformations.

The second controversy is about how the effects of RS on the cavernoma can be proven since MRI keeps showing the persistence of a treated CCM on follow-up imaging.

The third controversy concerns the real benefit of RS upon natural history of CM.

The fourth controversy is how the annual hemorrhage risk before RS must be calculated in order to compare with the annual risk after RS.

Moreover, controversies do exist concerning the risk of adverse radiation effects (clinical events associated with radioinduced parenchymal changes on MRI) that was very high (up to 59%) and unacceptable in the early series.

The last controversy regards the dose to be delivered to a CCM when using RS.

44.2 Rationale for Radiation Therapy

In arteriovenous malformations (AVM), the goal of RS is to achieve obliteration of the AVM through endothelial proliferation, hyalinization and calcification of the vessel's wall, thrombosis, and necrosis of the vessels [25]. Cavernomas do not have such vessels but still have very thin afferences and efferences, as seen during surgical procedures, and are therefore low-flow vascular malformations. Pathological findings after RS of CM showed endothelial cell destruction in the cavernous sinusoid vessels, with marked fibrosis and hyaline degeneration, and obliterated vessels in the irradiated samples. However, granulation tissues with newly formed thin-walled channels were observed as well [26, 27]. These data may suggest a real vascular effect of RS which may lead to partial or total obliteration of CM.

44.3 Comparison of Hemorrhage Risk Before and After RS

Whereas angiography can prove the complete obliteration of an AVM after RS, the situation is quite different for CM. If there is no complication from RS, MRI follow-up shows little changes of the CM after RS, mainly related to resorption of the hematoma from previous bleedings [25, 28–32]. The disappearance of the CM on MRI is very rare. And since CM are angiographically occult, it is difficult to prove its definitive cure. The first series reporting RS of cavernoma compared actuarial hemorrhage risk before and after RS [33, 34]. They showed a progressively decreasing risk, with a "latency interval" of 2 years. The risk became very low after the latency interval. Since then, many series have been published with the same methods [25, 28, 30, 31, 33–60], but hemorrhage risk before RS was calculated either since birth or from the last hemorrhage, leading to considerable differences in the evaluation of this risk (Table 44.1). Moreover, some series revealed that the natural history of untreated CM showed similar decrease of the hemorrhage risk with a latency interval of 2 years [18, 61]. Despite these concerns, recent meta-analyses comparing the effects of RS with natural history [18, 62, 63] seem to

Table 44.1 Main series published on radiosurgery of cerebral cavernomas

Author	Year	NB Pts	Location	Mean marginal dose (Gy)	Mean follow-up (months)	Annual hemorrhage rate			Seizures free	Seizures improved	SRS type	ARE	Calculated HR from
						Pre-SRS	Post SRS LI	Post-SRS after LI					
STEA	1994	12	All locations	21.67	27 median		8.3% global				LINAC	16.6%	
KONDZIOLKA	1995	47	All locations	16	43.2	32%	8.8%	1.1%			GK	26%	Hemorrhage
KIDA	1995	20	All locations	15–20							GK	25%	
AMIN-HANJANI	1998	95	All locations	15	65	17.3%	NA	4.5%			PBR	20.6%	
CHANG	1998	57	All locations	13.5–20	90	9.4%	9.4%	1.6%			LINAC	7%	
KARLSSON	1998	22	All locations	18	78	NA	10.0%	5.0%			GK	27%	
LISCAK	2000	26	Brainstem	14.76 median	24 median	4%	6.8%				GK	28% (8% permanent)	Birth
POLLOCK	2000	17	All locations	18	51	40.1%	8.8%	2.9%			GK	59% (41% permanent)	Hemorrhage
HASEGAWA	2002	82	All locations	16	60	33.9%	12.3%	0.76%			GK	13.4%	
REGIS	2004	49	All locations	19.17	23.66		3.8% global		53% (Engel I)	20% (Engel II)	GK	3.8%	
LIU	2005	125	All locations	12.1	64	NA	10.3%	3.3%		53% (Engel I et II)	GK	13.1%	
HUANG	2006	30	All locations	16	59.9	NA	0.67%	0.67%	61.5%		LINAC	6.7%	
MONACO	2010	68	Brainstem	15.84	62	32.38%	8.22%	1.37%			GK	11.8%	Hemorrhage
NAGY	2010	113	Brainstem/deep	12/13/15[a]	48	2.2–2.9%[b]	5.1%/15%	1.3/2.4%			GK	7.3% permanent	Birth
BLAMEK	2010	23	All locations	8–28 Gy	27.3						Linac	30.4%	
LUNSFORD	2010	103	All locations	16	67.8	32.48%	10.8%	1.6%			GK	11.65%	Hemorrhage

(continued)

Table 44.1 (continued)

Author	Year	NB Pts	Location	Mean marginal dose (Gy)	Mean follow-up (months)	Annual hemorrhage rate			Seizures free	Seizures improved	SRS type	ARE	Calculated HR from
						Pre-SRS	Post SRS LI	Post-SRS after LI					
LEE	2012	49	Brainstem	11	40.6	31.3%	4.29%	3.64%			GK	4.1%	Hemorrhage
FUETSCH	2012	14	Brainstem	13.9	85.2	121.8%	12.5%	1.8%			LINAC	16.7%	
PARK	2013	21	Brainstem	13	38.9	39.5%	8.2%	0%			GK	5% permanent	Hemorrhage
LISCAK	2013	112	All locations	16	48 median	2%	3.2%	0.5%		45% (Engel I and II)	GK	14.6% (0.9% permanent)	
SAGER	2014	52	All locations	15 (10–20)	62	39%	NA	1.21%			LINAC	17%	Hemorrhage
KIM	2014	39	Brainstem	13	49	33.6%	8.1%	2.4%			GK	10.3%	Hemorrhage
FRISCHER	2014	38	Brainstem	12	62.4	47.6%	2.6%	0.6%			GK	NA	Hemorrhage
AZIMI	2015	100	All locations	13	42	4.1%	NA	1.9%			GK	12%	
FERDORCZAK	2015	51	All locations	NA	NA	21.7%	4%	0%			GK	NA	Hemorrhage
LIU	2016	43	Brainstem	11.9	36	25%	3.92%	1.85%			GK	2.32%	Hemorrhage
ABOUKAIS	2016	24	Brainstem	14.8	51.2	27.31%	2.46%	2.46%			GK	0%	Hemorrhage
LOPEZ-SERRANO	2017	95	All locations	11.87	78	3.06%	1.4%	0.16%			GK	7%	Birth
SHEEN	2018	95	All locations	13.7 (9–20)	79		7.5% global	7.5% global			GK	16.3%	
KEFELI	2019	82	Brainstem	12	50	55.7%	0.87%	0.87%			GK	0%	Hemorrhage
YAPRAK	2019	19	All locations	15	82					75%	CyberKnife		

SRS stereotactic radiosurgery, *ARE* Adverse radiation effects, *PBR* Proton beam radiosurgery, *Pre-SRS* before radiosurgery, *Post SRS LI* after radiosurgery during the latency interval (2 years), *Post-SRS* after LI after radiosurgery after the latency interval (2 years), *HR* hemorrhage risk, *GK* Gamma Knife, *LINAC* Linear accelerator

[a]12 Gy brainstem, 13 Gy basal ganglia, 15 Gy basal ganglia high risk

[b]Low risk-high risk

show a real benefit of RS in selected patients with CM located in the brainstem or deep/functional areas who had previous hemorrhage.

44.4 Dose and Toxicity

In the first series of RS in cavernoma, doses were as high as for the treatment of AVMs. But complication rates of RS, especially adverse radiation effects (ARE), were unacceptably high: up to 59% with 41% permanent deficits [38]. This first led to performing a moratory [28, 38, 64] and then a progressive decrease of dose given to the CCMs. Recent series showed a dramatic decrease of complications by decreasing the dose. In 2013, Liscak proposed doses between 14 and 15 Gy that could represent a compromise between the higher risk of edema with the increasing marginal dose and the higher risk of rebleeding with the lower marginal dose [50]. The most recent series proposed doses as low as 11 Gy or 12 Gy [46, 54, 55, 57, 59]. Such doses can be safely delivered to the brainstem and functional areas. Currently, accepted doses may vary between 11 and 13 Gy as a function of location and evaluation of the hemorrhage risk of the CM.

Since the therapeutic dose is around 12 Gy, which is the dose tolerated in one single fraction by the brainstem, it is now possible to treat a brainstem CCM from the first hemorrhage.

44.5 Effects of Radiosurgery on Seizures

The risk of developing an epileptic seizure after an incidental CCM diagnosis is relatively low (0.9%/person/year). In patients with CCM presenting with their first seizure without hemorrhage, the 5-year risk of epilepsy is, on the contrary, high (94%) [19]. However, the risk of developing intractable seizures remains quite low [65]. In case of medically refractory seizures, their control may be often achieved by microsurgical resection of the CCM, if the CCM is solitary and if there is a good correlation with the electroclinical pattern [4].

Few series emphasized the role of RS in treating seizures. Regis published a multicentric series of 49 patients with a long duration and drug-resistant epilepsy presumably related to CCM and treated by RS with high doses (mean marginal dose 19.17 Gy). He reported a 53% rate of seizure-free cases (Engel class I) and an improvement in 20% of cases (Engel class II) with a low rate of ARE (2%) in spite of such a high dose [66].

Hsu reported a rate of 64.3% of seizure-free cases (Engel class I) with doses varying from 20 Gy for non-eloquent locations and <16 Gy for eloquent brain areas [67].

Huang reported a rate of 61.5% of seizure-free patients (median marginal dose 16 Gy) with a rate of 6.7% ARE [42].

Shih reported a rate of 25% seizure-free patients (Engel class I) after RS with low dose (mean marginal dose 13.3 Gy) for patients presenting with seizures [68], while Liu reported a rate of 53% improvement (Engel class I and II) with low dose (margin dose 12.1 Gy). But in most of these series, there were treatment selection bias, incomplete data about epilepsy severity, and insufficiently precise outcome data.

Nevertheless, in the case of severe intractable seizures due to a cavernoma in an eloquent region (where the surgical risk is high), RS appears to be a reasonable option with curative perspectives [69].

44.6 Single-Fraction Radiosurgery

Only radiosurgical (single-fraction) series were published for treatment of CCM. Fractionated radiotherapy was not proposed with a therapeutic aim, but conversely, a correlation between radiation therapy and cavernomas occurrence is well-known since 1994, in adults as well as children [70, 71]. Even radiosurgery may induce cavernoma occurrence [72, 73].

44.7 Image-Guided Radiosurgery

Most published series are based on gamma-unit and LINAC radiosurgery (Table 44.1). One series has recently been published on CyberKnife® radiosurgery [60]. The type of radiation does not affect results on efficacy and toxicity.

44.8 Procedure

A thermoplastic mask was used to immobilize the patient's head. Then, a computerized tomography scan and magnetic resonance imaging (T1w with gadolinium enhancement) with 1-mm-thick slices were performed in the treatment position.

The images were imported in the Accuray CyberKnife® Multiplan TPS software and then fused.

The target volume was defined on MRI as the area of mixed-signal included inside the hyposignal hemosiderin ring. No additional margin was given, so CTV = PTV.

Then, inverse planning was performed on TPS. The maximum doses given to eloquent areas were, respectively, 14 Gy for pyramidal tract, 12 Gy for the brainstem, 8 Gy for optical tracts, 4 Gy for cochlea, and 8 Gy for the fornix.

A representative treatment plan is shown in Figs. 44.1 and 44.2.

Follow-up included both clinical and imaging data (MRI with T1w, T2w or FLAIR, T1w with gadolinium enhancement, and protons weighted sequences such as T2*) in order to detect small cavernoma and hemorrhage. An example of follow-up imaging is shown in Fig. 44.3.

44.9 Toxicity

For brain cavernoma, toxicity related to RS has been well documented.

As for other lesions treated by RS, parenchymal changes associated with neurological deficits defined "adverse radiation effects" (ARE). On MRI, imaging semeiology has been described and is the same after RS for brain AVM: hypersignal on T2w sequences, contrast enhancement with a cocarde "necrosis-like" aspect [74] showing a grading between dose and ARE occurrence.

It is now clear that dose to give to CCM is much lower than for AVM. First series showed unacceptable rates of complications, while recent series with low doses seem to show comparable results in terms of efficacy but a much lower rate of complications.

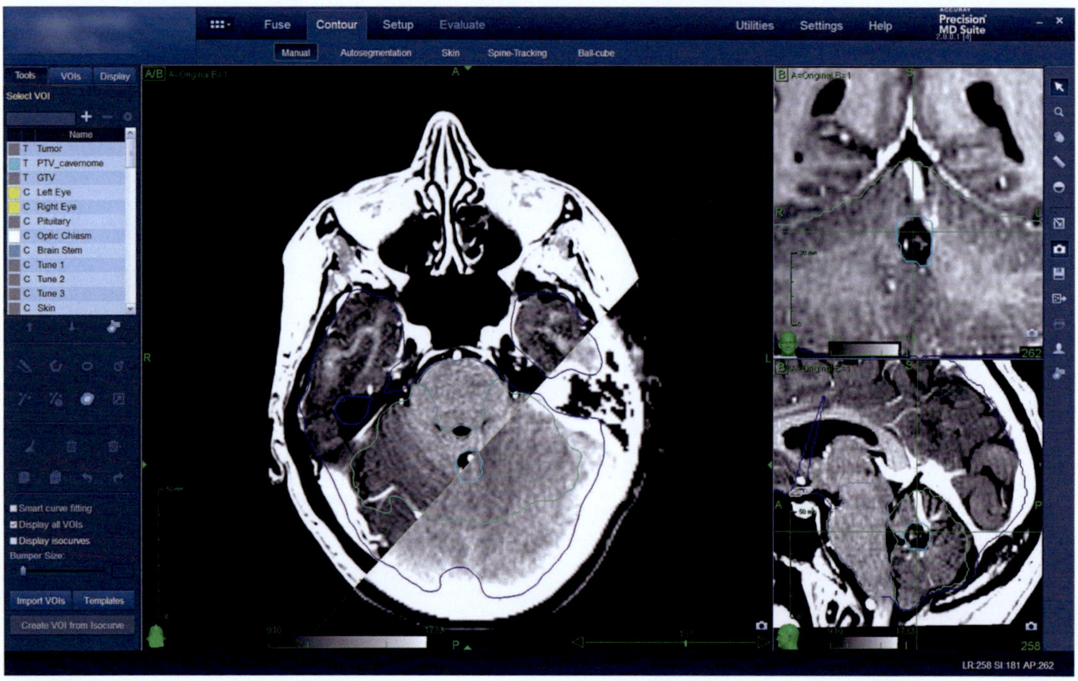

Fig. 44.1 Cerebellar cavernoma. Target planning on Multiplan TPS. Target includes all signal anomalies into the hyposignal

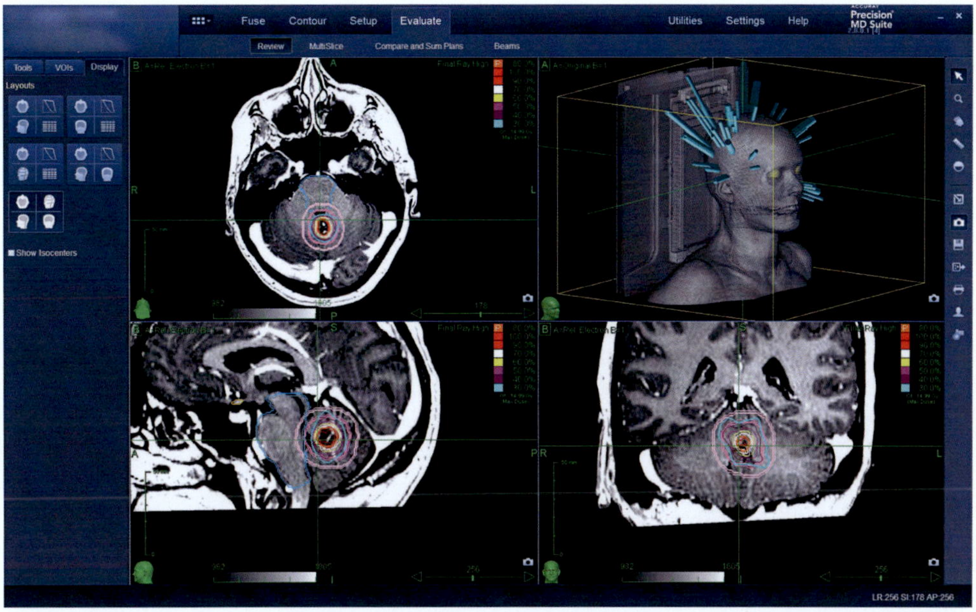

Fig. 44.2 Cerebellar cavernoma. Dosimetry with planning beams. Margin dose: 12 Gy in a single fraction

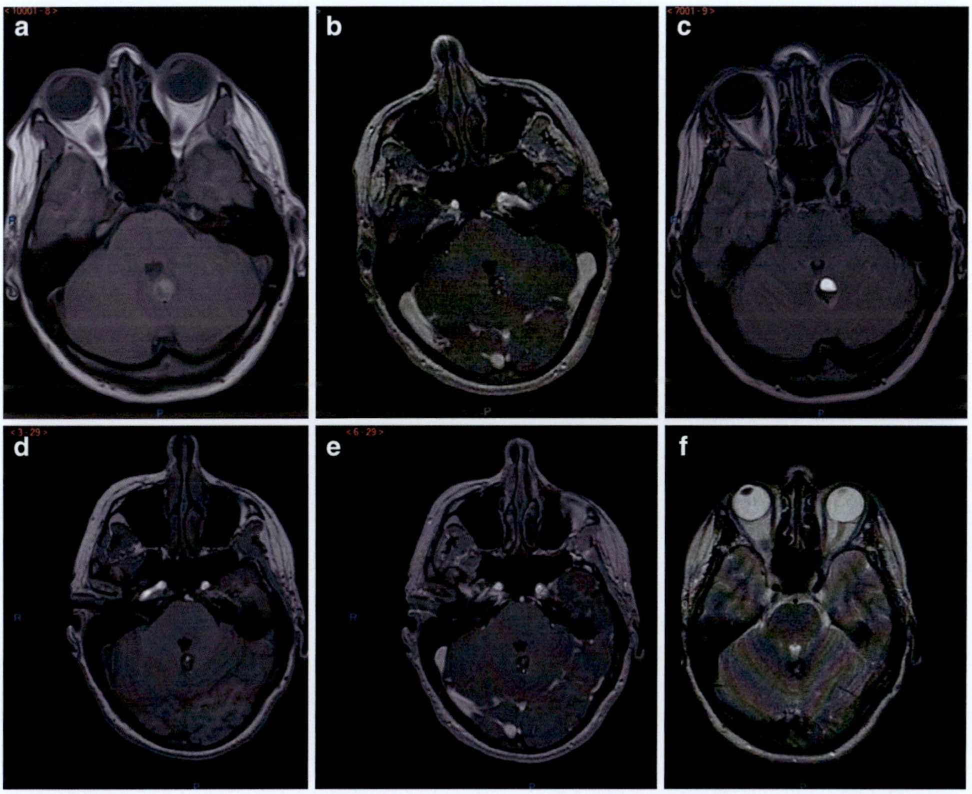

Fig. 44.3 Cerebellar cavernoma that bled 3 times in 3 years. MRI before treatment (at third hemorrhage) (**a–c**) and 1 year after treatment (CyberKnife radiosurgery) (**d–f**). No parenchymal changes were visible, and cavernoma image remained unchanged. No hemorrhage and no radio-induced changes

However, the optimal dose of CMs for a positive response and minimum side effects is controversial and may depend on volume and location.

44.10 Conclusion

The main goal of RS for CCM is to dramatically decrease the hemorrhage risk without completely eliminating that risk.

Despite numerous controversies about the use of RS in the treatment of CCM, consensus guidelines gradually have appeared after reviews and meta-analyses of literature [75–77]. RS may be considered in solitary CCM with previous symptomatic hemorrhage if the CCM lies in eloquent areas involving an unacceptably high surgical risk (Class IIb, Level B).

RS is not recommended for asymptomatic CCMs, for CCMs that are surgically accessible, or in familiar CCM because of the concern about de novo CCM genesis (Class III, Level C).

The dose is still a matter of debate, but there is a trend to give low doses such as 12 Gy on brainstem CCM, and higher doses between 12 and 16 Gy may be given as a function of location and volume.

44.11 Practical Guide

Selection of Patients

- Patients with a sporadic cavernoma with at least one hemorrhage, deep-seated, in an eloquent region or the brainstem, with an unacceptable surgical risk
- Discuss for progressive neurologic deterioration or intractable seizures.

Procedure

- Explanation of the rationale, benefits, and risks of the treatment
- Immobilization with a thermoplastic mask
- CT scan with 1-mm-thick slices with mask
- MRI scan volumetric acquisition 3D spT1 with gadolinium enhancement

Target Planning

- Import CT and MRI scan
- Fusion CT-MR
- Delineating the target: on Multiplan TPS (Accuray CyberKnife®) always inside the hyposignal T1, including all mixed-signal changes, no additional margin

Dose Selection

- 12 Gy could be the reference dose in the brainstem (margin dose) and even other locations.
- If high-risk CM, one can raise up to maximum 13–16 Gy as margin dose in case of locations other than the brainstem, but this may increase the risk of ARE.

Follow-Up

- MRI and clinical evaluation at 3, 6, and12 months and then annually.
- Long time follow-up is required.

References

1. Frischer JM, Pipp I, Stavrou I, Trattnig S, Hainfellner JA, Knosp E. Cerebral cavernous malformations: congruency of histopathological features with the current clinical definition. J Neurol Neurosurg Psychiatry. 2008;79(7):783–8.
2. Raychaudhuri R, Batjer HH, Awad IA. Intracranial cavernous angioma: a practical review of clinical and biological aspects. Surg Neurol. 2005;63(4):319–28; discussion 328.
3. Stapleton CJ, Barker FG. Cranial cavernous malformations: natural history and treatment. Stroke. 2018;49(4):1029–35.
4. Awad IA, Polster SP. Cavernous angiomas: deconstructing a neurosurgical disease. J Neurosurg. 2019;131(1):1–13.
5. Robinson JR, Awad IA, Masaryk TJ, Estes ML. Pathological heterogeneity of angiographically occult vascular malformations of the brain. Neurosurgery. 1993;33(4):547–54; discussion 554–555.
6. Rigamonti D, Spetzler RF. The association of venous and cavernous malformations. Report of four cases and discussion of the pathophysiological, diagnostic, and therapeutic implications. Acta Neurochir. 1988;92(1–4):100–5.
7. Clatterbuck RE, Eberhart CG, Crain BJ, Rigamonti D. Ultrastructural and immunocytochemical evidence that an incompetent blood-brain barrier is related to the pathophysiology of cavernous malformations. J Neurol Neurosurg Psychiatry. 2001;71(2):188–92.

8. Rigamonti D, Drayer BP, Johnson PC, Hadley MN, Zabramski J, Spetzler RF. The MRI appearance of cavernous malformations (angiomas). J Neurosurg. 1987;67(4):518–24.
9. Al-Shahi Salman R, Whiteley WN, Warlow C. Screening using whole-body magnetic resonance imaging scanning: who wants an incidentaloma? J Med Screen. 2007;14(1):2–4.
10. Del Curling O, Kelly DL, Elster AD, Craven TE. An analysis of the natural history of cavernous angiomas. J Neurosurg. 1991;75(5):702–8.
11. Morris Z, Whiteley WN, Longstreth WT, Weber F, Lee Y-C, Tsushima Y, et al. Incidental findings on brain magnetic resonance imaging: systematic review and meta-analysis. BMJ [Internet]. 17 août 2009 [cité 27 déc 2019];339. https://www.bmj.com/content/339/bmj.b3016.
12. Otten P, Pizzolato GP, Rilliet B, Berney J. [131 cases of cavernous angioma (cavernomas) of the CNS, discovered by retrospective analysis of 24,535 autopsies]. Neurochirurgie. 1989;35(2):82–3, 128–31.
13. Washington CW, McCoy KE, Zipfel GJ. Update on the natural history of cavernous malformations and factors predicting aggressive clinical presentation. Neurosurg Focus. 2010;29(3):E7.
14. Horne MA, Flemming KD, Su I-C, Stapf C, Jeon JP, Li D, et al. Clinical course of untreated cerebral cavernous malformations: a meta-analysis of individual patient data. Lancet Neurol. 2016;15(2):166–73.
15. Labauge P, Denier C, Bergametti F, Tournier-Lasserve E. Genetics of cavernous angiomas. Lancet Neurol. 2007;6(3):237–44.
16. Nimjee SM, Powers CJ, Bulsara KR. Review of the literature on de novo formation of cavernous malformations of the central nervous system after radiation therapy. Neurosurg Focus. 2006;21(1):e4.
17. Al-Shahi Salman R, Berg MJ, Morrison L, Awad IA, Angioma Alliance Scientific Advisory Board. Hemorrhage from cavernous malformations of the brain: definition and reporting standards. Angioma Alliance Scientific Advisory Board. Stroke. 2008;39(12):3222–30.
18. Salman RA-S, Hall JM, Horne MA, Moultrie F, Josephson CB, Bhattacharya JJ, et al. Untreated clinical course of cerebral cavernous malformations: a prospective, population-based cohort study. Lancet Neurol. 2012;11(3):150–6.
19. Josephson CB, Leach J-P, Duncan R, Roberts RC, Counsell CE, Al-Shahi Salman R, et al. Seizure risk from cavernous or arteriovenous malformations: prospective population-based study. Neurology. 2011;76(18):1548–54.
20. Taslimi S, Modabbernia A, Amin-Hanjani S, Barker FG, Macdonald RL. Natural history of cavernous malformation: systematic review and meta-analysis of 25 studies. Neurology. 2016;86(21):1984–91.
21. Abla AA, Lekovic GP, Turner JD, de Oliveira JG, Porter R, Spetzler RF. Advances in the treatment and outcome of brainstem cavernous malformation surgery: a single-center case series of 300 surgically treated patients. Neurosurgery. 2011;68(2):403–14; discussion 414–415.
22. Garrett M, Spetzler RF. Surgical treatment of brainstem cavernous malformations. Surg Neurol. 2009;72 Suppl 2:S3–9; discussion S9–10.
23. Ferroli P, Sinisi M, Franzini A, Giombini S, Solero CL, Broggi G. Brainstem cavernomas: long-term results of microsurgical resection in 52 patients. Neurosurgery. 2005;56(6):1203–12; discussion 1212–1214.
24. Wang C, Liu A, Zhang J, Sun B, Zhao Y. Surgical management of brain-stem cavernous malformations: report of 137 cases. Surg Neurol. 2003;59(6):444–54; discussion 454.
25. Chang SD, Shuster DL, Steinberg GK, Levy RP, Frankel K. Stereotactic radiosurgery of arteriovenous malformations: pathologic changes in resected tissue. Clin Neuropathol. 1997;16(2):111–6.
26. Nyáry I, Major O, Hanzély Z, Szeifert GT. Pathological considerations to irradiation of cavernous malformations. Prog Neurol Surg. 2007;20:231–4.
27. Gewirtz RJ, Steinberg GK, Crowley R, Levy RP. Pathological changes in surgically resected angiographically occult vascular malformations after radiation. Neurosurgery. 1998;42(4):738–42; discussion 742–743.
28. Karlsson B, Kihlström L, Lindquist C, Ericson K, Steiner L. Radiosurgery for cavernous malformations. J Neurosurg. 1998;88(2):293–7.
29. Kim DG, Choe WJ, Paek SH, Chung HT, Kim IH, Han DH. Radiosurgery of intracranial cavernous malformations. Acta Neurochir. 2002;144(9):869–78; discussion 878.
30. Lunsford LD, Khan AA, Niranjan A, Kano H, Flickinger JC, Kondziolka D. Stereotactic radiosurgery for symptomatic solitary cerebral cavernous malformations considered high risk for resection. J Neurosurg. 2010;113(1):23–9.
31. Kim BS, Yeon JY, Kim J-S, Hong S-C, Lee J-I. Gamma knife radiosurgery of the symptomatic brain stem cavernous angioma with low marginal dose. Clin Neurol Neurosurg. 2014;126:110–4.
32. Yoon PH, Kim DI, Jeon P, Ryu YH, Hwang GJ, Park SJ. Cerebral cavernous malformations: serial magnetic resonance imaging findings in patients with and without gamma knife surgery. Neurol Med Chir (Tokyo). 1998;38(Suppl):255–61.
33. Kondziolka D, Lunsford LD, Flickinger JC, Kestle JR. Reduction of hemorrhage risk after stereotactic radiosurgery for cavernous malformations. J Neurosurg. 1995;83(5):825–31.
34. Stea RA, Schicker L, King GA, Winfield JA. Stereotactic linear radiosurgery for cavernous angiomas. Stereotact Funct Neurosurg. 1994;63(1–4):255–65.
35. Kida Y, Kobayashi T, Tanaka T. Treatment of symptomatic AOVMs with radiosurgery. Acta Neurochir Suppl. 1995;63:68–72.
36. Amin-Hanjani S, Ogilvy CS, Candia GJ, Lyons S, Chapman PH. Stereotactic radiosurgery for cavernous

malformations: Kjellberg's experience with proton beam therapy in 98 cases at the Harvard Cyclotron. Neurosurgery. 1998;42(6):1229–36; discussion 1236–1238

37. Liscák R, Vladyka V, Simonová G, Vymazal J, Novotny J. Gamma knife radiosurgery of the brain stem cavernomas. Minim Invasive Neurosurg. 2000;43(4):201–7.
38. Pollock BE, Garces YI, Stafford SL, Foote RL, Schomberg PJ, Link MJ. Stereotactic radiosurgery for cavernous malformations. J Neurosurg. 2000;93(6):987–91.
39. Hasegawa T, McInerney J, Kondziolka D, Lee JYK, Flickinger JC, Lunsford LD. Long-term results after stereotactic radiosurgery for patients with cavernous malformations. Neurosurgery. 2002;50(6):1190–7; discussion 1197–1198.
40. Régis J, Bartolomei F, Rey M, Hayashi M, Porcheron D, Chauvel P, et al. [Gamma knife radiosurgery for the treatment of severe epilepsy]. Rev Neurol (Paris). 2002;158(4):405–11.
41. Liu-Li A, Wang C, Dai K. [Gamma knife radiosurgery for cavernous malformations]. Zhongguo Yi Xue Ke Xue Yuan Xue Bao. 2005;27(1):18–21.
42. Huang Y-C, Tseng C-K, Chang C-N, Wei K-C, Liao C-C, Hsu P-W. LINAC radiosurgery for intracranial cavernous malformation: 10-year experience. Clin Neurol Neurosurg. 2006;108(8):750–6.
43. Monaco EA, Khan AA, Niranjan A, Kano H, Grandhi R, Kondziolka D, et al. Stereotactic radiosurgery for the treatment of symptomatic brainstem cavernous malformations. Neurosurg Focus. 2010;29(3):E11.
44. Nagy G, Razak A, Rowe JG, Hodgson TJ, Coley SC, Radatz MWR, et al. Stereotactic radiosurgery for deep-seated cavernous malformations: a move toward more active, early intervention. Clinical article. J Neurosurg. 2010;113(4):691–9.
45. Blamek SC, Idasiak A, Larysz D, Rudnik A, Ficek K, Miszczyk L, et al. Linac-based stereotactic radiosurgery for brain cavernomas. Int J Radiat Oncol Biol Phys. 2010;78(3):S291–2.
46. Lee C-C, Pan DH-C, Chung W-Y, Liu K-D, Yang H-C, Wu H-M, et al. Brainstem cavernous malformations: the role of Gamma Knife surgery. J Neurosurg. 2012;(117 Suppl):164–9.
47. Fuetsch M, El Majdoub F, Hoevels M, Müller RP, Sturm V, Maarouf M. Stereotactic LINAC radiosurgery for the treatment of brainstem cavernomas. Strahlenther Onkol. 2012;188(4):311–6.
48. Park S-H, Hwang S-K. Gamma knife radiosurgery for symptomatic brainstem intra-axial cavernous malformations. World Neurosurg. 2013;80(6):e261–6.
49. Liscak R. Radiosurgery of brain cavernomas—long-term results. Prog Neurol Surg. 2013;27:147–56.
50. Liscak R, Urgosik D, Simonova G, Vymazal J, Semnicka J. Gamma knife radiosurgery of brain cavernomas. Acta Neurochir Suppl. 2013;116:107–11.
51. Sager O, Beyzadeoglu M, Dincoglan F, Uysal B, Gamsiz H, Demiral S, et al. Evaluation of linear accelerator (LINAC)-based stereotactic radiosurgery (SRS) for cerebral cavernous malformations: a 15-year single-center experience. Ann Saudi Med. 2014;34(1):54–8.
52. Azimi P, Shahzadi S, Bitaraf MA, Azar M, Alikhani M, Zali A, et al. Cavernomas: outcomes after gamma-knife radiosurgery in Iran. Asian J Neurosurg. 2015;10(1):49.
53. Fedorcsák I, Nagy G, Dobai JG, Mezey G, Bognár L. [Radiosurgery of intracerebral cavernomas—current hungarian practice]. Ideggyogy Sz. 2015;68(7–8):243–51.
54. Frischer JM, Gatterbauer B, Holzer S, Stavrou I, Gruber A, Novak K, et al. Microsurgery and radiosurgery for brainstem cavernomas: effective and complementary treatment options. World Neurosurg. 2014;81(3–4):520–8.
55. Liu HB, Wang Y, Yang S, Gong FL, Xu YY, Wang W. Gamma knife radiosurgery for brainstem cavernous malformations. Clin Neurol Neurosurg. 2016;151:55–60.
56. Aboukais R, Estrade L, Devos P, Blond S, Lejeune J-P, Reyns N. Gamma knife radiosurgery of brainstem cavernous malformations. Stereotact Funct Neurosurg. 2016;94(6):397–403.
57. López-Serrano R, Martínez NE, Kusak ME, Quirós A, Martínez R. Significant hemorrhage rate reduction after gamma knife radiosurgery in symptomatic cavernous malformations: long-term outcome in 95 case series and literature review. Stereotact Funct Neurosurg. 2017;95(6):369–78.
58. Sheen JJ, Lee DH, Lee DH, Song Y, Kwon DH. Long-term outcome of gamma knife radiosurgery for brain cavernoma: factors associated with subsequent de novo cavernoma formation. World Neurosurg. 2018;120:e17–23.
59. Kefeli AU, Sengoz M, Peker S. Gamma knife radiosurgery for hemorrhagic brainstem cavernomas. Turk Neurosurg. 2019;29(1):14–9.
60. Yaprak G, Ozen A, Demir H, Tuğrul F, Karabulut S, Isık N. Stereotactic radiosurgery in brain cavernomas: single-center experience. 2019.
61. Barker FG, Amin-Hanjani S, Butler WE, Lyons S, Ojemann RG, Chapman PH, et al. Temporal clustering of hemorrhages from untreated cavernous malformations of the central nervous system. Neurosurgery. 2001;49(1):15–24; discussion 24–25.
62. Lu X-Y, Sun H, Xu J-G, Li Q-Y. Stereotactic radiosurgery of brainstem cavernous malformations: a systematic review and meta-analysis. J Neurosurg. 2014;120(4):982–7.
63. Wen R, Shi Y, Gao Y, Xu Y, Xiong B, Li D, et al. The efficacy of gamma knife radiosurgery for cavernous malformations: a meta-analysis and review. World Neurosurg. 2019;123:371–7.
64. Steiner L, Karlsson B, Yen C-P, Torner JC, Lindquist C, Schlesinger D. Radiosurgery in cavernous malformations: anatomy of a controversy. J Neurosurg. 2010;113(1):16–21; discussion 21–22.

65. Kondziolka D, Lunsford LD, Kestle JR. The natural history of cerebral cavernous malformations. J Neurosurg. 1995;83(5):820–4.
66. Régis J, Bartolomei F, Kida Y, Kobayashi T, Vladyka V, Liscàk R, et al. Radiosurgery for epilepsy associated with cavernous malformation: retrospective study in 49 patients. Neurosurgery. 2000;47(5):1091–7.
67. Hsu P-W, Chang C-N, Tseng C-K, Wei K-C, Wang C-C, Chuang C-C, et al. Treatment of epileptogenic cavernomas: surgery versus radiosurgery. Cerebrovasc Dis. 2007;24(1):116–20; discussion 121.
68. Shih Y-H, Pan DH-C. Management of supratentorial cavernous malformations: craniotomy versus gamma knife radiosurgery. Clin Neurol Neurosurg. 2005;107(2):108–12.
69. Mouchtouris N, Chalouhi N, Chitale A, Starke RM, Tjoumakaris SI, Rosenwasser RH, et al. Management of cerebral cavernous malformations: from diagnosis to treatment. Sci World J. 2015;2015:808314.
70. Burn S, Gunny R, Phipps K, Gaze M, Hayward R. Incidence of cavernoma development in children after radiotherapy for brain tumors. J Neurosurg. 2007;106(5 Suppl):379–83.
71. Jain R, Robertson PL, Gandhi D, Gujar SK, Muraszko KM, Gebarski S. Radiation-induced cavernomas of the brain. Am J Neuroradiol. 2005;26(5):1158–62.
72. Nagy G, McCutcheon BA, Giannini C, Link MJ, Pollock BE. Radiation-induced cavernous malformations after single-fraction meningioma radiosurgery. Oper Neurosurg (Hagerstown). 2018;15(2):207–12.
73. Wang Q, Zhang S, Hui X. Cavernous malformation induced by stereotactic radiosurgery: a report and literature review. Neurol India. 2018;66(2):515–8.
74. Nataf F, Ghossoub M, Missir O, Merienne L, Roux FX, Meder JF, et al. Parenchymal changes after radiosurgery of cerebral arteriovenous malformations. Preliminary report of a proposed classification. Stereotact Funct Neurosurg. 1997;69(1–4 Pt 2):143–6.
75. Flemming KD, Lanzino G. Stereotactic radiosurgery for cavernous malformations: natural history or treatment effect? Neurology. 2019;93(21):921–2.
76. Niranjan A, Lunsford LD. Stereotactic radiosurgery guidelines for the management of patients with intracranial cavernous malformations. Prog Neurol Surg. 2013;27:166–75.
77. Akers A, Al-Shahi Salman R, A Awad I, Dahlem K, Flemming K, Hart B, et al. Synopsis of guidelines for the clinical management of cerebral cavernous malformations: consensus recommendations based on systematic literature review by the Angioma Alliance Scientific Advisory Board Clinical Experts Panel. Neurosurgery. 2017;80(5):665–80.

45 Dural Arteriovenous Fistulas

Juhana Frösen and Antti Lindgren

45.1 Introduction

Dural arteriovenous fistulas (DAVFs) are pathological, direct transcranial connections from extracranial arteries (i.e., meningeal, ethmoidal, occipital artery) to intracranial venous sinuses or veins (Fig. 45.1).

DAVFs lead to abnormally high flow and abnormally high pressure in the venous sinuses or veins to which they drain, leading in specific cases to a retrograde flow from the DAVF draining site to the cortical veins. This cortical venous reflux leads to supraphysiological flow and pressure in the cerebral cortical veins that may subsequently rupture causing intracerebral, subarachnoidal, or subdural hemorrhages.

Treatment options for DAVFs include (1) occlusion through endovascular embolization with liquid embolic agents or coils, (2) occlusion through microsurgical ligation, or (3) stereotactic radiosurgery (SRS). This book chapter discusses the pathophysiological basis of SRS for DAVFs, its indications, results, and limitations.

Carotid-cavernous fistulas (CCFs) represent a subgroup of DAVFs located in the cavernous sinuses. These lesions have an angiographical presentation somewhat similar to true DAVFs and with similar clinical problems caused primarily by hypertension in the intracranial veins due to the arteriovenous shunting of the fistula. Moreover, the therapeutic options for CCFs, including SRS, are similar to those of non-cavernous sinus DAVFs [1]. Radiotherapy of CCFs is therefore also discussed in this context.

45.2 Pathophysiology and Untreated Clinical Course of DAVFs

Why and how DAVFs form is not known. What seems clear, however, is that they are mostly acquired lesions that develop during lifetime [2–5]. The only known clinical risk factor for DAVFs is a history of sinus thrombosis [2, 3, 5]. Interestingly, a case of familial DAVFs associated with prothrombotic mutations has been reported [6] as well as the association of sporadic DAVFs with other prothrombotic mutations and abnormalities of coagulation [7–10]. This supports the hypothesis that sinus thrombosis plays a key role in the development of DAVFs. DAVFs do not have a clear gender predominance, and the average age at diagnosis is approximately 60 years [11]. The incidence of new DAVF diagnosis has been estimated to be around 0.3/100,000/year [12].

J. Frösen (✉)
Department of Neurosurgery, Tampere University Hospital, Tampere, Finland
e-mail: juhana.frosen@tuni.fi

A. Lindgren
Department of Neurosurgery, Kuopio University Hospital, Kuopio, Finland

© Springer Nature Switzerland AG 2020
A. Conti et al. (eds.), *CyberKnife NeuroRadiosurgery*,
https://doi.org/10.1007/978-3-030-50668-1_45

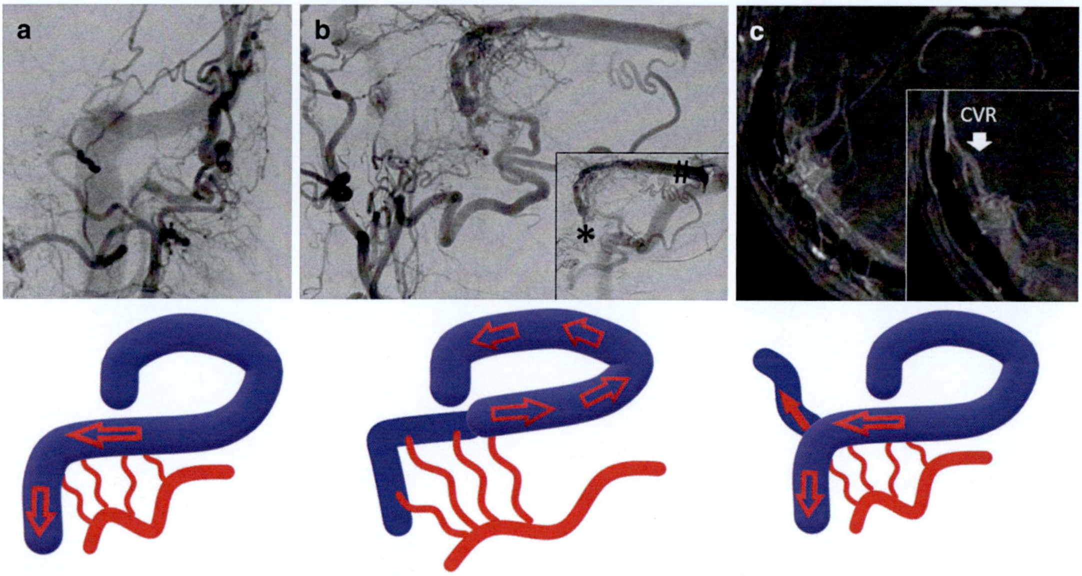

Fig. 45.1 Digital subtraction angiography (DSA) examples of DAVFs draining to the transverse and sigmoid sinus with antegrade filling of the sinus (Cognard 1) (**a**, coronal view. Note filling of the ipsilateral transverse and sigmoid sinus) and with retrograde flow to the transverse sinus (Cognard 2a) (**b**, sagittal view. Note that the ipsilateral sigmoid sinus (*) is not filling, but instead the drainage is to the contralateral sigmoid sinus through retrograde filling of the transverse sinus (#)). DAVF of the sinus with cortical venous reflux (CVR) (Cognard 2a + b) is demonstrated in with MRA in (**c**). Schematic illustrations of the corresponding DAVFs are provided in the lower row

DAVFs are dynamic lesions that may grow or regress once formed [13–16] (Fig. 45.2). This is also illustrated in their clinical presentation which may vary over time in the same patient with the same DAVF. The symptoms that lead to the DAVF diagnosis are (1) annoying pulsating tinnitus (bruit of retrosigmoid high-flow fistula), which should lead to a DSA when unilateral (approximately 30% were caused by DAVFs [17]); (2) decrease in visual acuity (caused by increased intracranial venous pressure) [18]; (3) otherwise unexplained sinus thrombosis (symptoms ranging from headache to neurological deficits) [5]; (4) intracranial hemorrhage from rupture of a cortical draining vein [19–26]; and (5) in DAVFs of the medulla oblongata and spinal cord, progressive myelopathy caused by venous congestion [27] and producing tetra-, para-, or monoparesis with or without bladder dysfunction depending on the affected level. DAVFs can also be asymptomatic, at least initially, and be diagnosed as incidental findings during neuroradiological investigations.

45.3 Risk of Hemorrhage from a DAVF and the Indications for Treatment

The risk of intracranial hemorrhage from a DAVF is dependent on the presence and type of cortical venous reflux [19–26]. In the classical series by Cognard et al., 40% of fistulas with a direct connection to the cortical veins presented with rupture [20]. In the same series, 10% of fistulas that drained to the cerebral sinus and had cortical venous reflux (CVR) presented with rupture, while none of the fistulas that had no cortical venous reflux had ruptured [20]. Later follow-up studies confirmed the role of CVR as a predictor of hemorrhagic risk [21–24]. Unruptured DAVFs presenting with CVR had a 1.5–1.7% annual risk of hemorrhage in studies by Söderman et al. and Strom et al. [22, 23], which adds up to significant cumulative risk of hemorrhage already in a 5-year period. Of note is the study by van Dijk et al. that reports an

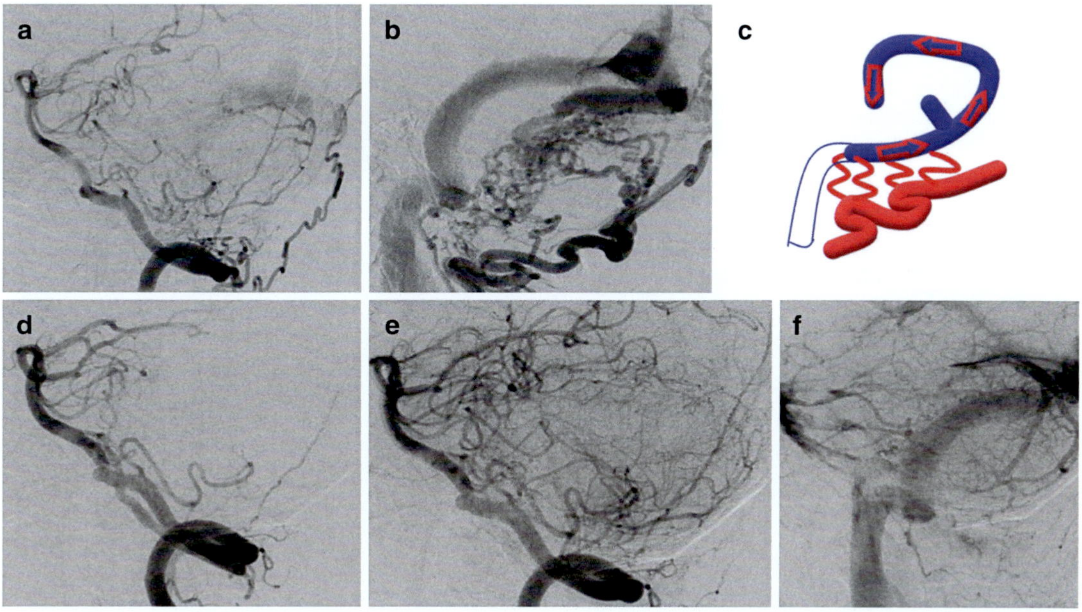

Fig. 45.2 Digital subtraction angiography (DSA) examples of a DAVF that initially presented with a pulsating bruit without a cortical venous reflux was managed conservatively and progressed during follow-up (**a** DSA at baseline, **b** DSA after 2 years' follow-up, **c** a schematic illustration of the DAVF after 2 years' follow-up). The DAVF was treated with CyberKnife radiosurgery, and in control DSA 2 years after the treatment, the DAVF was completely occluded (**d** early arterial phase, **e** late arterial phase, **f** venous phase)

annual hemorrhage rate of 8% for Borden II or III grade DAVFs (DAVFS with CVR) [21]. Due to this high risk of rupture, DAVFs with cortical venous reflux should be considered for prophylactic occlusion to prevent rupture. On the other hand, DAVFs without cortical venous reflux may be treated if they cause disabling symptoms, the most common of which is pulsating tinnitus when the DAVF is located retrosigmoidally in proximity of the cochlea. It should be noted, however, that the risk of hemorrhage for unruptured DAVFs without CVR was 0% in follow-up studies [22, 23]. Accordingly, any risk of disabling complication should be avoided and the indications for invasive treatments carefully considered and discussed with the patient when the only indication to treat is annoying symptoms such as bruit (Table 45.1).

Risk of rebleeding from a previously ruptured DAVF (by definition with CVR) exceeded 7%/year in the follow-up study by Söderman et al. [22]. DAVFs that present with rupture should thus be treated to prevent rehemorrhage.

DAVFs that cause symptoms due to venous hypertension (visual deficits, sinus thrombosis, myelopathy) are usually also treated urgently in order to relieve the disabling symptoms and for the risk that prolonged waiting time may lead to permanent deficits. Whether a DAVF causing only tinnitus should be treated or not is a more ambiguous clinical problem. Although the tinnitus can be severe enough to disturb sleep and reduce the quality of life, it needs to be remembered that the endovascular and microsurgical interventions do carry a risk of procedure-induced morbidity and even mortality [28–31]. While SRS seems to carry fewer risks than the other DAVF treatment options, it is not completely devoid of risks or side effects as will be discussed later. Furthermore, in our clinical experience, the DAVF-related tinnitus may spontaneously resolve even if the fistula has in fact not been occluded. The possibility of spontaneous disappearance of the tinnitus should also be considered when planning imaging follow-up for DAVFs, since loss of bruit may signify rerouting of shunting flow, which may be associated

Table 45.1 Borden [53] and Cognard classifications [20] and recommended treatment modalities based on our experience and published literature

Cognard	Borden	Drainage	Flow	Cortical venous reflux	Rupture risk	Treatment	
						Primary alternative	Second alternative
1	I	Dural sinus	Antegrade	No	No	STT RTx	Endovascular emboliz
2a	II	Dural sinus	Antegrade	No	No	STT RTx	Endovascular emboliz
2b	II	Dural sinus	Retrograde	Yes	Yes	Endovascular emboliz.	STT RTx
2a + b	II	Dural sinus	Retrograde	Yes	Yes	Endovascular emboliz.	STT RTx
3	III	Subarachnoid veins	Directly to veins	Yes (direct fistula)	Yes	Surgical	Endovascular emboliz
4	III	Subarachnoid veins	Directly to veins	Yes (direct fistula + venous ectasia)	Yes	Surgical	Endovascular emboliz
5	III	Subarachnoid veins	Directly to veins	Spinal	Yes	Surgical	Endovascular emboliz

with the development of CVR and thus a new risk of rupture. As stated above, DAVFs are dynamic lesions that may evolve as a function of time, and subsequently the associated symptoms also evolve. In practice, this means that DAVFs that initially present without cortical venous reflux and risk of rupture may, with time, develop CVR and become prone to rupture. It also means that without verification with imaging studies, it cannot be concluded simply based on evolution of symptoms that a DAVF would have regressed or remained stable. Thus, imaging follow-up needs to be considered when complete occlusion is not achieved or attempted.

45.4 Indications for Stereotactic Radiotherapy or Radiosurgery of DAVFs

There are two clear indications for SRS in the treatment of DAVFs: (1) preventing intracranial hemorrhage from a DAVF with cortical venous reflux and (2) relieving intolerable tinnitus.

Unlike endovascular or microsurgical occlusion of DAVFs that provide immediate therapeutic effect, the therapeutic effect of SRS on the DAVFs comes with a delay ranging from several months to several years, subsequently exposing the patient to the risk of hemorrhage [32]. SRS is therefore not the primary choice of treatment modality if the DAVF presents with rupture or with neurological deficits related to venous hypertension/congestion. However, sometimes SRS may be the best and most appropriate treatment for unruptured DAVFs with rupture risk due to CVR or DAVFs with deficits related to venous hypertension, if more immediate occlusion (by surgery or embolization) involves a significant risk. It is worth noting that the efficacy and risk of SRS for DAVFs is similar in older patients than in younger ones [33], while the risk of surgery or embolization is higher in the elderly and more fragile patient [28, 31].

For those DAVFs that only present with tinnitus or that are completely asymptomatic, SRS seems to be the most appropriate primary treatment choice given its low risk and relatively high efficacy in long-term follow-up (Table 45.2). In such DAVFs, however, conservative follow-up without any intervention may also be an option.

Table 45.2 Systematic review of published literature on stereotactic radiotherapy/stereotactic radiosurgery (SRS) for DAVFs. Studies with >20 patients are described in the table below, while smaller studies are included in references [54, 56–59, 65, 67, 68]

Study	Patient number	Margin dose	Type of SRS	Types of fistulas	Occlusion rate	STT RTX induced neurological deficit
Starke et al. 2019 [47]	114	Mean 21.8 Gy	Gamma knife	Cognard I $n = 38$ 33% Cognard IIa $n = 9$ 8% Cognard IIb $n = 6$ 5% Cognard IIa + b $n = 10$ 9% Cognard III $n = 6$ 5% Cognard IV $n = 20$ 18% Cognard V $n = 24$ 21%	68.4% obliteration at mean FU of 4 years	Post-SRS hemorrhage 4 patients (0.9% annual risk), SRS-induced imaging changes in 10% of patients
Mohammed et al. 2019 [33]	96	14–33 Gy	Gamma knife	Cognard I 39 Cognard II 18 Cognard III 4 Cognard IV 10 Cognard V 15	56% obliteration	Upfront SRS: 8% adverse SRS effect, no hemorrhages Salvage SRS: 16% post SRS hemorrhage
Chen et al. 2018 [55]	41	Mean 18.9 Gy	Gamma knife	Borden 2–3	Complete obliteration 63%	Permanent SRS-related deficit: 23%
Tonetti et al. 2017 [56]	61	Mean 20.0 Gy	Gamma knife	dAVFS without cortical venous reflux	11 with FU: 9 (82%) complete obliteration	None
Park et al. 2017 [35]	30	Median 17 Gy	Gamma knife	Cognard I $n = 5$ 16% Cognard IIa $n = 6$ 20% Cognard IIb $n = 2$ 7% Cognard IIa + b $n = 14$ 47% Cognard III $n = 3$ 10%	43% obliteration at 1 year 79% at 2 years 95% at 5 years	None
Park et al. 2016 [60]	31	Mean 16.8 Gy	Gamma knife	Cognard I $n = 3$ 15% Cognard IIa $n = 5$ 25% Cognard IIa + b $n = 10$ 50% Cognard III $n = 1$ 5% Cognard IV $n = 1$ 5%	90% obliteration	1 adverse SRS effect

(continued)

Table 45.2 (continued)

Study	Patient number	Margin dose	Type of SRS	Types of fistulas	Occlusion rate	STT RTX induced neurological deficit
Söderman et al. 2013 [61]	65	"Most commonly 20–26 Gy to 40–60% isodose"	Gamma knife	Cognard I, IIa $n = 20$ Cognard IIa + b, IIb $n = 19$ Cognard III N16 Cognard IV $n = 12$	59% obliteration 27% regression 14% unchanged	2 hemorrhages 5 "minor adverse radiation effects"
Hanakita et al. 2012 [62]	22	Median 20.0 Gy	Gamma knife	Cognard 1 $n = 3$ 14% Cognard 2 $n = 11$ 50% Cognard III $n = 3$ 14% Cognard IV $n = 5$ 23%	12 patients complete obliteration (median 25 months) 10 patients incomplete obliteration (median 42 months)	None
Pan et al. 2012 [37]	115 non-cavernous dAVFs	Mean 17.2 Gy	Gamma knife	Cognard I $n = 25$ 22% Cognard IIa $n = 38$ 33% Cognard IIb $n = 9$ 8% Cognard IIa + b $n = 26$ 23% Cognard III $n = 6$ 5% Cognard IV $n = 8$ 7% Cognard V $n = 3$ 3%	Mean FU 28 months 59% obliteration 37% regression 2% unchanged 1% progression 1% death	n.a.
Yang et al. 2010 [63]	40	Median 21 Gy	Gamma knife	Cognard I $n = 24$ 55% Cognard IIa $n = 6$ 14% Cognard IIb $n = 6$ 14% Cognard IIa + b $n = 8$ 17%	70% total obliteration at median FU of 45 months	1 death by post-SRS hemorrhage 4 transient worsening of symptoms
Cifarelli et al. 2010 [64]	55	Mean 21 Gy	Gamma knife	Borden 1 16 29% Borden 2 12 22% Borden 3 27 49%	46 patients with FU: 30 (65%) obliteration rate	3 (5%) post-SRS hemorrhage 12% post-SRS MRI changes
Söderman et al. 2006 [66]	49 patients with 52 dAVFs	Mean 22 Gy	Gamma knife	Borden 1 $n = 16$ 31% Borden 2–3 $n = 36$ 69%	41 with 2-year FU: 68% obliteration 28% regression	2 post-SRS hemorrhages 1 transient cranial nerve palsy 1 late radiation reaction

Table 45.2 (continued)

Study	Patient number	Margin dose	Type of SRS	Types of fistulas	Occlusion rate	STT RTX induced neurological deficit
Pan et al. 2002 [69]	20	16.5–19 Gy	Gamma knife	Cognard I $n = 4$ Cognard IIa $n = 7$ Cognard IIb $n = 2$ Cognard IIa + b $n = 7$	58% obliteration 16% subtotal obliteration 26% regression	None
Friedman et al. 2001 [70]	25	Median 18 Gy	Gamma knife	Cognard I $n = 12$ 52% Cognard IIa $n = 7$ 30% Cognard III $n = 4$ 17%	17 patients with FU: 65% total obliteration, regression 35%	None
Pollock et al. 1999 [71]	20	20 Gy	Gamma knife	Cognard I $n = 1$ 5% Cognard IIa $n = 15$ 75% Cognard IIa + b $n = 3$ 15% Cognard III $n = 1$ 5%	65% total obliteration 5% subtotal obliteration Median 12 months	None

PubMed was searched (January 2020) with the following search terms: dural arteriovenous fistula (DAVF) *and* (radiosurgery OR radiotherapy). Only original studies published in English and with abstracts were included. The studies were screened according to their abstracts, and studies reporting the outcome of a clinical series of DAVFs treated with stereotactic radiotherapy were selected. A summary of the selected studies with ≥20 patients included is presented in the table below

45.5 Results of Stereotactic Radiotherapy for DAVFs

The use of SRS to treat DAVFs was first reported by Barcia-Salorio et al. [34] in 1982. Despite approximately 40 years of clinical use, there are not many large published clinical series of SRS-treated DAVFs. A summary of the published clinical series with >20 patients (non-CCFs) treated with SRS is presented in Table 45.2. Overall, SRS leads to complete occlusion of the DAVF in approximately 60% of cases (median rate of complete occlusion in the published literature 63%, range 40–95%, follow-up ranging from 1 to 5 years (Table 45.2 and Fig. 45.3)). What seems clear from the published literature is that the rate of occlusion increases with longer follow-up (with up to 95% occlusion rate at 5 years post SRS reported by Park et al. [35], but a significant number of DAVFs are actually already occluded 1 year post SRS (40% in the series by Park et al. [35]).

Literature on SRS in the treatment of carotid-cavernous fistulas is limited, but the few studies published suggest that low-flow CCFs respond well to SRS (Table 45.3), perhaps even better than DAVFs in other locations [36]. The rate of complete occlusion reached 59%, 70%, and 91%, respectively, in low-flow fistulas of the type Barrow B–D in the three published studies with >20 treated patients (Table 45.3). High-flow CCFs (Barrow A) that are often posttraumatic seem to have a low occlusion rate (1/3) [37] after SRS and should thus preferentially be treated primarily by other modalities.

Although all the SRS studies with >20 DAVF patients reported in the literature have been performed with Gamma Knife, the results are consistent with those obtained in our own practice with

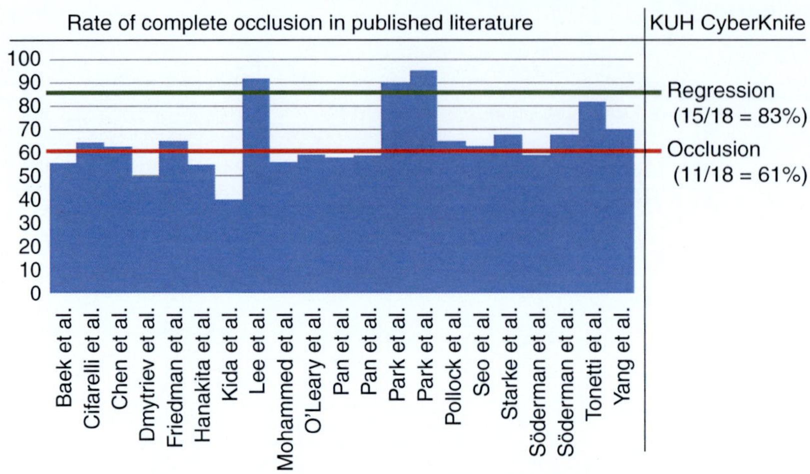

Fig. 45.3 Results of stereotactic radiotherapy of DAVFs treated with CyberKnife in Kuopio University Hospital compared with results of Gamma Knife-based treatment reported in the literature. PubMed was searched (January 2020) with the following search terms: dural arteriovenous fistula (DAVF) *and* (radiosurgery *or* radiotherapy). Only original studies published in English and with abstracts were included. The studies were screened according to their abstracts, and studies reporting the outcome of a clinical series of DAVFs treated with stereotactic radiotherapy were selected. A summary of the selected studies with ≥20 patients included is presented in Table 45.2

Table 45.3 Systematic review of published literature on stereotactic radiotherapy for CCFs

Study	Patient number	Margin dose	Type of STT RTx	Types of fistulas	Occlusion rate	STT RTX-induced neurological deficit
Wu et al. 2019 [72]	123	16.6–17.5 Gy	Gamma knife	Barrow B $n = 9$ Barrow C $n = 3$ Barrow D $n = 111$	59% complete obliteration at a median of 13 months	1 post-SRS hemorrhage
Pan et al. 2012 [37]	206	Mean 17.2 Gy	Gamma knife	Barrow B $n = 19$ 9% Barrow C $n = 12$ 6% Barrow D $n = 175$ 85%	70% obliteration 30% regression	None
Barcia-Salorio J. et al. 1994 [34]	25	30–40 Gy	?	Barrow B $n = 11$ Barrow C $n = 4$ Barrow D $n = 7$ Traumatic $N = 3$	Non-traumatic fistulas: 91% complete obliteration in a mean of 7.5 months	None

PubMed was searched (January 2020) with the following search terms: carotico-cavernous fistula (CCF) *and* (radiosurgery OR radiotherapy). Only original studies published in English and with abstracts were included. The studies were screened according to their abstracts, and studies reporting the outcome of a clinical series of CCFs (at least 20 cases) treated with stereotactic radiotherapy were selected. A summary of the selected studies is presented in the table below

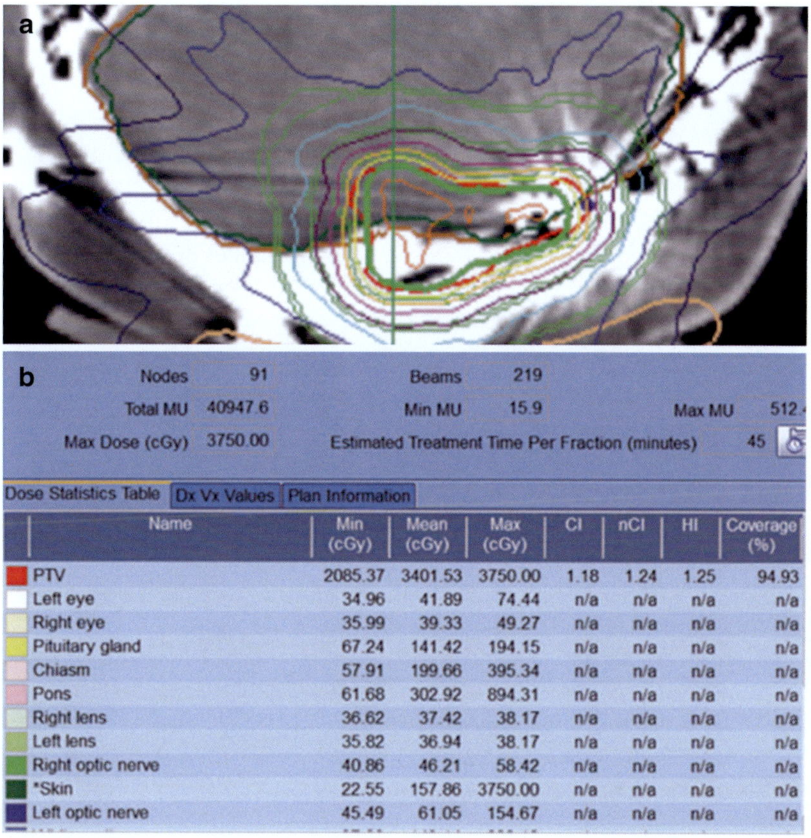

Name	Min (cGy)	Mean (cGy)	Max (cGy)	CI	nCI	HI	Coverage (%)
PTV	2085.37	3401.53	3750.00	1.18	1.24	1.25	94.93
Left eye	34.96	41.89	74.44	n/a	n/a	n/a	n/a
Right eye	35.99	39.33	49.27	n/a	n/a	n/a	n/a
Pituitary gland	67.24	141.42	194.15	n/a	n/a	n/a	n/a
Chiasm	57.91	199.66	395.34	n/a	n/a	n/a	n/a
Pons	61.68	302.92	894.31	n/a	n/a	n/a	n/a
Right lens	36.62	37.42	38.17	n/a	n/a	n/a	n/a
Left lens	35.82	36.94	38.17	n/a	n/a	n/a	n/a
Right optic nerve	40.86	46.21	58.42	n/a	n/a	n/a	n/a
*Skin	22.55	157.86	3750.00	n/a	n/a	n/a	n/a
Left optic nerve	45.49	61.05	154.67	n/a	n/a	n/a	n/a

Fig. 45.4 CyberKnife radiotherapy plan for a retrosigmoid/occipital DAVF. (**a**) shows the dose distribution with isodose lines, and (**b**) shows the dose to the DAVF (designated as PTV, or primary target volume) and the adjacent intracranial structure. Note that the doses are in centiGy (cGy)

CyberKnife and conventional LINAC (Figs. 45.3 and 45.4). We have aimed at a margin dose of 18–20 Gy given in one single fraction if the DAVF volume and location allow it. In DAVFs with a volume larger than 10 cm^3, we have considered fractionated SRS with a dosage of 5Gy/ fraction × 6 fractions, or volume-staged SRS in which subvolumes of the DAVF are treated with a single fraction of 20–18Gy sequentially. In large DAVFs, for which a single-fraction SRS is not indicated, the possibility of downsizing the DAVF with embolization and treating only the remaining residual DAVF with SRS should be considered. It seems, however, based on clinical series of brain arteriovenous malformations (bAVM) treated with SRS, that prior embolization compromises the efficacy of SRS [38]. While this remains to be demonstrated in DAVFs, it should be considered since the embolization material may affect both the visibility of small fistulous target vessels in the angiogram used for SRS planning, as well as cause dispersion of the radiation. Thus, rather than treating post-embolization residuals with SRS, downsizing the large DAVF with SRS followed by embolization of the residual could be considered.

Why the fistulous connections of DAVFs regress with radiotherapy while the feeding artery and the draining vein usually remain patent despite being exposed at least focally to the same radiation dose remains to be elucidated. Radiation damages the endothelium of vessels [39], which predisposes to thrombosis. In addition, it induces a fibrotic reaction in the vessel wall [40]. This has been thought to explain why the pathological nidal vessels of many arteriovenous malformations of the brain (bAVMs) regress after radiotherapy [41, 42]. While the majority of bAVMs are explained by presence of angiogenesis activating somatic mutations (KRAS) in the endothelial cells [43], the pathogenesis and underlying biological aberration in

DAVFs is likely very different. Nevertheless, radiotherapy induces occlusion of the fistulous vessels of DAVFs (Tables 45.2 and 45.3) similarly to bAVMs, perhaps because DAVF vessels are in an immature and active state of angiogenetic remodelling.

45.6 Limitations and Contraindications for Stereotactic Radiotherapy of DAVFs

As mentioned above, volume of the target lesion limits the use of SRS. Targets with a size larger than 3–3.5 cm or a volume larger than 10 cm^3 are usually not considered for single-fraction SRS because of increased risk of radiation-induced injury [44]. In these lesions, dose-fractionated (multiple consecutive fractions with smaller doses delivered in 2–3 day intervals [45]) or volume-fractionated (large target volume divided into subvolumes treated with sequentially single-fraction SRS [44]) can, however, be considered.

Another important limitation for SRS is the radiosensitivity of neurological or other anatomical structures adjacent to the target volume. Typically, these involve the cochlea or brain stem in the DAVFs of the posterior fossa extending to the sigmoid sinus (Fig. 45.5), the spinal cord in spinal DAVFs, or the optic nerve and the pituitary gland in CCFs. DAVFs involving the transverse sinus or the superior sagittal sinus are also fairly superficial, with the fistulous target vessels arising on the external surface of the skull in the subcutis. Thus, the radiation dose to which the skin is exposed also needs to be considered and may limit the dose given to the DAVF.

Side effects from SRS of DAVFs include transient headache and nausea that usually arise on the day of treatment and resolve in 1–2 days. In our experience, per oral corticosteroids are efficient in alleviating these symptoms. Radiotherapy to the brain may cause radionecrosis, even with a delay of several years [46]. This radionecrosis may cause new neurological symptoms, the clinical presentation of which varies depending on the structure affected. Of the 22 published clinical series with >20 DAVF patients treated with SRS, only 9 report the frequency of transient radiotherapy-induced symptoms or deficits which ranges from 3% to 15% with half of those studies reporting a rate close (±1) to the median rate of 7% (Table 45.2). A large international meta-analysis of bAVMs treated with Gamma Knife delivered SRS reports a rate of 5.2% for radiotherapy-induced transient new neurological symptoms and a rate of 3.5% for radiotherapy-induced permanent neurological symptoms [47]. These results seem comparable to the published

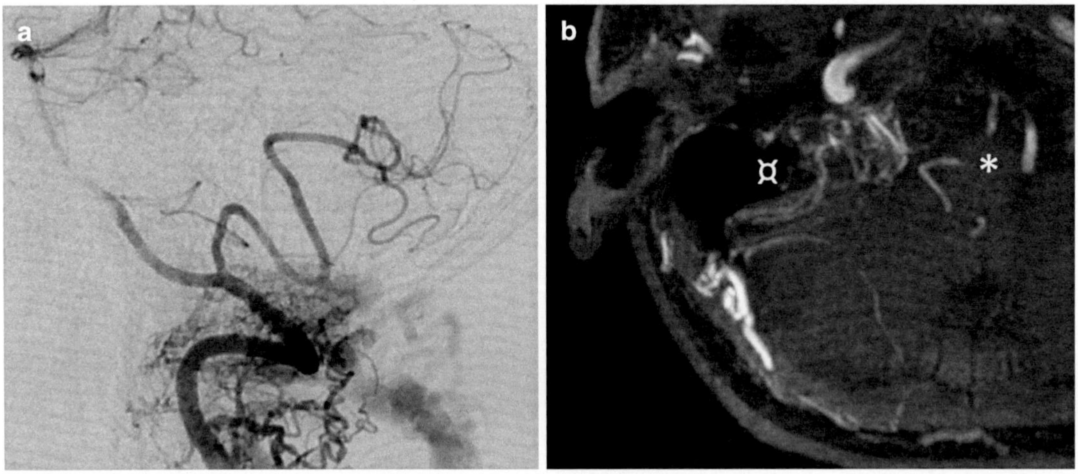

Fig. 45.5 Examples of a DAVF near radiosensitive structure such as the brain stem (*) and the cochlea (¤) in the posterior fossa (digital subtraction angiography, **a**, and MRI, **b**)

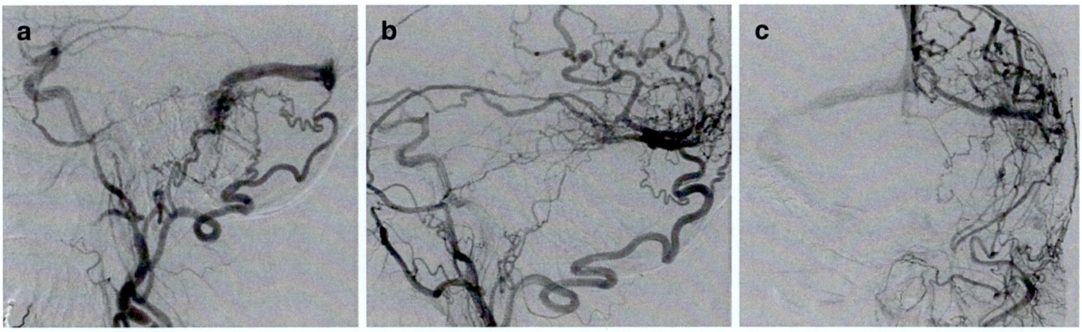

Fig. 45.6 Example of a DAVF that progressed despite radiotherapy. Digital subtraction angiography (DSA) prior to radiotherapy (**a**) showing a simple retrosigmoid DAVF with antegrade draining to the transverse sinus and no cortical venous reflux. A single fraction of radiotherapy with 20Gy margin dose was given. In a control DSA 3 years after the radiotherapy, the DAVF has progressed with now retrograde flow and cortical venous reflux (**b** lateral view, **c** coronal view)

literature on SRS of DAVFs (Table 45.2), as well as our own experience in treating 24 DAVF patients with CyberKnife or other LINAC (rate of new but transient neurological symptoms 2 cases in 24 patients, or 8%).

Radiotherapy for non-malignant, i.e., non-cancerous, lesions often raises the concern of potentially inducing secondary malignancies with the radiotherapy and, when the lesion is in the brain, of causing radiation-induced cognitive decline which is a known side effect of whole brain radiotherapy [48]. While there is no literature specific for SRS of DAVFs on the topic, follow-up studies of bAVM patients treated with similar single radiation doses [49, 50] or fractionated doses [51] demonstrate that these concerns are negligible in the treatment of bAVMs or DAVFs with SRS and need not to be worried, especially when compared with the hemorrhage risk of DAVFs with CVR [22, 23].

A clinically more valid concern is the time delay between the SRS and the DAVFs, which ranges in average from 1 to 3 years or more and during which the DAVF retains its risk of hemorrhage – if initially presenting with CVR. Considering the overall high rebleeding rates of ruptured DAVFs (7.4% per year; Söderman et al. 22), and the high risk of early rebleeding reported in the literature (up to 35% within 2 weeks in the study by Duffau et al. [52]), the risk of rebleeding during a 2–3-year latency period after SRS of a ruptured DAVF seems noticeably high. Therefore, SRS should not be considered as the treatment of choice for ruptured DAVFs unless endovascular or microsurgical occlusion is unfeasible or are associated with risks higher than that of rebleeding during the latency period. Furthermore, DAVF response to SRS or permanent occlusion cannot be achieved in all cases (Table 45.2 and Fig. 45.4), and some DAVFs may even progress after SRS (Fig. 45.6). For these lesions, other treatment modalities should be considered.

References

1. Barrow DL, Spector RH, Braun IF, Landman JA, Tindall SC, Tindall GT. Classification and treatment of spontaneous carotid-cavernous sinus fistulas. J Neurosurg. 1985;62:248–56.
2. Hacein-Bey L, Konstas AA, Pile-Spellman J. Natural history, current concepts, classification, factors impacting endovascular therapy, and pathophysiology of cerebral and spinal dural arteriovenous fistulas. Clin Neurol Neurosurg. 2014;121:64–75.
3. Wanke I, Rüfenacht DA. The dural AV-fistula (DAVF), the most frequent acquired vascular malformation of the central nervous system (CNS). Clin Neuroradiol. 2015;25(Suppl 2):325–32.
4. Ohshima T, Tamari Y, Yamamoto T, Goto S, Ishikawa K. Midterm follow-up of 20 consecutive patients with nonaneurysmal subarachnoid hemorrhage of unknown origin in a single-center: two cases of de novo development of dural arteriovenous fistula. J Stroke Cerebrovasc Dis. 2017;26:2788–92.
5. Preter M, Tzourio C, Ameri A, Bousser MG. Long-term prognosis in cerebral venous thrombosis. Follow-up of 77 patients. Stroke. 1996;27:243–6.

6. Orina JN, Daniels DJ, Lanzino G. Familial intracranial dural arteriovenous fistulas. Neurosurgery. 2013;72:E310–33.
7. Kraus JA, Stüper BK, Berlit P. Association of resistance to activated protein C and dural arteriovenous fistulas. J Neurol. 1998;245:731–3.
8. Aiello G, Rinaldo L, Marshall AL, Vine RL, Lanzino G. Incidence of hereditary thrombophilia in patients with cranial dural arteriovenous fistulae. J Clin Neurosci. 2020;S0967-5868(19):32044–2. [Epub ahead of print].
9. Izumi T, Miyachi S, Hattori K, Iizuka H, Nakane Y, Yoshida J. Thrombophilic abnormalities among patients with cranial dural arteriovenous fistulas. Neurosurgery. 2007;61:262–8.
10. Gerlach R, Yahya H, Rohde S, Böhm M, Berkefeld J, Scharrer I, Seifert V, Raabe A. Increased incidence of thrombophilic abnormalities in patients with cranial dural arteriovenous fistulae. Neurol Res. 2003;25:745–8.
11. Hiramatsu M, Sugiu K, Hishikawa T, Haruma J, Tokunaga K, Date I, Kuwayama N, Sakai N. Epidemiology of dural arteriovenous fistula in Japan: analysis of Japanese registry of neuroendovascular therapy (JR-NET2). Neurol Med Chir (Tokyo). 2014;54(Suppl 2):63–71.
12. Kuwayama N. Epidemiologic survey of dural arteriovenous fistulas in Japan: clinical frequency and present status of treatment. Acta Neurochir Suppl. 2016;123:185–8.
13. Kannath SK, Rajan JE, Mukherjee A, Sarma PS. Factors predicting spontaneous thrombosis of aggressive cranial dural arteriovenous fistulas. World Neurosurg. 2017;103:821–8.
14. Yen CP, Khaled MA, Schwyzer L, Vorsic M, Dumont AS, Steiner L. Early draining vein occlusion after gamma knife surgery for arteriovenous malformations. Neurosurgery. 2010;67:1293–302.
15. Shah MN, Botros JA, Pilgram TK, Moran CJ, Cross DT 3rd, Chicoine MR, Rich KM, Dacey RG Jr, Derdeyn CP, Zipfel GJ. Borden-shucart type I dural arteriovenous fistulas: clinical course including risk of conversion to higher-grade fistulas. J Neurosurg. 2012;117:539–45.
16. Cognard C, Houdart E, Casasco A, Gabrillargues J, Chiras J, Merland JJ. Long-term changes in intracranial dural arteriovenous fistulae leading to worsening in the type of venous drainage. Neuroradiology. 1997;39:59–66.
17. In 't Veld M, Fronczek R, de Laat JA, Kunst HPM, Meijer FJA, Willems PWA. The incidence of cranial arteriovenous shunts in patients with pulsatile tinnitus: a prospective observational study. Otol Neurotol. 2018;39:648–53.
18. Cognard C, Casasco A, Toevi M, Houdart E, Chiras J, Merland JJ. Dural arteriovenous fistulas as a cause of intracranial hypertension due to impairment of cranial venous outflow. J Neurol Neurosurg Psychiatry. 1998;65:308–16.
19. Brown RD Jr, Wiebers DO, Nichols DA. Intracranial dural arteriovenous fistulae: angiographic predictors of intracranial hemorrhage and clinical outcome in nonsurgical patients. J Neurosurg. 1994;81:531–8.
20. Cognard C, Gobin YP, Pierot L, Bailly AL, Houdart E, Casasco A, et al. Cerebral dural arteriovenous fistulas: clinical and angiographic correlation with a revised classification of venous drainage. Radiology. 1995;194:671–80.
21. van Dijk JM, ter Brugge KG, Willinsky RA, Wallace MC. Clinical course of cranial dural arteriovenous fistulas with long-term persistent cortical venous reflux. Stroke. 2002;33:1233–6.
22. Söderman M, Pavic L, Edner G, Holmin S, Andersson T. Natural history of dural arteriovenous shunts. Stroke. 2008;39:1735–9.
23. Strom RG, Botros JA, Refai D, Moran CJ, Cross DT III, Chicoine MR, et al. Cranial dural arteriovenous fistulae: asymptomatic cortical venous drainage portends less aggressive clinical course. Neurosurgery. 2009;64:241–8.
24. Satomi J, Ghaibeh AA, Moriguchi H, Nagahiro S. Predictability of the future development of aggressive behavior of cranial dural arteriovenous fistulas based on decision tree analysis. J Neurosurg. 2015;123:86–90.
25. Gross BA, Albuquerque FC, McDougall CG, Jankowitz BT, Jadhav AP, Jovin TG, Du R. A multi-institutional analysis of the untreated course of cerebral dural arteriovenous fistulas. J Neurosurg. 2018;129:1114–9.
26. Fassett DR, Rammos SK, Patel P, Parikh H, Couldwell WT. Intracranial subarachnoid hemorrhage resulting from cervical spine dural arteriovenous fistulas: literature review and case presentation. Neurosurg Focus. 2009;26:E4.
27. Atkinson JL, Miller GM, Krauss WE, Marsh WR, Piepgras DG, Atkinson PP, Brown RD Jr, Lane JI. Clinical and radiographic features of dural arteriovenous fistula, a treatable cause of myelopathy. Mayo Clin Proc. 2001;76:1120–30.
28. Hiramatsu M, Sugiu K, Hishikawa T, Nishihiro S, Kidani N, Takahashi Y, Murai S, Date I, Kuwayama N, Satow T, Iihara K, Sakai N. Results of 1940 embolizations for dural arteriovenous fistulas: Japanese registry of neuroendovascular therapy (JR-NET3). J Neurosurg. 2019;28:1–8.
29. Sugiyama T, Nakayama N, Ushikoshi S, et al. Complication rate, cure rate, and long-term outcomes of microsurgery for intracranial dural arteriovenous fistulae: a multicenter series and systematic review. Neurosurg Rev. 2020. [published online ahead of print, 2020 Jan 2]. https://doi.org/10.1007/s10143-019-01232-y.
30. Wachter D, Hans F, Psychogios MN, Knauth M, Rohde V. Microsurgery can cure most intracranial dural arteriovenous fistulae of the sinus and non-sinus type. Neurosurg Rev. 2011;34:337–45.

31. Dützmann S, Beck J, Gerlach R, Bink A, Berkefeld J, du Mesnil de Rochement R, Seifert V, Raabe A. Management, risk factors and outcome of cranial dural arteriovenous fistulae: a single-center experience. Acta Neurochir. 2011;153:1273–81.
32. Starke RM, McCarthy DJ, Chen CJ, et al. Hemorrhage risk of cerebral dural arteriovenous fistulas following Gamma Knife radiosurgery in a multicenter international consortium. J Neurosurg. 2019;1–9. [published online ahead of print, 2019 Mar 15]. https://doi.org/10.3171/2018.12.JNS182208.
33. Mohammed N, Hung YC, Xu Z, Starke RM, Kano H, Lee J, Mathieu D, Kaufmann AM, Grills IS, Cifarelli CP, Vargo JA, Chytka T, Janouskova L, Feliciano CE, Mercado RR, Lunsford LD, Sheehan JP. A propensity score-matched cohort analysis of outcomes after stereotactic radiosurgery in older versus younger patients with dural arteriovenous fistula: an international multicenter study. World Neurosurg. 2019;125:e1114–24.
34. Barcia-Salorio JL, Soler F, Barcia JA, Hernández G. Stereotactic radiosurgery for the treatment of low-flow carotid-cavernous fistulae: results in a series of 25 cases. Stereotact Funct Neurosurg. 1994;63:266–70.
35. Park SH, Park KS, Kang DH, Hwang JH, Hwang SK. Stereotactic radiosurgery for intracranial dural arteriovenous fistulas: its clinical and angiographic perspectives. Acta Neurochir. 2017;159:1093–103.
36. Hung YC, Mohammed N, Kearns KN, Chen CJ, Starke RM, Kano H, Lee J, Mathieu D, Kaufmann AM, Wang WG, Grills IS, Cifarelli CP, Vargo J, Chytka T, Janouskova L, Feliciano CE, Rodriguez-Mercado R, Lunsford LD, Sheehan JP. Stereotactic radiosurgery for cavernous sinus versus noncavernous sinus dural arteriovenous fistulas: outcomes and outcome predictors. Neurosurgery. 2020;86(5):676–84.
37. Pan DH, Wu HM, Kuo YH, Chung WY, Lee CC, Guo WY. Intracranial dural arteriovenous fistulas: natural history and rationale for treatment with stereotactic radiosurgery. Prog Neurol Surg. 2013;27:176–94.
38. Xu F, Zhong J, Ray A, Manjila S, Bambakidis NC. Stereotactic radiosurgery with and without embolization for intracranial arteriovenous malformations: a systematic review and meta-analysis. Neurosurg Focus. 2014;37:E16.
39. Baselet B, Sonveaux P, Baatout S, Aerts A. Pathological effects of ionizing radiation: endothelial activation and dysfunction. Cell Mol Life Sci. 2019;76:699–728.
40. Fernández-Alvarez V, López F, Suárez C, Strojan P, Eisbruch A, Silver CE, Mendenhall WM, Langendijk JA, Rinaldo A, Lee AWM, Beitler JJ, Smee R, Alvarez J, Ferlito A. Radiation-induced carotid artery lesions. Strahlenther Onkol. 2018;194:699–710.
41. Saunders WM, Winston KR, Siddon RL, Svensson GH, Kijewski PK, Rice RK, Hansen JL, Barth NH. Radiosurgery for arteriovenous malformations of the brain using astandard linear accelerator: rationale and technique. Int J Radiat Oncol Biol Phys. 1998;15:441–7.
42. Graffeo CS, Sahgal A, De Salles A, et al. Stereotactic Radiosurgery for Spetzler-Martin Grade I and II Arteriovenous Malformations: International Society of Stereotactic Radiosurgery (ISRS) Practice Guideline. Neurosurgery. 2020;nyaa004. [published online ahead of print, 2020 Feb 17].
43. Nikolaev SI, Vetiska S, Bonilla X, Boudreau E, Jauhiainen S, Rezai Jahromi B, Khyzha N, DiStefano PV, Suutarinen S, Kiehl TR, Mendes Pereira V, Herman AM, Krings T, Andrade-Barazarte H, Tung T, Valiante T, Zadeh G, Tymianski M, Rauramaa T, Ylä-Herttuala S, Wythe JD, Antonarakis SE, Frösen J, Fish JE, Radovanovic I. Somatic activating KRAS mutations in arteriovenous malformations of the brain. N Engl J Med. 2018;378:250–61.
44. Pollock BE, Kline RW, Stafford SL, Foote RL, Schomberg PJ. The rationale and technique of staged-volume arteriovenous malformation radiosurgery. Int J Radiat Oncol Biol Phys. 2000;48:817–24.
45. Zhong J, Press RH, Olson JJ, Oyesiku NM, Shu HG, Eaton BR. The use of hypofractionated radiosurgery for the treatment of intracranial lesions unsuitable for single-fraction radiosurgery. Neurosurgery. 2018;83:850–7.
46. Statham P, Macpherson P, Johnston R, Forster DM, Adams JH, Todd NV. Cerebral radiation necrosis complicating stereotactic radiosurgery for arteriovenous malformation. J Neurol Neurosurg Psychiatry. 1990;53:476–9.
47. Starke RM, McCarthy DJ, Chen CJ, Kano H, McShane B, Lee J, Mathieu D, Vasas LT, Kaufmann AM, Wang WG, Grills IS, Patibandla MR, Cifarelli CP, Paisan G, Vargo JA, Chytka T, Janouskova L, Feliciano CE, Rodriguez-Mercado R, Tonetti DA, Lunsford LD, Sheehan JP. Evaluation of stereotactic radiosurgery for cerebral dural arteriovenous fistulas in a multicenter international consortium. J Neurosurg. 2019;132(1):114–21.
48. Warrington JP, Ashpole N, Csiszar A, Lee YW, Ungvari Z, Sonntag WE. Whole brain radiation-induced vascular cognitive impairment: mechanisms and implications. J Vasc Res. 2013;50:445–57.
49. Steinvorth S, Wenz F, Wildermuth S, Essig M, Fuss M, Lohr F, Debus J, Wannenmacher M, Hacke W. Cognitive function in patients with cerebral arteriovenous malformations after radiosurgery: prospective long-term follow-up. Int J Radiat Oncol Biol Phys. 2002;54:1430–7.
50. Riva D, Pantaleoni C, Devoti M, Lindquist C, Steiner L, Giorgi C. Radiosurgery for cerebral AVMs in children and adolescents: the neurobehavioral outcome. J Neurosurg. 1997;86:207–10.
51. Murray AL, Dally M, Jeffreys A, Hwang P, Anderson JF. Neuropsychological outcomes of stereotactic radiotherapy for cerebral arteriovenous malformations. J Clin Neurosci. 2014;21:601–6.
52. Duffau H, Lopes M, Janosevic V, Sichez JP, Faillot T, Capelle L, et al. Early rebleeding from intracranial

dural arteriovenous fistulas: report of 20 cases and review of the literature. J Neurosurg. 1999;90:78–84.
53. Borden JA, Wu JK, Shucart WA. A proposed classification for spinal and cranial dural arteriovenous fistulous malformations and implications for treatment. J Neurosurg. 1995;82:166–79. (Erratum in J Neurosurg 82:705–706, 1995).
54. Baek HG, Park SH, Park KS, Kang DH, Hwang JH, Hwang SK. Stereotactic radiosurgery for dural arteriovenous fistulas involving the transverse-sigmoid sinus: a single center experience and review of the literatures. J Korean Neurosurg Soc. 2019;62: 458–66.
55. Chen CJ, Buell TJ, Diamond J, Ding D, Kumar JS, Taylor DG, Lee CC, Sheehan JP. Stereotactic radiosurgery for high-grade intracranial dural arteriovenous fistulas. World Neurosurg. 2018;116:e640–8.
56. Seo Y, Kim DG, Dho YS, Kim JW, Kim YH, Park CK, Chung HT, Paek SH. A retrospective analysis of the outcomes of dural arteriovenous fistulas treated with gamma knife radiosurgery: a single-institution experience. Stereotact Funct Neurosurg. 2018;96:46–53.
57. Tonetti DA, Gross BA, Jankowitz BT, Atcheson KM, Kano H, Monaco EA, Niranjan A, Lunsford LD. Stereotactic radiosurgery for dural arteriovenous fistulas without cortical venous reflux. World Neurosurg. 2017;107:371–5.
58. Lee CC, Chen CJ, Chen SC, Yang HC, Lin CJ, Wu CC, Chung WY, Guo WY, Hung-Chi Pan D, Shiau CY, Wu HM. Gamma Knife surgery for clival epidural-osseous dural arteriovenous fistulas. J Neurosurg. 2018;128:1364–71.
59. Dmytriw AA, Schwartz ML, Cusimano MD, Mendes Pereira V, Krings T, Tymianski M, Radovanovic I, Agid R. Gamma knife radiosurgery for the treatment of intracranial dural arteriovenous fistulas. Interv Neuroradiol. 2017;23:211–20.
60. Park KS, Kang DH, Park SH, Kim YS. The efficacy of gamma knife radiosurgery alone as a primary treatment for intracranial dural arteriovenous fistulas. Acta Neurochir. 2016;158:821–8.
61. Söderman M, Dodoo E, Karlsson B. Dural arteriovenous fistulas and the role of gamma knife stereotactic radiosurgery: the Stockholm experience. Prog Neurol Surg. 2013;27:205–17.
62. Hanakita S, Koga T, Shin M, Shojima M, Igaki H, Saito N. Role of Gamma Knife surgery in the treatment of intracranial dural arteriovenous fistulas. J Neurosurg. 2012;117(Suppl):158–63.
63. Yang HC, Kano H, Kondziolka D, Niranjan A, Flickinger JC, Horowitz MB, et al. Stereotactic radiosurgery with or without embolization for intracranial dural arteriovenous fistulas. Neurosurgery. 2010;67:1276–85.
64. Cifarelli CP, Kaptain G, Yen CP, Schlesinger D, Sheehan JP. Gamma knife radiosurgery for dural arteriovenous fistulas. Neurosurgery. 2010;67:1230–5.
65. Kida Y. Radiosurgery for dural arteriovenous fistula. Prog Neurol Surg. 2009;22:38–44.
66. Söderman M, Edner G, Ericson K, Karlsson B, Rähn T, Ulfarsson E, et al. Gamma knife surgery for dural arteriovenous shunts: 25 years of experience. J Neurosurg. 2006;104:867–75.
67. Koebbe CJ, Singhal D, Sheehan J, Flickinger JC, Horowitz M, Kondziolka D, et al. Radiosurgery for dural arteriovenous fistulas. Surg Neurol. 2005;64:392–9.
68. O'Leary S, Hodgson TJ, Coley SC, Kemeny AA, Radatz MW. Intracranial dural arteriovenous malformations: results of stereotactic radiosurgery in 17 patients. Clin Oncol (R Coll Radiol). 2002;14:97–102.
69. Pan DH, Chung WY, Guo WY, Wu HM, Liu KD, Shiau CY, et al. Stereotactic radiosurgery for the treatment of dural arteriovenous fistulas involving the transverse-sigmoid sinus. J Neurosurg. 2002;96:823–9.
70. Friedman JA, Pollock BE, Nichols DA, Gorman DA, Foote RL, Stafford SL. Results of combined stereotactic radiosurgery and transarterial embolization for dural arteriovenous fistulas of the transverse and sigmoid sinuses. J Neurosurg. 2001;94:886–91.
71. Pollock BE, Nichols DA, Garrity JA, Gorman DA, Stafford SL. Stereotactic radiosurgery and particulate embolization for cavernous sinus dural arteriovenous fistulae. Neurosurgery. 1999;45:459–67.
72. Wu CA, Yang HC, Hu YS, Wu HM, Lin CJ, Luo CB, Guo WY, Lee CC, Liu KD, Chung WY. Venous outflow restriction as a predictor of cavernous sinus dural arteriovenous fistula obliteration after gamma knife surgery. J Neurosurg. 2019;25:1–8. [Epub ahead of print].

46 Cavernous Sinus Hemangioma

Enmin Wang

46.1 Introduction

Cavernous sinus hemangiomas are an extremely rare benign tumor that have been reported to account for less than 3% of all benign tumors occurring in the cavernous sinus and 2% of all tumors, both benign and malignant [1–3]. Cavernous sinus hemangiomas may present with a variety of neurological features, for example, headache, seizure, hemorrhage, or neurological deficits. The incidence is higher in females in the fifth and sixth decade of their lives. Cavernous sinus hemangiomas can be diagnosed by their characteristic imaging appearance. The optimum treatment strategy is still controversial. Current treatment modalities for symptomatic cavernous sinus hemangiomas include microsurgical resection, embolization, fractionated radiation therapy, and stereotactic radiosurgery. Complete resection of cavernous sinus hemangioma is potentially curative but may be complicated by severe intraoperative hemorrhage and the complicated neurovascular structures. The reported incidences of postoperative complications have varied from 8 to 80%, whereas complete excision can be accomplished in only 30–60% of cases [4–15]. Thus, treatment of these lesions remains a therapeutic challenge. Because of the high possibility of profuse bleeding during surgical intervention combined with potential for long-term cranial nerve deficits, some practitioners have focused on radiotherapy and stereotactic radiosurgery for treatment of these lesions. Patients with cavernous sinus hemangiomas often respond well to treatment involving ionizing radiation [16–19]. In 1999, Iwai et al. reported the first cavernous sinus hemangioma case treated with Gamma Knife radiosurgery. The tumor had previously been partially removed, but removal was complicated by severe bleeding. Radiosurgery was performed as adjuvant therapy and dramatically decreased tumor size [20]. Since then, more reports have been published, all showing good response. In most reports, radiosurgery is used as adjuvant therapy after partial tumor removal. Radiosurgery has also been used as the primary treatment modality with good results for patients who had small- to medium-sized CSHs [21–32].

However, the lesion size and location make single-fraction radiosurgery for large CS hemangiomas very challenging. Although the targeting accuracy and dose fall-off of radiosurgery are excellent, single-fraction radiosurgery may not be ideal for large or giant tumors adjacent to optic pathways [33, 34]. As the target lesion size increases, so does the area of normal brain and other critical anatomic structures, such as optical nerves, that is irradiated, thereby increasing the risk of radiation complications [35].

E. Wang (✉)
Department of Neurosurgery, Huashan Hospital, Fudan University, Shanghai, China
e-mail: wangem@fudan.edu.cn

© Springer Nature Switzerland AG 2020
A. Conti et al. (eds.), *CyberKnife NeuroRadiosurgery*,
https://doi.org/10.1007/978-3-030-50668-1_46

Recently, multisession CyberKnife radiosurgery has been shown to be effective, even in control of large tumors, with excellent functional preservation [36–38]. The CyberKnife system (Accuray, Sunnyvale, CA) allows convenient dose fractionation with a frameless delivery that can comfortably treat patients on consecutive days in the outpatient setting without reacquisition of imaging or reapplication of rigid head frame fixation. To achieve favorable local control while maintaining low optic pathway toxicity for large CSHs, we started protocol-based multisession stereotactic radiosurgery with the CyberKnife system for large CSHs in 2007 [36]. This article addresses the safety and efficacy of Gamma Knife radiosurgery and multi-session CyberKnife radiosurgery for CSHs.

46.2 Presentations

Cavernous sinus hemangiomas have an overwhelming predilection for women. In our center, of the 186 cases, 42 (22.5%) have occurred in men, and 144 (77.5%) have occurred in women. Mean age at diagnosis was 50 years (range, 22–80 years). Clinical signs and symptoms and clinical tumor course cannot distinguish cavernous sinus hemangiomas from other cavernous sinus neoplasms. The most common symptoms in these patients were cranial neuropathy including diplopia, blurred vision, facial numbness, and abducens palsy. The second most common symptoms were headache and dizziness. Less frequent signs or symptoms include endocrinopathy, exophthalmos, trigeminal neuralgia, and nausea, vomiting, and gait instability when the tumor was giant. About 5% of cases were an asymptomatic incidental MRI examination finding. Duration of symptoms was from 1 week to 10 years.

46.3 Radiographic Appearance

Cavernous sinus hemangiomas characteristically are not calcified and demonstrate bony erosion or remodeling rather than hyperostosis. The MR images obtained in these patients showed well demarcated, low- to iso-signal mass on T1-weighted images, extremely high signal on T2-weighted images (as bright as cerebrospinal fluid signal), high signal on FLAIR-weighted (fluid attenuated inversion recovery) images, and strong homogeneous or heterogeneous enhancement after Gd-DTPA injection. When the CSHs were small to medium-sized tumors, they usually showed sharply delineated and intensely enhanced sellar masses without "dural tail sign." But when the CSHs were large or giant in size, some tumors showed heterogeneous enhancement and delayed homogeneous enhancement [40–44]. We found the CSHs have no metabolism on 18F- FDG PET CT. That means the SUV value was lower (very low) compared with normal brain tissue. The radiological characteristics of a representative case of CSH are shown in Fig. 46.1.

For patients with atypical manifestations, digital subtraction angiography (DSA) was performed to differentiate CSH from meningioma, neurilemmoma, and other cavernous sinus tumors. In patients with CSH, there is no tumor staining on the cerebral angiography with internal carotid artery DSA. But there is little flecked tumor enhancement on external carotid artery DSA. No intracavernous carotid artery stenosis was ever seen despite the complete involvement of the cavernous sinus in these patients (Fig. 46.2).

46.4 Therapeutic Option

Treatment options include the wait-and-see approach with serial images, microsurgery, single-fraction SRS and multisession SRS. If patients have asymptomatic small-sized tumors, a wait-and-see approach with serial images may be a reasonable treatment option, because CSHs are generally slow-growing tumors. Microsurgery was a common treatment before the advent of SRS. Complete resection is an ideal treatment, but it is not so easily achievable without any complications. Although recently skull base surgery has contributed to an increased rate of complete resection, there is no doubt that this is one of the most invasive treatments. SRS is a less invasive

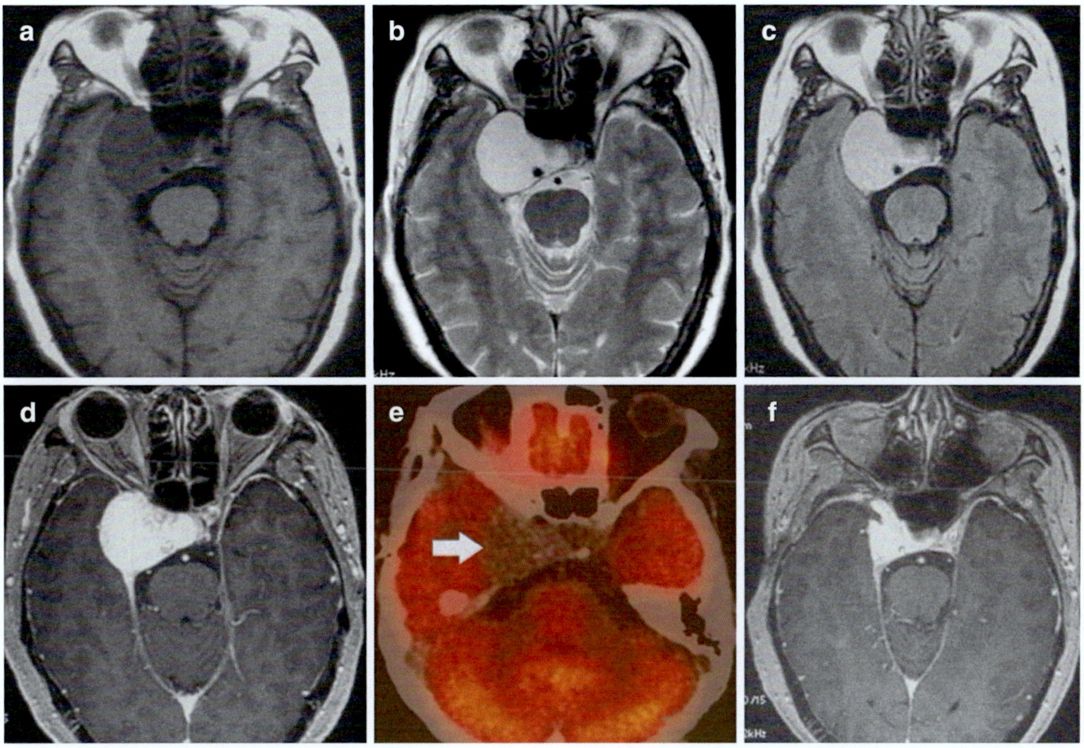

Fig. 46.1 MR images of a representative case showing radiological characteristics of CSH. A 50-year-old male patient suffered from slight headache and went for an MR scan, which showed a sharply delineated iso- or hypo-signal mass lesion on T1-weighted images (**a**), extremely high signal on T2-weighted images (**b**) and FLAIR sequence images (**c**), and intensely enhanced sellar mass (**d**). He had an operation for the cavernous sinus hemangioma and CSH was confirmed by histopathology. 18-F FDG PET-CT showed that CSH had no metabolism with lower SUV (**e**). During the operation only a small piece of tumor was removed because of bleeding. Then the patient was treated with CyberKnife radiosurgery. The CSH reduced in volume at 6 months post CyberKnife (**f**)

treatment option for small- to medium-sized CSHs, and multisession CyberKnife radiosurgery is a treatment option for large or giant CSHs, with good tumor control as well as improving neurological deficits.

46.5 Clinical Outcomes

Contemporary series of Gamma Knife radiosurgery and multisession CyberKnife radiosurgery for CSHs are shown in Table 46.1.

Our colleague Dr. Wang published a systematic review and meta-analysis of 59 cases of CSHs treated with Gamma Knife radiosurgery (GKR) [46]. The results suggest that GKR not only achieves good tumor control and symptom improvement but also avoids the complications associated with embolization, biopsy, and attempted microsurgical resection. However, most of the patients underwent GKR as an adjuvant treatment after open surgery, and the number of cases using GKR as a primary treatment was limited.

At the author's institution, 53 patients harboring CSHs were treated using Leksell Gamma Knife radiosurgery. Of the 53 patients, 15 with definitive histopathologic diagnoses after surgery, 38 were diagnosed based on their MR imaging findings. There were 15 male and 38 female patients with a mean age of 52 (range 25–76) years old. The mean volume of the tumors was 13.2 ± 8.2 cm^3 (range 1–41 cm^3). A mean marginal dose of 13.3 Gy (range 8–15 Gy) was directed to the 49–64% isodose line (mean 53%). The mean radiological and clinical follow-up

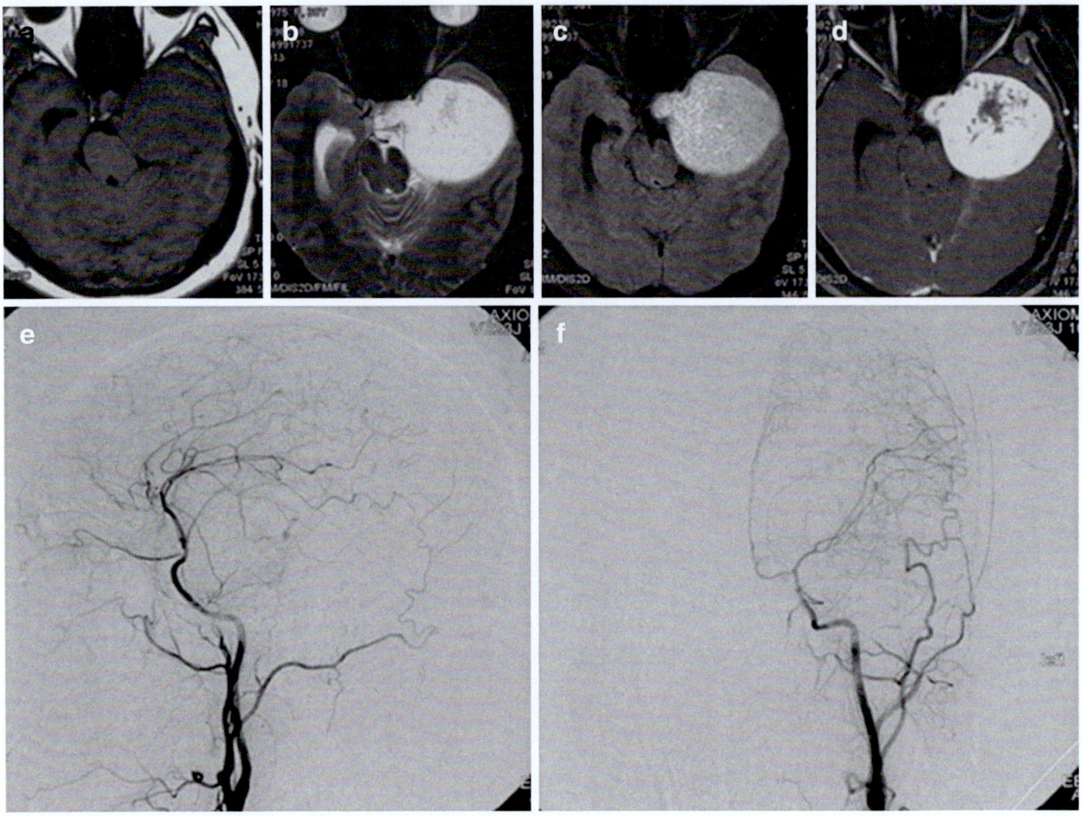

Fig. 46.2 A 32-year-old female had headache and vomiting and MRI examination was performed. The MRI showed low- to iso-signal giant mass on T1-weighted images (**a**), extremely high signal on T2-weighted images (as bright as cerebrospinal fluid signal) (**b**), high signal on FLAIR-weighted (fluid attenuated inversion recovery) images (**c**) and strong homogeneous or heterogeneous enhancement after Gd-DTPA injection (**d**). The DSA demonstrates elevation of the left middle cerebral artery and medial deviation of the left internal carotid artery (**e**). No tumor staining was seen on artery phase angiography although the CSH is a hypervascular lesion by nature (**f**). The patient had an operation and CSH was confirmed by histopathology

Table 46.1 Stereotactic radiosurgery series of cavernous sinus hemangioma

Authors	No. of patients	Methods	Tumor volume [Range] (cm^3)	Marginal dose [Range] (Gy)	Follow-up (months)	Tumor control rate (%)	Improvement of clinical symptoms/ signs (%)	New or worsening symptoms (%), no.
Huang [37]	*12*	CK	24.3 [11–96]	19.5–30	16.3	100	100	No
Wang [38]	31	CK	64.4 [40–145.3]	18–22	30	100	100	1 (3%)
Wang [48]	32	GK	30.5 [2.5–78.6]	14.2 [11–16]	30.2	97	97	No
Lee et al [32]	31	GK	9.3 [1.5–42.1]	12.6 [12–19]	54	100	97	1 (3%)
Xu [45]	7	GK	12.5 [5.3–33.2]	14 [10–15]	20	100	100	No
Park [39]	13	GK	2.7 [1.3–8.0]	12.9 [12–14]	90	100	100	No
Tang [46]	*53*	GK	13.2 [1–41]	13.3 [8–15]	34	100	94	3 (5.6%)

Table 46.1 (continued)

Authors	No. of patients	Methods	Tumor volume [Range] (cm^3)	Marginal dose [Range] (Gy)	Follow-up (months)	Tumor control rate (%)	Improvement of clinical symptoms/ signs (%)	New or worsening symptoms (%), no.
Anqi [47]	15	GK	29.3 [8.5–138]	13.4 [10–16]	13	100	100	No
Song [31]	19	GK	6.1	14.5 [11.5–16]	37	100	100	No
Li [30]	16	GK	30.4 [1.5–61.5]	13.3 [11–14]	21.5	100	100	No
Chou [28]	7	GK	7.6 [2.9–23.1]	12.5	22	100	100	No
Yamamoto [29]	30	GK	11.5 [1.5–51.4]	13.8	53	100	97	1 (3%)

time of this study was 24 (range 2–67 months) and 34 months (range 2–73 months), respectively. The tumor control rate was 100%. The mean tumor volume reduction was 79.5% (range 16.5–100%) compared with the pre-GKR volume. A typical case of CSH is shown in Fig. 46.3. Neurologically, only two of these patients showed clinical deterioration, and the other 51 patients demonstrated an obvious improvement in symptoms.

Similarly, Lee and colleagues reported the results of GKR in 31 patients with CSHs. The median radiosurgery target volume was 9.3 cm^3 and the median marginal dose was 12.6 Gy. At a mean follow-up period of 54 months (range 6–200 months), the tumor control rate was 100%. The average tumor reduction at 12 months was 64%; at 24 months, 73%; at 36 months, 79%; at 48 months, 82%; and at 60 months, 84%. No recurrence was found.

46.6 Multi-session CyberKnife Radiosurgery

We first reported the effectiveness and safety of multisession CyberKnife radiosurgery for controlling large CSHs. A Phase II study to substantiate the role of CyberKnife radiosurgery was conducted in our department in which we evaluated patients with a large CSH that had a clear "geographic" separation between the tumor and the optic apparatus on MRI. We found that a multi-fraction SRS dose of 21 Gy delivered in three fractions was effective in reducing the tumor volume without causing any new neurological deficits [36].

46.7 Rationale for Multisession CyberKnife Radiosurgery

Our dose not only was based on our and others' previous experience with SRS and radiotherapy, but also previous studies concerning the dose tolerance of the optic apparatus and brainstem. Although the exact α/β ratio that corresponds to CSHs has not been established, we assumed a value of 3 Gy and used the linear-quadratic model to estimate the biologically equivalent dose to a single-dose SRS of 10–13 Gy (a dose that yields remarkable tumor shrinkage for CSHs after radiosurgery) for a 3- or 4-fraction course of CyberKnife SRS (listed in Table 46.2). The dose prescription was based on intent to cover the entire tumor with a higher dose while ensuring dose limitation to the visual pathways and brainstem. General guidelines for dose limitations to normal structures included the following. The maximum permissible point dose to the optic nerves and chiasm was 14.1 Gy in 3 fractions (4.7 Gy per fraction) or 15.6 Gy in 4 fractions (3.9 Gy per fraction), similar to a single-fraction SRS of 9 Gy. The maximum permissible point dose to the brainstem was 21 Gy in 3 fractions (7 Gy per fraction) or 23.6 Gy in 4 fractions (5.9 Gy per fraction), similar to a single-session SRS dose of 13 Gy.

Thirty-one patients harboring giant CSHs were treated with multisession CyberKnife radio-

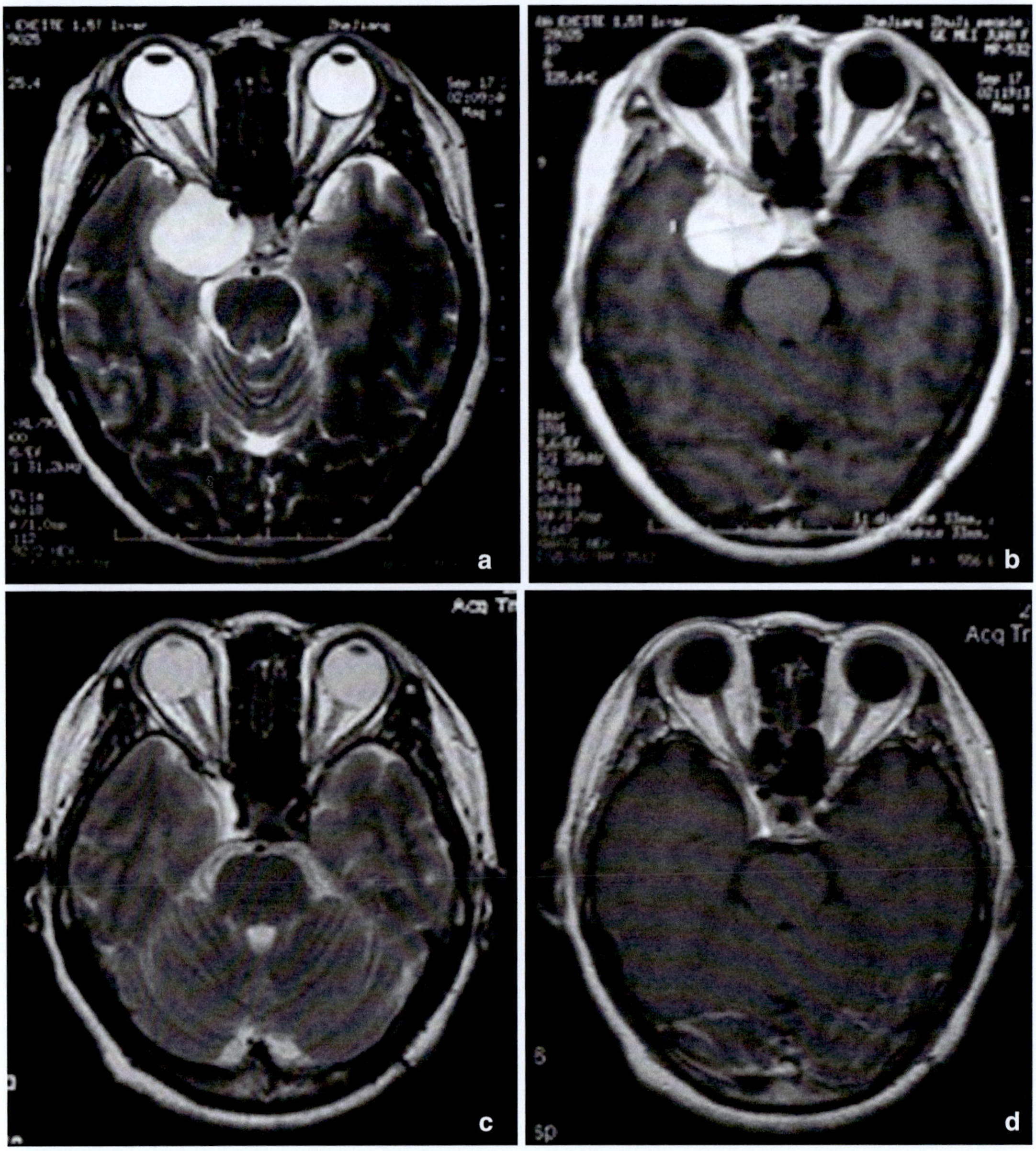

Fig. 46.3 A 47-year-old female patient underwent GKR primarily (maximal dose of 23.3 Gy and a margin dose of 14 Gy) (**a**, **b**). Her diplopia and blurred vision disappeared 1 month after GKR and the tumor exhibited evidence of shrinkage 45 months after treatment (**c**, **d**)

surgery at the author's institution, among whom 27 underwent multisession CyberKnife radiosurgery as the initial treatment. Four patients had a histological diagnosis at the time of a prior resection. In those four patients, three underwent a biopsy and one underwent partial resection. There were 8 male and 23 female patients with a mean age of 54 (range 22–80) years old. The median CSH volume was 64.4 cm^3 (range 40.9–145.3 cm^3). Three or four sessions of CyberKnife radiosurgery were used with a prescription dose based on the intent to cover the entire tumor with a higher dose while ensuring dose limitation to the visual pathways and brainstem. The median marginal dose to the tumor was 21 Gy (range 19.5–21 Gy) in 3 fractions for 11 patients and 22 Gy (range 18–22 Gy) in 4 fractions for 20 patients. During a median follow-up of 30 months (range 6–78 months), all patients achieved good tumor control. Follow-up MRI scans revealed a median tumor volume reduction of 88.1% (62.3–99.4%) at last examination compared with the pretreatment volume. Ten patients developed new or aggravated temporary headache, and 5 experienced vomiting during the treatment. These acute symptoms were relieved completely after steroid administration. Among the 30 patients with symptoms observed before treatment, 19 achieved complete symptomatic remission, and 11 had partial remission. One patient reported seizures, which were controlled after antiepileptic drug administration. No radiation-induced neurological deficits or delayed complications were reported during the follow-up period [38]. A typical case of giant CSH is shown in Fig. 46.4.

Table 46.2 Summary of dosing schedules, biologically equivalent doses, and single-dose equivalents

Total dose (Gy)	No. of fractions	Biologically equivalent dose (α/β = 3)	Single-dose equivalent (Gy)	No. of patients (%)
Prescribed dose				
18	4	45.00	10.22	5
20	4	53.33	11.23	2
20.8	4	56.9	11.65	2
22	4	62.33	12.25	11
19.5	3	61.75	12.20	1
20.4	3	66.60	12.71	1
21	3	70	13	9

Huang and colleagues documented 12 patients with high volume cavernous sinus hemangio-

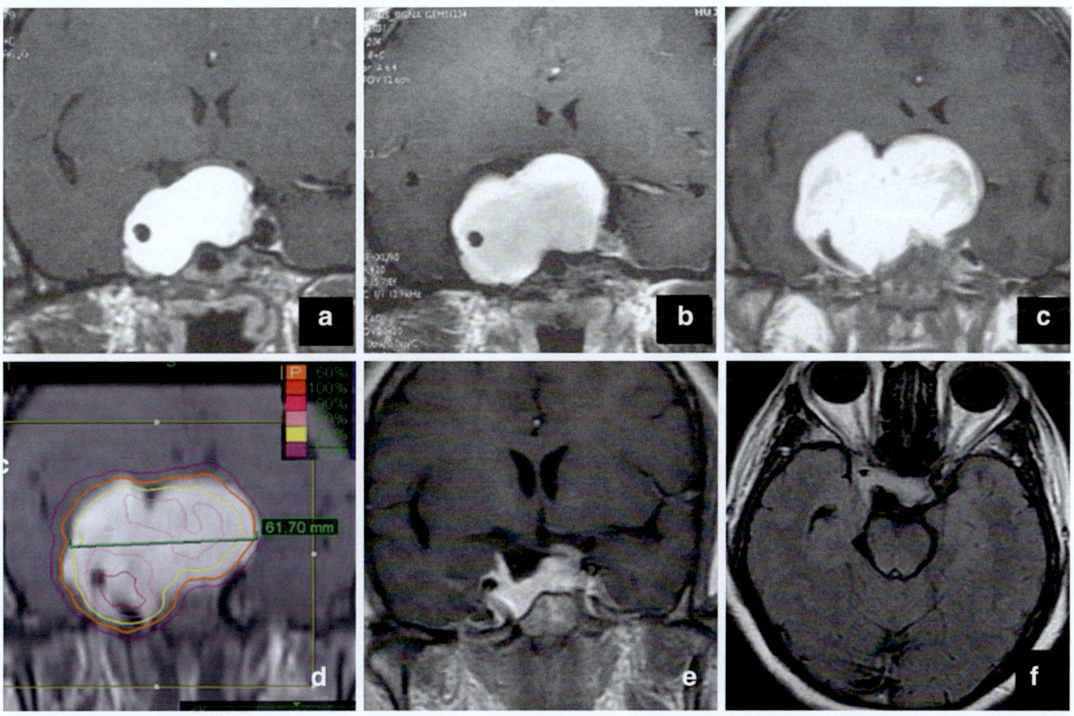

Fig. 46.4 A 43-year old woman had a visual field defect, blurred vision, and ipsilateral abducens nerve palsy. MRI found a cavernous sinus hemangioma on the right side in 1998 (**a**). The patient underwent resection and experienced severe intraoperative bleeding. Only a piece of tumor was removed during the operation. The tumor progressed slowly (**b**) and in 2011 the tumor became a giant tumor (**c**). The pretreatment tumor volume was 76.5 cm^3 (**d**). Three and half years post multisession CyberKnife radiosurgery (21 Gy in 3 fractions), the tumor shrank to 2 cm^3 (**e**, **f**). Clinical symptoms also improved

mas treated with hypofractionated stereotactic CyberKnife [37]. Initial tumor volumes ranged from 11.8 to 96.6 cm^3 with a median of 24.3 cm^3. Irradiation doses were 19.5 Gy with 3 fractions in 2 patients, 21 Gy with 3 fractions in 8 patients, 25 Gy with 5 fractions in 1 patient, and 30 Gy with 3 fractions in 1 patient. Follow-up ranged from 3 to 54 months, with a mean follow-up of 16.3 months. All tumor volumes decreased (28.6–94.1%) and symptoms improved (including blurred vision, visual field defects, diplopia, headaches, and facial numbness) after therapy. A patient experienced radiotherapy-related cerebral edema, which resolved after 5 days of mannitol and dexamethasone.

46.8 Summary

Gamma Knife radiosurgery is a safe and effective treatment for patients with small- to medium-sized cavernous sinus hemangiomas. When the tumor is relatively large (e.g., compressing the optic nerves or chiasm), multisession CyberKnife radiosurgery would be an alternative treatment option. Multisession CyberKnife radiosurgery delivery of 18–22 Gy in 3–4 fractions is effective in reducing tumor volume and improving neurological symptoms without causing any new complications. Until now microsurgery has been a common treatment option for cavernous sinus tumors, but, even with recent refinement of microsurgical techniques, it is unable to avoid the risk of complications such as cranial nerve injury and avoid partial resection of the tumors. Accordingly, single-fraction or multisession stereotactic radiosurgery can be a reasonable alternative to surgical resection not only as adjuvant treatment but also as the initial treatment, with respect to preservation of cranial nerve function.

References

1. Linskey ME, Sekhar LN. Cavernous sinus hemangiomas: a series, a review, and an hypothesis. Neurosurgery. 1992;30(1):101–8.
2. Gonzalez LF, Lekovic GP, Eschbacher J, Coons S, Porter RW, Spetzler RF. Are cavernous sinus hemangiomas and cavernous malformations different entities? Neurosurg Focus. 2006;21(1):e6.
3. Gliemroth J, Missler U, Sepehrnia A. Cavernous angioma as a rare neuroradiologic finding in the cavernous sinus. J Clin Neurosci. 2000;7(6):554–7.
4. Zhou LF, Mao Y, Chen L. Diagnosis and surgical treatment of cavernous sinus hemangiomas: an experience of 20 cases. Surg Neurol. 2003;60(1):31–6.
5. Shi J, Hang C, Pan Y, Liu C, Zhang Z. Cavernous hemangiomas in the cavernous sinus. Neurosurgery. 1999;45(6):1308–13; discussion 1313–1314.
6. Ohata K, El-Naggar A, Takami T, Morino M, El-Adawy Y, El-Sheik K, Inoue Y, Hakuba A. Efficacy of induced hypotension in the surgical treatment of large cavernous sinus cavernomas. J Neurosurg. 1999;90(4):702–8.
7. Bansal S, Suri A, Singh M, Kale SS, Agarwal D, Sharma MS, Mahapatra AK, Sharma BS. Cavernous sinus hemangioma: a fourteen year single institution experience. J Clin Neurosci. 2014;21(6):968–74.
8. Cure GC, Mejía JA, Roldan NG. Giant cavernous hemangioma of the cavernous sinus. Case report and review of literature. Rev Chil Neurocirugía. 2010;34:66–71.
9. Dou Y, Meng Q, Yan Z, Xu J, Che S, Jiao Y, Wu Z. Diagnosis and microsurgical treatment of cavernous sinus hemangioma. Artif Cells Blood Substit Immobil Biotechnol. 2010;38(2):109–12.
10. Fraser JF, Mass AY, Brown S, Anand VK, Schwartz TH. Transnasal endoscopic resection of a cavernous sinus hemangioma: technical note and review of the literature. Skull Base. 2008;18(5):309–15.
11. Goel A, Muzumdar D, Sharma P. Extradural approach for cavernous hemangioma of the cavernous sinus: experience with 13 cases. Neurol Med Chir (Tokyo). 2003;43(3):112–8; discussion 119.
12. Suzuki Y, Shibuya M, Baskaya MK, Takakura S, Yamamoto M, Saito K, Glazier SS, Sugita K. Extracerebral cavernous angiomas of the cavernous sinus in the middle fossa. Surg Neurol. 1996;45(2):123–32.
13. Suri A, Ahmad FU, Mahapatra AK. Extradural transcavernous approach to cavernous sinus hemangiomas. Neurosurgery. 2007;60(3):483–8; discussion 488–489.
14. Hashimoto M, Yokota A, Ohta H, Urasaki E. Intratumoral injection of plastic adhesive material for removal of cavernous sinus hemangioma. Technical note. J Neurosurg. 2000;93(6):1078–81.
15. Kim IM, Yim MB, Lee CY, Son EI, Kim DW, Kim SP, Sohn CH. Merits of intralesional fibrin glue injection in surgery for cavernous sinus cavernous hemangiomas. Technical note. J Neurosurg. 2002;97(3):718–21.
16. Jamjoom AB. Response of cavernous sinus hemangioma to radiotherapy: a case report. Neurosurg Rev. 1996;19(4):261–4.
17. Yamamoto Y, Weining Z, Ohashi T. Intracavernous cavernous hemangioma: dynamic CT findings and effectiveness of irradiation—case report. Neurol Med Chir (Tokyo). 1992;32(2):93–5.

18. Maruishi M, Shima T, Okada Y, Nishida M, Yamane K, Okita S. Cavernous sinus cavernoma treated with radiation therapy—case report. Neurol Med Chir (Tokyo). 1994;34(11):773–7.
19. Miserocchi G, Vaiani S, Migliore MM, Villani RM. Cavernous hemangioma of the cavernous sinus. Complete disappearance of the neoplasma after subtotal excision and radiation therapy. Case report. J Neurosurg Sci. 1997;41(2):203–7.
20. Iwai Y, Yamanaka K, Nakajima H, Yasui T. Stereotactic radiosurgery for cavernous sinus cavernous hemangioma: case report. Neurol Med Chir (Tokyo). 1999;39(4):288–90.
21. Seo Y, Fukuoka S, Sasaki T, Takanashi M, Hojo A, Nakamura H. Cavernous sinus hemangioma treated with gamma knife radiosurgery: usefulness of SPECT for diagnosis: case report. Neurol Med Chir (Tokyo). 2000;40(11):575–80.
22. Thompson TP, Lunsford LD, Flickinger JC. Radiosurgery for hemangiomas of the cavernous sinus and orbit: technical case report. Neurosurgery. 2000;47(3):778–83.
23. Kida Y, Kobayashi T, Mori Y. Radiosurgery of cavernous hemangiomas in the cavernous sinus. Surg Neurol. 2001;56(2):117–22.
24. Nakamura N, Shin M, Tago M, Terahara A, Kurita H, Nakagawa K, Ohtomo K. Gamma knife radiosurgery for cavernous hemangiomas in the cavernous sinus. Report of three cases. J Neurosurg. 2002;7(5):477–80.
25. Peker S, Kilic T, Sengoz M, Pamir MN. Radiosurgical treatment of cavernous sinus cavernous hemangiomas. Acta Neurochir. 2004;146:337–41.
26. Ivanov P, Chernov M, Hayashi M, Nakaya K, Izawa M, Murata N, Kubo O, Ujiie H, Muragaki Y, Nakamura R, Iseki H, Hori T, Takakura K. Low-dose gamma knife radiosurgery for cavernous sinus hemangioma: report of 3 cases and literature review. Minim Invasive Neurosurg. 2008;51(3):140–6.
27. Khan AA, Niranjan A, Kano H, Kondziolka D, Flickinger JC, Lunsford LD. Stereotactic radiosurgery for cavernous sinus or orbital hemangiomas. Neurosurgery. 2009;65(5):914–8.
28. Chou CW, Wu HM, Huang CI, Chung WY, Guo WY, Shih YH, Lee LS, Pan DH. Gamma knife surgery for cavernous hemangiomas in the cavernous sinus. Neurosurgery. 2010;67(3):611–6.
29. Yamamoto M, Kida Y, Fukuoka S, Iwai Y, Jokura H, Akabane A, Serizawa T. Gamma knife radiosurgery for hemangiomas of the cavernous sinus: a seven-institute study in Japan. J Neurosurg. 2010;112(4):772–9.
30. Li P, Ren H, Zhang S, Wang W. Clinical results of Gamma Knife surgery for cavernous sinus hemangiomas. J Neurosurg. 2012;117 Suppl:89–95.
31. Song SW, Kim DG, Chung HT, Paek SH, Han JH, Kim YH, Kim JW, Kim YH, Jung HW. Stereotactic radiosurgery for cavernous sinus hemangiomas. J Neurooncol. 2014;118(1):163–8.
32. Lee CC, Sheehan JP, Kano H, Akpinar B, Martinez-Alvarez R, Martinez-Moreno N, Guo WY, Lunsford LD, Liu KD. Gamma knife radiosurgery for hemangioma of the cavernous sinus. J Neurosurg. 2017;126(5):1498–505.
33. Liu AL, Wang C, Sun S, Wang M, Liu P. Gamma knife radiosurgery for tumors involving the cavernous sinus. Stereotact Funct Neurosurg. 2005;83(1):45–51.
34. Kuo JS, Chen JC, Yu C, Zelman V, Giannotta SL, Petrovich Z, MacPherson D, Apuzzo ML. Gamma knife radiosurgery for benign cavernous sinus tumors: quantitative analysis of treatment outcomes. Neurosurgery. 2004;54(6):1385–93; discussion 1393–1394.
35. Mayo C, Martel MK, Marks LB, Flickinger J, Nam J, Kirkpatrick J. Radiation dose-volume effects of optic nerves and chiasm. Int J Radiat Oncol Biol Phys. 2010;76(3):S28–35. J Neurooncol (2012) 107:239–245.
36. Wang X, Liu X, Mei G, Dai J, Pan L, Wang E. Phase II study to assess the efficacy of hypofractionated stereotactic radiotherapy in patients with large cavernous sinus hemangiomas. Int J Radiat Oncol Biol Phys. 2012;83(2):e223–30.
37. Huang L, Sun L, Wang W, Cui Z, Zhang Z, Li J, Wang Y, Wang J, Yu X, Ling Z, Qu B. Therapeutic effect of hypofractionated stereotactic radiotherapy using CyberKnife for high volume cavernous sinus cavernous hemangiomas. Technol Cancer Res Treat. 2019;18:1–7.
38. Wang X, Zhu H, Knisely J, Mei G, Liu X, Dai J, Mao Y, Pan L, Qin Z, Wang E. Hypofractionated stereotactic radiosurgery: a new treatment strategy for giant cavernous sinus hemangiomas. J Neurosurg. 2018;128(1):60–7.
39. Park CK, Choi SK, Kang IH, Choi MK, Park BJ, Lim YJ. Radiosurgical considerations for cavernous sinus hemangioma: long-term clinical outcomes. Acta Neurochir. 2016;158(2):313–8.
40. Jinhu Y, Jianping D, Xin L, Yuanli Z. Dynamic enhancement features of cavernous sinus cavernous hemangiomas on conventional contrast-enhanced MR imaging. AJNR Am J Neuroradiol. 2008;29(3):577–81.
41. Mendonça JL, Viana SL, Matsumine M, Silva RF, Viana MA, Freitas FM. Cavernous angioma of the cavernous sinus: imaging findings. Arq Neuropsiquiatr. 2004;62(4):1004–7.
42. Salanitri GC, Stuckey SL, Murphy M. Extracerebral cavernous hemangioma of the cavernous sinus: diagnosis with MR imaging and labeled red cell blood pool scintigraphy. AJNR Am J Neuroradiol. 2004;25(2):280–4.
43. Sohn CH, Kim SP, Kim IM, Lee JH, Lee HK. Characteristic MR imaging findings of cavernous hemangiomas in the cavernous sinus. AJNR Am J Neuroradiol. 2003;24(6):1148–51.
44. Tannouri F, Divano L, Caucheteur V, Hacourt A, Pirotte B, Salmon I, Balériaux D. Cavernous haemangioma in the cavernous sinus: case report and review of the literature. Neuroradiology. 2001;43(4):317–20.

45. Xu Q, Shen J, Feng Y, Zhan R. Gamma knife radiosurgery for the treatment of cavernous sinus hemangiomas. Oncol Lett. 2016;11(2):1545–8.
46. Tang X, Wu H, Wang B, Zhang N, Dong Y, Ding J, Dai J, Yu T, Pan L. A new classification and clinical results of gamma knife radiosurgery for cavernous sinus hemangiomas: a report of 53 cases. Acta Neurochir. 2015;157(6):961–9.
47. Anqi X, Zhang S, Jiahe X, Chao Y. Cavernous sinus cavernous hemangioma: imaging features and therapeutic effect of gamma knife radiosurgery. Clin Neurol Neurosurg. 2014;127:59–64.
48. Wang Y, Li P, Zhang XJ, Xu YY, Wang W. Gamma Knife surgery for cavernous sinus hemanginoma: a report of 32 cases. World Neurosurg. 2016;94:18–25.

Part VIII

Functional Disorders

Trigeminal Neuralgia

47

Pantaleo Romanelli and Alfredo Conti

47.1 Introduction

Trigeminal neuralgia (TN) is the most common cranio-facial pain syndrome, with an incidence of up to 5 cases per 100,000 inhabitants. Primary trigeminal neuralgia is linked to a contact between the trigeminal nerve and an arterial or venous vessel in the cisternal segment of the nerve adjacent to the inlet zone into the brainstem, namely, the root entry zone (REZ) [1–4]. Focal demyelination disturbing the regular transmission and processing of inputs through the trigeminal nerve and ephaptic transmission between afferent unmyelinated axons and partially damaged myelinated axons are involved in the pathogenesis of the sudden lancinating pain characteristic of TN [5–8]. Aside from peripheral injury or disease of the trigeminal nerve with increased afferent firing in the nerve, a failure of the central inhibitory mechanisms is also likely to play a role in the genesis of TN. Secondary neuralgias (less than 20% of trigeminal neuralgia) are in general a consequence of nerve damage induced by tumor compression (schwannomas, brain stem tumors, meningiomas) but can also be associated, more rarely, with vascular lesions. Furthermore, there is an incidence of 1–2% of trigeminal neuralgia in patients with multiple sclerosis, in which symptoms are often atypical or bilateral.

The complications associated to long-term intake of medical therapy have promoted the development and wide application of several surgical techniques to treat TN [9]. Microvascular decompression [10] aims to resolve the cause of the TN by an intracranial and direct approach to resolve the neurovascular conflict [11, 12]. Other techniques aim to interrupt the trigeminal nociceptive pathways either by percutaneous lesioning of the Gasserian ganglion or by irradiation of the nerve using stereotactic radiosurgery techniques.

47.2 Stereotactic Radiosurgery for Trigeminal Neuralgia

47.2.1 Target Area

The application of stereotactic irradiation in the treatment of trigeminal neuralgia was pioneered by Lars Leksell in 1951 [13]. In his original description, Leksell performed a "radiogangliotomy" on a patient with facial neuralgia with the goal to block pain transmission in the trigeminal ganglion. Later on, in the early 1990s, the target was moved to REZ with the goal of

P. Romanelli (✉)
Brain Radiosurgery, Cyberknife Center, CDI, Milan, Italy
e-mail: pantaleo.romanelli@cdi.it

A. Conti
Alma Mater Studiorum University of Bologna, Bologna, Italy

IRCCS ISNB Istituto delle Scienze Neurologiche di Bologna, Bologna, Italy

Department of Neurosurgery, Charité Faculty of Medicine, Berlin, Germany
e-mail: alfredo.conti2@unibo.it

© Springer Nature Switzerland AG 2020
A. Conti et al. (eds.), *CyberKnife NeuroRadiosurgery*,
https://doi.org/10.1007/978-3-030-50668-1_47

reducing complications and to increase mid- and long-term treatment efficacy [14, 15]. Assuming a greater radiosensitivity of the central myelin, this proximal portion of the nerve embracing the transition between central myelin (oligodendrocytes) and peripheral myelin (Schwann cells) represented an ideal target [10, 16]. The major limitation of using the REZ as target is the higher dose of radiations delivered to the brainstem and the consequent risk of complications.

Regis et al. suggested to target the retrogasserian portion of the nerve, also called the triangular plexus [17, 18]. Other authors have subsequently adopted this approach but, considering the great variability of the length of the cisternal portion of the trigeminal nerve, proposed to place the isocenter at a fixed distance from the nerve emergence (5–8 mm) [19, 20].

Flickinger et al. in 2001 analyzed the use of two isocenters, separated by 3–5 mm, on the retrogasserian trigeminal nerve in the hypothesis that increasing the length, and thus the volume of irradiated nerve, could improve pain control [21]. This approach, however, had no advantages in terms of pain control but increased the complications.

47.2.2 Results of Radiosurgery

The majority of data available originates from the use of Gamma Knife. Two large cohorts of patients who underwent Gamma Knife radiosurgery have been recently reported [17, 22]. The group of Marseille illustrated the results of 497 patients with primary TN. Ninety-one percent of patients were pain-free within a median time of 10 days. Pain recurred in 34.4% of cases [22]. Marshall et al. reported the data of 448 patients describing satisfactory results in terms of pain control in 86% and recurrence rate in 28% of patients [17]. Using the Novalis (Brainlab, Munich, Germany), De Salles et al. published in 2011 the results obtained in a cohort of 179 patients treated with a dose of 70–90 Gy delivered at the REZ [23, 24]. 79.3% of the patients treated had significant pain relief at a mean of 28.8 months after the treatment [24]. Actually, Tuleasca et al. [25] reported a comprehensive systematic review of SRS in the treatment of TN. According to the data reported, the freedom from pain response with or without medication was (mean and median [range]) 84.8% and 85.6% (66.6–100%) for Gamma Knife SRS and 87.3% and 88.5% (75%–100%) for LINAC SRS [25].

47.2.3 Dose Selection for Radiosurgery

As mentioned above, there are essentially two different approaches in target selection: one, according to the group from Pittsburgh, is to place the target at the nerve emergence, aiming to obtain post-irradiation hypoesthesia, likely as a result of higher doses to the brainstem, that is usually associated to higher probability of long-term pain control [23, 26, 27]. The alternative is the use of a retrogasserian target. This may reduce the risk of bothersome complications, but requires a higher maximum dose to the nerve to obtain optimal results [28–31].

With regard to clinical studies, Kondziolka and coll. published in 1996 a multi-institutional series of 50 patients irradiated with the Gamma Knife Model G and 4 mm collimator [32]. Thirty-two patients had undergone a surgical treatment before radiosurgery. The prescribed dose was 60–90 Gy, depending on the institution, on a target represented by the REZ. With a median follow-up of 18 months, 56% of patients had an excellent response and 32% a good response [19]. Median time to response was 1 month. The efficacy of treatment was better in patients irradiated between 70 and 90 Gy, compared to those treated with a lower dose. The dose of treatment was the key factor to control the pain, and the tolerance was satisfactory, so that author suggested a dose of 70 Gy for treatment [32]. This study set the bases for the practice of radiosurgery in TN, introducing important concepts of target and dose escalation.

These results were later confirmed by several other experiences using the Gamma Knife. Maesawa et al. reported that analgesic results were excellent or good in 70% of cases at 1 year and 55.8% of patients had pain under control 5 years later. With a dose escalation from 50 to

90 Gy (median 80 Gy), Sheehan et al. reported excellent results in 47% of patients 1 year after treatment, 45% at 2 years, and 34% at 3 years [33]. Régis et al. have reported the results of a prospective series of 100 patients with 88% control rate at 1 year and 58% at 2 years, using a dose of 85 Gy (70–90 Gy) and a retrogasserian target [31]. Massager et al. reported the results of an irradiation of 90 Gy in a similar target [34]. Pain control was excellent in 59% of patients and excellent or good in 71% of patients at the 42-month follow-up. On this basis, Pollock et al. began a comparison of two dose levels for patients treated at the REZ [35]. Twenty-seven patients received a dose of 70 Gy at the isocenter and 41 patients a dose of 90 Gy. At the last follow-up examination, 11 (41%) of the 27 patients with low-dose radiosurgery remained pain-free compared with 25 (61%) of the 41 patients with high-dose radiosurgery ($P = 0.17$). High-dose radiosurgery was associated with an increased rate of permanent trigeminal nerve dysfunction (54% vs. 15%, $P = 0.003$) [35]. Kim et al. reported 104 patients were with either a dose of 80 Gy or a dose of 85 Gy to the REZ. Pain control at 1 and 3 years were 75.0% and 61.2% for the 80 Gy group and 65.9% and 60.3% for the 85 Gy group, with no statistically significant difference. Only the response time was shorter for patients receiving 85 Gy [36]. According to these results, analgesic control (no pain and no medication) can be achieved in 35–87% of patients and satisfactory results, grades I–III according to the Barrow Neurological Institute (BNI) Pain Intensity Scale, in 44–100% in the first year. Importantly, results are strongly dependent on the dose delivered to the nerve. The time necessary for pain response varies, being reported quickly after treatment procedure, a few days later, and up to 6 months after the procedure, and also appears depending on the dose [18].

47.2.4 Long-Term Efficacy and Complications

In 2006, Pollock et al. analyzed 13 clinical series. The control rate, no pain and no drugs, ranged 16–61% at 2 years [37]. At 3 years, the rate of BNI pain score grade I was 60–62.5% for doses of 80–90 Gy, whereas the pain control was achieved, with I–III grade, in 48% for doses of 75 Gy [24, 38, 39]. At 5 years with the reported doses of 75–90 Gy, 20–32% of patients had still grade I; 46–58% had a BNIPS grades I–III [40–44].

Furthermore, Dhople et al. reported that 30% of patients had grades I–III 7 years after treatment, and Kondziolka et al. [41] reported 26% of patients with BNI pain grades I–III at 10 years. It has also been suggested that pain relapses are related to the radiation dose. Thus, 13–26% relapse has been reported at 18 months and 52.2–55% at 5 years for doses of less than 80 Gy. For doses >80 Gy, it was found a relapse rate of 18% at 14 months, 23% at 18 months, and 29% at 24 months with 26% patients with pain under control at 10 years in series targeting the REZ [41, 45, 46]. In series in which a retrogasserian target was used, with doses of 90 Gy, the relapse rates were 17% at 12 months and 34% between the 1-and 72-month follow-up (median 10 months) [30, 31].

Beyond the dose delivered to the isocenter, other factors have influenced results. The literature review found unfavorable prognostic factors: a diagnosis of multiple sclerosis, atypical neuralgia, diabetes, and history of previous surgery on the same side (pain control rate decreases from 81 to 61% at 1 year, from 53 to 35% at 2 years, and 41 to 15% in 5 years depending on whether or not there is a history of surgical procedure) [17, 20, 31, 38, 41, 43, 46–49].

Hypoesthesia or facial numbness is frequently observed after high-dose trigeminal irradiation, and this has even been considered an efficacy endpoint of the procedure [37]. The objective is, however, to avoid painful anesthesia, which can cause a discomfort that is even more bothersome than neuralgia.

In the literature, the occurrence of sensitive trigeminal disturbances in patients treated at the REZ has rates ranging between 6 and 54%, with a rate of 0–17% of bothersome numbness [35, 41, 44]. The prescribed dose is correlated with the incidence of sensitive trigeminal disturbances [28, 35], but there are other parameters impacting tolerance and rate of hypoesthesia. One main factor is the dose delivered to the brainstem.

With regard to the adopted dose limits, in the series of Massager et al. [34], the dose of 90 Gy was delivered to a target lying at a distance of 5–8 mm from the brainstem, and treatment was planned aiming to keep a dose constraint on the brainstem of 13–15 Gy to a volume of 1 mm^3 and 10–12 Gy to 10 mm^3 [34]. In series in which the target was the REZ, the reported dose to the brainstem varies based on prescribed doses. It is possible to summarize it as follows: 21–28 Gy for 70 Gy delivered to the target, 24–40 Gy for 80 Gy, and 18–57.6 Gy for 90 Gy [17, 20, 24, 35, 41, 43, 45, 49, 50]. The incidence of numbness increased with the dose to the brainstem. At 90 Gy, with the 30% isodose line lying on the brainstem, the rate of grade IV–V numbness was 10% and 17% when the brainstem was embraced by the 50% isodose line [24, 51].

Another parameter to be considered is the length of the nerve treated. Flickinger et al. conducted a prospective study on 87 patients, using a dose of 75 Gy with one or two isocenters. In the group of patients who were treated with a single isocenter, the length of nerve included in the 50% prescription isodose was 5.4 ± 0.4 mm, whereas it was 8.7 ± 1.1 mm in the group treated with two isocenters. At 2 years, a higher incidence of facial numbness and moderate or severe paresthesia have been observed in the group treated with two isocenters [51].

It has also been suggested that irradiating a long portion of the nerve using a non-isocentric technique may cause a higher rate of sensitive disturbances without increasing functional results [52, 53]. Further side effects, different from a simple hypoesthesia, have been described. Eye drought was described in 1–8.3% of cases [24, 42, 54]. Anesthesia dolorosa has been rarely reported, occurring for high doses in patients who had undergone multiple treatments or re-irradiation [35, 44, 53]. Cases of lockjaw and weakness of the mandible, diplopia, and hearing loss have rarely been reported [53].

47.2.5 Re-irradiation

As previously described, a significant part of patients may suffer a relapse after few months to several years after radiosurgery. This has raised the question of a second irradiation. In 2004, Herman et al. described 18 patients retreated using a dose of 70 Gy to the REZ; the cumulative dose was 145 Gy [51]. The 50% isodose covered the brainstem. Fifty percent of patients had excellent results and 11% presented new facial numbness [51]. Authors conclude that repeated SRS is more efficacious for those patients who experienced longer periods of pain relief after the initial SRS and that the incidence of complications is not significantly different from that observed for initial treatment.

The study by Dvorak et al. provided results of 28 patients, retreated with 45 Gy after a first treatment of 80 Gy, with a dose of 20 Gy to the brainstem. Results were excellent in 29%, good in 32%, and poor in 39%. Authors described 26% incidence of moderate trigeminal dysfunction including 11% of trigeminal numbness, 11% paresthesia, 4% dysesthesia, and 4% weaknesses of the mandible [55].

Huang et al. treated 28 patients with a dose of 40–76 Gy at the REZ and a cumulative dose of 110–152 Gy [56]. The recorded pain relief in 68% of patients and 36% of new sensory deficit were described as moderate [56]. The incidence of facial numbness was associated to a cumulative dose of >115 Gy.

Gellner et al. re-irradiated 22 patients using a median dose of 74.3 Gy and provided a longer follow-up. Their results showed a long-term persistence of the therapeutic response (more than 5 years). However, their treatment (>140 Gy cumulative dose) resulted in a rate of sensory deficit of 73.7%, described however as not bothersome.

Pollock et al. delivered a cumulative dose of 163.1 Gy in 19 patients. They observed an excellent response in 71% and 61% of patients at 1 and 2 years after radiosurgery, respectively [57]. Nonetheless, 11 patients (58%) had sensory deficits including facial paresthesia ($n = 3$), numbness ($n = 5$), or dysesthesias ($n = 3$). Two patients (11%) developed corneal numbness. Authors suggested to reduce the cumulative dose to <150 Gy [57].

Finally, Aubuchon et al. have published a series of 37 patients for persistent pain or recurrence of painful symptoms. The mean retreatment dose was 84.4 Gy (range, 60–90), and the cumulative dose was 171.5 Gy delivered at an anterior target. The cumulative doses to the brainstem surface and the REZ were 105.9 Gy and 80.6 Gy, respectively.

Of the 37 patients, 81% achieved a >50% pain relief response to repeat treatment, and 57% experienced some form of trigeminal dysfunction. Two patients (5%) experienced clinically significant toxicity: one with bothersome numbness and one with corneal dryness requiring tarsorrhaphy. A REZ dose >26.6 Gy at the retreatment predicted for treatment success (61% vs. 32%, $p = 0.0716$). A cumulative dorsal root entry zone dose of >84.3 Gy (72% vs. 44%, $p = 0.091$) and a cumulative pons surface dose of >108.5 Gy (78% vs. 44%, $p = 0.018$) predicted for post-treatment numbness. The presence of any new sensory deficit predicted the >50% decrease in pain intensity (100% vs. 60%, $p = 0.0015$). In the CyberKnife series by Romanelli et al. [58], all but one patient, who developed a BNI numbness scale grade III (somewhat bothersome) hypoesthesia, reported such complication exclusively following re-irradiation. This corresponds to 4.3% of the entire cohort of 138 patients and 18.2% of the 33 patients who received a second treatment.

In essence, re-irradiation appears to be a safe treatment option for recurrent trigeminal neuralgia, but the global dose delivered needs to be considered carefully. Results are satisfactory in the majority of patients, but there is an increased risk of sensory complications and nerve dysfunction. Treating with Gamma Knife, the target could be placed anteriorly with respect to the first one so that the radiosurgical volume at the second procedure overlaps only partially (by 50% approximately) with the first one. It can also be advisable to use lower radiation doses (50–60 Gy) for the second procedure to keep the cumulative dose to less than 145 Gy, controlling the doses to the brainstem as much as possible.

47.3 Frameless Radiosurgery for TN

The use of frameless non-isocentric stereotactic radiosurgery for the treatment of TN was introduced at Stanford by John Adler, the inventor of CyberKnife, and first reported by his fellow, Pantaleo Romanelli, in a study that was the first clinical demonstration of the accuracy and safety of frameless image-guided TN radiosurgery [14]. Almost immediate pain relief (within days) was found in this first cohort of patients following the delivery of a prescribed dose ranging from 65 to 70 Gy to a nerve segment up to 11 mm. The irradiation of such a long nerve segment, however, caused a high rate of bothersome numbness and prompted a reduction of dose and length of the nerve treated, leading to the treatment protocol reported here [52, 59, 60]. During early CyberKnife treatments for TN, a CT cisternography was a version of the system. CT cisternography has been abandoned after the preliminary Stanford experience because later versions of MultiPlan-TPS provided a rather accurate CT-MR fusion capability. Also, bony landmarks, indicating the entrance of the trigeminal nerve root into Meckel's cave, are easily recognizable directly on a bone CT scan. The identification of these points greatly supports a precise co-registration of the CT with the MR sequence. Further experience was, therefore, based on MR-CT fusion targeting [53, 60–65].

In 2008, Villavicencio et al. published data from a multicenter study illustrating the results of 95 patients who underwent CyberKnife radiosurgery [53]. This heterogeneous study included patients treated with widely different modalities (isocentric and non-isocentric), as well as variable doses and treatment volumes. The median dose used was 75 Gy. Certain variables were predictive of stable pain relief over pain recurrence, including the median maximum dose (77.5 vs. 65 Gy), median minimum dose (64 vs. 52 Gy), and median nerve length treated (4 mm vs. 6 mm). After 2 years, 50% of the population had excellent results, but 47% suffered new facial numbness. An update from the Stanford series reported on 46 patients receiving a treatment delivered over a 6-mm segment of the nerve, with a mean marginal prescription dose of 58.3 Gy and a mean maximal dose of 73.5 Gy [59]. Symptoms disappeared completely in 39 patients (85%). After a mean follow-up period of 14.7 months, patient-reported outcomes were excellent in 33 (72%), good in 11 (24%), and poor in 2 patients (4%). Ipsilateral bothersome facial numbness (grade III on the BNI numbness scale) was found in seven patients (15%). Further studies focusing on the general treatment of TN1, TN2, and multiple sclerosis (MS)-related TN provided similar outcomes [63–67].

Our group has recently published a prospective study reporting the long-term outcomes of a cohort of 138 TN patients treated with frameless image-guided radiosurgery [58]. Median follow-up was 52.4 months; median target length 5.7 mm; median target volume 40 mm^3; and median prescription dose 60 Gy (to the 80% isodose line). The actuarial pain control rate scored using the BNI classification at 6 months, 12 months, 24 months, and 36 months was, respectively, 93.5%, 85.8%, 79.7%, and 76%. A second treatment, due to primary treatment failure or recurrent pain, was offered to 33 patients (24%). Sensory complications have also been assessed using BNI scores. Overall, the rate of sensory disturbances was 18.1%. BNI grade II (not bothersome) hypoesthesia was reported by 18 patients (13%), with 10 patients receiving 2 treatments and 1 patient having 3 treatments. Six patients (4.3%) developed BNI grade III (somewhat bothersome) hypoesthesia, all after retreatment. One patient (0.7%) developed BNI grade IV dysfunction. The average delay for the appearance of sensory complications was 16.4 months after irradiation. Nerve length (<6 mm vs. >6 mm), smaller nerve volume (<30 mm^3 vs. >30 mm^3), and lower prescription dose (<58 vs. >58 Gy) were found to be associated with treatment failure. Re-irradiation independently predicted sensory disturbance ($P < 0.001$). The pain control rates became stable after 3 years and remained so for the following 3 years.

A more recent report [68] substantially confirmed the earlier findings on a larger number of patients (343 participants) with a follow-up of 3 years. Frameless image-guided targeting of a 6 mm segment of the trigeminal nerve with a prescribed dose of 60 Gy and a maximum dose not superior to 75 Gy is a safe and effective technique, achieving stable pain control with minimal risks. These results are consistent with those reported by similar studies reporting about frame-based radiosurgery [8, 17, 18, 20, 25, 31, 33, 34, 43, 45, 50, 57]. The two studies reporting about the largest patient populations show a high rate of pain control [17, 22]. The Marseille group illustrated the results of 497 patients with primary TN after 1 year of follow-up. Of these patients, 91% achieved pain freedom in a short time. Pain recurred in 34.4% of patients [22]. In their report on 448 patients, Marshall et al. described satisfactory pain control in 86% and recurrence in 28% after 3 months of follow-up [17].

Sensory complications have been found in 20.1% of our patients, which is also in line with the literature. Regis et al. have recently updated their original study of 497 patients, reporting the long-term results of this large group of patients. The rate of new sensory disturbances was 20.4% [31]. In our series, only 21 patients (6.1%) developed bothersome hypoesthesias, with a very substantial number (18 out of 21) having been treated twice. Therefore, the risk of bothersome sensory deficit after a single treatment can be calculated as less than 1% following frameless CyberKnife radiosurgery using a retrogasserian target and specific dose/volume constraints. The dose constraints we prescribe to the brainstem and Gasserian ganglion cannot exceed 15 and 25 Gy, respectively. Frame-based irradiation using the Gamma Knife can be associated with higher brainstem doses and higher rates of sensory complications, especially when maximum doses exceed 80 Gy.

Careful planning by an expert physician is crucial to enhance treatment safety and to provide the best clinical outcomes. Detailed knowledge of the involved anatomy is extremely important in order to avoid complications. Critical structures drawn by the treating physicians include the brainstem, Gasserian ganglion, cranial nerves VII and VIII, cochlea, and labyrinth. The offending vessel, if present, is also drawn, thus avoiding the inclusion within the target volume. Special care is taken in the identification of the affected trigeminal nerve, which is often compressed and distorted by the offending vessel or atrophic due to previous invasive treatments. The motor root of the trigeminal nerve lies medial and slightly superior to the sensory root and can be identified and drawn as a separate structure, thus receiving lower doses than the sensory root. Figure 47.1 shows the treatment planning developed over the anatomy shown in Fig. 47.2.

47.3.1 Summary of Efficacy

We reviewed efficacy of CyberKnife radiosurgery for TN considering four factors: speed of pain relief, response rate for pain relief, duration of pain relief, and pain recurrence rate.

47.3.1.1 Speed of Pain Relief

Romanelli et al. [69] recorded an almost immediate pain relief in the first cohort of treated patients following delivery of a prescribed dose ranging from 65 to 70 Gy to a nerve segment up to 11 mm. The cohort included 10 patients, of which 5 achieved pain relief within the first 72 h after the treatment. These five patients received doses between 66 and 70 Gy. Three of five patients reported overnight onset of pain relief and maximal effect within 48 h. One patient had onset of pain relief after 48 h with complete relief within 72 h, and another had onset of pain relief after 72 h and complete relief within 7 days after the treatment. Similarly, Tang et al. [70] reported significant pain relief (a decrease in VAS score of >5 points) in 11 of 14 patients within 3 days with a median prescription dose of 66 Gy and a median volume of 59.2 mm^3.

On the other hand, high rate of bothersome paresthesia that developed over time in patients suggested reducing doses and, particularly, the

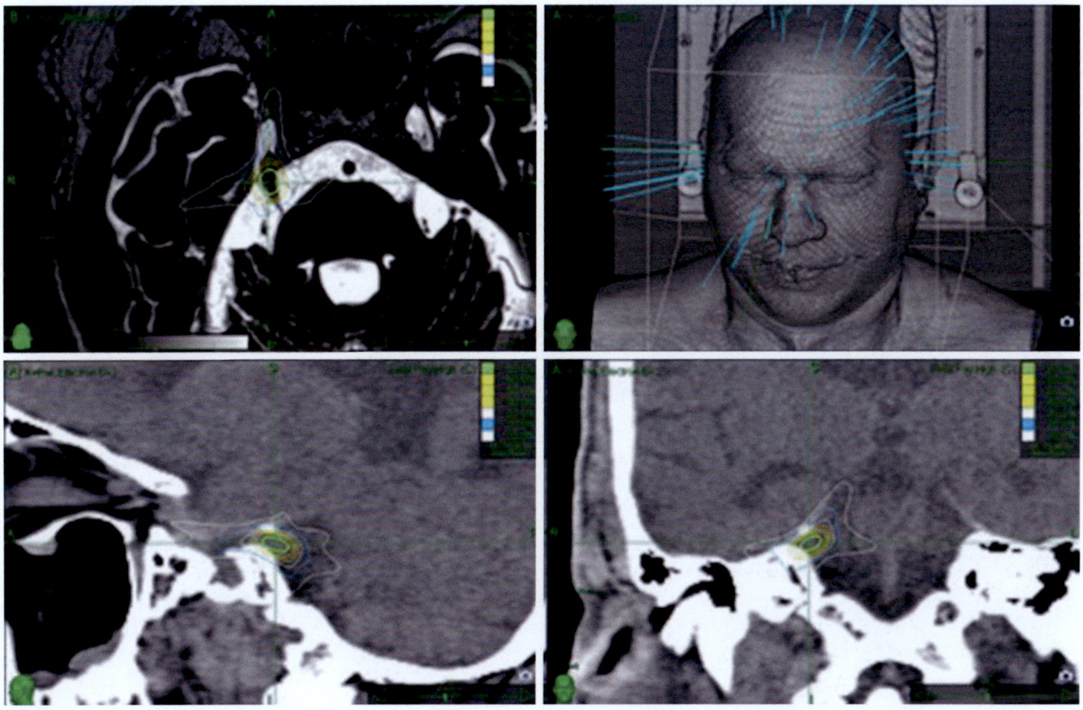

Fig. 47.1 CyberKnife radiosurgery for trigeminal neuralgia. The retrogasserian section of the trigeminal nerve was targeted, excluding the REZ for an elongated segment of about 6 mm (30 mm^3). The marginal dose was 64 Gy prescribed at the 85% isodose line. The brainstem was kept outside the 20% isodose line

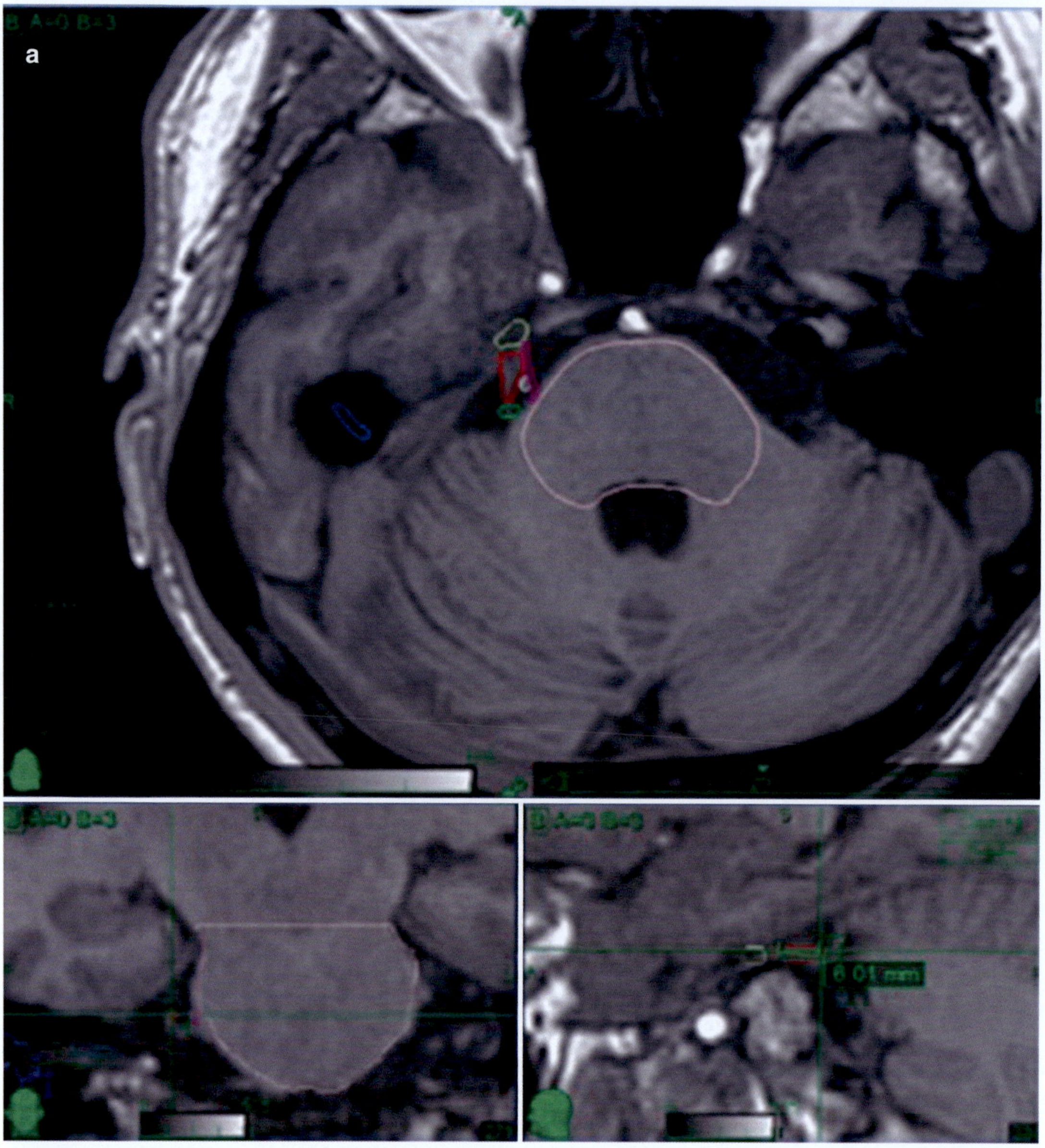

Fig. 47.2 (**a**) 3D reconstruction of the target (sensory root of the affected trigeminal nerve) and critical structures nearby (brainstem, Gasserian ganglion, trigeminal motor root-offending vessels). The sensory root is drawn using a red line, the motor root with a purple line, the brainstem with a pink line, and the Gasserian ganglion with a yellow line. There are two offending vessels, one located at the root entry zone (green line) and one dissecting the sensory and motor roots (white line). Axial view is above; coronal and sagittal views are below, respectively, on the left and right side. Crosshairs are placed at the proximal end of the target. As shown on the sagittal view, the target includes a 6.01 mm segment of the sensory root. (**b**) This plan shows how the careful identification of the anatomic structures involved can grant a certain degree of dose sparing on the trigeminal motor roots and the offending vessels as well as on the Gasserian ganglion and brainstem while preserving the intended prescribed dose to the sensory root

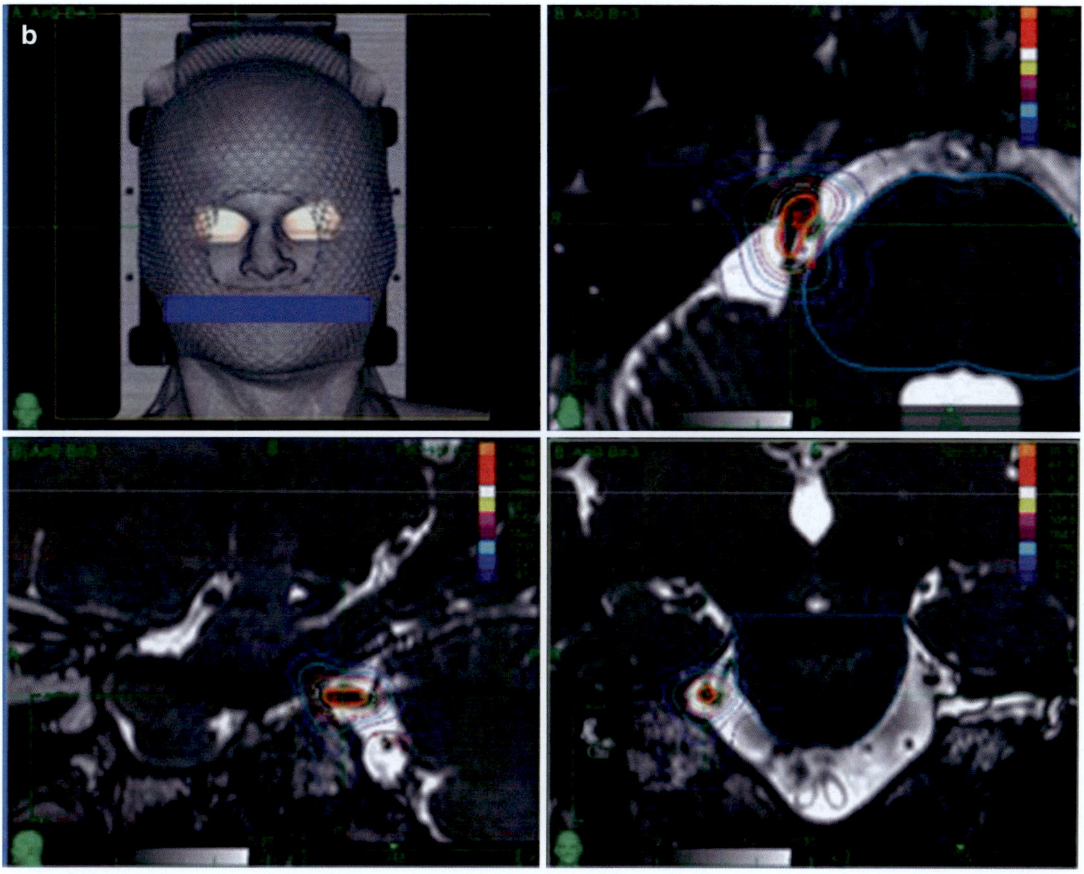

Fig. 47.2 (continued)

length of the nerve segment treated. Following a reduction of the prescribed doses down to 60 Gy and length of nerve segment treated down to 6 mm, a reduction in the rate of sensory loss and paresthesia was observed. However, the time needed to appreciate pain relief was substantially increased, becoming comparable to that following Gamma Knife treatment, namely, weeks to months. In most series, the interval ranges between 3 weeks and 6 months [52, 53, 58].

47.3.1.2 Response Rate for Pain Relief

Pain relief data were available in 15 out of 16 studies and for 757 out of 780 patients (see Table 47.1). Pain relief was defined as controlled pain with or without medications (Barrow Neurological Institute [BNI] pain scale I-III). Pain relief so defined was achieved in a percent of treated patients ranging between 67% [53] and

Table 47.1 Response rates in the different series

Authors	Year	Response rate	Percentage
Romanelli et al. [69]	2003	7/10	70%
Lim et al. [52]	2005	38/41	89.6%
Lim et al. [71]	2006	26/29	89.6%
Patil et al. [65]	2007	4/7	57%
Villavicencio et al. [53]	2008	64/95	67%
Adler et al. [59]	2009	33/46	72%
Borchers et al. [60]	2009	45/46	97.8%
Fariselli et al. [63]	2009	22/33	67%
Tang et al. [70]	2011	11/14	78.6%
Lazzara et al. [72]	2013	15/17	88%
Karam et al. [64]	2014	18/25	72%
Singh et al. [73]	2016	142/163	87.1%
Conti et al. [67]	2017	23/27	85.2%
Romanelli et al. [58]	2018	129/138	93.5%
Zhang et al. [74]	2018	55/66	83.3%
Romanelli et al. [68]	2019	323/343	94.2%
Pooled		955/1100	86.8%

97.8% [60]. Considering pooled data, pain relief was achieved in 632 out of 757 available patients (83.5%). The response rates in the different series are reported here:

Duration of Pain Relief

A follow-up longer than 12 months was available in 9 of 16 reviewed studies. Even though it is not possible to pool data in precise time intervals, it is clear that the response rate declines with time. Actually, the 12-month response rate is 82.6% in Borchers et al. [60] and 85.8% in Romanelli et al. [58]. In Zhang et al. [74], a "complete" pain relief was recorded in 39% of patients who received a standard dose to brainstem (45 Gy max point dose) and in 55% of patients who received a reduced brainstem dose (25 Gy max point dose). The 2-year response rate was 49.5% in Villavicencio et al. [53] and 33% in Fariselli et al. [63]. Longer follow-up data were available in Karam et al. [64] (28 months and actuarial response rate 56%) and Conti et al. [67] with a mean follow-up of 37 months and a 48-month response rate of 44%. Furthermore, Romanelli et al. [58] presented a large cohort of patients (138) with a minimum follow-up of 36 months (mean 52.4 months). In the series by Romanelli et al. [58], the response rate 3 years after the treatment was 76% and remained stable in later follow-up evaluation.

Pain Recurrence Rate

These data could be extracted clearly from 12 of 16 studies. Table 47.2 summarizes results of data analysis. Actually, the relapse rate ranged between 0% in Tang et al. [70] and 66% in Conti et al. [67]. It is reasonable that the relapse rate increases along with the length of the follow-up. In the series with longer follow-up, namely, Romanelli et al. [58] and Berti et al. [61], the pain relapse rates were 18.6% and 17.4%, respectively, whereas in the series of Conti et al. [67], the relapse rate was much higher (66%) as a probable consequence of the specific composition of this patient population including only subjects affected by multiple sclerosis. The relapse rate for pooled data was 23.9%.

47.3.2 Summary of Safety

The purpose of this section is to present a critical analysis of the clinical data pertinent to the safety of CyberKnife in the treatment of trigeminal neuralgia.

Table 47.3 lists the types of complications by the number of cases reported in the 16 studies reviewed, and summarizes these data. A total of 780 patients were treated in the 16 studies.

Table 47.2 Summary of data on recurrence rate

Authors	Year	Follow-up (months)	Relapse/responders	Percentage
Lim et al. [52]	2005	11	6/38	15.8%
Lim et al. [71]	2006	10	4/26	15.4%
Villavicencio et al. [53]	2008	23.5	17/64	26.6%
Adler et al. [59]	2009	14.7	1/45	2.2%
Fariselli et al. [63]	2009	23	11/22	50%
Tang et al. [70]	2011	20.4	0/14	0%
Lazzara et al. [72]	2013	11.8	4/14	28.6%
Karam et al. [64]	2014	28	2/20	10%
Conti et al. [67]	2017	37	15/23	66%
Romanelli et al. [58]	2018	52.4	24/129	18.6%
Berti et al. [61]	2018	2–13 years	4/23	17.4%
Zhang et al. [74]	2018	25 (standard) 19 (reduced)	25/55	45.4%
Romanelli et al. [68]	2019	36	24/316	7.6%
Pooled			137/789	17.4%

Table 47.3 Complications with CyberKnife treatment

Complications	Number of patients (%)
Facial numbness (not bothersome)	104 (13.3%)
Facial numbness (bothersome)	70 (9%)
Decreased corneal reflex	11 (1.4%)
Increase in trigeminal distribution anesthesia	7 (0.9%)
Dysesthesia (tolerable and bothersome)	6 (0.8%)
Generalized pain	5 (0.6%)
Masticator weakness	5 (0.6%)
Anesthesia dolorosa	5(0.6%)
Nausea	4 (0.5%)
Trismus	3 (0.4%)
Diplopia	2 (0.3%)
Urticaria	2 (0.4%)
Hearing loss	3 (0.1%)
Paresis	1 (0.1%)
Other not specified	9 (4.1%)

Specific types of complications reported in the 16 studies:

- Facial numbness (tolerable and bothersome) in 13 studies (Lim et al. [52, 71]; Patil et al. [58]; Villavicencio et al. [53]; Adler et al. [59]; Borchers et al. [60]; Tang et al. [70]; Lazzara et al. [72]; Karam et al. [64]; Singh et al. [73]; Conti et al. [67]; Romanelli et al. [58]; Berti et al. [61]; Zhang et al. [74]
- Dysesthesia (tolerable and bothersome) in five studies (Romanelli et al. [69]; Villavicencio et al. [53]; Fariselli et al. [63]; Patil et al. [58]; Lazzara et al. [72])
- Decreased corneal reflex in three studies (Lim et al. [52, 71]; Villavicencio et al. [53])
- Masticator weakness in three studies (Lim et al. [52, 71]; Villavicencio et al. [53])
- Anesthesia dolorosa in two studies (Lim et al. [71]; Villavicencio et al. [53])
- Diplopia in two studies (Lim et al. [71]; Villavicencio et al. [53])
- Trismus in two studies (Lim et al. [52]; Villavicencio et al. [53])
- Increase in trigeminal distribution anesthesia in one study (Lim et al. [71])
- Paresis in one study (Villavicencio et al. [53])
- Generalized pain in one study (Singh et al. [73])
- Nausea in one study (Singh et al. [73])
- Hearing loss in one study (Singh et al. [73])
- Urticaria in one study (Singh et al. [73])

47.3.2.1 Occurrence of Facial Numbness

Facial numbness is the most frequently cited specific complication of CK treatment reported by 174 of 780 patients (22.3%) in 14 studies after CK treatment. Nevertheless, bothersome facial numbness was reported only in 9% of patients.

47.3.2.2 Occurrence of Dysesthesia

Dysesthesia is the fourth most cited specific complication of CK treatment reported by 6 of 780 patients (0.8%) after decreased corneal reflex (1.4%) and increase in trigeminal distribution anesthesia (0.9%).

47.3.2.3 Other Adverse Events

Other types of complications are reported such as generalized pain in 0.6%, masticator weakness in 0.6%, anesthesia dolorosa in 0.5%, nausea in 0.5%, trismus in 0.4%, diplopia in 0.3%, and urticaria in 0.3%. The types of complications that are least observed are paresis (0.1%) found only in one study [53] and hearing loss (0.1%) also found in one study [73] (Table 47.3).

47.4 Conclusions

Stereotactic radiosurgery is emerging as a valid first-line treatment option for TN. Patients undergoing frameless radiosurgery using the CyberKnife show that this technique is safe and effective. Using our constraints for dose, volume of the nerve, and dose to the brainstem and Gasserian ganglion, the incidence of bothersome sensory complications was low, whereas a durable pain control was achieved in 76% of patients. The rarity of bothersome complications and the fact that frameless radiosurgery represents the less invasive technique for the surgical treatment of TN provide a particularly favorable profile to this technique, as compared with other systems, to deliver homogeneous irradiation to an extended length of the trigeminal nerve.

References

1. Jannetta PJ. Arterial compression of the trigeminal nerve at the pons in patients with trigeminal neuralgia. J Neurosurg. 1967;26(1 Suppl):159–62.
2. Jannetta PJ. Observations on the etiology of trigeminal neuralgia, hemifacial spasm, acoustic nerve dysfunction and glossopharyngeal neuralgia. Definitive microsurgical treatment and results in 117 patients. Neurochirurgia (Stuttg). 1977;20(5):145–54.
3. Jannetta PJ. Microsurgical management of trigeminal neuralgia. Arch Neurol. 1985;42(8):800.
4. Jannetta PJ. Trigeminal neuralgia. Neurosurgery. 1986;18(5):677.
5. Love S, Hilton DA, Coakham HB. Central demyelination of the Vth nerve root in trigeminal neuralgia associated with vascular compression. Brain Pathol. 1998;8(1):1–11; discussion 11–2.
6. Love S, Coakham HB. Trigeminal neuralgia: pathology and pathogenesis. Brain. 2001;124(Pt 12):2347–60.
7. Hilton DA, Love S, Gradidge T, Coakham HB. Pathological findings associated with trigeminal neuralgia caused by vascular compression. Neurosurgery. 1994;35(2):299–303; discussion.
8. Kondziolka D, Perez B, Flickinger JC, Habeck M, Lunsford LD. Gamma knife radiosurgery for trigeminal neuralgia: results and expectations. Arch Neurol. 1998;55(9865796):1524–9.
9. Sweet WH. The history of the development of treatment for trigeminal neuralgia. Clin Neurosurg. 1985;32(3905141):294–318.
10. McLaughlin MR, Jannetta PJ, Clyde BL, Subach BR, Comey CH, Resnick DK. Microvascular decompression of cranial nerves: lessons learned after 4400 operations. J Neurosurg. 1999;90(10413149):1–8.
11. Zakrzewska JM, Akram H. Neurosurgical interventions for the treatment of classical trigeminal neuralgia. Cochrane Database Syst Rev. 2011;(9):CD007312.
12. Zakrzewska JM. Treatment for trigeminal neuralgia. Controlled long term study of all surgical options is being planned. BMJ. 1997;314(7079):520.
13. Leksell L. Sterotaxic radiosurgery in trigeminal neuralgia. Acta Chir Scand. 1971;137(4948331):311–4.
14. Lindquist C, Kihlstrom L, Hellstrand E. Functional neurosurgery—a future for the gamma knife? Stereotact Funct Neurosurg. 1991;57(1725560):72–81.
15. Rand RW, Jacques DB, Melbye RW, Copcutt BG, Levenick MN, Fisher MR. Leksell Gamma Knife treatment of tic douloureux. Stereotact Funct Neurosurg. 1993;61 Suppl 1(8115760):93–102.
16. De Ridder D, Moller A, Verlooy J, Cornelissen M, De Ridder L. Is the root entry/exit zone important in microvascular compression syndromes? Neurosurgery. 2002;51(12182781):427–33.
17. Marshall K, Chan MD, McCoy TP, Aubuchon AC, Bourland JD, McMullen KP, et al. Predictive variables for the successful treatment of trigeminal neuralgia with gamma knife radiosurgery. Neurosurgery. 2012;70(21849918):566–72.
18. Riesenburger RI, Hwang SW, Schirmer CM, Zerris V, Wu JK, Mahn K, et al. Outcomes following single-treatment gamma knife surgery for trigeminal neuralgia with a minimum 3-year follow-up. J Neurosurg. 2010;112(19780644):766–71.
19. Guclu B, Sindou M, Meyronet D, Streichenberger N, Simon E, Mertens P. Cranial nerve vascular compression syndromes of the trigeminal, facial and vago-glossopharyngeal nerves: comparative anatomical study of the central myelin portion and transitional zone; correlations with incidences of corresponding hyperactive dysfunctional syndromes. Acta Neurochir. 2011;153(21947457):2365–75.
20. Tawk RG, Duffy-Fronckowiak M, Scott BE, Alberico RA, Diaz AZ, Podgorsak MB, et al. Stereotactic gamma knife surgery for trigeminal neuralgia: detailed analysis of treatment response. J Neurosurg. 2005;102(15796377):442–9.
21. Flickinger JC, Pollock BE, Kondziolka D, Phuong LK, Foote RL, Stafford SL, et al. Does increased nerve length within the treatment volume improve trigeminal neuralgia radiosurgery? A prospective double-blind, randomized study. Int J Radiat Oncol Biol Phys. 2001;51(11567820):449–54.
22. Tuleasca C, Carron R, Resseguier N, Donnet A, Roussel P, Gaudart J, et al. Patterns of pain-free response in 497 cases of classic trigeminal neuralgia treated with gamma knife surgery and followed up for least 1 year. J Neurosurg. 2012;117(Suppl):181–8.
23. Gorgulho AA, De Salles AAF. Impact of radiosurgery on the surgical treatment of trigeminal neuralgia. Surg Neurol. 2006;66(17015103):350–6.
24. Smith ZA, Gorgulho AA, Bezrukiy N, McArthur D, Agazaryan N, Selch MT, et al. Dedicated linear accelerator radiosurgery for trigeminal neuralgia: a single-center experience in 179 patients with varied dose prescriptions and treatment plans. Int J Radiat Oncol Biol Phys. 2011;81(21236592):225–31.
25. Tuleasca C, Regis J, Sahgal A, De Salles A, Hayashi M, Ma L, et al. Stereotactic radiosurgery for trigeminal neuralgia: a systematic review. J Neurosurg. 2018;130(3):733–57.
26. Kondziolka D, Flickinger JC, Lunsford LD, Habeck M. Trigeminal neuralgia radiosurgery: the University of Pittsburgh experience. Stereotact Funct Neurosurg. 1996;66 Suppl 1(9032878):343–8.
27. Rogers CL, Shetter AG, Fiedler JA, Smith KA, Han PP, Speiser BL. Gamma knife radiosurgery for trigeminal neuralgia: the initial experience of The Barrow Neurological Institute. Int J Radiat Oncol Biol Phys. 2000;47(10863073):1013–9.
28. Massager N, Murata N, Tamura M, Devriendt D, Levivier M, Regis J. Influence of nerve radiation dose in the incidence of trigeminal dysfunction after trigeminal neuralgia radiosurgery. Neurosurgery. 2007;60(17415205):681–7.
29. Regis J. High-dose trigeminal neuralgia radiosurgery associated with increased risk of trigeminal nerve dysfunction. Neurosurgery. 2002;50(12051192):1401–2.

30. Regis J, Arkha Y, Yomo S, Murata N, Roussel P, Donnet A, et al. [Radiosurgery in trigeminal neuralgia: long-term results and influence of operative nuances]. Neurochirurgie. 2009;55(19339026):213–22.
31. Regis J, Metellus P, Hayashi M, Roussel P, Donnet A, Bille-Turc F. Prospective controlled trial of gamma knife surgery for essential trigeminal neuralgia. J Neurosurg. 2006;104(16776335):913–24.
32. Kondziolka D, Lunsford LD, Flickinger JC, Young RF, Vermeulen S, Duma CM, et al. Stereotactic radiosurgery for trigeminal neuralgia: a multiinstitutional study using the gamma unit. J Neurosurg. 1996;84(8847587):940–5.
33. Sheehan J, Pan H-C, Stroila M, Steiner L. Gamma knife surgery for trigeminal neuralgia: outcomes and prognostic factors. J Neurosurg. 2005;102(15796376):434–41.
34. Massager N, Lorenzoni J, Devriendt D, Desmedt F, Brotchi J, Levivier M. Gamma knife surgery for idiopathic trigeminal neuralgia performed using a far-anterior cisternal target and a high dose of radiation. J Neurosurg. 2004;100(15070111):597–605.
35. Pollock BE, Phuong LK, Foote RL, Stafford SL, Gorman DA. High-dose trigeminal neuralgia radiosurgery associated with increased risk of trigeminal nerve dysfunction. Neurosurgery. 2001;49(11440460):58–62.
36. Kim YH, Kim DG, Kim JW, Kim Y-H, Han JH, Chung H-T, et al. Is it effective to raise the irradiation dose from 80 to 85 Gy in gamma knife radiosurgery for trigeminal neuralgia? Stereotact Funct Neurosurg. 2010;88(20431328):169–76.
37. Pollock BE. Radiosurgery for trigeminal neuralgia: is sensory disturbance required for pain relief? J Neurosurg. 2006;105 Suppl(18503340):103–6.
38. Dellaretti M, Reyns N, Touzet G, Sarrazin T, Dubois F, Lartigau E, et al. Clinical outcomes after Gamma Knife surgery for idiopathic trigeminal neuralgia: review of 76 consecutive cases. J Neurosurg. 2008;109 Suppl(19123905):173–8.
39. Petit JH, Herman JM, Nagda S, DiBiase SJ, Chin LS. Radiosurgical treatment of trigeminal neuralgia: evaluating quality of life and treatment outcomes. Int J Radiat Oncol Biol Phys. 2003;56(12829153):1147–53.
40. Han JH, Kim DG, Chung H-T, Paek SH, Kim YH, Kim C-Y, et al. Long-term outcome of gamma knife radiosurgery for treatment of typical trigeminal neuralgia. Int J Radiat Oncol Biol Phys. 2009;75(19515510):822–7.
41. Kondziolka D, Zorro O, Lobato-Polo J, Kano H, Flannery TJ, Flickinger JC, et al. Gamma knife stereotactic radiosurgery for idiopathic trigeminal neuralgia. J Neurosurg. 2010;112(19747055):758–65.
42. Little AS, Shetter AG, Shetter ME, Bay C, Rogers CL. Long-term pain response and quality of life in patients with typical trigeminal neuralgia treated with gamma knife stereotactic radiosurgery. Neurosurgery. 2008;63(19005382):915–23.
43. Maesawa S, Salame C, Flickinger JC, Pirris S, Kondziolka D, Lunsford LD. Clinical outcomes after stereotactic radiosurgery for idiopathic trigeminal neuralgia. J Neurosurg. 2001;94(11147887):14–20.
44. Verheul JB, Hanssens PEJ, Lie ST, Leenstra S, Piersma H, Beute GN. Gamma Knife surgery for trigeminal neuralgia: a review of 450 consecutive cases. J Neurosurg. 2010;113 Suppl(21121797):160–7.
45. Jawahar A, Wadhwa R, Berk C, Caldito G, DeLaune A, Ampil F, et al. Assessment of pain control, quality of life, and predictors of success after gamma knife surgery for the treatment of trigeminal neuralgia. Neurosurg Focus. 2005;18(5):E8.
46. Longhi M, Rizzo P, Nicolato A, Foroni R, Reggio M, Gerosa M. Gamma knife radiosurgery for trigeminal neuralgia: results and potentially predictive parameters—part I: idiopathic trigeminal neuralgia. Neurosurgery. 2007;61(18162905):1254–60.
47. Brisman R. Gamma knife surgery with a dose of 75 to 76.8 Gy for trigeminal neuralgia. J Neurosurg. 2004;100(15137604):848–54.
48. Dhople AA, Adams JR, Maggio WW, Naqvi SA, Regine WF, Kwok Y. Long-term outcomes of gamma knife radiosurgery for classic trigeminal neuralgia: implications of treatment and critical review of the literature. Clinical article. J Neurosurg. 2009;111(19326987):351–8.
49. Fountas KN, Smith JR, Lee GP, Jenkins PD, Cantrell RR, Sheils WC. Gamma Knife stereotactic radiosurgical treatment of idiopathic trigeminal neuralgia: long-term outcome and complications. Neurosurg Focus. 2007;23(6):E8.
50. Park S-H, Hwang S-K. Outcomes of gamma knife radiosurgery for trigeminal neuralgia after a minimum 3-year follow-up. J Clin Neurosci. 2011;18(21371892):645–8.
51. Herman JM, Petit JH, Amin P, Kwok Y, Dutta PR, Chin LS. Repeat gamma knife radiosurgery for refractory or recurrent trigeminal neuralgia: treatment outcomes and quality-of-life assessment. Int J Radiat Oncol Biol Phys. 2004;59(15093906):112–6.
52. Lim M, Villavicencio AT, Burneikiene S, Chang SD, Romanelli P, McNeely L, et al. CyberKnife radiosurgery for idiopathic trigeminal neuralgia. Neurosurg Focus. 2005;18(5):E9.
53. Villavicencio AT, Lim M, Burneikiene S, Romanelli P, Adler JR, McNeely L, et al. Cyberknife radiosurgery for trigeminal neuralgia treatment: a preliminary multicenter experience. Neurosurgery. 2008;62(18425011):647–55.
54. Chen JCT, Greathouse HE, Girvigian MR, Miller MJ, Liu A, Rahimian J. Prognostic factors for radiosurgery treatment of trigeminal neuralgia. Neurosurgery. 2008;62(18580781):53–60.
55. Dvorak T, Finn A, Price LL, Mignano JE, Fitzek MM, Wu JK, et al. Retreatment of trigeminal neuralgia with gamma knife radiosurgery: is there an appropriate cumulative dose? Clinical article. J Neurosurg. 2009;111(19326978):359–64.
56. Huang C-F, Chuang J-C, Tu H-T, Lin L-Y. Repeated Gamma Knife surgery for refractory trigeminal neuralgia. J Neurosurg. 2006;105 Suppl(18503339):99–102.

57. Pollock BE, Foote RL, Link MJ, Stafford SL, Brown PD, Schomberg PJ. Repeat radiosurgery for idiopathic trigeminal neuralgia. Int J Radiat Oncol Biol Phys. 2005;61(15629611):192–5.
58. Romanelli P, Conti A, Bianchi L, Bergantin A, Martinotti A, Beltramo G. Image-guided robotic radiosurgery for trigeminal neuralgia. Neurosurgery. 2018;83(5):1023–30.
59. Adler JR Jr, Bower R, Gupta G, Lim M, Efron A, Gibbs IC, et al. Nonisocentric radiosurgical rhizotomy for trigeminal neuralgia. Neurosurgery. 2009;64(2 Suppl):A84–90.
60. Borchers JD 3rd, Yang HJ, Sakamoto GT, Howes GA, Gupta G, Chang SD, et al. Cyberknife stereotactic radiosurgical rhizotomy for trigeminal neuralgia: anatomic and morphological considerations. Neurosurgery. 2009;64(2 Suppl):A91–5.
61. Berti A, Ibars G, Wu X, Sabo A, Granville M, Suarez G, et al. Evaluation of CyberKnife radiosurgery for recurrent trigeminal neuralgia. Cureus. 2018;10(5):e2598.
62. Conti A, Pontoriero A, Ricciardi GK, Granata F, Vinci S, Angileri FF, et al. Integration of functional neuroimaging in CyberKnife radiosurgery: feasibility and dosimetric results. Neurosurg Focus. 2013; 34(4):E5.
63. Fariselli L, Marras C, De Santis M, Marchetti M, Milanesi I, Broggi G. CyberKnife radiosurgery as a first treatment for idiopathic trigeminal neuralgia. Neurosurgery. 2009;64(2 Suppl):A96–101.
64. Karam SD, Tai A, Snider JW, Bhatia S, Bedrick EJ, Rashid A, et al. Refractory trigeminal neuralgia treatment outcomes following CyberKnife radiosurgery. Radiat Oncol. 2014;9:257.
65. Patil CG, Veeravagu A, Bower RS, Li G, Chang SD, Lim M, et al. CyberKnife radiosurgical rhizotomy for the treatment of atypical trigeminal nerve pain. Neurosurg Focus. 2007;23(6):E9.
66. Berti A, Granville M, Wu X, Huang D, Schwade JG, Jacobson RE. Delayed development of trigeminal neuralgia after radiosurgical treatment of a tentorial meningioma. Cureus. 2017;9(8):e1628.
67. Conti A, Pontoriero A, Iati G, Esposito F, Siniscalchi EN, Crimi S, et al. Frameless stereotactic radiosurgery for treatment of multiple sclerosis-related trigeminal neuralgia. World Neurosurg. 2017;103:702–12.
68. Romanelli P, Conti A, Redaelli I, Martinotti AS, Bergantin A, Bianchi LC, et al. Cyberknife radiosurgery for trigeminal neuralgia. Cureus. 2019;11(10):e6014.
69. Romanelli P, Heit G, Chang SD, Martin D, Pham C, Adler J. Cyberknife radiosurgery for trigeminal neuralgia. Stereotact Funct Neurosurg. 2003;81(1–4):105–9.
70. Tang CT, Chang SD, Tseng KY, Liu MY, Ju DT. CyberKnife stereotactic radiosurgical rhizotomy for refractory trigeminal neuralgia. J Clin Neurosci. 2011;18(11):1449–53.
71. Lim M, Cotrutz C, Romanelli P, Schaal D, Gibbs I, Chang SD, et al. Stereotactic radiosurgery using CT cisternography and non-isocentric planning for the treatment of trigeminal neuralgia. Comput Aided Surg. 2006;11(1):11–20.
72. Lazzara BM, Ortiz O, Bordia R, Witten MR, Haas JA, Katz AJ, et al. Cyberknife radiosurgery in treating trigeminal neuralgia. J Neurointerv Surg. 2013;5(1):81–5.
73. Singh R, Davis J, Sharma S. Stereotactic radiosurgery for trigeminal neuralgia: a retrospective multi-institutional examination of treatment outcomes. Cureus. 2016;8(4):e554.
74. Zhang M, Lamsam LA, Schoen MK, Mehta SS, Appelboom G, Adler JK, et al. Brainstem dose constraints in nonisometric radiosurgical treatment planning of trigeminal neuralgia: a single-institution experience. World Neurosurg. 2018;113:e399–407.

48 Movement Disorders

Marcello Marchetti, Alfredo Conti, Angelo Franzini, and Laura Fariselli

48.1 Introduction

Tremor is the most common movement disorder affecting up to five million people in the USA. It can be classified as essential tremor (ET), tremor associated to Parkinson disease (PD), and tremor related to multiple sclerosis (MS). Tremor may significantly impair patient performances, so remarkably affecting daily activities and overall quality of life (QoL). Medical treatment of tremor has limited efficacy and, usually, it decreases over time. Thus, surgical treatment remains the principal option for patients with tremor. Deep brain stimulation (DBS) or ablative procedures using radiofrequency thermocoagulation (RFT) are conventionally used to produce a neuromodulation of the extrapyramidal motor network. Nevertheless, morbidity and mortality have been reported for both these procedures [1, 2].

Recently MR-guided high-intensity focused ultrasound (MRgFUS) has been approved for ET (and Parkinson disease-related tremor in Europe). Preliminary data are promising, but long-term efficacy and evidence of a favorable cost/benefit ratio are still awaited [3–7]. Stereotactic radiosurgery (SRS) represents a minimally invasive management option for surgical treatment of tremor. The lesioning of thalamus and/or basal ganglia for the treatment of tremor is a well-known procedure which, before the introduction of DBS, was usually performed using stereotactic surgical procedures [8].

SRS is usually offered for the elderly patients, patients with surgical contraindications, and patients who failed either DBS or RFT.

Published experience in functional radiosurgery using the Gamma Knife (GK) or linear accelerators (LINAC) has been consistently positive [9–20]. As a matter of fact, stereotactic lesioning of the thalamus and basal ganglia for radiosurgery on invisible targets to treat movement disorders and intractable pain is still the domain of frame-based procedures because of the need for a solid reference system registered to the anterior commissure-posterior commissure (AC-PC) line, which allows the use of stereotactic atlases.

M. Marchetti · L. Fariselli
Department of Neurosurgery, Radiotherapy Unit, Fondazione IRCCS Istituto Neurologico "C. Besta", Milan, Italy
e-mail: marcello.marchetti@istituto-besta.it; laura.fariselli@istituto-besta.it

A. Conti (✉)
Alma Mater Studiorum University of Bologna, Bologna, Italy

IRCCS ISNB Istituto delle Scienze Neurologiche di Bologna, Bologna, Italy

Department of Neurosurgery, Charité Faculty of Medicine, Berlin, Germany
e-mail: alfredo.conti2@unibo.it

A. Franzini
Department of Neurosurgery, Fondazione IRCCS Istituto Neurologico "C. Besta", Milan, Italy
e-mail: bsvjf@tin.it

© Springer Nature Switzerland AG 2020
A. Conti et al. (eds.), *CyberKnife NeuroRadiosurgery*,
https://doi.org/10.1007/978-3-030-50668-1_48

Table 48.1 Study reporting results of Gamma Knife thalamotomy for essential tremor

Study	Patients (No.)	Level of evidence	Target	Follow-up (months)	Tremor reduction (%)
Friedman et al. (1999) [24]	17	4	Vim	3	68
Kondziolka et al. (2008) [25]	26	4	Vim	36	54
Lim et al., (2010) [26]	14	4	Vim	7–30	7
Young et al. (2010) [27]	119	4	Vim	44	40
Ohye et al. (2012) [28]	72	4	Vim	24	48
Kooshkabadi et al. (2013) [29]	86	4	Vim	23	45
Cho et al. (2015) [30]	7	4	Vim	12	44
Witijas et al. (2015) [23]	50	4	Vim	3–8	54
Tuleascca et al. (2017) [31]	38	4	Vim	12	63
Niranjan et al. (2017) [22]	73	4	Vim	28 (6–152)	68

48.2 Results of Radiosurgery for Tremor

Gamma Knife radiosurgery has recently become a new and well-defined treatment paradigm to treat many movement disorders [21–23]. The literature available on GK thalamotomy concerns 350 unique patients from 7 centers worldwide that were all retrospectively reviewed. Typically, patients are placed in a stereotactic frame and a central dose averaging 140 Gy is delivered using a 4 mm collimator. The ventralis intermedius (Vim) nucleus is the target in all patients (Table 48.1). Plans are optimized to reduce the dose to the internal capsule [17, 22, 23, 25, 26, 28, 29, 31]. Overall, reduction of tremor in the cohort of treated patients ranges from 7 to 61% and 3 to 68% at 3–9 months and last follow-up (median 7–152 months) respectively (Table 48.1).

Notably, the results from the two studies [23, 26] with the best outcome assessment (blinded) are contrasting. Witjas et al. [23] report an upper limb tremor score improvement by 54.2%. Activities of daily-living improve by 72.2%. Cognitive functions remain unchanged. The median delay of improvement is 5.3 months. On the other hands, Lim et al. [26] found that radiosurgery provided only modest anti-tremor efficacy.

Finally, a recent systematic review by the ISRS (International Stereotactic Radiosurgery Society) suggests that radiosurgery to the unilateral thalamic ventral intermediate nucleus, with a dose of 130–150 Gy, is a well-tolerated and effective treatment for reducing medically refractory tremor, recommendation level IV [32].

48.3 Treatment related adverse events

Permanent side effects (most frequently hemiparesis, followed by paraesthesia, dysphasia, and dysphagia) were relatively rare complications (median 0%; range 0–18%), and no deaths were reported. Side effects were observed years after SRS in two patients: a haemorrhagic stroke occurred in the irradiated area in patients taking anticoagulants for atrial fibrillation [26]. One case of delayed, complex involuntary movements has been reported. However, transient side effects including paraesthesia, dysphasia, dysphagia, and hemiparesis were more common (median 2%; range 0–9%). The main complications of SRS are reported in Table 48.2.

48.4 CyberKnife Thalamotomy for Tremor

The CyberKnife, compared to frame-based radiosurgery, is a pain-free procedure which offers the potential advantage of a better patient compliance by avoiding the pain and discomfort of the rigid frame. However, the subtle but clear differences in the 3D dose distribution and the dose fall-off features between Gamma Knife and

Table 48.2 Studies reporting complication after radiosurgery of essential tremor

Study	Paresthesia (%)	Gait/ataxia (%)	Hemiparesis (%)	Dysarthria (%)
Friedman et al. (1999) [24]	0	17	8	
Kondziolka et al. (2008) [25]	0	0	0	0
Lim et al. (2010) [26]	9	0	9	9
Young et al. (2010) [27]	2	0	5	3
Ohye et al. (2012) [28]	0	0	0	0
Kooshkabadi et al. (2013) [29]	1	0	1	1
Cho et al. (2015) [30]	0	0	0	0
Witijas et al. (2015) [23]	0	0	0	0
Tuleascca et al. (2017) [31]	0	0	0	0
Niranjan et al. (2017) [22]	0	0	0	0

CyberKnife make investigations on the effectiveness and safety of CK for this high-dose treatment mandatory.

We reported previously on two patients, treated with 70 and 90 Gy, who obtained good tremor control although one did not develop a lesional radiological necrosis. Based on this finding we postulated that relatively low, even sublesional doses could be effective. Thus, we began to look for a minimal effective dose. We investigated the efficacy and safety of 75, 80, and 90 Gy single-session radiosurgery. None of the current patients treated to 75 or 80 Gy demonstrated tremor control, neither did they show MRI changes. Both of the patients treated with 90 Gy (for whom follow-up is available), however, showed good tremor control (unpublished data). The clinical improvement became evident 18 and 26 months posttreatment concurrently with the MRI appearance of radiation necrosis-like images.

48.5 Target Definition and Stereotactic Atlas Registration of the CT Images

Our preferred target was always the ventralis oralis anterior-ventralis oralis posterior (VoA-VoP) complex of the thalamus. The choice of this target instead of the more frequently targeted ventral intermediate (VIM) nucleus was intended to lessen the risks. Indeed, this target is relatively far from the motor fibres running in the posterior limb of the internal capsule and far from other "eloquent" nuclei. The different target seemed to be safe based on the absence of serious toxicity, but it may also necessitate a higher dose to yield a therapeutic effect and evidence of necrosis and result in a longer latency between these effects and radiosurgery.

The relative stereotactic coordinates were 12–13 mm lateral to the midline, 2 mm posterior to the midcommissural point, and 0–2 mm superior to the commissural plane.

Targeting methods were reported previously. The assumption of the procedure is that if the head is immobile, the CT gantry can be treated as a solid reference system with fixed relationships to the brain structures.

During CT scanning, it is critical that the patient's head remains in a fixed position in order to avoid movement artefacts. For the patients reported here, high-quality images were obtained by restraining the patient's head using a standard thermoplastic mask and acquiring the images very rapidly, always taking less than 40 s, with the CT equipment at our institution (Light Speed Ultra, General Electric, Fairfield, CT). Mild sedation may be necessary for some patients with head tremor. In more severe cases, in which body and/or head movements could prevent an optimal image acquisition, administration of a low dosage of midazolam under anesthesiologic control is mandatory. If the head is immobile, the CT gantry behaves like a solid reference system with fixed relationships to the brain structures. In other words, the CT screen may be seen as a bidimensional stereotactic frame, and each pixel of the CT screen represents a discrete part of the brain identified by X lateral and Y anteroposterior coordinates in relation to the screen origin. The slice containing the anterior commissure (AC) is arbitrarily assigned to depth = 0 (Z coordinate); the depth of each slice is calculated relative to this point (the slices are 1.25 mm thick). In this system, we can calculate the AC X, Y, and Z coordinates (Z = 0) and the coordinates of the posterior commissure (PC) where Z is the distance in mm from slice zero. In cases in which the AC and PC lie on the same slice, AC and PC Z coordinates are both equal to zero, and the calculations are easier.

Then, the values in pixels are converted into millimeters based on the matrix/FOV ratio of the CT screen. In other words, the X and Y values of each pixel of the brain image on the CT screen are obtained, and Z is derived as the depth of the slice measured as the vertical distance from slice 0.

Finally, we calculate the coordinates of the AC-PC midpoint, which is the origin of the stereotactic atlas, and a simple roto-translation between the origin of the screen and the origin of the stereotactic atlas allows us to obtain atlas-registered X, Y, and Z coordinates in millimeters on the CT axial brain slices.

Target coordinates of the VoA/VoP complex derived from the stereotactic digital atlases registered to the midcommissural point are easily transposed onto the corresponding CT slice, and the target is drawn on the treatment planning system (Multiplan, Accuray Inc.). In other words, the roto-translation of the axes between the CT screen and the commissural system of the patient allows the use of atlas-derived stereotactic coordinates to make the invisible functional target detectable. The CT images may be fused with MRI to obtain more details about the anatomical structures surrounding the estimated target.

High quality control of CT couch movements is of course mandatory for the above-described procedure, and possible undesired movements of the CT couch during the examination could affect the precision of the Z coordinate, even if the VoA/VoP complex is relatively close to the slice containing the anterior commissure (slice 0), possible errors must be taken into account.

48.6 Dose Definition

After the target volume is identified and the critical healthy structures contoured, including the internal capsule (Fig. 48.1), in collaboration with a medical physicist, the 3D dose distribution is defined using an inverse planning algorithm. The aim of this procedure is to cause a lesion confined to the estimated target. Reports on frame-based radiosurgery show that, while a mean dose of 140 Gy has been effectively used, toxicity increases with higher doses. Nevertheless, the minimal effective dose has never been reported [10, 18, 24, 33]. Because the 3D dose distribution of the GK varies noticeably from the CyberKnife, we cautiously chose to test a lower prescription dose during our first attempts to treat movement disorders. When we started our experience, no previous CyberKnife treatments had been reported. Moreover, the first two patients treated at our institution with relatively low doses (70 and 90 Gy) had remarkably positive results, leading us to consider the real relative effectiveness of low doses.

Treatment plan evaluation was always based on dose-volume histogram (DVH) analyses of the PTV and critical healthy structures. The investigated doses were 75, 80, and 90 Gy prescribed to the 100% isodose line, delivered in a single session (unpublished data).

48.7 Patients Assessment and Tremor Response Criteria

The clinical assessment commonly includes a detailed neurological examination and assessment of the Fahn, Tolosa, Marin Tremor Rating Scale (FTMTRS). Patients are evaluated before the treatment, 2 months after treatment, and then every 4 months. Tremor control is based on the quantitative/qualitative analysis of the pre- and posttreatment FTMTRS scores. The Jancovich spiral drawing of the FTMTRS scores are assessed visually. Anatomical changes due to radiosurgery are evaluated on posttreatment MRI. Adverse events are assessed

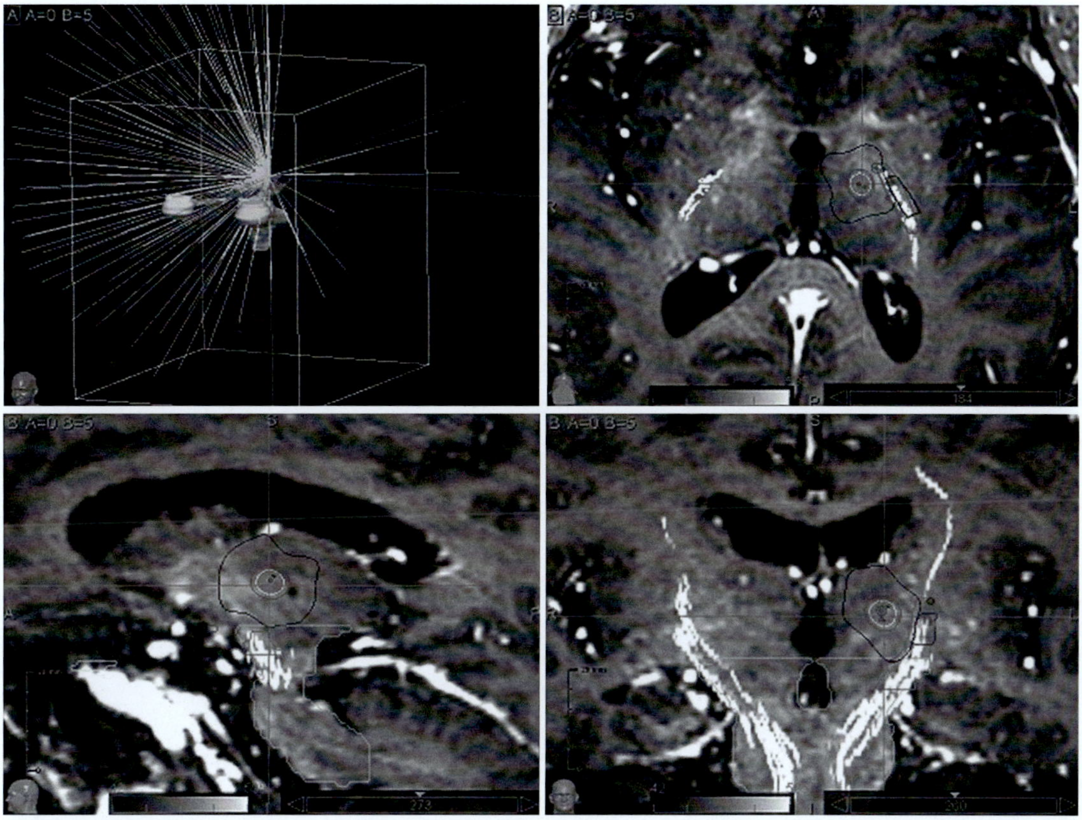

Fig. 48.1 Multimodality imaging, including CT, MRI, and tractography (not mandatory), are utilized to optimize the target definition, the dose distribution and the organs at risk sparing

using the CTCAE version 3.0. Treatment failure is defined as the lack of an effect on tremor or the absence of MRI changes 26 months after the treatment.

48.8 Results of CyberKnife Thalamotomy for Essential Tremor

According to our experience, the mean contoured treatment volume averaged 12.9 mm^3 ± 10.2 (median 11 mm^3; range 1.9–25.7 mm^3). Patients were treated with 75/80/90 Gy. The mean $V_{90\%}$, $V_{80\%}$, $V_{70\%}$, $V_{50\%}$, $V_{10\%}$ were, respectively, 14.2 ± 1.9, 39.2 ± 6.2, 78.2 ± 10.9, 206.6 ± 31.7, and 3376.0 ± 822.2 mm^3. The mean maximum point to the posterior half of the posterior arm of the capsula was 12.3 ± 5.4 Gy (median 11.4 Gy; range 7.3–21.6 Gy). The isodose overlapping the posterior half of the posterior arm of the capsula never exceeded the 20%.

48.9 Clinical and Radiological Response and Precision of Treatment Delivery

Two out of the three patients receiving a 90 Gy treatment developed MRI evidence of radiation necrosis at the 18th and 24th month post-radiotherapy and experienced a significant tremor relief. The fusion of the plans and the posttreatment images confirmed that, when present, radiation necrosis occurred where it has been expected. The radiation necrosis developed entirely inside the 70% isodose line, corresponding to the 63 Gy isodose line. The treatment precision was confirmed by the fusion of the post-necrosis MRI images with the treatment plan dose distribution (Fig. 48.2).

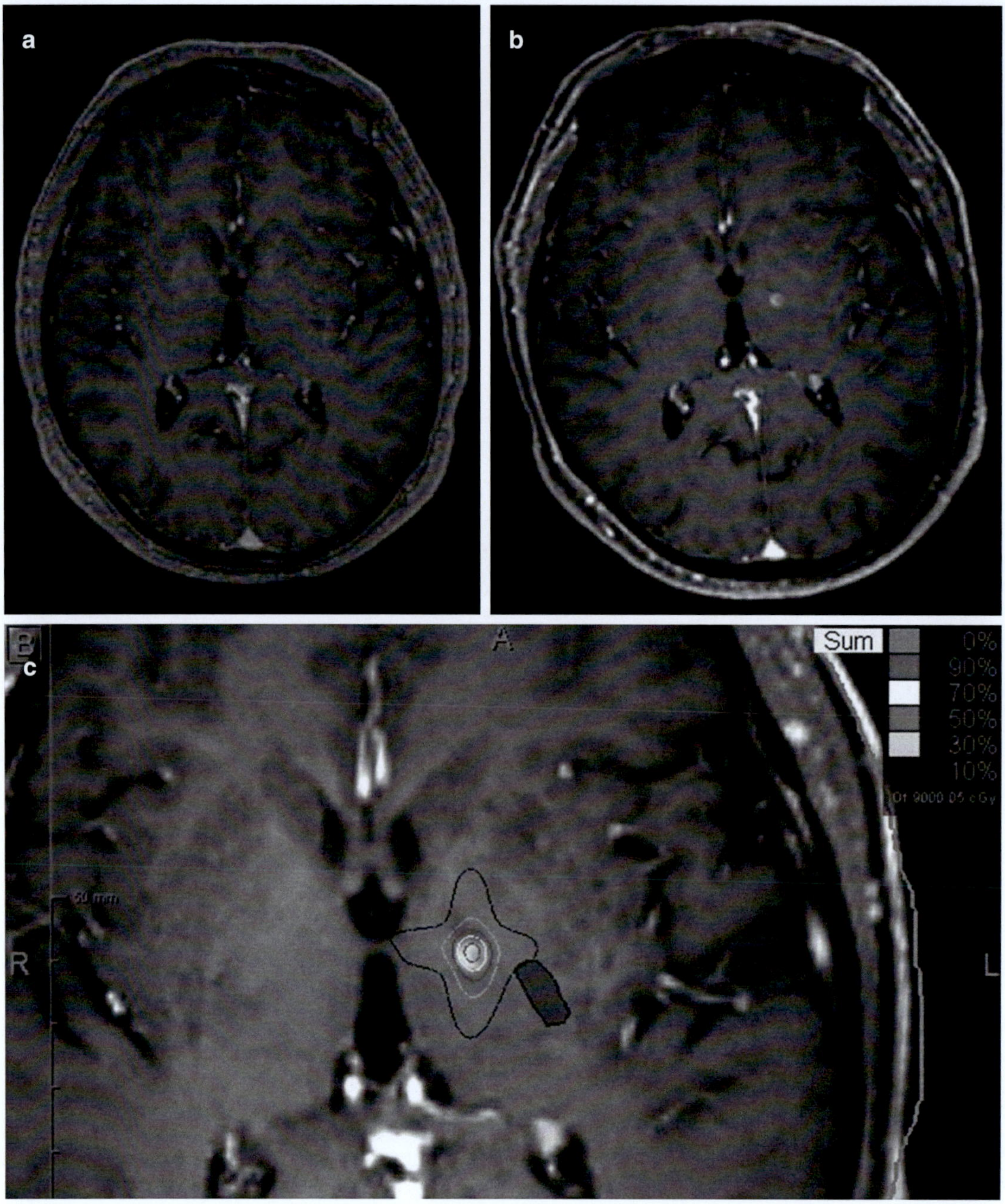

Fig. 48.2 A pre-treatment image (**a**) together with an 18 months post-treatment image (**b**) are represented here. The post-treatment MRI fusion with the treatment plan image confirms the accuracy of the treatment delivery. In this experience the treatment-related ring-shaped lesion never exceeded the 63 Gy isodose line

48.10 Toxicity

One patient developed a postural tremor 18 months after the treatment. This symptom appeared when intentional and rest tremors improved. This antigravity tremor completely recovered 6 months later, without therapy. No other complications have been observed.

48.11 Conclusions

In our preliminary experience frameless radiosurgery appears to be a safe alternative to treat tremor. Doses of 75 and 80 Gy were inadequate to provide tremor relief or radiological thalamotomy. The 90 Gy dose was generally effective, but the latency between the treatment and the tremor control was longer than expected. This long latency, in the authors' opinion, presents an ethical dilemma. Indeed, it is not easy to determine whether a treatment with such a delayed effect should be recommended. In conclusion, considering the results on the tremor, the long latency to the effects, but also the complete absence of permanent treatment-related side effects, and the potential for good patient compliance, we think that further dose-escalation studies are warranted.

References

1. Kumar R, Lozano AM, Sime E, Lang AE. Long-term follow-up of thalamic deep brain stimulation for essential and parkinsonian tremor. Neurology. 2003;61(11):1601–4.
2. Niranjan A, Jawahar A, Kondziolka D, Lunsford LD. A comparison of surgical approaches for the management of tremor: radiofrequency thalamotomy, gamma knife thalamotomy and thalamic stimulation. Stereotact Funct Neurosurg. 1999;72(2–4):178–84.
3. Chang JW, Park CK, Lipsman N, Schwartz ML, Ghanouni P, Henderson JM, et al. A prospective trial of magnetic resonance-guided focused ultrasound thalamotomy for essential tremor: results at the 2-year follow-up. Ann Neurol. 2018;83(1):107–14.
4. Chang WS, Jung HH, Kweon EJ, Zadicario E, Rachmilevitch I, Chang JW. Unilateral magnetic resonance guided focused ultrasound thalamotomy for essential tremor: practices and clinicoradiological outcomes. J Neurol Neurosurg Psychiatry. 2015;86(3):257–64.
5. Elias WJ, Huss D, Voss T, Loomba J, Khaled M, Zadicario E, et al. A pilot study of focused ultrasound thalamotomy for essential tremor. N Engl J Med. 2013;369(7):640–8.
6. Elias WJ, Lipsman N, Ondo WG, Ghanouni P, Kim YG, Lee W, et al. A randomized trial of focused ultrasound thalamotomy for essential tremor. N Engl J Med. 2016;375(8):730–9.
7. Ravikumar VK, Parker JJ, Hornbeck TS, Santini VE, Pauly KB, Wintermark M, et al. Cost-effectiveness of focused ultrasound, radiosurgery, and DBS for essential tremor. Mov Disord. 2017;32(8):1165–73.
8. Oh MY, Hodaie M, Kim SH, Alkhani A, Lang AE, Lozano AM. Deep brain stimulator electrodes used for lesioning: proof of principle. Neurosurgery. 2001;49(2):363–9.
9. Alvarez L, Macias R, Guridi J, Lopez G, Alvarez E, Maragoto C, et al. Dorsal subthalamotomy for Parkinson's disease. Mov Disord. 2001;16(1):72–8.
10. Duma CM, Jacques DB, Kopyov OV, Mark RJ, Copcutt B, Farokhi HK. Gamma knife radiosurgery for thalamotomy in parkinsonian tremor: a five-year experience. J Neurosurg. 1998;88(6):1044–9.
11. Frighetto L, De Salles A, Wallace R, Ford J, Selch M, Cabatan-Awang C, et al. Linear accelerator thalamotomy. Surg Neurol. 2004;62(2):106–13.
12. Lindquist C, Kihlström L, Hellstrand E. Functional neurosurgery—a future for the gamma knife? Stereotact Funct Neurosurg. 1991;57(1–2):72–81.
13. Ohye C. From selective thalamotomy with microrecording to gamma thalamotomy for movement disorders. Stereotact Funct Neurosurg. 2006;84(4):155–61.
14. Ohye C, Shibazaki T, Ishihara J, Zhang J. Evaluation of gamma thalamotomy for parkinsonian and other tremors: survival of neurons adjacent to the thalamic lesion after gamma thalamotomy. J Neurosurg. 2000;93(supplement_3):120–7.
15. Ohye C, Shibazaki T, Sato S. Gamma knife thalamotomy for movement disorders: evaluation of the thalamic lesion and clinical results. J Neurosurg. 2005;102:234–40.
16. Svennilson E, Torvik A, Lowe R, Leksell L. Treatment of parkinsonism by stereotactic thermolesions in the pallidal region. A clinical evaluation of 81 cases. Acta Psychiatr Scand. 1960;35(3):358–77.
17. Young RF, Jacques S, Mark R, Kopyov O, Copcutt B, Posewitz A, et al. Gamma knife thalamotomy for treatment of tremor: long-term results. J Neurosurg. 2000;93(supplement_3):128–35.
18. Young RF, Shumway-Cook A, Vermeulen SS, Grimm P, Blasko J, Posewitz A, et al. Gamma knife radiosurgery as a lesioning technique in movement disorder surgery. J Neurosurg. 1998;89(2):183–93.
19. Young RF, Vermeulen SS, Grimm P, Posewitz AE, Jacques DB, Rand RW, et al. Gamma knife thalamotomy for the treatment of persistent pain. Stereotact Funct Neurosurg. 1995;64(1):172–81.

20. Yu C, Main W, Taylor D, Kuduvalli G, Apuzzo MLJ, Adler JR, et al. An anthropomorphic phantom study of the accuracy of CyberKnife spinal radiosurgery. Neurosurgery. 2004;55(5):1138–49.
21. Higuchi Y, Matsuda S, Serizawa T. Gamma knife radiosurgery in movement disorders: indications and limitations. Mov Disord. 2016;32(1):28–35.
22. Niranjan A, Raju SS, Kooshkabadi A, Monaco E 3rd, Flickinger JC, Lunsford LD. Stereotactic radiosurgery for essential tremor: retrospective analysis of a 19-year experience. Mov Disord. 2017;32(5):769–77.
23. Witjas T, Carron R, Krack P, Eusebio A, Vaugoyeau M, Hariz M, et al. A prospective single-blind study of gamma knife thalamotomy for tremor. Neurology. 2015;85(18):1562–8.
24. Friedman DP, Goldman HW, Flanders AE, Gollomp SM, Curran WJ. Stereotactic radiosurgical pallidotomy and thalamotomy with the gamma knife: MR imaging findings with clinical correlation—preliminary experience. Radiology. 1999;212(1):143–50.
25. Kondziolka D, Ong JG, Lee JYK, Moore RY, Flickinger JC, Lunsford LD. Gamma knife thalamotomy for essential tremor. J Neurosurg. 2008;108(1):111–7.
26. Lim S-Y, Hodaie M, Fallis M, Poon Y-Y, Mazzella F, Moro E. Gamma knife thalamotomy for disabling tremor. Arch Neurol. 2010;67(5):584–8.
27. Young RF, Li F, Vermeulen S, Meier R. Gamma knife thalamotomy for treatment of essential tremor: long-term results. J Neurosurg. 2010;112(6):1311–7.
28. Ohye C, Higuchi Y, Shibazaki T, Hashimoto T, Koyama T, Hirai T, et al. Gamma knife thalamotomy for Parkinson disease and essential tremor. Neurosurgery. 2012;70(3):526–36.
29. Kooshkabadi A, Lunsford LD, Tonetti D, Flickinger JC, Kondziolka D. Gamma knife thalamotomy for tremor in the magnetic resonance imaging era. J Neurosurg. 2013;118(4):713–8.
30. Cho KR, Kim HR, Im YS, Youn J, Cho JW, Lee JI. Outcome of gamma knife thalamotomy in patients with an intractable tremor. J Korean Neurosurg Soc. 2015;57(3):192–6.
31. Tuleasca C, Witjas T, Najdenovska E, Verger A, Girard N, Champoudry J, et al. Assessing the clinical outcome of Vim radiosurgery with voxel-based morphometry: visual areas are linked with tremor arrest! Acta Neurochir. 2017;159(11):2139–44.
32. Martínez-Moreno NE, Sahgal A, De Salles A, Hayashi M, Levivier M, Ma L, Paddick I, Régis J, Ryu S, Slotman BJ, Martínez-Álvarez R. Stereotactic radiosurgery for tremor: systematic review. J Neurosurg. 2018;1:1–12.
33. Mathieu D, Kondziolka D, Niranjan A, Flickinger J, Lunsford LD. Gamma knife thalamotomy for multiple sclerosis tremor. Surg Neurol. 2007;68(4):394–9.

Image-Guided Robotic Radiosurgery for the Treatment of Drug-Refractory Epilepsy

49

Pantaleo Romanelli and Alfredo Conti

49.1 Introduction

Epilepsy is the most common neurological disorder, with an incidence reaching 0.5% of the population. Approximately one third of patients with epilepsy are affected by medically refractory seizures: seizures are unaffected or only partially improved by prolonged treatment with multiple drugs at therapeutic or even toxic doses. Medically refractory seizures expose the patient to the risk of death. Severe traumatic, metabolic, and neuropsychological sequelae can be suffered by patients that also often stand social stigma. Epilepsy surgery aiming to resect the epileptic focus is a valid option for medically refractory patients. However, the surgical risk of severe neurological and neuropsychological deficits becomes significant when the epileptic focus is adjacent or interspersed with eloquent brain areas or if the dominant hippocampus is involved.

Stereotactic radiosurgery (SRS) is an emerging treatment option for selected cases of medically refractory epilepsy [1, 2]. A noninvasive treatment such as SRS is possible for those cases in which invasive monitoring is not required to locate the epileptogenic focus. Deep-seated epileptogenic foci requiring complex and extensive neurosurgical procedures can also benefit from SRS. Seizure control is obtained through the delivery of high doses of radiations to the seizure focus, with longstanding edema being the most common complication after treatment. The main limit of the radiosurgical treatment of epilepsy, as compared to surgical resection, is the delay (sometimes up to 2 years) between treatment and seizure control. During this time, the patient remains exposed to the seizures and to the effects of antiepileptic drugs.

The exact mechanism of seizure abolition after radiosurgery is unknown. Depending on the target volume, radiosurgery can induce necrosis and consequent destruction of the epileptic focus and its pathways of spread. Suppression of epileptic activity by a neuromodulation effect using subnecrotizing doses has been suggested as a possible mechanism [3–5]. Doses of less than 20 Gy to volumes <7 mL seemingly do not generate radionecrosis, but impact neuronal density and perivascular sclerosis as documented is hippocampal specimens resected after the procedure [6].

Satisfactory epilepsy control has been reported in hippocampal sclerosis, brain tumors, arteriovenous malformations, cavernomas, and deep-seated epileptogenic lesions such as hypothalamic hamartomas (for a review, see [1, 2]). Most of these reports are from Gamma Knife series.

P. Romanelli (✉)
Brain Radiosurgery, Cyberknife Center, CDI, Milan, Italy
e-mail: pantaleo.romanelli@cdi.it

A. Conti
Alma Mater Studiorum University of Bologna, Bologna, Italy

IRCCS ISNB Istituto delle Scienze Neurologiche di Bologna, Bologna, Italy

Department of Neurosurgery, Charité Faculty of Medicine, Berlin, Germany
e-mail: alfredo.conti2@unibo.it

© Springer Nature Switzerland AG 2020
A. Conti et al. (eds.), *CyberKnife NeuroRadiosurgery*,
https://doi.org/10.1007/978-3-030-50668-1_49

We here review the current applications of CyberKnife in selected cases of refractory epilepsy from a medical point of view and highlight the use of frameless image-guided radiosurgery. A brief description of evolving research on the treatment of epileptogenic foci located over eloquent cortex using micro-radiosurgical transections will be provided as well.

49.2 Stereotactic Radiosurgery for Mesial Temporal Lobe Epilepsy

Radiosurgical amygdalohippocampal for mesial temporal lobe epilepsy was introduced by Jean Régis and coworkers in 1993 [4, 7–11]. Selection criteria for Gamma Knife radiosurgery were the same as those adopted for microsurgical amygdalohippocampectomy (i.e., the presence of hippocampal sclerosis and the absence of space-occupying lesions). The target was a volume of approximatively 7 mL, including the head and body of the hippocampus, the anterior part of the parahippocampal gyrus, and the basolateral region of the amygdaloidal complex (sparing the upper and mesial part) receiving 25 Gy to the 50% isodose line. At the follow-up, MR often showed transitory hippocampal swelling, with an increased T2 signal followed by the development of a contrast-enhancing ring demarcating the 50% isodose line and a diffusely increased T2 signal spreading from the hippocampus to the temporal lobe and adjacent white matter. Image changes appeared at 12 months (8–15 months) and were sometimes symptomatic (i.e., headaches, nausea, and vomiting) but sensitive to corticosteroid treatment. All abnormal MRI findings resolved within 24 months post-treatment [4, 7–11]. After 2 years, 65% (13 of 20 patients) of this cohort of patients treated in three different centers were seizure-free, with a reduction of the median number of seizures per month from 6.2 to 0.3 [10]. Ten patients out of 20 (50%) developed visual field deficits consisting of a quadrantanopia (8 cases), hemianopia (1 case), or a mixed deficit (1 case) [10]. These results have been confirmed by a recent prospective multicenter pilot trial delivering radiosurgery to epileptic patients with mesial temporal sclerosis. The overall seizure remission rate was 69% during the third follow-up year after treatment, a result that can be compared to those reported in resective temporal lobectomy [12].

The efficacy of radiosurgery in cases of mesial temporal tumors associated with longstanding epilepsy has been specifically analyzed in a retrospective study of 19 cases treated by Gamma Knife radiosurgery (GKRS) [13]. All tumors were within the mesial temporal structures, and the histology (biopsy or tumor resection) included 15 (79%) low-grade astrocytomas, 3 (16%) gangliogliomas, and 1 (5%) cavernous angioma. GKRS was performed in order to obtain local growth control and alleviation of epilepsy. The latter aim was achieved by irradiating the epileptic foci placed over the gray matter located immediately outside the tumor volume. The mean 50% isodose volume surrounding the tumor was 6.2 mL (range 1.1–18 mL). The mean marginal dose was 17.3 Gy (range 12–30 Gy). After a follow-up of 1.7–9.7 years (mean, 6.5 years), 11 patients (57.9%) were significantly improved (Engel I and Engel II), 7 patients (36.8%) had worthwhile improvement (Engel III), and 1 patient (5.3%) was unchanged.

Barbaro et al. [14] reported results of a pilot multicenter trial describing seizure freedom in 77% of 13 patients who received high-dose (24 Gy) treatment and in 59% of 17 individuals who received low-dose (20 Gy) therapy at 1 year. Verbal memory impairment was described in 15% of patients, although none declined on more than one measure, while verbal memory improvement was seen in 12% of individuals [15, 16]. Side effects were reasonable in most cases, including headache and visual field deficits requiring a brief period of steroid administration. Only one patient suffered from malignant edema after treatment, including severe headaches, visual field deficit, and papilledema not responsive to steroids, and this patient eventually underwent temporal lobectomy [15].

We found a similar case following CyberKnife radiosurgery on a 36-year-old patient with mesial temporal sclerosis and medically refractory epilepsy, characterized by complex multiple daily seizures. A massive hemispheric edema developed 9 months following the delivery of 20 Gy prescribed to the 74% isodose with a maximum dose of 27 Gy.

The edema subsided after intravenous administration of a 100 mg/4 mL dose of bevacizumab (Avastin). Figure 49.1 shows treatment planning and T1 post-contrast MR follow-up 13 months after treatment in which a radionecrotic response can be appreciated within the treated volume. Subsequent scans showed progressive healing of the radionecrotic scar, which was barely visible after 21 months. This patient became seizure-free (Enga grade 1a) 7 months after the procedure and remained so after 4 years.

Barbaro et al. have reported results of a prospective randomized trial on SRS versus open anterior temporal lobectomy (Radiosurgery or Open Surgery for Epilepsy [ROSE] trial) for treatment of pharmacoresistant unilateral mesial temporal lobe epilepsy (MTLE) [17].

Adult patients were eligible for open surgery among 14 centers in the USA, the UK, and India.

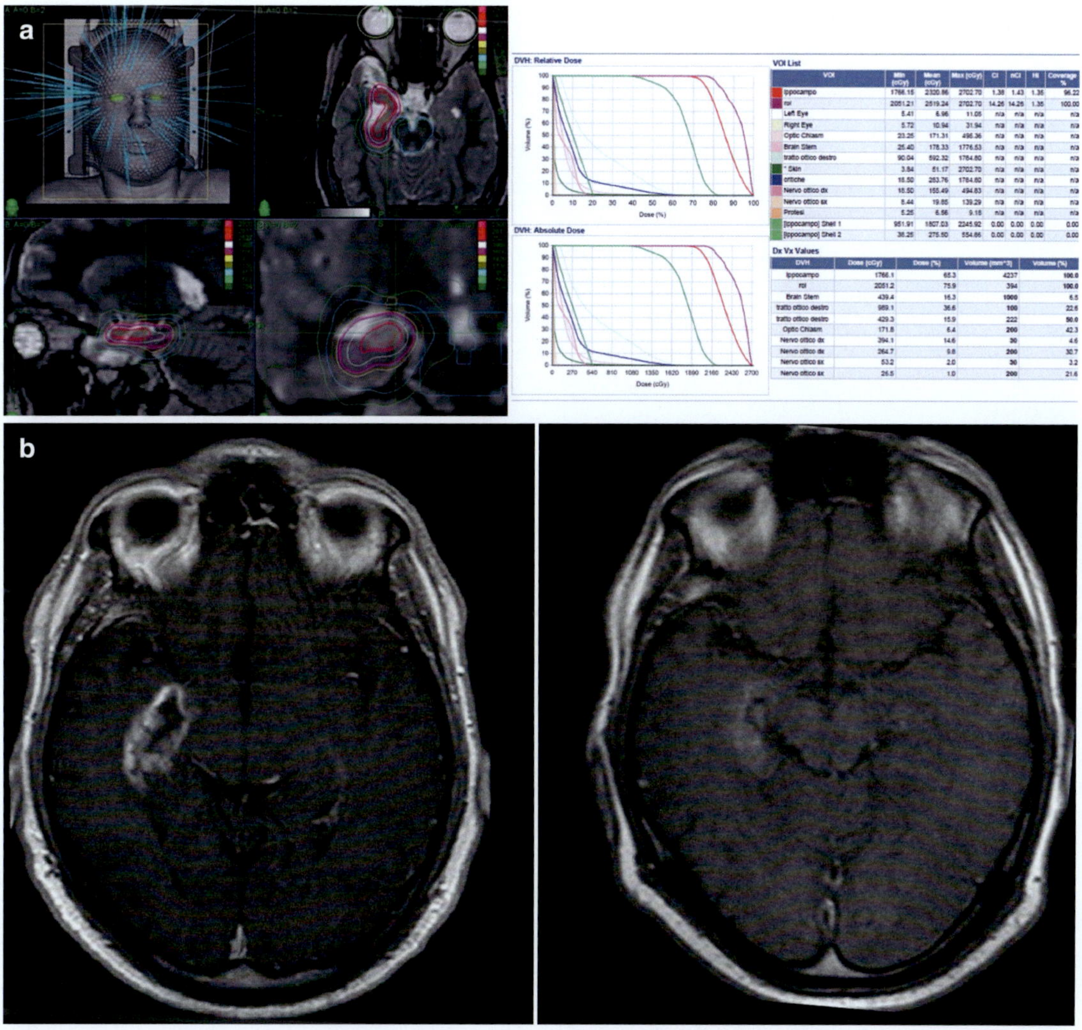

Fig. 49.1 (**a**) CyberKnife radiosurgery treatment plan for mesial temporal sclerosis. (**b**) (Left) MRI showing the development of a radionecrosis in the target area at 13 months. (Right) the MRI shows the resolution of the radionecrosis at 21 months after radiosurgery

Treatment was either SRS using 24 Gy to the 50% isodose targeting mesial structures with a volume ranging between 5 and 7.5 cm^3 or standardized arterial temporal lobectomy (ATL). Outcomes were seizure remission (absence of disabling seizures between 25 and 36 months), verbal memory (VM), and quality of life (QOL) at 36 months. A total of 58 patients (31 in SRS, 27 in ATL) were treated. Sixteen (52%) SRS and 21 (78%) ATL patients achieved seizure remission (difference between ATL and SRS = 26%, upper one-sided 95% confidence interval = 46%, *P* value at the 15% noninferiority margin = 0.82). Noninferiority of SRS compared to ATL was therefore not demonstrated.

Mean VM changes from baseline for 21 English-speaking, dominant-hemisphere patients did not differ between groups. Consistent worsening occurred in 36% of SRS and 57% of ATL patients. Adverse events were anticipating cerebral edema and related symptoms for some SRS patients and cerebritis, subdural hematoma, and others for ATL patients without a statistical difference between the two cohorts. The authors concluded that ATL has an advantage over SRS in terms of proportion of seizure remission, and both SRS and ATL appear to have effective and reasonable safety as treatments for MTLE. Accordingly, radiosurgery is an alternative treatment to ATL for patients with contraindications for or with reluctance to undergo open surgery [17]. Unlike resection, the beneficial effects of SRS on seizures are typically delayed up to 12 months or more after treatment. Chang and colleagues found that MRI characteristics during the first year following SRS might serve as a predictor of seizure outcome at 3 years after therapy [18]. Specifically, T2 hyperintensity volumes 9 months after the procedure were found to be highly related to seizure remission and were more pronounced in patients who received 24 Gy SRS compared to treatment with a lower dose of 20 Gy [12, 15].

In the ROSE trial, patients treated with SRS demonstrated a gradual increase in the proportion of seizure remission: only 2 (6%) patients had no seizures within 3 months post-radiosurgery, whereas 22 (81%) ATL patients experienced remission in the same period. By the final follow-up period (months 34–36), 23 (74%) of SRS patients and 23 (85%) of ATL patients experienced short-term seizure remission [17].

Currently, there are no published reports dedicated to the use of CyberKnife radiosurgery for the treatment of epilepsy caused by mesial temporal sclerosis. Figure 49.1 illustrates the case of a 12-year-old boy with right mesial temporal sclerosis treated with CyberKnife radiosurgery.

This child was affected by multiple daily complex partial seizures that were refractory to medical therapy. The MRI was remarkable for atrophy of the right anterior temporal lobe and mesial temporal sclerosis. A frameless image-guided CyberKnife treatment prescribing 23 Gy to the 81% isodose was delivered using 181 nonisocentric beams. The maximum dose was 28.4 Gy. Treatment volume was 6.49 cm^3 and included the head and body of the hippocampus, the anterior part of the parahippocampal gyrus, and the amygdala. Treatment delivery required 56 min and was uneventful. The patient received an intramuscular injection of dexamethasone (8 mg) immediately after the treatment and was discharged. A temporary increase of the seizures was observed 6 weeks after the treatment. Oral dexamethasone was given, with immediate seizure resolution.

The patient became seizure-free after this episode, remaining so during the last 7 years.

49.3 Stereotactic Radiosurgery for Neocortical Epilepsy

Neocortical seizures caused by arteriovenous malformations and cavernous angiomas have been treated by radiosurgery. Radiosurgery also represents a useful tool for ablation of seizure foci located in eloquent or surgically challenging brain regions, if surgical resection is associated with an unacceptably high risk of complications.

The combination of noninvasive seizure focus localization with radiosurgery represents an attractive alternative to traditional resection. It is a thoroughly noninvasive approach to map a cortical epileptic focus that is available. Magnetoencephalography (MEG) can be used to guide stereotactic irradiation in refractory seizures arising from eloquent cortical areas [14, 18, 19].

Current developments toward MEG ictal [20] and interictal mapping [21] as well as the use of epileptic network analysis [22] are likely to further implement the application of radiosurgery for epilepsy. Figure 49.2 illustrates a MEG-driven CyberKnife procedure performed on a 23-year-old patient with clinically severe refractory seizures (secondarily generalized tonic-clonic seizures) originating from the right frontal opercular region.

One of the seminal papers on radiosurgery to treat neocortical epilepsy was published by Barcia and coworkers [23–25]. Eleven patients with epileptic foci localized by neuroimaging and invasive electrode recordings underwent SRS for ablation. Nine subjects were treated with a ^{60}Co source receiving doses ranging from 10 to 20 Gy, while the remaining two patients received an estimated dose of 10 Gy delivered through a 10–15 MeV single beam betatron. At a mean follow-up of 102.5 months, four patients were off medication and seizure-free, five had a marked reduction in seizure frequency (75–98%), while two had no response (supposedly related to inaccuracy of seizure focus localization). No complications were reported.

A large number of data are available on seizure outcomes following radiosurgery of arteriovenous malformations (AMVs). Pollock and coworkers [26] retrospectively studied 67 patients with small AVMs, Spetzler-Martin Grade I or II, who refused to undergo an open surgical procedure and elected to receive a radiosurgical treatment. Thirty-one patients had experienced seizures prior to radiosurgery, which was performed with a 201 source Gamma Knife system. Mean AVM volume was 3.1 mL, with a mean marginal dose of 21 Gy (maximum dose: 36 Gy). Sixteen patients (52%) had seizure frequency reduced to less than 1 seizure per year, while 15 had no change in seizure control.

No patients developed new seizures following treatment. There was a 7.7% record of hemorrhage and a 3% mortality rate (caused by hemorrhage) within 8 months following radiosurgery. However, there was no risk of bleeding if there was total obliteration or subtotal obliteration with patency of early draining veins only.

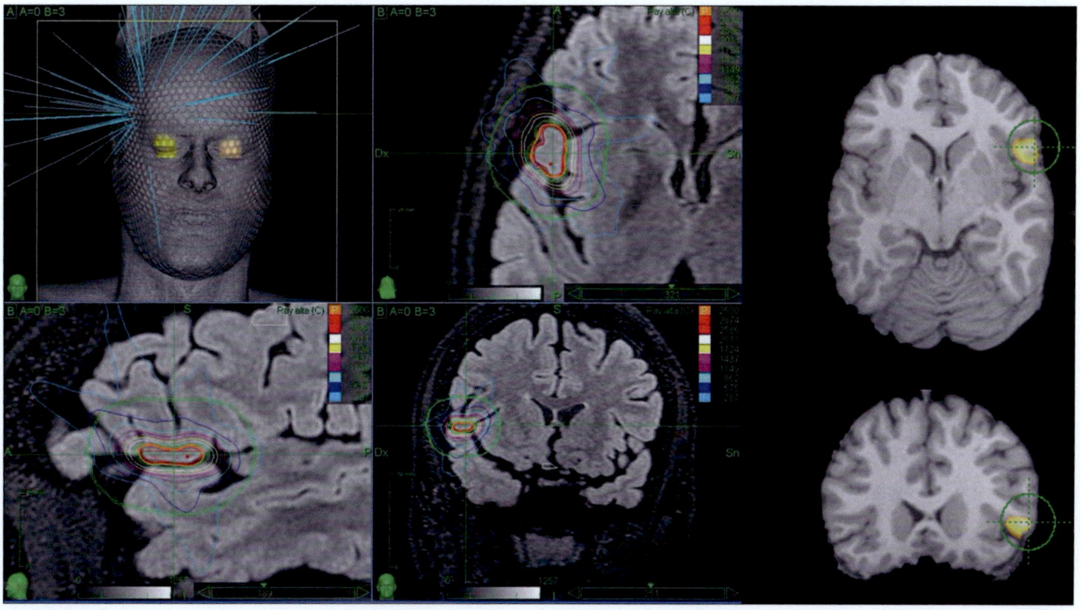

Fig. 49.2 (Left) CyberKnife radiosurgery treatment plan. (Right) The treatment was performed on neocortical seizure focus identified through MEG imaging in a patient with severe medically refractory seizures who refused invasive monitoring

Thirty-three patients with AVMs of the precentral gyrus were selected for retrospective study from a group of 770 patients who had been treated with Gamma Knife radiosurgery for AVMs [27]. Median AVM volume was 3 cm^3, and the median dose to the margin was 20 Gy. Twenty-seven (87%) of the patients had presented with seizures. After a mean follow-up of 54 months, there was a 63% seizure-free rate. The remaining 37% continued to experience seizures at a frequency no greater than before radiosurgery.

A retrospective review of 40 pediatric cases of AVMs treated by a multimodality approach and followed for a mean of 38.7 months has been reported by Hoh et al. [28]. Ten patients with AVM-related seizures were treated with proton beam radiotherapy if lesions were located in eloquent areas or had a particular pattern of venous drainage. Mean dose was 15.9 Gy to a mean volume of 9.9 cm^3. Nine out of ten patients (90%) became seizure-free following treatment. It should be noted that AVM embolization was also performed concurrently, and seizure outcome was not analyzed in detail to assess the respective weight of radiosurgery versus embolization in the outcome of seizure freedom. The authors reported no radiosurgery-related complications or morbidity. One patient had a hemorrhage after radiosurgery before the AVM had been obliterated.

A further retrospective study published by the same group in 2002 reported on 141 patients with AVMs and seizures, representing 33% of 424 patients treated for AVM over an 8-year period [29]. Follow-up data have been available for 110 patients out of 141.These patients were treated with a multimodality, multidisciplinary approach, including various combinations of surgery, radiosurgery, and embolization. The mean follow-up period was 34.8 months. Those who received radiosurgery were treated with proton beam therapy with a mean dose of 15.5 Gy to a mean volume of 7.7 cm^3. These investigators identified the following pretreatment risk factors for the development of symptomatic seizures with AVM: male gender, age less than 65 years, AVM size larger than 3 cm, and temporal lobe AVM location. Of the 110 patients treated with the multiple modalities, 73 (66%) were seizure-free (Engel Class I), 11 (10%) were Class II, 1 (0.9%) was Class III, and 22 (20%) were Class IV. Treatment-specific analysis revealed that surgery had the highest number of patients with Class I outcome (81%), followed by embolization (50%), and then radiosurgery (43%). However, if the AVM was completely obliterated, then all treatments yielded the same percentage of patients with a Class I outcome. The following factors were associated with a Class I outcome: short seizure history, associated intracranial hemorrhage, generalized tonic-clonic seizure type, deep and posterior fossa AVM location, surgical resection, and complete AVM obliteration. In addition, 5.7% of patients who did not have pretreatment symptomatic seizures developed seizures.

Seizure improvement has also been reported after SRS for cavernous malformations (CMs). Ninety-five patients were treated for CMs by proton beam therapy at the Massachusetts General Hospital [30]. Eighteen of the subjects had seizures prior to treatment. There was a significant improvement in seizure control after treatment, and no patients developed new-onset seizures or intractable epilepsy after treatment. Regis et al. [31] published a retrospective multicenter report of 49 patients treated for CMs with Gamma Knife radiosurgery. All patients had drug-resistant epilepsy and were followed for longer than 12 months after treatment. Mean marginal dose and volume were 19.17 Gy and 2.4 cm^3, respectively. Seizure outcome was favorable: 53% became seizure-free (Engel Class I), 20% experienced a significant decrease in number of seizures, and 26% had little to no improvement. The average time to seizure remission was 4 months. Five of the patients who failed to improve following radiosurgery were treated with microsurgery: three of these later became seizure-free, and one had rare seizures. One patient experienced no change in seizure control. Seven patients developed major post-treatment edema but fully recovered. Better outcome was associated with simple partial seizures compared with complex partial ones. Mesiotemporal location was associated with a poor outcome, while laterotemporal and central locations were associated with a good outcome. Based upon their results, Regis et al. recommended that Gamma Knife radiosurgery be considered for CMs associated with seizures arising from eloquent cortex surrounding the lesion.

Seizures associated with tuberous sclerosis can also respond well to radiosurgery, as illustrated by a report on a patient with intractable seizures related to a frontal subependymal nodule who did not improve after subtotal resection but experienced seizure freedom following radiosurgical treatment of the residual lesion [32].

49.4 Stereotactic Radiosurgery for Hypothalamic Hamartomas

Hypothalamic hamartomas (HHs) are epileptogenic developmental malformations that grow inside the hypothalamus (sessile or intrahypothalamic) or mostly within the third ventricle (pedunculated or parahypothalamic). Their size is commonly less than 2 cm but larger or even giant lesions can be found as well. HHs can be associated with a wide range of neurological or endocrine manifestations: pedunculated HHs are occasionally associated with endocrine deficits, while intrahypothalamic hamartomas are often associated with gelastic seizures (GSs) and severe medically refractory epilepsy [33]. Early seizure onset in newborns and young children is often associated with catastrophic epilepsy, including drop attacks, gelastic and generalized seizures, and cognitive and behavioral deterioration leading to mental retardation [34]. This early onset form is aggressive and poorly responsive to medical therapy, requiring timely surgical intervention to prevent severe neuropsychological sequelae. Seizure onset later in life is typically associated with a milder course.

Surgical approaches include microsurgical resection through the transcallosal interforniceal, pterional, or subfrontal translamina terminalis routes, microsurgical disconnection, endoscopic resection or disconnection, radiofrequency ablation, laser thermal ablation, and interstitial brachytherapy [35].

SRS provides an excellent treatment option for small- to medium-sized HHs causing catastrophic epilepsy and is best performed in the early years of childhood before the development of secondarily generalized epilepsy, developmental delay, and behavioral problems. Best results of seizure control are associated with prescribed doses equal or superior to 16 Gy [36, 37]. Temporary worsening of seizures is often observed weeks to months after SRS and is a good predictor of the final success of the procedure [36, 37]. There are no severe neurological complications following SRS for HHs, and a remarkable improvement not only of seizures but also of learning, memory, behavior, and sleep is frequently observed. The main limit of radiosurgery is its delayed effect, since seizures start to decrease 3–6 months after the procedure in most patients and with great variability in the timing of response.

Gamma Knife, Novalis, and CyberKnife radiosurgery provide safe and effective treatment options for HHs, including for small children [36–39].

Regis et al. have recently published a prospective study on 57 patients with HH and drug-refractory epilepsy associated with severe cognitive and psychiatric comorbidities [40]. Follow-up longer than 3 years was available for 48 patients.

Twenty-eight patients (58.3%) required a second treatment due to poor results after the first irradiation. Engel Class I outcome rate was 39.6%, Engel Class II was 29.2%, and Engel Class III was 20% after 3 years. Overall, a 68.8% rate of complete or near-complete seizure control (Class I and Class II) was achieved. Global psychiatric comorbidity was considered cured in 28%, improved in 56%, and stable in 8%, while further worsening was seen in the remaining 8%. No permanent neurological side effects were reported (in particular, no memory deficit). Non-disabling transient poikilothermia was observed in three patients (6.2%).

Frameless radiosurgery using the CyberKnife may further facilitate the use of radiosurgery in children, offering a totally noninvasive option devoid of major complications [1, 37]. Surgical and radiosurgical treatments can be optimally integrated in patients with large HHs. In such cases, a surgical debulking procedure can be followed by SRS delivered to the unresectable epileptogenic intrahypothalamic component. Figure 49.3 illustrates a frameless CyberKnife treatment performed uneventfully and with full seizure resolution in a 9-year-old child with an intrahypothalamic hamartoma associated with catastrophic epilepsy.

49.5 Radiosurgical Callosotomy

Callosotomy is performed as a palliative procedure in patients with severe generalized epilepsy who are not candidates for seizure focus resection. The aim of callosotomy is to prevent the propagation of seizures from one hemisphere to the other and control drop attacks caused by generalized tonic or atonic seizures, but a benefit has also been reported for secondarily generalized tonic-clonic seizures, myoclonic seizures, and simple and complex partial seizures. The adoption of surgical callosotomy has decreased since the introduction of vagal nerve stimulation

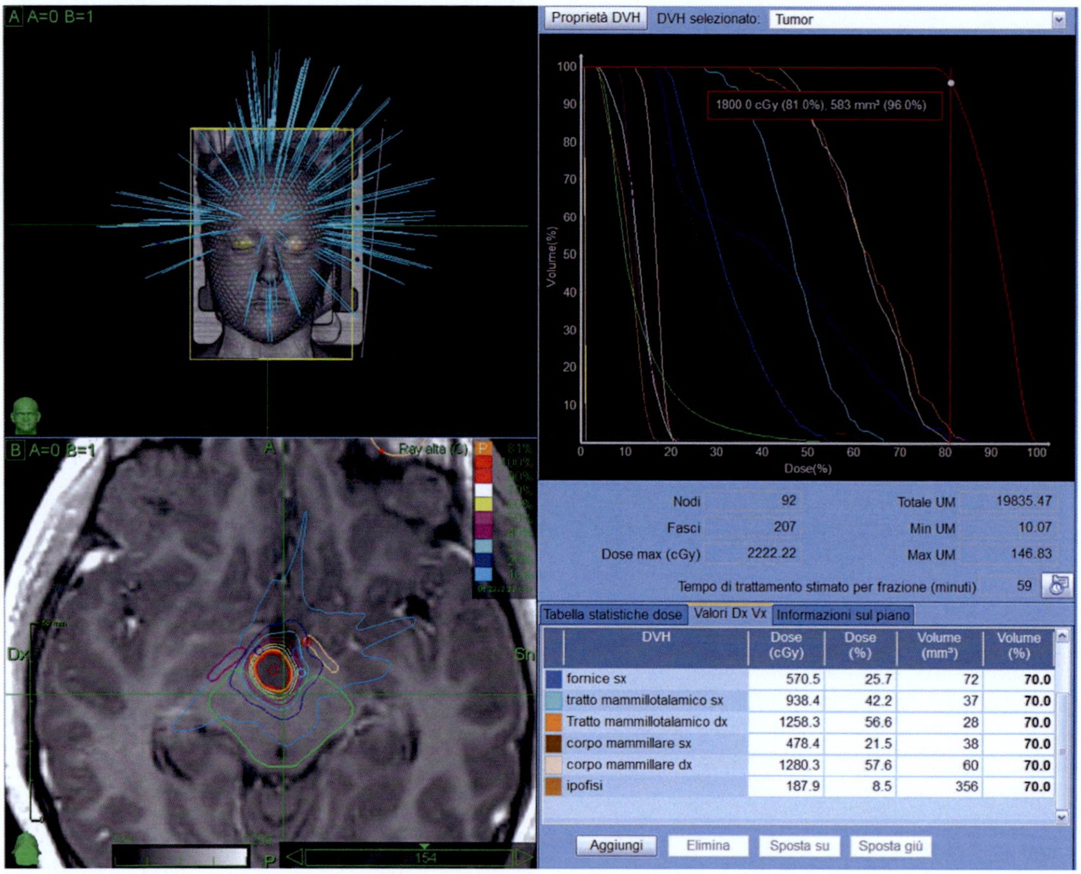

Fig. 49.3 CyberKnife treatment planning for a medium-sized sessile HH located in the interpeduncular fossa extending into the interpeduncular cistern. The lesion is adjacent to the optic chiasm, posterior to the optic tracts, and attached to the mammillary bodies which are compressed, distorted, and displaced posteriorly. Optic tracts, brainstem, and hippocampi outside the 30% isodose, thus being spared by high-dose irradiation. The mammillary bodies, fornix, and mammillary-thalamic tract were contoured to reduce their direct irradiation. Near-seizure freedom (Engel grade Ib) was achieved after 1 year, and complete seizure freedom (Engel grade Ia) was achieved after 2 years

(VNS), which has been demonstrated to prevent tonic and atonic seizures, although controversies remain about which procedure has the best efficacy/risk profile [41]. Radiosurgical callosotomy induces a slow and progressive axonal degeneration of white matter fibers as a consequence of neuronal and/or axonal injury. Diffusion tensor imaging (DTI) acquired 3 and 9 months after radiosurgery showed a progressive decrease of the fractional anisotropy in the irradiated region, indicating a progressive disconnection of callosal fibers [42]. Feichtinger and coworkers [43] have described the long-term results of Gamma Knife callosotomy on eight patients with severe generalized epilepsy and drop attacks. Six patients underwent anterior callosotomy (involving the rostral third of the corpus callosum), while two patients received a posterior callosotomy following hemispherotomy. In one patient, a second procedure was performed involving the middle third of the corpus callosum. The treatment was performed using a 4 mm collimator and placing three to five isocenters along the selected region (anterior, middle, or posterior third of the corpus callosum). The maximum dose ranged from 110 to 170 Gy. The prescribed 50% isodose ranged from 55 to 85 Gy. General anesthesia was required in five patients. Target volume ranged from 0.1 to 0.7 mL. Three patients experienced a complete disappearance of drop attacks, while two more experienced a 60% reduction. Generalized tonic-clonic seizures disappeared in two patients, while two others experienced a 50% and 60% decrease. Subacute transient headache and nausea appeared 4–6 months after the treatment in two patients. These symptoms were related to mild radio-induced edema in one case and to radionecrosis within the target region associated with bifrontal edema (this case received 55 Gy prescribed in the 50% isodose) in the second. Steroid administration induced symptomatic remission in both cases. Later MRI controls showed edema regression, while the radionecrosis remained limited to the callosal region. On the basis of this experience, the authors recommend prescribing a 50% isodose delivering 45–50 Gy to the selected callosal region and then, if needed, adding a further segment to the treatment.

Figure 49.4 shows a CyberKnife callosotomy administered to a 22-year-old patient with daily drop attacks plus other seizure types (multiple daily complex partial seizures and tonic-clonic generalized seizures twice a week). Presurgical evaluation could not identify a resectable seizure focus: prolonged video EEG monitoring was remarkable for showing a bilateral interictal activity localized over the parietal and occipital lobes. No ictal onset could be identified. A dedicated MRI study failed to show structural lesions. The target region (posterior third of the corpus callosum) received 40 Gy prescribed to the 82% isodose. Target volume was 0.4 mL. Maximum dose was 48.76 Gy. Total number of beams was 193. Beam delivery was nonisocentric, with a large number of beams penetrating around or below the orbito-meatal line. Treatment duration was 76 min. The patient was discharged home uneventfully immediately after the treatment. A single dose of dexamethasone (8 mg intramuscularly) was administered before discharge. This patient developed mild brain edema associated with worsening of the seizures 3 months after the treatment. Steroid administration was needed to improve seizure control. Six months after the treatment, no further drop attacks and tonic-clonic generalized seizures occurred.

49.6 Synchrotron-Generated Cortical and Hippocampal Transections

Microscopic arrays of radiograph beams (microbeams) originating from a synchrotron source can induce the equivalent of a microsurgical neocortical or hippocampal incision by delivering very high doses of radiation to tissue slices of microscopic thickness.

Neurons, glia, and axons along the penetration path receive peak doses of up to 1000 Gy and die immediately, while the immediately adjacent tissue is exposed to much lower doses of the valley (<6 Gy) which are unable to induce damage to histologically evident tissues [44].

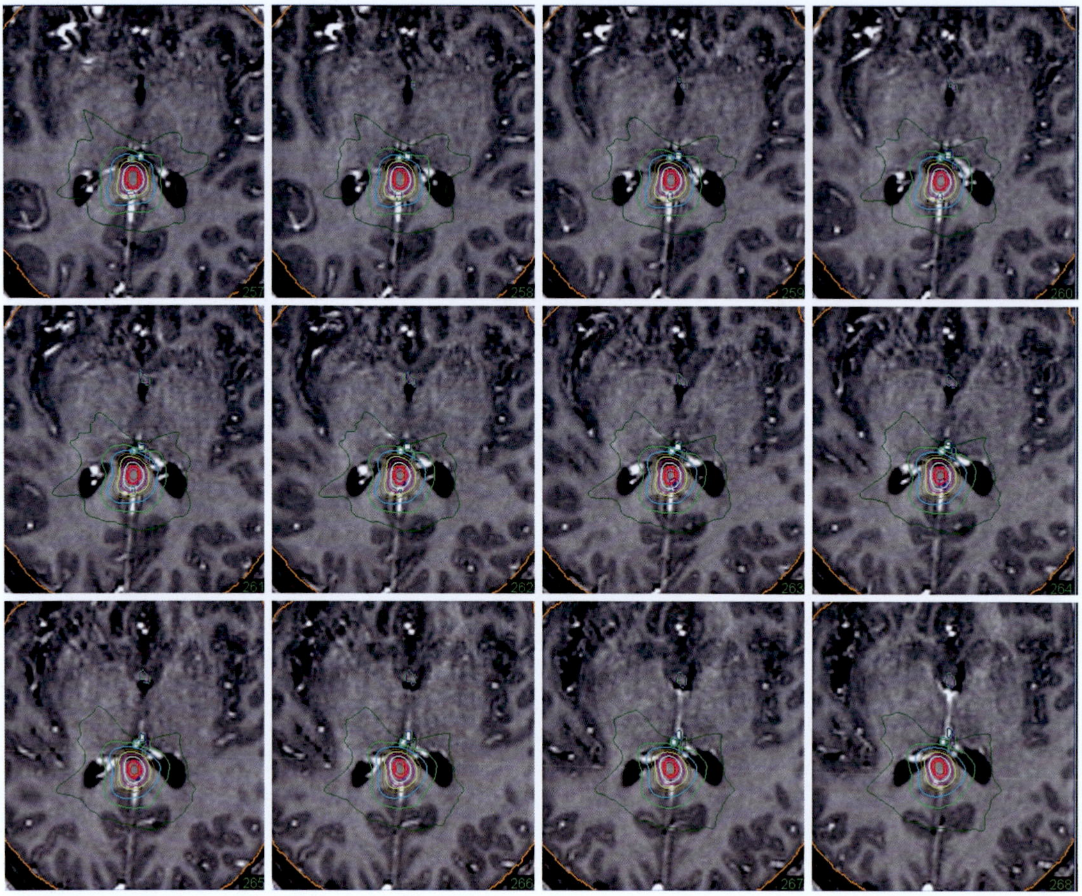

Fig. 49.4 CyberKnife posterior callosotomy: the absence of a stereotactic frame facilitates a wide array of beam trajectories. Venous structures surrounding the splenium are protected from radiation injury by the tight dosimetry

In essence, synchrotron-generated cortical transections provide a microradiosurgical equivalent of multiple subpial transections (MSTs), a non-resective surgical technique developed to treat patients with medically refractory epilepsy involving eloquent cortex [45–47].

This technique requires the placement of vertical incisions through the epileptic cortex in order to cut the horizontal axons responsible for the propagation of seizures while preserving the vertical axons subserving neurologic functions. The vertical columns working as the basic unit of cortical function are disconnected but not injured by MSTs, allowing the treatment of epileptic foci located over the sensorimotor or language cortex not amenable to surgical resection.

In the first experimental experiences, microbeam transections were performed on an epileptogenic focus located in the sensorimotor cortex, with almost immediate cessation of seizures and excellent conservation of motor function [48–50].

Hippocampal transection has been investigated as well in a rat model (Fig. 49.5), and further studies are ongoing to characterize the ability of hippocampal transections to control seizures originating from this region. These results suggest further investigations aimed at assessing the potential of microbeam transections to modulate cortical functions and to treat focal epilepsy.

The microbeam transection, either placed over neocortical seizure foci or through the hippocam-

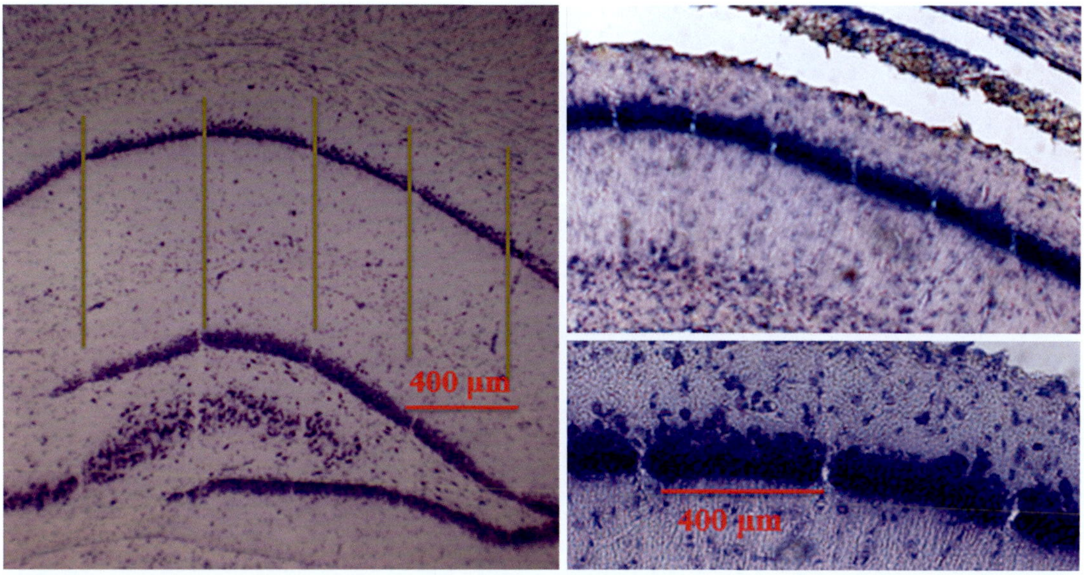

Fig. 49.5 Synchrotron-generated hippocampal transection shown on Nissl staining obtained 3 months after delivery of an array of nine parallel microbeams to a healthy Wistar rat. Microbeam size was 75 µm, and spacing across the beams was 400 µm. Incident dose was 600 Gy. The treatment was well tolerated, causing no radionecrosis or edema or evident behavioral or cognitive deficits

pus, could prove to be an excellent tool to add to the current radiosurgical techniques used to control seizures. The development of clinical devices delivering submillimetric beams able to generate cortical transections might add a powerful new tool to the clinical treatment of epilepsy and, more generally, to modulate cortical functions in a wide variety of neuropsychiatric disorders (Fig. 49.5).

49.7 Conclusions

CyberKnife radiosurgery is an option for selected cases of medically refractory epilepsy. It provides frameless and minimally invasive treatment, offering an attractive alternative to surgical resection for a wide variety of patients who refuse or are not candidates for conventional surgery.

References

1. Romanelli P, Anschel DJ. Radiosurgery for epilepsy. Lancet Neurol. 2006;5(7):613–20.
2. Romanelli P, Striano P, Barbarisi M, Coppola G, Anschel DJ. Non-resective surgery and radiosurgery for treatment of drug-resistant epilepsy. Epilepsy Res. 2012;99(3):193–201.
3. Regis J, Bartolomei F, Hayashi M, Chauvel P. Gamma knife surgery, a neuromodulation therapy in epilepsy surgery! Acta Neurochir Suppl. 2002;84:37–47.
4. Regis J, Carron R, Bartolomei F, Chauvel P. Seeking new paradigms in epilepsy: stereotactic radiosurgery. Clin Neurosurg. 2012;59:59–69.
5. Regis J, Carron R, Park M. Is radiosurgery a neuromodulation therapy? : a 2009 Fabrikant award lecture. J Neurooncol. 2010;98(2):155–62.
6. Srikijvilaikul T, Najm I, Foldvary-Schaefer N, Lineweaver T, Suh JH, Bingaman WE. Failure of gamma knife radiosurgery for mesial temporal lobe epilepsy: report of five cases. Neurosurgery. 2004;54(6):1395–402; discussion 402–4.
7. Regis J, Bartolomei F, Metellus P, Rey M, Genton P, Dravet C, et al. Radiosurgery for trigeminal neuralgia and epilepsy. Neurosurg Clin N Am. 1999;10(2):359–77.
8. Regis J, Bartolomei F, Rey M, Genton P, Dravet C, Semah F, et al. Gamma knife surgery for mesial temporal lobe epilepsy. Epilepsia. 1999;40(11):1551–6.
9. Regis J, Bartolomei F, Rey M, Hayashi M, Chauvel P, Peragut JC. Gamma knife surgery for mesial temporal lobe epilepsy. J Neurosurg. 2000;93(Suppl 3):141–6.
10. Regis J, Rey M, Bartolomei F, Vladyka V, Liscak R, Schrottner O, et al. Gamma knife surgery in mesial

temporal lobe epilepsy: a prospective multicenter study. Epilepsia. 2004;45(5):504–15.
11. Regis J, Semah F, Bryan RN, Levrier O, Rey M, Samson Y, et al. Early and delayed MR and PET changes after selective temporomesial radiosurgery in mesial temporal lobe epilepsy. AJNR Am J Neuroradiol. 1999;20(2):213–6.
12. Chang EF, Quigg M, Oh MC, Dillon WP, Ward MM, Laxer KD, et al. Predictors of efficacy after stereotactic radiosurgery for medial temporal lobe epilepsy. Neurology. 2010;74(2):165–72.
13. Schrottner O, Unger F, Eder HG, Feichtinger M, Pendl G. Gamma-knife radiosurgery of mesiotemporal tumour epilepsy observations and long-term results. Acta Neurochir Suppl. 2002;84:49–55.
14. Kurita H, Suzuki I, Shin M, Kawai K, Tago M, Momose T, et al. Successful radiosurgical treatment of lesional epilepsy of mesial temporal origin. Minim Invasive Neurosurg. 2001;44(1):43–6.
15. Barbaro NM, Quigg M, Broshek DK, Ward MM, Lamborn KR, Laxer KD, et al. A multicenter, prospective pilot study of gamma knife radiosurgery for mesial temporal lobe epilepsy: seizure response, adverse events, and verbal memory. Ann Neurol. 2009;65(2):167–75.
16. Quigg M, Broshek DK, Barbaro NM, Ward MM, Laxer KD, Yan G, et al. Neuropsychological outcomes after gamma knife radiosurgery for mesial temporal lobe epilepsy: a prospective multicenter study. Epilepsia. 2011;52(5):909–16.
17. Barbaro NM, Quigg M, Ward MM, Chang EF, Broshek DK, Langfitt JT, et al. Radiosurgery versus open surgery for mesial temporal lobe epilepsy: the randomized, controlled ROSE trial. Epilepsia. 2018;59(6):1198–207.
18. Stefan H, Hummel C, Grabenbauer GG, Muller RG, Robeck S, Hofmann W, et al. Successful treatment of focal epilepsy by fractionated stereotactic radiotherapy. Eur Neurol. 1998;39(4):248–50.
19. Heers M, Rampp S, Stefan H, Urbach H, Elger CE, von Lehe M, et al. MEG-based identification of the epileptogenic zone in occult peri-insular epilepsy. Seizure. 2012;21(2):128–33.
20. Tovar-Spinoza ZS, Ochi A, Rutka JT, Go C, Otsubo H. The role of magnetoencephalography in epilepsy surgery. Neurosurg Focus. 2008;25(3):E16.
21. Wu XT, Rampp S, Buchfelder M, Kuwert T, Blumcke I, Dorfler A, et al. Interictal magnetoencephalography used in magnetic resonance imaging-negative patients with epilepsy. Acta Neurol Scand. 2013;127(4):274–80.
22. Stefan H, Lopes da Silva FH. Epileptic neuronal networks: methods of identification and clinical relevance. Front Neurol. 2013;4:8.
23. Barcia JA, Barcia-Salorio JL, Lopez-Gomez L, Hernandez G. Stereotactic radiosurgery may be effective in the treatment of idiopathic epilepsy: report on the methods and results in a series of eleven cases. Stereotact Funct Neurosurg. 1994; 63(1–4):271–9.
24. Barcia-Salorio JL, Barcia JA, Hernandez G, Lopez-Gomez L. Radiosurgery of epilepsy. Long-term results. Acta Neurochir Suppl. 1994;62:111–3.
25. Barcia-Salorio JL, Barcia JA, Roldan P, Hernandez G, Lopez-Gomez L. Radiosurgery of epilepsy. Acta Neurochir Suppl (Wien). 1993;58:195–7.
26. Pollock BE, Lunsford LD, Kondziolka D, Maitz A, Flickinger JC. Patient outcomes after stereotactic radiosurgery for "operable" arteriovenous malformations. Neurosurgery. 1994;35(1):1–7; discussion 7–8.
27. Hadjipanayis CG, Levy EI, Niranjan A, Firlik AD, Kondziolka D, Flickinger JC, et al. Stereotactic radiosurgery for motor cortex region arteriovenous malformations. Neurosurgery. 2001;48(1):70–6; discussion 6–7.
28. Hoh BL, Ogilvy CS, Butler WE, Loeffler JS, Putman CM, Chapman PH. Multimodality treatment of nongalenic arteriovenous malformations in pediatric patients. Neurosurgery. 2000;47(2):346–57; discussion 57–58.
29. Hoh BL, Chapman PH, Loeffler JS, Carter BS, Ogilvy CS. Results of multimodality treatment for 141 patients with brain arteriovenous malformations and seizures: factors associated with seizure incidence and seizure outcomes. Neurosurgery. 2002;51(2):303–9; discussion 9–11.
30. Amin-Hanjani S, Ogilvy CS, Candia GJ, Lyons S, Chapman PH. Stereotactic radiosurgery for cavernous malformations: Kjellberg's experience with proton beam therapy in 98 cases at the Harvard Cyclotron. Neurosurgery. 1998;42(6):1229–36; discussion 36–38.
31. Regis J, Bartolomei F, Kida Y, Kobayashi T, Vladyka V, Liscak R, et al. Radiosurgery for epilepsy associated with cavernous malformation: retrospective study in 49 patients. Neurosurgery. 2000;47(5):1091–7.
32. Romanelli P, Verdecchia M, Rodas R, Seri S, Curatolo P. Epilepsy surgery for tuberous sclerosis. Pediatr Neurol. 2004;31(4):239–47.
33. Arita K, Kurisu K, Kiura Y, Iida K, Otsubo H. Hypothalamic hamartoma. Neurol Med Chir (Tokyo). 2005;45(5):221–31.
34. Romanelli P. CyberKnife(R) radiosurgery as first-line treatment for catastrophic epilepsy caused by hypothalamic hamartoma. Cureus. 2018;10(7):e2968.
35. Calisto A, Dorfmuller G, Fohlen M, Bulteau C, Conti A, Delalande O. Endoscopic disconnection of hypothalamic hamartomas: safety and feasibility of robot-assisted, thulium laser-based procedures. J Neurosurg Pediatr. 2014;14(6):563–72.
36. Regis J, Scavarda D, Tamura M, Villeneuve N, Bartolomei F, Brue T, et al. Gamma knife surgery for epilepsy related to hypothalamic hamartomas. Semin Pediatr Neurol. 2007;14(2):73–9.
37. Romanelli P, Muacevic A, Striano S. Radiosurgery for hypothalamic hamartomas. Neurosurg Focus. 2008;24(5):E9.
38. Selch MT, Gorgulho A, Mattozo C, Solberg TD, Cabatan-Awang C, DeSalles AA. Linear accelerator stereotactic radiosurgery for the treatment of gelas-

tic seizures due to hypothalamic hamartoma. Minim Invasive Neurosurg. 2005;48(5):310–4.
39. Unger F, Schrottner O, Feichtinger M, Bone G, Haselsberger K, Sutter B. Stereotactic radiosurgery for hypothalamic hamartomas. Acta Neurochir Suppl. 2002;84:57–63.
40. Regis J, Lagmari M, Carron R, Hayashi M, McGonigal A, Daquin G, et al. Safety and efficacy of gamma knife radiosurgery in hypothalamic hamartomas with severe epilepsies: a prospective trial in 48 patients and review of the literature. Epilepsia. 2017;58(Suppl 2):60–71.
41. Englot DJ, Birk H, Chang EF. Seizure outcomes in nonresective epilepsy surgery: an update. Neurosurg Rev. 2017;40(2):181–94.
42. Moreno-Jimenez S, San-Juan D, Larraga-Gutierrez JM, Celis MA, Alonso-Vanegas MA, Anschel DJ. Diffusion tensor imaging in radiosurgical callosotomy. Seizure. 2012;21(6):473–7.
43. Feichtinger M, Schrottner O, Eder H, Holthausen H, Pieper T, Unger F, et al. Efficacy and safety of radiosurgical callosotomy: a retrospective analysis. Epilepsia. 2006;47(7):1184–91.
44. Romanelli P, Bravin A. Synchrotron-generated microbeam radiosurgery: a novel experimental approach to modulate brain function. Neurol Res. 2011;33(8):825–31.
45. Devinsky O, Romanelli P, Orbach D, Pacia S, Doyle W. Surgical treatment of multifocal epilepsy involving eloquent cortex. Epilepsia. 2003;44(5):718–23.
46. Morrell F, Whisler WW, Bleck TP. Multiple subpial transection: a new approach to the surgical treatment of focal epilepsy. J Neurosurg. 1989;70(2):231–9.
47. Morrell F, Whisler WW, Smith MC, Hoeppner TJ, de Toledo-Morrell L, Pierre-Louis SJ, et al. Landau-Kleffner syndrome. Treatment with subpial intracortical transection. Brain. 1995;118(Pt 6):1529–46.
48. Romanelli P, Fardone E, Battaglia G, Brauer-Krisch E, Prezado Y, Requardt H, et al. Synchrotron-generated microbeam sensorimotor cortex transections induce seizure control without disruption of neurological functions. PLoS One. 2013;8(1):e53549.
49. Anschel DJ, Bravin A, Romanelli P. Microbeam radiosurgery using synchrotron-generated submillimetric beams: a new tool for the treatment of brain disorders. Neurosurg Rev. 2010;34(2):133–42.
50. Romanelli P, Fardone E, Bucci D, Battaglia G, Brauer-Krisch E, Requardt H, et al. Microradiosurgical cortical transections generated by synchrotron radiation. Phys Med. 2015;31(6):642–6.

50 Behavioral/Psychiatric Disorders

Sait Sirin and Kaan Oysul

50.1 Introduction

Obsessive-compulsive disorder (OCD) is a chronic, severe, and frequent disorder with an estimated lifetime prevalence of 2.3% in the United States [1]. OCD is characterized by repetitive and intrusive thoughts and behaviors that cause clinically significant distress or impairment [2].

It has a poor outcome, with a remission rate of just 53% [3]. OCD typically runs a chronic course, with sequential periods of remission and relapse, and is associated with disabling comorbidities, including major depressive disorder (15%), social anxiety disorder (14%), generalized anxiety disorder (13%), persistent depressive disorder (13%), tic disorder (12.5%), body dysmorphic disorder (8.71%), and self-harming behavior (7.43%) [4].

Approximately 40–60% of patients with OCD fail to satisfactorily respond to standard treatments, including serotonin reuptake inhibitors and cognitive behavioral therapy. In some OCD patients (less than 1% of treatment-seeking individuals), the condition is severe and considered "intractable" [5]. These patients are potential candidates for neurosurgical intervention (Table 50.1).

S. Sirin (✉) · K. Oysul
Department of Radiation Oncology, CyberKnife Radiosurgery Center, Medicana International Ankara Hospital, Ankara, Turkey
e-mail: ssirin@medicana.com.tr; kaan@oysul.com

© Springer Nature Switzerland AG 2020
A. Conti et al. (eds.), *CyberKnife NeuroRadiosurgery*,
https://doi.org/10.1007/978-3-030-50668-1_50

In some cases, neurosurgery might be the only viable therapeutic option. Table 50.2 describes the current selection criteria for neurosurgical candidacy [11, 12, 17, 18]. Types of interventions include target brain lesioning or electrical brain stimulation. Lesioning can be performed using radiofrequency or radiation energy [19]. Since the discovery that lesion of the anterior limb of the internal capsule may alleviate the symptoms of OCD, other neurosurgical targets for lesioning and deep brain stimulation (DBS) for OCD have since been investigated [20]. Indeed, the ventral striatum, the nucleus accumbens, the anterior cingulate gyrus, the substantia innominata, and the subthalamic nucleus have been proposed targets for OCD [21, 22]. While DBS avoids the creation of a permanent lesion, it requires lifelong follow-up and carries the risk of infection and intracerebral hemorrhage [23].

50.2 History of Radiosurgery for OCD

The development of frame-based stereotactic procedures in the late 1940s enabled neurosurgical lesions to be created in a relatively precise, reproducible way, unlike the broadly destructive frontal lobotomies which have been disparaged for indiscriminate application and occasionally dramatic adverse effects [24, 25].

Table 50.1 Summary of patients with OCD who had anterior capsulotomy with Gamma Knife. Adapted from Martinez-Alvarez and improved with more series. Results were given as the mean

Authors	Patient, number	Age, years	Follow-up, months	Dose (Gy)	Preop Y-BOCS	Postop Y-BOCS	Success, *n* (%)	Side effects, %
Ruck et al. [6]	9	43.9	136.8	180–200	33.4	14.2	5 (55)	44
Lopes et al. [7]	5	35	48	180	32	24	3 (60)	40
Gouvea et al. [8]	1	34	12	180	37	0	–	0
Kondziolka et al. [9]	3	44	42	140–150	37.3	16.3	2 (67)	0
Sheehan et al. [10]	4	38	26	140–160	32	17	3 (80)	0
Lopes et al. [11, 12]	12	33.9	55.2	180	33.6	17.3	7 (58.3)	13
Peker et al. [13]	10	NA	9	140–150	38	16	7 (70)	NA
Rasmussen et al. [14]	SSR 15 DS 40	35.8 32.8	36 36	180 180	33.2 34.1	19.2 16.7	7 (46.6) 30 (75)	5
Martinez-Alvarez [15]	10	41.2	40.2	120	32.7	14.4	7 (70)	0

SSR single shot repeated, *DS* double shot

Table 50.2 Current selection criteria for neurosurgery for intractable OCD [16]

Inclusion criteria
• Main diagnosis of OCD (if comorbid Axis I or II disorders are present, OCD symptoms should be the most troublesome)
• Y-BOCS OCD severity rating of 28 or higher (extremely ill) or 14 if only obsessions or only compulsions are present. In any potential candidate, OCD must be extremely time-consuming or impairing
• ≥5 years of severe OCD symptoms despite adequate treatment trials
• Refractoriness, as evidenced by insufficient response to the following: – 3 trials with an SRI (selective or not), at least one of which should be with clomipramine. All trials should have a minimum duration of 12 weeks, at the maximum tolerated dose – 2 augmentation strategies, such as the use of antipsychotic drugs (typical or atypical) or clomipramine, with adequate duration and dose – 20 h of OCD-specific BT (i.e., ERP). Participation for shorter times may be permitted if nonadherence is due to symptom severity rather than to noncompliance
• Independent confirmation of the above refractoriness criteria with previous mental health providers
• Age 18–75 years (increasing age is a relative contraindication)
• Ability to provide informed consent
• Appropriate expectations of the outcomes of surgery
Exclusion criteria
• Comorbid psychiatric disorder that may interfere with treatment (e.g., severe personality disorder or psychosis)
• Clinically significant condition affecting brain function or structure
• Cognition in the low range
• Past history of head injury, with posttraumatic amnesia
• Current substance use disorder
• Recent suicide attempt or active, formed suicidal ideation

OCD obsessive-compulsive disorder, *Y-BOCS* Yale-Brown Obsessive-Compulsive Scale, *SRI* serotonin reuptake inhibitor, *BT* behavior therapy, *ERP* exposure and response prevention

In 1949, Jean Talairach proposed treating psychiatric disorders by using stereotactic RF thermocoagulation to create lesions in the anterior limb of the internal capsule (ALIC) [26]. In addition to the ALIC, targets have included the anterior cingulate cortex (ACC) and subcaudate white matter. The procedures developed were capsulotomy (targeting the ALIC), anterior cingulotomy (targeting the ACC), subcaudate tractotomy (targeting the subcaudate white matter), and a fourth procedure, limbic leucotomy, combining the last two [26–33].

The first case of anterior capsulotomy was performed with stereotactic radiosurgery using a 300-kV industrial X-ray tube by Lars Leksell in 1953, 4 years after the first description of RF capsulotomy. Proton beam-based radiosurgery was used by Lars Leksell to perform anterior capsulotomy in 1960. In 1967, Leksell presented a radiosurgical apparatus intended for research and routine clinical use, equipped with sources of a radioactive isotope of cobalt (^{60}Co) emitting high-energy gamma rays with a half-life of 5.27 years. The "Gamma Knife" (GK) employs many ^{60}Co sources, arranged in a hemispherical or conical configuration within a helmet-like part of the device [34]. Since 1976, Gamma Knife radiosurgery for anterior capsulotomy was performed by Lars Leksell and coworkers. The first case of gamma ventral capsulotomy (GVC) was treated as proposed by Steve Rasmussen in 1993 [16]. Thereafter, other specialized centers started using GK for intractable OCD.

50.3 Anatomy, Physiopathology, and Target Selection

The ALIC carries ascending and descending fibers connecting the prefrontal cortex (PFC) to deep gray matter including the thalamus and basal ganglia. These cortical areas are associated with control over emotion, motivation, and cognition and are linked to psychiatric illnesses including OCD, major depressive disorder, schizophrenia, addiction, and several others [16]. Appreciation of this complex anatomy has largely been obtained from nonhuman primate tract-tracing studies [16]. The PFC is considerably larger and more complex in humans than in nonhuman primates, complicating comparisons. Methodological limitations preclude detailed studies of the anatomy and topography of the human ALIC, because neuronal tracing methods are not suitable for human use. In vivo MRI-based techniques, in particular diffusion tensor imaging (DTI), have been the mainstay of this effort.

Imaging studies in obsessive-compulsive disorder observe hypermetabolic changes in the orbital frontal cortex, the anterior cingulate region, the caudate nucleus, and the thalamus [21, 22]. Current studies support cortico-striato-thalamo-cortical dysfunction models, providing a basis for modulation or effects on this circuitry [35]. The ventral anterior internal capsule should be an appropriate target for such modulation.

Target optimization involves strategies to improve efficacy and reduce side effects. An attempt to find optimal lesion placement regarding treatment response used a retrospective analysis of postoperative MRI data. They found bilateral clusters of voxels in the ventral portion of the ALIC (in the coronal plane), approximately near the posterior putaminal border (in the axial plane) that were statistically related to responder status, suggesting that lesions including this region are more likely to produce a clinical response [36].

50.4 Gamma Knife Experience in OCD

50.4.1 Target, Number of Shots, and Dose

Between 1976 and 1979, Leksell, Backlund, and Rylander at the Karolinska Institute in Stockholm, Sweden, treated 12 patients with OCD using the first prototype of the Gamma Knife and the 3 × 5-mm collimator and a maximum dose of 160–180 Gy [37] (Fig. 50.1a). Then, the Karolinska group started using three, 4-mm bilateral shots with GK Model B and 200 Gy [38] (Fig. 50.1b).

In the United States, Rasmussen and coworkers treated 15 patients with a single 4-mm bilateral shot at 180 Gy using the GK Model U, resulting in spheroidal isocenters that are slightly prolate. The shots were located centrally in the internal capsule (midcapsule), 1/3 of the distance up from the base of the IC. Of those 15 patients, 13 underwent a second procedure, receiving another 180 Gy, using a 4-mm bilateral shot immediately ventral to the previous midpoint shot, bordering the ventral striatum. A refined technique combining both of the "shots" at one

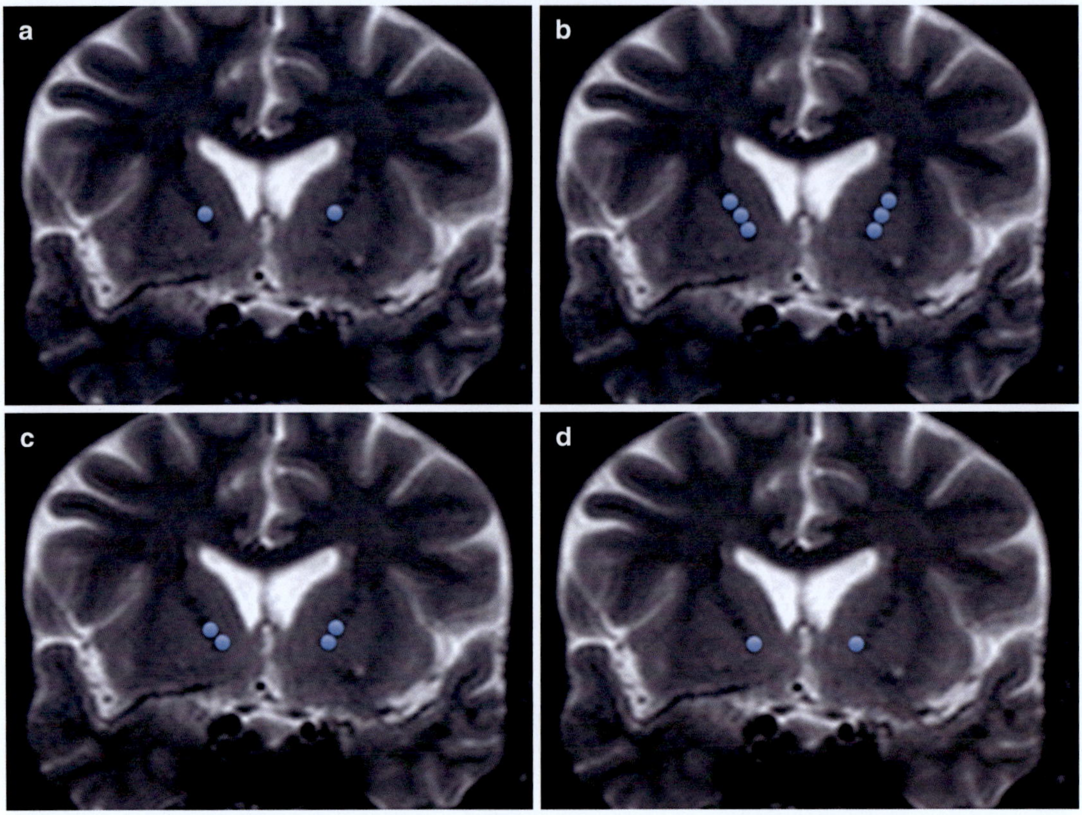

Fig. 50.1 Evolution of Gamma Knife capsulotomy. (**a**) Bilateral single-shot target for GK I, (**b**) bilateral triple shots for GK B, (**c**) bilateral double-shot GVC for GK C, (**d**) bilateral single-shot GVC for GK Perfexion

time ("double shot") was subsequently used in 40 patients, 22 treated with the Model U, and 18 with the Model C. The term gamma ventral capsulotomy (GVC) was coined to describe this "double-shot" procedure, with a 4-mm collimator targeting the ventral ALIC and bordering the ventral striatum [14] (Fig. 50.1c).

A group at the University of São Paulo (USP) treated five patients, reproducing all parameters of the 180-Gy double-shot GVC technique with a GK Model B. Following that pilot study, the same group conducted a double-blind, sham-controlled, randomized trial involving 16 patients [11, 12] (Fig. 50.1c). The University of Pittsburgh group has proposed reducing the radiation dose while maintaining ventrally focused targets in the ALIC. The Pittsburgh group treated three patients using GK Model C and 4C with 140–150 Gy and bilateral double 4-mm shot [9] (Fig. 50.1c). The UVA group treated five patients with a single bilateral ventral shot (140–160 Gy), using the newer GK model Perfexion, which differs from previous models in its source geometry but with isodose distributions similar to models B and C, creating oblate spheroidal isocenters [10] (Fig. 50.1d). Spatola et al. from Madrid treated patients with refractory OCD using GK model 4C and Perfexion. They adopted GVC with 120 Gy [39].

50.4.2 Gamma Knife Outcomes

Early studies from the Karolinska Institute reported some degree of clinical benefit in 36–56% of patients treated with GK without standardized patient selection criteria and validated OCD rating scale [16]. Several adverse events were observed. Of the nine patients, five (56%) exhibited severe frontal lobe edema at

12 months, with symptoms including headache, apathy, fatigue, loss of initiative, and disinhibition. Two of the patients improved over time, but three (33%) remained symptomatic.

The most commonly used symptom scale for OCD, the Yale-Brown Obsessive-Compulsive Scale (Y-BOCS), was developed in 1989 [40]. Using Y-BOCS in combination with other standardized scales to measure symptom severity, Lippitz et al. reported a ≥50% improvement in the scores in 70% of ten patients undergoing GK capsulotomy [41]. The most common modern criterion for a full treatment response in OCD patients is a 35% decrease in the Y-BOCS score. In a series of nine patients with the same criterion, Rück and coworkers from Karolinska reported that 56% of patients undergoing GKC responded [6].

In the later Karolinska cohort, patients also developed frontal lobe dysfunction, associated with radiation dose. Patients had adverse events who received higher radiation doses from three 4-mm shots at 200 Gy or who were treated more than once.

Rasmussen et al. started to treat their first cohort of 15 patients with bilateral single mid-capsule shots. After observing insufficient response except one patient, they retreated 13 of the cohort by adding more ventral bilateral shots. At 3-year follow-up, 7/13 (54%) were full responders, and 2/13 (15%) were partial responders. With this experience, the Brown University group treated another 40 patients with double-shot GVC. At 36 months (using last observation carried forward), 30 (75%) were full responders and 5 (12.5%) were partial responders [14]. In this series, three patients (5%) developed radionecrotic cysts 3–5 years after GVC with double-shot 180 Gy with the GK Model C (oblate spheroidal isocenters). One of three patients required surgical decompression. Another four patients were treated with corticosteroids [14].

The University of São Paulo (USP) group selected GVC in their five OCD patients, and at 48 months, three (60%) were complete responders, and one (20%) was a partial responder [7]. By observing those promising results, the USP group conducted a double-blind, sham-controlled randomized trial involving 16 patients with OCD. Eight patients were randomized to active GVC, and the other eight to a well-executed sham procedure that included the same head frame placement as the active procedure, but with a sham attachment on the GK device. In the double-blinded period of 12 months, only 2/8 were responders in GVC group using both Y-BOCS and Clinical Global Impressions-Improvement (CGI-I) scale. In the 54-month open-label period, there were three more responders rated at months 14, 18, and 24. During that period, 4/8 of the sham group accepted open-label treatment, and 2/4 of them had complete response at months 6 and 36. Therefore, 7/12 (58%) of the patients who underwent GVC were complete responders after long-term follow-up in the open-label period of the study [11, 12]. If only the Y-BOCS criterion had been applied in the USP trial, two additional patients during the open-label follow-up would also have been labeled as responders, for a total of 9/12 (75%) overall [11, 12]. Of the 12 patients who received GVC in the USP randomized trial, two patients with a history of hypomania experienced manic episodes that were successfully treated pharmacologically, and one patient with no history of drug abuse subsequently developed drug dependence [87]. One patient (8%) developed symptoms of delirium, confabulation, and visual hallucinations 8 months after treatment. An MRI showed peri-lesional edema, and the patient was treated with corticosteroids, with resolution of the symptoms 5 months later [11, 12].

The Pittsburgh group published three OCD patients with Y-BOCS scores superior to 24 underwent GVC with double shots and 140–150 Gy. If the 35% reduction in Y-BOCS score response criterion was applied, two (67%) of the three patients would be categorized as complete responders. The score of the third patient decreased from 35 to 24 (a 29% reduction) after 55 months [9].

The Virginia group treated five patients with OCD and Y-BOCS score more than 24 using GVC with bilateral single ventral shots and 140–160 Gy.

At a median follow-up of 24 months, four of the five patients (80%) had a reduction in Y-BOCS score ranging between 59 and 62% which meets the conventional ≥35% response criterion [10]. There were no adverse events in the Pittsburgh and Virginia series with relatively low-dose GVC. The Madrid group also treated ten patients using GVC with 120 Gy, and they achieved 70% success in reducing Y-BOCS score more than 35%, and at the last follow-up, none of the patients had experienced any significant adverse neuropsychological effects or personality changes [15].

Complete elimination of OCS is not to be expected after surgery, and this expectation should be clearly explained to all involved. A more realistic goal is to aim for levels of improvement that enhance the effects of conventional therapies, engendering possible synergistic effects between surgical and nonsurgical treatments [42]. Therefore, pharmacological and psychotherapeutic regimens are always maintained after GVC, being reduced only when clinical improvement occurs and persists. Medications are rarely discontinued after surgery. Given the possibility of delayed side effects (e.g., swelling or cyst formation), patients should be followed for years. A recent report from the Karolinska group took advantage of the national health registry system in Sweden to provide very long-term follow-up information (from 13 to 43 years) on 70 patients who had undergone capsulotomy [43]. A notable finding was that among the patients who were still alive, 75% were still being prescribed at least two psychiatric medications, most commonly antidepressants.

50.5 CyberKnife Experience in OCD

Although there are many studies investigating the results of CyberKnife radiosurgery for trigeminal neuralgia, there is only one study on radiosurgery for mental illnesses using CyberKnife [44]. Kim and coworkers reported management of 11 patients with OCD and 4 patients with depression using CyberKnife radiosurgery. Computed tomography (CT) was used in patients for the localization of the target. The 80% isodose line was prescribed in a conformal fashion to a 7-mm diameter of the target. The authors started 75 Gy with 10-mm collimator at the 80% isodose line, but the necrotic lesion volume was larger than their expectations, and the dose parameter was reduced to 50 Gy with 7-mm collimator at the 80% margin dose line. In four patients with depression who underwent CKRS, the median score in Hamilton Depression Rating Scale (HAM-D) declined from 34 to 12, and three patients returned to previous social life. With follow-up in 11 patients with OCDs after CKRS, the median score in Y-BOCS of 6 patients declined from 37 to 23 after 10 months, and clinical improvement was observed. There was no operative mortality after CKRS and no significant morbidity except one patient with fatigue and malaise [44].

At our center, we treated 15 patients with intractable OCD between 2014 and 2019. Patient selection criteria were the same as for GVC patients. Y-BOCS, Beck Anxiety Scale, and Beck Depression Scale were used for patients to rate the severity of the symptoms prior to the treatment and at the follow-ups. Navigation CT, T2W, and T1W with contrast MR images were obtained and co-registered using MultiPlan.

Bilateral target volumes were contoured in the anterior limb of internal capsule at midputaminal region on axial plane and reaching the base of IC on the coronal plane. Target volume delineation simulates double-shot GVC (Figs. 50.2 and 50.3). Oblique distance on coronal plane varied 7–9 mm and each target volume varied. In the first five patients, we prescribed 70 Gy at the 80–85% isodose line; in the second five patients, 80 Gy at the 80% isodose line; and for the last five patients, 95 Gy at the 79–83% isodose line. In the 70-Gy group, two patients had a repeated radiosurgery with 60 Gy after 7 and 10 months. In the 80-Gy group, one patient was retreated

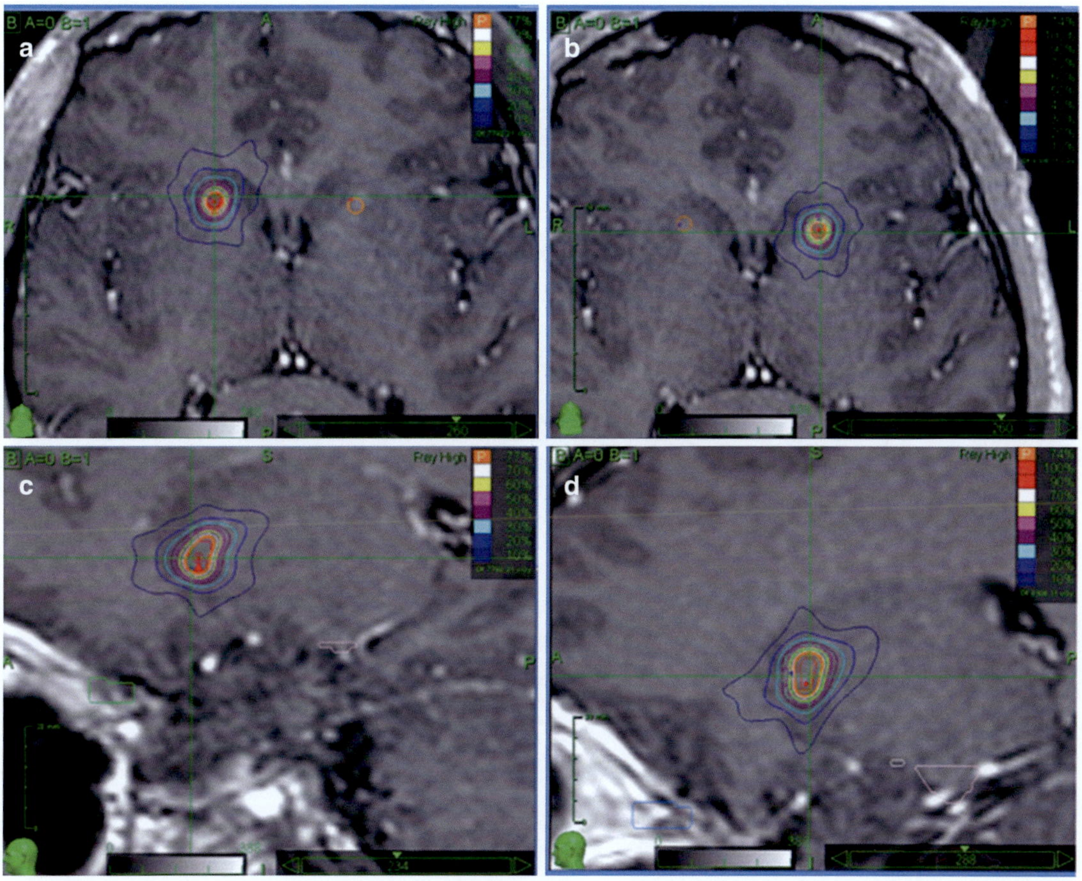

Fig. 50.2 CyberKnife radiosurgery plan for both sides. Target contouring is similar to bilateral double-shot GVC. Prescription dose is 95 Gy at the 83% isodose and maximum dose is 115 Gy, and dark blue represents the 10% isodose line. (**a**) Right-side axial plane, (**b**) left-side axial plane, (**c**) right-side sagittal plane, (**d**) left-side sagittal plane

with 60 Gy after 8 months. Retreatment decision was done due to lack of symptom relief and absence of bilateral capsulotomy lesions in the follow-up MR images. Median follow-up time is 28 months (range 2–68 months). Nine of the 15 patients (60%) had a reduction of more than 35% in Y-BOCS score. Typical MR features of the patients with good response are small bilateral lesions in ALIC which are hyperintense on T2W images and rim-like contrast enhancement on T1W images (Fig. 50.4). Patients had no acute adverse effect after the treatment. One patient in the 95-Gy group experienced headache and mental slowness after 5 months of treatment. MR images showed unanticipated bilateral radionecrosis that required medical treatment including steroid and hyperbaric oxygen treatment. While the symptoms of side effect totally resolved, Y-BOCS score reduced by more than 35% after the treatment.

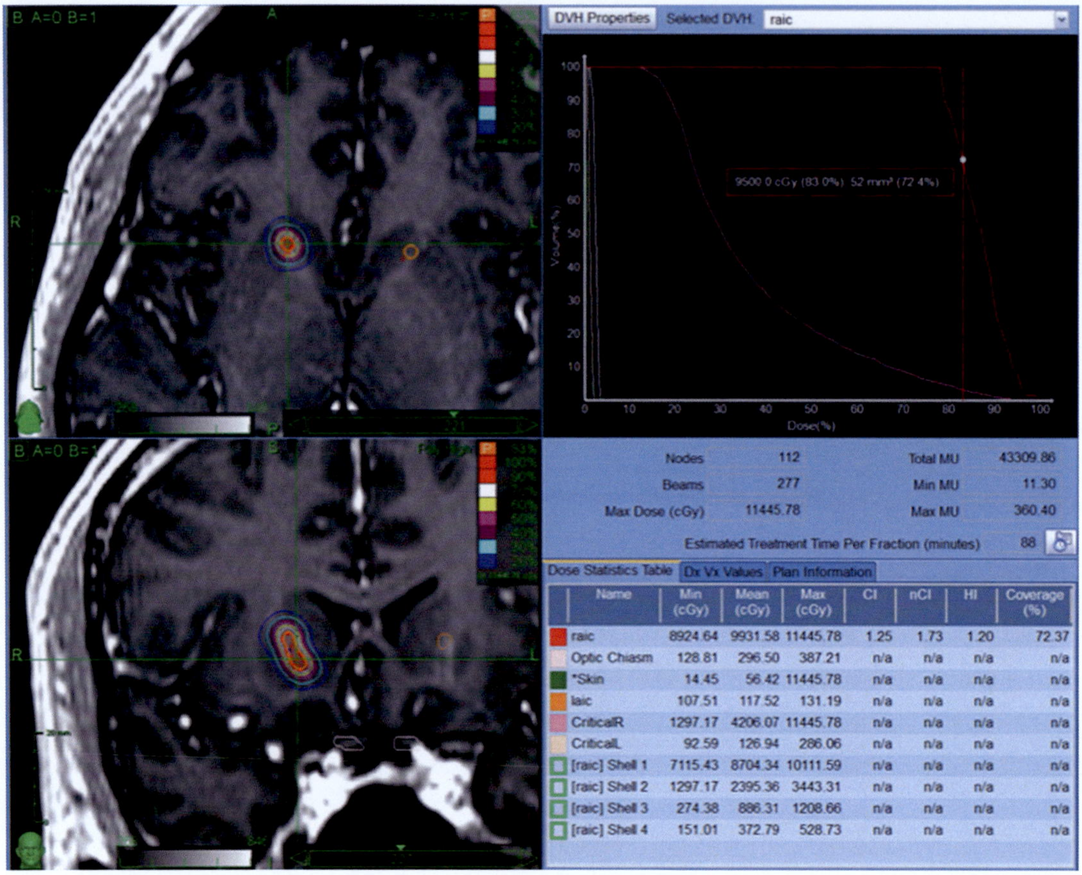

Fig. 50.3 CyberKnife radiosurgery plan for the right-side lesion in axial and coronal plane, prescription dose is 95 Gy at the 83% isodose line. High dose gradient protects midline structures and optic pathways

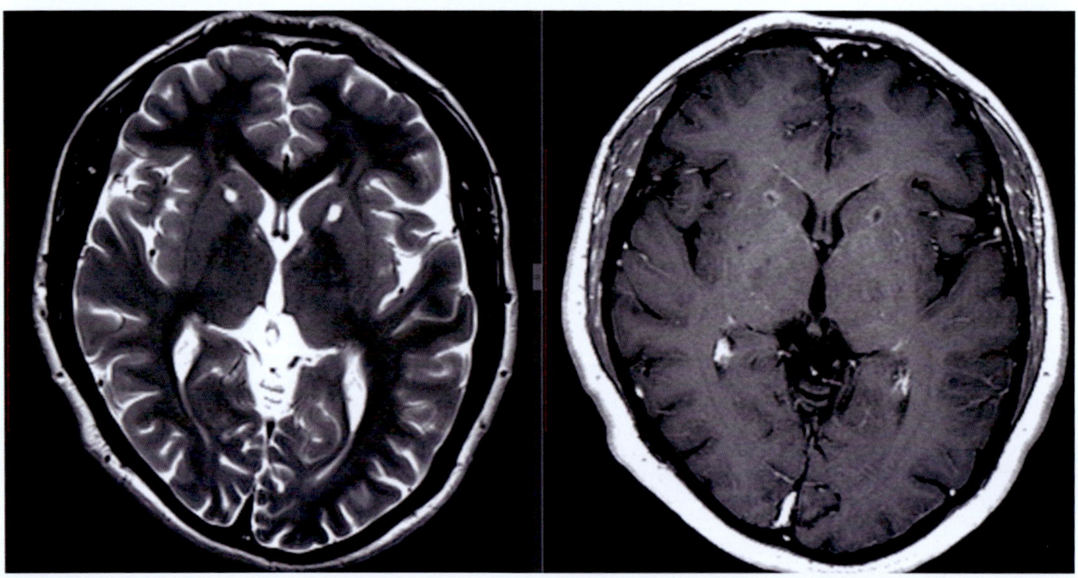

Fig. 50.4 Four-year follow-up axial T2W and contrast-enhanced T1W MR images of the patient with intractable OCD show well-defined bilateral ventral capsulotomies, hyperintense on T2W image and rim-like contrast enhancement on T1W image

50.6 Conclusion

Performing ventral capsulotomy with CyberKnife radiosurgery is safe and effective for patients with a refractory obsessive disorder. There is more to study for the optimal dose, target position, target volume, and preoperative and postoperative imaging characteristics for successful intervention. As in many radiosurgical indications such as benign tumors, there is a tendency to reduce the dose to ventral capsulotomy.

References

1. Ruscio AM, Stein DJ, Chiu WT, Kessler RC. The epidemiology of obsessive-compulsive disorder in the National Comorbidity Survey Replication. Mol Psychiatry. 2010;15(1):53–63.
2. American Psychiatric Association. Diagnostic and statistical manual of mental disorders: DSM-5. Washington, DC: American Psychiatric Association; 2013.
3. Lochner C, Fineberg NA, Zohar J, van Ameringen M, Juven-Wetzler A, Altamura AC, et al. Comorbidity in obsessive-compulsive disorder (OCD): a report from the International College of Obsessive-Compulsive Spectrum Disorders (ICOCS). Compr Psychiatry. 2014;55(7):1513–9.
4. Sharma E, Thennarasu K, Reddy YC. Long-term outcome of obsessive-compulsive disorder in adults: a meta-analysis. J Clin Psychiatry. 2014;75(9):1019–27.
5. Garnaat SL, Greenberg BD, Sibrava NJ, Goodman WK, Mancebo MC, Eisen JL, et al. Who qualifies for deep brain stimulation for OCD? Data from a naturalistic clinical sample. J Neuropsychiatry Clin Neurosci. 2014;26:81–6.
6. Ruck C, Karlsson A, Steele JD, Edman G, Meyerson BA, Ericson K, Nyman H, Asberg M, Svanborg PL. Capsulotomy for obsessive-compulsive disorder: long- term follow-up of 25 patients. Arch Gen Psychiatry. 2008;65:914–21.
7. Lopes AC, Greenberg BD, Noren G, Canteras MM, Busatto GF, de Mathis ME, Taub A, D'Alcante CC, Hoexter MQ, Gouvea FS, Cecconi JP, Gentil AF, Ferrão YA, Fuentes D, de Castro CC, Leite CC, Salvajoli JV, Duran FL, Rasmussen S, Miguel EC. Treatment of resistant obsessive-compulsive disorder with ventral capsular/ventral striatal gamma capsulotomy: a pilot prospective study. J Neuropsychiatry Clin Neurosci. 2009;21:381–92.
8. Gouvea F, Lopes A, Greenberg B, Canteras M, Taub A, Mathis M, Miguel E. Response to sham and active gamma ventral capsulotomy in otherwise intractable obsessive-compulsive disorder. Stereotact Funct Neurosurg. 2010;88:177–82.
9. Kondziolka D, Flickinger JC, Hudak R. Results following gamma knife radiosurgical anterior capsulotomies for obsessive compulsive disorder. Neurosurgery. 2011;68:28–32.
10. Sheehan JP, Patterson G, Schlesinger D, Xu Z. Gamma knife surgery anterior capsulotomy for severe and refractory obsessive-compulsive disorder. J Neurosurg. 2013;119:1112–8.
11. Lopes AC, Greenberg BD, Canteras MM, Batistuzzo MC, Hoexter MQ, Gentil AF, Pereira CA, Joaquim MA, de Mathis ME, D'Alcante CC, Taub A, de Castro DG, Tokeshi L, Sampaio LA, Leite CC, Shavitt RG, Diniz JB, Busatto G, Norén G, Rasmussen SA, Miguel EC. Gamma ventral capsulotomy for obsessive-compulsive disorder: a randomized clinical trial. JAMA Psychiatry. 2014a;71:1066–76.
12. Lopes AC, Greenberg BD, Canteras MM, Batistuzzo MC, Hoexter MQ, Gentil AF, et al. Gamma ventral capsulotomy for obsessive-compulsive disorder: a randomized clinical trial. JAMA Psychiatry. 2014b;71:1066–76.
13. Peker S, Yilmaz M, Sengoz M, Ulku N, Ogel K. Gamma knife radiosurgical capsulotomy for obsessive compulsive disorder. J Neurosurg. 2016;124:A1185.
14. Rasmussen SA, Norén G, Greenberg B, Marsland R, McLaughlin N, Malloy P, et al. Gamma ventral capsulotomy in intractable obsessive-compulsive disorder. Biol Psychiatry. 2018;84(5):355–64.
15. Martinez-Alvarez R. Radiosurgery for behavioral disorders. In: Niranjan A, Lunsford LD, Kano H, editors. Leksell Radiosurgery. Prog Neurol Surg, vol. 34. Basel: Karger; 2019. p. 289–97.
16. Miguel EC, Lopes AC, McLaughlin NCR, Norén G, Gentil AF, Hamani C, Shavitt RG, Batistuzzo MC, Vattimo EFQ, Canteras M, De Salles A, Gorgulho A, Salvajoli JV, Fonoff ET, Paddick I, Hoexter MQ, Lindquist C, Haber SN, Greenberg BD, Sheth SA. Evolution of gamma knife capsulotomy for intactable obsessive-compulsive disorder. Mol Psychiatry. 2019;24(2):218–40.
17. Greenberg BD, Rauch SL, Haber SN. Invasive circuitry-based neurotherapeutics: stereotactic ablation and deep brain stimulation for OCD. Neuropsychopharmacology. 2010;35:317–36.
18. Sheth SA, Neal J, Tangherlini F, Mian MK, Gentil A, Cosgrove GR, et al. Limbic system surgery for treatment-refractory obsessive-compulsive disorder: a prospective long-term follow-up of 64 patients: clinical article. J Neurosurg. 2013;118:491–7.
19. Baxi N, Bruswick A, Mazel E, Kodziolka D. Gamma Knife radiosurgery for obsessive-compulsive disorder. In: Lunsford LD, Sheehan J, editors. Intracranial stereotactic radisurgery. New York: Thieme Publishers; 2016. p. 177–82.
20. Mindus P, Rasmussen SA, Lindquist C. Neurosurgical treatment for refractory obsessive-compulsive disorder: implications for understanding frontal lobe function. J Neuropsychiatry Clin Neurosci. 1994;6(4):467–77.

21. Lipsman N, Neimat JS, Lozano AM. Deep brain stimulation for treatment-refractory obsessive-compulsive disorder. Neurosurgery. 2007a;61(1):1–13.
22. Lipsman N, Neimat JS, Lozano AM. Deep brain stimulation for treatment- refractory obsessive-compulsive disorder: the search for a valid target. Neurosurgery. 2007b;61(1):1–13.
23. Gupta A, Shepard MJ, Xu Z, Maiti T, Martinez-Moreno N, Silverman J, Iorio-Morin C, Martinez-Alvarez R, Barnett G, Mathieu D, Borghei-Razavi H, Kondziolka D, Sheehan JP. An international radiosurgery research foundation multicenter retrospective study of gamma ventral capsulotomy for obsessive compulsive disorder. Neurosurgery. 2019;85(6):808–16.
24. Spiegel EA, Wycis HT, Marks M, Lee A. Stereotaxic apparatus for operations on the human brain. Science. 1947;106:349–50.
25. Moniz E. Essai d'un traitement chirurgical de certaines psychoses. Bull Acad Natl Med. 1936;115:385–92.
26. Talairach J, Hecaen H, David M. Lobotomie préfrontale limitée par électrocoagulation des fibres thalamo-frontales à leur émergence du bras antérieur de la capsule interne. Rev Neurol (Paris). 1949;83:59.
27. Whitty C, Duffield J, Tow P, Cairns H. Anterior cingulectomy in the treatment of mental disease. Lancet. 1952;259:475–81.
28. Ballantine HT Jr, Cassidy WL, Flanagan NB, Marino R Jr. Stereotaxic anterior cingulotomy for neuropsychiatric illness and intractable pain. J Neurosurg. 1967;26:488–95.
29. Knight G. The orbital cortex as an objective in the surgical treatment of mental illness. The results of 450 cases of open operation and the development of the stereotactic approach. Br J Surg. 1964;51:114–24.
30. Knight G. Stereotactic tractotomy in the surgical treatment of mental illness. J Neurol Neurosurg Psychiatry. 1965;28:304.
31. Kelly D, Richardson A, Mitchell-Heggs N, Greenup J, Chen C, Hafner R. Stereotactic limbic leucotomy. Br J Psychiatry. 1973a;123:141–8.
32. Kelly D, Richardson A, Mitchell-Heggs N. Stereotactic limbic leucotomy: neurophysiological aspects and operative technique. Br J Psychiatry. 1973b;123:133–40.
33. Foltz EL, White LE Jr. Pain "relief" by frontal cingulumotomy. J Neurosurg. 1962;19:89–100.
34. Leksell L. Stereotaxis and radiosurgery: an operative system. Springfield: Thomas; 1971.
35. Rauch SL, Dougherty D, Malone D, et al. A functional neuroimaging investigation of deep brain stimulation in patients with obsessive-compulsive disorder. J Neurosurg. 2006;104(4):558–65.
36. McLaughlin NCR, Nanda P, Banks GP, Miguel EC, Sheehan J, Lopes AC, et al. Gamma knife capsulotomy for intractable OCD: impact of lesion size and location. Biol Psychiatry. 2017;81:S276.
37. Kihlström L, Hindmarsh T, Lax I, Lippitz B, Mindus P, Lindquist C. Radiosurgical lesions in the normal human brain 17 years after gamma knife capsulotomy. Neurosurgery. 1997;41:396–402.
38. Kihlström L, Guo W-Y, Lindquist C, Mindus P. Radiobiology of radiosurgery for refractory anxiety disorders. Neurosurgery. 1995;36:294–302.
39. Spatola G, Martinez-Alvarez R, Martínez-Moreno N, Rey G, Linera J, Rios-Lago M, Sanz M, Gutiérrez J, Vidal P, Richieri R, Régis J. Results of Gamma Knife anterior capsulotomy for refractory obsessive-compulsive disorder: results in a series of 10 consecutive patients. J Neurosurg. 2018;131(2):376–83.
40. Goodman WK, Price LH, Rasmussen SA, Mazure C, Fleischmann RL, Hill CL, et al. The Yale-Brown obsessive compulsive scale: I. Development, use, and reliability. Arch Gen Psychiatry. 1989;46:1006–11.
41. Lippitz BE, Mindus P, Meyerson BA, Kihlström L, Lindquist C. Lesion topography and outcome after thermocapsulotomy or gamma knife capsulotomy for obsessive-compulsive disorder: relevance of the right hemisphere. Neurosurgery. 1999;44:452–8.
42. Spofford CM, McLaughlin NCR, Penzel F, Rasmussen SA, Greenberg BD. OCD behavior therapy before and after gamma ventral capsulotomy: case report. Neurocase. 2014;20:42–5.
43. Rück C, Larsson JK, Mataix-Cols D, Ljung R. A register-based 13- year to 43-year follow-up of 70 patients with obsessive–compulsive disorder treated with capsulotomy. BMJ Open. 2017;7:e013133.
44. Kim MC, Lee TK. Stereotactic lesioning for mental illness. Acta Neurochir Suppl. 2008;101:39–43.

The manufacturer's authorised representative in the EU is Springer Nature Customer Service Centre GmbH, Europaplatz 3, 69115 Heidelberg, Germany. If you have any concerns regarding our products, please contact ProductSafety@springernature.com

Printed and bound by CPI Group (UK) Ltd, Croydon, CR0 4YY
30/09/2026
02224058-0003